EXPLORING THE UNKNOWN

NASA SP-4407

EXPLORING THE UNKNOWN

Selected Documents in the History of the
U.S. Civil Space Program
Volume IV: Accessing Space

John M. Logsdon, Editor
with Ray A. Williamson, Roger D. Launius, Russell J. Acker.
Stephen J. Garber, and Jonathan L. Friedman

The NASA History Series

National Aeronautics and Space Administration
NASA History Division
Office of Policy and Plans
Washington, D.C.   1999

Library of Congress Cataloguing-in-Publication Data

Exploring the Unknown: Selected Documents in the History of the U.S. Civil Space Program/John M. Logsdon, editor ... [et al.]
p. cm.—(The NASA history series) (NASA SP: 4407)

    Includes bibliographical references and indexes.
    Contents: v. 1. Organizing for exploration
    1. Astronautics—United States—History.   I. Logsdon, John M., 1937–
               II. Series.   III. Series   V. Series: NASA SP: 4407.

TL789.8.U5E87   1999   96-9066
387.8'0973–dc20   CIP

Dedicated to the Memory of
Robert H. Goddard,
Hermann J. Oberth,
and
Konstantin E. Tsiolkovskiy

The Pioneers of Rocketry

Contents

Acknowledgments xix

Introduction xxi

Biographies of Volume IV Contributors xxv

Glossary xxvii

Chapter One

Essay: "Access to Space: Steps to the Saturn V," by Ray A. Williamson 1

Documents

I-1 and I-2   Hugh J. Knerr, Major General, USA, Deputy Commanding General,
              U.S. Strategic Air Forces in Europe, Memorandum for Commanding
              General, U.S. Strategic Air Forces in Europe, June 1, 1945; and
              Memorandum to the Director of Research and Development, DC/S,
              Material, Attn: General Craigie, "Utilization of German Scientists by
              U.S.S.R. and U.S.," March 22, 1948 32

I-3   H. Julian Allen and A.J. Eggers, Jr., NACA Research Memorandum,
              "A Study of the Motion and Aerodynamic Heating of Missiles
              Entering the Earth's Atmosphere at High Supersonic Speeds,"
              RM A53D28, August 25, 1953, pp. 1–3, 26–29 34

I-4   Homer J. Stewart, Chairman, Ad Hoc Advisory Group on Special
              Capabilities, Report to Donald A. Quarles, Assistant Secretary of
              Defense (Research and Development), August 4, 1955, pp. iii, 1, 3–8. . . . 38

I-5   Reuben B. Robertson, Jr., Deputy Secretary of Defense,
              Memorandum for the Secretaries of the Army, Navy, and Air Force,
              "Technical Program for NSC 5520 (Capability to Launch a Small
              Scientific Satellite During IGY)," September 9, 1955 43

I-6   Joseph C. Myers, Deputy Secretary, Advisory Group on Special
              Capabilities, Office of the Assistant Secretary of Defense,
              "Memorandum for Members, Advisory Group on Special
              Capabilities," December 1, 1955, with attached: Homer J. Stewart,
              Chairman, Advisory Group on Special Capabilities, Memorandum
              to Assistant Secretary for Defense (R&D), "Activities of the
              Advisory Group on Special Capabilities," December 1, 1955 45

I-7 and I-8   Colonel A.J. Goodpaster, "Memorandum for Record," June 7, 1956;
                 and E.V. Murphree, Special Assistant for Guided Missiles,
                 Memorandum for Deputy Secretary of Defense, "Use of the
                 JUPITER Re-entry Test Vehicle as a Satellite," July 5, 1956. 49

I-9 and I-10   Brigadier General A.J. Goodpaster, "Memorandum of Conference
                 With the President, October 8, 1957, 8:30 AM," October 9, 1957;
                 and Robert Cutler, Special Assistant to the President, Memorandum
                 for the Secretary of Defense, "U.S. Scientific Satellite Program
                 (NSC 5520)," October 17, 1957 51

I-11   Donald Quarles, Memorandum for the President, "The Vanguard-
                 Jupiter C Program," January 7, 1958. 54

I-12   James C. Hagerty, "Memorandum on Telephone Calls Between
                 Brigadier General Andrew J. Goodpaster in Washington and James C.
                 Hagerty in Atlanta, Georgia, Friday Afternoon and Evening,
                 January 31, 1958 and Saturday Morning, February 1, 1958" 56

I-13   Brigadier General Bernard A. Schriever, USAF Commander,
                 Western Development Division, Air Research and Development
                 Command, Memorandum for Lt. General [Thomas] Power, "Air Frame
                 Industries vs. Air Force ICBM Management," February 24, 1955. 64

I-14   Brigadier General A.J. Goodpaster, "Memorandum of Conference
                 With the President, March 10, 1958—10:20 AM," March 11, 1958 69

I-15   Hugh L. Dryden, Deputy Administrator, NASA, Memorandum for
                 the President, "Use of Solid Propellants in the U.S. Space Program,"
                 April 7, 1961 70

I-16   NASA, News Release, "Mercury Redstone Booster Development Test,"
                 Release No. 61-57, March 22, 1961 73

I-17   Assistant Deputy Chief of Staff, Development, U.S. Air Force, to
                 AFCCS, "Convair Analysis of Atlas Booster Space Launches," with
                 attached: J.V. Naish to Dr. T. Keith Glennan, Administrator, NASA,
                 December 21, 1960, and A.D. Mardel, Senior Flight Test Group
                 Engineer, Convair Astronautics, General Dynamics Corporation, to
                 Distribution, "Short Summary of Atlas Space Boosters," EM-1691,
                 December 17, 1960. 75

I-18   George E. Mueller, "NASA Learning From Use of Atlas and Titan
                 for Manned Flight," with attached: "Summary Learning From the
                 Use of Atlas and Titan for Manned Flight," December 21, 1965 81

I-19   Staff of Aerophysics Laboratory, North American Aviation, "Feasibility of Nuclear Powered Rockets & Ram Jets," Report No. NA-47-15, February 11, 1947, pp. 2, 11–14 85

I-20   AEC-NASA Press Kit, "Nuclear Rocket Program Fact Sheet," March 1969 88

I-21 and I-22 Senator Howard Cannon to the President, October 19, 1971; and James Fletcher to Senator Howard Cannon, January 24, 1972 91

I-23   NASA, in consultation with the Advanced Research Projects Agency, "The National Space Vehicle Program," January 27, 1959, pp. 1–7 94

I-24   Development Operations Division, Army Ballistic Missile Agency, "Proposal: A National Integrated Missile and Space Vehicle Development Program," Report No. D-R-37, December 10, 1957, pp. 1–7 97

I-25   Abe Silverstein, Chairman, Source Selection Board, Memorandum for the Administrator, "Recommendations of the Source Selection Board on the One Million Pound Thrust Engine Competition," December 12, 1958 105

I-26   Roy W. Johnson, Director, Advanced Research Projects Agency, to Commanding General, U.S. Army Ordnance Missile Command, Huntsville, Alabama, ARPA Order No. 14-59, August 15, 1958 109

I-27   F.C. Schwenk, Memo for Record, "Visit to ABMA on June 16–17, 1959," June 24, 1959 111

I-28   Abraham Hyatt, Deputy Director, Launch Vehicle Programs, to Wernher von Braun, Army Ballistic Missile Agency, January 22, 1960 . . . 115

I-29   Saturn Vehicle Team, "Report to the Administrator, NASA, on Saturn Development Plan," December 15, 1959, pp. 1–4, 7–9 116

I-30   Robert R. Gilruth, Director, Space Task Group, to Dr. N.E. Golovin, Director, DOD-NASA Large Launch Vehicle Planning Group, September 12, 1961 119

I-31 and I-32 Milton Rosen, Director, Launch Vehicles and Propulsion, Office of Manned Space Flight, Memorandum to Brainerd Holmes, Director of Manned Space Flight, "Large Launch Vehicle Program," November 6, 1961; and Milton Rosen, Director, Launch Vehicles and Propulsion, Office of Manned Space Flight, Memorandum to Brainerd Holmes, Director of Manned Space Flight, "Recommendations for NASA Manned Space Flight Vehicle Program," November 20, 1961, with attached: "Report of Combined Working Group on Vehicles for Manned Space Flight" 122

I-33   Future Projects Design Branch, Structures and Mechanics Division,
George C. Marshall Space Flight Center, NASA, Huntsville, Alabama,
"NOVA Preliminary Planning Document," August 25, 1961, pp. 1–6 . . 129

I-34   A.O. Tischler, Chief, Liquid Fuel Rocket Engines, NASA, to
David Aldrich, Program Engineer, Rocketdyne, July 29, 1959 134

I-35 thru I-37 D. Brainerd Holmes, Director of Manned Space Flight, to Wernher
von Braun, Director of Marshall Space Flight Center, "Combustion
Instability of F-1 Engine," January, 26, 1963; A.O. Tischler, Assistant
Director for Propulsion, to Milton Rosen, Director, Launch Vehicles
and Propulsion, "First monthly report of F-1 instability problems,"
February 15, 1963; and Wernher von Braun, Director of Marshall
Space Flight Center, to D. Brainerd Holmes, Director of Manned
Space Flight, "Response to Letter of January 26, 1963," March 11, 1963. . . 135

I-38   George E. Mueller, Deputy Associate Administrator for Manned
Space Flight, NASA, to the Directors of the Manned Spacecraft
Center, Launch Operations Center, and Marshall Space Flight
Center, "Revised Manned Space Flight Schedule," October 31, 1963 . . . 142

I-39   Wernher von Braun, "The Detective Story Behind Our First Manned
Saturn V Shoot," *Popular Science*, November 1968, pp. 98–100, 209. . . . 144

I-40 and I-41 Kurt H. Debus, Director, Launch Operations Center, NASA, to
Captain John K. Holcomb, Office of Manned Space Flight, NASA,
"Reference draft DOD/NASA Agreement dated 20 December 1962
regarding management of Merritt Island and AMR," January 2, 1963,
pp. 1–2; and Robert S. McNamara, Secretary of Defense, and
James E. Webb, Administrator, NASA, "Agreement between the
Department of Defense and the National Aeronautics and Space
Administration Regarding Management of the Atlantic Missile
Range of DoD and the Merritt Island Launch Area of NASA,"
January 17, 1963 147

I-42   "Minutes of the Management Council," Office of Manned Space
Flight, May 29, 1962 154

I-43 thru I-46   James E. Webb, Administrator, NASA, Memorandum to Associate
                 Administrator for Manned Space Flight, NASA, "Termination of the
                 Contract for Procurement of Long Lead Time Items for Vehicles 516
                 and 517," August 1, 1968; W.R. Lucas, Deputy Director, Technical,
                 Marshall Space Flight Center, NASA, Memorandum to Philip E.
                 Culbertson, NASA Headquarters, "Long Term Storage and Launch
                 of a Saturn V Vehicle in the Mid-1980's," May 24, 1972; George M.
                 Low, Deputy Administrator, NASA, Memorandum to Associate
                 Administrator for Manned Space Flight, NASA, "Leftover Saturn
                 Hardware," June 2, 1972; and Dale D. Myers, Associate Administrator
                 for Manned Space Flight, NASA, Memorandum to Administrator,
                 "Saturn V Production Capability," August 3, 1972. 155

Chapter Two

Essay: "Developing the Space Shuttle," by Ray A. Williamson 161

Documents

II-1 and I-2   Ad Hoc Subpanel on Reusable Launch Vehicle Technology, "Report
                 for presentation to the Supporting Space Research and Technology
                 Panel," September 14, 1966, pp. 1–8; and Supporting Space Research
                 and Technology Panel, "Final Report, Ad Hoc Subpanel on Reusable
                 Launch Vehicle Technology," submitted to the Aeronautics and
                 Astronautics Coordinating Board, September 2, 1966, pp. 7–10 194

II-3   Dr. George Mueller, Associate Administrator for Manned Space
                 Flight, NASA, "Honorary Fellowship Acceptance," address delivered
                 to the British Interplanetary Society, University College, London,
                 England, August 10, 1968, pp. 1–10, 16–17 202

II-4   NASA, Space Shuttle Task Group Report, Volume II, "Desired
                 System Characteristics," revised, June 12, 1969. 206

II-5   Charles J. Donlan, Acting Director, Space Shuttle Program, NASA,
                 to Distribution, "Transmittal of NASA paper 'Space Shuttle Systems
                 Definition Evolution,'" July 11, 1972. 211

II-6   Maxime A. Faget and Milton A. Silveira, NASA Manned Spacecraft
                 Center, "Fundamental Design Considerations for an Earth-Surface-
                 to-Orbit Shuttle," presented at the XXIst International Congress of
                 the International Astronautical Federation, Constance, German
                 Federal Republic, October 4–10, 1970 215

II-7   Office of Management and Budget, "Documentation of the Space
                 Shuttle Decision Process," February 4, 1972. 222

II-8   George M. Low, Deputy Administrator, NASA, to Donald B. Rice, Assistant Director, Office of Management and Budget, November 22, 1971, with attached: "Space Shuttle Configurations" 231

II-9   Charles J. Donlan, Acting Director, Space Shuttle Program, to Deputy Administrator, "Additional Space Shuttle Information," December 5, 1971 234

II-10   Mathematica, "Economic Analysis of the Space Shuttle System," Executive Summary, prepared for NASA, January 31, 1972 239

II-11   James C. Fletcher, Administrator, NASA, to Caspar W. Weinberger, Deputy Director, Office of Management and Budget, December 29, 1971 245

II-12   Arnold R. Weber, Associate Director, Office of Management and Budget, Memorandum for Peter Flanigan, "Space Shuttle Program," June 10, 1971, with attached: "NASA's Internal Organization for the Space Shuttle Project" and "NASA's Space Shuttle Program" 249

II-13   James C. Fletcher, Administrator, NASA, to Caspar W. Weinberger, Deputy Director, Office of Management and Budget, January 4, 1972 . . 252

II-14   James C. Fletcher, Administrator, NASA, to Caspar W. Weinberger, Deputy Director, Office of Management and Budget, March 6, 1972 . . . 253

II-15   James C. Fletcher, Administrator, NASA, to Senator Walter F. Mondale, April 25, 1972 256

II-16   James C. Fletcher, Administrator, George M. Low, Deputy Administrator, and Richard McCurdy, Associate Administrator for Organization and Management, NASA, Memorandum for the Record, "Selection of Contractor for Space Shuttle Program," September 18, 1972 262

II-17   The Comptroller General of the United States, Decision in the Matter of Protest by Lockheed Propulsion Company, File B-173677, June 24, 1974, pp. 1, 18–23 268

II-18   Gerald J. Mossinghoff, Assistant General Counsel for General Law, Memorandum for the Record, "Classification of the Space Shuttle as a 'Space Vehicle' and not an 'Aircraft,'" September 25, 1975 272

II-19   John F. Yardley, Associate Administrator for Space Transportation Systems, NASA, to Director, Public Affairs, NASA, Memorandum, "Recommended Orbiter Names," May 26, 1978, with attached: "Recommended List of Orbiter Names" 274

II-20   Eugene E. Covert, Chairman, Ad Hoc Committee for Review of
             the Space Shuttle Main Engine Development Program, National
             Research Council, Statement before the Subcommittee on
             Science, Technology, and Space, Committee on Commerce,
             Space, and Transportation, U.S. Senate, February 22, 1979 275

II-21   John F. Yardley, Associate Administrator for Space Transportation
             Systems, NASA, to Director, Space Shuttle Program, NASA,
             "Study of TPS Inspection and Repair On-Orbit," June 14, 1979 282

II-22   William A. Anders, Consultant, to Dr. Robert Frosch, Administrator,
             NASA, September 19, 1979 283

II-23   James C. Fletcher, Administrator, NASA, to James T. Lynn, Director,
             Office of Management and Budget, October 22, 1976 286

II-24   James T. McIntyre, Jr., Acting Director, Office of Management and
             Budget, to Robert A. Frosch, Administrator, NASA, December 23, 1977. 288

II-25   Robert A. Frosch, Administrator, NASA, to President Jimmy Carter,
             November 9, 1979 290

II-26   Brigadier General Robert Rosenberg, National Security Council,
             "Why Shuttle Is Needed," undated but November 1979 292

II-27   Office of Management and Budget, Background Paper, "Meeting
             on the Space Shuttle," November 14, 1979 294

II-28   Robert A. Frosch, Administrator, NASA, Special Announcement,
             "Examination of the Shuttle Program," August 18, 1980 305

II-29   NASA, Lyndon B. Johnson Space Center, "Major Safety Concerns:
             Space Shuttle Program," JSC 09990C, November 8, 1976, Preface
             and pp. 1-1–2-3, 5-1–A-6 306

II-30   Associate Administrator for Space Transportation Systems, NASA,
             to Administrator, NASA, "STS-1 Mission Assessment," May 12, 1981,
             with attached: "Postflight Mission Operation Report,"
             No. M-989-81-01, Foreword and pp. 1–3, 6–7, 9, 11, 13–14 323

II-31   Allen J. Lenz, Staff Director, National Security Council,
             Memorandum for Martin Anderson, Assistant to the President for
             Policy Development, et al., "Space Shuttle Policy," July 17, 1981,
             with attached: "Presidential Directive: Space Transportation Policy"
             and "Space Policy Review: Terms of Reference" 329

II-32   The White House, National Security Decision Directive 8, "Space
             Transportation System," November 13, 1981 333

II-33   President Ronald Reagan, "Remarks on the Completion of the
Fourth Mission of the Space Shuttle *Columbia*, "July 4, 1982,
pp. 869–872 334

II-34   NASA, "National Space Transportation System: Analysis of Policy
Issues," August 1982, pp. 5–12 337

II-35   Chester Lee, Director, STS Operations, to Manager, Space Shuttle
Payload Integration and Development Program Office, Johnson
Space Center, and Manager, STS Projects Office, Kennedy Space
Center, "Guidelines for Development of the Flight Assignment
Baseline," November 20, 1978 344

II-36   National Security Council, Senior Interagency Group (Space),
"Issue Paper on the Space Transportation System's (STS) Fifth
Orbiter," late 1982 347

II-37   James C. Fletcher, Consultant, Memorandum to Al Lovelace,
"Personal Concern about the Launch Phase of Space Shuttle,"
July 7, 1977 352

II-38   President Ronald Reagan, Executive Order 12546, "Presidential
Commission on the Space Shuttle *Challenger* Accident,"
February 3, 1986 354

II-39   Presidential Commission on the Space Shuttle *Challenger*
Accident, "Report at a Glance," June 6, 1986 356

II-40   Richard H. Truly, Associate Administrator for Space Flight,
NASA, to Distribution, "Strategy for Safely Returning the Space
Shuttle to Flight Status," March 24, 1986 375

II-41   John W. Young, Chief, Astronaut Office, to Director, Flight Crew
Operations, "One Part of the 51-L Accident—Space Shuttle
Program Flight Safety," March 4, 1986, with attached: "Examples
of Uncertain Operational and Engineering Conditions or Events
Which We 'Routinely' Accept Now in the Space Shuttle Program" 378

II-42 and II-43 President Ronald Reagan, "Statement by the President,"
August 15, 1986; and The White House, Fact Sheet, NSDD-254,
"United States Space Launch Strategy," December 27, 1986 382

II-44   H. Guyford Stever, Panel on Redesign of Space Shuttle Solid
Rocket Booster, Committee on NASA Scientific and Technological
Program Reviews, National Research Council, to James C. Fletcher,
Administrator, NASA, Seventh Interim Report, September 9, 1986 . . . 385

II-45   Office of Technology Assessment, "Shuttle Fleet Attrition if Orbiter
Recovery Reliability is 98 Percent," August 1989, p. 6 393

II-46   Dale D. Myers, Deputy Administrator, NASA, to Robert K. Dawson,
Associate Director for Natural Resources, Energy and Science,
Office of Management and Budget, January 20, 1988, with
attachment on the benefits of the Shuttle-C, December 1987 394

II-47   "Report of the Space Shuttle Management Independent Review
Team," February 1995, pp. iii–iv, vii–x, A-1–A-2 399

Chapter Three

Essay: "Commercializing Space Transportation," by John M. Logsdon and Craig Reed. . 405

Documents

III-1   NASA, *We Deliver,* brochure, 1983. 423

III-2   Space Launch Policy Working Group, "Report on Commercialization
of U.S. Expendable Launch Vehicles," April 13, 1983, pp. 1–4, 34. . . . 426

III-3   National Security Decision Directive 94, "Commercialization of
Expendable Launch Vehicles," May 16, 1983 428

III-4   Commercial Space Launch Act of 1984, Public Law 98–575,
98 Stat. 3055, October 30, 1984 431

III-5   Craig L. Fuller, Memorandum for the President, "Determining the
Lead Agency for Commercializing Expendable Launch Vehicles,"
November 16, 1983 441

III-6   Gerald J. Mossinghoff, Assistant Secretary and Commissioner of
Patents and Trademarks, to Elizabeth Hanford Dole, Secretary of
Transportation, letter regarding recommendations of the Commercial
Space Transportation Advisory Committee, October 31, 1984, with
attached: committee recommendations, October 23, 1984 442

III-7   Robert C. McFarlane, Memorandum for The Honorable Elizabeth H.
Dole, Secretary of Transportation, "STS Pricing Issue," June 21, 1984 . . 444

III-8   James M. Beggs, Administrator, NASA, to President Ronald Reagan,
September 17, 1984. 446

III-9   Lawrence F. Herbolsheimer, Memorandum for Craig L. Fuller,
"OMB's study on U.S. ELV competitiveness," November 13, 1984 448

III-10   Gilbert D, Rye, National Security Council, Memorandum for
              Robert C. McFarlane, "Corporate Letters on the Shuttle Pricing
              Issue," July 1, 1985. 449

III-11   Robert C. McFarlane, Memorandum for the President, "Shuttle
              Pricing for Foreign and Commercial Users," July 27, 1985 450

III-12   The White House, National Security Decision Directive Number 181,
              "Shuttle Pricing for Foreign and Commercial Users," July 30, 1985. . . 452

III-13   U.S. Department of Transportation, Office of Commercial Space
              Transportation, "Federal Impediments to Development of a Private
              Commercial Launch Industry," report submitted to Congress,
              July 1985, pp. 1–2 453

III-14   President Ronald Reagan, Memorandum for the United States Trade
              Representative, "Determination Under Section 301 of the Trade Act
              of 1974," July 17, 1985. 454

III-15   Alfred H. Kingon, Assistant to the President, Cabinet Secretary, to
              James C. Fletcher, Administrator, NASA, Memorandum, "Space
              Commercialization," September 25, 1986 457

III-16   Commercial Space Launch Act Amendments of 1988, Public Law
              100–657, H.R. 4399, November 15, 1988 458

III-17   Shellyn G. McCaffrey, The White House, Through Eugene G.
              McAllister, Memorandum for Nancy J. Risque, "Space Launch
              Insurance," July 1, 1987 465

III-18   The White House, National Space Policy Directive 2, "Commercial
              Space Launch Policy," September 5, 1990. 466

III-19   The White House, National Space Policy Directive 4, "National
              Space Launch Strategy," July 10, 1991 467

III-20   Elizabeth Dole, Secretary of Transportation, Letter to the President,
              September 30, 1987. 471

III-21   Richard E. Brackeen, Chairman, COMSTAC, and President, Martin
              Marietta Commercial Titan, Inc., to James H. Burnley, Secretary of
              Transportation, January 29, 1988. 472

III-22   Office of Commercial Space Transportation, U.S. Department of
              Transportation, "Office of Commercial Space Transportation;
              Licensing Regulations," Final Rule (Preamble), *Federal Register* 53
              (No. 64 / Friday), April 4, 1988, pp. 11004–11011 476

III-23   Samuel Skinner, Secretary of Transportation, Letter to Member of
          Congress transmitting a study by the Office of Commercial Space
          Transportation on the scheduling of commercial launch operations
          at Government launch sites, June 1, 1989, with attached: "Executive
          Summary," pp. iii–vii 490

III-24   "Memorandum of Agreement Between the Government of the
          United States of America and the Government of the People's
          Republic of China Regarding International Trade in Commercial
          Launch Services," January 26, 1989 497

Chapter Four

Essay: "Exploring Future Space Transportation Possibilities," by Ivan Bekey 503

Documents

IV-1   The White House, Fact Sheet, "National Space Strategy," National
         Security Decision Directive 144, August 15, 1984 513

IV-2   NASA and the Department of Defense, "National Space Strategy—
         Launch Vehicle Technology Study," December 1984 517

IV-3   NASA/DOD Joint Steering Group, "National Space Transportation
         and Support Study, 1995–2010," May 1986, pp. ii–iii, 1–9, 21–24 521

IV-4   Department of Defense, NSDD-261 Report, "Recommendations for
         Increasing United States Heavy-Lift Space Launch Capability,"
         April 29, 1987, pp. iii–xvi 529

IV-5   James Fletcher, Administrator, NASA, and Frank Carlucci, Secretary
         of Defense, and approved by Ronald Reagan, President, "Advanced
         Launch System (ALS) Report to Congress," January 14, 1988 537

IV-6   Vice President's Space Policy Advisory Board, "The Future of the
         U.S. Space Launch Capability," November 1992, pp. 3–11, 29–40 542

IV-7   Darrell R. Branscome, Director, Advanced Program Development
         Division, Office of Space Flight, NASA, "The Next Manned
         Spacecraft . . . Which Path to Follow?," November 17, 1988 553

IV-8   Secretary of Defense and NASA Administrator, "Memorandum of
         Understanding Between the Department of Defense and the
         National Aeronautics and Space Administration for the Conduct
         of the National Aero-Space Plane Program (Revision B),"
         August 31, 1988 557

IV-9   Department of Defense, "Report of the Defense Science Board
Task Force on the National Aerospace Plane (NASP),"
September 1988, pp. 2–25 561

IV-10   The White House, Office of the Press Secretary, "Statement by the
Press Secretary," July 25, 1989 571

IV-11   Maxwell W. Hunter, "The Opportunity," April 26, 1987 (revised) 572

IV-12   Gary Hudson, Pacific American, Memo to Thomas L. Kessler,
General Dynamics/Space Systems Division, "Comments on SSTO
Briefing and a Short History of the Project," December 17, 1990 575

IV-13   Department of Defense, Strategic Defense Initiative Organization,
"Solicitation for the SSTO Phase II Technology Demonstration,"
June 5, 1991 577

IV-14   Office of Space Systems Development, NASA Headquarters,
"Access to Space Study—Summary Report," January 1994,
pp. i–ii, 1–6, 59–72 584

IV-15   Department of Defense, "Space Launch Modernization Plan—
Executive Summary," May 1994, frontmatter and pp. 1–18, 23–30.... 604

IV-16   The White House, Office of Science and Technology Policy,
"Fact Sheet—National Space Transportation Policy," August 5, 1994.. 626

IV-17   NASA, "A Draft Cooperative Agreement Notice—X-33 Phase II:
Design and Demonstration," December 14, 1995, pp. A-2–A-4....... 631

Biographical Appendix... 635

Index... 667

The NASA History Series... 681

Acknowledgments

This volume is the fourth in a series that had its origins more than a decade ago. The individuals involved in initiating the series and producing the initial three volumes have been acknowledged in those volumes—Volume I: *Organizing for Space* (1995), Volume II: *External Relationships* (1996), Volume III: *Using Space* (1998), and this Volume IV: *Accessing Space* (1999). Those acknowledgments will not be repeated here.

We owe thanks to the individuals and organizations that have searched their files for potentially useful materials, and for the staff members at various archives and collections who have helped us locate documents. Without question, first among them is Lee D. Saegesser, retired from the History Office at NASA Headquarters in 1997, who has helped compile the NASA Historical Reference Collection that contains many of the documents selected for inclusion in this work. All those who in the future will write on the history of the U.S. space program will owe a special debt of thanks to Lee; those who have already worked in this area realize his tireless contributions. We also thank his successors in the office, Terese K. Ohnsorg and Jane H. Odom, who have carried on Lee's traditional of exceptional support for researchers. We also thank Colin Fries and Mark Kahn, contract archivists in the NASA History Office, who have also been instrumental in tracking down documents and checking facts.

There are numerous other people at NASA associated with historical study, technical information, and the mechanics of publishing who helped in a myriad of ways in the preparation of this documentary history. NASA History Office interns Nicole Garrera, Brian Norton, and Madeleine Short compiled the biographical appendix. Stephen J. Garber helped immensely in the final proofing and organization and is a co-editor of this volume. M. Louise Alstork undertook the work of editing and proofreading the manuscript and preparing the index for the final book. Nadine J. Andreassen managed the production of the book for the History Office. We also thank the staffs of the NASA Headquarters Library, the Scientific and Technical Information Program, and the NASA Document Services Center for providing assistance in locating and preparing for publication the documentary materials in this work. The NASA Headquarters Printing and Design Office developed the layout and handled printing. Specifically, we wish to acknowledge the work of Jane E. Penn, Patricia M. Talbert, and Kelly L. Rindfusz for their consultation and design work. Indeed, Jonathan L. Friedman of this organization provided such substantive editorial assistance that we deemed it appropriate to include him as a co-editor of the volume. In addition, Michael Crnkovic, Stanley Artis, and Jeffery Thompson saw the book through the publication process. Thanks are due them all.

At the Space Policy Institute, research associate Russell J. Acker made so many contributions to the organization of material for this volume that he deservedly has been listed as co-editor. Former graduate students Bridget Ziegelaar and Tracy VanDerBeck also played key roles in assembling, scanning, and editing the many documents included.

My thanks go to all those mentioned above, and again to those who helped get this effort started almost a decade ago.

John M. Logsdon, George Washington University

Introduction

One of the most important developments of the twentieth century has been the movement of humanity into space with machines and people. The underpinnings of that movement—why it took the shape it did; which individuals and organizations were involved; what factors drove a particular choice of scientific objectives and technologies to be used; and the political, economic, managerial, and international contexts in which the events of the space age unfolded—are all important ingredients of this epoch transition from an Earthbound to a spacefaring people. This desire to understand the development of spaceflight in the United States sparked this documentary history series.

The extension of human activity into outer space has been accompanied by a high degree of self-awareness of its historical significance. Few large-scale activities have been as extensively chronicled so closely to the time they actually occurred. Many of those who were directly involved were quite conscious that they were making history, and they kept full records of their activities. Because most of the activity in outer space was carried out under government sponsorship, it was accompanied by the documentary record required of public institutions, and there has been a spate of official and privately written histories of most major aspects of space achievement to date. When top leaders considered what course of action to pursue in space, their deliberations and decisions often were carefully put on the record. There is, accordingly, no lack of material for those who aspire to understand the origins and evolution of U.S. space policies and programs.

This reality forms the rationale for this series. Precisely because there is so much historical material available on space matters, the National Aeronautics and Space Administration (NASA) decided in 1988 that it would be extremely useful to have a selective collection of many of the seminal documents related to the evolution of the U.S. civilian space program that was easily available to scholars and the interested public. While recognizing that much space activity has taken place under the sponsorship of the Department of Defense and other national security organizations, within the U.S. private sector, and in other countries around the world, NASA felt that there would be lasting value in a collection of documentary material primarily focused on the evolution of the U.S. government's civil space program, most of which has been carried out since 1958 under the agency's auspices. As a result, the NASA History Office contracted with the Space Policy Institute of George Washington University's Elliott School of International Affairs to prepare such a collection. This is the fourth volume in the documentary history series; two additional ones detailing programmatic developments with respect to space science and human spaceflight will follow.

The documents collected during this research project were assembled from a diverse number of both public and private sources. A major repository of primary source materials relative to the history of the civil space program is the NASA Historical Reference Collection of the NASA History Office located at NASA Headquarters in Washington, D.C. Project assistants combed this collection for the "cream" of the wealth of material housed there. Indeed, one purpose of this series from the start was to capture some of the highlights of the holdings at headquarters. Historical materials housed at the other NASA installations, at institutions of higher learning, and at presidential libraries were other sources of documents considered for inclusion, as were papers in the archives of individuals and firms involved in opening up space for exploitation.

Copies of more than 2,500 documents in their original form collected during this project (not just the documents selected for inclusion), as well as a database that provides a guide to their contents, will be deposited in the NASA Historical Reference Collection. Another complete set of project materials is located at the Space Policy Institute at George Washington University. These materials in their original form are available for use by researchers seeking additional information about the evolution of the U.S. civil space program or wishing to consult the documents reprinted herein in their original form.

The documents selected for inclusion in this volume are presented in four major chapters, each covering a particular aspect of access to space and the manner in which it has developed over time. These chapters focus on the evolution toward the giant Saturn V rocket, the development of the Space Shuttle, space transportation commercialization, and future space transportation possibilities. Volume I in this series covered the antecedents to the U.S. space program, as well as the origins and evolution of U.S. space policy and of NASA as an institution. Volume II addressed the relations between the U.S. civil space program and the space activities of other countries, between the U.S. civil program and national security space and military efforts, and between NASA and industry and academic institutions. Volume III provided documents on satellite communications, remote sensing, and the economic of space applications. As mentioned above, the remaining two volumes of the series will cover space science and human spaceflight.

Each chapter in this volume is introduced by an overview essay, prepared by individuals who are particularly well qualified to write on the topic. In the main, these essays are intended to introduce and complement the documents in the chapter and to place them, for the most part, in a chronological and substantive context. Each essay contains references to the documents in the chapter it introduces, and many also contain references to documents in other chapters of the collection. These introductory essays are the responsibility of their individual authors, and the views and conclusions contained therein do not necessarily represent the opinions of either George Washington University or NASA.

The project team, in concert with the essay writer, chose the documents included in each chapter from those assembled by the research staff for the overall project. The contents of this volume emphasize primary documents or long-out-of-print essays or articles and material from the private recollections of important actors in shaping space affairs. Key legislation and policy statements are also included. The contents of this volume thus do not comprise in themselves a comprehensive historical account; they must be supplemented by other sources, those both already available and to become available in the future. Indeed, a few of the documents included in this collection are not complete; some portions of them were still subject to security classification as the volume went to print.

The documents included in each chapter are generally arranged chronologically; sometimes the flow of the essay's content necessitated that some documents be placed a little out of chronological order. Each document is assigned its own number in terms of the chapter in which it is placed. As a result, the first document in Chapter Three of this volume is designated "Document III-1." Each document is accompanied by a headnote setting out its context and providing a background narrative. These headnotes also provide specific information about people and events discussed. We have avoided the inclusion of explanatory notes in the documents themselves and have confined such material to the headnotes.

The editorial method we adopted for presenting these documents seeks to preserve spelling, grammar, paragraphing, and use of language as in the original. We have sometimes changed punctuation where it enhances readability. We have used ellipses (". . .") to note where sections of a document have not been included in this publication, and we have avoided including words and phrases that had been deleted in the original document unless they contribute to an understanding of what was going on in the mind of the writer in making the record. Marginal notations on the original documents are inserted into the text of the documents in brackets, each clearly marked as a marginal comment. When deletions to the original document have been made in the process of declassification, we have noted this with a parenthetical statement in brackets. Except insofar as illustrations and figures are necessary to understanding the text, those items have been omitted from this printed version. Page numbers in the original document are noted in brackets internal to the document text. Copies of all documents in their original form, however, are available for research by any interested person at the NASA History Office or the Space Policy Institute of George Washington University.

We recognize that there are certain to be quite significant documents left out of this compilation. No two individuals would totally agree on all documents to be included from the more than 2,500 that we collected, and surely we have not been totally successful in locating all relevant records. As a result, this documentary history can raise an immediate question from its users: why were some documents included while others of seemingly equal importance were omitted? There can never be a fully satisfactory answer to this question. Our own criteria for choosing particular documents and omitting others rested on three interrelated factors:

- Is the document the best available, most expressive, most representative reflection of a particular event or development important to the evolution of the space program?
- Is the document not easily accessible except in one or a few locations, or is it included (for example, in published compilations of presidential statements) in reference sources that are widely available and thus not a candidate for inclusion in this collection?
- Is the document protected by copyright, security classification, or some other form of proprietary right and thus unavailable for publication?

As general editor of this volume, I was ultimately responsible for the decisions about which documents to include and for the accuracy of the headnotes accompanying them. It has been an occasionally frustrating but consistently exciting experience to be involved with this undertaking. My associates and I hope that those who consult it in the future find our efforts worthwhile.

John M. Logsdon
Director
Space Policy Institute
Elliott School of International Affairs
George Washington University

Biographies of Volume IV Contributors

Russell J. Acker is a graduate student in George Washington University's Science, Technology, and Public Policy program, where he is the recipient of the Lockheed Martin Graduate Fellowship. His research interests lie at the intersection of space, information technology, and environmental policy, while his prior education includes an MBA and a BBA in information systems. He previously worked in both the energy and software industries, most recently with PeopleSoft, Inc., in the San Francisco Bay area.

Ivan Bekey is an internationally known advanced space systems engineering consultant, providing services to a number of large and small established aerospace industry firms, entrepreneurial space ventures, and government entities as President of Bekey Designs, Inc. He is equally at home in national security, civil, and commercial space systems and applications, as well as the exploitation of new technologies. He is best known for innovative long-term thinking and conception of bold new technology applications. Bekey retired from NASA Headquarters in 1997 as Director of Advanced Concepts. For nineteen years at NASA, he directed the planning, conception, definition, advocacy, development, and flight-test demonstration of advanced programs across many areas of manned and unmanned civilian space and transportation activity. For the previous eighteen years, he was at The Aerospace Corporation, directing system engineering, advanced technology applications, and concept formulation activities on military space and missile systems in support of the U.S. Air Force. In prior positions, he worked at RCA in airborne radar countermeasures and at Douglas Aircraft on surface-guided missiles.

Jonathan L. Friedman is a technical writer for RS Information Systems, Inc., of McLean, Virginia, currently working under contract at NASA Headquarters in the Printing and Design Office. Prior to his five years at NASA, he spent ten years writing and editing for other federal government and state and local agencies on such diverse topics as environmental protection for the military, hazardous and solid waste management, resource recovery, forestry, energy, and foreign disaster assistance. Prior to that, he worked in health care and special education publishing and as a reporter and photographer for a suburban Boston newspaper. He holds a bachelor's degree in English from the University of Arizona.

Stephen J. Garber is a policy analyst in the NASA History Office, in Washington, D.C. He received a bachelor's degree in politics from Brandeis University and a master's in public and international affairs from the University of Pittsburgh's Graduate School of Public and International Affairs. He is currently a graduate student at the Virginia Polytechnic Institute and State University in the field of science and technology studies. He has written on such aerospace history topics as the congressional cancellation of NASA's Search for Extraterrestrial Intelligence program, President John F. Kennedy's attitudes toward space, and the design of the Space Shuttle.

Roger D. Launius is NASA's Chief Historian, located at NASA Headquarters in Washington, D.C. He has produced several books and articles on aerospace history, including *Innovation and the Development of Flight* (Texas A&M University Press, 1999); *NASA & the Exploration of Space* (Stewart, Tabori, & Chang, 1998); *Frontiers of Space Exploration* (Greenwood Press, 1998); *Organizing for the Use of Space: Historical Perspectives on a Persistent Issue* (Univelt, Inc., AAS History Series, Volume 18, 1995), editor; *NASA: A History of the U.S. Civil Space Program* (Krieger Publishing Co., 1994); *History of Rocketry and Astronautics: Proceedings of the Fifteenth and Sixteenth History Symposia of the International Academy of Astronautics* (Univelt, Inc., AAS History Series, Volume 11, 1994), editor; *Apollo: A Retrospective Analysis* (Monographs in Aerospace History No. 3, 1994); and *Apollo 11 at Twenty-Five*, an electronic picture book issued on computer disk by the Space Telescope Science Institute, Baltimore, Maryland, 1994.

John M. Logsdon is Director of the Space Policy Institute of George Washington University's Elliott School of International Affairs, where he is also Professor of Political Science and International Affairs and Director of the Center for International Science and Technology Policy. He holds a bachelor of science degree in physics from Xavier University and a Ph.D. in political science from New York University. He has been at George Washington University since 1970, and he previously taught at The Catholic University of America. He is also a faculty member of the International Space University and Director of the District of Columbia Space Grant Consortium. He is an elected member of the International Academy of Astronautics and the Board of Trustees of the International Space University and Chair of the Advisory Council of The Planetary Society. Dr. Logsdon has lectured and spoken to a wide variety of audiences at professional meetings, colleges and universities, international conferences, and other settings, and he has testified before Congress on numerous occasions. He is frequently consulted by the electronic and print media for his views on various space issues. He has been a Fellow at the Woodrow Wilson International Center for Scholars and was the first holder of the Chair in Space History of the National Air and Space Museum. He is a Fellow of the American Association for the Advancement of Science and an Associate Fellow of the American Institute of Aeronautics and Astronautics. In addition, he is North American editor for the journal *Space Policy*.

Craig R. Reed is Director of Business Development for Lockheed Martin Special Programs, an operating agent of Lockheed Martin Corporation, in Fairfax, Virginia. He completed a Ph.D. dissertation, "U.S. Commercial Launch Policy Implementation, 1986–1992," in political science from George Washington University in 1998. His contribution to this volume is drawn from that dissertation.

Ray A. Williamson is a Research Professor of Space Policy and International Affairs at George Washington University's Elliott School of International Affairs, focusing on the history, programs, and policy of Earth observations, space transportation, and space commercialization. He joined the Space Policy Institute in 1995. Previously, he was a Senior Associate and Project Director in the Office of Technology Assessment (OTA) of the U.S. Congress. He joined OTA in 1979. While at OTA, Dr. Williamson was Project Director for more than a dozen reports on space policy, including: *Russian Cooperation in Space* (1995), *Civilian Satellite Remote Sensing: A Strategic Approach* (1994), *Remotely Sensed Data: Technology, Management, and Markets* (1994), *Global Change Research and NASA's Earth Observing System* (1994), and *The Future of Remote Sensing from Space: Civilian Satellite Systems and Applications* (1993). He has written extensively about the U.S. space program. He holds a bachelor of arts degree in physics from Johns Hopkins University and a Ph.D. in astronomy from the University of Maryland. He spent two years on the faculty of the University of Hawaii studying diffuse emission nebulae and ten years on the faculty of St. John's College in Annapolis, Maryland. He is a member of the faculty of the International Space University and of the editorial board of *Space Policy*.

Glossary

AA . . . . Associate Administrator
AACB . . . . Aeronautics and Astronautics Coordinating Board
ABE . . . . airbreathing engine
ABMA . . . . Army Ballistic Missile Agency
AC . . . . Alternate current
ACRV . . . . Assured Crew Return Vehicle
AEC . . . . Atomic Energy Commission
AEDC . . . . Arnold Engineering Development Center
AFAL . . . . Air Force Astronautics Laboratories
AFB . . . . Air Force Base
AFMTC . . . . Air Force Missile Test Center
AGC . . . . Aerojet-General Corporation
AIA . . . . Aircraft Industries Association
ALS . . . . Advanced Launch System
ALT . . . . Approach and Landing Test
AMLS . . . . Advanced Manned Launch System
AMR . . . . Atlantic Missile Range
AOMC . . . . Army Ordnance Missile Command
APU . . . . Auxiliary power unit
ARDC . . . . Air Research and Development Command
ARPA . . . . Advanced Research Projects Agency
ASD . . . . Assistant Secretary of Defense
ASIS . . . . Abort sensing and implementation system
ASRM . . . . Advanced Solid Rocket Motor
ASSET . . . . Aerothermodynamic-Elastic Structural Systems Environment
                                 Tests (Air Force project)
AT&T . . . . American Telephone and Telegraph
ATV . . . . Automated Transfer Vehicle
BOB . . . . Bureau of the Budget
BUR . . . . Bottom-Up Review
CAN . . . . Cooperative Agreement Notice
CCAFS . . . . Cape Canaveral Air Force Station
CDR . . . . Critical Design Review
CELV . . . . Complementary Expendable Launch Vehicle
CFD . . . . Computational fluid dynamics
CLV . . . . Crewed Launch Vehicle
CNES . . . . Centre Nationale d'Études Spatiales (French space agency)
COCOM . . . . Coordinating Committee for Multinational Export Control
COMSTAC . . . . Commercial Space Transportation Advisory Committee
CSM . . . . Command and Service Module
CSOC . . . . Consolidated Satellite Operations Center
CTRV . . . . Crew Transfer and Return Vehicle
CY . . . . Calendar Year
DAB . . . . Defense Acquisition Board
DARPA . . . . Defense Advanced Research Projects Agency
DDT&E . . . . Design, development, test, and evaluation
DEW . . . . Directed energy weapon
DNA . . . . Defense Nuclear Agency
DOD/DoD . . . . Department of Defense
DOT . . . . Department of Transportation

EELV . . . . .Evolved expendable launch vehicle
ELV . . . . .Expendable launch vehicle
E-MAD . . . . .Engine Maintenance, Assembly and Disassembly (Building)
EPC . . . . .Economic Policy Council
ESA . . . . .European Space Agency
ESMC . . . . .Eastern Space and Missile Center
EST . . . . .Eastern Standard Time
ESTPD . . . . .Economics, Science, and Technology Program Division (OMB)
ET . . . . .External tank
ETR . . . . .Eastern Test Range
ETS . . . . .Engine Test Stand
EVA . . . . .Extravehicular activity
FAA . . . . .Federal Aviation Administration
FAR . . . . .Federal Acquisition Regulations
FCC . . . . .Federal Communications Commission
FEWS . . . . .Follow-on Early Warning System
FMOF . . . . .First Manned Orbital Flight
FOC . . . . .Full operational capability
FOF . . . . .First Operational Flight
FSO . . . . .Functional supplementary objective
FTO . . . . .Functional test objective
FY . . . . .Fiscal Year
FYDP . . . . .Future Years Defense Program
GAEC . . . . .Greenbelt Aerospace Engineering Corporation
GALCIT . . . . .Guggenheim Aeronautical Laboratory of the California
                               Institute of Technology
GAO . . . . .General Accounting Office
GE . . . . .General Electric
GEO . . . . .Geosynchronous Earth orbit
GET . . . . .Ground elapsed time
GN&C . . . . .Guidance, navigation, and control
GPC . . . . .General purpose computer
GSE . . . . .Ground support equipment
GTO . . . . .Geosynchronous transfer orbit
ICBM . . . . .Intercontinental ballistic missile
IGY . . . . .International Geophysical Year
ILC . . . . .Initial launch capability
ILRV . . . . .Integral Launch and Reentry Vehicle
INSAT . . . . .Indian National Satellite
INTELSAT/
Intelsat . . . . .International Telecommunications Satellite (consortium)
IOC . . . . .Initial operational capability
IR&D . . . . .Independent Research and Development
IRBM . . . . .Intermediate-range ballistic missile
IUS . . . . .Inertial Upper Stage
IVA . . . . .Intravehicular activity
JPL . . . . .Jet Propulsion Laboratory (formerly GALCIT)
JPO . . . . .Joint Program Office
JSC . . . . .Johnson Space Center

KEW . . . . .Kinetic energy weapon
KSC . . . . .Kennedy Space Center
LASL . . . . .Los Alamos Scientific Laboratory
L/D . . . . .Lift-to-drag (ratio)
LEO . . . . .Low-Earth orbit
LeRC . . . . .Lewis Research Center
LH_2 . . . . .Liquid hydrogen
LOC . . . . .Launch Operations Center
LOX/LO_2 . . . . .Liquid oxygen
MCR . . . . .Master change record
MDAC . . . . .McDonnell Douglas Aerospace Corporation
MDS . . . . .Malfunction Detection System
MECO . . . . .Main engine cutoff
MILA . . . . .Merritt Island Launch Area
MLV . . . . .Medium launch vehicle
MOL . . . . .Manned Orbiting Laboratory
MOR . . . . .Mission Operation Report
MosGIRD . . . . .Moscow Group for the Study of Reactive Motion
MOU . . . . .Memorandum of Understanding
MR . . . . .Mercury-Redstone
MSFC . . . . .Marshall Space Flight Center
MTCR . . . . .Missile Technology Control Regime
NACA . . . . .National Advisory Committee for Aeronautics
NASP . . . . .National Aerospace Plane
NATO . . . . .North Atlantic Treaty Organization
NERVA . . . . .Nuclear Engine for Rocket Vehicle Application (program)
NIO . . . . .NASP Inter-Agency Office
NLS . . . . .National Launch System
NOAA . . . . .National Oceanic and Atmospheric Administration
NRDS . . . . .Nuclear Rocket Development Station
NRL . . . . .Naval Research Laboratory
NRO . . . . .National Reconnaissance Office
NRX . . . . .NERVA Reactor Experiment
NSC . . . . .National Security Council
NSDD . . . . .National Security Decision Directive
NSIA . . . . .National Security Industries Association
NSPD . . . . .National Space Policy Directive
NSTL . . . . .National Space Technology Laboratories
NSTS . . . . .National Space Transportation System
NTS . . . . .Nevada Test Site
OAO . . . . .Orbiting Astronomical Observatory
OASD . . . . .Office of the Assistant Secretary of Defense
OAST . . . . .Office of Aeronautics and Space Technology
OCST . . . . .Office of Commercial Space Transportation
ODDR&E . . . . .Office of the Directorate of Defense Research and Engineering
OFPP . . . . .Office of Federal Procurement Policy
OFT . . . . .Orbital Flight Test
OMB . . . . .Office of Management and Budget
OMI . . . . .Operational Maintenance Inspection
OMRSD . . . . .Operational Maintenance Readiness Support Document
OMS . . . . .Orbital Maneuvering System
OMSF . . . . .Office of Manned Space Flight

OMV ....Orbital Maneuvering Vehicle
OSD ....Office of the Secretary of Defense
OSHA ....Occupational Safety and Health Administration
OSTP ....Office of Science and Technology Policy
OTS ....Orbit transfer system
OTV ....Orbital Transfer Vehicle
OV ....Orbiter Vehicle
PAM ....Payload Assist Module
PASS ....Primary Ascent Software System
PCIN ....Program change identification number
PDR ....Preliminary Design Review
PFB ....Pressure-fed booster
PFRT ....Preliminary Flight Rating Test
PLS ....Personnel Launch System
PM ....Program Manager
PMP ....Program Management Plan
PMR ....Pacific Missile Range
PRCB ....Program Requirements Control Board
PRC ....People's Republic of China
PRIME ....Precision Recovery Including Maneuvering Entry (Air Force project)
PSAC ....President's Science Advisory Committee
QFD ....Quality Function Deployment
R&D ....Research and Development
RAO ....Rocket assisted orbiter
RCC ....Rough combustion cut-off
RCS ....Reaction Control System
RDT&E ....Research, development, test, and evaluation
RFP ....Request for Proposals
RID ....Review item disposition
RIFT ....Reactor-in-Flight Tests (program)
RLV ....Reusable launch vehicle
R-MAD ....Reactor Maintenance Assembly and Disassembly (Building)
RSA ....Range Standardization and Automation
RSRM ....Redesigned Solid Rocket Motor
S&M ....Structures and Mechanics (Division)
S&T ....Science and technology
SAMSO ....Space and Missile Systems Organization
SANAC ....State-Army-Navy-Air Coordinating Committee
SDI ....Strategic Defense Initiative
SDIO ....Strategic Defense Initiative Organization
SE&I ....Systems Engineering and Integration
SEB ....Source Evaluation Board
SECDEF ....Secretary of Defense
SEI ....Space Exploration Initiative
SIG ....Senior Interagency Group (Space)
SLBM ....Submarine-launched ballistic missile
SLC ....Space launch complex
SLIIP ....Space Launch Infrastructure Investment Plan
SLMP ....Space Launch Modernization Plan
SLV ....Standard Launch Vehicle
SNPO ....Space Nuclear Propulsion Office
SNTP ....Space Nuclear Thermal Propulsion (program)

SR&QA . . . . .Safety, Reliability, and Quality Assurance
SRB . . . . .Solid rocket booster
SRM . . . . .Solid rocket motor
SRMU . . . . .Solid Rocket Motor Upgrade
SSO . . . . .Space Shuttle Office
SSME . . . . .Space Shuttle main engine
SSTO . . . . .Single-stage-to-orbit
SSX . . . . .Space Ship Experimental
START . . . . .Strategic Arms Reduction Treaty
STAS . . . . .Space Transportation Architecture Study
STG . . . . .Space Task Group
STL . . . . .Space Technology Laboratories
STME . . . . .Space Transportation Main Engine
STS . . . . .Space Transportation System
SWNCC . . . . .State-War-Navy Coordinating Committee
TAOS . . . . .Thrust Assisted Orbiter Shuttle
TARS . . . . .Three axis reference system
TCI . . . . .Transpace Carriers, Inc.
TDRS . . . . .Tracking and Data Relay Satellite
TDRSS . . . . .Tracking and Data Relay Satellite System
TMP . . . . .Technology Maturation Program
TOR . . . . .Terms of Reference
TOS . . . . .Transfer Orbit Stage
TPS . . . . .Thermal protection system
TVA . . . . .Tennessee Valley Authority
UCLA . . . . .University of California at Los Angeles
UCV . . . . .Unmanned cargo vehicle
U.S./US . . . . .United States
USAF . . . . .U.S. Air Force
USD(A) . . . . .Under Secretary of Defense for Acquisition
USD(A&T) . . . . .Under Secretary of Defense for Acquisition and Technology
USG/U.S.G. . . . . .U.S. Government
U.S.S.R./USSR . . . . . .Union of Soviet Socialist Republics
USTR . . . . .U.S. Trade Representative
VAB . . . . .Vertical Assembly Building (later, Vehicle Assembly Building)
VAFB . . . . .Vandenberg Air Force Base
VfR . . . . .Verein für Raumschiffahrt (German Rocket Society)
VTOHL . . . . .Vertical takeoff and horizontal landing
VTOL . . . . .Vertical takeoff and landing
VTV . . . . .Vanguard test vehicle
WTR . . . . .Western Test Range
XE . . . . .Experimental engine
XLR . . . . .Experimental Liquid Rocket

Chapter One

Access to Space: Steps to the Saturn V

by Ray A. Williamson

Building the Technology Base for Launch Systems

Prior to the creation of the huge national space programs that have marked the latter half of the twentieth century, individuals and small, privately funded groups in the United States and abroad confronted the challenges of spaceflight and developed the theoretical and experimental rudiments of rocket technology. By the late 1930s, experimenters in Germany, Russia, and the United States had successfully flown liquid-fueled rockets of various types and capacities. Many experimenters belonged to rocket societies, which assisted the progress of rocket development by developing new technological approaches and by creating broad interest in rocketry.[1]

The rocket societies often had strong connections with science fiction writers, who helped keep the dream of interplanetary travel in the forefront of people's imaginations.[2] In the United States, for example, the American Interplanetary Society was started in 1930 by several science fiction writers, including G. Edward Pendray and Hugo Gernsback, editor of *Science Wonder Stories*.[3] Members of the American Interplanetary Society, which in 1934 became the American Rocket Society, successfully experimented with liquid fuel rockets throughout the 1930s. In December 1941, just as the United States was entering World War II, four members of the American Rocket Society formed Reaction Motors, Inc., the first U.S. firm to build liquid-fuel rockets. Using ideas on cooling originally learned from reading one of Eugen Sänger's[4] papers, the Reaction Motors team developed a regeneratively cooled rocket engine[5] that circulated liquid oxygen (LOX) in a cooling jacket around the engine.[6] In 1947, the Army used this engine in the Bell X-1, the first aircraft to penetrate the sound barrier.[7] In the Soviet Union, several groups emerged to study rocketry, the most important of which was the Moscow Group for the Study of

---

1. See, for example, a book by a captain in the Austrian Army, Hermann Noordung (pseudonym of Herman Potočnik), *The Problem of Space Travel: The Rocket Motor* (Washington, DC: NASA Special Publication (SP)-4026, 1995). This book examines many technical aspects of space travel, including space stations. It was originally published in Berlin in 1929. For a discussion of the origins of many of the ideas regarding space travel, see John M. Logsdon, gen. ed., with Linda J. Lear, Jannelle Warren-Findley, Ray A. Williamson, and Dwayne A. Day, *Exploring the Unknown: Selected Documents in the History of the U.S. Civil Space Program, Volume I, Organizing for Exploration* (Washington, DC: NASA SP-4407, 1995), Chapter One.

2. See Frank H. Winter, *Prelude to Space Age: The Rocket Societies, 1926–1940* (Washington, DC: Smithsonian Institution, 1983), for a detailed examination of this period in the development of rocketry.

3. *Ibid.*, p. 73.

4. Eugen Sänger was an Austrian scientist, whose ideas about reusable spacecraft were commemorated in a German design in the 1980s for a two-stage launch system that carries his name. See E. Sänger, *Raketenflugtechnik*, 1933, whose English version is *Rocket Flight Engineering* (Washington, DC: NASA TT F-223, 1965).

5. Wernher von Braun, Frederick I Ordway III, and Dave Dooling, *Space Travel: A History of Rocketry and Space Travel*, rev. 3rd ed. (New York: Harper and Row, 1975), p. 82.

6. The advantage of regenerative cooling is that the propellant, while cooling the combustion chamber, is also preheated to make it more efficient in the burning cycle.

7. Known as the 6000C4, this engine was capable of generating 6,000 pounds of thrust. The Bell X-1 flew on nineteen contractor demonstration flights and fifty-nine Air Force test flights.

Reactive Motion (MosGIRD), led by Sergei P. Korolev, who until his death in 1966 led the Soviet rocket program.[8]

As early as 1921, Robert H. Goddard, the first American rocket engineer, had begun to work on liquid-fuel engines after first experimenting with solid-fuel rockets.[9] On March 16, 1926, he successfully launched the world's first liquid-fueled rocket[10] along a trajectory that took it to an altitude of forty-one feet and a distance of 184 feet.[11] This was a remarkable achievement. Yet the feat, which might have been publicly heralded, was lost to history for another decade because of Goddard's penchant for secrecy.[12] The test took place at his aunt's farm outside of Auburn, Massachusetts, but only three people besides himself witnessed it.[13] Goddard preferred to work alone. For example, at one point Goddard was asked by the American Interplanetary Society to assist its efforts, but he refused.[14] In doing so, he failed to reap the potential benefits that an association with such a group might have yielded in terms of greater appreciation and funding for his experiments.

By 1929, Goddard had completed four successful flights. The last one was the first liquid-fueled launch to carry measuring instruments—a thermometer, a barometer, and even a camera to record the dials in flight. After reaching a height of ninety feet, the rocket crashed and exploded. The powerful noise greatly disturbed his neighbors and, in Goddard's view, brought unwanted headlines in the local paper.[15] Soon after, Goddard moved to Roswell, New Mexico, a sparsely populated desert town, where he could more readily continue his experimentation beyond the watchful eyes of nervous neighbors.

Goddard's New Mexico work, which was supported at the suggestion of his friend Charles Lindbergh in part by the Guggenheim Fund for the Promotion of Aeronautics, was extremely fruitful. There, he tested thirty-one rockets, one of which attained an altitude of 7,500 feet; another reached a speed more than 700 miles per hour.[16] From 1941 until his death in 1945, Goddard worked for the Navy, helping it to develop liquid-fueled rockets for jet-assisted takeoff to assist heavily laden aircraft lift off a runway or a deck of an aircraft carrier.

In his Massachusetts work and in his later experimentation in New Mexico, Goddard contributed an impressive list of firsts to the world of rocketry and several important technical advances. He gained 214 patents for his efforts.[17] He even tested (in March 1923, years before the American Rocket Society did so) the principle of regenerative cooling. However, because of his desire for secrecy and the relative lack of interest from those who might have put his discoveries to work, his experiments contributed relatively little to the development of modern launch vehicles. Working without knowledge of Goddard's activities, government-supported experimenters in Germany eventually duplicated most of Goddard's discoveries and soon surpassed his rockets in size and lift capacity.

8. See James Harford, *Korolev: How One Man Masterminded the Soviet Drive to Beat America to the Moon* (New York: John Wiley & Sons, 1997).

9. See Documents I-7 and I-8 in Logsdon, *Exploring the Unknown*, 1: 86–133.

10. It was fueled by liquid oxygen and gasoline. Goddard chose liquid oxygen for the same reason later rocket designers used it—the liquid form can be transported relatively easily and can be stored in a relatively small volume.

11. Reported in Robert H. Goddard, *Liquid-propellant Rocket Development*, Smithsonian Miscellaneous Collections, Vol. 95, No. 3 (Washington, DC: Smithsonian Institution Press, 1936). See Document I-9 in Logsdon, gen. ed., *Exploring the Unknown*, 1: 134–40, for an extensive excerpt of this report.

12. Frank Winter, *Rockets Into Space* (Cambridge, MA: Harvard University Press, 1990), p. 31.

13. Robert H. Goddard, "Liquid-Propellant Rocket Development," March 16, 1936, in *The Papers of Robert H. Goddard* (New York: McGraw-Hill, 1970).

14. Apparently Goddard regarded them as amateurs, unworthy of his time. Winter, *Prelude to Space Age*, pp. 74–78.

15. Winter, *Rockets Into Space*, p. 33.

16. *Ibid.*, p. 34. Also, see Document I-9 in Logsdon, gen. ed., *Exploring the Unknown*, 1: 134–40.

17. Of these 214 patents, 131 were granted after his death in 1945. Winter, *Rockets Into Space*, p. 34.

Developing the Vengeance Weapon 2 (V-2)

Modern rocketry is a legacy of World War II and its aftermath, the Cold War. During World War II, France, Germany, Japan, the United Kingdom, and the United States attempted to build rockets in support of the war effort. Of these, only Germany was successful in building large rockets. Beginning in 1932, within about a decade, a team of scientists and engineers led by Captain Walter R. Dornberger designed, built, and tested the V-2 rocket. Starting in September 1944, the German army used the V-2 as an early ballistic missile to terrorize Allied military troops and civilian populations.

Experiments by members of the German Rocket Society (Verein für Raumschiffahrt, or VfR), founded July 5, 1927, provided the technical basis for Germany's early success with the V-2. Society members also gained valuable experience designing, building, and testing rockets and rocket components. Wernher von Braun started working for the German army in 1932, specifically to conduct secret research on rockets; he was the first of several VfR members recruited by Dornberger, head of the German army's research program.

They quickly went to work designing and testing a workable liquid-fuel engine. By December 1934, the team had succeeded in building a motor powered by liquid oxygen and alcohol, which it used to send two small, gyroscopically controlled rockets about 6,500 feet high.[18] The team designated this design Aggregat-2, or A-2. The team's success attracted the interest of the German air force, which desired to use rocket engines to assist propeller-driven aircraft at takeoff and to power aircraft and missiles. Out of this interest came a joint army-air force establishment centered at Peenemünde, an island in the Baltic Sea.

By 1936, the experimenters had arrived at the basic design of the A-4, the vehicle that a few years later became the V-2 missile. Further design and testing produced an engine capable of generating the remarkable (for the time) thrust of 59,500 pounds for sixty-eight seconds.[19] This engine, which was regeneratively cooled, operated at 750 pounds per square inch pressure. Kerosene fuel and LOX were fed to the combustion chamber at rates of fifty gallons or more per second by steam-driven centrifugal pumps.[20] The A-4 stood nearly fifty feet high and was just under five and a half feet in diameter. Fully loaded with fuel and a payload of 2,310 pounds, it weighed 28,229 pounds and was capable of flying up to 3,500 miles per hour. The A-4 had a range of 190 miles and could reach an altitude of sixty miles. After the first two test flights ended in failure, the A-4 was successfully flown on October 3, 1942. Twenty-three months and some 65,000 technical alterations later, the A-4 became the operational Vengeance Weapon-2 (known as V-2), the name given to the missile by Hitler (Figure 1–1). By early 1945, when Allied troops first entered the country, the German army had fired 3,225 warhead-carrying V-2 rockets, most of them toward London and Antwerp.

On May 2, 1945, Wernher von Braun, Dornberger, and 116 other rocket specialists surrendered to American officials in the Austrian Tyrol town of Reutte, just south of Bavaria.[21] Several months later, they were taken to the United States, along with about 100 V-2 rockets, many rocket components, and truckloads of scientific documents. This "rocket team" formed one of the foundations of U.S. progress in missiles and rocket development for several decades to come. [I-1, I-2]

18. These were named Max and Moritz after the Katzenjammer Kids of the popular comic strip of the day.
19. Compare the 6,000-pound thrust of the Reaction Motors engine used in the Bell X-1 a decade later.
20. In searching for a manufacturer of pumps with the right specifications, von Braun made the interesting discovery that his needs could be satisfied by pumps very similar in pressure, rate, and size to those used by firefighters.
21. Von Braun and Dornberger feared being captured by the Russians and calculated that they would have a better chance of pursuing their rocket research on acceptable terms in the United States than in the Soviet Union. Hence, in February, after seeing the way the war was going, they led most of the upper echelon of German rocket scientists south to Bavaria to meet the Americans and avoid being captured by the Russians. See Frederick Ordway III and Mitchell Sharpe, *The Rocket Team* (New York: Crowell, 1979), pp. 254–75.

Early Missile and Upper Atmosphere Research

By the mid-1930s, U.S. interest in rocket research had spread to several centers. In 1936, staff members of the Guggenheim Aeronautical Laboratory of the California Institute of Technology (GALCIT), which was directed by the noted aeronautical theoretician Theodore von Kármán, formed a rocket research group to work on both liquid and solid rocket motors. Among these experimenters was Frank Malina, a physics student at Caltech.[22] During World War II, the group's expertise was in high demand to develop small sounding rockets and jet-assisted takeoff solid-fuel rockets to provide additional takeoff boost for heavily loaded aircraft.[23]

GALCIT, which operated under the sponsorship of the U.S. Army, eventually was renamed the Jet Propulsion Laboratory (JPL). Among other rocket technologies, JPL developed slow-burning rocket propellant and storable liquid propellants that proved extremely useful after World War II. After the war, JPL developed a small sounding rocket called the WAC Corporal. This

Figure 1–1. The V-2 being launched from Peenemünde toward the end of World War II. The V-2 was the brainchild of German rocket expert Wernher von Braun and the first operational ballistic missile. (NASA photo)

rocket was powered by an engine using storable hypergolic fuels—red fuming nitric acid and aniline. The WAC Corporal made its first flight at White Sands Proving Grounds in New Mexico on October 11, 1945, attaining an altitude of forty-five miles.[24]

Intensive U.S. launch vehicle research and development essentially began with the testing of German V-2s on American soil following World War II. Nowhere can the close bonds between the development of weapon-carrying missiles and Earth-to-orbit launch vehicles be seen more clearly than in the use of these missiles to jumpstart U.S. rocket development. The United States employed them not only to catch up to the conquered Germans in missile development, but also to push the boundaries of spaceflight for scientific purposes. The V-2 technologies served as foundations for the development of U.S. sounding rockets and provided a vehicle for the first U.S. space science efforts, under the guidance of James van Allen, who directed the government's Upper Atmosphere Rocket

22. Malina and his colleague A.M.O. Smith published the first scholarly article on rocket research: Frank J. Malina and A.M.O. Smith, "Flight Analysis of the Sounding Rocket," *Journal of the Aeronautical Sciences* 5 (1938): 199–302. This article appears as Document I-11 in Logsdon, gen. ed., *Exploring the Unknown*, 1: 145–53.

23. *Ibid.*, plus Document I-12 in Logsdon, gen. ed., *Exploring the Unknown*, 1: 153–76, which is Theodore von Kármán, "Memorandum on the Possibilities of Long-Range Rocket Projectiles," and H.S. Tsien and F.J. Malina, "A Review and Preliminary Analysis of Long-Range Rocket Projectiles," Jet Propulsion Laboratory, California Institute of Technology, November 20, 1943.

24. Frank J. Malina, "Is the Sky the Limit?," *Army Ordnance* (July–August 1946), pp 43–53.

Research Panel.[25] The investigation of V-2 technologies by the government and U.S. industry strengthened bonds that had begun in World War II. New firms were formed and old ones strengthened and enhanced by the partnership.

The U.S. Army set up launch facilities at White Sands and hired the General Electric Company (GE) to carry out a long series of tests with the V-2s. The Army and GE test-flew sixty-seven V-2s between 1946 and 1951, most of them at White Sands. Under Project Hermes, as the test program was called, GE and the Army also developed several different missiles. These included a series of launchers called the Bumper, which used the WAC Corporal as a second stage.[26] Although the weight and propellant advantages of using several rocket stages, in which progressively smaller rockets took over after the previous stage had expended its propellant and fallen back to Earth, were well known, this method had not been tried in a large rocket. Earlier experimenters faced the technical difficulties of igniting an upper stage in space and of separating the two while controlling the upper stage, as well as the lack of a reliable upper stage rocket. The WAC Corporal had proved sufficiently reliable as a sounding rocket. The testing of the Bumper was undertaken in part to reach high altitudes and in part to test the various techniques needed to control the ignition, separation, and control of a second stage. On February 24, 1949, one of these two-stage rockets reached into outer space at an altitude of 244 miles, an altitude record that stood for several years. Bumper 8, the last of the series, was the first rocket to be launched from Cape Canaveral, Florida, on July 24, 1950. During these tests, the Army and GE experimented with developing a tactical missile using radio-inertial guidance.[27]

By making copies of the V-2 engines beginning in 1949, North American Aviation, Inc., was able to gain valuable experience in rocket motor design and construction that the company soon used to good effect in developing larger and more powerful rocket engines. By March 1950, North American was able to build and conduct successful tests on a LOX-alcohol engine that generated 75,000 pounds of thrust (the Experimental Liquid Rocket 43, or XLR43). By January 1956, Rocketdyne, North American's newly named rocket division,[28] had produced a version containing three firing chambers that generated a then-astounding 415,000 pounds of thrust, burning LOX-kerosene (the XLR83).[29] Rocketdyne's engine was originally destined for incorporation into the experimental Navaho cruise missile, a development program begun by the U.S. Army Air Forces in 1946.[30] In July 1957, in a budget-cutting measure, the Army cancelled Project Navaho. However, the effort that had gone into developing the XLR83 resulted in a powerful engine that, in various modifications, served as the basis for many of America's future missiles and space launch vehicles.[31] For example, the lessons learned in building the Navaho engine were later put to good use for the very large F-1 engine, which powered the first stage of the Saturn V. The Navaho program produced a number of other technical advances, including the development of chemical milling for reducing structural weight

---

25. The panel had representatives from the Army Signal Corps, the Johns Hopkins University Applied Physics Laboratory, the Army Air Forces, the Naval Research Laboratory, Princeton, Harvard, Caltech, and the University of Michigan. See John P. Hagen, "Viking and Vanguard," in Eugene Emme, ed., *The History of Rocket Technology: Essays on Research, Development, and Utility* (Detroit: Wayne State University Press, 1964), p. 123.

26. Using a WAC Corporal as a second stage was suggested by engineer Frank Malina, who had a major role in developing the WAC Corporal. See Malina, "Is the Sky the Limit?," p. 45.

27. Radio-inertial guidance is a form of guidance in which the launch vehicle or missile is tracked by radar and commands are issued by radio to change attitude as the flight progresses. It is a technique that was used on the Titan launch vehicles until recently.

28. Rocketdyne was formed as a separate division of North American Aviation in 1955.

29. Compare the 350,000 pounds of thrust from the Space Shuttle main engines (at sea level).

30. The version of this engine actually destined for the Navaho generated 120,000 pounds of thrust.

31. Julius H. Braun, "Development of the JUPITER Propulsion System," IAA-91-673, 42nd Congress of the International Astronautical Federation, Montreal, Canada, October 1991.

while retaining strength and the use of a titanium skin. It also developed an inertial guidance device that used the first transistorized launch vehicle computer.[32]

The Cold War tensions of the 1950s, the development of nuclear weapons, and the Korean War spawned several additional missile-building programs.[33] Among them was the Redstone rocket, which originated in the Hermes C project. In July 1950, the Army chief of ordnance asked the Ordnance Guided Missile Center at Redstone Arsenal in Huntsville, Alabama, to study the feasibility of building a missile with a range of 500 miles. Wernher von Braun's team of scientists and engineers, which the Army had just moved from Texas to the Redstone Arsenal, was given the task. The team decided to use the XLR43, the engine from the Navaho test missile, and an inertial guidance system using a stabilized platform and accelerometers, because they were "simple, reliable, accurate— and available."[34] The engine also employed many other features taken from the V-2. The pressures of the Korean War soon resulted in a redirection of the Hermes program to the development of a single-stage, surface-to-surface ballistic missile having only 200-mile range, but with high mobility, allowing field deployment. Christened the Redstone, the new missile first flew successfully on August 20, 1953, on a test flight of 8,000 yards. Between 1953 and 1958, the Arsenal fired thirty-seven Redstone test vehicles.[35] It was the first large ballistic missile developed in the United States and the first U.S. missile to use an inertial guidance system.

While the Redstone was under development, the Army and the Navy began a joint project to build an intermediate-range ballistic missile (IRBM) that could be launched at sea as well as on land. The Jupiter missile, as it was called, was developed in two versions, both using the Redstone as a basis. Jupiter A was an IRBM designed to carry a warhead. North Atlantic Treaty Organization (NATO) forces deployed it in Europe until 1963, after the Cuban Missile Crisis. Jupiter C, with the official name Jupiter Composite Re-entry Test Vehicle, was a vehicle primarily designed to test reentry technology. Before the United States could build and successfully operate a ballistic missile, it had to solve the difficult problem of reentry into the atmosphere. Opinions differed on how best to protect the nose cone of a nuclear warhead reentering the atmosphere from overheating and destroying the warhead before it reached its target. In 1953, H. Julian Allen, a scientist with the National Advisory Committee for Aeronautics (NACA), had postulated that a blunt rather than a sharp nose would more readily survive reentry. [I-3] The Jupiter C nose cone was not only blunt, but was coated with a fiberglass material that ablated, or burned off, as the surface of the nose cone heated up, thereby keeping the contents of the nose cone cool.[36]

JPL supplied the second and third upper stages for the Jupiter C. On August 8, 1957, the launch team used a Jupiter C to fire a warhead 600 miles high and 1,200 miles down-range, where it was recovered from the Atlantic by U.S. Navy teams. The reentry nose cone on this flight was the first object crafted by humans to be recovered from space.[37] The

---

32. Dale D. Myers, "The Navaho Cruise Missile: A Burst of Technology," IAA-91-679. 42nd Congress of the International Astronautical Federation, Montreal, Canada, October 1991.

33. Technologies from sounding rockets, for example, were incorporated into medium-range missiles. Jacob Neufeld, *Ballistic Missiles in the United States Air Force 1945–1960* (Washington, DC: U.S. Government Printing Office, 1989).

34. Wernher von Braun, "The Redstone, Jupiter, and Juno," in Emme, *The History of Rocket Technology*, p. 109.

35. Twenty-five of these were essentially Jupiter A missiles.

36. Wernher von Braun, "The Redstone, Jupiter, and Juno," p. 113–14.

37. The von Braun team did not accept gracefully the 1955 decision to assign the satellite launch mission to the Naval Research Laboratory team and the Vanguard rocket. Throughout 1956, it kept pushing for a reconsideration of this decision and permission to attempt a satellite launch sometime in 1957. After review within the Pentagon, this suggestion was rejected, but still there was some sense that the Army team would try to launch a satellite without top-level permission. Thus, for this launch, the upper stage was loaded with sand to prevent it from orbiting Earth.

success of this test, along with further refinements, later led to the incorporation of this technology into the design of the Mercury, Gemini, and Apollo capsules, and it made possible the return of astronauts from space. As discussed below, this launch, or perhaps even one earlier in the test series, might have been able to launch an initial U.S. satellite, months before Sputnik 1. History might then have been rather different.

Vanguard, Juno, and the First American Satellite

The ultimate goal of many of the early rocket researchers was to reach orbit. In the early 1950s, sounding rocket and balloon research on the upper atmosphere and growing interest in geophysics and radio propagation led to serious interest among scientists in launching a scientific research satellite. In 1954, meetings of the International Scientific Radio Union and the International Union of Geodesy and Geophysics passed resolutions calling for the launch of a scientific satellite during the International Geophysical Year (IGY), which had been set for 1957–58, when scientists expected peak sunspot activity. The United States and the Soviet Union in 1955 both announced their intentions to orbit a satellite sometime during the IGY.

A committee within the Department of Defense (DOD) picked the launch vehicle for this satellite from among three proposals: the Atlas intercontinental ballistic missile (ICBM), which was still in the development stage; the Jupiter, using several upper stages; and an unnamed vehicle that would use the Viking as a first stage, the Aerobee as a second stage, and a new solid-fuel third stage. The Viking and the Aerobee were liquid-fueled sounding rockets with proven launch records. The Viking-Aerobee combination, which had been proposed by the Naval Research Laboratory (NRL), had an advantage, because it used available sounding rockets and thus would not compete with the development of the higher priority Atlas ICBM program. Furthermore, the Viking rocket used gimbaled engines for control and had some growth potential. The resulting program, which was managed by the NRL, was called Project Vanguard. The first and second stages used storable nitric acid and dimethylhydrazine as fuel. On September 9, 1955, DOD authorized the Navy to proceed with Project Vanguard. As John P. Hagen, the director of Project Vanguard, has noted, "The letter from the Secretary of Defense stated clearly that what was needed was a satellite (i.e., one) during the I.G.Y. which was in no way to interfere with the on-going military missile programs."[38] [I-4, I-5, I-6]

The role of the aerospace industry as contractor in the construction of launch vehicles was an important though not entirely easy one. For example, the NRL contracted with the Martin Company, the developer of Viking, to build the first stage of the Vanguard rocket and to oversee the vehicle's assembly. During the negotiations, the NRL and Martin had protracted discussions about which organization should have responsibility for overall systems design and engineering. Despite strong arguments to the contrary from Martin, the NRL maintained systems responsibility. As would become very evident ten years later in the development of the Saturn V, such a division of labor sometimes led to friction between the contractor and the government office overseeing launch vehicle development.

Project Vanguard selected Cape Canaveral, Florida, where there was already a missile test range, for its launch site, and it established a worldwide tracking network using the NRL's Minitrack system to maintain control over the launcher after it left the Cape. The Minitrack system also served to collect data from the orbiting satellite.

The debate over which of these satellite-launching proposals best served the nation's interests involved a good measure of interservice rivalry as well as rivalry among rocket

---

38. Hagen, "Viking and Vanguard," p. 123.

teams. As Wernher von Braun has written, with a detachment that understates the strong feelings prevailing among the engineers at the Army Ballistic Missile Agency (ABMA): "While Project Vanguard was the approved U.S. satellite program, we at Huntsville knew that our rocket technology was fully capable of satellite application and could quickly be implemented."[39] It is indeed likely that the von Braun team could have launched a simple satellite in 1957 (barring a launch failure), but it was prohibited from doing so after a 1956 Washington and White House review of that option. [I-7, I-8] Von Braun's team got its chance only after the first attempt at launching a satellite with a Vanguard launch vehicle resulted, on December 6, 1957, in an embarrassing launch pad explosion.

The embarrassment came about, in part, from President Dwight D. Eisenhower's decision to announce the attempted liftoff well in advance. In October 1957, shortly after the surprise launch of Sputnik, Eisenhower was briefed on the situation and told of a planned Vanguard test launch in December. The test was to be the first launch of all three stages and the first launch of the second stage. The Vanguard team and the nation got their first taste of the political sensitivity of the space program when on October 9, 1957, President Eisenhower announced in a news conference that "the satellite project was assigned to the Naval Research Laboratory as Project Vanguard. . . . The first of these test vehicles is planned to be launched in December of this year." [I-9, I-10] This put the launch team in the unenviable, and untenable, position of attempting in public the launch of an untried rocket—the three-stage Vanguard had never been tested as a unit. On December 6, the Project Vanguard team and the United States watched in dismay as the engines of the first stage ignited, then exploded in a fiery exhibition, while the world looked on. The press had a field day with the incident: "Vanguard was Kaputnik, Stayputnik, or Flopnik, and Americans swilled the Sputnik Cocktail: two parts vodka, one part sour grapes."[40]

It was the first and last failure of the first stage in the Vanguard program, but it set a tone that carried through the early days of the U.S. space program. Not only had the Soviets been first into space, but the United States was not even a near second. The Vanguard failure heightened the perception that U.S. engineers were space bunglers, and it stiffened U.S. resolve to best the Soviets. As some U.S. policymakers (but never President Eisenhower) saw it, winning the space race would demonstrate to the world, and to the nation, the superiority of the U.S. political and economic system. But first, rocket engineers had to launch a satellite.

A month before the Vanguard failure, after receiving White House permission to proceed with an alternative to Vanguard, DOD ordered the Army team at Redstone Arsenal to prepare its Jupiter launch vehicle for a satellite launch. [I-11] The Army team quickly made itself ready. Adding an additional upper stage to the Jupiter C gave the vehicle the ability to reach orbit with a small satellite. When the order came to the ABMA to attempt a satellite launch, the Jupiter C with the fourth upper stage became the Juno I, which on January 31, 1958, lifted the first U.S. satellite, Explorer I, into space (Figure 1-2). [I-12] Because Juno I's lift capacity was limited, Explorer I weighed only eighteen pounds, but it carried instruments that made possible the discovery of one of Earth's natural radiation belts, now known as the Van Allen belts.

The Vanguard rocket, too, finally achieved success on March 17, 1958, when it launched the Vanguard I satellite into orbit (Figure 1-3). Although Vanguard was quickly superceded by other, more powerful rockets, its components, especially its Aerobee second stage and its solid-fuel third stage, had important roles in the later success of the Scout and Delta launchers.

---

39. Wernher von Braun, "The Redstone, Jupiter, and Juno," in Emme, *History of Rocket Technology*, p. 114.
40. Walter A. McDougall, . . . *the Heavens and the Earth: A Political History of the Space Age* (New York: Basic Books, 1985), p. 154.

Figure 1-3. A test of the Viking rocket used to launch the Vanguard satellite, TV 3 BU. This satellite was part of the U.S. Earth satellite program to place in Earth orbit the first American satellite on February 5, 1958. The satellite would measure atmospheric density and conduct geodetic measurements. After 57 seconds of flight, connection units of the first-stage control system failed. At 20,000 feet, the rocket veered off course and broke apart. Not until March 1958 did Vanguard I successfully enter Earth orbit. (NASA photo VAN 9A)

Figure 1-2. The launch of Explorer I, on January 31, 1958, 10:48 p.m. Eastern Standard Time, atop the Jupiter-C rocket originally developed by Wernher von Braun as part of the ballistic missile program at the Redstone Arsenal in Huntsville, Alabama. (U.S. Army photo)

In these early days of the U.S. space program, relatively small modifications to the launch vehicles that were already available enabled designers to create launchers for ever more demanding projects. For example, by the end of April 1961, Juno II, which derived directly from the Jupiter IRBM and was essentially a larger version of the Redstone, carried the deep space probes Pioneer III and Pioneer IV toward the Moon and Explorers VII, VIII, and XI into Earth orbit to return data about the physical characteristics of near-Earth space.

Missile Development

Until the early 1950s, missile designers had focused on the eventual development of large ICBMs produced to carry the massive nuclear warheads that the United States had developed immediately after World War II. The U.S. strategic doctrine of the period depended on large bombers to carry nuclear warheads over the Soviet Union should hostilities between the two superpowers reach the flash point, and only a few dreamers expected ICBMs to gain ascendancy much before the middle of the 1960s. By 1953, however, scientists discovered how to make a relatively lightweight thermonuclear weapon, and U.S. officials discovered that the Soviet Union had made considerable progress in developing a long-range missile. These events led to a reevaluation of the U.S. approach to ICBM development. In 1954, the Air Force Strategic Missiles Evaluation Committee, chaired by mathematician John von Neumann, urged the development of a relatively small ICBM, capable of launching the newly developed weapons toward the Soviet Union. It also recommended the creation of a special development group with sufficient funding and authority to proceed with dispatch. The Air Force created the Western Development Division, which later became the Air Force Ballistic Missile Division. The Space Technology Laboratories of the Ramo-Wooldridge Corporation, the precursor to the Aerospace Corporation, provided systems engineering and technical direction for the Ballistic Missile Division. [I-13][41]

During the mid-1950s, the Air Force started work on two major ICBM systems, both of which still play a major role in U.S. space transportation efforts—Atlas and Titan. It also started work on the Thor IRBM, which employed related technologies. These projects came to fruition in the late 1950s, adding to U.S. strength in the missile race and, soon after, to its ability to place satellites in orbit.

The Atlas ICBM—a project that had originally started in 1945 as a classified Air Force effort (Project MX-774) and had died in 1947—was reborn in 1951 as a five-engine missile generating a takeoff thrust of 650,000 pounds. In 1954, it was redesigned and reduced to using three engines based on those originally developed for the Navaho. The Atlas incorporated several new design features, but one of the most important was the introduction of a pressurized stainless steel fuel tank, designed to carry some of the structural burden. This innovation, introduced by Convair engineer Karel J. Bossart, reduced the need for stiffeners and made the Atlas much lighter for a given thrust than earlier designs. Bossart's team also introduced gimbaled thrust nozzles and a warhead that separated from the missile after burnout. The first successful flight of the Atlas occurred in December 1957, after a series of both major and minor development problems. Its first use as a space launcher occurred on December 18, 1958, when an Atlas booster launched into orbit a communications payload weighing sixty-eight kilograms.

Out of technical conservatism and a desire to reduce the risk of depending on single industrial sources for the Atlas, the Air Force contracted with other firms to develop alternative approaches for the major subsystems. After assuring themselves that the Atlas design was on a sound track, in April 1955, Air Force officials approved the incorporation of several of the alternative subsystems, which involved more sophisticated technology development, into an alternate Titan missile, which was to be built by the Martin Corporation. Unlike the Atlas, the Titan missile had a monocoque airframe, in which the aluminum skin absorbed much of the stress of flight, and a more sophisticated guidance system. It also had a different first-stage engine, built by Aerojet, which burned LOX-kerosene fuel instead of LOX-alcohol.[42] The Titan was also a true two-stage missile designed to be launched from a hardened, underground launch silo. The Titan I missile, guided by a combination radio-inertial guidance mechanism, had its first full test in February 1959 and was declared operational in 1962.

During the early 1950s, many Air Force officers had become convinced that the United States needed an IRBM, and in January 1955, the Scientific Advisory Committee of the Office of the Secretary of Defense recommended that the Air Force proceed. However, the Army, which was developing the Jupiter, objected, as did the Navy, which also wanted its own program. The Joint Chiefs of Staff compromised by recommending to Secretary of Defense Charles Wilson in November 1955 that the Air Force develop the Thor, while the Army and the Navy worked jointly on the Jupiter. The Western Development Division got the Thor assignment a month later.[43]

Thor was undertaken as a high-risk program, having the express goal of achieving flight within the shortest possible time. Using engines originally developed for the Atlas, the Thor had its first complete launch pad test in January 1957 and a full range flight test in September of that year. By December 16, 1958, the Strategic Air Command successful-

---

41. Note that the documents at the end of this chapter are not necessarily in chronological order.

42. In keeping with its desire to maintain more than one supplier for critical launch technology, the Air Force chose Aerojet to build the engine for the Titan I. Aerojet used the same rocket motor technology used in the Atlas missile and originally developed for the Navaho missile. The Martin Company built the structure.

43. Neufeld, *Ballistic Missiles*, pp. 146–47.

ly launched a Thor from Vandenberg Air
Force Base in California. The test marked
the passage from development to initial military readiness.[44] On February 28, 1959, a
Thor missile combined with an Agena second stage launched the first Air Force satellite, Discoverer I, into low-Earth orbit.[45]
Under NASA's control, the Thor, using its
Delta upper stage and numerous detailed
modifications, later evolved into the highly
successful Delta launch vehicle, one of the
standard vehicles used to launch NASA's scientific payloads and commercial communications satellites. The Delta evolved from
the original model capable of placing a few
hundred pounds into low-Earth orbit to one
(Delta II 7925) that by the mid-1990s was
capable of launching payloads weighing
3,965 pounds to geostationary transfer orbit
(Figure 1–4).

The V-2, Redstone, Jupiter, and Atlas
missiles were all propelled by LOX-alcohol
or LOX-kerosene. LOX—liquid oxygen—
has the serious drawback of requiring cooling and special handling. It therefore cannot
be stored for long periods and must be
loaded immediately prior to launch. Hence,
LOX is ultimately unsuitable for use in military missiles, when speed in launching could
be critical. In the late 1950s, missile designers spent considerable effort to develop storable liquid propellants and solid fuels.

*Figure 1–4. This photograph of a long-tank Delta no. 73 rock-
et was taken on August 22, 1969, at Kennedy Space Center
in Florida. It shows the launch configuration for the Pioneer
E, which would have been called Pioneer 10. When launched
on August 27, however, the launch vehicle malfunctioned
and was destroyed 8 minutes and 3 seconds into powered
flight by the range safety officer. (NASA photo 69-H-1442)*

The desire to operate from a "hard-
ened" launch site, below ground and solidly encased in concrete, and to be ready to
launch with only a few minutes' notice, also acted to speed up the development of hypergolic, storable fuels. The Air Force embarked on the development of the Titan II missile,
which later became a modest-capacity launch vehicle.[46] It used a mixture of unsymmetri-
cal dimethylhydrazine and hydrazine, oxidized by nitrogen tetroxide. The two-stage Titan
II represented a major leap in technology development over the Titan I. Not only did it
use storable propellants, it also had all-inertial guidance. NASA chose the Titan II to
launch the Gemini spacecraft into orbit.[47]

---

44. Robert L. Perry, "The Atlas, Thor, Titan, and Minuteman," in Emme, *The History of Rocket Technology,*
p. 151.

45. This was the first of many satellites in the Corona series of spy satellites. See Dwayne A. Day, John M.
Logsdon, and Brian Latell, *Eye in the Sky: The Story of the Corona Spy Satellites* (Washington, DC: Smithsonian
Institution Press, 1998).

46. The Titan II was decommissioned as an ICBM between 1982 and 1987. Fourteen were refurbished as
launchers and, during the 1980s and 1990s, have been used to launch a variety of automated payloads, including the National Oceanic and Atmospheric Administration (NOAA) series of polar-orbiting launch vehicles.

47. See, especially, Barton C. Hacker and James M. Grimwood, *On the Shoulders of Titans: A History of Project
Gemini* (Washington, DC: NASA SP-4302, 1970).

The Atlas, Thor, and Titan were all developed according to the management technique called "concurrency," in which all major systems and subsystems were developed in parallel. This technique called for the planning and construction of industrial production facilities and operational bases even before initial flight testing began. It put great pressure on the development team to oversee each step of development very closely. It also meant that (1) both the authority and responsibility for decisions had to be located within the same agency, (2) program managers had to have a high degree of technical competence, and (3) "funding and programming decisions outside of the authority of the program director had to be both timely and firm."[48]

All three of these programs achieved their objectives relatively quickly. As Perry has noted, "The *management* of technology became the pacing element in the Air Force ballistic missile program. Moreover—as had not been true of any earlier missile program—technology involved not merely the creation of a single high-performance engine and related components in a single airframe, but the development of a family of compatible engines, guidance subsystems, test and launch site facilities, airframes, and a multitude of associated devices."[49]

Missile development also led to one other major technology advance that is now a common element of modern launch vehicles—the creation of rocket motors propelled by solid fuels. Solid propellants are composed of an oxidizer such as ammonium perchlorate, a fuel such as aluminum powder, and an organic binder to create a mixture capable of being cast in a rocket motor casing. When ignited, the mixture continues to burn without benefit of an external source of oxygen. The advantages of using solid rocket fuel for a military missile are enormous. Rockets loaded with solid propellants can be built and stored for long periods, and they can be moved around readily. As noted above, JPL[50] developed solid-fuel jet-assisted takeoff rockets during World War II, and its small solid rocket motors were later used as upper stages in the Jupiter C.[51] However, the difficulties of mixing and casting solid propellant in motors large enough to carry a nuclear weapon, and the absence of a satisfactory igniter, had prevented its use in missiles. Ammonium perchlorate, the oxidizer of choice, is hard to handle in large quantities and difficult to mix evenly with an organic binder. In addition, Air Force scientists and engineers needed to develop methods for controlling the fuel's burn, its rate of thrust, and its direction, as well as ways of constructing high-strength, lightweight engine cases.

By October 1957, the Air Force had made substantial progress toward building rocket motors large enough to propel a nuclear weapon, but it had no solid-fuel missile development project in place. Although solid-fuel missiles had been considered for tactical deployment, they had not reached a level of reliability and thrust sufficient to serve as ICBMs. However, the perceived crisis of responding to Sputnik, coupled with the technical progress made in the 1950s, injected a new urgency into U.S. plans for developing a solid-fuel ICBM.[52] [I-14] Studies developed the concept for Weapon System Q, a three-stage, solid-fueled ICBM, which would be deployed in large quantity in hardened missile pads. In September 1957, this was named Minuteman. By the end of 1957, the Office of the Secretary of Defense Ballistic Missile Command recommended that the Air Force

48. Perry, "The Atlas, Thor, Titan, and Minuteman," p. 148.

49. *Ibid.*, p. 150. See also O. J. Ritland, "Concurrency," *Air University Quarterly Review* 12 (Winter–Spring 1960–61): 57–62.

50. JPL, which was established in 1944 as a U.S. Army facility, was transferred to NASA on December 3, 1958. The California Institute of Technology (Caltech) operates it under contract to NASA.

51. The Jupiter C used eleven solid-fuel Baby Sergent rockets for its second stage, six of them for stage three. To place Explorer I in orbit in 1958, the ABMA employed a single Baby Sergent rocket as a fourth stage.

52. Neufeld, *Ballistic Missiles*, p. 227.

begin a program to develop the Minuteman, and in June 1958, Secretary of Defense Neil H. McElroy approved the request. Although the Air Force did not complete the selection of the contractors for the rocket's stages and other major systems until July 1958, the first flight test took place only two and one half years later on February 1, 1961. It was highly successful. The developers used the same concurrency process that had worked well for the development of the Atlas, Thor, and Titan. Of these four missiles, only the Minuteman has not yet been upgraded and made into a working Earth-orbit launch vehicle, although there have been moves in this direction as Minutemen have become excess to security requirements as a result of arms limitation agreements.

In March 1956, the Navy had also gained permission to start its own missile program, which eventually led to the Polaris missile, launched from a submarine below the surface of the ocean. Like the Minuteman, the Polaris depended on a solid rocket motor for propulsion for much the same reasons that Air Force officials were drawn to it for the Minuteman—solid rocket motors can be fired nearly immediately, and they can be stored for long periods without degrading. They are also much easier to handle than liquid motors, making them especially attractive for launching from submarines. The Lockheed Aircraft Corporation was the prime contractor for the Polaris. It conducted the first successful test of an inertially guided Polaris missile on January 7, 1960, from Cape Canaveral. On July 20 of that same year, the nuclear submarine *George Washington* conducted the first undersea firing of a Polaris.[53]

The experience gained in manufacturing solid rocket motors for the Minuteman and the Navy's Polaris[54] missile programs enabled NASA to develop the solid-fueled Scout small launch vehicle. The Scout was first completed and launched in July 1960.

In the late 1950s, while developing rocket motors for missiles, rocket engineers also began to work on ever larger solid rocket motors in hopes of creating a space booster capable of placing moderate-sized payloads into orbit. Rockets based solely on solid propellants require motors capable of generating several million pounds of thrust for durations of 100 seconds or more. Rocket designers faced the major problem of achieving a sustained, even burn, rather than igniting the entire mass of propellant at once. Among other things, this involved developing the means to disperse an oxygen-rich compound, commonly ammonium perchlorate, uniformly in an organic binder that would provide the fuel. It also involved building high-strength, lightweight engine cases. After considerable testing, they finally mastered the technique of casting solid propellant in large sizes and with internal shapes capable of sustaining an even burn rate. Nevertheless, it was clear that the enormous sizes (diameter and length) needed to develop millions of pounds of thrust would create difficult construction and transportation problems. However, if the rocket motors could be built in segments and bolted together on the launch pad, they would be much easier to construct and to transport to the launch site.

Starting in 1957 with funding from the U.S. Air Force, Aerojet General Corporation, which had manufactured jet-assisted takeoff units during World War II, demonstrated that the concept was feasible by first cutting a twenty-inch-diameter Regulus II booster rocket into three pieces, filling the pieces with propellant, reattaching them, and firing the segmented rocket motor. Following a successful test in early 1959, Aerojet attempted the same procedure with a sixty-five-inch-diameter Minuteman rocket motor, which also fired successfully. On February 17, 1960, Aerojet successfully test-fired a three-segment, 100-inch-diameter rocket motor more than 400 inches long that produced an average of 534,000 pounds of thrust for nearly ninety seconds.[55] The test program concluded in

53. Von Braun, Ordway, and Dooling, *Space Travel*, pp. 130–32.

54. Wyndham D. Miles, "The Polaris," in Emme, *The History of Rocket Technology*, p. 162–75.

55. K. Klager, "Segmented Rocket Demonstration: Historical Development Prior to their Use as Space Boosters," IAA-91-687, 42nd Congress of the International Astronautical Federation, Montreal, Canada, October 1991.

October 1962 after achieving partial success with tests of two longer 100-inch-diameter motors.[56] These developments demonstrated that reliable segmented solid-fuel rockets could be built and fired in ground tests. Such experience enabled the Air Force and NASA to develop the large segmented solid rocket boosters that were later used to power both the Titan III and IV launchers and the Space Shuttle.

During the early development of the Saturn liquid-fueled booster, proponents of solid rocket motors suggested their use in that program. NASA had explored the potential of solid rockets and decided that, while advantageous for some tasks, such as launching scientific payloads, they had not yet reached the level of development that would make them suitable for launching people into space. [I-15] In the immediate aftermath of President John F. Kennedy's May 1961 announcement that the United States would send people to the Moon, a joint NASA-Department of Defense team examined the possible use of solid-fueled rockets in accomplishing that mission; however, NASA decided to stand by its earlier position. Hence, NASA carried out relatively little development work on solid rocket motors until they were under consideration for the Space Shuttle.

Launching People: Mercury-Redstone,
Mercury-Atlas, and Gemini-Titan

Launching people into orbit introduced another set of considerations into booster design and manufacture. Although the armed services and NASA were concerned about launch vehicle reliability because of the costs involved in replacing an expensive payload, they had little concern about safety beyond the obvious issues of possible launch pad and range damage. Once the many tons of steel, aluminum, and propellants were on their way to space, the loss of the vehicle primarily meant extra costs and the loss of the payload and research results. However, the loss of human life was another matter, the costs of which could not be reckoned in dollars alone. The creation of Project Mercury, a high-visibility, U.S. human spaceflight program, led to the need to reduce the risks of spaceflight, not only to protect the astronauts, but to protect the space program itself from cancellation. Astronauts were not merely test pilots; they were highly visible manifestations of U.S. technological and political accomplishments, and they soon became American icons. NASA began to institute different procedures for designing, building, and launching the rockets destined to carry humans. Because the Redstone had previously demonstrated relatively high reliability and flight stability, NASA requested eight Redstone launchers for the suborbital portion of Project Mercury. These boosters were modified to allow additional propellant to increase their lift capacity and to add an abort-sensing system to increase their safety.

"Man-rating" the Redstone also meant additional verifications of the reliability of launcher hardware and launch software and extensive testing for electronic and mechanical compatibility with the Mercury spacecraft payload. After an initial launch test to assure that all the systems and subsystems performed together, the first flight with a live passenger occurred on January 31, 1961, when the second Mercury-Redstone mission (MR-2) carried the chimpanzee Ham briefly into space and back on a parabolic trajectory. However, the Redstone boosted the Mercury capsule to a greater height than planned, and thus the capsule landed much further downrange than had been planned. The cause of the booster malfunction was quickly identified and remedied, but von Braun and his associates insisted on an additional test flight before committing an astronaut to a mission atop the Redstone. That additional flight took place on March 23, 1961, and was totally successful. If it had not been inserted into the Mercury schedule, the March flight could

---

56. The test failures were related to malfunctions of the motors' nozzle assembly, not the segment joints.

well have carried an astronaut, and it would have been an American, not the Soviet cosmonaut Yuri Gagarin, who would have been first into space (though not into orbit). [I-16]

On May 5, 1961, Navy Lieutenant Commander Alan B. Shepard, Jr., did become the first American human in space aboard *Freedom 7* (Figure 1–5). His flight was followed by the second and last crewed Mercury-Redstone flight on July 21, which carried Air Force Captain Virgil I. "Gus" Grissom into space and back aboard *Liberty Bell 7*. A more powerful rocket would be needed to place an astronaut into orbit.

For the orbital launch of the Mercury capsule, NASA officials decided at the start of the program to use the Atlas launcher, which was capable of carrying about 3,000 pounds into a 150- by 100-mile elliptical orbit. Using the Atlas to carry people required upgrading the launcher to increase its safety margins; there was concern that there had been frequent failures during the use of the Atlas for unmanned space launches. [I-17] In all, NASA procured nine Atlas D launchers from the Air Force for the task, of which four carried astronauts into orbit. NASA successfully completed the first orbital Mercury flight ten months after Gagarin first circled the globe. On February 20, 1962, astronaut John Glenn orbited Earth three times in *Friendship 7*, landing in the Atlantic Ocean southeast of Bermuda (Figure 1–6).

The Titan II became the second and last modified ICBM to be used for launching humans to orbit; it was employed in launching all ten spacecraft in the two-astronaut Gemini program. The extra payload capacity of the Titan II compared to the Atlas made it possible to launch a heavier capsule, large enough to accommodate two individuals. Gemini was designed to develop the astronauts' skills in orbital rendezvous and docking as a precursor to the Apollo lunar program (Figure 1–7). It was also used to extend NASA's experience with spaceflight to a duration long enough to reach the Moon and return and to test extravehicular activities. [I-18]

Figure 1–5. The launch of the first American into space, astronaut Alan Shepard, on the Mercury-Redstone 3 space vehicle from the Cape Canaveral launch site on May 5, 1961. (NASA photo 61-MR3-72A)

Figure 1–6. The launch of the first American to orbit the Earth, astronaut John Glenn, on the Mercury-Atlas 6 space vehicle from the Cape Canaveral launch site on February 22, 1962. The Atlas was the first ICBM developed by the Air Force in the 1950s. (NASA photo 62-MA6-111)

During the early 1960s, as military and national security payloads quickly grew in weight, it became clear that the Air Force would need a booster larger than the Titan II to lift its planned payloads to orbit. Hence, it modified the Titan II by adding an additional stage and solid "strap-on" booster rockets and designated the new rocket Titan III. The first Titan IIIA, carrying a third "Transtage," successfully flew on September 1, 1964. Shortly thereafter, the Air Force used an Agena upper stage to create the Titan IIIB, capable of carrying 3,300 kilograms into low-Earth orbit. Because still greater lift was needed to launch the Air Force's largest satellites, the Air Force added segmented solid-fuel rockets to create the Titan IIIC. It employed two strap-on boosters made up of five ten-foot-diameter segments that extended eighty-six feet in height. The boosters were developed and manufactured by United Technology Center, using techniques it developed in the late 1950s.[57] The first test flight took place in June 1965. The Titan IIIC was capable of lifting 13,100 kilograms into low-Earth orbit. For even more massive loads, the Air Force contracted with Martin Marietta to build the Titan IIID and Titan IIIE, both of which used the solid rocket boosters from the Titan IIIC but had more powerful upper stages. The rocket combination with the greatest lift capacity was the IIIE, which employed a cryogenic upper stage called the Centaur, first designed for use on an Atlas rocket.[58] In the 1970s, NASA used the Titan IIIE with a Centaur upper stage to launch the two Mars Viking landers (Figure 1–8). The two successful flights are particularly notable for occurring with-

Figure 1–7. The launch of the first piloted mission of the Gemini program, Gemini 3, atop the sturdy and reliable Titan launch vehicle. The Titan was originally developed as part of the Air Force's ICBM program in the late 1950s. This launch took place on March 23, 1965, with astronauts Gus Grissom and John Young aboard. (NASA photo 65-H-448)

Figure 1–8. The launch of the Viking space probe to Mars in 1974 atop the Titan III launch system. Significantly modified, and thrust-enhanced over time, the Titan family of launchers has enjoyed enormous success as a vehicle that can place in orbit, and en route to other planets, a variety of spacecraft engaged in all manner of applications. (NASA photo)

---

57. Winter, *Rockets Into Space*, p. 92.
58. The term "cryogenic" refers to the low temperatures required to create and store liquid oxygen and liquid hydrogen.

in only three weeks of one another. Viking 1 was launched on August 20, 1975, followed on September 9 by Viking 2.[59] They were launched from the Air Force-maintained Titan launch pads at Cape Canaveral, Florida.

Convair, with funding from the Advanced Research Projects Agency (ARPA) and the Air Force, developed the Centaur upper stage, which was successfully test-flown on an Atlas rocket on November 27, 1963. The Centaur, which is still in use, employs two Pratt & Whitney RL-10 engines, and it was the first liquid oxygen-liquid hydrogen (LOX-hydrogen) engine to demonstrate the capability to restart in space.[60] The Atlas-Centaur rocket has launched spacecraft to Mercury, Venus, Mars, Jupiter, and Saturn, as well as many communications satellites to geosynchronous orbit. The development of the Centaur provided the Air Force and NASA with significant experience with the problems encountered in using liquid hydrogen for propulsion, which assisted in the later development of the cryogenic engines used in the Saturn space booster program.

Nuclear Propulsion

One of the more interesting aspects of rocket development was the partnership between NASA and the Atomic Energy Commission (AEC) in developing nuclear rocket engines. Seen strictly from the standpoint of available power for rocket thrust, a nuclear rocket generating heat from fission is much more efficient than chemical propulsion, allowing much higher thrust. Nuclear rockets have been of particular interest for interplanetary spaceflight, because they could markedly shorten trips to the planets. However, they also present formidable engineering and safety challenges. The notion of using atomic energy as a fuel source was briefly explored by Konstantin E. Tsiolkovskiy, Robert H. Goddard, and others, but these early theoreticians and experimenters were daunted by the problem of controlling the enormous potential for explosive, rather than controlled, releases of energy. It was not until after controlled nuclear fission had been achieved in 1942 and after World War II that the technology began to receive serious attention in rocketry.

North American Aviation completed the first detailed (classified) study of the issue in 1947. [I-19] It concluded that a nuclear rocket would be feasible if some serious technical hurdles could be overcome. Beginning in the late 1940s, the AEC also experimented with the use of nuclear power in aircraft, which generally contributed to the government's technical expertise in nuclear propulsion. Robert W. Bussard, who worked on the nuclear aircraft program at the Oak Ridge National Laboratory, became interested in the challenge of nuclear rocketry and published an important report in 1953 that influenced the Air Force in its decision to start up a nuclear rocket program.[61] His work convinced officials that nuclear rockets might be feasible alternatives to chemical propulsion for ballistic missiles.[62]

Both the Los Alamos and Lawrence Livermore National Laboratories established small programs to investigate nuclear propulsion technologies in detail. In November

---

59. Edward C. Ezell and Linda N. Ezell, *On Mars: Exploration of the Red Planet 1958–1978* (Washington, DC: NASA SP-4212, 1984), pp. 325–26.

60. This test took place on October 26, 1966. The capability to start in the near vacuum of space was extremely important to the success of the Apollo program. See John L. Sloop, "Technological Innovation for Success: Liquid Hydrogen Propulsion," in Frederick C. Durant, ed., *Between Sputnik and Shuttle: New Perspectives on American Astronautics* (Washington, DC: American Astronautical Society, 1985), pp. 225–39.

61. R. W. Bussard, "Nuclear Energy for Rocket Propulsion," Oak Ridge National Laboratory, ORNL CF-53-6-6, July 2, 1953. This was published in *Reactor Science and Technology*, December 1953, pp. 79–170. This secret publication was declassified on November 4, 1960.

62. A historical summary of the early research in nuclear rockets appears in Robert W. Bussard, "Nuclear Rocketry—The First Bright Hope," *Astronautics* (December 1962): 32–35.

1955, the Air Force and the AEC formally started Project Rover, with the goal of harnessing the enormous power of nuclear fission for spaceflight. Some Air Force officials felt that nuclear power would be of use in powering ICBMs. Livermore was directed to focus on nuclear ramjets under Project Pluto, leaving Los Alamos to develop a nuclear reactor for a rocket engine. In 1957, program officials had chosen the area at the Nevada Test Site called Jackass Flats to conduct engine tests. Los Alamos developed the Kiwi experimental nuclear reactor, testing several versions at Jackass Flats between 1959 and 1964.[63] These tests demonstrated the use of carbide coatings to prevent hydrogen erosion of the graphite and established numerous crucial details about reactor design and control. The testing of the first version, KIWI-A, established the technical feasibility of creating a nuclear rocket. Nevertheless, it soon became clear that solid rocket propulsion was of much greater use for ballistic missiles than nuclear engines. Among other things, nuclear bomb engineers had managed to create nuclear warheads of much reduced mass, thereby relaxing the lift requirements for missiles.

Soon after the creation of NASA, the Eisenhower administration transferred the Air Force's responsibility for nuclear rocketry to NASA. NASA and the AEC created the NASA-AEC Space Nuclear Propulsion Office in August 1960. During his May 25, 1961, speech titled "Urgent National Needs," President Kennedy urged a speed-up of the Rover nuclear rocket program, proposing a threefold increase in funding.[64]

Soon after, the Space Nuclear Propulsion Office began the Nuclear Engine for Rocket Vehicle Application (NERVA) program, with the eventual goal of flight-testing the NERVA engine on a Saturn rocket. Aerojet-General and the Westinghouse Electric Corporation were awarded a contract to develop the NERVA engine, which was to be derived from the KIWI-B test engine then undergoing tests at Jackass Flats. In a program called Reactor-in-Flight Tests (RIFT), NASA planned to use a flight-rated version of the NERVA engine to power the third stage of a Saturn V.[65] A few NASA officials contemplated that it might serve as a second or third stage on the even larger Nova vehicle for which some NASA engineers had been arguing. In the spring of 1962, NASA selected Lockheed Missiles and Space Company as the prime contractor for the nuclear stage. As planned, the RIFT test vehicle was to consist of Saturn IC and Saturn II stages, topped by the Saturn N nuclear stage. In its lunar flight configuration, it would launch a crewed spacecraft and lunar lander. After the first two stages carried the spacecraft beyond Earth's atmosphere, the nuclear engine would be started to carry the crew to the Moon.

By the end of 1963, the nuclear rocket effort was already in decline as NASA focused on making the Apollo program a success using more conventional rocket engines. Budget reductions forced NASA and the AEC to terminate the RIFT project. They converted the NERVA project to a technology effort using ground tests of nuclear engines and components. Between May 1964 and March 1969, the NERVA project tested thirteen reactors, essentially completing the technology phase. The KIWI series was followed by a 5,000-megawatt reactor named Phoebus, designed to achieve higher temperatures and longer operating times at lower specific weights. NASA planned to use a flight-rated version of Phoebus for space travel. [I-20]

---

63. The name "kiwi" for the reactor derives from the name of the flightless bird native to New Zealand.

64. In this speech, Kennedy announced that "an additional $23 million, together with $7 million already available, will accelerate development of the ROVER nuclear rocket. This gives promise of some day providing a means for even more exciting and ambitious exploration of space, perhaps beyond the moon, perhaps to the very ends of the solar system itself." See Document III-12 in Logsdon, gen. ed., *Exploring the Unknown*, 1: 453–54.

65. W. Scott Fellows, "RIFT," *Astronautics* (December 1963): 38–47.

In 1968, the project initiated work on a 75,000-pound thrust flight-rated engine having a specific impulse of 850 seconds, but the program was nearing its end.[66] As work on the proposed Space Shuttle increased, program officials even proposed that the Shuttle would transport a NERVA engine into orbit for testing. Yet that effort fell on deaf ears, in part because the Nixon administration and Congress continued to decrease NASA's budget, reducing the need for propulsion to support interplanetary travel, but also because of mounting opposition to nuclear power. In 1972, Project Rover was terminated. [I-21, I-22]

The nuclear rocket program had been quite ambitious, and it showed the technical feasibility of nuclear propulsion. As Dr. Glenn T. Seaborg, chairman of the AEC, stated in 1970:

> *Lest you get the impression that the development of such a nuclear rocket is as simple as its principle sounds, let me point out what is involved in it. What we must do is build a flyable reactor, little larger than an office desk, that will produce the 1500 megawatt power level of Hoover Dam and achieve this power in a matter of minutes from a cold start. During every minute of its operation, high-speed pumps must force nearly three tons of hydrogen, which has been stored in liquid form at 420°F below zero, past the reactor's white-hot fuel elements which reach a temperature of 4,000°F. And this entire system must be capable of operating for hours and of being turned off and restarted with great reliability.[67]*

Although the nuclear program had been relatively successful from a technical standpoint, and it still had many proponents within NASA and the AEC, it could not survive the funding competition with programs that carried less technical risk, especially given the diminished prospects for interplanetary flight involving large payloads. Nuclear propulsion interested mission planners once again in the late 1980s and early 1990s after President George Bush announced on July 20, 1989, a plan to send humans to Mars and back by 2019. However, that effort, which became known as the Space Exploration Initiative, was very short lived. Congressional proponents of NASA's other programs became worried that such a public effort, requiring many billions of dollars of investment, would use up funds planned for other NASA programs, including the long-planned International Space Station.

Saturn and the Race to the Moon

Meeting President Kennedy's 1961 challenge to put people on the Moon before 1970 required much larger launch vehicles; in many ways, the race to the Moon was a rocket-building competition. Because planning for such large vehicles had been initiated by the von Braun team and others even before NASA was officially opened in 1958, the space agency was able to respond quickly. [I-23] Among other things, NASA sped up work on technologies that led to the Saturn I and to the huge Saturn V, which in its final form was capable of lifting 260,000 pounds to low-Earth orbit. Although the roots of the design of the Saturn V ultimately trace back to the V-2, the Saturn evolved along a different developmental path from the Redstone, Titan, Thor-Delta, and other launchers that were originally designed as missiles to carry nuclear warheads efficiently. The Saturn family was the first designed as pure space boosters.

Well before Kennedy's speech to Congress, von Braun's team at the ABMA had begun to consider building a large multi-stage rocket capable of launching large objects into space. [I-24] Von Braun and many of his engineering team had the Moon and Mars as

---

66. Compare, for example, the Space Shuttle main engine's superior specific impulse of 450 seconds.

67. Glenn T. Seaborg, "A Nuclear Space Odyssey." Remarks to the Commonwealth Club of California, San Francisco, CA, July, 24, 1970. U.S. Atomic Energy Commission release number S-27-70.

their ultimate goals, but they also had in mind an orbiting space station.[68] U.S. officials had been astonished by the lift capacity of the initial Soviet rocket. Although U.S. intelligence had known that the Soviets were building rockets based on the V-2, the United States was unprepared for the scope of this effort. The Eisenhower administration decided that the United States might need a much larger U.S. vehicle than was available—one capable of launching large military payloads and perhaps humans and the gear to support them. Building and testing a successful high-power engine were the most difficult of the many tasks planners faced in developing such a vehicle. Hence, engineers began to tackle the difficult problem of providing the propulsion to propel a large payload into space. [I-25]

By late 1957, they had settled on a launch design that would employ a first stage propelled by a cluster of eight powerful engines based on the S-3D engine from the Jupiter IRBM. In August 15, 1958, the newly created ARPA, which was organizing the U.S. military space effort, issued orders to begin work on a new large launcher. [I-26] Increasing the S-3D's thrust by 14 percent made it possible to achieve 1.5 million pounds of thrust in the cluster of eight engines. The engine was named the H-1; the launcher was tentatively named Juno V. Team members adopted a clustered approach out of necessity because building a brand new, high-thrust engine would have been too expensive.[69] ARPA officials were forcing the von Braun team to live on low budgets and encouraging it to use off-the-shelf hardware wherever possible. As a result, the Juno V's designers became quite inventive.[70] Although engine clusters raise many technological challenges, by meeting them at this early stage, the team was able to provide a firm base for the development of later engine clusters.[71] ARPA conceived of the Juno V as a static test vehicle, but von Braun's team had clearly intended that it serve as the basis for a new launcher, which von Braun and his associates called "Saturn."[72] Shortly after ARPA gave the ABMA the green light to proceed with the Saturn, NASA came into being officially, and the issue of transferring the ABMA to NASA began to be discussed in earnest. [I-27] On November 2, 1959, President Eisenhower approved that transfer.

By late December 1959, NASA and DOD had already made many of the initial technology decisions that would lead first to the Saturn IB launch vehicle and then to the huge Saturn V (Figure 1-9). [I-28] NASA, working with DOD, organized a Saturn Vehicle Team, chaired by Abe Silverstein of NASA. The Silverstein Committee made three important technological choices that set the stage for later Saturn developments. They decided to (1) use liquid hydrogen (LH$_2$) as the fuel for the upper stages of the Saturn booster, (2) develop a series of multi-stage rockets, and (3) follow an evolutionary path for growth in which each succeeding vehicle used the proven stages of the preceding one. The Silverstein Committee saw three primary functions for the Saturn family: (1) lunar and deep-space missions with an escape payload of 9,900 pounds; (2) geostationary orbit payloads of 4,950 pounds; and (3) missions carrying humans into low-Earth orbit) in the Dyna-Soar program, an Air Force human spaceflight effort. These choices, while they introduced

---

68. Wernher von Braun, "Crossing the Last Frontier," *Collier's*, March 22, 1952, pp. 23–29, 72–73. This was reprinted as part of Document I-13 in Logsdon, gen. ed., *Exploring the Unknown*, 1: 179–88.

69. As it was, the ABMA team ran into difficulties uprating the S-3D engine from 165,000 pounds of thrust to 188,000 pounds because the more powerful engine developed a combustion instability that threatened to destroy the engine. This led to a costly redesign.

70. "The dire need made us more inventive, and we bundled the containers to be loaded with propellants." As quoted by William A. "Willy" Mrazek in Roger E. Bilstein, *Stages to Saturn: A Technological History of the Apollo/Saturn Launch Vehicles*, rep. ed. (Washington, DC: NASA SP-4206, 1996), p. 30.

71. ABMA engineers also clustered the propellant tanks by using eight Redstone tanks, which alternately held RP fuel (a form of kerosene) and LOX, surrounding a single large Jupiter tank in the center that carried RP fuel. *Ibid.*, p. 82.

72. In writing about it a few years later, von Braun noted that the "Juno V was, in fact, an infant Saturn." See von Braun, "The Redstone, Jupiter, and Juno," p. 120.

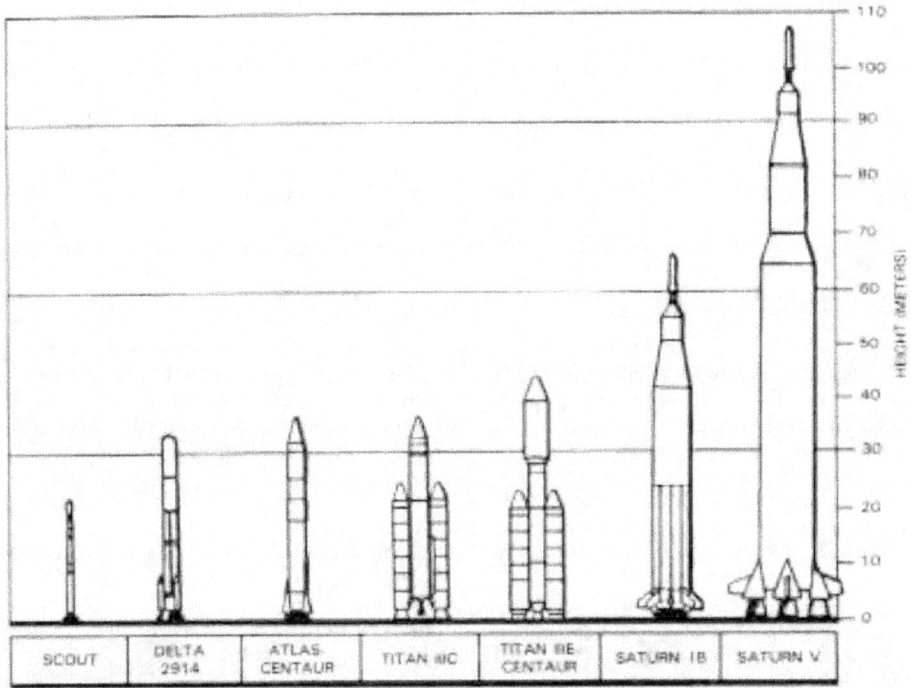

Figure 1-9. Expendable launch vehicles (1974).

some serious technical hurdles, were the backbone of the Saturn's ultimate success as a launch vehicle. [1-29]

The decision to use high-energy LH_2 as a fuel was the most controversial of the three. It was also the crucial one in allowing the program to develop efficient boosters. Just after the turn of the century, Tsiolkovskiy and Goddard had determined that using liquid hydrogen as a fuel in a liquid oxygen environment would provide superior specific impulse.[73] In 1923, Hermann Oberth even suggested that the LOX-hydrogen combination would be especially appropriate for the upper stages of rockets.[74] Yet liquid hydrogen, which requires cooling to −423 degrees Fahrenheit, is hard to handle and causes the imbrittlement of many metals. Nevertheless, with von Braun in concurrence, Silverstein was able to convince the other committee members to accept LH_2 as a fuel, despite its handling problems. As Sloop has noted: "It was a very bold and crucial decision to stake the success of the entire manned space program on a relatively new high-energy fuel, but subsequent developments proved it to be a sound decision and a key one in the success of the Saturn V and the Apollo missions."[75]

---

73. Konstantin Tsiolkovskiy, in A.A. Blagonravov, ed., *Collective Works of K.E. Tsiolkovskiy, Volume 2, Reactive Flying Machines* (Washington, DC: NASA Technical Translation (TT) F-237, 1965), pp., 78–79; Robert H. Goddard, in Esther Goddard and G. Edward Pendray, eds., *The Papers of Robert H. Goddard*, three volumes (New York: McGraw-Hill, 1970).

74. Hermann Oberth, *Rockets in Planetary Space* (Washington, DC NASA TT F-9227, 1965).

75. Sloop, "Technological Innovation for Success."

As explained above, the Centaur upper stage used two hydrogen-fueled Pratt & Whitney RL-10 engines that developed 15,000 pounds of thrust apiece. By clustering these engines in a group of six, NASA planned to build a powerful second stage for the Saturn I, called the S-IV. It was to be the first major Saturn stage to be built under contract by industry, rather than developed within the ABMA.

The decision to let the S-IV contract to the Douglas Aircraft Company illustrates the importance of subjective factors in NASA's choice of contractors. Two companies— Convair and Douglas—placed well above the other nine that submitted proposals. In choosing between them, NASA officials considered not only technical competence, but also their judgment of the firms' ability to manage a large, complex contract and the firms' business acumen. Convair, which was developing the Centaur upper stage, placed slightly higher on technical competence, but lower in the latter two categories. NASA Administrator T. Keith Glennan felt that Douglas's proposal was more imaginative. He was also concerned that giving the S-IV contract to Convair would inadvertently create a monopoly in the development of cryogenic upper stages.[76] NASA officials were well aware of the need to develop a broad, competitive contractor base from which to choose, especially in building systems that required the development of new, untried technologies. As a result, NASA announced the choice of Douglas on May 26, 1960. The closeness of the decision, and the subjective reasons for the selection of Douglas, caused some concern within Congress, which directed the General Accounting Office (GAO) to investigate. The GAO report, however, generally sustained NASA's decision.[77]

Modern launch systems consist of hundreds of interacting systems, each of which is itself composed of thousands of smaller subsystems and parts. Designing and successfully launching moderate and small-sized launchers pose a major challenge. For systems the size and complexity of the Saturn I, its descendent the Saturn IB, and the Saturn V Moon rocket, the task seemed daunting. [I-30] Building the Saturn vehicles forced NASA and the aerospace industry to solve numerous practical problems, including the handling of large structures, flawless welding, and the testing and tracking of millions of components. It also required the development of new manufacturing methods. For example, Douglas Aircraft and NASA had to overcome a panoply of obstacles to manufacture the S-IV to a standard sufficient to carry people reliably and safely to space. To build a rocket stage of requisite size and strength, the designers decided to carry the two propellants in only two tanks, one above the other, and to give them a common bulkhead. The size of the S-IV and the decision to use large propellant tanks brought their own production problems. The tanks' welded seams needed to be flawless. New machinery needed to be developed to handle the large tanks. New fabrication methods had to be invented to create the common bulkhead. In addition, Douglas also had to build special facilities to handle components the size of the tanks. Historian Roger E. Bilstein has commented that the development of Saturn hardware "frequently came down to a question of cut-and-try."[78] This approach, of course, made it extremely difficult to estimate the developmental cost of any of the launchers.

The Saturn I was a research and development project designed to gather data and experience with large launch vehicles. NASA made the first flight to test its first stage on

76. James Webb, who became the second NASA administrator, noted in 1963 hearings before the House of Representatives that "one of the principal factors cited in the selection of the Douglas Aircraft Company was that the addition of the company would broaden the industrial base in the hydrogen technology field." NASA Authorization Hearings, U.S. House of Representatives, 87th Cong., 2d sess., Part 2, 1963, p. 825.

77. Controller General of the United States to Overton Brooks, Chairman, Committee on Science and Astronautics, June 22, 1960.

78. Roger Bilstein, "The Saturn Launch Vehicle Family," in *Apollo: Ten Years Since Tranquility Base* (Washington, DC: Smithsonian Institution Press, 1979), p. 117.

October 27, 1961, carrying only a dummy S-IV stage. The first flight of an operating S-IV second stage was made on January 29, 1964. On July 30, 1965, the Saturn I made its last flight, having prepared the way for the more powerful Saturn IB. During its ten flights, the Saturn I had been used in a variety of engineering experiments in low-Earth orbit and had given NASA's engineers valuable insights into the complexities of building and launching a large cryogenic rocket.

The launch requirements considered by the Silverstein Committee demanded an even larger propulsion stage than the S-IV, and instead of uprating the RL-10, or somehow adding more of them to the cluster, the Silverstein Committee began to look toward a much larger, more powerful single engine that would generate 200,000 pounds of thrust. On June 1, 1960, a source evaluation board chose the Rocketdyne Division of North American Aviation to build a high-thrust cryogenic rocket engine called the J-2. Marshall Space Flight Center developed the concept and monitored the contractor's work, while Rocketdyne attempted to bend metal around Marshall's ideas.

From the beginning, the Saturn IB rocket was designed to carry humans. Hence, the final engine contract, which was awarded to Rocketdyne in September 1960, contained the important phrase "to insure maximum safety for manned flight."[79] In other words, although reliability had been an important ingredient of earlier designs, for the first time, a contract specified that a rocket engine was to be designed with human safety as part of the initial specifications.[80] Because the Saturn IB was intended to carry humans, each stage of the design and manufacturing process was closely scrutinized for high reliability and each part tested individually as well as in concert with other parts. Rocketdyne engineers faced serious problems finding appropriate metals and other materials that would work properly in a liquid hydrogen environment. They also had to trace down every leak in great detail, for a small amount of gaseous hydrogen in the wrong place could lead to a devastating explosion. After pursuing a number of intermediate short-duration tests for approximately nine previous months, Rocketdyne successfully ran the first model of the J-2 in a 250-second test on October 4, 1962.[81]

When this contract was let in 1960, NASA had not yet decided which vehicle would use the powerful engine. Outside of NASA, there was relatively little interest in pursuing a program that would require the lift capacity of an upper stage that used the J-2 rocket. However, President Kennedy's May 1961 decision to "shoot for the Moon" dramatically changed the situation. By July 1962, NASA settled on proceeding with the uprated S-IV, called the S-IVB, which it planned to use as the second stage of the Saturn IB; the stage would be powered by a single J-2 engine. The Saturn IB would loft an Apollo spacecraft to low-Earth orbit as part of the sequence of tests that would lead to a landing on the Moon. This powerful launcher, capable of placing 41,000 pounds into an orbit 110 miles above Earth, had an important role in the execution of the Apollo program. Not only did it carry the first Apollo spacecraft into orbit during the test phases of the Apollo program of the mid-1960s; it also served to ferry astronauts to Skylab in the 1970s and was the launch vehicle used in the Apollo-Soyuz mission of 1975.

The Saturn IB made its first flight two years after the first flight of the Saturn I, on February 26, 1966, using the S-IVB second stage. On October 11, 1968, it carried the first Apollo capsule containing astronauts into orbit for a ten-day, twenty-hour flight—

79. Bilstein, *Stages to Saturn*, p. 141.

80. One of the best ways to ensure safety to astronauts and their launch crew is to build a highly reliable vehicle. However, even a vehicle of high reliability may not have adequate safety margins for human flight if it fails catastrophically and does not provide some means to protect its passengers. Conversely, a vehicle meeting a lower reliability rating could, in principle, be safer for humans if it incorporated sufficient means to ensure the crew's ability to survive a failure.

81. Bilstein, *Stages to Saturn*, p. 143.

Apollo 7 (Figure 1–10). That was also the last Apollo flight for the Saturn IB. It was followed two months later by the first Saturn V to carry astronauts, when the Apollo 8 mission launched astronauts Frank Borman, James A. Lovell, Jr., and William A. Anders into space for the first human flight around the Moon.

The Saturn V dwarfed even the powerful Saturn I and Saturn IB launchers. Standing 363 feet high and weighing 500,000 pounds unfueled, the Saturn V was capable of launching more than 200,000 pounds to low-Earth orbit. It was built to place three astronauts on the Moon and allow them to take sufficient fuel and equipment to return to Earth. The Saturn V had three stages: the S-IC first stage (powered by five F-1 LOX-kerosene engines), the S-II second stage (propelled by five J-2 LOX-LH$_2$ engines, and the S-IVB third stage (with one J-2 engine).

Rocketdyne, which, as noted above, NASA later chose to build the J-2 engine, had in January 1959 received the contract to build the F-1.[82] The contract for the giant power plant, which would employ RP kerosene fuel and LOX, stipulated that it should develop 1.5 million pounds of thrust, nearly four times the thrust of the Navaho missile engine from which it was derived.[83] Rocketdyne's experience in the Navaho program made it the logical candidate for the task. In awarding this contract, NASA was betting that it could bypass the more common evolutionary approach to engine development and make a revolutionary jump to this enormous engine.

Figure 1–10. The launch stack of the Apollo 7 mission sits on Launch Complex 34 at Kennedy Space Center, on September 16, 1968. The launch vehicle, the Saturn IB, would power the crew of astronauts Wally Schirra, Donn Eisele, and Walter Cunningham into Earth orbit on October 11, 1968, for a checkout of the Apollo command and service modules in preparation for flights to the Moon. (NASA photo 68-H-920)

At the time of this decision, the United States had no program to attempt a Moon landing. NASA did not make a final decision about the configuration of the first stage of the Saturn Moon rocket until January 10, 1962, after it had already chosen the Boeing Aircraft Company to build it. That choice was based on the recognition that a booster with five F-1 engines in its first stage might be able to accomplish the lunar landing mission in a single launch, if NASA were to adopt the controversial lunar orbital rendezvous approach to a lunar landing. With this decision, the large booster was named the Saturn V.[84] [I-31, I-32]

---

82. The F-1 design had its origins in earlier studies at the ABMA. See David E. Aldrich, "The F-1 Engine," *Astronautics* (February 1962): 40. Also see Document I-24 at the end of this chapter.

83. The Navaho engine developed 415,000 pounds of thrust. The original Air Force goal had been an engine of 1 million pounds of thrust, but the company was not able to reach that goal until March 1959, when it fired up a "boilerplate" thrust chamber and injector that achieved the goal. By then, the program had been transferred to NASA. Bilstein, *Stages to Saturn*, p. 106.

84. NASA chose Boeing in December 1961.

As planning for the lunar landing mission had proceeded at an intense pace following Kennedy's May 1961 speech, there had been serious consideration of the need for a
booster even larger than that which became the Saturn V; that booster, named Nova,
would use eight F-1 engines in its first stage. [I-33] The difficulty of a rapid evolution to
such a gigantic rocket was one of the reasons that mission planners, in the second half of
1961, began to converge on some kind of rendezvous approach to accomplishing the
lunar mission; adopting such an approach would allow the use of one or several Saturn
rockets. A rocket the size of Nova—and all of the very large ground facilities required to
launch it—would not be needed. Although studies of the Nova continued until 1963, the
development of that vehicle was never initiated.

Although the difficulties of working with LOX-LH$_2$ on the J-2 engine created novel
complications for rocket engine designers, the sheer size and thrust of the LOX-RP F-1
engine also presented formidable challenges. David E. Aldrich, the manager of Marshall
Space Flight Center's Engine Program Office, acknowledged that "the development of the
F-1 engine, while attempting to stay within the state of the art, did, by size alone, require
major facilities, test equipment, and other accomplishments which had not been attempted prior to F-1 development."[85] The F-1 was a gimbaled engine, whose bell-shaped expansion nozzle was regeneratively cooled by liquid oxygen. Although nearly every subsystem
brought its own technological hurdles and special challenges, the F-1 injector, which controlled the flow and pattern of both fuel and oxygen into the thrust chamber, turned out
to be the stiffest challenge of all. The injector forced fuel through 3,700 orifices into the
combustion chamber to meet oxygen that entered from 2,600 additional openings. The
injector had to endure greater heat and pressure than in any previous engine.
Unfortunately, it proved impossible simply to scale-up previous designs to the required
size. Initial tests with early models of the F-1 injector led to unacceptable combustion
instability that could not be stopped short of cutting off the flow of fuel. [I-34] New
designs, based on tests with scale models and the use of high-speed photography in a specially designed test chamber, looked promising, but when scaled up to F-1 size and tested,
they also failed. On June 28, 1962, one of these tests resulted in the loss of an F-1 test
engine. By early 1963, NASA Associate Administrator Robert C. Seamans, Jr., was quite
concerned; he told the head of the Apollo program, Brainerd Holmes, that "as you and
Wernher [von Braun] know, I feel that F-1 instability may seriously delay the MLLP
[manned lunar landing program] and consequently is the technological pacing item."[86]

Eventually, after empirical "cut-and-try" redesign and exhaustive testing, coupled with
an intensive theoretical attack on injector combustion instability, Rocketdyne and NASA
engineers developed an injector that would pass muster. [I-35, I-36, I-37] Rocketdyne,
working closely with NASA and university researchers, came up with a flight-rated model
by January 1965. Still, despite the satisfaction of having developed a working engine, the
Rocketdyne engineers noted that "the causes of such instability are still not completely
understood."[87]

The decision to build a cryogenic second stage for the launch vehicle that became the
Saturn V was also rooted in the Silverstein Committee report of 1959. Committee members knew that an extremely powerful second stage would be needed to launch humans
to the Moon. Soon after the committee issued its report, NASA designers had begun to
define the general outlines of the S-II stage that would eventually become the second

85. Aldrich, "The F-1 Engine," p. 40.

86. Memorandum from R.C. Seamans, Jr., to Brainerd Holmes, January 23, 1963

87. William Brennan, "Milestones in Cryogenic Liquid Propellant Rocket Engines," AIAA Paper 67-978,
October 1967, p. 9. Quoted in Bilstein, *Stages to Saturn*, p. 116.

stage of the Saturn V. Before he left NASA in January 1961, Administrator T. Keith Glennan wrote: "The Saturn program is left in mid-stream if the S-II stage is not developed and phased in as the second stage of the C-2 launch vehicle."[88] (The C-2 was a proposed Saturn version using two F-1 engines in its first stage.)

Soon after James Webb was sworn in as the new NASA Administrator, Marshall Space Flight Center started contractor selection. The S-II, at that point to be propelled by four J-2 engines, was to be the largest rocket project to be given to U.S. industry. Although thirty interested contractors attended the first meeting, only seven submitted bids, of which four survived source evaluation board scrutiny.[89] Those remaining—Aerojet, Convair, Douglas, and North American Aviation—faced the difficult task of attempting to bid on a contract that was still largely undefined. NASA had not yet decided the size of the S-II stage, nor had it settled on any of the myriad other specifications of the project. Yet NASA was under considerable pressure to get the design process under way, and it needed to choose a contractor as soon as possible.[90] NASA was attempting to build a team, and it needed to select a contractor that would manage the construction well. Wilbur Davis, of Marshall's Procurement and Contacts Office, stated, "I wish to emphasize at this point that the important product that NASA will buy in this procurement is the efficient management of a stage system."[91]

NASA chose North American Aviation for the job on September 11, 1963—a decision that raised some eyebrows outside of NASA. North American already had the contract for building the Apollo capsule and was not considered to be a major player in the launch vehicle area. However, North American had built the highly successful X-15 rocket plane, and because NASA emphasized the importance of a strong management team, it selected North American largely on that basis.

The relationships between NASA and North American on the S-II contract provide important insights into the development of NASA as an institution. Although NASA was determined here, as in other contracts, to use the intellectual capacity and manufacturing experience of American industry, because NASA retained its own cadre of engineers and other specialists, it often found itself second-guessing North American. In an area in which both NASA and the contractor were "pushing the envelope" of the state of the art, misunderstandings and disagreements over the best way to proceed inevitably arose, which led to tensions between individuals and sometimes whole departments in the two organizations. In addition, North American's approach was one of an aircraft company, used to building high-performance flying machines. By contrast, Marshall Space Flight Center, NASA's lead center for the Saturn V, had little experience in aircraft procedures, but much experience in building hefty rockets that looked more like boilers than aircraft. Marshall, for example, had designed and built the first stage of the Saturn I. Added to these tensions was the fact that the U.S. Navy had oversight over the construction of new government facilities for building the S-II. The coordination among the three entities was not always smooth.

---

88. T. Keith Glennan, "Memorandum for the Administrator," January 19, 1961.

89. The seven proposals came from Aerojet General Corporation, Chrysler Corporation, Convair, Douglas Aircraft Corporation, Lockheed Aircraft Corporation, Martin Company, and North American Aviation.

90. The Marshall project team felt considerable pressure to move as quickly as possible on the design of this crucial stage. "You can see that we have a whole lot of doubt in what we say here, and there are a lot of conflicting problems. We are presently trying to resolve them. We could have asked you not to come here today and could have taken, say, six weeks time to resolve these problems internally, in which case we would have lost six weeks on the S-II contract." Marshall spokesperson, Marshall "Minutes of the Phase II Pre-Proposal Conference for Stage S-II Procurement on June 21, 1961," Johnson Space Center files.

91. Quoted in *ibid*.

In many respects, the final S-II stage resembled an aircraft component more than a rocket. It was much more efficient than the first stage, not only because it carried more efficient rocket engines, but also because it was lighter for a given strength factor. During the design and manufacturing process in the early 1960s, NASA continually asked North American to shed pounds on the S-II, as the Apollo payload relentlessly gained weight. To add one additional kilogram of payload delivered to orbit, NASA had to cut fourteen kilograms from the S-IC first stage, nearly five kilograms from the S-II second stage, or only one kilogram from the S-IVB. Thus, it would have been most effective to remove weight from the third, the S-IVB stage. However, the S-IVB was already in production at McDonnell Aircraft by the time NASA began to experience the most severe weight problems. Taking fourteen kilograms from the S-IC for each kilogram of overweight in the Apollo stage was also not very feasible. The burden fell on the S-II stage—and on the NASA-North American Aviation team.

The need to find innovative ways in which to shave weight from the stage, and the requirements for innovative manufacturing processes, took their toll on both Marshall and North American Aviation. By September 29, 1965, when an S-II test article failed catastrophically during a test designed to simulate forces that the S-II would experience at the end of the first-stage burn, concern within NASA over North American's management reached major proportions. The Apollo program's manager, General Samuel C. Phillips of NASA Headquarters, was appointed head of a so-called "Tiger Team" to investigate the problems. The Tiger Team effort served two purposes—it helped NASA investigate the problems and recommend solutions, and it put North American on notice that NASA considered the perceived problems extremely serious. The resulting report, sent to J. Leland "Lee" Atwood, president of North American Aviation, on December 19, 1965, was extremely critical of North American's management of the S-II and also on its handling of spacecraft development.[92] It was a wake-up call to North American. By this time, problems with the S-II threatened to hold up the first launch of the Saturn V.

Stung by the criticism, the company responded to this crisis by reorganizing its management team and rethinking how it organized the work on the S-II. Among other things, it brought in new top managers and improved the sharing of information on the progress and problems experienced by the company's engineering teams. Still, despite making significant progress,[93] North American continued to experience problems. On May 28, 1966, a second S-II stage (S-II-T) failed during a pressure check of the LH_2 tank. The loss once again indicated poor management control. The tank exploded as technicians filled it with helium during a test for leaks. Unfortunately, they were unaware that other technicians had previously disconnected the pressure sensors and relief switches that would have prevented an explosion. The accident injured five individuals. To add to the problem, the team investigating the S-II-T failure found tiny cracks and other problems in the test article. Nevertheless, North American continued to make progress, and by January 9, 1967, Phillips could report that the company had markedly improved its management and test procedures.[94]

Unfortunately, on January 27, 1967, a deadly fire broke out in the Apollo command module during a test with crew aboard, killing three astronauts. Because North American

92. This report appears in Document III-17 in John M. Logsdon, gen. ed., with Dwayne A. Day and Roger D. Launius, *Exploring the Unknown: Selected Documents in the History of the U.S. Civil Space Program, Volume II: External Relationships* (Washington, DC: NASA SP-4407, 1996), 2: 527–35.

93. George Mueller, NASA's Associate Administrator for Manned Space Flight, even commented to Lee Atwood, "Your recent efforts to improve the stage schedule position have been most gratifying and I am confident that there will be continuing improvement." George E. Mueller to Lee Atwood, February 23, 1966.

94. Samuel Phillips to Associate Administrator, "S-II-T Failure Corrective Action," January 9, 1967.

also had responsibility for building the command module, NASA asked for and got a further reorganization of the company's top management. Harrison Storms, president of the Space and Information Systems Division, and the highest official directly overseeing command module and S-II construction, was replaced and moved to a slot as corporate vice president.[95] NASA continued to follow closely the development of the S-II. On November 9, 1967, after numerous delays caused not only by problems with the S-II stage, but also by other aspects of the Saturn V development, the SA-501 launch vehicle successfully carried the crewless Apollo 4 capsule into orbit for tests.

The Apollo 4 flight was the first "all up" test of the Saturn V launch vehicle. For NASA, it was also a major risk, because not only had the stages never been launched together, neither the S-IC nor the S-II had flown at all. The S-IVB stage, the command module, and the instrument unit that provided inertial guidance and avionics to the vehicle had been tested on Saturn IB flights. As one writer put it, "The all-up concept is, in essence, a calculated gamble, a leap-frogging philosophy that advocates compression of a number of lunar landing preliminaries into one flight. It balances the uncertainties of a number of first-time operations against a 'confidence factor' based on the degree of the equipment reliability achieved through the most exhaustive ground-test program in aerospace history."[96]

This all-up test was the result of an earlier decision by NASA's Associate Administrator for Manned Space Flight, George E. Mueller. In September 1963, when Mueller took his post, NASA was beginning to feel the enormity of meeting Kennedy's deadline for reaching the Moon. It was also experiencing the first hint of a shrinking yearly budget. After he succeeded Kennedy, President Lyndon B. Johnson was under considerable pressure to keep NASA's 1964 budget under $5 billion, versus a $5.75 billion budget request to Kennedy earlier in the year. Mueller notified the directors of the Manned Spacecraft Center in Houston, the Launch Operations Center in Cocoa Beach, Florida, and Marshall Space Flight Center in Huntsville, Alabama, that the first Saturn IB flight and the first Saturn V flight would be made with all stages operating. Both should also carry complete spacecraft. [I-38]

Although Mueller's directive caused considerable debate among the highly conservative staff at the centers, particularly at Marshall, eventually even the Marshall team came around.[97] Nevertheless, everyone recognized the risks, and it was with considerable trepidation that the launch team prepared the first Saturn V for launch. Yet at 7 a.m. EST, November 9, 1967, the first Saturn V, AS-501, lifted off the pad, carrying the Apollo 4 command module and performing nearly flawlessly. The risk had paid off. With one exception, the remainder of the Saturn V launches were also highly successful. In all, there were thirteen Saturn Vs launched between November 1967 and May 1973.

The one troublesome launch was AS-502, or Apollo 6. NASA had planned to fly both it and AS-503 without a crew. However, on November 16, in light of the success of AS-501, Phillips decided that tests were going so well that if the flight of Apollo 6 proved successful, NASA would proceed directly to human flights with AS-503.[98] As it turned out, Phillips's optimism was short lived. AS-502 lifted off from Launch Complex 39 on April 4, 1958. All went well until about 125 seconds into the flight, near the end of first-stage burn, when the launcher began to experience strong longitudinal oscillations that created a "pogo" effect for nearly ten seconds. Despite the pogo, the separation and ignition of the second stage occurred normally, but after four and a half minutes of operation, its

---

95. For a journalistic and rather biased (in favor of Storms) account of this relationship and its development, see Mike Gray, *Angle of Attack* (New York: W.W. Norton & Company, 1992).

96. James J. Haggerty, "Apollo 4: Proof Positive," *Aerospace* (Winter 1967): 3. Quoted in Bilstein, *Stages to Saturn*, p. 348.

97. Bilstein, *Stages to Saturn*, pp. 348–51.

98. General Sam Phillips to NASA centers, teletype, November 15, 1967.

number-two engine shut down, followed a second later by a shutdown of the number-three engine. The instrument unit, which performed the vehicle's guidance and control, compensated by steering the rocket into a new trajectory and causing the three remaining J-2 engines to fire longer than planned. Following a normal third-stage firing and shutdown of its single J-2 engine, Apollo 6 coasted into Earth orbit. After waiting two orbits, NASA flight controllers signaled the J-2 third-stage engine to restart to complete as much of the flight plans as possible. Despite many attempts, it failed to function. Flight controllers finally gave up and managed to separate the command service module and command module from the third stage. They then finally resorted to using the smaller engine on the command service module to position the command module for a successful reentry.

The flight was successful in proving that even with two second-stage J-2 engines out, the command module could still reach orbit and return safely. The pogo phenomenon had been experienced on Gemini-Titan and other launches, but not with such intensity. Although the vibrations were apparently not severe enough to harm the vehicle or the astronauts directly, it would have caused extra stress to the astronauts, and NASA officials decided they should not risk the possibility of stronger vibrations with people aboard. An intensive investigation by a specially constituted pogo task force composed of representatives of NASA, industry, and the universities disclosed that the F-1's thrust chamber and combustion chamber vibrated at about five and a half hertz during burning. The vehicle as a whole vibrated with a variable frequency. When the vehicle vibrations reached five and a half hertz, the two effects combined to produce the pogo effect. The pogo team was able to devise a repair that involved "de-tuning" the F-1 engine to change its frequency of vibration.

The J-2 problem was much more serious, in part because NASA had no idea what might have gone wrong on the two engines. Fortunately, the second stage was extremely well instrumented; one of the thermocouples showed a temperature drop about seventy seconds into second-stage burn, indicating a leak of cold gas. Then, just before engine shutdown, another thermocouple registered a suddenly higher temperature, suggesting that there had been an eruption of hot gas, probably from the igniter fuel line. With these data in hand, NASA and Rocketdyne engineers began to perform extensive tests on the J-2 fuel lines. At sea-level temperatures and pressures, they could not reproduce the failure. However, by pumping liquid hydrogen through eight separate lines in a vacuum chamber, thereby simulating operational conditions, they were able to cause every one of them to fail about 100 seconds into the test. Once the engineers had reproduced the failure in the laboratory, they were then able to devise a suitable repair, and the Apollo program was back on track.[99] [I-39]

Launch Operations

The relatively mundane tasks of assembling all of the launch vehicle's parts and preparing the vehicle for liftoff are easily overlooked when examining the development of large, powerful launch vehicles. However, a well-organized manufacturing and logistics chain and smooth running launch operations are absolutely crucial to a successful launch. The manufacturing, assembly, preparation, and launch of the completed Saturn V constituted an engineering and organizational marvel. It required huge machines for handling the Saturn V's three stages and the barges, specially modified aircraft, and trucks for transporting them to the launch site at Cape Canaveral. It involved a logistics chain that stretched across the United States, fed by major manufacturing sites along the east, west, and gulf coasts.

99. Bilstein, *Stages to Saturn*, pp. 360–62.

Marshall Space Flight Center had responsibility for launch vehicle construction; the Manned Spacecraft Center in Houston was responsible for the spacecraft and for mission control once liftoff had occurred; and the Launch Operations Center in Florida was in charge of ensuring a safe, successful launch. Merritt Island near Cape Canaveral was chosen for the Launch Operations Center in part because it offered ready access by barge from manufacturing sites on the west and gulf coasts. Its location on the Atlantic Ocean simplified the safety precautions during launch. In addition, the Air Force already maintained a launch facility immediately southeast of NASA's launch range. After some rather tense negotiations, an agreement was reached on sharing responsibilities for range communications and other facilities between NASA and the Air Force. [I-40, I-41]

Early in the Saturn program, officials realized they would need massive facilities in which to erect the massive Saturn V and prepare it for launch. After some discussion regarding whether to construct the vehicle and mate it with its payload on the launch pad, NASA engineers decided that the most efficient operation would result from keeping vehicle construction and payload integration separate from launch operations. [I-42] Hence, they conceived of a large enclosed structure capable of holding the entire vehicle and its payload. Building engineers designed a large Vehicle Assembly Building that covered eight acres. The high bay stands 441 feet high. NASA officials anticipated a high launch rate and designed the building to accommodate four fully assembled Saturn V launch vehicles, to minimize their time on the launch pad.[100] Each Saturn V was erected on a massive device called a mobile launcher, which supported the launcher from the initial assembly through the launch.

For safety purposes, the Vertical Assembly Building (VAB) had to be located far away from the launch pad. Originally, NASA had explored the possibility of building a shallow canal between the VAB and the launch pads and floating the assembled launcher to the pad on a barge. However, tests at the Navy's David Taylor Basin near Washington, D.C., soon showed that the mobile launcher's huge gantry and the launcher would act like an enormous sail, making steering such a mammoth contraption impossible. After considering a rail line and rejecting it because of the enormous forces the rails and their bedding would have had to sustain, NASA settled on a large tractor built by the Marion Power Shovel Company, which had built similar tractors for strip-mining coal.[101] After the launcher was assembled in the VAB, a huge crawler-transporter lifted the assembled vehicle and atop its mobile launcher, which together weighed nearly 12 million pounds, and it slowly crawled the three and a half miles to the launch pad (Figure 1–11).[102]

This system was used thirteen times to bring a Saturn V and its payload to the launch pad. There were two test launches carrying Apollo spacecraft without crews. The first launch with a crew aboard was the Apollo 8 mission to lunar orbit in December 1968; this was followed by the Apollo 9 Earth-orbital test of the lunar module in February 1969 and the Apollo 10 "dress rehearsal" in May. Then there were seven launches of crews to the Moon, beginning of course with the July 1969 Apollo 11 mission (Figure 1–12) and ending, prematurely in terms of the original plans, with the Apollo 17 mission in December 1972. A final launch in May 1973 carried the Skylab space station, which was in fact a modified S-IVB Saturn V upper stage, to Earth orbit.

With the exception of its second flight, discussed above, the Saturn V performed almost flawlessly in each of its missions. The Saturn launch system was truly a triumph of U.S. organizational, management, and technological capabilities.

---

100. However, ultimately it was equipped to handle only three.

101. The Vertical Assembly Building, renamed the Vehicle Assembly Building, Launch Pads 39A and 39B, the mobile launchers, and the crawler-transporters are still in use in the Space Shuttle program, although they have been modified to accommodate the rather different shape, size, and load requirements of the Shuttle system.

102. Walter Flint, "Operational Support for Apollo," in *Apollo: Ten Years Since Tranquility Base*, pp. 109–14.

Figure 1–11 The massive size of the Saturn V and the Vertical Assembly Building (later renamed the Vehicle Assembly Building) are shown in this 1969 aerial photograph with the launch complex more than three miles in the background. Delivering 7.5 million pounds of thrust in its first stage and standing 363 feet tall, it was the most powerful rocket ever successfully built and flown with astronauts aboard. (NASA photo)

However, as the United States, after achieving the goal of a successful lunar landing, shifted priorities away from human space exploration beyond Earth orbit, there were no new missions requiring the booster's power. NASA's original order was for fifteen Saturn V rockets. As early as 1968, NASA faced the issue of whether it would need more Saturn Vs and decided to wait before ordering the long-lead-time components involved. [I-43] Later, it became clear in 1970 that there would be no early approval of a large space station. As NASA in 1971 and 1972 struggled with getting approval to develop a new space transportation system, the Space Shuttle, NASA officials reluctantly decided that they had no choice but to give up hopes of preserving the two remaining Saturn V boosters for future use and of maintaining production capabilities for additional vehicles. [I-44, I-45, I-46]

Thus the two remaining Saturn V rockets became museum pieces, reminders of a time when the United States pioneered the space frontier beyond Earth orbit. They may be seen today at NASA's Kennedy Space Center and Johnson Space Center. Those who actually saw them in use, not as they exist today, were indeed fortunate.

Figure 1–12. The mighty Saturn V launch vehicle was the booster that allowed the United States to go to the Moon in the late 1960s and early 1970s. This photo shows the launch of Apollo 11, the first lunar landing mission, lifting off from Kennedy Space Center's Launch Complex 39 on July 16, 1969. (NASA photo 69-H-1111)

Document I-1

Document title: Hugh J. Knerr, Major General, USA, Deputy Commanding General, U.S. Strategic Air Forces in Europe, Memorandum for Commanding General, U.S. Strategic Air Forces in Europe, June 1, 1945.

Document I-2

Document title: Memorandum to the Director of Research and Development, DC/S, Material, Attn: General Craigie, "Utilization of German Scientists by U.S.S.R. and U.S.," March 22, 1948.

Source: Both in NASA Historical Reference Collection, NASA History Office, NASA Headquarters, Washington, D.C.

At the conclusion of World War II, U.S. commanders gave considerable thought to the problem of how best to use German technical staff to promote U.S. aerospace capabilities. Not only did U.S. officials want to prevent capable German personnel from reviving Germany's war potential, they also wanted to benefit from their technical capabilities. On June 1, 1945, one of those most involved, General Hugh J. Knerr, formally put forth the bold plan to bring qualified German scientists and engineers and their families to the United States to work for the U.S. scientific and defense establishments. This and other efforts by farsighted U.S. military personnel led to Project Paperclip (originally called Operation Overcast). Of great concern to military planners was the relative disposition of German scientists and engineers among the United States, the United Kingdom, France, Germany, and especially Russia. The United States tended to focus on skimming the cream of German scientists, while Russia centered on removing whole laboratories and factories, including their operating personnel, back to Russian soil.

Document I-1

HEADQUARTERS
UNITED STATES STRATEGIC AIR FORCES IN EUROPE
Office of the Deputy Commanding General

1 June 1945

MEMORANDUM FOR: Commanding General, U.S. Strategic Air Forces in Europe

It is suggested that the program for exploiting German scientific and aeronautical research include a plan for transporting to the U.S. the key German personnel associated with each subject. Due to the political and economic factors involved in uprooting these scientists, it is considered essential that their immediate dependent families accompany them. Such a realistic arrangement will guarantee willing cooperation and maximum contribution to the program of aeronautical development that we must expedite if we are to come abreast of and attempt to surpass those of other countries.

It is considered feasible to assemble such a party, place it in charge of a project officer for transportation to Wright where it can be established as a unit in a block of houses set aside for the purpose in the adjacent Osborn housing project. This location is considered essential in order that full use may be made of the laboratory equipment of Wright Field. Also in view of the fact that this is a military enterprise, none of this personnel should be dispersed to the uses of civil activities.

.If undertaken, these men should be paid a good salary and in nowise treated as pris-oners or slave workers. The scientific mind simply does not produce under duress. Control can be easily exercised by a clear understanding that violation of the privilege of living in the U.S. for several years will result in prompt deportation to the area from which the individual came.

Occupation of German scientific and industrial establishments has revealed the fact that we have been alarmingly backward in many fields of research. If we do not take this opportunity to seize the apparatus and the brains that developed it and put the combina-tion back to work promptly, we will remain several years behind while we attempt to cover a field already exploited. Pride and face-saving have no place in national insurance.

> HUGH J. KNERR
> Major General, USA
> Deputy Commanding General

Document I-2

[originally stamped "SECRET"]
[1]

AFOIR-CO/Capt
Macken/nc/6282

22 Mar 48

MEMORANDUM FOR: Director of Research and Development, DC/S, Material
Attn: General Craigie
SUBJECT: Utilization of German Scientists by U.S.S.R. and U.S.

1. Project Paperclip is a program for the employment of certain outstanding German and Austrian scientists and technicians in connection with the research and development programs of the Army, Navy and Air Force. It was authorized by the State-War-Navy Coordinating Committee (SWNCC), now the State-Army-Navy-Air Coordinating Committee (SANAC), as a procedure to provide the military services with the technical advances which the Germans had made in those fields in which they were admittedly ahead of us.

2. A procedure for the admission of these scientists under normal methods of entry into the United States had been blocked. The Department of State gave the War and Navy Departments to understand that under existing regulations visas could not be granted to the scientists. A year having passed since the inauguration of the program, it was mean-while becoming imperative that some steps be taken to get the scientists whom we need-ed out of Germany, since they were constantly being contacted by Russian agents who made them attractive offers.

3. Under these circumstances, SWNCC adopted the policy of bringing the scientists to the United States in military custody, with the intention that the legalizing of their sta-tus as immigrants would be accomplished after their arrival here. It was in implementa-tion of this policy that the present Paperclip Program is operating.

4. After their arrival in the United States, the project was further affected by the housing shortage, which made it impossible for a long time to bring the families of the sci-entists to this country. During this period the scientists in the United States reported to their families in Germany difficulties of this sort which were encountered, and were in turn informed by their relations of the promises, at least, which were made to their friends by the Russians.

5.   These difficulties, reported from one side to the other, worked to our disadvantage, in that they served to prevent a number of important scientists from signing contracts with the United States, and drove them into the arms of other powers.

[2] 6.   Information which has reached us through intelligence channels indicates the effect of the evacuations by the various powers upon the over-all exploitation of German experience. German scientists (not counting engineers and production technicians) seem to be apportioned as follows:

Russian Service	17%
British Service	11%
French Service	11%
U.S. Service	6%
German Service	10%

The rest can be assumed to be engaged outside of their profession or to be fully idle. The main interest seems to be concentrated on every kind of war technique which the Russians did not use in the recent war due to lack of experience and skill in those fields.

7.   There is a considerable difference between the Russian and the American basis of selection of the scientists. The Russians seem to have aimed at taking large numbers of "working" scientists, together with technicians and even laborers working upon specific projects. They have been known to remove from Germany entire factories and laboratories, together with their equipment and the personnel of these establishments down to and including the lowest "skilled labor" classes. These plants and laboratories have been moved to Russia and have been set up there with the same staffs which originally operated them in Germany.

8.   Reliable information indicates that under "operation Ossawakim," for example, the number of people affected would run into the hundred of thousands. Unquestionably, a considerable number of the skilled German labor involved in this operation volunteered to move with their families to Russia. In all probability, a majority of the leading scientists and executives evacuated under the project have also gone voluntarily.

9.   While it can generally be said that the United States obtained a large number of the cream of German scientists and have a few others of some number still available (they are at present allocated to the British, under an arrangement made with them), the Russians did obtain a much larger number of personnel, and have appreciably boosted their scientific experience and skill-level by the wholesale evacuations to their territory which they have carried out.

Document I-3

Document title: H. Julian Allen and A.J. Eggers, Jr., NACA Research Memorandum, "A Study of the Motion and Aerodynamic Heating of Missiles Entering the Earth's Atmosphere at High Supersonic Speeds," RM A53D28, August 25, 1953, pp. 1–3, 26–29.

Source: NASA Historical Reference Collection, NASA History Office, NASA Headquarters, Washington, D.C.

One of the most difficult problems faced in early ballistic missile programs involved the high temperatures generated during atmospheric reentry. In 1953, H. Julian Allen, who worked for the Ames Aeronautical Laboratory of the National Advisory Committee for Aeronautics (NACA), produced a report noting that blunt nose cones were much less likely to overheat than were sharp ones. The Air Force did not include Allen's principle in missile design until 1956. NASA later incorporated this idea, for which Allen won NASA's Distinguished Service Medal, into the designs of the Mercury, Gemini, and Apollo spacecraft.

[typed: "August 25, 1953," "Declassified April 8, 1957"]

RM A53D28

NACA Research Memorandum

A Study of the Motion and Aerodynamic Heating of Missiles Entering the Earth's Atmosphere at High Supersonic Speeds

By H. Julian Allen and A.J. Eggers, Jr.

[1]   SUMMARY

A simplified analysis is made of the velocity and deceleration history of missiles entering the earth's atmosphere at high supersonic speeds. It is found that, in general, the gravity force is negligible compared to the aerodynamic drag force and, hence, that the trajectory is essentially a straight line. A constant drag coefficient and an exponential variation of density with altitude are assumed and generalized curves for the variation of missile speed and deceleration with altitude are obtained. A curious finding is that the maximum deceleration is independent of physical characteristics of a missile (e.g., mass, size, and drag coefficient) and is determined only by entry speed and flight-path angle, provided this deceleration occurs before impact. This provision is satisfied by missiles presently of more usual interest.

The results of the motion analysis are employed to determine means available to the designer for minimizing aerodynamic heating. Emphasis is placed upon the convective-heating problem including not only the total heat transfer but also the maximum average and local rates of heat transfer per unit area. It is found that if a missile is so heavy as to be retarded only slightly by aerodynamic drag, irrespective of the magnitude of the drag force, then convective heating is minimized by minimizing the total shear force acting on the body. This condition is achieved by employing shapes with a low pressure drag. On the other hand, if a missile is so light as to be decelerated to relatively low speeds, even if acted upon by low drag forces, then convective heating is minimized by employing shapes with a high pressure drag, thereby maximizing the amount of heat delivered to the atmosphere and minimizing the amount delivered to the body in the deceleration process. Blunt shapes appear superior to slender shapes from the standpoint of having lower maximum convective heat-transfer rates in the region of the nose. The maximum average heat-transfer rate per unit area can be reduced by [2] employing either slender or blunt shapes rather than shapes of intermediate slenderness. Generally, the blunt shape with high pressure drag would appear to offer considerable promise of minimizing the heat transfer to missiles of the sizes, weights, and speeds presently of interest.

INTRODUCTION

In the design of long-range rocket missiles of the ballistic type, one of the most difficult phases of flight the designer must cope with is the re-entry into the earth's atmosphere, wherein the aerodynamic heating associated with the high flight speeds of such missiles is intense. The air temperature the boundary layer may reach values in the tens of thousands of degrees Fahrenheit which, combined with the high surface shear, promotes very great convective heat transfer to the surface. Heat-absorbent material must therefore be provided to prevent destruction of the essential elements of the missile. It is a characteristic of long-range rockets that for every pound of material which is carried to "burn-out," many pounds of fuel are required in the booster to obtain the flight range. It is clear, therefore, that the amount of material added to protect the warhead from excessive aerodynamic heating must be minimized in order to keep the take-off weight to a practicable value. The importance of reducing the heat transferred to the missile to the least amount is thus evident.

For missiles designed to absorb the heat within the solid surface of the missile shell, a factor which may be important, in addition to the total amount of heat transferred, is the rate at which it is transferred since there is a maximum rate at which the surface material can safely conduct the heat within itself. An excessively high time rate of heat input may promote such large temperature differences as to cause spalling of the surface, and thus result in loss of valuable heat—absorbent material, or even structural failure as a result of stresses induced by the temperature gradients.

For missiles designed to absorb the heat with liquid coolants (e.g., by "sweat cooling" where the surface heat-transfer rate is high, or by circulating liquid coolants within the shell where the surface heat-transfer rate is lower), the time rate of heat transfer is similarly of interest since it determines the required liquid pumping rate.

These heating problems, of course, have been given considerable study in connection with the design of particular missiles, but these studies are very detailed in scope. There has been need for a generalized heating analysis intended to show in the broad sense the means available for minimizing the heating problems. Wagner . . . [3] made a step toward satisfying this need by developing a laudably simple motion analysis. This analysis was not generalized, however, since it was his purpose to study the motion and heating of a particular missile.

It is the purpose of this report to simplify and generalize the analysis of the heating problem in order that the salient features of this problem will be made clear so that successful solutions of the problem will suggest themselves.

A motion analysis, having the basic character of Wagner's approach, precedes the heating analysis. The generalized results of this analysis are of considerable interest in themselves and, accordingly, are treated in detail. . . .

|26| DESIGN CONSIDERATIONS AND CONCLUDING REMARKS

In the foregoing analysis and discussion, two aspects of the heating problem for missiles entering the atmosphere were treated. The first concerned the total heat absorbed by the missile and was related to the coolant required to prevent its disintegration. It was found that if a missile were relatively light, the least required weight of coolant (and hence of missile) is obtained with a shape having a high pressure drag coefficient, that is to say, a blunt shape. On the other hand, it was found that if the missile were relatively heavy the least required weight of coolant, and hence of missile, is obtained with a shape having a low skin-friction drag coefficient, that is to say, a long slender shape.

The second aspect of the heating problem treated was concerned with the rate of heat input, particularly with regard to thermal shell [27] stresses resulting therefrom. It was seen that the maximum average heat-input rate and, hence, maximum average thermal stress could be decreased by using either a blunt or a slender missile, while missiles of intermediate slenderness were definitely to be avoided in this connection. The region of highest local heat-transfer rate and, hence, probably greatest thermal stress was reasoned to be located at the forward tip of the missile in most cases. This was assumed to be the case and it was found that the magnitude of this stress was reduced by employing a shape having the largest permissible tip radius and over-all drag coefficient; that is to say, the blunt, high drag shape always appears to have the advantage in this respect.

These results provide us with rather crude, but useful, bases for determining shapes of missiles entering the atmosphere which have minimized heat-transfer problems. If the over-all design considerations of payload, booster, et al[.], dictate that the re-entry missile be relatively heavy in the sense of this report, then it may be most desirable to make this missile long and slender, especially if the entry speed is very high (say 20,000 ft/sec or greater). Perhaps the slender conical shape is appropriate for such a missile. It seems clear, too, that the tip of this missile should be given the largest practicable nose radius in order to minimize the maximum local heat-transfer rate and hence maximum local shell stress problem. Even then it may be necessary to employ additional means to minimize the heat-transfer rate and, hence, thermal stress encountered in this region (e.g., by sweat cooling).

Let us now consider the case where the over-all design conditions dictate that the re-entry missile be relatively light in the sense of this report. This case is believed to be of more immediate importance than the one just considered since the lower sizes, weights, and entrance speeds to which it applies are more nearly in line with those presently of interest. The relatively light re-entry missile will therefore be treated at greater length.

A shape which should warrant attention for such missile application is the sphere, for it has the following advantages:

1. It is a high drag shape and the frictional drag is only a few percent of the total drag.
2. It has the maximum volume for a given surface area.
3. The continuously curved surface is inherently stiff and strong.
4. The large stagnation-point radius significantly assists in reducing the maximum thermal stress in the shell.
[28] 5. Aerodynamic forces are not sensitive to attitude and, hence, a sphere may need no stabilizing surfaces.
6. Because of this insensitivity to attitude, a sphere may purposely be rotated slowly, and perhaps even randomly[11] during flight, in order to subject all surface elements to about the same amount of heating and thereby approach uniform shell heating.

On the other hand, the sphere, in common with other very high drag shapes[,] may be unacceptable if:

1. The low terminal speed permits effective countermeasures.
2. The lower average speed of descent increases the wind drift error at the target.
3. The magnitude of the maximum deceleration is greater than can be allowed.

The first two of these disadvantages of the sphere might be minimized by protruding a flow-separation-inducing spike from the front of the sphere to reduce the drag coefficient to roughly half. . . . Stabilization would now be required but only to the extent required to counterbalance the moment produced by the spike. Special provision would have to be made for cooling the spike.

These possible disadvantages of very high drag shapes may also be alleviated by another means, namely, using variable geometry arrangements. For example, an arrangement which suggests itself is a round-nosed shape with conical afterbody of low apex angle employing an extensible skirt at the base. . . . With the skirt flared, the advantages of high drag are obtained during the entry phase of flight. As the air density increases with decreasing altitude, the skirt flare is decreased to vary the drag so as to produce the desired deceleration and speed history. If the deceleration is specified in the equation of motion (see motion analysis), the required variation of drag coefficient with altitude can be calculated and, in turn, the heating characteristics can be obtained.

---

14. Note that if rotation is permitted, slow, random motion may be required in order to prevent Magnus forces from causing deviation of the flight path from the target. It should also be noted that at subsonic and low supersonic speeds gun-fired spheres, presumably not rotating, have shown rather large lateral motions in flight. . . . It is not known whether such behavior occurs at high supersonic speeds.

[29] The examples considered, of course, are included only to demonstrate some of the means the designer has at hand to control and diminish the aerodynamic heating problem. For simplicity, this problem has been treated, for the most part, in a relative rather than absolute fashion. In any final design, there is, of course, no substitute for step-by-step or other more accurate calculation of both the motion and aerodynamic heating of a missile.

Even from a qualitative point of view, a further word of caution must be given concerning the analysis of this paper. In particular, throughout, we have neglected effects of gaseous imperfections (such as dissociation) and shock-wave boundary-layer interaction on convective heat transfer to a missile, and of radiative heat transfer to or from the missile. One would not anticipate that these phenomena would significantly alter the conclusions reached on the relative merits of slender and blunt shapes from the standpoint of heat transfer at entrance speeds at least up to about 10,000 feet per second. It cannot tacitly be assumed, however, that this will be the case at higher entrance speeds. . . . Accurate conclusions regarding the dependence of heat transfer on shape for missiles entering the atmosphere at extremely high supersonic speeds must await the availability of more reliable data on the static and dynamic properties of air at the high temperatures and pressures that will be encountered.

Ames Aeronautical Laboratory
National Advisory Committee for Aeronautics
Moffett Field, Calif., Apr. 28, 1953

Document I-4

Document title: Homer J. Stewart, Chairman, Ad Hoc Advisory Group on Special Capabilities, Report to Donald A. Quarles, Assistant Secretary of Defense (Research and Development), August 4, 1955, pp. iii, 1, 3–8.

Source: National Archives, Washington, D.C.

President Dwight D. Eisenhower announced on July 29, 1955, that the United States would launch satellites as part of the U.S. contribution to the International Geophysical Year (IGY). Although the United States could have modified an existing military rocket, such as the Jupiter or Atlas, which were then under development, Eisenhower and his advisors were concerned about siphoning off energy from attention to the development of ballistic missiles. Donald A. Quarles, Assistant Secretary of Defense for Research and Development, asked the Ad Hoc Advisory Group on Special Capabilities, chaired by Homer J. Stewart, to advise on the best course of action. The resulting report from the so-called Stewart Committee provided the basis for President Eisenhower to choose the as-yet-undeveloped rocket based on Viking technology, rather than run the risk of diverting resources from the development of the Atlas or Jupiter missiles.

|iii|

The Honorable Donald A. Quarles
Assistant Secretary of Defense   4 August 1955
(Research and Development)

Dear Mr. Quarles:

I have the honor to transmit the attached report of the Ad Hoc Advisory Group on Special Capabilities pursuant to your directive dated 13 July 1955.

The Group has reviewed the earth satellite plans and programs of the military departments and presents its conclusions, recommendations and observations in the report, together with summaries of the programs proposed.

The Group has interpreted the National Security Council directive and its own charter as implying the highest and broadest national interest and urgency for prestige and political as well as scientific reasons, limited only by the military necessities. The prestige and political elements have been considered, however, only to the extent necessary in weighing the relative merits of certain phases of the technical problems of placing the satellite on orbit.

The opinion of the Group was unanimous on the conclusions, and on three of the four recommendations. The differences of view on Recommendation 3 recorded in the report are mainly due to differences in judgment on the practicability of making the proposed modification required in either alternative in time to ensure the minimum objective within the IGY period.

Sincerely yours,
Homer J. Stewart
Chairman . . .

[1]   **CONCLUSIONS**

1.   There is a reasonable assurance that the United States can have the capability to put up a small scientific satellite during 1958 on an orbit having a minimum (perigee) attitude of 150 to 200 statute miles and carrying a payload on the order of 5 to 50 pounds; however, none of the existing proposals will provide this capability without considerable development work.

2.   Any use of current military programs to accomplish the objective within the International Geophysical Year (IGY) period will run some risk of interference with such military programs, if only indirectly in the drain on skills and facilities; but, if such a program is properly carried out, it can result in long-term benefits to the military programs.

3.   In addition to any program intended to fulfill the immediate needs of the IGY, there should also be a continuing program of geophysical observations that could be provided adequately by means of a number of small scientific satellites having a payload of approximately 50 pounds, launched from time to time on different orbits. Such a continuing program would be useful, even after a large satellite has been developed.

4.   If attainment of the objective of the National Security Council directive is to be ensured, clear and undivided administrative responsibility in the Department of Defense must be promptly defined, assigned and ordered. Great caution is imperative to ensure that existing techniques, existing contractors, group skills and facilities be used. Diversion from this policy must be strictly controlled at the highest level; otherwise, additional and unnecessary delays will be inevitable.

5.   The immediate and direct cost of such a satellite program is likely in any event to be of the order of twenty million dollars, but it will be much larger unless full advantage is taken of existing programs, facilities and reasonable logistic support. . . .

[3]   **RECOMMENDATIONS**

Recommendation 1.
The development of scientific satellite vehicles should be carried out in two phases as follows:

Phase I. An immediate program designed for maximum assurance of placing at least a small payload (5 to 10 pounds), including a small radio transmitter, in an orbit having a minimum (perigee) altitude of at least 150 miles during 1958.

There should be a concurrent effort to improve the components used in this program which should have as its goal a somewhat larger payload (on the order of 30 to 50 pounds) with eventual full attitude control. Some part of this development might also be accomplished in time to be of value in the program of the International Geophysical Year.

The detailed planning for Phase I must provide enough flexibility to permit launching the most useful satellite practicable within the then current state of development.

Phase II. A program to launch a satellite vehicle capable of carrying a significantly larger payload (up to 2,000 pounds) or of achieving a significantly higher orbit. In this phase also, flexibility is desirable in the planning, so that advantage can be taken quickly of any advances in technology.

Recommendation 2.

The use of the ICBM booster would unquestionably give the greatest performance margin and therefore the highest probability of placing a useful payload in a long-duration orbit. Whether or not this could be accomplished during the IGY period would depend on (a) the degree of interference with the ICBM program that might be tolerated and (b) the degree of certainty that can be assigned to the ICBM schedule.

On the assumption that any effort which is needed from the ICBM program will be made available, that this effort will not seriously interfere with the ICBM program, that the ICBM program will be on schedule and that only a single satellite program can be approved, the Group would unanimously favor a program using the ICBM booster. The question raised in the assumptions of Recommendation 2 involve points of national policy outside the competence of the Group. For this reason, the Group considered two alternative methods for accomplishing Phase I.

[4] Recommendation 3.

(a) The use of an improved VIKING as a booster with a liquid-propellant second stage based on the AEROBEE-HI and a solid-propellant third stage. For the initial part of the program, characteristics similar to those of the present AEROBEE-HI and the scale SERGEANT motor were assumed; however, it is expected that both these stages would be improved during the course of the development and that the last stage would eventually be attitude-stabilized.

(b) The use of the REDSTONE missile, as currently being modified for use in re-entry tests, as a booster, either with three additional solid-propellant stages or with one liquid-propellant stage based on AEROBEE-HI and two additional solid propellant stages. For the initial part of the program, it is assumed that a cluster of seven scale SERGEANT motors would be used for the second stage and one scale SERGEANT motor for each of the third and fourth stages. Concurrent development would include replacing the present REDSTONE motor with a liquid-oxygen-gasoline motor, better proportioning of the stages, the possible substitute of a liquid-propellant second stage and attitude stabilization of the final stage.

Five members of the Group recommend alternative (a), and two members support alternative (b).

Recommendation 4.

Regardless of the course of action taken on Phase I, Phase II should make use of an ICBM booster and should be made a responsibility of the Air Force. The work should be carefully coordinated with the ICBM program and any military satellite program which may be undertaken. If possible, an attempt should be made to use some of the same components as are used in Phase I.

[5]   **DISCUSSION**

Factors Considered.
The Ad Hoc Advisory Group on Special Capabilities has considered three proposals:
(1) An Army proposal based on a modified REDSTONE missile with three additional stages. . . .
(2) A Navy proposal based on a modified VIKING rocket with two additional stages. . . .
(3) An Air Force proposal based on the ICBM booster with one additional stage. . . .
The findings of the Group set forth in this report are based largely on the presentations by the Army, Navy and Air Force representatives on 6–9 July 1955 in Washington, D.C. and Redstone Arsenal and on additional data and information obtained by the Group from authoritative sources.
The Group has weighed the several proposals and alternatives against the primary question:
What program will be most certain of placing the most useful satellite vehicle on an orbit of at least 150 to 200 statute miles' perigee (minimum altitude) within the International Geophysical Year period and with minimum interference with priority military programs?
The factors considered by the Group in developing the recommendations and conclusions include:
(1)   The practicability of putting up any satellite within the time period;
(2)   The minimum payload and altitude for something useful;
(3)   The duration of the orbit;
(4)   Tracking requirements;
(5)   The growth potential of the equipments [sic] proposed;
(6)   Maximum use of available facilities and skills;
(7)   Minimum delay of priority projects;
(8)   Maximum scientific utility;
(9)   Broad national interest;
(10) Over-all economy for about a 5-year period.

[6] Basis for Opinions.
The majority of the Group (five members) supports as first preference using a program along the lines of alternative (a) of Recommendation 3, using the VIKING rocket as a booster, for the following reasons:
(1) Despite its smaller size, the proposed VIKING booster offers better performance and more reserve margin than the REDSTONE with its permanent 75,000-pound-thrust engine.
(2) As a result, the VIKING requires only two additional stages, whereas the REDSTONE requires three or four, at least one of which is a multiple cluster in the current proposals.
(3) The necessary modifications to the VIKING for a minimum program seem to be well within demonstrated engineering capability, and it appears that the facilities required, including those for the GE X-400 rocket engine, could be made available without any interference from, or with, existing weapons projects.
(4) There is at least a finite probability that the objectives of the minimum program will not only be met but exceeded to such an extent that essentially the full objectives of Phase I above will be achieved during the IGY period. It seems less likely that this result would be accomplished if the REDSTONE booster should be used, unless immediate steps are taken to replace the present REDSTONE rocket motor with the ICBM motor.
(5) One single agency, the Naval Research Laboratory [NRL], has had an extensive experience with the VIKING rocket, with the AEROBEE and with upper atmosphere

research equipment. This agency has an excellent reputation for on-time accomplishment of its objectives and is not primarily involved in any high-priority weapons project.

(6) Because of its smaller size, the proposed VIKING would require less logistic support and would be more suitable and more economical for a continuing small satellite program.

(7) The associated AEROBEE-HI development would undoubtedly be used in more advanced steps of the satellite program.

(8) The improved VIKING components might also be used eventually to make a larger second stage for use with the ICBM booster to achieve still greater performance.

(9) The fact that the VIKING project has been declassified would simplify the handing of security problems in any international collaborative project and would increase the amount of technical data that could be released.

[7] On the other hand, a minority of the Group (two members) maintains that a program along the lines of alternative (b) should be adopted. The minority lists the following reasons:

(1) The REDSTONE is larger than the VIKING, has more flexibility in application to a satellite program and therefore has more chance of achieving success during IGY than VIKING.

(2) The VIKING proposal retires each stage to meet predicted values within narrow margins to reach the goal of the program, and the development problems are so great that they might make it impossible to meet the objective within the IGY period.

(3) The REDSTONE proposal, in comparison, has fewer development problems.

(4) The REDSTONE missile is an active weapon program now entering intensive flight testing and thus will have the benefit of many tests prior to the first effort to launch a satellite; this will also reduce the costs of additional satellite vehicles.

(5) Range facilities planned for the REDSTONE missile as a weapon can be used for the satellite based on REDSTONE, resulting in less interference on facilities.

Need for Extensive Tests.

Although the configuration proposed by NRL is basically sound and, in the opinion of the majority of the Group, with appropriate modifications has an adequate prospect for success, it is believed that the program proposed by NRL should be expanded to include more check-out runs of components and trial runs of complete assemblies.

For example, the NRL has proposed three flight tests of the M10; this seems insufficient. The combined second and third stages should have trial runs as a unit before the satellite launchings. As soon as possible, tests should be run on ignition and other problems connected with the launching of a spinning solid rocket from the nose of an AEROBEE-HI. Tests of this sort might be started with existing AEROBEE-HIs and solid rockets. An existing AEROBEE-HI, with a 7-inch scale SERGEANT motor in vertical firings carrying a 5-pound payload, would rise to nearly 800 miles and stay out of the atmosphere 15 minutes or more; so preliminary data might be obtained on the operation of beacons at high altitudes and possibly on their degradation due to the effect of comic rays, direct sunshine, ionization at high altitudes and so on. Similar considerations would apply to any alternative second- and third-stage combinations that might be developed. Before the first complete satellite launching, it might also be desirable to launch several complete vehicles with full communication [8] equipment on a near-vertical trajectory in order to obtain the most realistic system tests possible before launching an actual satellite.

Selection of Orbit.

The relative merits of various orbits with respect to equatorial, inclined or polar, were considered by the Group in connection with tracking, and the following comments are made:

The smaller the vehicle that is to be placed on orbit, the more important it becomes to ensure radio tracking. Otherwise, the risk of losing contact with a small object as a result of the marginal capabilities of the optical methods described to the Group becomes too great.

An orbit inclined about 30 degrees and symmetrical to the equator would seem to offer the greatest immediate scientific utility. It would seem to give adequate initial observations for establishing the orbital elements; it would, in effect, yield data over an equatorial band 60 degrees in width, and it would provide increased data on geophysical phenomena which are functions of latitude.

Other reasons for favoring the inclined orbit are that it will be simpler to place on orbit, will reduce logistic costs and preparation make maximum use of United States installations, offer wide opportunity for international collaboration, provide more opportunity to use skilled observers in numerous astronomical observatories (where special instruments might be needed, this would afford the possibility of enlisting United States mutual scientific aid at a low cost for operation) and afford extended possibilities of enlisting radio amateurs and amateur astronomers as observers in many countries, thereby increasing popular interest and support from other nations.

On the other hand, an equatorial orbit, although providing more opportunities for observations from any one point in a given period of time, offers limited opportunities for international collaboration, generates more serious logistic problems and greatly reduces the potential number of observers. If the slight improvement in the performance margin which would result from launching on an equatorial orbit as compared with an orbit inclined about 30 degrees is necessary to achieve orbiting conditions, the proposed performance is considered to be too speculative.

For the preceding reasons, the Group recommends the inclined orbit and recommends that a radio transmitter be carried in the satellite vehicle, regardless of which program is activated. . . .

Document I-5

Document title: Reuben B. Robertson, Jr., Deputy Secretary of Defense, Memorandum for the Secretaries of the Army, Navy, and Air Force, "Technical Program for NSC 5520 (Capability to Launch a Small Scientific Satellite During IGY)," September 9, 1955.

Source: NASA Historical Reference Collection, NASA History Office, NASA Headquarters, Washington, D.C.

This memorandum implemented the decision by the Eisenhower administration to focus on the development of a new rocket based on the existing Viking (booster), Aerobee (second stage), and Sargent (third stage) rockets. This was later termed the Vanguard rocket.

[no pagination]

THE SECRETARY OF DEFENSE
WASHINGTON

September 9, 1955

MEMORANDUM FOR: THE SECRETARY OF THE ARMY
THE SECRETARY OF THE NAVY
THE SECRETARY OF THE AIR FORCE

SUBJECT: Technical Program for NSC 5520 (Capability to Launch a Small Scientific Satellite During IGY)

References: (a) NSC Action 1408, meeting of 26 May 1955, pertaining to NSC Report 5520
           (b) Memo from DepSecDef to Multiple Addressees, 8 June 1955, subject:
                NSC 5520

1. The National Security Council Report 5520 provides for a program to launch a scientific satellite during the period of the International Geophysical Year (July 1957–December 1958). The implementing directive charges the Secretary of Defense with the over-all responsibility of the scientific satellite program as delineated in NSC 5520, and the Assistant Secretary of Defense [ASD] (R&D) has been assigned the responsibility for coordinating the implementation of the scientific satellite program within the Defense Department by reference b.

2. In carrying out the technical program preliminary to launching the satellite, the following course of action is approved:

   a. A joint three-service program [will] be established to produce and launch a small scientific satellite based on the Navy proposal involving the improved Viking (booster), Aerobee-Hi (second stage), solid-propellant modified Sergeant (third stage).

   b. The Navy Department will manage the technical program with policy guidance from the Assistant Secretary of the Defense (R&D) and will provide the funds required to implement the action in a above with the understanding that reimbursement will be made as soon as funds can be made available from other sources.

   c. The Departments of the Army and Air Force will participate in the prosecution of the technical program and will assign appropriate priorities to permit attainment of the schedule to be established by the Navy for such work. Any major interference resulting from such priorities will be brought to the attention of the Assistant Secretary of Defense (R&D).

   d. The Assistant Secretary of Defense (R&D) will continue the Technical Advisory Group already established to advise the ASD (R&D) and the military departments on the technical program.

3. Any departmental interest or requirement in connection with the scientific program of observation after satellite launching will be programmed by the military departments in accordance with existing policies and procedures.

4. It is requested that the addressees, as appropriate, provide for the immediate implementation of the action above. The Secretary of the Navy is also requested to advise the ASD (R&D) as soon as practicable of the detailed plan for undertaking the technical program and for coordination of that program with the other military departments.

5. In order to provide for the coordination of inter-agency matters and the exchange of information on this program with other government agencies, separate action is being taken to establish a coordinating group under the chairmanship of the Assistant Secretary of Defense (R&D) with membership to be invited from State, Central Intelligence Agency, National Science Foundation, and National Academy of Sciences.

6. The international scientific purposes, the classified military-related rocketry, and the political and propaganda aspects of this program pose special problems with regard to security classification and information release. The following principles apply:

   a. The classification of equipment and techniques pertaining to the launching and rocketry which are common to military weapons systems will be governed by the security classification of the military weapons.

   b. Information regarding the satellite itself, any inclosed [sic] instrumentation, the orbit and other items relating to the scientific program will be unclassified, at least by time of launching.

   c. All information material intended for public release relating to this project will be submitted to the Office of Security Review. In this regard the Department of Defense is operating under the specific guidance of the Operations Coordination

Board. Information on <u>military participation</u> [handwritten underlining] in the <u>program and possible relationship to military programs</u> [handwritten underlining] will be kept to a minimum.

> Reuben B. Robertson, Jr.
> Deputy

Document I-6

Document title: Joseph C. Myers, Deputy Secretary, Advisory Group on Special Capabilities, Office of the Assistant Secretary of Defense, "Memorandum for Members, Advisory Group on Special Capabilities," December 1, 1955, with attached: Homer J. Stewart, Chairman, Advisory Group on Special Capabilities, Memorandum to Assistant Secretary of Defense (R&D), "Activities of the Advisory Group on Special Capabilities," December 1, 1955.

Source: NASA Historical Reference Collection, NASA History Office, NASA Headquarters, Washington, D.C.

The attached memorandum, prepared just four months after the decision to proceed with the development of the upgraded Viking rocket, refers back to the earlier Stewart Report. The memorandum notes the importance of developing a backup plan as outlined in the Phase II recommendation of the Stewart Report, in the event that development delays would make it impossible for the Naval Research Laboratory (NRL) and the Martin Company to meet their schedule. The approved schedule is also included in the memo.

OFFICE OF THE ASSISTANT SECRETARY OF DEFENSE
WASHINGTON 25, D.C.

1 December 1955

DEPARTMENT OF DEFENSE
RESEARCH AND DEVELOPMENT

MEMORANDUM FOR MEMBERS, ADVISORY GROUP ON SPECIAL CAPABILITIES
Attachment (A)

1. The attached memorandum, drafted by the Advisory Group Chairman and edited by the Staff, delineates for the Assistant Secretary of Defense (R&D) the results of the Group's action of the 15 November 1955 meeting and its future plans.
2. For reasons of expedition this memorandum is to be an agenda item for the Policy Council at its 15 December meeting.
3. If the Advisory Group members have any disagreements of a substantive nature they are requested to inform the Staff by Friday, 9 December.

> JOSEPH C. MYERS
> Deputy Secretary
> Advisory Group on Special Capabilities

****

[attachment]

[1]

OFFICE OF THE ASSISTANT SECRETARY OF DEFENSE
WASHINGTON 25, D.C.

1 December 1955

DEPARTMENT OF DEFENSE
RESEARCH AND DEVELOPMENT

MEMORANDUM TO ASSISTANT SECRETARY OF DEFENSE (R&D)
SUBJECT: Activities of the Advisory Group on Special Capabilities

Attachment (A)

1. The Advisory Group on Special Capabilities held its Fifth Meeting 18 October 1955 in the Pentagon, Washington, D.C. Those in attendance are given on the attached list. The morning session, at which all attendees were present, was devoted to a progress report by:
(a) Mr. Samuel Clements of OASD [Office of the Assistant Secretary of Defense] (R&D) on OSD [Office of the Secretary of Defense] developments
(b) Representatives of the NRL on Navy progress in contractual arrangements, tri-service cooperation, and propulsion control and end item detail.

2. The afternoon session was executive, with Mr. Milton Rosen of NRL and Messrs. Samuel Clements and Al Waggoner of OASD for brief periods of consultation.

3. After this consultation the following comments and recommendations on the progress to date were formulated.

4. Comments and Recommendations
(a) In the Advisory Group's earlier consideration of the best way to carry out the satellite program, a paramount issue was the possible interference at Martin if that company became involved in the ICBM program. The probability of this action occurring was considered to be so small that a majority of the group recommended approval of the basic programs proposed by Martin. Now that Martin is heavily involved in the ICBM program, it is inevitable that the participation of the senior [2] Martin technical personnel in the satellite program is, and will continue to be, severely reduced. The very compressed schedules required by the satellite program now will require the Navy to maintain unusually close supervision to ensure early recognition of potential difficulties due to technical management dilution.
This situation gives the Advisory Group particular concern that there is as yet no formally approved program aimed toward providing an emergency back-up in case severe development delays should occur in the NRL-Martin program. The Advisory Group understands that the OASD (R&D) action dated 9 September 1955 which formally approved the NRL-Martin program is also based on a decision not to implement at this time the Air Force program designed to carry out the Phase II recommendation in our 6 August 1955 report (RD 263/9) which might, in its early stages, have provided such a back-up. In response to an OASD (R&D) request, NRL is preparing a report summarizing their satellite launching system for presentation to the Policy Council referred to in a 19 September 1955 memorandum from the Assistant Secretary of Defense (R&D) to the Air Force and relating to the Air Force Phase II

back-up proposal. This same memorandum requested that the other services present their considerations of alternate systems which could provide this emergency back-up. The Advisory Group deems it extremely important that such plans be made. Even if it is not practical to provide funds to implement such proposals fully, it should be possible to authorize preliminary engineering studies (unconnected with ICBM personnel) to provide completely defined plans and to determine the time at which implementation would be necessary if they were in fact to provide a back-up.

(b) NRL and Martin have completed the review of proposals for the development of a propulsion system for Stage II, and a preliminary contract, dated 14 November 1955, has been placed with Aerojet-General Corporation. Two systems were proposed by Aerojet, one based on gas-pressurization with highly stressed 410 steel tanks and the other based on an adaptation of a turbine pumping system from one of their assist take-off developments. In view of the very short time schedule (the delivery [3] date for the first unit is 1 November 1955) [and] the development problems inherent in either system, NRL is proposing to authorize Aerojet to proceed with both systems until the time when one can be shown to be satisfactory or superior.

It is recommended that the NRL proposed course of action be followed.
(c) NRL and Martin have obtained and are evaluating proposals from five different organizations for the development of the spinning, solid-propellant, Stage III. Again, in view of the development problems and the short time scale, NRL plans that two of these proposals be accepted and pursued until the time when one can be shown to be satisfactory or superior.

It is recommended that this course of action be followed.
(d) NRL has prepared a preliminary test schedule and this is being discussed with the Air Force Missile Test Center (Patrick Air Force Base). In view of the already active missile programs at Patrick and of the two ICBM and the two IRBM programs which should be active with high priority near the time of the NRL program, we are requesting NRL to present to us at our next meeting a review of the range scheduling problems.

(e) A particular range scheduling problem concerns the supply of liquid oxygen. We have noted that the supply is becoming critical in several places at this time. In view of the large quantity usage of this propellant by all the high priority ballistic missile programs, it is important that an adequate supply at Patrick Air Force Base for the NRL-Martin program be assured.

We are requesting NRL also to present a review of this problem at our next meeting.
(f) We note with concern that the system for controlling the trajectory and altitude of the satellite launching vehicle is not yet determined. Various systems for both Stage I and Stage II are still under discussion. In view of the well known development problems associated with such devices, we deem it imperative that this fluid situation not be unduly prolonged.

[4] We are requesting that Martin present to us at our next meeting a complete review of the trajectory control and attitude stabilization problems and plans.
(g) Probably the most difficult development problems associated with the satellite launching systems are those of propulsion. These items require the longest lead time and thus affect most strongly the overall schedule. GE [General Electric] has made considerable progress in re-activating the Malta test station and they expect to arrive in the near future at the point when a motor combustion chamber can be tested. Although the delivery of the first complete propulsion system for flight test is

scheduled for October, 1956, no detailed development test schedule is yet available. Similarly, no detailed development schedules for Stage II or Stage III propulsion systems are yet available.

We are recommending Martin to present a review of the propulsion system development schedules at our next meeting.

(h) In order to summarize progress on the project simply, a brief comparison of the test schedule as presented on 22 August 1955 and the current schedule is useful.

Vanguard Schedule

Schedule Submitted by NRL 18 November 1955		Schedule Presented by NRL 22 August 1955
1. Delivery by GE of first stage propulsion system	Oct. '56	Aug. '56
2. Firing Viking No. 13 to test Vanguard instrumentation	Oct. 1, '56	
3. Firing Viking No. 14 to test ignition and burning performance of Vanguard 2d stage	Dec. 1, '56	
4. Fire first Vanguard test vehicle [VTV] No. 1 to test first stage performance and third stage start	Feb. '57	Feb. '57
5. Fire 2nd VTV to test first stage performance and third stage stabilization	April 1, '57	
[5] 6. Fire 3rd VTV to test complete performance of first two stages	June 1, '57	
7. Fire 4th VTV to test complete performance of complete 3-stage vehicle	Aug. 1, '57	
8. Fire 1st complete Vanguard in attempt to get a satellite in orbit	Oct. '57	May '57

During test firing programs three reserve vehicles are to be available, and any necessary repeat firings would be made between scheduled firings.

(i) In general the time interval allowed for the propulsion system developments is unchanged but delayed by the time which has been required for contractual negotiations. Martin expects to be able to make up this loss by shortening the time interval between delivery of the first Stage I propulsion system and the first Stage I flight. The other changes are caused by introducing a fourth preliminary Stage I flight and by lengthening the basic flight test interval to two months (with a spare vehicle which can be flown in the intervening month in case of failure). In addition, the firing of Viking rounds 13 and 14 have been canceled as such and these rounds will now be incorporated in the tests preceding the first Stage I flight.

Since the various development projects are now just starting, it is too early to make a new evaluation of the realism of the current schedule except to note that a two-month firing interval should be much easier to attain than a one-month firing interval and to note that the scheduled delivery of the first Stage II propulsion system is now one month late for its scheduled incorporation in the Viking 14 test.

5. The Advisory Group will hold its next meeting 19 December 1955 at the Glenn L. Martin plant in Baltimore for further surveillance of that contractor's effort in the program; the Group also plans a visit to Patrick Air Force Base sometime in January for a check on the test facility program as it bears on the test program of Project Vanguard.

HOMER J. STEWART
Chairman
Advisory Group on Special Capabilities

Document I-7

Document title: Colonel A.J. Goodpaster, "Memorandum for Record," June 7, 1956.

Document I-8

Document title: E.V. Murphree, Special Assistant for Guided Missiles, Memorandum for Deputy Secretary of Defense, "Use of the JUPITER Re-entry Test Vehicle as a Satellite," July 5, 1956.

Source: Both in Dwight D. Eisenhower Library, Abilene, Kansas.

Although there had been a decision to assign the International Geophysical Year scientific satellite project to the Naval Research Laboratory team and its Project Vanguard, Wernher von Braun and his associates of the Army Ballistic Missile Agency (ABMA) at Redstone Arsenal in Huntsville, Alabama, continued to argue that they could launch a satellite well before the first scheduled Vanguard launch. This claim was brought to the attention of President Eisenhower's assistant, Colonel Andrew J. Goodpaster, who consulted the president on how to respond. Inquiries on the issue were made to the Department of Defense. A quick review of the situation was conducted within the Department of Defense. The result was a recommendation that the approved plan not be changed, and an order was given to the Army Ballistic Missile Agency that it should not plan for, or attempt, a satellite launch.

Document I-7

[stamped "UNCLASSIFIED" over "SECRET"]

June 7, 1956

MEMORANDUM FOR RECORD

On May 28th Secretary Hoover called me over to mention a report he had received from a former associate in the engineering and development field regarding the earth satellite project. The best estimate is that the present project would not be ready until the end of '57 at the earliest, and probably well into '58. Redstone had a project well advanced when the new one was set up. At minimal expense ($2–5 million) they could have a satellite ready for firing by the end of 1956 or January 1957. The Redstone project is one essentially of German scientists, and it is American envy of them that has led to a duplicate project.

I spoke to the President about this to see what would be the best way to act on the matter. He asked me to talk to Secretary Wilson. In the latter's absence, I talked to Secretary Robertson today and he said he would go into the matter fully and carefully to try to ascertain the facts. In order to establish the substance of this report, I told him it had come through Mr. Hoover (Mr. Hoover had said I might do so if I felt it necessary).

A.J. Goodpaster
Colonel, CE, US Army

Document I-8

[stamped "UNCLASSIFIED" over "CONFIDENTIAL"]
[no pagination]

July 5, 1956

MEMORANDUM FOR DEPUTY SECRETARY OF DEFENSE

SUBJECT: Use of the JUPITER Re-entry Test Vehicle as a Satellite

I have looked further into the matter of the use of the JUPITER re-entry test vehicle as a possible satellite vehicle in order to obtain an earlier satellite capability as we discussed recently. I find that there is no question but that one attempt with a relatively small effort could be made in January 1957. [handwritten underlining] Also, an earlier attempt in September of this year is theoretically possible, although a decision to do this would clearly delay the JUPITER program. However, there are certain other aspects [handwritten underlining] of the matter which must be considered and which, in my judgment, are overriding. [handwritten underlining]

The proposal for making an attempt at a satellite is not new and, in fact, has been raised on several occasions during the history of the VANGUARD program. This may be explained by the fact that the original REDSTONE satellite and re-entry test vehicle proposals resulted from a common study, the results of which indicated that essentially the same vehicle could accomplish either task. Moreover, the first two flights of JUPITER re-entry test vehicles are scheduled primarily for propulsion system tests and could continue to serve a major part of their purpose in the over-all JUPITER test program even if they were used to carry the satellite vehicle. There is, however, room for serious doubt that two isolated flight attempts would result in achieving a successful satellite [handwritten underlining], and the dates of such flights would be prior to the Geophysical Year for which a satellite capability is specifically required, and prior to the time when tracking instrumentation will be available.

These facts were well known at the time that competing proposals were reviewed in the Office of the Secretary of Defense for undertaking the satellite program and the decision to assign this program to the Navy VIKING group, i.e., the Glenn L. Martin Company, under the code name VANGUARD, was made with the Army test vehicle possibilities taken into full consideration. That decision was based largely on a conviction that the VANGUARD proposal offered the greater promise of success. The history of increasing demands for funds for this program confirms the conviction that this is not a simple matter. I know of no new evidence available to warrant a change in that decision at this time. [handwritten underlining]

While it is true that the VANGUARD group does not expect to make its first satellite attempt before August 1957, whereas a satellite attempt could be made by the Army Ballistic Missile Agency as early as January 1957, little would be gained by making such an early satellite attempt as an isolated action with no follow-up program. In the case of VAN-GUARD, the first flight will be followed up by five additional satellite attempts in the ensuing year. It would be impossible for the ABMA group to make any satellite attempt that has a reasonable chance of success without diversion of the efforts of their top-flight scientific personnel from the main course of the JUPITER program, and to some extent, diversion of missiles from the early phase of the re-entry test program. There would also be a problem of additional funding not now provided.

For these reasons, I believe that to attempt a satellite flight with the JUPITER re-entry test vehicle without a preliminary program assuring a very strong probability of its success would most surely flirt with failure. Such probability could only be achieved through the application of a considerable scientific effort at ABMA. The obvious interference with the progress of the JUPITER program would certainly present a strong argument against such diversion of scientific effort.

On discussing the possible use of the JUPITER re-entry test vehicle to launch a satellite with Dr. Furnas, he pointed out certain objections to such a procedure. He felt there would be a serious morale effect on the VANGUARD group to whom the satellite test has been assigned. Dr. Furnas also pointed out that a satellite effort using the JUPITER re-entry test vehicle may have the effect of disrupting our relations with the non-military scientific community and international elements of the IGY group.

I don't know if I have a clear picture of the reasons for your interest in the possibility of using the JUPITER re-entry test vehicle for launching the satellite. I think it may be helpful if Dr. Furnas and I discuss this matter with you, and I'm trying to arrange for a date to do this on Monday.

E.V. MURPHREE
Special Assistant for
Guided Missiles

Copy furnished:
ASD (R&D)

Document I-9

Document title: Brigadier General A.J. Goodpaster, "Memorandum of Conference With the President, October 8, 1957, 8:30 AM," October 9, 1957.

Document I-10

Document title: Robert Cutler, Special Assistant to the President, Memorandum for the Secretary of Defense, "U.S. Scientific Satellite Program (NSC 5520)," October 17, 1957.

Source: Both in Dwight D. Eisenhower Library, Abilene, Kansas.

Five days after the launch of Sputnik 1, President Eisenhower met with a number of his advisers to assess the significance of the Soviet achievement and to consider how to respond. He was told that it would have been possible for the United States to have launched a satellite well before the Soviet Union. Eisenhower decided that it was best to proceed with the Vanguard program as it was planned; he announced later that day that the first Vanguard test launch was scheduled for December. Eisenhower also insisted that the program go forward on its current schedule, rather than be delayed to improve the instrumentation on the initial satellites.

Document I-9

[stamped "SECRET," declassified May 7, 1979]

[1]   October 9, 1957

MEMORANDUM OF CONFERENCE WITH THE PRESIDENT
October 8, 1957, 8:30 AM

                              Others present: Secretary Quarles
                                               Dr. Waterman
                                               Mr. Hagen
                                               Mr. Holaday
                                               Governor Adams
                                               General Persons
                                               Mr. Hagerty
                                               Governor Pyle
                                               Mr. Harlow
                                               General Cutler
                                               General Goodpaster

Secretary Quarles began by reviewing a memorandum prepared in Defense for the President on the subject of the earth satellite (dated October 7, 1957). He left a copy with the President. He reported that the Soviet launching on October 4th had apparently been highly successful.

The President asked Secretary Quarles about the report that had come to his attention to the effect that Redstone could have been used and could have placed a satellite in orbit many months ago. Secretary Quarles said there was no doubt that the Redstone, had it been used, could have orbited a satellite a year or more ago. The Science Advisory Committee had felt, however, that it was better to have the earth satellite proceed separately from military development. One reason was to stress the peaceful character of the effort, and a second was to avoid the inclusion of materiel, to which foreign scientists might be given access, which is used in our own military rockets. He said that the Army feels it could erect a satellite four months from now if given the order—this would still be one month prior to the estimated date for the Vanguard. The President said that when this information reaches the Congress, they are bound to ask why this action was not taken. He recalled, [2] however, that timing was never given too much importance in our own program, which was tied to the IGY and confirmed that, in order for all scientists to be able to look at the instrument, it had to be kept away from military secrets. Secretary Quarles pointed out that the Army plan would require some modification of the instrumentation in the missile.

He went on to add that the Russians have in fact done us a good turn, unintentionally, in establishing the concept of freedom of international space—this seems to be generally accepted as orbital space, in which the missile is making an inoffensive passage.

The President asked what kind of information could be conveyed by the signals reaching us from the Russian satellite. Secretary Quarles said the Soviets say that it is simply a pulse to permit location of the missile through radar direction finders. Following the meeting, Dr. Waterman indicated that there is some kind of modulation on the signals, which may mean that some coding is being done, although it might conceivably be accidental.

The President asked the group to look ahead five years, and asked about a reconnaissance vehicle. Secretary Quarles said the Air Force has a research program in this area and gave a general description of the project.

Governor Adams recalled that Dr. Pusey had said that we had never thought of this as a crash program, as the Russians apparently did. We were working simply to develop and transmit scientific knowledge. The President thought that to make a sudden shift in our approach now would be to belie the attitude we have had all along. Secretary Quarles said that such a shift would create service tensions in the Pentagon. Mr. Holaday said he planned to study with the Army the back up of the Navy program with the Redstone, adapting it to the instrumentation.

There was some discussion concerning the Soviet request as to whether we would like to put instruments of ours aboard one of their satellites. He said our instruments would be ready for this. Several present pointed out that our instruments contain parts which, if made available to the Russians, would give them substantial technological information.

A.J. Goodpaster
Brigadier General, USA

Document I-10

[stamped "SECRET," declassified February 27, 1986]

October 17, 1957

MEMORANDUM FOR THE SECRETARY OF DEFENSE

Subject: U.S. Scientific Satellite Program (NSC 5520)

I am writing this memorandum to you as Secretary of Defense because the Department of Defense is the responsible executive agency for carrying out the U.S. scientific satellite program in accordance with NSC 5520.

At a recent meeting of the National Security Council the President made very plain that the overriding objective of the IRBM and the ICBM programs is the successful achievement of these ballistic missiles with the necessary range and reasonable accuracy, in priority over related problems.*

Although recent Council action has not reflected a similar expression by the President with reference to the U.S. scientific satellite, the President's concern in this regard is no less clear. As you know, the President issued a statement to the press on October 9 that the first satellite test vehicle was planned to be launched in December, and that the first fully-instrumented satellite vehicle would be launched in March, 1958.

In line with this statement the President said yesterday that he wanted to be sure that the launching of the U.S. scientific satellite proceed as planned and scheduled. He is, of course, conscious of the understandable desire of the scientists to perfect the instrumentation that goes into the satellite. Nevertheless, he made very plain that any efforts further to perfect such scientific instrumentation should not be permitted to delay the planned launching schedule.

In order that there might be no ambiguity, I thought it advisable to send this memorandum to you as head of the responsible executive agency, with a copy to the Director of the National Science Foundation.

Robert Cutler
Special Assistant
to the President

cc: Director, National Science Foundation
General Goodpaster

---

* (NSC Action No. 1800-c)

Document I-11

Document title: Donald Quarles, Memorandum for the President, "The Vanguard-Jupiter C Program," January 7, 1958.

Source: Anne Whitman File, Dwight D. Eisenhower Library, Abilene, Kansas.

Although the rocket team at the Army Ballistic Missile Agency in Huntsville, Alabama, had been campaigning for several years to be allowed to attempt a satellite launch, it was only after the October 4, 1957, launch of Sputnik 1 that it was given permission to prepare to do so. Even then, the Army's Jupiter C launch vehicle was treated only as a backup to the Vanguard launcher. Only after the failure of the first Vanguard test launch in December 1957 was the Jupiter C satellite effort accelerated to aim at a late January 1958 launch.

[original marked "SECRET," crossed out by hand]
[1]

THE SECRETARY OF DEFENSE
WASHINGTON

JAN 7 1958

MEMORANDUM FOR THE PRESIDENT

SUBJECT: The VANGUARD-JUPITER C Program

The documents which led to the authorization on 8 November 1957 of the JUPITER-C back-up for the VANGUARD program and those which describe the present program are summarized as follows:

7 October 1957

The Secretary of the Army by memorandum to the Secretary of Defense stated that the success of the third JUPITER-C re-entry firing had solved the JUPITER re-entry problem leaving 8 remaining JUPITER-Cs in various stages of assembly as excess to direct JUPITER needs and that JUPITER-Cs could be readily modified to provide an early satellite capability. The Army estimated it would require four months from a decision date to the first launching and recommended a program based on launching six satellite vehicles requiring a total of $12,752,000 of non-Army funds.

14 October 1957

The Secretary of Defense by memorandum to the Secretary of the Army advised that it was planned to continue the VANGUARD program along the current scientific lines. The Army was asked to restudy its proposal and suggest means appropriate for a back-up of VANGUARD directed toward the launching of the 21-lb. sphere, a part of the VANGUARD scientific program. The suggestion was made that the possibility of component assistance to VANGUARD as well as the possibility of an independent Army launching program be covered together with estimates of time required and the cost of the project.

23 October 1957

The Secretary of the Army by memorandum to the Secretary of Defense stated the Army believed it could place a VANGUARD sphere in orbit in June 1958 by using the JUPITER as the first stage and the JUPITER-C three stage solid propellant cluster as the upper three stages. This program, estimated to cost $16.2 million, called for four launchings. To give the proposed approach a high assurance of success, the Army recommended the launching of a JUPITER-C cylindrical satellite in February and another in [2] April to provide the basic knowledge to help to place a VANGUARD sphere in orbit in June. A second JUPITER/JUPITER-C/VANGUARD sphere vehicle which could be launched in September was proposed as additional assurance. In this memorandum the Army advised that it would be possible to package in the JUPITER-C cylindrical satellite instrumentation which would directly support scientific experiments which are a part of the VANGUARD program. Other approaches were considered by the Army but in the opinion of the Army offered little chance of success.

29 October 1957

The Special Assistant for Guided Missiles by memorandum to the Secretary of Defense advised him of the program recommended to him by the Advisory Group on Special Capabilities to provide maximum assurance of success for the VANGUARD satellite program. This group recommended the use of two JUPITER-C type vehicles to be used to carry two scientific satellites on orbit. The Army's estimated cost of converting REDSTONES to JUPITER-Cs and providing for the launching was given to be $3.5 million with the cost of the REDSTONE missiles to be absorbed by the Army. The Special Assistant for Guided Missiles approved this back-up program and recommended that it be called to the attention of the President for his approval.

8 November 1957

The Secretary of Defense by memorandum to the Secretary of the Army acknowledged the Army offer (memorandum of the Secretary of the Army to the Secretary of Defense dated 23 October 1957) to help assure that the U.S. IGY scientific satellite would maintain the announced schedule. In this memorandum, the Army was requested to provide the capability of launching a satellite containing scientific instrumentation by the use of a modified JUPITER-C test vehicle. The Army was authorized to proceed with the necessary preparation to attempt two launchings during March 1958, the actual dates to be determined later. Funds in the amount of $3.5 million were authorized to support the program.

22 November 1957

The Director of Guided Missiles by memorandum to the Secretary of the Army notified the Army of the assignment to the JUPITER-C vehicle of the cosmic ray experiment originally scheduled for VANGUARD and disapproved the provision of additional microlock telemetry receiving stations in connection with the program.

[3] 3 December 1957

The Director of Guided Missiles by memorandum advised the Director of Research and Development, Department of the Army, that disapproval of additional telemetering ground receiving facilities did not limit the Army from using whatever ground equipment is required in conjunction with missile-borne instrumentation to assure success during the launching phase of the flight.

21 December 1957

By letter to Maj. Gen. D.N. Yates, Commanding General, Air Force Missile Test Center, the Director of Guided Missiles advised Gen. Yates of the schedule for a launching that would be complied with as nearly as possible and furnished the same information to the Army, the Navy and the U.S. National Committee for the IGY.

TV-3 BU	January 18, 1958
JUPITER C-1	January 29, 1958
TV-4	February 10, 1958
TV-5	March 3, 1958
JUPITER C-2	March 5, 1958
SLV-1	March 24, 1958

Possible adjustment in the schedules for TV-5 and JUPITER C-2 may be made as follows:

a.　If TV-5 launches a successful satellite at its scheduled time (Mar 3, 1958), JUPITER C-2 will be delayed until about March 8, 1958, in order to provide adequate time for geophysical data gathering and reduction on TV-5 satellite.

b.　If TV-5 satellite is unsuccessful, JUPITER C-1 launching attempt will be made on March 5, 1958 or as soon thereafter as possible.

c.　Should the scheduled launching date for TV-5 be delayed beyond March 3, 1958, the following will apply as appropriate:

(1) Launch JUPITER C-2 on the 5th of March or immediately thereafter. If flight is successful, TV-5 may be scheduled for 7 days later.

(2) If JUPITER C-2 fails, TV-5 would be launched at the earliest possible date.

[signature only] Donald A. Quarles

Document I-12

Document title: James C. Hagerty, "Memorandum on Telephone Calls Between Brigadier General Andrew J. Goodpaster in Washington and James C. Hagerty in Atlanta, Georgia, Friday Afternoon and Evening, January 31, 1958 and Saturday Morning, February 1, 1958."

Source: Dwight D. Eisenhower Library, Abilene, Kansas.

Particularly after the December 1957 public failure of the first attempt to launch a Vanguard satellite, getting a U.S. satellite into orbit became a matter of great interest to President Dwight Eisenhower. The January 31, 1958, attempt to launch the Explorer I satellite was not announced in advance, as had been the case with the Vanguard launch attempt. The White House decided to announce the launch only after the satellite was already in orbit. James Hagerty was President Eisenhower's press secretary; General Andrew Goodpaster was one of his senior staff assistants.

[1]

Memorandum on Telephone Calls Between Brigadier General Andrew J. Goodpaster in Washington and James C. Hagerty in Augusta, Georgia, Friday Afternoon and Evening, January 31, 1958 and Saturday Morning, February 1, 1958

2:30 P.M.   On arrival in Augusta, I called General Goodpaster at the White House and he told me that high winds in the stratosphere were still postponing the Jupiter-C shooting. I relayed this information to the President in his cottage at the Augusta National Golf Club.

5:30 P.M.   General Goodpaster called me at my room at the Bon Air Hotel to tell me that the weather was improving at Cape Canaveral and that the Army was going to try to shoot the Jupiter-C tonight at 10:30 P.M. plus four minutes.

                 I immediately drove out to the Augusta National. The President was in the living room playing bridge with Barry Leithead, Cliff Roberts, and Clarence Schoo. He was playing a game bid which he made in four hearts. At the conclusion of his hand, he walked with me to the opposite corner of the room, and I told him of the message from General Goodpaster.

                 The President was immediately very interested and said that he certainly hoped that if the weather were right, the shot would be made tonight. I told him that General Goodpaster said he would call again at 8:30, and the President left it this way: If at 8:30 I had additional news that the launching was still on, I would come out and tell him. If, however, it was scrubbed out, I would merely call John Moaney and ask him to tell the President, "Nothing doing."

                 I then left to return to the Bon Air Hotel.

[2] 8:30 P.M.   General Goodpaster called me again at this time and told me that the weather had improved to the point that it was acceptable as of now, that the Army was planning to go ahead and that they were beginning fueling of the rocket as of 8:30.

                 I drove out again to the President's cottage and told him the news. He asked me when I thought the launching would occur, and I told him it was now scheduled for 10:34 P.M. He told me to keep in touch with him.

                 I then returned to the Bon Air Hotel.

9:50 P.M.   General Goodpaster called again to say that the launching was definitely on and that he was leaving his home at this time to go to the office. He said that they were still four minutes behind schedule and that the launching was scheduled for 10:34 P.M.

General Goodpaster said he would call me after getting to the office, and after he had a chance to check once again on the launching.

10:25 P.M.   General Goodpaster called from his White House office to tell me that the launching was now running 10 to 15 minutes behind the schedule and that it looked as if the launching now would be held between 10:40 and 10:45.

10:30 P.M.   I called the President at his cottage at the Augusta National and informed him of the additional delay. He again urged me to keep in touch with him and let him know when the launching was made.

[3] 10:40 P.M.   General Goodpaster called again to say that he was going to keep the White House line from his office to my office at the Bon Air open, that he also had a direct line into the Telecommunications Center at the Pentagon where he was receiving reports. The Telecommunications Center had a direct line to Canaveral.

10:42 P.M.   General Goodpaster reported that he had nothing definite yet on any launching, but that it was expected soon. He asked me if I had told the President about the delay, and I said I had, and that the President had said, "I'll be here. Call me as soon as you get anything."

10:43 P.M.   General Goodpaster told me that he had just received word that X minus 7 was at 10:41. (In other words, the reports he was receiving from the Telecommunications Room were running two minutes behind the actual events at Cape Canaveral.)

                  From then on, General Goodpaster gave me the countdown, which went as follows:

                  "X minus 6—10:42 P.M.
                  X minus 5—10:43 P.M.
                  X minus 4—10:44 P.M.
                  X minus 3—10:45 P.M.
                  X minus 2—10:46 P.M.
                  X minus 1—10:47 P.M.
                  X minus 20 seconds—10:47:40"

                  Twenty seconds after this, General Goodpaster said, "Jim, they have given the firing command at 10:48. It takes 16 seconds to start the rocket lifting off the ground. Here's the report.

[4]   "The main stage lifted off at 10:48:16.

                  "The program is starting O.K.

                  "They are putting it in the right attitude.

                  "It is still going, they say.

                  "It is still going at 55 seconds.

"It is still going and looks good at 90 seconds.

"Jupiter is on the way!

"It is through the jet stream—They were worried about that jet stream.

"115 seconds, it is going higher and higher.

"Everything is going all right at 145 seconds."

I interrupted at this time to say that I thought it would be a good idea if we were to call the President now and get him in on a three-way conversation between him, General Goodpaster and myself. General Goodpaster agreed, and I asked the Signal Corps operator to cut the President in.

Meanwhile, I had the radio turned on in my room, and the first bulletin on the launching came in at this time. It was a CBS station, and Chuck Von Fremd was doing the reporting.

General Goodpaster continued to relay to me the reports which he was receiving from the Telecommunications Center. They went like this:

[5]   "The first stage has been cut off O.K.

"180 seconds report—Everything going O.K.

"Everything O.K. at three and a half minutes after the launching."

I interrupted to say, "Andy, I am talking notes on this. I will dictate it when I get back to Washington. I am sure you will want a copy of this."

"You bet I will—Thanks," Goodpaster said.

10:56 P.M.   The President was cut in to the conversation, and General Goodpaster brought him up to date on the reports thus far.

10:58 P.M.   General Goodpaster said that the second stage ignition had gone off O.K.

The President asked Goodpaster when the announcement would come that it was in orbit and Goodpaster replied that that would take one and a half or two hours before they were definitely sure.

The President thanked Goodpaster for the information. He said that I was to let him know just as soon as it went in orbit.

[6] 11:03 P.M.   Goodpaster, still relaying information from the Telecommunications Center, said that the launching was completely successful.

I told the President that I would tell the press that the President was being kept informed, and he said that that was right. He then said he would cut out of this conversation, that if we had anything more of a major announcement to tell him within 30 minutes, we were to call back, but we were to call back anytime we heard the satellite went in orbit.

11:05 P.M.   The President cut out of the conversation.

11:06 P.M.   While we were awaiting further word, General Goodpaster and myself discussed the "color" that I was planning to give to the newsmen on how the President received the information. We both agreed that this would be a good thing to show the President was in close touch with the situation.

General Goodpaster told me he was in direct touch with the Telecommunications Center—"in the heart of the Army section."

He was receiving reports directly over the telephone from Major Nicholson Parker, USA, and Miss Jean Ferguson, a civilian receptionist.

11:10 P.M.   Goodpaster: "We are waiting for a little more information right now. All we need is some more information. We have been on the phone a little more than a half hour. Do you realize that?"

[7]   Hagerty: "No, I didn't. Time sure goes pretty fast, doesn't it?"

11:15 P.M.   Goodpaster: "It's been a long time now since we've received additional reports."

Hagerty: "What's the trouble?"

Goodpaster: "Nothing. Probably they have fired all stages by now, but they have got to be sure. I don't have anything on the last two stages. As a matter of fact, General Maderas has just sent word to the Telecommunications personnel to go out for a cup of coffee and sweat it out with him."

11:22 P.M.   Hagerty: "Can you tell me who is in the Telecommunications Room, besides Major Parker and Miss Ferguson?"

Goodpaster: (Talking to Telecommunications Room on the other phone)

"Who is there with you?"

Goodpaster to me:

Dr. Wernher von Braun is in the Telecommunications Room with Major Parker. So is Secretary Brucker. Also—

General Lyman Lemnitzer
Vice Chief of Staff

Dr. Herbert York
University of California
Director of Radiation Laboratory

[8]   Dr. William Pickering
                            Director, Jet Propulsion Laboratory

                            Dr. James A. Van Allen, who designed the cosmic ray experiment

                            Dr. Richard H. Porter, who is a member of the working group for
                            earth satellites of the National Academy of Sciences

                            and Murray Snyder

11:28 P.M.   Goodpaster told me that Orville Splitt would call in on a White House
                            phone from the National Academy of Sciences and that when the time
                            for the orbiting announcement became necessary, Splitt would hold until
                            I had a chance to call the President, and then Goodpaster would arrange
                            with them the time for the simultaneous announcement.

11:30 P.M.   Goodpaster was on the phone with the Telecommunications Room, and
                            reported to me as follows:

                            "Jim, it looks as if it will be a number of minutes, probably fifteen, before
                            we know anything further. Reports are presently being analyzed and stud-
                            ied, and we won't know anything for a little time. Maybe I had better call
                            you back."

                            Hagerty: "No, I'll hold on if you don't mind. After all, I can't tell the
                            President anything now anyway."

11:37 P.M.   Goodpaster: "This is secret. The first analysis that we have received is that
                            the satellite has passed over the first station, Antigua, on time. This is very
                            encouraging, and it tends to show that the [9] third and fourth stages
                            went off all right. Yes, I think it is a fair statement that the third and
                            fourth stages went off O.K."

11:40 P.M.   Goodpaster said that it would be at least a half hour more before we got
                            any word on whether it had gone into orbit, and I told him that I would
                            go to see the press and fill them in on some "color" as to how the
                            President was keeping in touch with the news from Canaveral.

                            Goodpaster agreed that that would be a good thing to do since it would
                            show to the world that even though the President is out of Washington,
                            he keeps in close touch with all important situations as they develop.

11:45 P.M.   I signed off temporarily with General Goodpaster and went to the press
                            room for a press conference.

                                            * * * * *

While I was having the press conference, Goodpaster tried to get me, but talked to the President directly, and it was agreed between the President and General Goodpaster that the President would stay up until there was definite word about whether the satellite was in orbit or not.

As a sidelight to this conversation, I had left Betty Allen in my room manning the phone while I went down to the press conference. The phone rang, and it was the President. He asked Betty whether we had heard any later word, and she said we had not. The President then started to ask her other questions about the launching. She had taken notes on all the conversations between Goodpaster and myself, but the notebook was in the other room. Just at this point, [10] General Goodpaster called in again, and Betty transferred him to the President.

I finished my press conference about 12:05 A.M. and told the newspapermen that I would see them next when I had any definite word.

12:09 A.M.   I got back on the phone with General Goodpaster. He told me that the preliminary analyses were quite favorable. He also told me of his conversation with the President and the fact that the President had told Goodpaster—"Let's not make too great a hullabaloo on this."

12:11 A.M.   General Goodpaster got a call from Orville Splitt who told him that we would receive the scientific word from Dr. J. Wallace Joyce, Head of the International Geophysical Year office of the National Science Foundation, in place of Dr. Alan Waterman, the Director, as our original announce-ment had contemplated. Dr. Waterman had left Washington earlier in the day after it looked as if bad weather would postpone the launching Friday evening. I changed the advance statement to make it read: "Dr. J. Wallace Joyce, Head of the International Geophysical Year office of the National Science Foundation, informed me . . ." (meaning the President).

12:28 A.M.   Goodpaster: (who was now working three phones—one to the Telecommunications Room in the Pentagon, one to me in Augusta, and one to Orville Splitt at the National Academy of Sciences)

                  Goodpaster told me:

[11]   "Governor Brucker has just told me that everything is going good, that the Army has an open line to Pasadena and that they expect the satellite to pass over San Diego fairly soon. As soon as that happens, it will be final proof that it is in orbit and it will be O.K. to announce."

12:32 A.M.   Goodpaster said: "Governor Brucker says they are expecting it over San Diego very shortly now and that they should be hearing from Pasadena within four minutes."

12:42 A.M.   I could hear General Goodpaster say over the phone to the Pentagon: "Yes . . . Yes . . . You say it's in orbit? Good! That's fine!"

12:43 A.M.   General Goodpaster to me: "Jim, it's in orbit. You can call the President."

12:44 A.M.	I asked the Signal Corps to ring the President. They did so, and he answered immediately. I said, "Mr. President, it's in orbit. General Goodpaster has just received the official word."

The President replied, "That's wonderful. That's wonderful. Are you going to put out my statement now?"

I told him that I was, and he replied, "That's wonderful. I sure feel a lot better now."

12:45 A.M.	I came back on the phone with Goodpaster and said, "Andy, is it O.K. for me to make the announcement now?"
[12]	He said, "Yes, how much time do you want and what should we tell Orville Splitt?"

I said, "Give me five minutes. That's all I need so that the President's announcement can get out of Augusta as the first official word of the orbiting."

I hung up and dashed down to the press room to make the announcement.

1:10 A.M.	I returned from my press conference and called Andy again and said, "That cleans us up for the night unless you have anything further to add."

He said he had not, and at 1:12 we ended the conversation for the evening.

(Later in the morning)

8:30 A.M.	I called General Goodpaster, and he suggested that the President send a message to Dr. Alan Waterman, which read as follows:

"My congratulations to you and your colleagues. May I ask you to extend my personal congratulations to all—in whatever capacity they participated—who have been working in the development of satellites for scientific purposes. Would you also extend my congratulations to the personnel who took part in the successful orbiting of our satellite last night."

[13]	General Goodpaster, in answer to questions from me, also said they would try with the Vanguard, weather permitting, as soon as possible anytime from Monday morning on. The Vanguard, it was reported, was now back on the pad.

I went out to the Augusta National Golf Club and the President approved the message to Dr. Alan Waterman.

10:00 A.M.	I called back General Goodpaster and told him that the message was O.K., that I would send a telegram directly from Augusta to Dr. Waterman, and would also send him a copy on the teletype so he could get it immediately to Dr. Waterman.

Document I-13

Document title: Brigadier General Bernard A. Schriever, USAF Commander, Western Development Division, Air Research and Development Command, Memorandum for Lt. General [Thomas] Power, "Air Frame Industries vs. Air Force ICBM Management," February 24, 1955.

Source: Professor Stephen Johnson, Department of Space Studies, University of North Dakota, Grand Forks, North Dakota.

As its space efforts took shape in the late 1950s, access to Earth orbit and beyond for the United States came from three lines of development. One was the work of Wernher von Braun and his "rocket team" working at the Army Ballistic Missile Agency in Huntsville, Alabama. Another was the development, under the management of the Naval Research Laboratory, of the Vanguard booster designed specifically to launch the first U.S. scientific satellite. The third was the adaptation of various Air Force ballistic missiles, including the Thor-Delta intermediate range ballistic missile and the Atlas and Titan intercontinental ballistic missiles (ICBMs), for use as space launchers.

In the mid-1950s, Air Force ICBM efforts were managed by the Western Development Division of the Air Research and Development Command; its commander was Brig. General Bernard A. Schriever, who was to become a strong advocate of the Air Force in a lead role for the national space program. Given the urgent nature of the ICBM program, Schriever adopted innovative management approaches, such as concurrent development of various system elements. He also placed the Air Force, and his organization, in the role of systems manager for ICBM development efforts. To assist the Western Development Division and Schriever in this systems management role, two individuals from Hughes Aircraft, Simon Ramo and Dean Wooldridge, formed the Ramo-Wooldridge Corporation (which eventually became TRW).

As this memorandum suggests, although there was industry resistance to this strong centralized systems management approach by a government agency, the Air Force pursued such an approach with significant success. As the United States organized its civilian space effort in 1958, subsequently, the new National Aeronautics and Space Administration adopted elements of the approach. In particular, the approach was important to the success of the Apollo program (see Documents I-43 through I-46), and several of those steeped in it (particularly George Mueller and Lt. General Sam Phillips) were key Apollo program managers.

[all pages stamped "SECRET," crossed out; stamped "CONFIDENTIAL," crossed out; stamped "DOWNGRADED AT 3 YEAR INTERVALS[.] DECLASSIFIED AFTER 12 YEARS. DOD DIR 5200.10"]

[1]   WESTERN DEVELOPMENT DIVISION
                              HEADQUARTERS
              AIR RESEARCH AND DEVELOPMENT COMMAND
                         Post Office Box 262
                         Inglewood, California

[stamped 24 Feb 55]

MEMORANDUM FOR LT GENERAL POWER

SUBJECT: Airframe Industries vs Air Force ICBM Management

1. In the 8 November 1954 issue of Aviation Week, the following item appeared—

    "MISSILE PROBLEMS
        Aircraft industry is growing increasingly uneasy over recent trends in the busi-
    ness pattern for new USAF missile developments. Aircraft Industries Assn. |AIA|
    is considering a strong protest to the Pentagon. Big battle on upper Pentagon lev-
    els looms now between the established missile contractors and the Johnny-come-
    latelies in the field."

2. In the last month, I have had conversations with several people which lead me to
    believe that the AIA may apply pressure top side against the Air Force ICBM man-
    agement approach. A straw in the wind was an invitation from General Baker to
    attend the AIA convention in Phoenix for the purpose of clarifying any questions
    the industry might have. I declined the invitation, but advised him by letter that I
    would gladly discuss the matter with appropriate company officials, if they
    desired, and a need-to-know existed.

|2| 3. To the above, can be added Frank Collbohm's letter and the knowledge that his
    views have been disseminated to a number of RAND personnel including the
    RAND Board of Trustees. Also, there is definite evidence that these views have
    been passed to some RAND visitors, who had no official connection with the pro-
    gram. I also have good reason to believe that this matter has been brought to the
    attention of some members of Congress, who I am sure, have not heard the offi-
    cial Air Force position.

4. Although to my knowledge, the AIA has not made a specific counter-proposal, it
    is fairly simple to speculate on the management approach they favor. It would cer-
    tainly be in the pattern of the past Air Force missile developments, in which an
    airframe company is designated prime contractor, with complete weapon system
    responsibility. They are naturally motivated by self-interests, which I believe to be
    as follows:
        a. Adherence to the prime contractor concept.
        b. Avoidance of strong Air Force system management control.
        c. The Ramo-Wooldridge Corporation as potential competitors.

    Adherence to the prime contractor concept—
        This concept has permitted a broad expansion by the airframe industry into
    component field, such as electronics, propulsion, inertial guidance and control
    (automatic pilot), etc.
|3| North American is perhaps the outstanding example of such post-war expan-
    sion. AFR 70-9, recently published, clearly indicates that the Air Force desires to
    reverse this trend with its inherent disadvantages which are:
        (1) narrowing the industrial base at the expense of existing component
        industries,
        (2) large scale proselytizing of scientific and engineering personnel.

(3) second-rate competence of airframe industry in component areas—certainly during the build-up stage and resulting program delay and higher cost.

The ICBM management approach is based on using the most competent component industries in each of the major technical areas. This will eliminate time-consuming build-up of staff and facilities and insure [sic] the broadest and most competent base for this complex program.

Avoidance of strong Air Force system management control—

While the airframe industry offers lip service to Air Force weapon systems management control—past performance clearly indicates that our management has not been too effective. The project office has been no match for the powerful pressure that industry can, and has, exerted at political and high military levels. As a result, industry has usually prevailed on major matters in their interest.

[4]   In the present management approach, a very high degree of authority has been vested in the Commander, [Western Development Division], and through the services of Ramo-Wooldridge, the organization possesses a high degree of technical competence for managing the program. This will permit the making of hard decisions, based on a rationalization of technical and military factors.

I think it is clear to the aircraft industry that our organization has the potential of exercising very strong weapon systems management control. In the past, the aircraft industry, as prime contractor, has usurped much of this control. This has resulted in the expansion of airframe companies mentioned in a. above. For example, in the NAVAHO program, North American, in addition to the airframe, is building the rocket motor booster, the inertial navigation and guidance system and most of the electronic system. CONVAIR, in the ATLAS program[,] was developing the radar tracker and communications links—despite the fact that every review of the program by competent and objective scientists and engineers concluded that the CONVAIR approach was wrong and that their competence in the electronic field was considerably below that of a number of first-line electronic companies. The history of Northrop's performance in the SHARK program follows the same pattern.

[5]   The Ramo-Wooldridge Corporation as potential competitors—

It is only natural that industry does not welcome a competitor with open arms. In the case of Ramo-Wooldridge, the feeling is perhaps stronger since they have an outstanding reputation in science and industry. Many knowledgeable individuals credit Dr. Ramo and Wooldridge as the major factor in the rapid rise of Hughes Aircraft. Of course, this is a point on which the industry cannot be articulate, but there can be little doubt that they do not relish Ramo-Wooldridge as future competitors.

5.   The above motives of the aircraft industry are camouflaged by a number of assortions [sic], which I have heard and will enumerate below.

a.   The Air Force is building up Ramo-Wooldridge and this is un-American. Discussions on this point take several forms and there is always the inference that Ramo and Wooldridge, in leaving Hughes, were unethical and therefore, not deserving of Air Force support.

First, in leaving Hughes, and forming a new company, they followed a time-worn pattern in U.S. industry in which the airframe companies are perhaps the leaders. Everyone in aviation knows the history of Douglas, Kindleberger, Bell and others and that most of our major companies were formed from a splinter of an existing company.

[6]   Furthermore, the airframe industry owes its existence and present affluence to Government support in contracts and Government furnished facilities. For example, the total North American Government-owned facilities amount to $61,684,722. An additional $24,017,015 has been approved by the Air Force for further construction, of which approximately $19,000,000 is in support of the ATLAS program. Douglas Aircraft have [sic] $108,050,000 worth of Government facilities at their disposal, and Lockheed has $22,000,000. As for the Ramo-Wooldridge Corporation, the Air Force has to date furnished them with an analogue computer. In summary, the assertion that the Air Force is building up Ramo-Wooldridge and the inference that they were unethical in leaving Hughes Aircraft is very much in the category of "the pot calling the kettle black." Air Force action in this case, is entirely in keeping with past practices.

b.   Perhaps the statement heard most often is that the ICBM system is ready for production engineering and should be turned over to an old-line company with a free hand technically, and the funds required to proceed.

The corollary of this statement is that Ramo and Wooldridge are scientists, surrounding themselves with scientists, and are optimizing to such an extent that the program will be delayed for several years.

[7]   The first statement is of course a matter of judgment and while no one quarrels with the technical feasibility, the Air Force is convinced that the technical complexities and the advances of the ICBM are each substantially greater than past development projects. For example, the project has many conspicuous firsts. No one had brought a vehicle of anything approaching this size, to a speed of 20,000 feet per second. No one has controlled the velocity, even much more crudely than this, at the ranges required. No one has made to fly stably [sic], a vehicle which changes its "primary autopilot" constants as it flies, by virtue of radical changes in weight, center of pressure and the like.

In this connection, it is also well to note the position taken by Douglas and North American in conversations with representatives of these companies late this summer. Douglas indicated the ICBM was too big a bite to take in one step and the development should be in series starting with the short range ballistic missile. North American indicated that they had constantly maintained interest in the ballistic missile but gave the impression that it was rather far out in the blue, and the NAVAHO ramjet approach was the correct one and much more realistic. [8] Aircraft industries' performance in the missile field has also not been impressive. For example, the NAVAHO program has slipped a total of 8.3 years and the SHARK and MATADOR over 4 years. All other missile programs have slipped varying amounts.

With respect to optimization by Ramo-Wooldridge, it should be noted that the Strategic Missile Committee recommended that a comparative analysis and technical study to be undertaken to establish a reoriented ICBM program. This was concurred in by the Air Force Council. The analysis and study had been underway for a number of months and while the objective has been to optimize, it is to optimize the approach which will lead to the earliest operational capability. We strongly feel that this is being accomplished. One outstanding example of this being the reorientation of the configuration. We are certain that a three-engine configuration weighing slightly over 200,000 lbs, can do the job and will replace the CONVAIR five-engine 450,000 lbs. configuration. In this connection, CONVAIR, as late as early September, was recommending that the Air Force approve the five-engine configuration and launch into an all-out program on this basis.

[9]   c.   Finally, the assertion which I have heard a number of times is that the present management approach eliminates competition. The fact is that the opposite is true. We are opening up the program for competition. The top electronics companies have been invited to compete for the development of the radar tracker, and the same applies to the computer and inertial guidance system. Likewise, in conjunction with the [Atomic Energy Commission], we are giving consideration to the outstanding companies for development of the nose cone. In other words, we are going to the industries where the greatest competence exists for each of the major components of the system. Compare this with the approach taken by the airframe companies in the development of NAVAHO, SHARK, and ATLAS. As I have already pointed out, in each instance, they established within their own company, departments for development of components, where component industries of great competence already existed.

6.   CONCLUSIONS:
        a.   The airframe industries based on self-interest apparently desire to upset the present Air Force ICBM management approach.
        b.   They probably favor the prime contractor approach along the pattern of NAVAHO, SHARK, and (ATLAS, prior to the establishment of [the Western Development Division]).
[10]   c.   The component industries are not organized on this matter but would probably support the Air Force management approach once it is entirely clear to them.
        d.   The assertions made by the airframe industries concerning the [Western Development Division]-Ramo-Wooldridge set up do not bear up under close inspection.
        e.   The Air Force management approach is sound in that it
             (1) provides the strongest possible weapon system management team, with control remaining in the Air Force,
             (2) insures [sic] that the most competent component industries participate,
             (3) is consistent with AFR 70-9, reversing the trend of airframe company expansion into component fields.
             (4) has the support of the scientific community,
             (5) is streamlined to permit crash operations and is most likely to convince higher authority that the Air Force is not pursuing this program on a "business-as-usual" basis.

7.   RECOMMENDATIONS:

        a.   That the Air Force remain firm on the ICBM management
[11]   b.   that the Secretary's level be advised of certain information contained in this memorandum to off-set any pressure which the industry may bring to bear at that level.

                                        BERNARD A. SCHRIEVER
                                        Brigadier General, USAF
                                        Commander, Western Development
                                        Division (ARDC)

Document I-14

Document title: Brigadier General A.J. Goodpaster, "Memorandum of Conference With the President, March 10, 1958—10:20 AM," March 11, 1958.

Source: Dwight D. Eisenhower Library, Abilene, Kansas.

As part of organizing the government's space and missile programs in the months following Sputnik, President Eisenhower, in the meeting recorded in this memorandum, took initial steps to begin a significant examination of solid fuels for missile and space use and to eliminate overlap in the intermediate range ballistic missile (IRBM) program. He also initiated consideration of giving a significant role in the space program to the von Braun rocket team based in Huntsville, Alabama. George B. Kistiakowsky would later replace James R. Killian, Jr., as Eisenhower's science advisor in July 1959.

[stamped "UNCLASSIFIED" over "SECRET," declassified May 5, 1987]
[1]

March 11, 1958

MEMORANDUM OF CONFERENCE WITH THE PRESIDENT
March 10, 1958—10:20 AM

> Others present: Dr. Killian
> Dr. Kistiakowsky
> General Goodpaster

Dr. Killian spoke from a memorandum, the original of which he handed to the President.

With regard to the proposal for a well-conceived basic research effort on solid propellants, the President strongly stressed that an overall group, such as ARPA [Advanced Research Projects Agency], should conduct this research. Other-wise, it would be done in bits and pieces. In fact, he thought that all research on fuels should be kept centralized, avoiding the wastes of duplicating effort. Dr. Kistiakowsky reported that there has really been very little support for, or interest in, a solid propellant develop-ment program. There have been many starts and stops, and the effort that has been devoted to these fuels has been very small. In the interest of economy of effort and continuity, he would agree with putting the program into ARPA. The President suggested that it might even be put in the civil agency now under consideration.

Dr. Killian stressed the need for a review by the President of proposals for "second generation" missiles. The President strongly agreed and asked that necessary directives be developed.

The President further agreed with the recommendation for a program of improve-ment on the TITAN missile, and for phasing out the ATLAS as soon as consistent with an adequate rate of buildup of total missile forces.

The President said that he conceived of the missile activity as separate and distinct from traditional air, ground, and sea operations. He would accept the logic of a decision by the Department of Defense to assign a submarine-based missile such as POLARIS to the Navy, but he saw no reason for the Air Force or for the Army to try to preempt the field. Instead, he would incline toward a single missile command. Specifically, he agreed that we should not rush into the proposed [2] Minuteman program: he asked that there be no approval along these lines until the matter had been much more carefully considered, and presented to him. Dr. Killian repeated his recommendation that Defense should not produce both THOR and JUPITER. The President said that so far as he is concerned there is no problem with dropping either of these. He asked what could be done with the team at Huntsville, which he understood was a group of outstanding ability. Dr. Killian said that they are working on the PERSHING missile family. He also said that this group is well suited to conducting space program activities, either under ARPA or NASA.

The President asked why Drs. Killian and Kistiakowsky thought that the THOR was a better missile than the JUPITER. Dr. Kistiakowsky said it is not better, but simply nearer to quantity production. He feels that the shift to industrial producers of the JUPITER (Chrysler, Ford Instrument, and Goodyear) would delay its availability. The President said that he would agree to closing out the JUPITER, but thought the Huntsville force should be promoted to space and similar activities. He thought consideration should be given to taking them out of their present assignment and assigning them to ARPA, or even to NASA. Dr. Kistiakowsky commented that the PERSHING is an excellent approach, and the President said that the Huntsville group could work on that project too.

The President asked Dr. Killian to prepare for him a series of decisions very tightly drafted and very positive in tenor to accomplish what had been recommended. He said he strongly agreed with the basic proposal to obtain centralized direction and thought this should be done soon.

Dr. Killian asked whether he should ask the Secretary of Defense to carry out studies to give effect to the proposals. The President said this would be all right, but that we should make clear what the scientific conclusions and recommendations are. Dr. Killian said he was prepared to do this.

> A.J. Goodpaster
> Brigadier General, USA

Document I-15

Document title: Hugh L. Dryden, Deputy Administrator, NASA, Memorandum for the President, "Use of Solid Propellants in U.S. Space Program," April 7, 1961.

Source: Documentary History Collection, Space Policy Institute, George Washington University, Washington, D.C.

Solid rocket motors had a place in the development of the U.S. space program from the very beginning, although they tended to find more favor within the Department of Defense than within NASA. This memorandum summarizes for President Kennedy the state of the use of solid rocket motors in the U.S. space program three months after Kennedy assumed office and just a month and a half before he made his decisive speech before Congress calling for landing a human on the Moon within the decade.

[1]
NATIONAL AERONAUTICS
AND SPACE ADMINISTRATION

OFFICE OF THE ADMINISTRATOR
1520 H STREET NORTHWEST
WASHINGTON 25, D.C.

April 7, 1961

MEMORANDUM FOR THE PRESIDENT

Subject: Use of Solid Propellants in U.S. Space Program

Solid propellant rockets have been used in the U.S. space program from its very beginning. Explorer I used a liquid fueled rocket for the first booster stage, but the three upper stages were made up of clusters of solid propellant rockets. Although the performance of solid propellants is somewhat inferior to that of liquid propellants, the simplicity of solid rockets makes their use extremely attractive. When NASA was established in October of 1958 a project was started to develop a space vehicle capable of placing 150 pounds in a 300-mile orbit about the earth based completely on the use of solid propellant rockets. An attempt was made to use the Polaris as the first-stage booster but at the time the Navy could not make this rocket available. As a substitute one of the rockets which had been used in the development of Polaris was selected. This vehicle now supports an important part of our space program, and is called "Scout."

The extension of the use of solid propellants to larger vehicles has been carefully studied. Even in the size used for the first stage of Scout it is necessary to transport the fully loaded stage as a unit by means of a special trailer accompanying each booster, provided with an electric heating blanket to maintain the temperature of the booster within certain limits. Until very recently the necessity of transporting the completely loaded stage as a unit has seemed to present rather formidable difficulties for still larger stages. It has [2] been proposed in at least one instance to manufacture the solid propellant at the launch site and to load the vehicle in place. This seems to be a rather impractical proposal to most of us. The recent development of segmented solid propellant rockets, i.e., those which are made in a number of separate pieces which can be bolted or otherwise fastened together at the launch site, seems to offer a way of overcoming these logistic difficulties. NASA has supported one group in developmental testing of this segmented approach and the Department of Defense has supported still another approach of the same general character. Another proposed solution is to cluster a number of smaller solid propellant rockets in the same fashion as liquid propellant rockets are clustered in the Saturn booster. In its planning NASA has studied the desirability of proceeding to an all-solid propellant space vehicle of larger size than the Scout, to ultimately replace those space vehicles based on the Thor booster, whose manufacture will ultimately be discontinued.

Because of differing technical characteristics of liquid propellant and solid propellant boosters, there is much confusion about the proper basis for comparison of their performance. In fact such comparisons can be based soundly only on detailed computations of the performance in specific missions. The principal differences arise in the rate of burning of the fuel, and the method of control of burning time and direction of thrust. In general solid propellants burn more rapidly and thus provide a larger thrust than a liquid fueled rocket containing a similar amount of fuel. However the thrust rating is not a measure of the effectiveness of the rocket in placing weight into orbit. The significant quantity is the total impulse, or approximately the product of the thrust by the time of burning. The solid propellant rocket accomplishes its job by a large thrust with a short time of burning and the liquid propellant rocket accomplishes the same job with a smaller thrust and a longer burning time. One result of this is that the solid propellant rocket usually gives a higher acceleration, which is partly beneficial and partly detrimental. The beneficial result is to reduce somewhat the penalties of atmospheric drag in the lower atmosphere; the detrimental effect is to impose higher accelerations on the apparatus and equipment which is carried.

[3] In the application to manned space flight, it will be necessary to carefully control the maximum accelerations imposed on the man. Further it will be necessary to study very carefully whether there are practical methods of forecasting impending difficulties with solid propellant rockets in time to enable a man to escape, if the booster is defective. However, at the present time, the fully developed solid propellant rockets are more reliable in performance than liquid fueled rockets, and many of their failures are non-catastrophic.

The number of firings required for the development of a solid propellant rocket is a matter of some controversy between experts. Some information is available from the experience with Polaris, and more will be available from the Minuteman program when many more firings have been made. Experience with the development of sounding rockets and with the development of the Scout components does not give much ground for such optimism as is expressed by estimates that a mere ten to fifteen firings will be sufficient.

Currently proposals are being made to move immediately from solid propellant rockets in the sizes now available to much larger rockets or to clusters of rockets to duplicate the performance of Saturn. Claims are made that this can be done in much shorter time, but analysis shows that the comparisons are made between the first firings of a first-stage booster and the use of a developed multi-stage space vehicle. There is no question that a structure could be built to hold a number of existing solid propellant rockets in a cluster within about eighteen months. In fact the Saturn is nothing but a cluster of eight existing liquid fueled engines. This first stage has been built and static-tested within less than two years and the first firing of this cluster as a first stage will be done during the current year. This firing of the Saturn cluster by no means constitutes a useful space vehicle, nor will a mere cluster of solid propellant rockets. It is necessary to develop a complete multi-stage vehicle with its guidance and control systems. The development of even the simplest multi-stage space vehicle assembled from existing components has invariably taken an additional eighteen to twenty-four months, and some of these assemblies have never been successfully fired. Because of the variability in performance of solid propellants, it is necessary to provide in the first stage solid propellant booster means for control of the direction of thrust and of the burning time.

[4] In the present U.S. space program, approximately one-third of the total funds are being expended in the development of larger vehicles than now available. The initiation of a large booster project using solid propellants would add another $500 million or so in the vehicle area to provide a complete multi-stage space vehicle. There is no reason to suggest that such a development could be completed prior to the Saturn development.

Hugh L. Dryden
Deputy Administrator

Document I-16

Document title: NASA, News Release, "Mercury Redstone Booster Development Test." Release No. 61-57, March 22, 1961.

Source: NASA Historical Reference Collection, NASA History Office, NASA Headquarters, Washington, D.C.

The second suborbital Mercury/Redstone (MR-2) test flight, launched on January 31, 1961, carried the chimpanzee Ham further downrange into the Atlantic Ocean than had been planned, and it subjected the ape to very high gravity loads. While an astronaut would have survived the flight, he would have been quite uncomfortable. The developers of the Redstone at the Marshall Space Flight Center, Wernher von Braun and his associates, quickly identified the cause of the flight anomaly as a valve that stuck in the open position, and they proposed a simple remedy. The Marshall team insisted that an unplanned test flight be inserted in the Project Mercury schedule to test the fix. This meant that the first suborbital flight by a U.S. astronaut was slipped six weeks. In the interim, Yuri Gagarin became the first human to go into space.

[1]

NEWS RELEASE
NATIONAL AERONAUTICS AND SPACE ADMINISTRATION
1520 H Street, Northwest, Washington 25, D.C.

FOR RELEASE: IMMEDIATE
March 22, 1961

Release No. 61-57

Mercury Redstone Booster Development Test

A special development flight test of a Mercury-Redstone launch vehicle will be conducted from Cape Canaveral, Florida, in the next few days.

The purpose of the test will be to provide engineers of the National Aeronautics and Space Administration with additional performance data on the Redstone vehicle which will lift the Astronaut-carrying Project Mercury spacecraft on short suborbital training flights down the Atlantic Missile Range.

The upcoming flight will be devoted exclusively to proving the modifications which have been incorporated in the rocket system as a result of earlier Mercury-Redstone flights. If the flight goes as planned, the Redstone, carrying a full scale boilerplate—or dummy—Mercury craft will reach a peak altitude of about 100 (statute) miles and land about 300 miles down range.

In this test, the dummy Mercury craft will not be separated from the Redstone launch vehicle. No recovery of either the spacecraft or the launch vehicle is planned. The Mercury craft will not contain any operating systems or instrumentation and has been included in the test to provide only the proper weight and aerodynamic factors for the flight. The Mercury escape rocket will be inert and not capable of removing the craft from the launch vehicle in case of malfunction.

[2] Two Redstone launch vehicles have been flown in earlier Mercury tests. The first successfully launched a heavily instrumented production Mercury spacecraft on a suborbital flight to verify the operation of the Mercury systems in the space environment. Conducted on December 19, 1960, this test was termed an unqualified success with regard to the overall Mercury mission.

On January 31, 1961, a second Mercury-Redstone combination was flown. The Mercury spacecraft carried a chimpanzee to check out the Mercury life support system in flight.

In that flight (MR-2), test results indicated the Redstone engine ran with the throttle literally "wide open." As a result, liquid oxygen was consumed at a higher rate than planned causing the engine to cut-off sooner than planned. The Mercury automatic abort sensing system (ASIS) properly activated the spacecraft's emergency escape system to pull the craft away from the launch vehicle. Firing the escape rocket added still further to the already greater range and altitude of the flight path.

"Ham," the animal passenger on the MR-2 flight, was recovered unharmed. The flight provided vital data on the performance of the animal and the operation of the Mercury spacecraft operating systems.

Analysis of previous Mercury-Redstone flights has revealed a control system vibration problem related to the greater length and altered mass distribution of the rocket. Corrective steps have been taken to prevent reaction of the attitude control vanes to vehicle body oscillations.

In the Redstone to be used in this test, an electrical filter has been installed in the attitude control system to damp out undesirable signals. In addition, a thick, vibration-reducing undercoating has been applied to the inner skin of the upper part of the instrument compartment of the Redstone.

[3]   The Redstone Launch Vehicle

The Redstone launch vehicle used in the Project Mercury flight program is 83 feet long, including the spacecraft and escape tower. This is compared to 69 feet for the standard earlier Redstone rockets. The vehicle is 70 inches in diameter and liftoff weight is 66,000 pounds including the spacecraft.

The basic Redstone rocket has been modified for the Mercury mission. Modifications include:

1.   Elongation of the tank section by about six feet to increase fuel and liquid oxygen capacity. The added fuel increases burning time by about 20 seconds. The Redstone was similarly elongated for its role in the launching of early Explorer satellites. That earlier version was known as the Jupiter C.

2.   The North American Rocketdyne engine to be used in this flight is of the latest Redstone design (A-7), modified for this application. Using alcohol and liquid oxygen, the thrust level of the engine in this launching will be 78,000 pounds. Modifications have been incorporated in the engine system to provide for the extra long burning time and for improvements in the peroxide system which drives the fuel and liquid oxygen pumps, and provides thrust control.

3.   The Mercury-Redstone, as compared to the earlier standard Redstone, has a less complex control system which is designed for simpler and more reliable operation. The system uses an autopilot in conjunction with carbon jet vanes in the exhaust of the propulsion unit and air rudders to maintain proper flight attitude.

4.   An automatic abort sensing and implementation system has also been built in to the Redstone for the Mercury mission. It is an electronic system which serves to give an advance warning of a possible impending launch vehicle malfunction. In the event any one of several deviations from planned launch vehicle performance occurs, the system gives an electric signal which terminates the launch vehicle thrust, separates the spacecraft from the launch vehicle, and activates the spacecraft's escape rocket to propel the craft a safe distance away within about one second.

[4] The abort system senses and is activated by such conditions as unacceptable deviations in the programmed attitude, excessive turning rates, loss of thrust, critical irregularities in thrust in the engine, or loss of electrical power.

In the MR-BD flight, the automatic abort sensing and implementation system will be flying "open loop." It will observe all of the functions and report its findings through telemetry but will not be capable of initiating the Mercury escape system.

Instruments installed in the Redstone launch vehicle will telemeter about 65 measurements surveying all aspects of the rocket behavior during the flight such as attitude, vibrations, accelerations, temperature, pressure and thrust level. Several tracking signals will also be telemetered by the vehicle.

Document I-17

Document title: Assistant Deputy Chief of Staff, Development, U.S. Air Force, to AFCCS, "Convair Analysis of Atlas Booster Space Launches," with attached: J.V. Naish to Dr. T. Keith Glennan, Administrator, NASA, December 21, 1960, and A.D. Mardel, Senior Flight Test Group Engineer, Convair Astronautics, General Dynamics Corporation, to Distribution, "Short Summary of Atlas Space Boosters," EM-1691, December 17, 1960.

Source: Library of Congress, Washington, DC.

The Atlas ICBM was modified to serve as a launch vehicle for a number of early space missions. Many of those missions experienced very visible failures during the launch phase. These failures were of concern to NASA Administrator T. Keith Glennan, because they cast public doubt on the ability of NASA to carry out its missions successfully and because the Atlas was scheduled to be the launch vehicle for the first U.S. human spaceflight effort, Project Mercury. The Atlas was manufactured by the Convair Division of General Dynamics. In the attached letter, Convair president J.V. Naish attempts to assure Glennan that the basic Atlas booster was reliable enough to be counted on as a space mission launcher.

[stamped "SECRET," declassified December 12, 1980]

DEPARTMENT OF THE AIR FORCE
HEADQUARTERS UNITED STATES AIR FORCE
WASHINGTON 25, D.C.

REPLY TO
ATTN OF:

SUBJECT: Convair Analysis of Atlas Booster Space Launches

TO: AFCCS

1. The attached copy of a letter and inclosure [sic] from Mr. J.V. Naish, President of Convair Division of General Dynamics Corporation, to Dr. Keith Glennan is forwarded as an item of interest to you in conjunction with the report on NASA/Air Force space project relations recently provided to you.

2. I consider the letter a very candid approach, with valid reasoning, as well as an understandable reaction on the part of Convair. The summary of the Atlas booster space launches has been reviewed and the information is factual in content and void of any bias on the part of the contractor.

                                                1 Atch—Ltr from Mr. J.V. Naish
VICTOR R. [signature illegible]   w/Atch EM-1691
Major General, USAF
Asst Deputy Chief of Staff,
Development

                            ****

                THIS DOCUMENT STANDING ALONE IS UNCLASSIFIED.

[1]

                                                December 21, 1960

Dr. T. Keith Glennan
Administrator
National Aeronautics and Space Administration
1520 H Street, N.W.
Washington 25, D.C.

Dear Keith:
    During our recent evening in Washington I was particularly concerned about the impressions you had regarding the specific reliability of Atlas as the booster for space vehicles.
    As soon as I returned, therefore, I had the attached analysis prepared, which examines the detail situation of the Atlas performance in its use to date as a space booster. In general, it shows that the Atlas has been used as a space booster in ten attempts so far and that in only one of these ten has the Atlas failed insofar as mission performance is concerned, and that failure was during a static test of Missile 9C, the first Atlas-Able. In the other nine operations Atlas has performed successfully as far as its mission is concerned or a failure has occurred which definitely cannot be isolated but is peculiar to actions of the upper stage on top of the Atlas.
    Since in a number of these cases, however, the immediately obvious result of the flight attempt is a spectacular explosion of the booster stage, it is frequently reported in the immediate press reaction as an Atlas explosion. Although this is true, it is also true that this is a secondary reaction.
    I am sending this material because we both know that Atlas has been programmed for a booster for a number of NASA and Air Force space programs and that it is important for public confidence in these programs that the Atlas performance be accurately stated in public discussions by all the people concerned in these programs. In fact, in the last two weeks we have had several calls from the press in which they mistakenly [2] ascribed failures in programs to the Atlas because, as stated above, the end results of failures were generally explosions in the booster stage.
    I certainly share with you not only the disappointment in program failures, but I fully agree that each of us as members of the team cannot find solace because the responsibility of any failure is attributable to any other member of the team. We at Convair will do everything possible to achieve the team success which the urgency of these programs demands.

                                Sincerely,

                                J.V. Naish

cc - Lt Gen B.A. Schriever

                            ****

[1]   COPY No. 5
CONVAIR ASTRONAUTICS
                                                                             EM-1691
GENERAL DYNAMICS CORPORATION   17 December 1960

To: Distribution

From: A.D. Mardel

Subject: Short Summary of Atlas Space Boosters

Ten Atlas missiles have been used as space boosters to date:

| | | | Mission Failure |
Project	Missile No.	Brief Comment	Responsibility
SCORE	10B	Completely satisfactory.	None
Mercury	10D	Atlas failed to stage.	None
Atlas-Able	9C	Bad plumbing, Atlas blew on FRF.	Atlas*
Atlas-Able	20D	Upper stages fell off at 47 seconds.	Upper stage
MIDAS	29D	Incident during Agena separation.	Upper stage
MIDAS	45D	Completely satisfactory.	None
Mercury	50D	Incident at 57 seconds.	?
Atlas-Able	80D	Incident during Able separation.	Upper stage
SAMOS	57D	Atlas autopilot only, Agena lost control gas.	Upper stage
Atlas-Able	91D	Incident at 66 seconds.	?

*Not a flight, but a ground firing.

The purpose of this memo is to summarize the performance of each of the ten boosters.

Missile 10B
    Missile 10B was launched from Complex 11 at AMR [Atlantic Missile Range] on
18 December 1958. The entire missile, minus the booster section, was placed into an orbit
around the earth in fulfillment of its primary objective. The missile carried two Signal
Corps packages for transmission of voice and teletype messages to and from the satellite.
The capability of this equipment was successfully demonstrated.
    Only one problem was evident during the flight. Tracking data indicated an excessive
azimuth error during the self-guided phase of flight. An 11° roll error was established
prior to 23 seconds. Despite the azimuth error, the guidance system satisfactorily
provided the proper steering commands to place the missile on the correct azimuth. The
cause of difficulty was attributed to a misalignment of the gyro canister in roll by 11
degrees.

References: Convair Reports ZC-7-208 and AE60-0103

Missile 29D
    Missile 29D was launched from Complex 14 at AMR on 28 February 1960. This was
the first Atlas missile designated to support the MIDAS Project. Performance of the Atlas
vehicle was completely satisfactory during powered phase. Shortly [2] after vernier cutoff,
a guidance discrete command was sent to separate the satellite vehicle. An incident
occurred shortly after firing of the retrorockets which affected both the booster and satellite vehicles. The primary objective of placing a MIDAS satellite, carrying an infrared
detection payload, into a circular orbit of 300 statute-miles altitude was not achieved.

The incident resulted in loss of Atlas lox [liquid oxygen] tank structural integrity and indeterminate damage to the satellite vehicle (satellite telemetry lost at the time of the incident). The final Lockheed report advanced the following theories:

    a.   Explosion of one or more retrorockets located in the Lockheed adapter, starting a chain of events leading to Atlas lox tank rupture.

    b.   Explosion of the Agena destruct charge, thereby rupturing the Atlas lox tank.

    c.   Explosion of the Atlas lox tank for unknown reasons.

The MIDAS Joint Flight Test Working Group Report states that the most probable cause of the incident was either inadvertent activation of the satellite premature separation destruct charge circuitry or a random failure of the satellite high pressure gas spheres which resulted in a hypergolic explosion of the satellite propellants.

On the next MIDAS flight, Missile 45D, Lockheed made many changes to their vehicle such as rewiring of electrical equipment and disabling of the Agena premature separation destruct system. No changes were made to the Atlas, and the flight was a complete success; therefore it is inferred that the cause of difficulty on the 29D flight was an inadvertent activation of the Agena destruct system.

References: Convair Report AZC-27-118
               Lockheed Reports LMSD-445912-08 and LMSD-445962-1

Missile 45D

Missile 45D was launched from Complex 14 at AMR on 24 May 1960. The primary objective of this flight was to place a MIDAS satellite in a circular orbit, approximately 261 nautical miles from earth, carrying an infrared detection payload. This objective was fully satisfied. The operation of the Atlas booster was completely satisfactory.

References: Convair Report AE60-0320, Lockheed Report LMSD-445912-07

Missile 57D

Missile 57D was launched from Pad 1 at PMR [Pacific Missile Range] on 11 October 1960. The primary objective of this flight was to place a SAMOS satellite in a circular orbit, approximately 261 nautical miles from earth. This objective was not satisfied [3] because of damage to the satellite vehicle at liftoff when an umbilical failed to release satisfactorily.

The operation of the Atlas booster was satisfactory in accomplishing its mission, despite a guidance system failure. Complete loss of the guidance track subsystem during booster stage prevented the generation of any commands, solely by the pre-programmed flight control system. The guidance loss resulted from a failure of the airborne pulse beacon or decoder or its associated waveguide. The exact cause remains unknown because no guidance system measurements were telemetered on this flight.

At liftoff the nitrogen control gas fitting in the Agena was broken off, causing control gas completion shortly after launch. Also, damage is believed to have occurred to the helium system. The lack of control gas prevented stabilization of the Agena satellite. During engine burning, the thrust was not developed along the flight path and the satellite failed to orbit. Engine performance was slightly subnominal due to low helium pressure.

References: Convair Report AE60-0749, Lockheed Report LMSD-445919-1

Missile 10D

Missile 10D was launched from Complex 14 at AMR on 9 September 1959. This was the first missile designated to support the Mercury Project. All flight objectives were not satisfied because the booster section failed to jettison 3 seconds after cutoff as planned. Because of the added weight of the booster section, fuel depletion occurred before the sustainer cutoff discrete was transmitted. As a result, the sustainer fuel and lox valves remained open, and residual thrust was indicated. This residual thrust prevented a normal capsule separation, with continuous or almost continuous coupling of the capsule and tank section indicated out to approximately 83 seconds after retrorocket fire. The capsule was recovered in a satisfactory condition.

The strongest possibility for the failure was that the electrical signal did not reach the squibs of the Conax valves. The plug connecting the Conax valve wiring to the wires from the programmer was relocated and rewired to increase accessibility. This task was accomplished by TVA [Tennessee Valley Authority] at AMR. Continuity checks made at the plug during preflight testing verified that the wiring was intact; however, it is possible that the wires were connected to the improper contacts when the plug was moved and that the continuity checks were also made on the wrong contacts.

References: Convair Reports AZC-27-077 and AE60-0103
            NASA Working Paper No. 107

Missile 50D
    Missile 50D was launched from Complex 14 at AMR on 29 July 1960. This was the second missile designated to support the Mercury Project and the first to boost a McDonnell capsule. Performance of the Atlas booster was completely [4] satisfactory until 57.60 seconds. At this time an incident occurred which culminated in loss of telemetry and destruction of the Atlas booster at 58.99 seconds.

At 57.60 seconds an impulse disturbance was registered by the missile axial accelerometer and the capsule high range longitudinal accelerometer. The data indicates that the capsule accelerated at approximately 22 g's while the missile was decelerated at approximately 2 g's. Available data does not permit detailed determination of the cause of flight failure; however, a logical explanation for the sequence of events is that static or dynamic loads were introduced into, and caused rupture of, the forward portion of the lox tank.

All evidence indicates that the capsule survived the disturbance without damage but was then destroyed upon impact with the surface of water.

References: Convair Report AE60-0323, NASA Working Paper No. 159

Missile 9C
    Missile 9C, assigned as the first stage of the four-stage Atlas-Able IV Lunar Satellite Project, was destroyed by fire and explosion during a flight readiness firing on 24 September 1959, at Complex 12 at AMR.

The loss of the missile followed a premature cutoff of the engines at 2.1 seconds. The cutoff was preceded by an unloading of the sustainer fuel pump and subsequent turbine overspeed, followed by rupture of the sustainer lox pump low pressure system. Liquid oxygen entering the engine compartment started a fire of such intensity that normal firex facilities were incapable of extinguishing the flame.

It was determined that sustainer fuel pump cavitation was caused by entrainment of helium in the fuel flowing to the pump. The helium entered the system when the vernier fuel tank vented into the main missile tank in the vicinity of the sustainer fuel outlet. Improper installation of the vent line to a port below the baffle was a result of poor engineering judgment. This modification (5-second tanks) was unique to 9C in the C Series. A similar modification performed earlier on Missiles 10B and 13B resulted in satisfactory performance. The extension of the modification to 9C was unsatisfactory due to a change in configuration between B and C Series.

Reference: Convair Report FTA 6182

Missile 20D

Missile 20D was launched from Complex 14 at AMR on 28 November 1959. This missile was the second missile to be assigned as the first stage of the Atlas-Able IV Lunar Satellite Project. Performance of the Atlas vehicle was satisfactory. Engine ignition and separation of the Able second stage from the [5] Atlas was effected satisfactorily at 261 seconds but mission objectives were not accomplished because of an upper stage failure at 47 seconds.

The upper stage failure at 47 seconds was reflected as a disturbance in Atlas rate gyro and accelerometer data, loss of second stage guidance signals, ultimate loss of second stage telemetry signals, and was observed both visually and photographically to be several objects falling away from the missile. Portions of the payload and payload adapter were recovered in the vicinity of the Cape. No STL Report is available stating what the cause of failure was. The Atlas missile was in no way implicated in the upper stage failure.

One minor problem was apparent in the Atlas booster. The initial operating level of the booster engines was somewhat reduced because of momentary faulty operation of the booster reference regulator. The temporary reduction in booster thrust, from before liftoff to 3 seconds after, had no overall adverse effect on Atlas performance. The transient condition in the regulator was the result of an out-of-tolerance manufacturing condition.

References: Convair Reports AZC027-080 and AEOO-0103

Missile 80D

Missile 80D was launched from Complex 12 at AMR on 25 September 1960. This was the first booster vehicle for the Atlas-Able V Lunar Satellite Project. The Atlas vehicle was successful in boosting the upper stage to the planned position and velocity. Ignition of the Able second stage engine occurred at the proper time; however, the thrust chamber pressure dropped abruptly to a lower level during separation with complete shutdown occurring prematurely. As a result the overall mission was not completed.

No STL Report is available stating what the cause of failure was. The Atlas missile was in no way implicated in the upper stage failure.

Three minor problems were apparent in the Atlas booster. None of these problems had any overall adverse effect on Atlas performance. The first problem was failure of the vernier engines to shut down when a command was generated 5 seconds after sustainer cutoff. The failure has been attributed to a short circuit in the vernier cutoff relay. The second problem was an abnormal pressure decay in the separation bottle; the pressure dropped from 3,135 to 2,590 psig between liftoff and booster cutoff. The exact cause of this pressure decay is unknown. The third problem was an excessive bending mode buildup starting at the time of commencement of the pitch program. Use of quadratic-lead, triple-lag stabilization filters incorporated in the square type autopilot packages resulted in insufficient attenuation near 24 cps for the Atlas-Able configuration.

Reference: Convair Report AE60-0748

[6] Missile 91D

Missile 91D was launched from Complex 12 at AMR on 15 December 1960. This was the second booster for the Atlas-Able V Lunar Satellite Project. All data indicates that operation of the Atlas booster was satisfactory until 66.680 seconds. At this time an incident occurred which culminated in destruction of the Atlas booster at 74 seconds.

At 66.680 seconds an impulse disturbance was registered by the missile axial accelerometer and the Able vehicle axial accelerometer. The data indicates that the Able vehicle accelerated while the booster was decelerated. A vibration measurement in the

Able second stage engine compartment showed a buildup in output starting approximately 15 milliseconds before the disturbance indicated on the Atlas booster axial accelerometer.

No films are yet available in San Diego and the data are in the process of being analyzed at this time.

Reference: Convair Memo EM-1689

> A.D. Mardel
> Senior Flight Test
> Group Engineer

Document I-18

Document title: George E. Mueller, "NASA Learning From Use of Atlas and Titan for Manned Flight," with attached: "Summary Learning From the Use of Atlas and Titan for Manned Flight," December 21, 1965.

Source: NASA Historical Reference Collection, NASA History Office, NASA Headquarters, Washington, D.C.

A converted Atlas missile was used for each of the four orbital flights in Project Mercury, and a converted Titan II ICBM was used in each of the ten flights in Project Gemini. This experience was crucial as NASA began to plan for the initial Apollo missions. When this memorandum was written in 1966 by NASA's Associate Administrator for Manned Space Flight, George Mueller, the Air Force was planning to launch military crews in the Manned Orbiting Laboratory (MOL) program.

[1]
A/Administrator   DEC 21 1965

M/Associate Administrator for Manned Space Flight

NASA Learning From Use of Atlas and Titan for Manned Flight

The attached summary of what NASA has learned in the last six or seven years through working with the Air Force in making use of Atlas and Titan for manned flight is submitted in response to your request.

For your convenience, following is a recapitulation of the salient knowledge we have acquired:

a.   The unique management procedures, techniques, philosophy and related experience acquired and developed by the Air Force during the course of the ballistic missile program. These have been adopted and adapted by NASA to meet our specific requirements.

b.   The difficult, detailed and productive process of converting selected operational military missiles to man rated boosters with the associated system reliability (redundancy, quality assurance and control, etc.) requirements.

c.   The detailed procedures, checkouts and operational techniques required for the successful integration and operation of a launch complex and the launching of manned vehicles.

d. A vast amount of ancillary technical-engineering knowledge and experience from the Air Force in essential areas such as guidance, performance, propellants, vehicle and spacecraft component design manufacture and procurement, etc.

The NASA, Air Force and Industry have learned together and have mutually benefited from working together in expanding on the technology of the Atlas and Titan vehicle systems. The sum of this experience [2] and acquired knowledge is being effectively applied to the NASA Apollo-Saturn and the Air Force MOL-Titan III manned flight programs.

George E. Mueller

****

Enclosure

[1]

Summary Learning From the Use of Atlas and Titan for Manned Flight

What we have learned in the past six or seven years through working with the Air Force in making use of the Atlas and Titan launch vehicles for manned flight is summarized in the following five sections.

I. Management (Procedures, Techniques, Philosophy and Personnel)

1. Probably the most important item that NASA learned was how to apply and directly benefit from the management techniques and the government-industry team approach established by the Air Force's ballistic missile program. NASA adopted and modified the Air Force system management concept. The direct experience, management and procurement know-how of the Air Force has been effectively transferred to NASA.

2. As a result of the joint effort, familiarity with the internal operations and organization of each agency was developed by both NASA and the Air Force as a basis for future cooperation and mutual support of manned space flight problems.

3. The Mercury and Gemini programs are largely responsible for a large number of Air Force Officers serving NASA at all levels on direct loan or in supporting Air Force efforts. This day-to-day working is and has been a productive learning process for members of both agencies.

4. NASA learned to work together with the management panels of the DOD [Department of Defense] resulting in the formation of the AACB [Aeronautics and Astronautics Coordinating Board], etc., joint panels and boards. The Mercury program resulted in the development of procedures and ground rules for manned space program interagency committees and was a major factor in the recognition of the need for and establishment of standardized Air Force NASA spacecraft standards. The Air Force was responsible for NASA's early recognition of the necessity for formalized procedures. In the early phases of the Mercury program, NASA was inclined toward very informal procedures.

[2] 5. NASA has learned that any particular contractor such as Martin or Convair must have only one specific "boss," either the Air Force or NASA, but not both at the same time on any one particular system or vehicle. This is often hard to learn on cooperative programs and early recognition of this fact was most important to manned space flight efforts.

6. NASA-Air Force-Industry learned and developed educational and unusual personnel handling techniques which highly motivated assembly line workers, technicians and clerical personnel to perform well above the routine level to insure [sic] success in the manned programs.

7.   NASA learned to use and benefit from the unusual expertise of specialized Air Force contractors such as the Aerospace Corporation and Space Technology Laboratories.

II. Man Rating of a Missile Booster
1.   The process of man rating the Atlas and Titan II taught both the Air Force and NASA the tremendous differences between an operational missile capability and a man rated booster.
2.   NASA and the Air Force learned firsthand the vast amount of additional procedures, time, effort and dollars necessary to achieve man rating standards. Atlas and Titan II were not designed to the mechanical limits and reliability criteria established for manned vehicles. Starting with Atlas, the most reliable booster avail-able, it took a large amount of modification, additional monitoring and checking beginning with design and parts procurement, then on thru full production line, and finally delivery and acceptance checkout at the Atlantic Missile Range (AMR). NASA and the Air Force learned, developed and established the significantly expanded detailed acceptance procedures, quality control efforts and rigid contractor control-supervision required for successful manned flight.
3.   NASA experienced the major advantages in time, dollars and confidence which resulted from starting with well engineered [3] mature hardware that had been flown repeatedly with high reliability compared to starting with a new untried booster.
4.   NASA and the Air Force learned and proved that the adaptation of military hardware to civilian space efforts can be accomplished successfully in the missile booster area as it has been in the adaptation of certain selected military aircraft to civilian transport.
5.   It is pertinent to review briefly the major modifications made in the Atlas D and Titan II vehicles to convert them from Air Force missiles to man rated boosters. Lists of the modifications to Atlas D and Titan II are attached. These modifications were the result of a large amount of study and effort by the NASA-Air Force-Industry team and represent a significant amount of technical learning and development.

III. Launch, Checkout & Operational Techniques
1.   The considerable Air Force past experience in the area of integrated launch complex-vehicle checkout and countdown was increased and crossfed to NASA in the process of launching the Atlas and Titan vehicles by the Air Force for NASA.
2.   NASA and the Air Force learned that unmanned and military launch procedures and checkouts, while useful as starting points, were inadequate for manned launches. Mercury and Gemini capsule interfaces with their respective launch vehicles posed significant additional checkout and complexity. Applying the experience from the complex interface problems associated with earlier Atlas-Agena launches, the Air Force and NASA developed and learned the significant additional and more stringent launch procedures with the extended more detailed countdowns required for manned launch.
3.   NASA learned the value and benefits of the formal certification procedures of the Flight Safety Review Board. This is a high-level Air Force and Industry board chaired for all manned flights by the Commander of the Air Force Space Systems Division. It has been [4] their collective responsibility to certify that each launch vehicle was indeed flight ready. This board continues to function in support of the Titan-Gemini Program.
4.   NASA and the Air Force learned and demonstrated that complex launch facilities as well as vehicles can be effectively used for both NASA and DOD launches with the resultant savings to the Nation by avoidance of duplicate facilities.
5.   From the Mercury launches the major effects of non-homogenous atmosphere on tracking accuracy at long range and low elevation were defined for both NASA and the Air Force. This resulted in the modification of approaches to vehicle tracking. Presently range and range rate are measured to determine vehicle position rather than attempting to measure elevation directly. Position accuracy was improved by two orders of magnitude.

IV. Ancillary Knowledge

1.   Beginning with Atlas-Mercury and continuing with Titan-Gemini, NASA drew heavily from the Air Force in the guidance and performance areas thus augmenting and further developing their own internal competence.

2.   The Air Force as a major user of electronic components and small parts had extensive experience with these items and with scattered bad lots. NASA was advised and able to readily apply this experience to their own designs, procurements and analogous problems. Similarly, working together on Mercury various NASA groups were made aware of many other technological traps which the Air Force had encountered and, thus, avoided the same blind alleys.

3.   In working with the Air Force on Atlas and Titan, NASA has learned much about the propellants involved, their sources, quality, handling and transfer characteristics. The Atlas vehicle was one of the earliest major users of cryogenics, i.e., liquid oxygen. [5] Experience with toxic storeable [sic] propellants in the Titan has been applied to use of these oxidants and fuels in Gemini and Apollo spacecraft systems for attitude control and main spacecraft propulsion subsystems. NASA also learned that technical specifications for these storeables [sic], while originally developed by the Air Force for missile use and apparently satisfactory for such purpose, had to be refined and rigidized [sic], particularly for use in the smaller spacecraft attitude control engines where orifices, etc. are much finer than in the larger Titan engines.

4.   The Standard Launch Vehicle (SLV) programs and the Aerospace Industry in general benefited by the quality control procedures which NASA-Air Force-Industry learned, developed and instituted during the Mercury program and further developed under Gemini.

V.   Conclusion

The NASA, the Air Force and Industry have learned a vast amount by working together and using the Atlas and Titan launch vehicles for manned flight. It has truly been a mutually productive and beneficial process.

The total and full import of what we have learned will probably never be completely identified. However, the total of this knowledge and experience is being effectively applied to the NASA Apollo-Saturn and the Air Force MOL-Titan III manned space flight programs.

[no page number] ATLAS D MAJOR MODIFICATIONS

1.   A new spacecraft adapter was installed.
2.   Wet start of the engines previously discarded in missile launches was used.
3.   Replaced the telemetry package with an all transistorized lightweight telemetry system.
4.   Removed the retro-rockets and vernier solo package.
5.   Insulated the LOX [liquid oxygen] dome.
6.   A three second delay was added to the range safety command destruct signals.
7.   The abort sensing and implementation system (ASIS) was added.
8.   The LOX boil off valve was changed from the weapon system valve to a type similar to that used in the Atlas C R&D [research and development] flight test program.
9.   Installed a modified autopilot. An all-electric transistorized programmer replaced the potentially unreliable Electro-Mechanical Programmer. A redundant rate gyro was added and system was "repackaged."
10. Installed a baffled injector in booster engines to eliminate traces of combustion instability.
11. Removed the insulation and the insulation bulkhead from inside the fuel tank. This reduced complexity and eliminated a problem with fuel seepage wetting the insulation.

12. Increased the skin thickness of the forward end of the LOX tank to provide adequate safety factors for heavy stress loads imposed by the spacecraft.

[no page number] TITAN II MAJOR MODIFICATIONS

1. Structural modifications were made to the transition section above the second stage for attaching the spacecraft to the launch vehicle.
2. The AC inertial guidance system was removed and the General Electric MOD III-G installed. In addition, a three AXIS reference system (TARS) was required for flight attitude.
3. By adding a tandem actuator system, a second hydraulic power supply, a second autopilot and redundant electrical power system, the failure probability in the flight control system was lowered by at least two orders of magnitude.
4. Weapon system batteries were replaced by rechargeable space system batteries in the electrical system.
5. A Malfunction Detection System (MDS) was installed to provide the astronauts with a detection system for noting malfunctions in order that an abort or escape action could be taken before a catastrophe occurs. Signals were provided in the spacecraft indicating pressure in fuel and oxidizer tanks, engine and thrust chamber pressure, staging signals, excessive attitude rate changes, and range safety officers' actions.
6. Since the spacecraft has its own maneuverable engines the vernier and retro-engines were removed.
7. The Titan II engine program was redirected to solve performance reliability and the longitudinal oscillation or "POGO" problem and combustion instability problems. The net result of this effort was an improved Titan II engine system that was man rated.

Document I-19

Document title: Staff of Aerophysics Laboratory, North American Aviation, "Feasibility of Nuclear Powered Rockets & Ram Jets," Report No. NA-47-15, February 11, 1947, pp. 2, 11–14.

Source: NASA Historical Reference Collection, NASA History Office, NASA Headquarters, Washington, D.C.

Before nuclear bomb designers figured out how to construct high-yield, low-mass nuclear warheads for delivery on missiles, they faced the problem of building high-thrust rockets. These considerations led to an exploration of nuclear-powered rockets. This formerly classified study is the first detailed examination of the potential of nuclear fission for propulsion. Only the text of the preface and abstract appear here.

NORTH AMERICAN AVIATION, INC.
AEROPHYSICS LABORATORY

DATE 2-11-47   REPORT NO. NA-47-15

Feasibility of Nuclear Powered Rockets & Ram Jets

PREPARED BY
STAFF OF AEROPHYSICS LABORATORY . . .

[2]   PREFACE

The nuclear powered rocket presented here is a single stage vehicle of about the same weight as a modern medium bomber. It is capable of escaping from the earth's gravitational field and travelling in interstellar space. With a bomb load of 8,000 pounds, it can orbit about the earth indefinitely, or deliver its payload to any point on the earth's surface.

The proposed nuclear ram jet has about the same weight as present day fighter planes. It is designed to carry an 8,000 pound bomb load and to cruise indefinitely in the stratosphere at the speed of a rifle bullet.

This report was prepared in accordance with Army Air Force Contract W33-038 ac-14191, Project MX-770, under the cognizance of the Guided Missiles Section, Air Materiel Command, Army Air Forces, Wright Field, Dayton, Ohio. . . .

[11]   ABSTRACT

This report examines the engineering feasibility of the application of nuclear energy to long range supersonic rockets and ram jets. Specifically, analysis indicates that a 10,000 mile rocket-missile, nuclear powered and hydrogen propelled, can be designed and constructed with a gross initial weight of about 100,000 pounds and a useful payload of 8,000 pounds. With slight modification, and without the payload, the rocket can escape from the gravitational field of the earth. The analysis further shows that a nuclear powered ram jet with an 8,000 pound payload has a gross weight of 17,600 pounds and can fly almost indefinitely at a speed of about 2,000 miles per hour.

The Rocket
The use of nuclear energy as a heat source permits the choice of rocket propellant to be free of the limitations of chemical combustion. Momentum considerations alone indicate that hydrogen, because of its small molecular weight, would be the best propellant. Unfortunately, the low density of liquid hydrogen requires large containers whose weight reduces the advantage gained by the low molecular weight of the gas. Therefore a denser propellant, liquid methane, was also investigated. Liquid ammonia may be used practically interchangeably with liquid methane. The study indicated that in spite of lower density, liquid hydrogen was the better propellant. The combined use of liquid hydrogen and liquid methane may have some advantages which are discussed in the report. The nuclear reactor in every case was a graphite assembly impregnated with uranium operated at about 5700°F (3160°C).

Alcohol-oxygen and hydrogen-oxygen multi-stage chemical rockets were compared with methane-propelled and hydrogen-propelled nuclear powered rockets. The study indicated that for ranges greater than 2,000 miles, the nuclear powered, hydrogen propelled rocket is roughly one-third the weight of a chemically powered hydrogen-oxygen multi-stage rocket. . . . The nuclear-hydrogen rocket, as a 10,000 mile missile or a satellite vehicle, would weigh less than 100,000 pounds. If the payload is removed and 500 pounds of instruments retained, a nuclear hydrogen-propelled escape vehicle (or lunar vehicle) would also weigh about 100,000 pounds.

A cost analysis of the chemical and nuclear rockets is summarized . . ., based on an estimated structure cost of fifty dollars [12] per pound and a suitable fuel cost. The fuel cost is always negligible compared to the total structure cost. The comparison between the various 10,000 mile missiles follows:

(PAYLOAD = 8,000 POUNDS)

PROPULSION SYSTEM	INITIAL GROSS WEIGHT (POUNDS)	ESTIMATED COST (MILLIONS OF DOLLARS)
Alcohol-Oxygen	680,000	5.1
Nuclear-Methane	428,000	3.6
Hydrogen-Oxygen	252,000	2.6
Nuclear-Hydrogen	93,000	1.5

The cost of uranium was not included in the estimates of the nuclear powered rockets since the cost of U-235 (enhanced in concentration) was not available. This report has considered only engineering aspects, without primary regard to uranium economy. Considerable saving of uranium can be achieved by the use of a reactor designed with economy as the principal criterion, at the expense of additional rocket weight. This is a subject for future study.

A detailed analysis of the components of the nuclear hydrogen-propelled 10,000 mile missile is presented here. The problems of structural arrangement, aerodynamic stability, steering control, turbines and pumps for propellants, and nuclear reactor design are all considered. An experimental study of the fabrication features of the graphite nuclear reactor is reported. Techniques for impregnating the graphite with uranium are described. In order to prevent chemical erosion of the graphite structure by reaction with hydrogen gas, a protective film of tantalum carbide has been developed and studied experimentally at high temperatures at the Aerophysics Laboratory of North American Aviation, Inc.

During flight the nuclear reactor develops heat at the rate of about 8 million horsepower. The problem of transferring this heat to the [13] gas stream involves theoretical considerations of an unusual magnitude. The analysis presented indicates that such heat transfer is feasible in a reactor of reasonable size. The experimental investigation of this problem is proceeding in the Aerophysics Laboratory.

The Ram Jet

A nuclear power plant can develop power for an extremely long time without regeneration. Since in a ram jet the propellant gas is provided by the air stream, the combination of nuclear power and ram jet action provides a vehicle that has an indefinitely long range. A nuclear reactor has been considered for the ram jet, to be fabricated of beryllium oxide, impregnated with uranium, and operated at about 3600°F (1980°C). Impregnation techniques have been developed at the Aerophysics Laboratory and are reported here. Such a reactor can be incorporated in a 3-foot diameter ram jet of only 14,000 pounds total weight, of which 8,000 pounds is bomb load. However, a minimum practical dimension for the war head requires a 5-foot diameter, 17,600 pound ram jet. Calculations show that successful operation of this ram jet could be anticipated in the range of Mach numbers from 2.5 to 3.5, and altitudes up to 50,000 feet.

The comparative performance of a chemical ram jet using gasoline as a fuel was investigated, both without booster . . . and with booster. . . . The results indicate clearly that only for short range use is the chemical ram jet of value, and that under the best possible circumstances such a ram jet has a maximum possible range of about 4,000 miles. For ranges greater than this, only nuclear power can be considered. The cooperative cost of the chemical and nuclear ram jets has also been evaluated without booster . . . and with booster. . . . The cost figures are based upon an estimated structure cost of fifty dollars per pound. The cost of the payload (bomb) is not included. The cost of uranium is not included in the case of the nuclear ram jet. As with the rocket, this cost is probably of the magnitude of a million dollars. The results indicate that a nuclear ram jet is probably economically justified for ranges greater than about 3,500 miles, and without uranium would cost about nine hundred thousand dollars with booster. This is only slightly less than the cost of the nuclear-hydrogen rocket.

[14] An engineering analysis of the nuclear powered ram jet is presented in the report. This study includes launching trajectories, aerodynamic stability, structural calculations, heat transfer analysis and determination of over-all ram jet performance. Since the ram jet must be launched at operative speeds, an acid-aniline rocket booster has been designed. The initial gross weight of the booster is 37,500 pounds, with a resulting launching weight of the combination of 55,100 pounds.

Conclusions and Recommendations

The conclusions reached in this report are that both the rocket and ram jet powered by nuclear energy are feasible from the engineering standpoint and are economically comparable to, or less costly than[,] the best chemically powered units for long range use. It is further concluded that vehicles having a useful payload and extremely long ranges, including "escape" or "space" vehicles, become practicalities only when propelled by nuclear power.

The detailed performance considerations that would permit a choice to be made between the ram jet and rocket can only come from continued development of these devices. It is therefore recommended that the development of both the nuclear powered ram jet and rocket be carried on in parallel. Since the nuclear reactor for both these devices is intimately related to the rest of the design, it is recommended that nuclear reactor development be part of the over-all program.

A program for the next five years, based upon these recommendations, is discussed in this report. A specific proposal for the next year is made. . . .

Document I-20

Document title: AEC-NASA Press Kit, "Nuclear Rocket Program Fact Sheet," March 1969.

Source: NASA Historical Reference Collection, NASA History Office, NASA Headquarters, Washington, D.C.

This document summarizes the experience of the joint nuclear rocket program of the Atomic Energy Commission (AEC) and NASA. In March 1969, the Nixon administration had just entered the White House with the intention to reduce the federal budget, and all programs were under review. In addition, the federal budget for fiscal year 1970 was under consideration in Congress, and the nuclear program was in jeopardy. This fact sheet enabled reporters to write about the program with greater knowledge about the components of the nuclear rocket development and test series.

[1]
[hand-dated "March 1969"]

Nuclear Rocket Program Fact Sheet

Nuclear Rocket Program: AEC-NASA program to develop nuclear rocket engine technology and systems for space exploration

Program Terminology:

NERVA Program: The program to develop the technology of nuclear rocket engines
and, based on that technology, a flight qualified engine called NERVA (Nuclear Engine
for Rocket Vehicle Application). The program work is being accomplished under a government contract (SNP-1) with the Aerojet-General Corporation (AGC). AGC's principal
subcontractor is the Westinghouse Electric Corporation. Westinghouse is responsible for
the development of the engine nuclear subsystem, which includes the reactor. The major
facilities used in the NERVA program are: the Aerojet-General Test Facility, Sacramento,
California; the Westinghouse Astronuclear Laboratory, Large, Pennsylvania; and the
Nuclear Rocket Development Station in Jackass Flats, Nevada.
NERVA Reactor Experiment (NRX): The name given to the series of Westinghouse
experimental reactors fabricated and tested as a part of the NERVA technology phase.
This portion of the technology effort was completed with the testing of the NRX-A6 reactor in December 1967.
NERVA Ground-Experimental Engine (XE): The ground-based, experimental
nuclear rocket engines designed, fabricated and tested by the Aerojet/Westinghouse contractor team as a part of the NERVA technology phase.
NERVA Engine: The 75,000 pound thrust engine being developed to flight qualification by the Aerojet/Westinghouse industrial contractor team.
KIWI: The name given to the series of non-flyable, ground-based, experimental reac-
tors and the Los Alamos Scientific Laboratory (LASL) project to develop basic graphite
reactor technology. The project was completed in 1964 with the testing of the LASL KIWI-
B4E reactor at the Nuclear Rocket Development Station in Nevada. The Los Alamos
Scientific Laboratory is operated by the University of California (UCLA) for the AEC.
Phoebus: The name given to the series of modified KIWI reactors, referred to as
Phoebus-1, and larger high-power reactors, called Phoebus-2, designed, fabricated and test-
ed by the Los Alamos Scientific Laboratory as a part of the effort to scale-up the basic reac-
tor technology developed under the KIWI project to higher powers and greater efficiency.
This LASL effort was completed in July 1968 with the testing of the Phoebus-2A reactor.
[2] Pewee: A small graphite reactor designed and assembled by the Los Alamos Scientific
Laboratory to evaluate the performance of fuel elements and other promising reactor
core components being considered for inclusion in NERVA.
Nuclear Rocket Development Station (NRDS): The national site for the ground testing of nuclear rocket reactors and engines. Comprises an area of approximately 90,000
acres (140 square miles) in the AEC Nevada Test Site (NTS). NRDS is located on U.S.
Highway 95, approximately 90 miles northwest of Las Vegas, Nevada.
The major NRDS test facilities are as follows:

Test Cell "C"—Facility for the testing of nuclear rocket reactors.

Engine Test Stand No. 1 (ETS-1)—Facility for the static testing of nuclear rocket
engines.

Engine Maintenance, Assembly and Disassembly Building (E-MAD)—A complex of "hot" cells and laboratories equipped with special remote handling equipment and devices for assembling, disassembling, servicing and examining nuclear rocket engines.

Reactor Maintenance Assembly and Disassembly Building (R-MAD)—A complex similar to the E-MAD building for assembling, disassembling, servicing and examining nuclear rocket reactors.

Space Nuclear Propulsion Office (SNPO): Joint office of the AEC and NASA which directs the nuclear rocket program. SNPO comprises a headquarters office located at the AEC in Germantown, Maryland, and three extension offices; the latter are located in Cleveland, Ohio; Jackass Flats, Nevada; and Albuquerque, New Mexico.

Budget: Total costs, cumulative through fiscal year 1968 (in millions): AEC, 692.8; NASA, 446.1; Both, 1138.9. FY 1969: AEC, 57.2; NASA, 32.0.

Chronology:

1955:	11/2, initiation of nuclear rocket program.
1957:	3/6, Nevada site for nuclear rocket tests authorized.
1959:	7/1, Kiwi-A reactor test.
1960:	7/8, Kiwi-A-Prime reactor test.
	8/31, SNPO established.
	10/19, Kiwi-A3 reactor test.
[3] 1961:	12/7, end of Kiwi-B1A reactor tests.
1962:	2/19, NRDS designated.
	7/16, Kiwi-B reactor "cold flow" tests (completion).
	9/1, reactor startup with liquid hydrogen, end of Kiwi-B1B reactor tests.
	11/30, Kiwi B4A reactor tests (completion).
1963:	5/15, Kiwi-B4A "cold flow" reactor test.
	7/12, Kiwi-B2A "cold flow" reactor tests (completion).
	8/21, Kiwi-B4B "cold flow" reactor tests.
1964:	2/13, Kiwi-B4D "cold flow" reactor test.
	4/16, NRX-A1 "cold flow" reactor test.
	5/13, Kiwi-B4D reactor power test.
	8/28, Kiwi-B4E reactor power test.
	9/10, Kiwi-B4E reactor restart.
	9/24, NRX-A2 reactor power test.
	10/15, NRX-A2 reactor restart.
1965:	1/12, Kiwi transient nuclear test.
	4/23, NRX-A3 reactor power test.
	5/20, NRX-A3 reactor restart.
	5/28, NRX-A3 reactor restart.
	6/25, Phoebus-1A reactor power test.
1966:	3/25, NRX EST (breadboard engine) power tests (completion).
	6/8, NRX-A5 reactor power test.
	6/23, NRX-A5 reactor restart.
1967:	2/23, Phoebus-1B reactor power test.
	7/12, Phoebus-2CF "cold flow" reactor tests.
	12/15, NRX-A6 reactor power test.
1968:	4/11, XE-CF "cold flow" engine tests (completion).
	6/26, Phoebus-2A reactor power test.
	7/18, Phoebus-2A reactor restart.
	12/11, Pewee-1 fuel-element test bed reactor tests (completion).

Manager, SNPO
Milton Klein

Deputy Manager, SNPO
David S. Gabriel

Propellant and Coolant: Hydrogen, carried in liquid form.

Reactor Fuel: Uranium, loaded in graphite fuel elements.

Other Nuclear Rocket Program Activities: Apart from NERVA, the nuclear rocket program includes a broad spectrum of supporting research and technology activities. Examples of these activities are: the work at the Y-12 Plant of [4] Oak Ridge and at LASL, which includes the Pewee reactor program, on improving reactor fuel elements and support hardware; the work at United Aircraft Research Laboratories on the gas core nuclear reactor; the work at Lewis Research Center (LeRC) on advances in component technology; and the in-house and contractual effort by Marshall Space Flight Center (MSFC) on nuclear stage technology.

Document I-21

Document title: Senator Howard Cannon to the President, October 19, 1971.

Document I-22

Document title: James Fletcher to Senator Howard Cannon, January 24, 1972.

Source: Both in NASA Historical Reference Collection, NASA History Office, NASA Headquarters, Washington, D.C.

As the civilian space program was reduced in priority and budget by the Nixon administration in the aftermath of the initial lunar landing, the future of the NASA-AEC nuclear rocket program was in obvious jeopardy. One of the strongest congressional supporters of the program was Senator Howard Cannon of Nevada; many of the program's tests were carried out in his state. His efforts to save the program were not successful, as explained in the letter from NASA Administrator James Fletcher.

Document I-21

[1]

United States Senate
WASHINGTON, D.C. 20510

October 19, 1971

The President
The White House
Washington, D.C.

Dear Mr. President:

It has come to my attention that the Administration's support of the nuclear rocket engine, NERVA, which is financed jointly by the AEC and NASA, once more is the subject of doubt.

I am advised that the [Office of Management and Budget] has been instructed to freeze $24 million authorized and appropriated by the Congress pending a decision on 1973 budget levels.

It seems to me, as I have written you many times, that this program continues to offer the nation's best chance to take the next logical step forward in space, and that the already staggering $1.4 billion investment in successful [research and development] would make continuation of the program not only desirable but mandatory, since we are so close to a flyable engine.

As one of the senior members of the Senate Space Committee, I am greatly disturbed that the space program in recent years has been progressively cut back. It seems to me that if we are to continue in space, the NERVA funding issue is critically important. On the other hand, if we are to cast aside our earlier desire to go forward in space and subject our investment to a less-than-starvation funding level, we are [2] only deceiving ourselves. I believe that rather than merely giving lip service to space, we ought to consider a total restructuring or delegation of NASA's role to the military.

I sincerely hope that my present assessment of our space posture is overly pessimistic and that you will recognize the opportunity and challenge which this deserving NERVA program presents to the Administration.

Since I believe that we are at a crossroad in deciding our space objectives, I bring this matter to your attention.

Sincerely,

HOWARD W. CANNON

Document I-22

[1]

Honorable Howard W. Cannon
United States Senate
Washington, D.C. 20510

Dear Senator Cannon:

I am writing to inform you in some detail of the decisions on the space nuclear propulsion program which have been made in connection with the President's FY 1973 Budget for NASA and AEC. As I advised you in my letter of September 29, 1971, we have been operating in this program during the first part of FY 1972 on the basis planned in the President's FY 1972 Budget, i.e., holding together a technical cadre of Government and contractor personnel so that development of the 75,000-lb. thrust NERVA engine could be resumed when it became timely to do so. I also advised you at that time that we anticipated that the decision on the future of the program would be made as a part of the FY 1973 budget decisions. This has now occurred.

As stated in the testimony on the FY 1972 budget, the reasons for suspending development of the NERVA engine in the FY 1972 budget were in part the fiscal constraints necessary in the budget for FY 1972 and succeeding years and in part the fact that the first missions using the NERVA engines would not take place before the middle or late 1980's. Therefore, the decision we presented to you in the FY 1972 budget was to suspend NERVA engine development and to endeavor to preserve the capability for resuming it at a later time when a development sequence—engine, stage, and payloads—leading to use of the nuclear engine in mission applications requiring its capabilities could be begun with a reasonable expectation of being carried to completion.

[2] In developing our FY 1973 budget we have given special attention to the problem of configuring the entire NASA program in such a way that it will not commit the nation to large increases in the total NASA budget in future years. This has meant some basic changes in our plans and another stretchout of the period over which our continuing and long range objectives in space exploration and space science will be achieved. Two major examples have been our decisions on the space shuttle and on the Grand Tour. We have

now selected a space shuttle configuration concept which will cost about half of what the configuration envisaged in our plans last year would have cost to develop; this decision serves to reduce substantially the peak funding required for shuttle development in any one year and thereby helps us avoid an increase in the total annual NASA budget. We have also decided to cancel plans for the Grand Tour missions which would have been possible only in the last half of the 1970's. This means that we will not be able to launch missions to explore the distant planets—Uranus, Neptune, and Pluto—until sometime in the 1980's, and that we will then have to depend on high efficiency propulsion stages to reach them.

By these and other actions we have been able to reconfigure our long range plans so that the total NASA program can be accomplished at an annual overall NASA budget at approximately the current level. The projections we will submit to Congress as required by the Legislative Reorganization Act will show that the estimated run-out costs of the total program do not rise above the FY 1972–1973 level. (By contrast, the run-out projections submitted last year with the FY 1972 budget rose to $4 billion.)

By properly phasing-in major new programs as we go along, we can maintain a viable and useful total NASA program in space and aeronautics at a total NASA annual budget level which (in 1971 dollars) can remain essentially at the FY 1972–1973 level for the indefinite future unless, of course, the President and Congress decide that the program should be expanded or accelerated. I strongly believe that this posture of a realistic long-term plan in which the nation's commitment is clearly limited to budgets of approximately the current size, is the proper one for NASA from the standpoint of responsible management and also is essential at this time to assure continued broad-based support for the NASA program.

[3] As we took the actions in the FY 1973 budget needed to establish a realistic long range plan for NASA, we had to take another look, of course, at the nuclear propulsion program. From the standpoint of holding our total plan within an acceptable total, it was clear that we could not afford to reinstate development of the 75,000-lb. thrust NERVA engine. The costs in the 1970's would be too high, and with the stretchout in our future plans the missions requiring this capability would be even farther in the future than the forecast a year ago. Under these circumstances and constraints, reinstatement of the NERVA 75,000-lb. thrust engine development could not be justified.

On the other hand, the cancellation of the Grand Tour missions introduced the possibility of a new class of future missions for which a much smaller nuclear engine appears to be needed and particularly well suited, namely, the first missions to explore the distant planets—Uranus, Neptune, and Pluto. Our preliminary analyses show that a small nuclear engine, in the 15,000–20,000-lb. thrust class, may provide the most practical means for the first mission to these planets sometime in the 1980's, in lieu of the cancelled Grand Tour missions, as well as perhaps providing at a later time, by clustering or staging several small engines, many of the capabilities the large 75,000-lb. thrust engine would have given us.

If these conclusions are confirmed in the studies we are now initiating and are proposing to carry on in FY 1973, we will be able to establish a firm and significant specific mission goal for the nuclear propulsion development program. This would be most significant. With a focused effort on a specific mission objective, the program could then proceed without the uncertainties and controversy that has characterized it in the past.

For the reasons outlined above, the FY 1973 budget reflects a decision to reorient the nuclear propulsion program. NASA and AEC will define a small nuclear rocket system in the 15,000–20,000-lb. thrust class. This effort will be a part of a broader program to define and make trade-off studies of alternative types of advanced propulsion systems, including chemical, solar-electric, nuclear-electric, and nuclear-rocket systems for possible future missions to the distant planets Uranus, Neptune, and Pluto. Supporting research and [4] component testing for nuclear systems will also be undertaken. Development of the NERVA 75,000-lb. thrust engine and the contractor effort directed at this goal are being terminated in favor of the program reoriented as above.

For the reoriented program, the FY 1973 budget includes a total of $16.5 million in budget authority, $8.5 million for NASA and $8 million for AEC. This funding will support engine definition work at the Los Alamos Scientific Laboratory (LASL), component development and test at LASL and the Nuclear Rocket Test Site, and some work in supporting research and advanced technology work. The development contracts with Aerojet and Westinghouse will be terminated in FY 1972, with termination costs to be met out of funds available in FY 1972.

I hope that the foregoing will give you an understanding of the reasons which have led to the decisions on the nuclear propulsion program which are reflected in the FY 1973 budget and which we will begin to implement in the remainder of FY 1972. I will be available to discuss this further with you at your convenience.

Sincerely,

James C. Fletcher
Administrator

Document I-23

Document title: NASA, in consultation with the Advanced Research Projects Agency, "The National Space Vehicle Program," January 27, 1959, pp. 1–7.

Source: NASA Historical Reference Collection, NASA History Office, NASA Headquarters, Washington, D.C.

In the months after it began operations on October 1, 1958, NASA assessed its launch needs as part of developing an initial ten-year plan. In this report prepared for the White House, NASA had the lead, but consulted with the Department of Defense's lead organization for space, the Advanced Research Projects Agency. This report was NASA's "Declaration of Independence" from the future use of Department of Defense missiles for meeting all of its launch needs; the space agency argued that there was a need to develop launch vehicles specifically for space applications. Of the vehicles proposed in this report for early development, the Vega was never approved. The following is the report's summary.

The National Space Vehicle Program

Prepared by the
National Aeronautics and Space Administration

In consultation with the
Advanced Research Projects Agency
of the
Department of Defense

[1]

SUMMARY

Under the National Space Act of 1958 (Public Law 85-568) the President of the United Sates is responsible for developing a continuing program of aeronautical and space activities to be conducted by agencies of the United States. The National Aeronautics and Space Administration presents in this report a National Space Vehicle Program. This program plan is a continuing effort to be reviewed annually and revised as needed.

The National Space Vehicle Program was formulated after discussion and consultation with agencies of the Department of Defense, principally the Advanced Projects Research Agency, the Department of the Air Force, and the Department of the Army. Existing and planned projects of the Department of Defense in this area, including those intended for military missions, have been taken into account with the purpose of avoiding any unnecessary duplication of effort.

The present generation of space flight vehicles is being used to place small payloads in close orbits around the earth and to propel very small instrument packages into space. The current group of booster vehicles, namely, Vanguard, Jupiter C, Juno II, and Thor-Able, were all hurriedly assembled under pressure of meeting the threat of Russian Sputniks and none of them possess the design characteristics required by future needs of the National Space Program. The Vanguard, which has the best basic design philosophy, has not yet demonstrated sufficient flight reliability. The Jupiter C, which has had the most flight success, has [2] a low load-carrying capability. The Juno II vehicle has a low injection altitude for satellite use, and requires that it be spun for stability. The Thor-Able booster that has been used in the Air Force moon shots has no attitude control system for the second stage during coast, so that the injection altitude for satellites is on the order of 150 miles. The Atlas-Able being prepared for one space mission has the best potential load-carrying capability but suffers, as do the others, from being designed for a specific mission.

Our approach up to this time has been much too diverse in that we fire a few vehicles of a given configuration, most of which have failed to achieve their missions, and then call on another vehicle to take the stage. In this situation no one type of vehicle is tested with sufficient thoroughness and used in enough firings to achieve a high degree of reliability.

The National Space Vehicle Program is directed toward avoiding past errors. The central idea is that one vehicle type, when fitted with guidance and payload appropriate to the mission, can serve for most of the space missions planned for a given 2 to 4 year period. By designing the vehicle with this purpose in view and by using it again and again for most of the space work, it appears inevitable that this one vehicle type will achieve a high degree of reliability. Therefore, this program presents a series of space-flight vehicles of increasing payload capability for successive periods of use. Each vehicle of the series will be useful for satellite work including low and high circular orbits, highly elliptical earth orbits, lunar exploration, planetary exploration, and deep space probing.

[3] In an attempt to achieve greater reliability in the existing vehicle area, NASA is sponsoring DELTA as an interim general purpose vehicle. DELTA is a more versatile version of Thor-Able, achieved by inserting a Vanguard design feature that had been deleted; namely, the coasting flight control system. Reliability rather than performance is to be emphasized by replacing or deleting those components of Vanguard and Thor-Able that have caused failures. It will be used for communication, meterological [sic] and scientific satellites and lunar probes during 1960 and 1961.

The first new general purpose vehicle of the National series is the VEGA. This is one of three vehicles based on the use of Atlas as a primary stage. The second stage is powered by the Vanguard first stage engine modified for high altitude operation. This engine has an excellent record of performance under Vanguard. The tanks are made up principally of standard Atlas parts, thus providing an early availability of the VEGA vehicle. When used for lunar or planetary missions, a third or terminal stage with solid or storable-liquid fuels will be employed. VEGA should see considerable use in the period from 1960 through 1964. It can boost two men into a close earth orbit with enough equipment to sustain them for several weeks. Its principal function, however, may be the exploration of the moon for which it is ideally suited. It should be possible in the next few years to take very high resolution photographs, first of the front or visible side of the moon and eventually of the back or heretofore unseen side. A close approach to a planet will require at least 1000 and probably 2000 pounds of equipment devoted principally to [4] guidance and communication. VEGA is the first vehicle that can carry payloads of this magnitude to the vicinity of Mars or Venus and should pave the way for the use of CENTAUR which is better adapted to the planet mission.

The second new general purpose vehicle of the National series is the CENTAUR which is well suited to be a successor to VEGA, because it requires no change in the Atlas booster. CENTAUR will be useable during the period from 1962 through 1966 for performing the same missions as VEGA but with from 50 to 100 percent more load-carrying capability. CENTAUR is the first vehicle to employ hydrogen as a fuel, and, if successful, should pave the way for use of this highest energy fuel in future vehicles of the National series. The payloads planned for SATURN and NOVA, more advanced vehicles of the National series, would have to be reduced if a lower energy fuel had to be substituted for hydrogen. There is every expectation, however, that CENTAUR will be successful, owing to the background of experience with hydrogen in industry and also within NASA.

ATLAS-HUSTLER is being developed by the Air Force. It should be available about six months prior to Vega but will have only about half of Vega's load-carrying capability. It could serve, however, as an interim version of the Atlas boosted series.

The third general purpose space vehicle of the National series is the SATURN, previously called JUNO V. Actually, JUNO V designates the first stage booster of a large multistage vehicle. This booster is being achieved by clustering eight ICBM-type engines and nine ballistic-missile-type tanks [5] to form a vehicle with a gross weight of about 3/4 million pounds. Second and third stages will have to be provided in order to make a complete vehicle of SATURN. The second stage is about the size of an ICBM, will use conventional fuels at first and will be designed for high altitude operation. The third stage is smaller, and may use conventional fuels at first, but is planned ultimately for hydrogen as a propellant. This vehicle will be capable of placing very large payloads (10–15 tons) in orbits around the earth. A typical mission would involve sending a crew of 5 men into orbit with enough facilities to sustain them for a long period of time, say several months, and the necessary equipment to permit them to perform experiments and make observations.

SATURN may well become the basic vehicle for orbital supply missions, involving the transport of food and supplies to crews in orbit, the exchange of crew members, and the transport of additional fuel and equipment to the orbiting vehicle. In order to perform these latter functions, techniques of navigation and rendezvous will have to be worked out. When used for lunar and planetary exploration, unmanned of course, the SATURN space vehicle has a load-carrying capability of between 1 and 4 tons. Starting about 1963, this vehicle should see use for at least 5 and perhaps 10 years and may, in time, become one of the most versatile vehicles in the National series.

The fourth general purpose vehicle of the National series is the NOVA, an entirely new vehicle based upon use of the one and one-half million pound thrust engine recently initiated. The earliest possible use of the large engine would come about by using a single unit to propel a [6] first stage booster. In this configuration, however, it would be about the same size as JUNO V and would be competitive to it. Therefore, the first use of the large engine is planned for NOVA; the first stage of which may employ a cluster of four of the large engines yielding a total thrust of six million pounds. The vehicle's second stage would be powered by a single million and one-half pound thrust engine and the third stage would be about the size of an ICBM but will use hydrogen as a fuel. As presently conceived, this vehicle would stand 260 feet high. NOVA is the first vehicle of the series that could attempt the mission of transporting a man to the surface of the moon and returning him safely to the earth without use of orbital supply operations.

With advances in the state-of-the-art which must surely occur over the next 5 to 10 years, it is conceivable that the NOVA would be improved to transport say 2 or 3 men on the earth-moon and return mission. Four additional. stages above the three already mentioned are required for the lunar return mission including the rockets for landing on the moon, taking off from the moon, and for re-entry into the earth's atmosphere. NOVA has the capability of transporting, if it is needed, very large payloads, on the order of 75 tons, into earth orbits.

NASA is now supporting Project ROVER in anticipation of using nuclear engines in the 1965 to 1975 period. Although it is too early to designate specific uses for nuclear rocket vehicles, they would probably be employed first as upper stages for Saturn and Nova.

A wide variety of low thrust engines and vehicles can be conceived for space travel. These are vehicles that do not land or take-off from [7] celestial bodies but are used as ferries, so to speak, between orbiting stations. The engines employ various combinations of nuclear, electrical and solar energy. Most of these engines are in early stages of development and would not see use in the near future. However, they hold promise, owing to their high efficiencies, of increased payload-carrying capabilities in the future.

Succeeding sections of this report are devoted to brief descriptions of existing vehicles and their capabilities and the plans for new vehicles and their missions. . . .

Document I-24

Document title: Development Operations Division, Army Ballistic Missile Agency, "Proposal: A National Integrated Missile and Space Vehicle Development Program," Report No. D-R-37, December 10, 1957, pp. 1–7.

Source: Dwight D. Eisenhower Library, Abilene, Kansas.

In the months following the launches of Sputniks 1 and 2, there was much activity as various groups attempted to stake out their roles in the emerging U.S. space and missile buildup. This report summarizes the arguments for a major role in launch vehicle development put forth by Wernher von Braun and his rocket team, who were working under the command of the Army Ballistic Missile Agency in Huntsville, Alabama. The following is the first seven pages of the report.

[original stamped "Secret," crossed out by hand]

Report No. D-R-37

Proposal

A National Integrated Missile and Space Vehicle Development Program

10 December 1957

DEVELOPMENT OPERATIONS DIVISION
ARMY BALLISTIC MISSILE AGENCY

[1] PURPOSE

The purpose of this report is to review U.S. missile programs in the light of known Soviet space flight capabilities and to propose an integrated national missile and space vehicle development program that will insure [sic] maximum security through appropriate expenditure of manpower, facilities and money.

The need for an integrated missile and space program within the United States is accentuated by the recent Soviet satellite accomplishments and the resulting psychological intimidation of the West. These facts demonstrate that we are bordering on the era of space travel and must very seriously consider the expansion of the principles of earth warfare to space warfare. A review and revision of our scientific and military efforts planned for the next ten years will insure [sic] that provisions for space exploration and warfare are incorporated into the overall development program.

Because of the short time available for preparation this report is preliminary in nature. It will be supplemented and revised as more possibilities are explored and more accurate information is available.

STATEMENT OF THE PROBLEM

To outline a feasible plan which allows the U.S. to catch up and ultimately overtake the Soviets in the race for scientific and military space supremacy without upsetting the Nation's economic stability, disrupting the manpower balance and draining national resources.

ASSUMPTIONS

1. The national objectives should include achievement of the following:
    a. Reaffirmation of national scientific and technological supremacy.
    b. Provision of adequate defense against the Soviet capability to engage in space warfare.
    c. Expansion of the national deterrent capability to include space warfare techniques.
    d. Evolution of a national capability for space exploration.
[2] 2. The development program should be conducted on a national basis devoid of the personal interests of any individual military or civilian group or organization.
3. Maximum use should be made of existing development teams and available hardware wherever possible.

DISCUSSION

1. AMERICAN vs[.] SOVIET SATELLITE AND MOON FLIGHT CAPABILITIES

The launching of SPUTNIK I on 4 Oct 1957 and SPUTNIK II on 3 Nov 1957 demonstrated clearly the Soviet capability in the field of long range rockets and orbital techniques. The U.S. satellite capabilities are inadequate in schedule and in satellite payload weights. Figure 1 shows the present and anticipated Soviet and U.S. satellite capabilities plotted against time. If these estimates are correct, the Soviet capabilities cannot be reached and surpassed before 1962 or 1963. This prediction is based on the assumption that immediate development of an orbital carrier with a booster stage of at least 1.5 million pounds of thrust will be initiated without delay. The Soviet lead is due largely to their early effort in developing large rocket engines in the 300,000 pound thrust class. A comparison of U.S. and Soviet moon flight capabilities is shown in Figure 2. The Soviet carriers are identified on these charts by the engine take-off thrust expressed in thousands of pounds ("K" equals one thousand pounds) for the individual stages. It is again very unlikely that the Soviet capabilities can be surpassed before 1963 because of their lead in basic transportation vehicles.

The key to rapid improvement of the U.S. capability for orbital and moon flight missions lies in an accelerated development of powerful booster stages. The overall impulse of the ICBM booster stage is insufficient for any large unmanned or manned space flight mission. A larger booster than the ICBM type booster is a mandatory requirement.

[no page number]

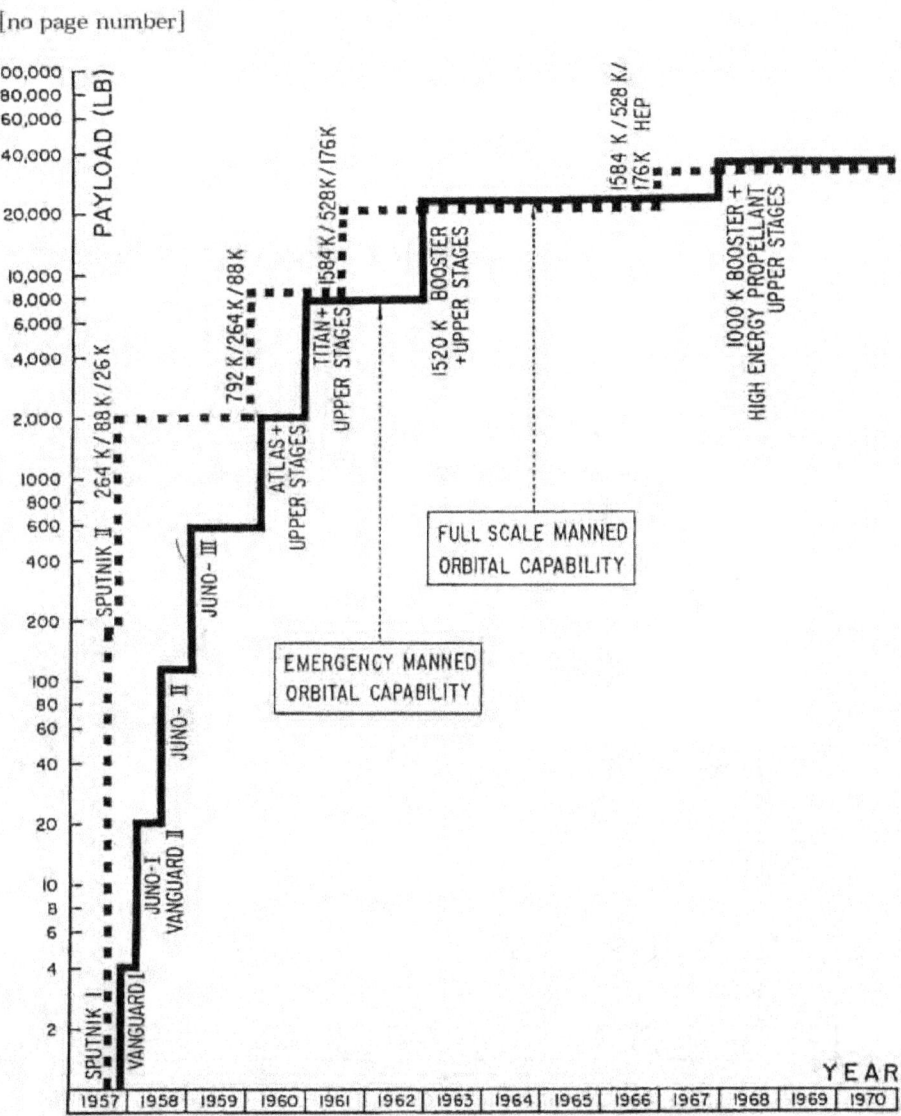

Figure 1. Present and Anticipated Satellite Payload Capabilities of Russian and American Satellite Carriers (One Way Mission)

[no page number]

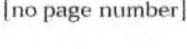

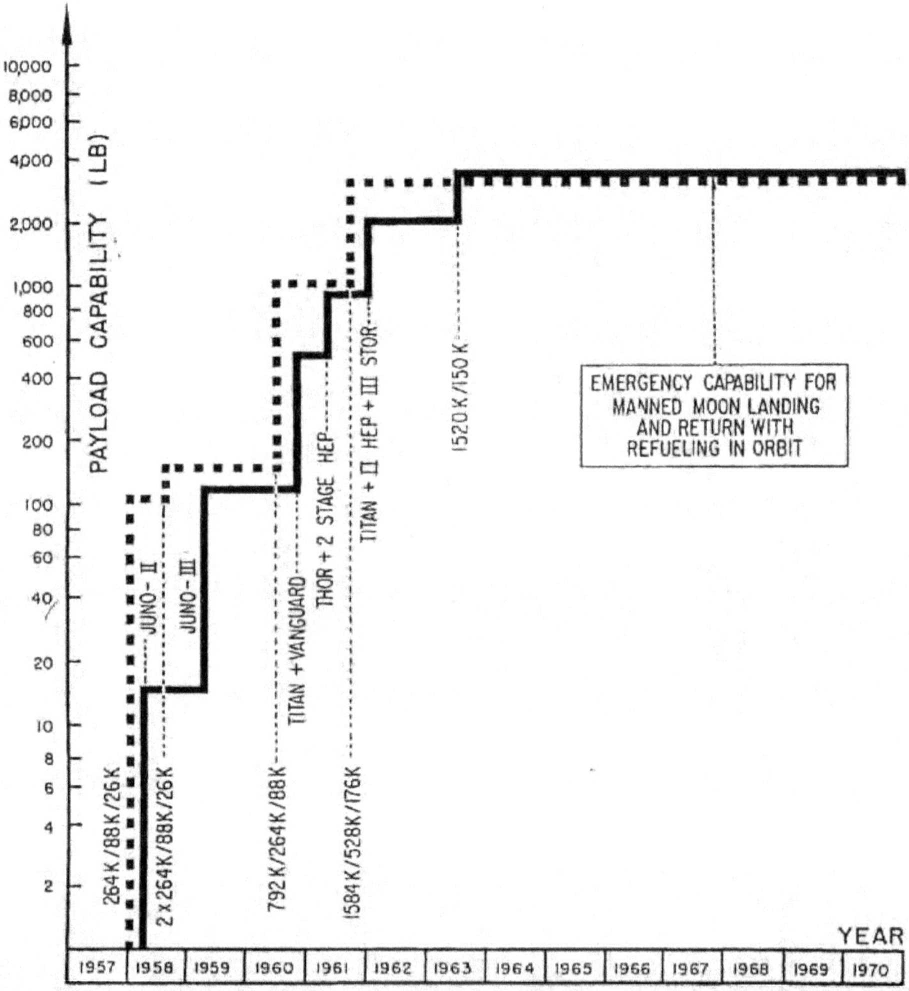

Figure 2. Present and Anticipated Moon Flight Payload Capabilities of Russian and American Carrier Vehicles (One Way Mission Without Refueling)

2.   BUILDING BLOCK SCHEME SUMMARY

A logical booster development sequence is portrayed in Figure 3 which depicts five basic families:
- a.   REDSTONE booster (Booster I; 78,000 pounds thrust).
- b.   JUPITER booster (IIa; 150,000 pounds thrust).
- c.   JUNO IV booster (IIb; 380,000 pounds thrust).

[no page number]

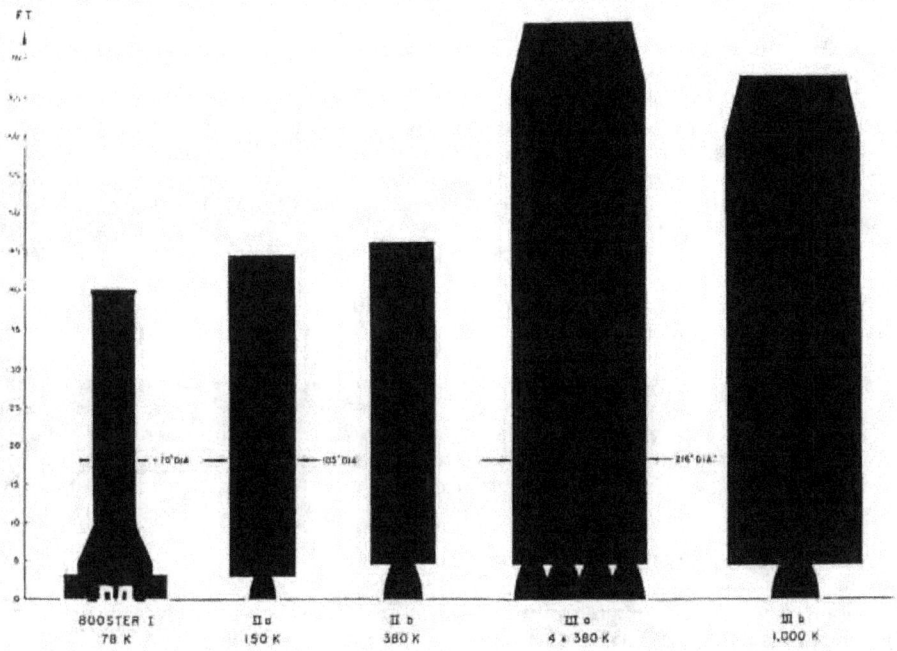

Figure 3. Typical Booster Family as Basic Transportation Elements

[3]   d.   SUPER JUPITER booster (IIIa; cluster of four 380,000 pound engines).
        e.   SUPER JUPITER, second generation booster (IIIb: 1,000,000 pounds thrust).
    The JUPITER booster appears in the sequence because of its availability in the desired time scale, and because detailed performance data were on hand. The study could as well be based upon other choices.
    Application of each of the five basic families to specific purposes is portrayed and discussed in Appendix A. Sufficient technical data are tabulated so that performance for each application can be indicated. The purpose of this portion of the study is to indicate the flexibility inherent within each family and the total program. The study also illustrates how work performed with one booster can contribute to the development of the next larger one or can provide an upper stage for a larger multi-stage missile. Interim and emergency capabilities can be readily achieved as an outgrowth from the basic booster development, in much the same way as a branch depends upon the trunk of a tree for its growth.

3. INTEGRATED MISSILE AND SPACE VEHICLE DEVELOPMENT PROGRAM
SUMMARY

A U.S. satellite capability of 20,000–30,000 pounds will be required by 1963. The
Soviets will be able to attain the necessary booster and stages with the existing 264,000
pound thrust engine or the 820,000 pound thrust engine reportedly in development.

It is imperative that the NAA [North American Aviation] E-1 380,000 pounds thrust
engine development be accelerated and that highest priority be given to a development
program incorporating this engine.

A logical short-cut development program to attain the 20,000–30,000 satellite by 1963
would be a booster of 4 x 380,000 pounds thrust, a second stage JUPITER booster with a
380,000 pounds engine and a third stage JUPITER booster-payload with existing
150,000 pounds thrust engine.

The space vehicle program should be organized into—

      a.   Orbit carriers for the transportation of cargo and personnel from the earth's
           surface into a selected orbit and back.

      b.   Scientific and military, unmanned and manned satellite vehicles for accom-
           plishing such missions as reconnaissance, satellite intercepts, scientific
           research, etc.

[4]   c.   Moon flight missions for purposes of scientific research with manned land-
           ings and return.

It is very important that every effort be made to accomplish a U.S. manned moon
landing prior to such a feat being done by the U.S.S.R. This is an extension of the manned
satellite and could be accomplished by approximately 1967.

Recommended development programs for orbital carriers, instrumented and
manned satellites and moon vehicles are tabulated in Appendix B. The development programs are treated by logical teams, time scales for development and operational phases,
payload capabilities and estimated project costs. In addition a recommended engine
development program is outlined as a critical component development requirement of
the overall plan.

4. COST AND SCHEDULE

Some actual and estimated overall costs for individual projects and missions have
been included in Tabs XV thru XVII. These were estimated on the basis of actual project
expenditures, estimated manpower requirements, team strength, number of flights
required for individual missions, available figures of cost per missile and missile launching and some relationships between cost and weight of components. In spite of the fact
that best available sources and judgement [sic] were used, the given figures are considered approximations only, especially for the larger satellite and moon flight projects anticipated to take place about 10 years from now. These estimates, however, serve to illustrate
the order of magnitude of effort or money involved and an average expenditure per year
to be expected for the program under discussion. It should be mentioned that the annual supporting costs for maintaining permanent manned satellites are not included in the
figures given in Tabs XV thru XVII. For a 50 man satellite, for example, maintenance costs
will be in the order of 100 to 200 million dollars per year; for a 20 man satellite approximately 50 to 100 million dollars per year. Thus, the overall cost for the space flight program proposed in Tabs XV thru XVII will be approximately as follows for the time period
of 1958 to 1971:

	Millions of $
Orbital Carrier Program	2,600
Satellite Program	2,500
Maintenance 20 Man Satellite 1966 thru 71	450
Maintenance 50 Man Satellite 1968 thru 71	600
Moon Flight Program	3,500
Component Development	2,800
[5] Additional Facilities	550
Ground Organization and Operation	1,000
5 Percent Inflation Rate	700
14 Year Space Flight Program Total Cost Estimated	$14,700 Mil

Average Expenditure Per Year (1958 thru 1971)	$1,050 Mil

Estimated Annual Supporting Research and Experimental Models	450 Mil

Total Per Year	$1,500 Mil

These expenditures would have to be spent in addition to the present military missile program. The development and production costs for the present Air-to-Air, Air-to-Surface, Surface-to-Air, and Surface-to-Surface missile projects, including the IRBM and ICBM programs, are not included in these figures—only the effort required for the modifications necessary for the space flight program under discussion.

The schedules given in Tabs XIII thru XX are derived on the basis of current experience and the assumption that a national missile and space flight development program will be established and authorized in early 1958. It is also assumed that the individual development teams obtain assignments with respect to their contribution to the overall program early enough to be able to carefully define the overall systems and to plan for a realistic operational date. With these principles as bases, the following U.S. accomplishments in the achievement of space superiority are attainable and should be strived [sic] for:

Jan	1958	1st 4 lb and/or 20 lb Satellite
Jun	1958	1st 100 lb Satellite
Jan	1959	1st 500 lb Satellite
Apr	1959	1st 100 lb Moon Flight (hard landing)
Spring	1960	1st 2000 lb Satellite
Fall	1960	1st 100 lb Moon Flight (soft landing)
Spring	1961	1st 5000 lb Satellite
Spring	1962	1st TV Instrumented Moon Circumnavigation
Fall	1962	1st Manned Satellite (1 to 2 man)
Spring	1963	1st 20,000 lb Orbital Capability
Fall	1963	1st Manned Moon Circumnavigation
Fall	1965	1st 20 Man Permanent Manned Satellite
Spring	1967	1st Manned Moon Landing & Return (3 man expedition)
Spring	1968	IOC 50-Man Permanent Manned Satellite
	1971	50-Man Moon Expedition

[6] CONCLUSIONS

1.   An integrated missile and space flight program is feasible and essential for national survival.

2.   It seems possible to overtake the Soviet capabilities provided an adequate long range space flight program is instituted immediately.

3.   The estimated annual cost of the program desceibed [sic] in this report (in addition to the present missile program) is 1.5 billon [sic] dollars.

4.   This U.S. space flight plan can be achieved without upsetting the nation's economic stability, manpower balance and other national resources if:

   a.   The plan makes maximum utilization of existing teams and hardware developed under existing missile programs.

   b.   The plan provides for adequate supplemental programs to develop essential hardware and techniques not provided in present programs. The most urgent of these is development of large boosters.

   c.   The program is closely coordinated with the military missile program and is based upon the same transport vehicles.

5.   The allocation of work loads to specific teams should take the following factors into account:

   a.   Matching of required techniques to skill, experience and facilities that the team possesses.

   b.   Availability of team capacity.

6.   Development of the large (1520 K-pounds thrust) booster is considered the key to space exploration and warfare.

RECOMMENDATIONS

1.   That the development of essential components and long lead items such as the 4 x 380,000 pound thrust power plant package be initiated immediately.

2.   That an integrated missile and space vehicle development program with immediate long range task assignment to individual teams be authorized without delay.

[7] 3.   That each development team be assigned system responsibility for a complete phase of the program to assure maximum economy and acceleration of the development. For example, the payload stage of the basic orbital carrier vehicle will carry the components of any manned satellite or space vehicle into the orbit and the design of these satellites and space vehicles should be carried out concurrently by the same team to insure [sic] maximum use of available components and to minimize effort.

4.   That the primary goal of the space flight program for the next 10–12 years be the accomplishment of a manned Moon landing and return to Earth.

5.   That maximum use be made of the transportation provided by the development program for all kinds of scientific exploration of the upper atmosphere, space environment and celestial bodies [such] as moons, planets and the sun. The ultimate use of space vehicles will be as carriers for men and instruments bound to resolve the laws and secrets of nature for the benefit and progress of mankind.

6.   That an early scientific space exploration program be developed parallel to the space vehicle program and coordinated with the individual development phases. This scientific space exploration program and allied military programs should be used as the basis of the integrated operational space program which will start as soon as the individual carriers become available. . . .

Document I-25

Document title: Abe Silverstein, Chairman, Source Selection Board, Memorandum for the Administrator, "Recommendations of the Source Selection Board on the One Million Pound Thrust Engine Competition," December 12, 1958.

Source: NASA Historical Reference Collection, NASA History Office, NASA Headquarters, Washington, D.C.

In 1958, soon after it began operation, NASA began the procurement of a new million-pound thrust liquid-fuel rocket engine designated the F-1. The program had been initiated by the Air Force, but it was transferred to NASA as part of the redistribution of space programs following the 1958 creation of the Advanced Research Projects Agency and NASA. The F-1 engine, as eventually developed, produced 1.5 million pounds of thrust and was used in a cluster of five on the first stage of the Saturn V. This memorandum from NASA's Source Selection Board documents the selection of the Rocketdyne Division of North American Aviation to develop the new engine. The two attachments mentioned in this memo are not included here.

[1]   Washington 25, D.C.
                                                                    December 12, 1958

MEMORANDUM for the Administrator

Subject: Recommendations of the Source Selection Board on the One Million Pound
           Thrust Engine Competition

INTRODUCTION

Extensive exploration of space beyond the sensible atmosphere will eventually require booster vehicles with several million pounds [of] thrust. A major step in this direction is the development of a million pound thrust engine which can be used singly or in clusters. To this end, the National Aeronautics and Space Administration has initiated a research and development type procurement for a million pound thrust, single chamber liquid fuel, rocket engine. It is expected that several years will be required to complete this project. The initial actions taken to date include the following:

October 14, 1956 —   Invitations to attend a briefing by NASA personnel at NASA
                           Hdqts on the proposed procurement were extended to seven
                           contractors—namely:

                           Rocketdyne, a Division of North American Aviation, Inc.
                           Aerojet-General Corporation
                           Aircraft Gas Turbine Division [of] General Electric
                           Pratt and Whitney Aircraft Division of United Aircraft Corporation
                           Reaction Motors Division of Thiokol Chemical Corporation
                           Wright Aeronautical Division, Curtiss-Wright Corporation
                           Bell Aircraft Corporation

October 21, 1958 —   Briefing of invited contractor.

October 23, 1958 —   NASA Specification HS-10 (Attachment A) sent to invited
                            contractors.

[2] November 24, 1958 — Six   contractors   submitted   proposals: Bell   Aircraft
                            Corporation declined to propose.

On November 24th, two assessment teams were organized for purposes of making a
thorough analysis of the six proposals submitted. On December 2nd a Source Selection
Board was appointed to evaluate the proposals. One of the assessment teams consisted of
scientific and technical specialists and the other of cost and management specialists. (The
membership of the two teams is listed on Attachment B). The teams were relieved from
all other work. For two weeks, the two teams conducted an intensive and exhaustive analysis and comparison of the Proposals and they prepared themselves to present their findings to the Source Selection Board.

On December 9th[,] 1958, the Source Selection Board was convened. The purpose of
the Board was to review and evaluate the entire matter and, thereupon, to recommend to
the Administrator the selection of a contractor for the development of the engine. The
Board consisted of:

Dr. A. Silverstein, Director of Space Flight Development[,] Chairman
Mr. J.W. Crowley, Director of Aeronautical and Space Research
Mr. A. Hyatt, Assistant Director for Propulsion
Mr. R.E. Cushman, Procurement and Supply Officer
Mr. Robert G. Nunn, Jr., Assistant General Counsel

The Board remained in continuous session during December 9th and 10th, and
reconvened again for several hours on December 11th. During this period the Board thoroughly reviewed the work of both assessment teams. All team members were available to
the Board for questioning.

The following main subjects were considered by the Board in the technical area:

Thrust Chamber and Injector
Turbo Pump Assembly
Controls
Overall System Design Features
Materials and Methods of Construction
Scheduling
Test Program
Technical Capability

[3] In the management area, the following main subjects were considered:

Availability of Facilities
Availability of Manpower
Realism of Programming
Cost Estimating
History of Past Performance
Management Capability

[4] EVALUATION

The following paragraphs set forth the factors supporting the recommended selection of a contractor with whom the procurement should, in the opinion of the Board, be placed.

Curtiss-Wright Corporation:

The Curtiss-Wright proposal fails to comply with three important requirements expressly contained in the NASA Specification HS-10 governing this competition. Specifically, Curtiss-Wright (a) fails to make any provision for a changeover to non-cryogenic propellants as required by HS-10; (b) fails to make any mention of the manner or possibility of up-rating the engine to 1.5 million pounds thrust, as required by HS-10; and (c) fails to base its proposal on the use of suitable test facilities to be located at Edwards Air Force Base, California as required in HS-10, but requested instead the development of a new test and support facility to be located in the vicinity of Reno, Nevada.

Because its proposal fails to comply with these three major requirements, the Curtiss-Wright Corporation was deemed not responsive.

General Electric Company:

The General Electric proposal is based upon a design which may be referred to as "the plug nozzle concept." No rocket engine of any size has as yet been built using this principle. One of the primary advantages claimed for it, namely, less likelihood of combustion instability, is of dubious validity. Moreover, the total heat to be removed from this engine is estimated to be about 60% greater than from conventional engines. The method of providing vector thrust control would present unusual and difficult design and development problems.

The General Electric Company also proposed to inaugurate a new department at Schenectady, New York, to execute this program. Their main effort in rocket engines now is centered at Evendale, Ohio.

Although the General Electric proposal is next to the lowest in estimated cost, its proposal lacks realism in that the test schedule and total test man-hours are considered too low. A more realistic test program would, of course, raise the cost estimate an indeterminate but substantial amount. In general, the General Electric proposal appears to be a high risk development program with insufficient compensatory advantages. It is altogether undesirable to undertake the development of [5] the million pound engine and at the same time attempt to develop the plug nozzle content of design.

Thiokol Corporation, Reaction Motors Division:

The Reaction Motors proposal is technically conventional but inferior in design in terms of the present state of the art. The engine is the heaviest and largest of all proposals received. In addition, Reaction Motors proposed to develop the thrust chamber in conjunction with the turbo-pump. Since the thrust chamber and the turbo-pump will initially be highly experimental devices this would mean that whenever a change on either the thrust chamber or turbo-pump was necessary, development testing on both would stop. Technically this approach is unsound and unacceptable. The scheduled number of engine tests and total engine test time is very low and unrealistic for the kind of development involved. Reaction Motors submitted the highest cost estimate received. It was almost twice as high as the nearest competitor. Reaction Motors does not have adequate physical facilities or technical capability within its organization to do the work required.

Pratt and Whitney:

Pratt and Whitney is one of the outstanding turbo–machinery companies in the world. However, they have no experience in the development of large liquid rocket engines. This fact is reflected in the sketchiness and incompleteness of the technical and other phases of their proposal. Serious difficulty was encountered with the Pratt and Whitney proposal as presented in that it did not adequately specify many major factors such as thrust chamber and injector method of construction weights, dimensions, and other details. This deficiency is apparent in their unrealistic cost estimation. They also lack proper facilities for this work. Pratt and Whitney states that they will build some facilities at company cost and in addition will require government furnished facilities to the extent of $6,297,000.

The cost estimate submitted is the lowest of any proposal received and clearly reflects a lack of appreciation of the magnitude of the tasks. A more realistic appreciation and programming of necessary tests would however raise the cost considerably.

[6] Aerojet and Rocketdyne:

Both the Aerojet and Rocketdyne proposals show a sound appreciation of the task. Both companies are believed to be capable of developing the one million pound engine. The Board has weighed all areas of these two proposals carefully and has determined that the Rocketdyne proposal is the superior proposal.

Particular points of difference between the proposals and some of the arguments for rating the Rocketdyne proposal superior to that of Aerojet are set forth below.

(1) The Rocketdyne development program shows a mature appreciation of the major technical problems and in addition is backed-up by alternate design concepts and hardware. The Aerojet proposal although containing a number of novel features is committed to a single design approach with almost no concept back-up.

(2) The Aerojet thrust chamber design cools the chamber with only 10 percent of the total fuel flow. While this concept provides for a light-weight thrust chamber, the decision to use only 10 percent of the fuel as coolant results in marginal cooling. Extension of thrust from 1.0 to 1.5 million pounds and conversion from cryogenic to storable propellants will most likely necessitate revision or redesign of the thrust chamber. Furthermore, the suggested method of cooling the combustion chamber, by the use of film cooling, is not only unproved in large-thrust engines but the method of accomplishing it is not clearly put forth in the proposal. Rocketdyne proposes a conventional thrust chamber cooled by all the fuel. This engine is accordingly heavier but should avoid the heat transfer problems of the Aerojet proposal.

(3) Both Aerojet and Rocketdyne turbo-pumps are direct drive arrangements, therefore, potentially more reliable than the geared arrangements used by the other contractors. Both use bi-propellant gas generators. They are equivalent in pump and turbine hydro-aero-dynamic and mechanical design. The Aerojet pump delivers propellant at a pressure only 300 psi above the chamber pressure. This low pressure differential offers little or no margin for correcting difficulties that might develop in the engine testing program such as, for example, additional pressure drop if the cooling of the thrust chamber must be increased. The Rocketdyne pump delivery pressure is very high, that is, over twice [the] chamber pressure on the fuel pump and 180 percent of chamber pressure on the oxidant pump. These high pressures require large horsepower from the [7] turbine with consequent large gas generator propellant consumption. While these pump outlet pressures may appear unnecessarily high, the approach is conservative and provides, at the cost of slightly lowered overall specific impulse, a wide pressure margin for controlling heat transfer processes and possible combustion driven oscillations.

(4) The Aerojet controls are simpler, more straightforward and, although experimental, may be regarded as superior to the Rocketdyne proposed controls. However, Rocketdyne proposes to use a control system which has previously been used.

(5) Both Rocketdyne and Aerojet have, in general, well balanced test programs. Rocketdyne is considered to have the more realistic schedule. By reason of an Air Force "feasibility" contract, Rocketdyne will have test facilities which will permit thrust chamber tests eight months sooner than Aerojet. Generally speaking, the thrust chamber development work will pace the engine development. Rocketdyne proposes eleven months more than Aerojet for full scale testing to PFRT (Preliminary Flight Rating Test).

(6) In the overall system arrangement Aerojet attaches the turbo-pump to the fixed portion of the engine mount. The result is that the high pressure (700 to 1000 psi) fuel and oxidant lines must have flexible joints. Rocketdyne on the other hand mounts the turbo-pump assembly on the movable portion of the engine. This arrangement permits the flexible lines to be on the low pressure (50 psi) side of the pumps, which is more desirable

(7) The Rocketdyne proposal is considered to be superior in the areas of facilities, manpower, and management. Rocketdyne has in existence more facilities and more available skilled manpower directly applicable to this program than any other company. It has more previous experience on large liquid rocket engines. These factors coupled with the excellent management concept in design approach and test scheduling extending over a longer period of time results in the conclusion that the Rocketdyne proposal is the most realistic of those submitted.

(8) The Aerojet and Rocketdyne cost estimates are within 5% of each other in total dollar amount. This small difference in cost and considering that the contract will most likely be a cost plus fixed fee type contract makes the weight to be assigned to the cost figure, in the overall evaluation, of relatively low importance.

[8] CONCLUSIONS

The Source Selection Boards, after a thorough evaluation of all factors relevant to this competition, has determined that the proposal of the Rocketdyne Division of North American Aviation is the best overall proposal submitted.

RECOMMENDATION

The Board recommends that the National Aeronautics and Space Administration undertake negotiations with the Rocketdyne Division of North American Aviation on a definitive contract to develop 1000K pound engine and that a letter be sent to that company initiating such negotiations.

Abe Silverstein,
Chairman, Source Selection Board

Document I-26

Document title: Roy W. Johnson, Director, Advanced Research Projects Agency, to Commanding General, U.S. Army Ordnance Missile Command, Huntsville, Alabama, ARPA Order No. 14-59, August 15, 1958.

Source: NASA Historical Reference Collection, NASA History Office, NASA Headquarters, Washington, D.C.

After witnessing the lift capacity of Soviet rockets, U.S. officials decided to develop a large booster capable of launching very heavy loads. The Advanced Research Projects Agency (ARPA), which was coordinating military space activities, authorized $5 million in 1958 for the Army Ordnance Missile Command (AOMC) to initiate the development of a booster with 1.5 million pounds of thrust. To save the time and money involved in developing a new engine, the booster was to achieve this thrust level by using a cluster of eight existing rocket engines. This booster eventually evolved into the Saturn I and Saturn IB vehicles used in the Apollo program. In fact, the Saturn IB launched the Apollo 7 flight into Earth orbit.

[1]   ADVANCED RESEARCH PROJECTS AGENCY
                                      Washington 25, D.C.

                                                                      ARPA Order No. 14-59
                                                                         August 15, 1958

TO: Commanding General
     U.S. Army Ordnance Missile Command
     Huntsville, Alabama

     1.   Pursuant to the provisions of the DoD Directive 5105.15, dated February 7, 1958, you are requested to proceed at once on behalf of the Advanced Research Projects Agency with the project specified below. Additional details and directives will be issued by ARPA from time to time and will become a part of this Order when so specified.
     2.   Initiate a development program to provide a large space vehicle booster of approximately 1.5 million pounds thrust based on a cluster of available rocket engines. The immediate goal of this program is to demonstrate a full-scale captive dynamic firing by the end of calendar year 1959.
     3.   You will submit, as soon as possible, for review and approval by the Advanced Research Projects Agency[,] a detailed development and related financial plan covering the program. These data shall include a time-phased schedule of work and estimates for work to be performed (a) by AOMC, (b) by contract, and (c) at other government facilities.
     4.   This Order makes available $5,000,000 under appropriation and account symbol "97X0113.001 Salaries and Expenses, Advanced Research Projects, Department of Defense" for obligation by the Army Ordnance Missile Command on behalf of the Advanced Research Projects Agency only for purposes necessary to accomplish the work specified herein. These funds are immediately available for direct obligation and for use in reimbursing the Army Ordnance Missile Command for costs incurred under this Order. Upon approval of development and financial plans, as required herein or in accordance with amendments to this Order, these funds will be increased as appropriate.
     5.   The Director, Advanced Research Projects Agency, will provide policy and technical guidance, either directly or through designated resident representatives. The Army Ordnance Missile Command [2] will be responsible for arranging for the detailed technical direction necessary to accomplish the specified objectives and to comply with ARPA policy and technical guidance. This general relationship may be specified in greater detail by amendment to this Order if such action is necessary.
     6.   The Director, Advanced Research Projects Agency, and the Office of the Secretary of Defense will be kept informed by such management, technical and accounting reports as may be prescribed pursuant to this Order.
     7.   The use of equipment and materials procured in connection with this project is subject to direction of ARPA and all reports, manuals, charts, data and information as may be collected or prepared in connection with the project shall be made available to ARPA prior to release to other agencies or individuals under procedures to be approved.

8. AOMC shall be responsible for preserving the security of this project in accor-dance with the security classification assigned and the security regulations and procedures of the Department of the Army.

9. Notwithstanding any other provisions of this Order, AOMC shall not be bound to take any action in connection with the performance of this work that would cause the amount for which the Government will be obligated hereunder to exceed the funds made available, and the obligation to the AOMC to proceed with the performance of this work shall be limited accordingly. AOMC shall be responsible for assuring that all commit-ments, obligations and expenditures fo [sic] the funds made available are made in accor-dance with the statutes and regulations governing such matters provided that whenever such regulations require approval of high authority such approvals will be obtained from or through the Director, ARPA, or his designated representative.

Roy W. Johnson
Director

cc: Secretary of the Army

[stamped "Classification Changed To: UNCLASSIFIED, By Authority of SCG-6, Date 5-6-70, By Lois F"]

Document I-27

Document title: F.C. Schwenk, Memo for Record, "Visit to ABMA on June 16–17, 1959," June 24, 1959.

Source: NASA Historical Reference Collection, NASA History Office, NASA Headquarters, Washington, D.C.

In late 1958, NASA had been unsuccessful in its hope to have the von Braun rocket development team transferred to it from the Army Ballistic Missile Agency (ABMA); the Army's leadership had resisted the change, and the Department of Defense or the White House had not overruled that resistance. Although NASA hoped that that decision might be reversed (as it was in late 1959), it also recognized that it might have to cooperate with the Army-led team to obtain the launchers needed for ambitious future missions, particularly human flights to the Moon. By mid-1959, such flights had already been identified as a long-term goal of NASA's human spaceflight program. Thus in mid-1959, a NASA delegation led by George M. Low, then Program Chief for Manned Space Flight at NASA Headquarters, made an initial visit to Huntsville so the group could better understand the potential of the Saturn family of launch vehicles.

[all pages formerly marked "CONFIDENTIAL"]
[1]

DPA (FCS:rlc)
June 24, 1959

Memo for Record

Subject: Visit to ABMA on June 16–17, 1959

1. NASA representatives attending these meetings were Messrs. Low, Disher, and Schwenk. The main purpose of the meeting was to discuss programs relating to the Saturn

system. In particular, we discussed the application of the Saturn system to a manned lunar landing mission.

2.   Saturn Development Program

According to the latest thinking at ABMA, the Saturn program will evolve along two major lines. The first or current line of development will use a thrust level of 1.5 million pounds in the first stage and low energy propellants in the 1st and 2nd stage. In the 2nd line of development, the Saturn will evolve as a vehicle having a take-off thrust of 2 million pounds. This thrust level is achieved with 8 LOX/RP engines. In addition, the 2nd generation of Saturn will use high energy propellants in the 2nd stage. This information regarding the 2nd generation of Saturn is different from what we had known the development program to be in the past.

a.   Saturn I is based on ICBM hardware in the upper stages. The 2nd stage is a modified Titan and the 3rd stage is a modified Centaur. The propellant loading in the 1st stage is 750,000 lbs.; in the 2nd stage, 200,000 lbs. and 50,000 lbs. in the 3rd stage. The Saturn I will provide the following payloads: 30,000 lbs. in 96 minute orbits; from 7500 to 8500 lbs. for an escape mission and 5,000 lbs. in a 24-hour equatorial orbit. Each of these payloads is an honest payload; that is 2500 lbs. of guidance and control have already been subtracted from the vehicle capability.

The Saturn I has an undesirable feature in that it will be very long. The resulting low bending frequencies due to the long length will create problems in the control system dynamics. The cluster also has a low characteristic frequency.

The basic Saturn I development program calls for 16 flight vehicles and one propulsion test vehicle. The propulsion test and the first four flight vehicles will use 8 engines of 165,000 lbs. of thrust. Starting with flight vehicle #5, the individual thrust rating of the engines will be raised to 188,000 lbs.

A 1 to 1 mock-up of the thrust mounts and engines is being constructed currently. The engines for the propulsion test vehicle and the first four flight vehicles have been ordered. The first hot test of the system will be run on December 21, 1959. The first flight of the booster will occur in October 1960.

[2] Flight vehicles Nos. 1 and 2 will be propulsion and flight tests of stage one, only. Major objectives of these tests will be to study booster performance, propellant depletion and booster recovery. Flight vehicles 3 and 4 will be propulsion testis of stages 1 and 2 combined which will place about 10,000 lbs. in a low altitude orbit. A recoverable satellite, much like a Jupiter nose cone, is planned for these vehicles. The satellite will contain engineering components and materials. According to the ABMA staff, there is [a] NASA-ARPA-Air Force ad hoc committee planning this engineering satellite.

Vehicles 5, 6, and 7 are not as yet ordered but are planned for the first half of 1962. These vehicles will fly all 3 stages and will be research and development flights. In the case of these 3 vehicles, the 3rd stage will be a standard Centaur insofar as the propellant volume is concerned. With one of these 3 vehicles, a lunar satellite could be planned and would provide an early test of guidance capabilities for the advanced lunar missions. Vehicles 1 through 7 will cost about $20 million each.

Vehicles 8 through 16 ($15 million each) will represent the complete prototype vehicle: that is with the full 1 1/2 million pound thrust on the booster and 3rd stage of 50,000 lbs. propellant capacity. Vehicle 8 could be readied by August, 1962: a date of a Mars opposition. Mr. Koelle suggested that NASA claim vehicle #8 if they are interested in achieving this early Mars shot with the vehicle having the capability (8000 lbs.) of the Saturn. The 16th flight vehicle is scheduled for September 1963.

There is an interesting use of the Saturn I vehicle as a space truck. A 2 1/2 stage version; that is with a 3rd stage based on a JPL 6K storable engine instead of the Centaur could place from 15,000 to 20,000 lbs. in a 300 nautical mile orbit. A preliminary design of a capsule of the Jupiter nose cone type shows that from 10 to 16 men could be taken into orbit and returned with this payload weight. With a Dyna Soar type re-entry vehicle, approximately 3 to 4 men could be carried to orbit and be returned. By means of attitude control of control flaps, the Jupiter nose cone can have a lift-to-drag ratio of 0.3. This lift-to-drag ratio affords sufficient control so that the landing area required is only 10 miles in diameter. Military missions may require the Dyna Soar type of re-entry vehicle, particularly if large numbers are to be used with limited recovery facilities; however, for non-military use, the Jupiter type of re-entry vehicle appears feasible if the landing area can be kept to this 10 mile diameter figure. Consequently, this 2 1/2 stage version of Saturn I appears to have sufficient payload capability to transfer men to orbit and return them for assembly or re-fueling or experimental operations.

b. Saturn II must be viewed as an entirely new vehicle. It will employ the same type of engine cluster and tank cluster as in Saturn I for the 1st stage; however, the 2nd stage will contain liquid-oxygen and liquid-hydrogen as will the 3rd stage.

The 1st stage will use the H-2 engine. This engine is similar in size to the H-1; however, it utilizes the Mark-14 turbo-pump (an [Air Force] development) which will allow a thrust rating from 250 to 300,000 lbs. The [3] H-2 engine does not use a gas generator for turbine power. It is planned to extract hot gases from the face of the injector plate. These eight H-2 engines will provide a take-off thrust of 2 million pounds.

A modified optimization* of the complete vehicle shows that the 2nd stage requires a thrust of about 1 million pounds (oxygen-hydrogen propellants). The 3rd stage requires a thrust of 200–250,000 lbs. and the 4th stage requires a thrust of about 100,000 lbs. The Saturn II (3 stages) can provide a payload of over 70,000 lbs. in the 96 minute orbit and can soft land about 8 or 9 thousand pounds on the surface of the moon. These payloads are conservative values based on adequate velocity assumptions and conservative structural weights. In addition, guidance and control weights have been subtracted.

3. <u>Manned Lunar Landing and Return</u>

According to ABMA estimates, a capsule weight of 8000 lbs. is required for this mission. This is the weight that houses the men and returns to the earth. If aerodynamic breaking and re-entry is employed, the required weight from the surface of the moon just prior to take-off is over 46,000 lbs. for a capsule weight of 8000 lbs. In order to place a payload of 46,000 lbs. on the surface of the moon, a weight of approximately 400,000 lbs. is required in a 96 minute orbit around the earth. These figures are based on the use of hydrogen for the transfer from the earth orbit to the moon and for landing on the moon; storable propellants are used for the lunar take-off. The calculations are also based on very conservative consumptions for velocity requirements so that those weight figures can be trusted. The question is: "How do we get a 400,000 lb. payload into an orbit around the earth?" If we use Saturn I, approximately 13 vehicles will be required to build up this payload and assembly in orbit will be a necessity. If we use Saturn II, then only 7 vehicles will be required and no assembly in orbit is needed, only re-fueling.

This technique of accomplishing any manned lunar landing requires many developments in the technology. However, there are peculiar developments associated with the Saturn approach to a manned lunar landing; that is the techniques of orbital rendezvous and techniques of construction or re-fueling must be developed. Not much definite can be said about the techniques of construction but there is a feeling that re-fueling should pose no serious problems if it is possible to have men there to oversee the operation. This, in turn, involves allowing the men to come out of the transporting capsules for purposes of making connections for the re-fueling operation. The opinions of the people at ABMA are that if a man is able to get out of a capsule and walk around on the surface of the moon, he should be able to leave an orbiting capsule to work on the refueling or assembling procedures.

---

*   In the modified optimization process, the burning time of the first stage is made long enough so that staging will occur after the vehicle passes through the high-Q region.

Orbital rendezvous has not been demonstrated either and there have been many comments about the difficulty of such operations as joining two vehicles in space. However, it is important to put the problem of rendezvous into the proper perspective. According to Dr. von Braun, two elements are necessary for a satisfactory rendezvous in orbit. 1—we must develop a capability of establishing accurate orbits; [4] 2—we must demonstrate the capability of launching our vehicles at a prescribed time. Dr. von Braun commented that the problem of establishing an accurate orbit is easier than the establishment of a 24-hour equatorial orbit which is planned rather seriously. Furthermore, the sending of a vehicle into an accurate orbit is also much easier than the problem posed for the Nike-Zeus anti-missile-missile. Compared to the guidance necessary for landing the vehicle on the moon, the guidance required for establishing accurate orbits appears to be a simple development. The ABMA personnel feel that the ability to meet a prescribed launch time has been demonstrated in two cases using rather simple vehicles, Juno II. The last two lunar probes launched with this vehicle were fired within a few seconds of the prescribed launch time.

During our meetings, one of the ABMA personnel described a promising technique for the final closure of the distance between two vehicles in orbit. One of the orbiting vehicles could be equipped with a net which could be deployed and which would have a source of infra-red radiation in its center. The approaching vehicle could fire a small rocket guided by a host seeker to carry a line to the net. Once the net captures the small rocket and the line, the two vehicles could easily be towed or pulled together.

The Saturn II vehicle appears to be the most reasonable one to use for this lunar mission since orbital assembly is not involved. Although the Saturn I may represent a cheaper approach from a vehicle cost standpoint, the costs of developing the techniques of orbital assembly may be overriding and make the development of the Saturn II doubly attractive.* On a very tight schedule, this mission could be accomplished in 1965 according to ABMA; however, 1966 seems to be a reasonable date. The development of the Saturn II vehicle, therefore, will require some early action on the development of high energy engines. If NASA undertakes the development of the hydrogen-oxygen engine of the 100–500,000 lbs. class in 1959, the development of a 500,000 lb. engine using oxygen and hydrogen should be initiated no later than the fall of 1960. Furthermore, we should take a close look at the hydrogen supplies, engine test facilities and launch sites for this mission.

The people at ABMA have invested a significant amount of time on their studies of the lunar mission; however, they have come to the point where it would be well for them to have some funding for further preliminary design studies of the entire lunar mission. These studies should be supported by the NASA and should encompass the use of both Saturn I and II for the program of landing a man on the moon.* We have the time for the study right now but if we delay too long then we will be forced into making some quick decisions a year from now if we ever hope to achieve this manned lunar landing by 1966 or in other words, if we ever hope to beat the Russians in their race to land a man on the moon. I visualize a study conducted by the staff of ABMA to last on the order of six months and which will cost at least [5] $500,000. I believe that we could find adequate justification for this study within our own propulsion group and from Mr. Low.

4.   Current Funding on the Saturn Program

In FY 1959, ABMA has added $25,000,000 to spend on the Saturn program. All but 2 million is employed for outside procurement of engines and materials. In order to keep the program moving, they need $145 million in FY 1960. The current ARPA request for

---

*   Studies that we have been doing here indicate that the Saturn I vehicle would also accomplish the lunar landing mission by orbital re-fueling only; consequently, the Saturn I vehicle would probably be used for the early launch landing mission.

ABMA for the Saturn program is only $50 million. Therefore, there is a deficit of $95 million. There is some hope for a supplemental appropriation for the Saturn program in January of 1960, particularly if a successful firing of the first propulsion test vehicle is accomplished in December of 1959. However, NASA help certainly could be used. Figures of from 10 to 20 million were mentioned as being a reasonable down payment for Saturn vehicles which might ultimately be used for the lunar mission.

F.C. Schwenk

Copy: Mr. Disher

Document I-28

Document title: Abraham Hyatt, Deputy Director, Launch Vehicle Programs, to Wernher von Braun, Army Ballistic Missile Agency, January 22, 1960.

Source: NASA Historical Reference Collection, NASA History Office, NASA Headquarters, Washington, D.C.

The decision to transfer the von Braun team from the Army to NASA was made on November 2, 1959. Although the transfer would not formally go into effect until mid-1960, from November 1959 on, NASA Headquarters was already dealing with von Braun as if he were part of the NASA team. NASA wished to centralize its rocket engine development efforts under the management of von Braun, including the 1.5-million-pound thrust F-1 engine program it had inherited from the Air Force in 1958 and the new upper stage cryogenic engine that would become the J-2. This management transfer put under Huntsville's control the two engines that would power the Saturn V moon rocket.

LD(AH:ad)
22 Jan 1960

Dr. Wernher von Braun
Army Ballistic Missile Agency
Huntsville, Alabama

Dear Dr. von Braun:

As a result of current policy determinations, it is the intent of the NASA Headquarters to transfer the administration and technical direction of certain development programs to the Huntsville Development Center. Those under consideration at this time are:

1.   The one and a half million pounds thrust rocket engine under development by the Rocketdyne Division of North American Aviation, Inc. (Headquarters Contract No. NASAw-16) and
2.   A new development of a 200 K thrust rocket liquid oxygen-liquid hydrogen engine.

It is desired that the transfer of these programs be accomplished in an efficient manner at the earliest practicable time. In order to accomplish this, it will be necessary to establish a mutually agreeable plan for the transfer of the responsibility to Huntsville.

It is requested that a plan be prepared and submitted to NASA Headquarters which outlines the manner in which your organization would propose to carry out the administration and technical direction of these programs. Rules for the groups involved and procedures to be followed should be indicated.

It is recommended that the people who are assigned to this task meet with Messrs. Elliot Mitchell and A. O. Tischler, who are the Headquarters cognizant personnel, to obtain information on the status and objectives of these projects.

Sincerely yours,

Abraham Hyatt
Deputy Director, Launch Vehicle Programs

Copy to:
Mr. Mitchell
Mr. Tischler

Document I-29

Document title: Saturn Vehicle Team, "Report to the Administrator, NASA, on Saturn Development Plan," December 15, 1959, pp. 1–4, 7–9.

Source: NASA Historical Reference Collection, NASA History Office, NASA Headquarters, Washington, D.C.

President Eisenhower approved the transfer of the Development Operations Division of the Army Ballistic Missile Agency to NASA on November 2, 1959. This meant that by mid-1960, Wernher von Braun and his rocket team would be part of NASA. In the interim, NASA assumed management responsibility for the Saturn launch vehicle through a working agreement with the Army. An immediate step was to form a "Saturn Vehicle Team" to advise NASA on the direction the Saturn program should take, particularly with respect to the vehicle's upper stages. The team was led by NASA Headquarters official Abe Silverstein, who was an advocate of the use of powerful but difficult-to-handle liquid hydrogen as a fuel for rocket engines. During the deliberations that led to this report, Silverstein was able to convince an initially skeptical von Braun that the upper stages of the Saturn vehicle should use engines employing liquid hydrogen and liquid oxygen. This decision set the stage for the creation of the S-II and S-IVB stages used in the Saturn V Moon rocket. The three tables, two figures, and three appendices referred to in this report do not appear here.

Report to the Administrator, NASA, on Saturn Development Plan

by
Saturn Vehicle Team

[stamped "Downgraded at 3 year intervals; declassified after 12 years"]

[1]   December 15, 1959

INTRODUCTION

The President of the United States, on 2 November 1959, announced his intention to transfer the Developmental Operations Division of the Army Ballistic Missile Agency (ABMA) and the Saturn project to NASA. In anticipation of this transfer, the NASA and

Department of Defense have established an interim working agreement that provides for immediate assumption by NASA of responsibilities for technical management of the Saturn vehicle development. On 17 November 1959, the Associate Administrator of NASA requested the Director of Space Flight Development to

"form a study group with membership from NASA, the Directorate of Defense Research and Engineering, ARPA, ABMA, and the Air Force from the Department of Defense to prepare recommendations for guidance of the development, and specifically, for selection of upper stage configurations.

Attention in the study should be directed toward
1. Missions and payloads,
2. Technical development problems,
3. Cost and time for development, and
4. Future growth in vehicle performance."

A Saturn vehicle team was established with the following membership:

Dr. Abe Silverstein, Chairman	NASA
Col. N. Appold	USAF
Mr. A. Hyatt	NASA
Mr. T. C. Muse	ODDR&E
Mr. G. P. Sutton	ARPA
Dr. W. von Braun	ABMA
Mr. E. Hall, Secretary	NASA

[2] The results and recommendations of the Saturn vehicle team are summarized in this report and the more detailed findings are presented in Appendices A, B, and C, which are attached.

The Saturn project was initiated on 15 August 1958 by an order from the Advanced Missile Command to develop a large booster vehicle of approximately 1.5 million pounds of thrust using available engines. Authorization was given for construction of test facilities, develop-ment and early captive firing of the first stage, launchings of three first stages with dummy upper stages, and one with a live upper stage. A brief chronology of important actions relative to the Saturn project are contained in Appendix A.

For the past several months technical studies have been conducted by ABMA, ARPA, and NASA to establish the performance characteristics of the Saturn vehicle with various upper stages. The results of these independent studies were in close agreement and form a basis for this evaluation.

Presentations were made to the Saturn vehicle team on missions for the Saturn vehicle by both NASA and the Department of Defense. The following missions, listed in their order of importance, were established for the Saturn vehicle (Appendix B).
   a. Lunar and deep space missions with an escape payload of about 10,000 pounds.
   b. Payloads of about 5,000 pounds in a 24-hour equatorial orbit.
   c. Manned spacecraft missions such as Dyna Soar, with a weight of about 10,000 pounds in a low orbit (two-stage launch vehicle).

These missions were established for the initial Saturn vehicle configuration. It is recognized that the initial Saturn configuration must provide for growth to permit increased pay-load capability in the lunar, deep space, and satellite missions. Early capability with an advanced vehicle and possibilities for future growth were accepted as elements of greatest importance in the Saturn vehicle development.

[3] The current Saturn first stage with eight engines giving a total thrust of nearly 1,500,000 pounds was reviewed. The many problems associated with its development and operation were discussed. Attention was given to alternate configurations for the first stage including the use of solid propellant rockets, a cluster of four 400,000-pound thrust engines, and a single engine of 1,500,000 pounds of thrust. The problems of clustered tanks as compared with those of a single large tank were also considered.

A wide variety of upper stages utilizing conventional and high-energy propellants and of various weights were compared on the basis of performance, technical feasibility, growth potential and probable time and cost to develop. Various tank configurations, including clusters of existing IRBM's, which were independently analyzed by ABMA and NASA, were also studied by the group. A discussion of the technical terms covered is contained in Appendix C.

[4] SUMMARY OF RESULTS

After a review of the many possible configurations of Saturn vehicles, the team reduced its detailed considerations to those shown in Table I.

The payload capabilities of the configurations shown in Table I for the most important missions listed in the Introduction are given in Table II.

Vehicle A-1, with upper stages consisting of a modified Titan stage 1 and Centaur upper stage, makes maximum utilization of existing hardware and would most likely have earliest flight availability and lowest cost. It fails, however, to meet the mission requirements for the lunar and 24-hour missions and, because of its slenderness (120-inch diameter upper stages), vehicle A-1 is a structurally marginal configuration. Development of a 160-inch diameter second stage similar in construction to the Titan first stage was reviewed and eliminated from detailed consideration because it limited the growth potential of the Saturn.

The A-2 vehicle, with a cluster of IRBM's as the second stage, is similar to the A-1 configuration in its use of existing hardware. Vehicle A-2 fails to meet the requirements for lunar and deep space missions and for the 24-hour equatorial orbit.

Vehicle B-1 meets the requirements of the missions, but requires the development of a new conventionally fueled second stage that is approximately twice the size of our current ICBM's. The cost and time to develop this large second stage which seemed to be interim in character for advanced missions raised doubts as to the desirability of developing this vehicle.

In examining vehicles A-1, A-2, B-1, and others, it became apparent that highest priority missions for the Saturn vehicle could not be accomplished in a reasonable design without the use of high-energy propellants in the top stages. If these propellants are to be accepted for the difficult top-stage applications, there seems to be no valid engineering reasons for not accepting the use of high-energy propellants for the less difficult . . . [7] application to intermediate stages. Of course, the maximum payload capability with the Saturn first stage booster will be achieved if high-energy propellants are used in all the upper stages. Current success in the Centaur engine program substantiates the choice of hydrogen and oxygen for the high-energy propellants.

The C-1 configuration (Tables I and II) is the first phase in the development of a vehicle using all hydrogen and oxygen upper stages (see figures 1 and 2). Succeeding phases are C-2 and C-3 with progressively increasing payload capability. As the development proceeds from phase to phase, a new stage is added to the vehicle. Stages developed for early phases continue to be used in all latter phases (see figure 2). Thus all developments lead to increased flight capability and reliability.

Configuration C-1 permits early flights and essentially meets the established mission requirements. The upper stages consist of a four engine hydrogen-oxygen second stage

(S IV) and a Centaur upper stage (S V) as a third stage. The engines for the second and third stage are the same. Uprating of the 15K Centaur engine to 20K is necessary for the second stage.

Configuration C-2 is adapted from C-1 by the addition of a new hydrogen-oxygen second stage (S III). The development of a 150K–200K pounds of thrust hydrogen-oxygen rocket engine is required to power the new stage.

Configuration C-3 increases the payload capability by adding a second stage (S II) with four 150K–200K pound thrust engines. The thrust of the first stage is also increased to over two million pounds. This thrust may be obtained by replacing the four center engines with one F-1 engine or by uprating all eight H-1 engines.

[8] RECOMMENDATIONS

It is recommended that:
1.   A long-range development plan for the Saturn vehicle be established that will provide, through a consecutive development of building-block upper stages, a substantial early payload capability and a final configuration that exploits the maximum capability of the Saturn first stage. Vehicle reliability will be emphasized in the building-block program through a continued use of each development stage in later vehicle configurations.
2.   All upper stages be fueled with hydrogen-oxygen propellants.
3.   The initial vehicle configuration, C-1, consists of the following:
     a.   The eight engine first stage currently under development at ABMA.
     b.   A newly developed second stage using four of the current Centaur engines uprated to 20,000 pounds of thrust.
     c.   The third stage using the current Centaur stage modified only as required for vehicle and payload attachments.
4.   The following developments be initiated immediately:
     a.   A 150–200K hydrogen-oxygen fueled rocket engine for stages S II and S III.
     b.   A design study of hydrogen-oxygen upper stages S II and S III using the 150–200K engines.
5.   The development schedule shown in Table III be adopted.

[9] Submitted by:

Abe Silverstein, NASA (Chairman)
Abraham Hyatt, NASA
George P. Sutton, ARPA
T.C. Muse, ODDR&E
Norman C. Appold, Col., USAF
Wernher von Braun, ABMA
Eldon Hall, NASA (Secretary)

Document I-30

Document title: Robert R. Gilruth, Director, Space Task Group, to Dr. N.E. Golovin, Director, DOD-NASA Large Launch Vehicle Planning Group, September 12, 1961.

Source: NASA Historical Reference Collection, NASA History Office, NASA Headquarters, Washington, D.C.

In designing the launch vehicle to take Apollo crews to and from the Moon, the relationship between launch vehicle reliability and crew safety was a critical concern. This document reviews several considerations related to this relationship as seen by the Space Task Group, which managed NASA's human spaceflight efforts. In particular, the letter notes the concern that a vehicle of less capacity than the proposed Nova superbooster might place astronauts at too great a risk. From the time he heard of President Kennedy's announcement until the end of the Apollo program, Robert Gilruth was concerned that the lunar landing mission was excessively risky. The figure referred to in the enclosure does not appear here.

[original marked "CONFIDENTIAL" on each page; classification change to "unclassified" by authority of Executive Order 11652, February 7, 1973]
[1]

NATIONAL AERONAUTICS AND SPACE ADMINISTRATION

SPACE TASK GROUP
Langley Field, Va.

September 12, 1961

Dr. N. E. Golovin, Director
DOD-NASA Large Launch Vehicle
    Planning Group
NASA Headquarters, Code AA-4
1520 H Street, Northwest
Washington 25, D.C.

Dear Dr. Golovin:

In answer to your request of August 16th, Space Task Group has prepared a list of mission criteria and other requirements for the launch vehicles to be used for Project Apollo missions. Project Apollo is aiming for a mission reliability goal of 0.90 and a safe crew return goal of 0.999. Necessarily, the launch vehicles must be designed and developed with better margins and more redundancy than missile boosters. In addition, the launch vehicle operation must cater to certain restrictions and provide the flight crew with failure information and some control capability.

Necessary and reasonable constraints and requirements upon the launch vehicle design and operation are listed. Most of these are self-explanatory, the others will be discussed later in this letter.

    a.   Mission Criteria
        (1) Launch longitudinal acceleration not to exceed 8g
        (2) Lateral acceleration at the spacecraft due to launch vehicle hard over control maneuvers shall not exceed 3g at any time
        (3) Vibration transverse and longitudinal shall not exceed those shown in the enclosure at any time
        (4) Maximum dynamic pressure no greater than 1000 psf
[2]   (5) Staging to be carried out in a noncritical environment
        (6) Minimum number of stages
        (7) Staging into parking orbit (parking orbit will be standard procedure) where performance penalty does not exceed 10 percent

      (8) All failure modes identified and either or both of the following provided
         (a) Failed systems overridden automatically
         (b) Failures and prospective failure activity signalled to crew with provision
            for crew to
             <u>1</u>   Switch to backup system or redundant mode
             <u>2</u>   Directly override failed system
             <u>3</u>   Shutdown launch vehicle and initiate escape sequence

b. Performance Required (Includes contingency for growth)
    (1) 35,000 pounds to escape for lunar orbit missions
    (2) 150,000 pounds to escape for lunar landing missions

c. Reliability Goal
    (1) 95 percent satisfactory insertion on translunar trajectory

d. Failure Sensing
    (1) 0.999 failures sensed prior to catastrophic condition

e. Structural Margin
    (1) The general structural factors of safety shall be 1.5 to ultimate and 1.1 to
    yield. The design limit load envelopes shall be established by superposition of
    rationally deduced critical loads for all flight modes. Loads envelopes shall
    recognize the cumulative effects of additive type loads. The structure shall be
    designed such that operation of the vehicle or any subsystem is unimpaired
    by structural deflection at limit load conditions.

[3] f. Reliability Demonstration
    (1) One successful operation of each complete launch vehicle system with a
    dynamically similar payload. Reliability demonstrated by analysis and compo-
    nent and subsystem tests. Reliability to be emphasized in design, production,
    and launch preparation procedures.

The Apollo spacecraft will be designed in a manner to place full control of the mission with the crew. This by no means implies that all control functions will be carried out by the crew. It may be desirable and sometimes necessary to utilize the speed and precision of automatic equipment. Likewise, certain repetitive tasks that would prove monotonous or time consuming may also better be done by automatic means. It is suggested that the launch vehicle design should also take advantage of the fact that there is a crew aboard in order to improve reliability. It is not obvious, however, that a general formula for crew participation is practical. This should be decided after careful and detailed study of the particular launch vehicle to be employed.

The Apollo missions are being planned on the basis of using a parking orbit as standard procedure. The operational flexibility provided is sufficient to justify this decision. The use of the parking orbit period for inflight checks may be of even more significance. After injection into the parking orbit, the spacecraft will have encountered its most severe launch stresses, and will come into contact with the space environment for the first time in the mission. By making status, functional, and operational checks of all equipment at this time the safety of the flight will be enhanced greatly. While in orbit the spacecraft can be brought back to the earth in a matter of minutes; on the other hand, if the start of these checks is delayed until translunar insertion, it will be hours before an earth return can be completed. Considering the duration of the mission that lies ahead it seems only reasonable to plan on spending some time in orbit in order to obtain assurance that all equipment may be relied upon.

Having established that a parking orbit as a checkout period has desirable features, one is led to consideration of launch vehicle operation relative to a parking orbit. The Space Task Group would like to see the launch vehicle designed so that staging naturally [4] occurs at orbital conditions. An obvious benefit is that the hazard of one start-up is eliminated. This is important both from a flight safety and a mission reliability standpoint. A less obvious benefit is that all launch stages are "single-burn." There should be only one stage required to accelerate from orbital conditions to the translunar trajectory. This stage will not have been used. By making proper status checks, the crew can be reassured that this stage is in the same condition as at launch. An example of degradation by a relight operation is cited. Consider that in the first shutdown a slow propellant leak, perhaps through a valve, develops. The stage may then be rendered incapable of providing adequate insertion velocity. Yet, there is at this time no known procedure for determination of propellant quantity in a partially-filled tank in a weightless environment.

I feel that it is highly desirable to develop a launch vehicle with sufficient performance and reliability to carry out the lunar landing mission using the direct approach. Therefore, I recommend that rocket motors larger than those presently under development be obtained for this program. Rendezvous schemes are and have been of interest to the Space Task Group and are being studied. However, the rendezvous approach itself will, to some extent, degrade mission reliability and flight safety. I am concerned that rendezvous schemes may be used as a crutch to achieve early planned dates for launch vehicle availability, and to avoid the difficulty of developing a reliable NOVA class launch vehicle. As you know, the mission most likely will not be attempted until a reasonable amount of confidence in completing the mission is established. Thus, from a program planning standpoint "system reliability in use" is more important than "hardware availability for use," even though earliest achievement of the mission is a primary goal.

Yours very truly,

Robert R. Gilruth
Director

Encl:
1 copy figure entitled
    "Vibration Limits"

Document I-31

Document title: Milton Rosen, Director, Launch Vehicles and Propulsion, Office of Manned Space Flight, Memorandum to Brainerd Holmes, Director of Manned Space Flight, "Large Launch Vehicle Program," November 6, 1961.

Document I-32

Document title: Milton Rosen, Director, Launch Vehicles and Propulsion, Office of Manned Space Flight, Memorandum to Brainerd Holmes, Director of Manned Space Flight, "Recommendations for NASA Manned Space Flight Vehicle Program," November 20, 1961, with attached: "Report of Combined Working Group on Vehicles for Manned Space Flight."

Source: Both in NASA Historical Reference Collection, NASA History Office, NASA Headquarters, Washington, D.C.

In the months following the May 1961 announcement by President Kennedy that the United States would send Americans to the Moon, there was intense activity examining various ways of achieving that mission and the overall acceleration of the national space effort that Kennedy had approved. In particular, a NASA-Department of Defense committee headed by Nicholas Golovin had spent the summer addressing launch vehicle requirements, but had become deadlocked. By the end of 1961, NASA needed decisions on what kind of launch vehicle would be needed for the Moon if it was to meet Kennedy's deadline of accomplishing the mission before the end of the decade. Milton Rosen, a veteran of the Vanguard program, was in charge of launch vehicle development for Apollo. The two-week study described in his November 6 memorandum and reported in his November 20 memorandum proposed that a Saturn vehicle using five F-1 engines in its first stage, rather than the two to four engines that had previously been under discussion, be developed. This recommendation was accepted by NASA leadership; the resulting vehicle soon became known as the Saturn V. When NASA in mid-1962 decided to use the lunar-orbit rendezvous approach to accomplishing the lunar landing mission, the additional power provided by the fifth first-stage engine meant that the mission could be carried out using a single launcher. Thus the key recommendation of Rosen's report was one of the significant enablers of meeting the Apollo deadline.

Document I-31

[1]
UNITED STATES GOVERNMENT
MEMORANDUM

TO: M—Mr. Holmes   DATE: November 6, 1961

FROM: ML—Mr. Rosen

SUBJECT: Large Launch Vehicle Program

Pursuant to discussions with you and Dr. Seamans, I have organized a working group consisting of members of my staff, augmented by representation from [Marshall Space Flight Center] and the Office of Spacecraft and Flight, to examine the reports of several committees and on the basis of these reports, and our judgment and analysis, to recommend to you a large launch vehicle program which will:
1. Meet the requirements of manned space flight, and
2. Have broad and continuing national utility (for other NASA and DOD missions)

Our principal background material will consist of the reports of the following groups:
1. The Large Launch Vehicle Planning Group (Golovin Committee)
2. The Fleming Committee
3. The Lundin Committee
4. The Heaton Committee
5. The Davis-Debus Committee

The following people are members of the working group:
    Launch Vehicles & Propulsion
        Mr. M. W. Rosen, Chairman
        Mr. R. B. Canright
        Mr. Eldon Hall
        Mr. Elliott Mitchell
        Mr. Norman Rafel
        Mr. Melvyn Savage
        Mr. A. O. Tischler

<u>Marshall Space Flight Center</u>
    Mr. Wm. Mrazek
    Mr. Hans Maus
    Mr. James B. Bramlet
<u>Spacecraft & Flight</u>
    Mr. John Disher

[2] Our approach is to start out by having sub-groups make critical evaluations of some of the most important problems. Having done this, we will be in a better position to formulate a recommended program. Some of the subjects we are considering are:

1. An assessment of the problems involved in orbital rendezvous
2. An evaluation of intermediate vehicles (C-3, C-4, C-5 class)
3. An evaluation of NOVA class vehicles
4. An assessment of the future course of large solid rocket motor development
5. An evaluation of the utility of TITAN-III for NASA missions
6. An evaluation of the realism of the spacecraft development program (schedules, weights, performances)

Preliminary discussions within the group as to our mode of operation and the scope of our work have taken place this week. This memorandum is the result of these discussions. We have set as a target having in your hands a recommended program, and an evaluation of the more critical factors affecting it, by November 20.

I need your help in the following areas:

1. Immediate access to the report of, and supporting data used by, the Golovin Committee.
2. The opportunity of completing our work before further decisions are made in the areas we are examining. Should the need arise for a critical decision prior to November 20, we will be available at any time on or after November 13 to give you an oral briefing of our up-to-date findings.

Milton W. Rosen
Director, Launch Vehicles & Propulsion
Office of Manned Space Flight

Document I-32

UNITED STATES GOVERNMENT
MEMORANDUM

TO: M—Mr. Holmes   DATE: November 20, 1961
                                               ML (MWR:pbm)
FROM: ML—Mr. Rosen

SUBJECT: Recommendations for NASA Manned Space Flight Vehicle Program

1. In accordance with my memorandum to you of November 6, I am presenting, for your consideration, a summary report prepared by the working group on vehicles for manned space flight. The members of the group were as stated in the November 6 memorandum, with the addition of Mr. David Hammock of the Space Task Group.

2. This report represents the distilled judgment of the group. No attempt was made to enforce or obtain unanimity. A small minority may differ with the wording of some of the

recommendations. The general approach of the report, as a whole, is supported by the group, as a whole, and in this sense represents a consensus. Differences of opinion arose in three areas: rendezvous vs. direct flight, solids vs. liquids, and the nature of the intermediate vehicle. These differences are in the nature of emphasis rather than content. This situation is best illustrated by the tape recording made during the final session of the group.

3.   The group had available the final recommendations of the Golovin Committee and preliminary drafts of several of the report chapters. We took the view that the Golovin Committee had opened doors to rooms which should be explored in order to formulate a program. Our report consists of a finer cut of the Golovin recommendations—it is more specific with regard to the content and emphasis of a program. We believe such closer definition is required in order to arrive at a 1963 budget.

4.   The program we are recommending to you is, in my opinion, the best we can offer at this time. It takes account not only of technical factors, but also of the realities of the budgetary and political situation. We are preparing a budget and schedule as an appendix to this document. I propose to have these in your hands by November 22. My gross estimate at this time is that the program recommended here can be funded by the Plan A budget ($4,238 million) recommended by Mr. Webb to the Director of the Budget. The Plan B ($3,699 million) budget would be inadequate. Should it develop that the Plan A budget is not obtainable, we are prepared to undertake a further condensation of the program to meet a lesser figure. It must be admitted, however, that such a step starts to eliminate some important alternative approaches.

5.   Those of us who participated in this intensive two-week effort feel that our work has been worthwhile in clarifying in our minds the very important issues that are the subject of this report.

Milton W. Rosen
Director, Launch Vehicles & Propulsion

Attachment: as stated

****

[each page of the attachment is stamped "FOR INTERNAL NASA USE ONLY"; no page number on first page]

Report of Combined Working Group on Vehicles for Manned Space Flight

Recommendations

1.   The United States should undertake a program to develop rendezvous capability on an urgent basis.
2.   To exploit the possibility of accomplishing the first manned lunar landing by rendezvous, an intermediate vehicle with five F-1 engines in the first stage and four or five J-2 engines in the second stage and one J-2 in the third stage should be developed. The vehicle should be so designed that it can be modified to produce a three engine first stage, if rendezvous is difficult to achieve. The three engine vehicle provides a better match with a large number of NASA and DOD requirements and earlier flights in support of the manned lunar program.
3.   The United States should place primary [crossed out and replaced with "major"] emphasis on the direct flight mode for achieving the first manned lunar landing. This mode gives greater assurance of accomplishment during this decade. In order to

implement the direct flight mode, a NOVA vehicle consisting of an eight F-1 first stage, a four M-1 second stage, and a one J-2 third stage should be developed on a top priority basis.

4. Large solid rockets should not be considered as a requirement for manned lunar landing. Should these rockets be developed for other purposes, the manned space flight program should support a solid first stage development in order to provide a backup capability for NOVA.

[2]

5. Development of the one J-2 engine S-IVB stage should be started, aiming toward flight tests on a Saturn C-1 in late 1964. It should be used as the third stage of both C-5 and NOVA, and also as the escape stage in the single earth orbit rendezvous mode.

6. NASA has no present requirement for the TITAN III vehicle. Should the TITAN III be developed by the DOD, NASA should maintain continuous liaison with the DOD development to ascertain if the vehicle can be used for future NASA needs.

[no page number]

Discussion

1. Rendezvous

The capability for rendezvous in space is essential to a variety of future space missions. These include crew rotation and resupply of orbiting laboratories and space stations, orbital assembly for future manned planetary missions, and rescue operations in orbit. For these reasons alone a vigorous high priority rendezvous development effort must be undertaken immediately.

The United States should undertake a program to develop rendezvous capability on an urgent basis.

Space rendezvous presents the possibility of accomplishing the initial manned lunar landing mission earlier than by other means and therefore should also be considered for that mission.

Several modes of rendezvous in space have been proposed for accomplishing the initial lunar landing mission. The favored modes are (1) a single rendezvous and docking in earth orbit, (2) a single rendezvous in lunar orbit by a lunar excursion vehicle which departs from a parent craft in lunar orbit, descends to the lunar surface and returns to the parent craft which remains in lunar orbit. The second alternative offers the possibility of mission accomplishment with only one earth launch of the same type launch vehicle of which two are required for the earth orbit rendezvous. It also offers the possibility of a smaller and simpler lunar landing vehicle for the initial landing attempt. However, the lunar orbit rendezvous operation entails [2] appreciably greater human risk than does earth orbit rendezvous because a missed rendezvous at the moon is fatal whereas a missed earth rendezvous simply aborts the mission. The lunar rendezvous vehicle also lacks substantial radiation protection and lands only a minimal payload on the moon with limited staytime and scientific equipment.

After comparing the advantages and disadvantages of the two rendezvous modes it has been concluded that the preferred rendezvous mode is the single rendezvous in earth orbit.

It is imperative to recognize that rendezvous offers only a possibility of carrying out the initial landing more rapidly than by other means. Because we will not have our first experimental indications of the difficulty of performing rendezvous until 1964 we will not until that time have a firm basis for estimating and scheduling the time required to develop high reliability space rendezvous, docking, and fuel transfer operations.

The Heaton Committee investigated the docking method for earth orbit rendezvous and concluded that the launch vehicle should have sufficient capability so that only one rendezvous would be required. About four rendezvous (5 vehicles) are required with the C-3. Hence, emphasis shifted from the C-3 to the C-4 vehicle. At that time it was believed that adequate capability could be obtained with two C-4 vehicles. A more detailed investigation indicates that the C-4, when designed and built with sufficient structural and flight margins for high confidence, [3] is inadequate with only one rendezvous for the desired allowable spacecraft weight. The C-5 has adequate margin with one rendezvous.

If several rendezvous in earth orbit are shown to be entirely feasible, the use of a C-3 class vehicle would be suitable with a fueling type of operation but not with a docking type because of the structural considerations of combining five vehicles. Two rendezvous maneuvers with three C-4 vehicles would be suitable with either docking or fueling. The C-5 vehicle is capable of performing the single earth orbit rendezvous mode without refueling and is also capable of performing the lunar orbit rendezvous mode as described above.

> To exploit the possibility of accomplishing the first manned lunar landing by rendezvous, an intermediate vehicle with five F-1 engines in the first stage and four or five J-2 engines in the second stage and one J-2 in the third stage should be developed. The vehicle should be so designed that it can be modified to produce a three engine first stage, if rendezvous is difficult to achieve. The three engine vehicle provides a better match with a large number of NASA and DOD requirements and earlier flights in support of the manned lunar program.

The working group examined rendezvous more intensively than any other subject in an attempt to understand the technical and operational problems involved. This effort led to the conclusion that the development of rendezvous, and its use for manned lunar landing, cannot be scheduled with any reasonable degree of assurance. We urge development [4] of rendezvous in its own right and so that a better assessment of its use for manned lunar landing can be made in the next year or two.

2. Direct Flight

In order to inject the Apollo spacecraft into a lunar trajectory without recourse to orbital assembly or refueling, a launch vehicle with capability equivalent to that provided by an 8 F-1 engine first stage is required. Such a launch vehicle presents no different order of technical problems than does a 5 F-1 engine first stage. Larger facilities are required for fabrication and test, and the first unit will take more man hours to build and test, but the problems are the same.

The group examined versions of NOVA suggested by the Golovin Committee. The chosen configuration places emphasis on achieving early manned lunar landing by direct flight, with sufficient margin for both spacecraft and vehicle contingencies, and in addition, offers potential for missions beyond manned lunar landing. This configuration consists of a first stage with 8 F-1 engines, a second stage with (4-1).* M-1 engines and an S-IVB third stage, the same as the third stage of the C-5 and the second stage of the C-IB Saturn. This version has growth potential and also offers the advantage that it could utilize the four 240-inch solid first stage if it were to be developed.

We have examined the feasibility of producing this NOVA vehicle and have concluded that it can be scheduled with a reasonable degree of assurance. An optimistic schedule would provide an earliest capability in late 1966; a pessimistic schedule would provide an

---

*   Four engines with one engine out capability

earliest lunar landing capability in 1968. It appears reasonable to plan on the availability of this type of NOVA vehicle in 1967 for the achievement of manned lunar landing. [5]

> The United States should place primary emphasis on the direct flight mode for achieving the first manned lunar landing. This mode gives greater assurance of accomplishment during this decade. In order to implement the direct flight mode, a NOVA vehicle consisting of an eight F-1 first stage, a four M-1 second stage, and a one J-2 third stage should be developed on a top priority basis.

3. Solid Rockets

The group examined the prospects for developing large solid rockets for first stages of the intermediate and NOVA vehicles. In particular, we examined the 156-inch segmented motor and the 240-inch monolithic motor. The group concluded that both of these versions could be developed, and that the elapsed time between now and the first motor test could be scheduled with reasonable assurance. There was considerable uncertainty as to the number of motor tests required to solve technical problems and to achieve a reasonable degree of reliability, to the number of stage tests which may be required and to the number of flight tests. On the other hand, success of the F-1 and J-2 engines must be assured if the program proposed here is to be undertaken at all. Since these engines must be developed to a high degree of reliability for the intermediate vehicle, it seems only sensible to use them in NOVA. These considerations led to the conclusion that the present program for manned lunar landing should be based on liquid propulsion, and that solid rockets should serve as a backup only. [6]

> Large solid rockets should not be considered as a requirement for manned lunar landing. Should these rockets be developed for other purposes, the manned space flight program should support a solid first stage development in order to provide a backup capability for NOVA.

4. Saturn Class Vehicles

As recommended by the Golovin Committee, development of Saturn C-1 should be continued to provide an early capability for orbital tests of Apollo.

A one J-2 engine top stage can serve the C-1, C-5, and NOVA. It also serves, with modification, as the escape tanker in the single earth orbit rendezvous operation. In other words, in any mode of operation recommended here, when the Apollo spacecraft is sent from orbit to escape, it uses the S-IVB. We have examined the development schedules of the S-IV and the S-IVB and have concluded that the S-IV leads the S-IVB by at least one year. Substitution of the S-IVB at this time would result in a year's delay in first flights of the Apollo spacecraft on Saturn. Since the Apollo orbital flights are to start with the Saturn C-1, using the S-IV, it may be prudent and desirable to continue this version of Saturn C-1 for all of the Apollo orbital tests. In this case, we recommend that two or three Saturn S-I's be devoted to vehicle tests of the S-IVB stage at an early date, in order to qualify the S-IVB for its future use on the C-5 and NOVA.

[7]

> Development of the S-IVB stage should be started, aiming toward flight tests on a Saturn S-I in late 1964, and use as the third stage of both C-5 and NOVA, and also as the escape stage in the single earth orbit rendezvous mode.

The group examined information available on the TITAN III, its performance, future availability and developmental problems.

The TITAN III and the Saturn C-1 are competitive in orbital performance. The TITAN III, alone, has some escape capability which is enhanced by addition of a fourth stage. The Saturn C-1 has an appreciable escape capability through the addition of a third stage. One major difference is that the TITAN III core has a 10-foot diameter and only with difficulty could carry large diameter payloads. The Saturn C-1, on the other hand, has an 18-foot diameter and could be provided with a third stage of similar diameter, for example, the following combination [S-I–S-IVB–S-IV]. Escape payloads presently planned by NASA for Centaur utilize the full 10-foot diameter of that vehicle. Future escape payloads, requiring greater launch vehicle capability, fall in the diameter class of 12 to 18 feet. Launch vehicle requirements for these payloads can be met by the Saturn C-1.

> NASA has no present requirement for the TITAN III vehicle. Should the TITAN III be developed by the DOD, NASA should maintain continuous liaison with the DOD development to ascertain if the vehicle can be used for future NASA needs.

Document I-33

Document title: Future Projects Design Branch, Structures and Mechanics Division, George C. Marshall Space Flight Center, NASA, Huntsville, Alabama, "NOVA Preliminary Planning Document," August 25, 1961, pp. 1–6.

Source: NASA Historical Reference Collection, NASA History Office, NASA Headquarters, Washington, D.C.

The notion that a more powerful booster than that originally designated Saturn would be required for ambitious future missions had been part of the planning of Wernher von Braun and his associates for some time. That vehicle, called the Nova, became part of NASA's future planning as early as 1959. With President Kennedy's 1961 decision to go to the Moon, planning for the Nova took on increased urgency; a vehicle of such capabilities was required for a direct flight to the Moon's surface and back to Earth by a crew-carrying spacecraft. Ultimately, a rendezvous approach to the lunar mission was adopted, and the Nova's extremely heavy-lift capabilities were not needed. Thus the vehicle never entered development, although it remained under study until 1963.

NOVA Preliminary Planning Document

August 25, 1961

Future Projects Design Branch
Structures and Mechanics Division
George C. Marshall Space Flight Center
National Aeronautics and Space Administration
Huntsville, Alabama

[no page number]

PREFACE

This document presents NOVA vehicle data generated to date by the Structures and Mechanics [S&M] Division of the George C. Marshall Space Flight Center [MSFC]. The data contained herein is preliminary and subject to change in the near future as more technical data and planning knowledge of the subject becomes available. The preliminary weights and performance data shown are particularly subject to change. Other divisions of the MSFC have not completed their inputs to the NOVA development plan, and the S&M Division will have changes to the data presented in this document. The NOVA Development Plan will be formalized by the MSFC in the near future.

The NOVA vehicle as presented in this document is a three stage launch vehicle which injects a payload into a lunar transfer orbit. The fourth, lunar landing, stage will be included in following documents. Data for this stage is not included here because it is not yet sufficiently refined.

[signed "Robert G. Voss for"]
W. B. Schramm
Chief, Future Projects Design Branch . . .

[1]
A. VEHICLE SYSTEM

1. Approach

The NOVA vehicle development will be aimed toward optimization for the three stage escape mission utilizing a first stage called the N-I, a N-II second stage, and a N-III third stage. The NOVA vehicle's objective is to provide a heavy weight lifting capability so that this nation's space exploration and manned lunar programs can be carried out. Among the several mission objectives slated for the NOVA, the prime objectives are:

    Manned Lunar Landing.
    Planetary Spacecraft Landing (such as Prospector).

Other missions for which the NOVA vehicle is needed are:

    350,000 lb. Max. Volume Orbiting Laboratory (96 Min Orbit).
    A booster vehicle (N-I & N-II) which is capable of boosting a nuclear reactor (NERVA) powered upper stage. . . .

True optimization of the NOVA vehicle for a three stage escape mission would demand ignoring the other possible missions to the end that the design of the N-I and N-II stages would not be easily modified. It is a basic intent that the NOVA vehicle shall be capable of easy modification to accommodate both two stage (orbital operations) and large, low density upper stages (nulcear [sic] powered stages). The choice of diameter is a good example of pre-planning for future growth. To plan ahead for "other missions," it was decided to have a diameter of 320 inches in the third stage so that the vehicle would be capable of boosting a large volume, low density, liquid hydrogen filled nuclear upper stage, thus "cashing in" on the extremely high specific impulse available from nuclear propulsion. Consideration was thus given, when this decision was made, for the possible growth potential.

[2] It can be expected that the development engineering to be conducted will lead to certain desirable changes to the vehicle system as known today. Where their worth is proven or unquestionable, they will be introduced into the vehicle system. Generally, introduction will be at block change points. However, mandatory changes discovered during ground or flight tests[,] and directly affecting mission reliability, will be introduced immediately upon discovery of the unsatisfactory condition. In all cases changes to the vehicle system will be limited to only those that will improve operational safety and mission reliability.

In the C-1 vehicle development program, the block concept of progressive development was necessarily spaced out over a long period because of the time differential between availability of the upper stage engines and the later initiation of the S-IV development with respect to stage S-I stage of development. In the case of NOVA, however, the N-I, N-II, and N-III stages are almost on an equal footing with respect to engine availability and the required development leadtime. As a consequence, the block concept for initial flight testing of these new stages will be accelerated.

2. Description
    a. General
    The NOVA vehicle is a three-stage general purpose space vehicle which will be greater than 280 feet in length, will weigh approximately 635,000 pounds when empty and 9,500,000 pounds when fueled. Its lift-off acceleration will be 1.25 g, reaching 5.37 g at cutoff of the first stage. Its initial thrust will be 12,000,000 pounds; it will be capable of placing 350,000 pounds of payload in a 96 minute orbit (300 n. mile) and will impart escape velocity to 180,000 pounds of payload.
    b. Stage N-I
    The N-I stage of the NOVA launch vehicle will consist of a cylindrical tank structure with propellants separated by a common bulkhead. The diameter will be 530 inches, and the length will be approximately 111 feet. It will be designed to load a capacity 7,030,000 pounds of usable mainstage propellants. The stage will be powered by eight Rocketdyne F-1 engines, each developing 1,500,000 pounds of thrust and using RP-1 for fuel and liquid oxygen as the oxidizer. Four engines will gimbal for vehicle control. Control signals commanding the engine control actuators originate in the NOVA vehicle instrument unit located forward of the N-III.
    c. Stage N-II
    The N-II stage is the second stage of NOVA. It will be a cylindrical tank, 396 inches diameter, and will be loaded with liquid hydrogen for fuel and liquid oxygen for the oxidizer. It will be powered [3] by eight Rocketdyne J-2 engines, each developing 200,000 pounds thrust, yielding a total thrust at altitude of 1,600,000 pounds. N-II will have a tank design capacity for loading a maximum of 1,333,000 pounds of mainstage propellant. Most of the design details are preliminary, pending evaluation and selection of an industrial contractor who will be responsible for the complete design, development and delivery of this stage system.
    d. Stage N-III
    The N-III stage is a cylindrical tank structure 320 inches in diameter. Two Rocketdyne J-2 engines of 200,000 pounds thrust each power the third stage, and both are

gimballed for control. This stage has a loading capacity of 440,724 pounds of propellant. Adapters at the forward and aft ends of the stage taper to connect to the mating surfaces of other vehicle components.

    e.   Instrument Unit

The instrument unit will house the primary NOVA vehicle guidance and control instrumentation. It should be located as "high" in the system as possible; that is, it should probably be located in the N-III stage or higher. Since the primary guidance and control equipment consists of a stabilized inertial platform, a guidance computer, and a control computer, the entire package should be placed as far away from high energy vibration as possible. Insofar as the commands to control surfaces and commands for separation are concerned, such commands should be relayed stage-to-stage downward from the control computer.

    f.   Payload

The payload will be determined by the mission, and the mission will be limited by the payload. The main mission will be manned lunar landing. One orbital payload should be a 350,000 pound orbiting laboratory, cylindrical, about 320 inches in diameter, orbiting in a 96 minute (300 n. mile) orbit. Other orbiting payloads should be spacecraft in 24-hour orbits. In the case of all orbiting payloads, they should be of minimum density (maximum volume). The NOVA vehicle has an escape capability of 180,000 pounds of payload, lending itself well for soft landing in investigative spacecraft on the surface of Mars or Venus.

3.   Ground Test Program

Prior to the flight testing, many ground tests of individual stage systems will be required to demonstrate assurance that high reliability can be maintained and progressively improved. A full and comprehensive ground test program is considered a mandatory requirement for NOVA development. It is axiomatic that the confidence NASA can place in the reliability of any stage, vehicle, or system, is directly proportional to the time and effort spent in hot-testing, evaluation, redesign, modification and retesting of all [4] components and systems. Since manned missions are planned for the SATURN vehicle, the design and development of the NOVA must result in a vehicle with an extremely high degree of reliability and assurance. To achieve this goal with so few vehicles, an intense design and testing program will be established in which safety, reliability, and quality take high precedence over most other considerations. Integrated with the design and testing programs must be a highly intensified inspection program. To fully demonstrate reliability and safety, an extensive flight test program would have to be performed; however, time and cost restrictions prohibit it. Therefore, a high level of confidence must be established through ground testing.

The outstanding factor associated with a comprehensive test program is the ability to perform early R&D propulsion (battleship) testing, followed by all-systems vehicle captive testing and finally full thrust and duration acceptance testing of development flight vehicles.

Equally important are the many detailed tests that must be performed on individual components and subsystems ranging from qualification of valves and switches to static loading of structural components and propellant tanks.

MSFC is fully aware of the benefits to [be] derived through testing, evaluation and redesign during the development phase of any space vehicle. Each of the proposing contractors will be evaluated, in part, based on the completeness of their proposed test program.

4.   Production

The fabrication of all three stages will occur at contractor plants. A hot test stand for the N-I and a hot test stand for the N-II will be fabricated. A dynamic test stand for the C-3 will be modified to accommodate the second stage N-II and higher stages as well as to independently have a capability for first stage N-I accommodation. This test stand in possession of MSFC will be available for NOVA testing by personnel of the NOVA prime contractor. The load testing facilities of MSFC, to be built for C-3 testing, are in this status capable of load testing NOVA stages. The importance here is that parallel dynamic and load testing is facilitated.

5. Reliability
As already mentioned in the above paragraphs, a high degree of reliability is required for all components and systems. Therefore, a comprehensive reliability program will be carried on by the NOVA prime contractors. MSFC will provide the detailed work statement for such a program which will encompass the following broad reliability areas: management and technical organization, program planning requirements, engineering design, subcontractor and vendor control, reliability goals and evaluations, testing programs, failure analysis and data collection, documentation and progress reports, and manufacturing and handling procedures. MSFC will monitor major control points in the program and will evaluate the program progress at definite detail and all stages in the NOVA program.

[5] 6. Possible Change in Configuration of N-III
In view of the recent decision to change the S-II diameter from 320 to 360 inches, it will be desirable to investigate a similar change for N-III. There are, however, some differences in the two stages which should be mentioned. One difference is in the flight trajectories: S-II is the injection stage in an orbital flight, whereas the N-III will be injection stage for escape missions. There is, also, a 1 to 4 weight ratio of payloads in orbital flights if the N-III is used for orbital injection. Another factor may be a probable difference in the number of engines. In view of these and other differences, it appears that little other than handling equipment and some tooling and internal parts will be interchangeable on the two stages. It still may be desirable, however, from the standpoint of test stands and other considerations, to make the N-III the same diameter as the S-II.

Performance-wise, there appear to be no major objections to either diameter except that payloads, guidance packages, and interface problems may be somewhat simplified by the use of a similar diameter.

7. Vehicle Description
General
The NOVA vehicle will be 283 feet in length; its first stage is 530 inches in diameter and approximately 111 feet long; its second stage is 396 inches in diameter and approximately 106 feet in length; the third stage is 320 inches in diameter and approximately 66 feet long. The first stage will be powered by eight Rocketdyne F-1 engines which will use LOX and RP-1 as propellants; each engine will develop 1.5 million pounds of thrust. The second stage will be powered by eight J-2 engines which will use LOX and LH_2 as propellants; each engine will develop 200,000 pounds of thrust. The third stage will be powered by two J-2 engines. . . .
Curves and charts appended to this report show the following characteristics of NOVA:
(1)   Distribution of Normal Force Coefficient
(2)   EI vs. Station
(3)   Weight and Propulsion Data
(4)   N-I, N-II, N-III Stage Weight Breakdown
(5)   [Control and Guidance] and Pitch Moment of Inertia vs. Burning Time
(6)   Normal Force Coefficient vs. Mach Number
(7)   Center of Pressure vs. Mach Number
[6] (8)   Vehicle Drag vs. Mach Number
(9)   Projected NOVA Reliability

(10) Trajectory Data
(11) Design and Expected Bending Moments vs. Station
(12) Design and Expected Shears vs. Station
(13) Longitudinal Force vs. Station . . .

Document I-34

Document title: A.O. Tischler, Chief, Liquid Fuel Rocket Engines, NASA, to David Aldrich, Program Engineer, Rocketdyne, July 29, 1959.

Source: NASA Historical Reference Collection, NASA History Office, NASA Headquarters, Washington, D.C.

The development of the powerful F-1 rocket engine was a technological challenge from the start. Nothing of similar scale had ever been attempted. Shortly after receiving the F-1 development contract from NASA in January 1959, Rocketdyne began full-scale injector and thrust chamber tests. It soon discovered a combustion instability problem—that is, the burning of the rocket fuel was not even across the full width of the injector plate. This concern was common to the early stages of most rocket engine development efforts, but given the size of the F-1, it could lead to major problems; shock waves from the instability could destroy an engine in milliseconds. The problem was not easily or quickly solved and, within three years, had become one of the pacing technological challenges of the Apollo program.

[1]

DPL (AOT:bw)
29 July 1959

Rocketdyne
A Division of North American Aviation, Inc.
6633 Canoga Avenue
Canoga Park, California

Attention: Mr. David Aldrich
             Program Engineer

Dear Dave:

As you know, the National Aeronautics and Space Administration is concerned about the occurrence of destructive combustion-driven oscillations in the experimental work on the F-1 engine. We feel that continued occurrence of combustion oscillations can jeopardize the development to a greater extent that any other single factor. Therefore, we are anxious to do all that can be done to eliminate combustion oscillations in order to assure expedient development of the engine. This must be accomplished in the face of a limitation of available funds: planned work must be consistent with funds.

In the assessment team meetings of June 13–15 your people reviewed your program on combustion-driven oscillations for our benefit. The assessment team's opinion of this review was that a more definitive step-by-step program aimed specifically at the F-1 engine development would be required. This is, I believe, in line with your own plans. Such a program may encompass model testing to develop empirical solutions as well as applied research into the more fundamental aspects of the problem. Since such as program requires well planned integration, your recent formation of a panel on combustion-driven oscillations will be valuable in putting together procedures aimed toward a solution. The program plan should explain what each test or each experiment is intended [2] to demonstrate, what the anticipated result will prove, what an opposite result would indicate, and what in either case should follow to carry your understanding progressively further.

In connection with this problem ARDC [the Air Research and Development Command] proposes to make a historical survey of methods and techniques used to circumvent the problem in the past. It would be appreciated if you would cooperate in discussing with and providing to ARDC people ([Ballistic Missile Division]) and the former [Western Air Development Command] WADC group, henceforth stationed at [Edwards Air Force Base], all technical information requested on this subject. Lt. Fred Anderson (now at [the Eastern Air Force Command requirements branch]) is expected to contact you shortly.

Since a better identification of what goes on in the chamber may yield valuable clues to the phenomena the assessment team favored more complete instrumentation of the test equipment in future operations. Records of valve opening and sequencing should be mated with chamber instrumentation records.

Concepts for attenuating the combustion oscillations before test hardware damage occurs must be considered. The RCC [rough combustion cut-off] device appears to detect the occurrence of oscillations but the slowly-operating valves prohibit shut down in time. Can some other faster system or method to attenuate the oscillations be used?

The assessment team observed that hydraulic simulation of the valving and flow operations would be valuable but that such simulation cannot be considered complete without a turbopumped fluid system to work with.

NASA sees manned vehicle application as a future requirement of the F-1 engine. While it is probably too early to consider what needs to be done to demonstrate a high degree of reliability in this engine the future need for such demonstration should be anticipated.

Yours truly,

[signed "Oscar Bessio for"]
A.O. Tischler
Chief, Liquid Fuel Rocket Engines

Document I-35

Document title: D. Brainerd Holmes, Director of Manned Space Flight, to Wernher von Braun, Director of Marshall Space Flight Center, "Combustion Instability of F-1 Engine," January, 26, 1963.

Document I-36

Document title: A.O. Tischler, Assistant Director for Propulsion, to Milton Rosen, Director, Launch Vehicles and Propulsion, "First monthly report on F-1 instability problems," February 15, 1963.

Document I-37

Document title: Wernher von Braun, Director of Marshall Space Flight Center, to D. Brainerd Holmes, Director of Manned Space Flight, "Response to Letter of January 26, 1963," March 11, 1963.

Source: All in NASA Historical Reference Collection, NASA History Office, NASA Headquarters, Washington, D.C.

NASA and Rocketdyne engineers had known for some time that addressing combustion instability would be a major problem in qualifying the F-1 engine for use in the Saturn/Apollo booster. They believed that they had the problem under control until an engine was destroyed during a June 1962 test. Attempts to address the problem during the rest of 1962 were not successful. D. Brainerd Holmes, NASA's Associate Administrator for Manned Space Flight, even considered abandoning the engine at one point. One concern was whether Wernher von Braun and his associates at the Marshall Space Flight Center, who were in charge of F-1 development, were being sufficiently responsive to suggestions coming from outside that center.

After a January 31, 1963, review of the situation, Holmes was persuaded not to start another engine development effort and to move forward on the assumption that the problem could be solved without threatening the overriding objective of meeting President Kennedy's "before the decade is out" objective for the lunar landing. He was also assured by von Braun that all good ideas, whatever their source, were being taken into account. Various ad hoc approaches to the problem were tried in succeeding months until a stable baffled injector design was developed. Even then, additional fixes had to be made to assist the engine in recovering from transient instability problems. Note that the enclosures with the Holmes letter to von Braun, as well as with Tischler's monthly report, do not appear here.

<hr>

Document I-35

[1]

January 26, 1963

Dr. Wernher von Braun, Director
George C. Marshall Space Flight Center
National Aeronautics and Space Administration
Huntsville, Alabama

Subject: Combustion Instability of F-1 Engine

Dear Wernher:

We have become increasingly concerned over the problem of combustion instability of the F-1 engine. In fact, the recent decision to limit test firings to fifteen seconds duration because of these instabilities was very disturbing to me. It is difficult to see that progress is being made, although I recognize that such development problems are not solved rapidly and often entail major hardware modifications even during the periods of experimentation.

We would, however, like to see the specific steps which are being planned in the analysis and experimentation to be programmed for the months ahead. I have asked Mr. Rosen to contact your propulsion people in order that I can be briefed by those intimately involved in the F-1 engine development concerning our plans in the handling of this matter. It is my understanding that this meeting is scheduled to be held in my office on January 31st.

As you know, Dr. Seamans has for some time believed that this problem is one of the most serious in our entire manned lunar landing development program. I have attached for your information three memoranda which I believe are self-explanatory. One is a memorandum from Mr. Dixon to Dr. Seamans concerning his view on the subject. The second is a letter to Dr. Seamans from Dr. [John C.] Evvard which references the third memorandum from Dr. [Richard] Priem [at Lewis Research Center]. As you will note, these memoranda give the impression that the suggestions of the Lewis people have been largely ignored. I do not know if this is a fact, but I do believe the problem has reached such serious proportions that we should all be very well aware of the specific steps being taken to endeavor to reach an early solution.

[2] I would most appreciate it if you would give this matter your personal and urgent attention and advise me at an early date of the specific actions which you judge we should undertake at this time.

Sincerely,

Dr. Brainerd Holmes
Director of Manned Space Flight

cc: Dr. Seamans
Mr. Rosen

Enclosures: (1) To Associate Administrator from Thomas F. Dixon, January 18, 1963, "Combustion Instability of the F-1 Engine"
(2) To Dr. Seamans from John C. Evvard, undated, same title
(3) To Deputy Associate Director for Research from Dr. Priem, dated December 12, 1962, "Combustion Instability with F-1 Engine"
(4) To ML/Mr. Rosen from M/Mr. Holmes, January 26, 1963

DBH:as

Document I-36

[1]

UNITED STATES GOVERNMENT   National Aeronautics and
MEMORANDUM   Space Administration

TO: ML/Milton Rosen
FROM: MLP/A. O. Tischler
SUBJECT: First monthly report on F-1 instability problems
DATE: February 15, 1963
          M-M L 4000.036

This report will discuss background of the F-1 instability problem, will review current theories of the oscillation mechanism, will survey design modifications in work and possible effect on F-1 engine program and will indicate supporting activities.

HISTORY

Combustion-driven oscillations were recognized at the onset of the F-1 engine development as the most critical problem facing the engine development. Project direction demanded intensive effort in this area and this effort, as planned by Rocketdyne, was reenforced [sic] after a number of occurrences in the early thrust chamber tests. Subsequently, repeated oscillation-free operation of the thrust chamber with one particular injector (5-U pattern) resulted in a tapering off of activity to examine engine stability. This injector furthermore permitted stable operation of engine tests during the early phases of engine testing. It is noted, however, that because of turbo-pump test failures on the turbo-pump stand many of the early engine tests were run at a derated thrust of about 1100k.

More recent testing of the engine at full thrust has, on occasion, resulted in main-stage combustion-driven oscillations leading to automatic termination of the tests by a device called the rough combustion cut-off (RCC). Eight such cases have occurred in

240 engine tests. Eighty-four of these engine tests have been at near-rated thrust levels. All of the full thrust tests have been made with either the 5U pattern injector, which has a flat face, or with a baffled version of this injector. The attached photos show the 5U and 5U Baffled injector patterns. (Enclosure 1)

Of the eight main-stage rough combustion cut-offs, seven have been at full thrust; one occurred below 1400k. It is noteworthy that, except for some early engine tests, the engine has shown remarkable stability through the start transient. Although starting often serves to trigger instabilities in liquid engines, the F-1 has been free of oscillatory troubles during start; all recent cases of rough combustion cut-off have occurred during steady thrust operation. In 168 engine tests with a flat-face 5U pattern injector, five cases of instability have been observed. This is an incidence rate of about 3%. In 15 tests with [2] the 5U Baffled injector, three cases of instability have occurred. This is an incidence rate of 20%. However, the severity of the oscillations is different for the 5U and 5U Baffled injectors. Instability with the 5U injector results in very rapid extensive damage to the injector face and, very often, to the combustion chamber walls as well. With the 5U Baffled pattern, however, cut-off can be initiated before damage becomes excessive. The chamber is generally operable. The injector face may be scorched and the baffles slightly bent but the injector is generally reusable

OSCILLATORY

The modes of instability of the 5U and 5U Baffled injector are different. Instrumentation traces of engine tests with a 5U injector show the characteristic frequency (670 cps) and wave form of the first tangential mode of oscillation in the chamber. Pressure amplitudes range from 1500-2000 psi peak-to-peak. This is the predominant combustion-driven mode that has destroyed hardware in other engine programs. Instrumentation traces of an engine run with a 5U Baffled injector show a frequency of about 350 cps at amplitudes of 700 to 900 psi peak-to-peak. The wave form, instead of being opposite in phase at opposite ends of a diameter of the chamber, is in this case in phase across the entire injector face. This is therefore not a transverse acoustical mode. A second form of instability with frequency of about 500 cps has appeared with one 5U Baffled injector during two tests. This instability had the phase relationship of a normal transverse mode although not the frequency. It is also of higher amplitude than the lower frequency mode. This may be a damped form of the transverse oscillation. Damping tends to depress frequencies.

The 670 cps corresponding to the first tangential mode of acoustical oscillation in the chamber also corresponds roughly to the wake frequency of the blades of the turbo-pump. The pressure pulses delivered by each blade have been measured just downstream of the turbo-pump. These pressure excursions are about 75 psi peak-to-peak. Both the fuel and the oxidant pump, which run at the same speed and have the same number of blades, cause such excursions. The coincidence of these frequencies is recognized as bad and the number of blades in both pumps is being changed from six blades to eight to mismatch the frequencies. In addition, the dome of the injector, which serves as a plenum chamber for the oxygen supply, has a characteristic "ring" under flow conditions of about 350 cps, which corresponds to the frequency observed with the baffled injectors. This dome is being redesigned to change its vibrational characteristic. Thus, the injection system of the chamber contains several driving forces which are potentially oscillatory and which can couple to produce the observed instabilities of the F-1 engine. It is clear that the engine is not likely to be cured of combustion-driven oscillations by injector [3] redesign changes alone. Both the contractor and NASA have recognized this. Steps are being taken to redesign and correct those coupling systems which appear to affect the oscillation tendency of the F-1 engine.

CONTRACTOR ACTION

Design actions taken by the contractor to suppress combustion-driven oscillations in the F-1 engine have been directed in courses. These are 1) attempts to isolate the feed system from the combustion chamber, and 2) injector modification intended to produce a stable injector pattern. The first course has included changes in the number of blades on the turbo-pumps to "detune" the systems already mentioned in this memorandum, use of dome modifications to provide improved feed system isolation, widening of certain restricted flow passages in the dome to prevent repeated acceleration and deceleration of the inlet (lox) flow. To date, a compartmented dome . . . which separates to prevent flow from the diametrically opposed oxygen inlets has been fabricated and is being subjected to test. It is believed that these opposed flows tended to generate a "flutter." After engine tests with this obstructed dome have been completed, it will be "bombed" with an explosive charge to determine whether it is dynamically stable.

Along the other course, a series of detailed injector designs have been laid out. Designs comprising approximately 14 different injectors are planned and eight are being fabricated for testing in a program continuing through most of calendar year 1963. The more conventional design injectors will be evaluated by the middle of June 1963. The production engine injector design release requirement is the end of June.

Among the injector patterns to be evaluated are some which depart from the usual variation in injector element patterns and explore the effects of grouping clusters of elements in a manner that will produce a non-uniform flame pattern within the combustor. Such concepts have not been applied heretofore to avoid combustion-driven oscillations. Preliminary results using scaled hardware (H-1 engine) indicate promise that such gross injector groupings will suppress oscillations. Another pattern will inject propellants through concentric tubes carried out into the combustion chamber at various distances to distribute the flame front axially.

The lead time for some of the major design changes, particularly those which involve the turbo-pump or the injector dome, is of the order of a half-year. Such changes, in work now for about three months, are still several months from experimental evaluation. Because of the importance of these experimental evaluations to the program progress and schedule, the normal hardware lead times have been greatly reduced by special handling on items affecting F-1 stability. Fabrication time on injector's hole pattern changes, for example, has been reduced from about five months to six weeks.

[4] In addition to these mainline courses, several other avenues are being explored by the contractor. These include the investigation of various additives to the fuel and to the oxidant.

RECENT TEST EXPERIENCE

Two advanced injector designs were tested early in February as part of the injector evaluation program. A triplet design, a radical departure from current F-1 pattern concepts, was tried in the thrust chamber stand and went unstable spontaneously as it went into main-stage. A splash ring injector (jets impinge on its surface and fan out) was tested and made one short run. The second run was "bombed" and it went unstable damaging a portion of the stand suction piping. In the meantime, a 5U flat faced injector with dams and baffles in the liquid passages feeding the injector face to isolate the feed system has been accumulating impressive running times without going unstable. Two short thrust chamber checkout runs were made with this injector before it was installed in engine #1 but it has not been "bombed." It has operated successfully seven times in the engine for a total of 670 seconds. One of these runs was for 151.3 seconds of duration at rated thrust. It is planned to continue this modified 5U injector in engine #9 for several additional runs. Then it will be removed and bombed in a thrust chamber test to see if it is capable of smoothing out the disturbance.

PERSONNEL

Rocketdyne has established a special development group within its R&D organization to attack this particular problem. This group is under Mr. Paul Castenholz and Dr. Daniel Klute. The group presently numbers 142 people. This relatively large group has autocratic authority within the program to take whatever action is deemed necessary to solve the combustion problem expeditiously. At present, there is no money deficiency. The higher rate of spending as a result of this group activity will generate a program money deficiency early in FY 64 unless money is forthcoming immediately. In FY 64, [the Office of Manned Space Flight] has projected a requirement for the F-1 program of which represents an increase over the original submission.

In carrying out the investigations on the cause of combustion oscillations Rocketdyne has had direct manpower support of NASA ([Marshall Space Flight Center]) personnel. Rocketdyne has employed nationally-known consultants to assist in the interpretation of the problem and the data records. In addition, [Marshall] has formed an ad-hoc combustion instability group under the chairmanship of Mr. Jerry Thomson. This group includes Dr. David Harjie of Princeton University and Dr. Richard Priem of [the Lewis Research Center]. [Marshall] is also buying some technical support in the form of additional contract work with Princeton University and General Electric. Some of the committee members are listed on the attached sheet. (Enclosure 4)

[5]   FACILITIES

To accommodate the additional development investigations by Rocketdyne, a second position of thrust chamber test stand 2-B is being activated. In addition, a high-liquid-flow-rate water bench is being constructed to test the hydraulics and dynamics of the injectors without combustion. Additional instrumentation suitable for measurement of high frequency phenomena will be employed in every test engine and chamber in order to obtain far fuller information about the combustion phenomena. The additional requirement being programmed for this purpose is reflected in an additional [Construction of Facilities] requirement of $3.33m during the current fiscal year. This amount does not include approximately $0.95m required as payment to the Air Force for not removing numerically controlled machine tool equipment from Rocketdyne's Canoga Park fabrication facility.

EFFECT ON LUNAR LANDING FLIGHT SCHEDULES

The present difficulties in the F-1 engine development do not jeopardize flight schedules. The PFRT [Preliminary Flight Rating Test] date is threatened and PFRT may be delayed to the end of the year to provide time to evaluate the several injectors which will be tested in June prior to final PFRT configuration selection. Such an occurrence will delay the delivery of the first complete S-5 set of ground test vehicle engines from January 64 to June 64. However, it is not intended to use all five engines in the earliest phase of the vehicle ground test programs. Accordingly, an April delivery of one F-1 engine would permit the accomplishment of planned ground test programs without delay to any flight schedules. Such a schedule, on the other hand, has the disadvantage of having taken up most of the "slack" in the ground testing program. The "fall-back" schedule proposed by [Marshall] personnel is shown on Enclosure 2. It should be noted that this schedule has not been reviewed by Dr. von Braun nor has any "fall-back" in schedule been sanctioned by [the Office of Manned Space Flight].

[OFFICE OF ADVANCED RESEARCH AND TECHNOLOGY] SUPPORT

A memorandum has also been prepared to encourage an intensified [Office of Advanced Research and Technology] program of support in the examination of the fundamental combustion processes driving these oscillations. A copy is attached. (Enclosure 3)

A. O. Tischler
Assistant Director for Propulsion

Enclosures:
Photos (C)
Schedule
Ltr to R. Bisplinghoff
Personnel Roster

Document I-37

[1]

NATIONAL AERONAUTICS AND SPACE ADMINISTRATION
GEORGE C. MARSHALL SPACE FLIGHT CENTER
HUNTSVILLE, ALABAMA

In reply refer to: MAR 11, 1963
M-DIR

Mr. D. Brainerd Holmes
Director
Office of Manned Space Flight
National Aeronautics and Space Administration
Washington 25, D.C.

Dear Brainerd:

In response to your letter of January 26, 1963, I want to reaffirm my personal concern and awareness of the problem confronting us regarding combustion instability of the F-1 engine. We at [Marshall Space Flight Center] have taken what we believe to be all the logical steps necessary to bring about a rapid and final solution.

Your letter contained a number of questions and comments which, I have been told, were adequately answered at NASA Headquarters during the [Marshall]/Rocketdyne presentation of January 31, 1963. However, I feel it necessary to re-emphasize some of the remarks made at that time.

As you are aware, the test limit of 15 seconds duration was imposed temporarily and voluntarily on engine runs with injector configurations proven to be risky and inadequate, and in the absence of any better known designs. This was done in an effort to conserve as much hardware as possible. At the same time it would permit us to run as many tests as feasible with hardware, which if permitted to run longer durations, would possibly fail. On the other hand, modified hardware incorporating the latest design changes would have no duration limit imposed since we are interested in exposing such new designs as realistically as possible to verify the validity of these modifications.

[2] At the present time, two engines are being tested at Edwards Air Force Base. Engine #009 with the new injector/dome hardware, having satisfactorily completed a series of tests (including eight long duration runs) since the first of February, is being replaced by

engine #008 (also with new injector/dome hardware). Engine #010, utilizing the older design injector/dome hardware, has been limited to 15 second tests. This engine will have as its primary objective the demonstration of the gimbal capability of the F-1 engine. I feel that the approach being taken on these engines is sound and reasonable.

Regarding your concern that suggestions from Lewis Research Center have been largely ignored, I am informed that this too was satisfactorily answered at the January 31, 1963, presentation. Suggestions made by the F-1 Stability Ad Hoc Committee, of which Lewis Research Center is a member, have been incorporated into the F-1 injector/dome program and have already resulted in hardware or are currently in design. As a matter of fact, the day of the presentation an injector configuration suggested by Lewis was component-tested with unsuccessful results.

I hope that as a result of the presentation on January 31, 1963, you have acquired the feeling that everything which can logically be done to bring about a rapid solution to this problem is being done. I also want to assure you that [Marshall] will continue to be responsive to constructive inputs from other areas, and that I will give my personal attention to the efforts on the F-1 program.

Yours very truly,

Wernher von Braun
Director

Copies to: NASA Headquarters
            Dr. Seamans, AA
            Mr. Low
            Capt. Freitag, ML
            Mr. Tischler, MLP
            Mr. King, MLPL
            Mr. Bessio, MLPL

Document I-38

Document title: George E. Mueller, Deputy Associate Administrator for Manned Space Flight, NASA, to the Directors of the Manned Spacecraft Center, Launch Operations Center, and Marshall Space Flight Center, "Revised Manned Space Flight Schedule," October 31, 1963.

Source: NASA Historical Reference Collection, NASA History Office, NASA Headquarters, Washington, D.C.

Finding ways to shorten the development time of the Saturn boosters was of considerable importance in achieving President Kennedy's goal of placing a human on the Moon within the decade. Based on his experience in managing the Minuteman ICBM program, George E. Mueller, NASA's new Deputy Associate Administrator for Manned Space Flight, proposed to accelerate the test flight schedule for the Saturn IB and Saturn V by testing all elements of the system together. The new schedule was approved after considerable debate, resulting in "all-up" test flights of the launch vehicles and spacecraft much earlier than had been originally planned. This acceleration of the test schedule was one of the crucial decisions leading to a 1969 lunar landing. The two figures referred to in this memorandum do not appear here.

[1]

NATIONAL AERONAUTICS AND SPACE ADMINISTRATION
WASHINGTON 25, D.C.

IN REPLY REFER TO:
M-C M 9330.186
OCT 31, 1963

TO: Director, Manned Spacecraft Center
           Houston 1, Texas
           Director, Launch Operations Center
           Cocoa Beach, Florida
           Director, Marshall Space Flight Center
           Huntsville, Alabama
FROM: Deputy Associate Administrator for Manned
           Space Flight
SUBJECT: Revised Manned Space Flight Schedule

Recent schedule and budget reviews have resulted in a <u>deletion of the Saturn I manned flight program and realignment of schedules and flight mission assignments on the Saturn IB and Saturn V programs</u> [handwritten underlining]. It is my desire at this time to plan a flight schedule which has a good probability of being met or exceeded. Accordingly, I am proposing that a flight schedule such as shown in Figure 1, with slight adjustments as required to prevent "stack-up," be accepted as the official launch schedule. Contractor schedules for spacecraft and launch vehicle deliveries should be as shown in Figure 2. This would allow actual flights to take place several months earlier than the official schedule. The period after checkout at the Cape and prior to the official launch date should be designated the "Space Vehicle Acceptance" period.

With regard to flight missions for Saturn I, [the Manned Spacecraft Center] should indicate when they will be in a position to propose a firm mission and spacecraft configuration for SA-10. [The Marshall Space Flight Center] should indicate the cost of a meteoroid payload for that flight. SA-6 through SA-9 missions should remain as presently defined.

[2] It is my desire that "all-up" spacecraft and launch vehicle flights be made as early as possible in the program. To this end, SA-201 and 501 should utilize all live stages and should carry complete spacecraft for their respective missions. SA-501 and 502 missions should be reentry tests of the spacecraft at lunar return velocity. It is recognized that the Saturn IB flights will have [Command Module/Service Module] and [Command Module/Service Module/Lunar Excursion Module] configurations.

Mission planning should consider that two successful flights would be made prior to a manned flight. Thus, 203 could conceivably be the first manned Apollo flight. However, the official schedule would show the first manned flight as 207, with flights 203–206 designated as "man-rating" flights. A similar philosophy would apply to Saturn V for "man-rating" flights with 507 shown as the first manned flight.

I would like your assessment of the proposed schedule, including any effect on resource requirements in FY 1964, 1965 and run-out by November 11, 1963. My goal is to have an official schedule reflecting the philosophy outlined here by November 25, 1963.

                        George M. Low [signed for]
                        George E. Mueller
                        Deputy Associate Administrator
                        for Manned Space Flight

Enclosures:
Figure 1
Figure 2

Document I-39

Document title: Wernher von Braun, "The Detective Story Behind Our First Manned Saturn V Shoot," *Popular Science*, November 1968, pp. 98–100, 209.

Source: NASA Historical Reference Collection, NASA History Office, NASA Headquarters, Washington, D.C.

The second flight of the Saturn V booster, launched on April 4, 1968, encountered several problems. Identifying them and introducing corrections were essential to maintaining a schedule that would put U.S. astronauts on the Moon before the end of 1969. Omitted here are photographs of author Wernher von Braun, the Saturn V, and the fuel line. Von Braun's original sketches have been redrawn for clarity.

[98]
The Detective Story Behind
Our First Manned Saturn V Shoot

By solving the mystery of what went awry last time, engineers give the giant moon rocket a "go" to carry astronauts on the next Apollo mission.

By DR. WERNHER VON BRAUN
Director of NASA's George C. Marshall Space Flight Center, Huntsville, Ala.
Sketches by the Author

A few weeks hence, at Cape Kennedy, the first manned Saturn V will thunder aloft—our 363-foot-high moon rocket. A triumph of detective work has cleared the way for astronauts to ride it.

So far, just two of the giant rockets have been launched, both unmanned. The first Saturn V flight went off flawlessly late last year. A string of mishaps, in contrast, beset the second one last April. But the diagnosis of these has been so conclusive, and the remedies so successful, that the unmanned trial will not need to be repeated. NASA has decided to go right ahead and fly the third Saturn V manned.

The story of how the second Saturn V flight's troubles were identified resembles a detective thriller. It illustrates, too, modern methods of shaking down a complex space vehicle.

The last flight. The second Saturn V's takeoff at the Cape was faultless. For two minutes everything looked like a repeat of the first Saturn V's textbook performance. Then came a little excitement in the launch control center when, around the 125th second, telemetered signals from accelerometers indicated an apparently mild "Pogo" vibration.

This is a lengthwise oscillation, named after the motion of a Pogo stick, which had caused no little concern with the earlier Titan-boosted Geminis. It makes a space vehicle lengthen and shorten like a concertina, several times a second. But [original placement of first figure] [99] the Pogo vibration disappeared at about the 132nd second.

The second stage's five J-2 engines, burning liquid hydrogen, ignited exactly on schedule. But engine No. 2 soon gave signs of trouble. After burning for almost 4 1/2 minutes, it suddenly lost thrust, and its low-thrust detection switch turned it off completely. Engine No. 3—which had performed perfectly up to this point—shut itself down a second later.

Deprived of two-fifths of its million-pound thrust, the second stage bravely fought on upward—with the trouble-sensing guidance system altering the climb path to help—and labored overtime before dropping off. The third stage's single J-2 engine started, and the bird arrived in a somewhat off-normal but stable parking orbit. When it had circled the

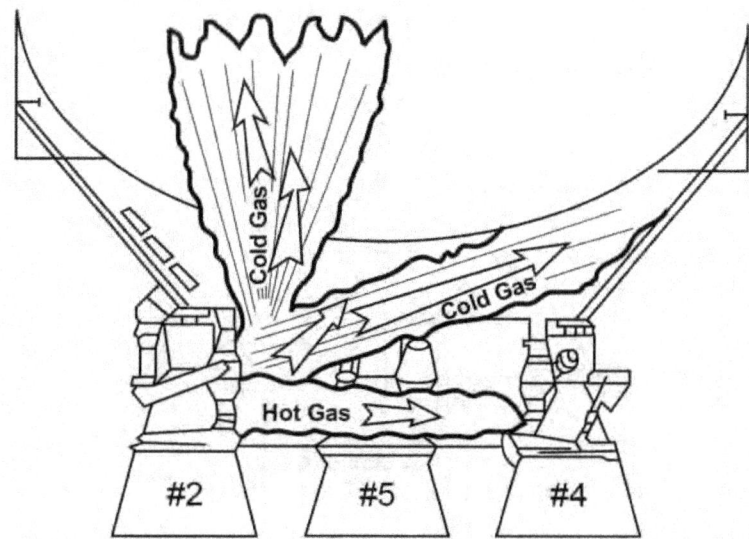

Clues to conking-out of second-stage engine No. 2, on last Saturn V, are sketched by Dr. von Braun. Thermocouples on rocket told of flow of cold gas, as liquid hydrogen leaked from igniter fuel line and vaporized; then, of hot blast, as fire gas from combustion chamber spurted like blowtorch from ruptured line. Losing thrust, engine shut itself off.

earth twice, a radio command to re-ignite the third stage was sent. But the J-2 engine failed to respond.

To get the most out of the rest of the flight, the Command and Service Module carried in the nose was commanded to separate from the disabled third stage. After two burns of the Service Module, the Command Module made its reentry and was successfully recovered.

Had the flight been manned, the astronauts would have returned safely. But the flight clearly left a lot to be desired. With three engines out, we just cannot go to the moon.

Despite the J-2's impressive reliability in tests, two of the engines had conked out in second-stage flight, and a third had balked in orbit. Why, suddenly, three failures on a single flight?

Sleuths find clues. A joint detective team of engineers from NASA's Marshall Space Flight Center and from Rocketdyne, the J-2's maker, went to work. Soon they discovered clues. Counting time from second-stage ignition, telemetered temperature readings of thermocouples in the second stage's tail told this story:

- At about the 70th second, a flow of cold gas was detected, which could come only from a liquid-hydrogen fuel leak. The flow pattern clearly located the leak in the upper part of engine No. 2.
- The cold flow seemed to be increasing from the 110th second on, the time when engine No. 2 began to falter.
- Between the 262nd and 263rd seconds, a sudden blast of very hot gas came from the same place—just a split second before engine No. 2 shut itself off.

This short hot blast before shutdown was the giveaway. Only the fuel line to the J-2's igniter could fail in just this way. (The igniter is a hydrogen-oxygen pilot flame that helps start the engine, and burns while it operates.)

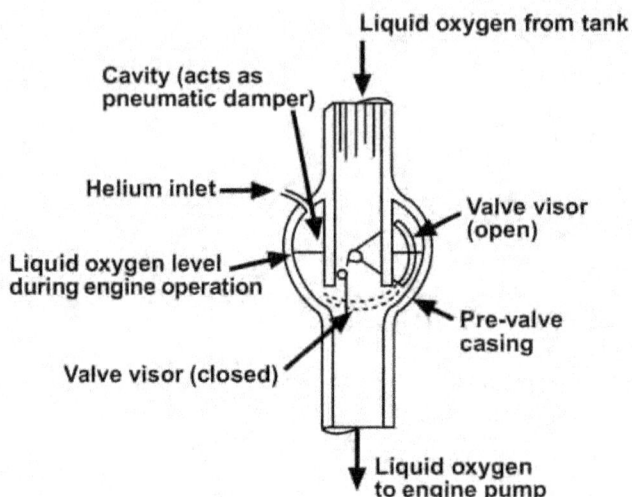

Successful cure for Saturn V's Pogo vibration, from slight pulsation in thrust of its mighty first stage, puts a shock-absorbing pneumatic damper of helium gas in each engine's oxygen prevalve.

A leaking igniter fuel line would spray the surrounding area with cold hydrogen, [original placement of second figure] [100] while it kept feeding some fuel into the igniter. But the moment the line failed completely, high-pressure fire gas from the rocket engine's combustion chamber would back up in it and rush out of the breach like a blow-torch, rapidly widening the uncooled opening. And when the engine's combustion pressure dropped below a certain point, the low-thrust sensing device would turn off the engine, by closing fuel and oxygen "prevalves" that control the propellants' flow to the engine pumps. That explained why engine No. 2 shut down.

But what made the healthy No. 3 engine quit an instant later? Embarrassingly, a plain human goof. Because of a mistake in wiring, the electrical signal intended to close engine No. 2's lox (liquid-oxygen) prevalve went instead to engine No. 3's prevalve. Thus engine No. 2, while it shut itself off by closing its own *fuel* supply, cut off engine No. 3's *lox* supply, too.

The third stage's J-2 engine shared the troubles of second-stage engine No. 2. During its first burn of 170 seconds there were the same telltale signs of leaking and rupture of the igniter fuel line, including the final hot blast. That put the engine out of commission—and so it could not be restarted.

What ailed the igniter fuel lines? Tortured in tests before, they went on the rack again. They proved immune to increased pressure and flow rates, and to a far more severe shaking up than in flight. Next came a study of resonant conditions: Did bellows sections in the lines which provided flexibility for expansion "buzz" at certain flow rates?

It turned out that they did—but it seemed impossible to make them fail as a result. Then eight lines were placed in a vacuum chamber. Liquid hydrogen flowed through them at the proper rate and pressure. Within 100 seconds, every line failed at the bellows section!

Movies made of bellows' tests solved the mystery. When the test chamber was *not* evacuated, surrounding air was liquefied by the extremely low temperature of the bellows (-350 to -400 F.) when liquid hydrogen flowed through it. The liquefied air, trapped by metal braid around the bellows, effectively damped its vibration at resonant points. Evacuate the chamber, and (as in space) the protective damping effect was gone.

Once this diagnosis of the engines' failures was made, the remedy was simple. New igniter fuel lines, with bends in the stainless-steel tubing for flexibility, eliminated the bellows sections—and that was all there was to it.

Then, the Pogo fix. The Marshall center set up a Pogo Task Force, too—supported by experts from other NASA centers, universities, and industry. The team studied the first-stage F-1 engines, made shake tests of parts of the Saturn V/Apollo structure, and reported:

Such is the nature of a rocket engine's operation that the F-1s' thrust and combustion chambers slightly pulsate, at a natural frequency of about 5 1/2 cycles a second. The entire Saturn V with a spacecraft in its nose has a natural frequency, too, at which it is especially susceptible to longitudinal (concertina-like) vibration. Increasing as propellants are consumed, this frequency also approaches 5 1/2 cycles a second at about 125 seconds after takeoff.

When the structure's responding frequency matches the engines' driving frequency, Pogo vibrations can occur.

While not necessarily destructive, it [209] undesirably imposes an extra, fluctuating fraction of a g load on the vehicle and crew. (Sitting atop the long Saturn V stack, the relatively light spacecraft is subjected to even higher Pogo-vibration loads than the engines at the other end that cause the problem.)

The Pogo team's solution: Detune the two frequencies by placing a pneumatic shock absorber in the liquid-oxygen line of each of the five F-1 engines.

Cavities in the engines' lox prevalves make this easy. Just fill them with helium gas—which doesn't condense at liquid-oxygen temperatures—and you have the desired shock absorbers. The first stage's ample supply of helium for pressurizing the fuel tank can be tapped to do it. Thus the Pogo fix was made.

Both a first stage with this shock-absorber modification, and a second stage with the new igniter fuel lines, were successfully test-fired last August at the Marshall center's Mississippi Test Facility. The two simple fixes qualify the Saturn V for manned flight.

New plans. Called Apollo 8, the first manned Saturn V flight will follow the initial manned Apollo mission, boosted by an Uprated Saturn I—Apollo 7, due to have taken place when this is read.

Apollo 8, likewise, will carry the Command and Service Module (CSM); contrary to earlier plan[s], it will not include the Lunar Module, whose debugging is taking longer than expected. Plans for the first manned Saturn V, and later missions, had therefore to be revised.

Apollo 8's new basic mission plan provides operations with the manned CSM in low earth orbit—and, after separation of the CSM, an unmanned orbital launch of the Saturn V's third stage into an escape trajectory possibly grazing the moon. However, if Apollo 7 has gone very well, possible options are under consideration for the Saturn V. It might launch the CSM several thousand miles into space. There is even a remote possibility of a spectacular swing around the moon by the manned spacecraft. That a mission as bold as the last is even considered, for the first Saturn V to be manned, bespeaks planners' confidence that all about it has been set aright.

Document I-40

Document title: Kurt H. Debus, Director, Launch Operations Center, NASA, to Captain John K. Holcomb, Office of Manned Space Flight, NASA, "Reference draft DOD/NASA Agreement dated 20 December 1962 regarding management of Merritt Island and AMR," January 2, 1963, pp. 1–2.

Document I-41

Document title: Robert S. McNamara, Secretary of Defense, and James E. Webb, Administrator, NASA, "Agreement between the Department of Defense and the National Aeronautics and Space Administration Regarding Management of the Atlantic Missile Range of DoD and the Merritt Island Launch Area of NASA," January 17, 1963.

Source: Both in NASA Historical Reference Collection, NASA History Office, NASA Headquarters, Washington, D.C.

To accomplish the lunar landing mission, NASA recognized that it would have to establish a new, quite large, launch operations complex. After examining several possible locations, NASA decided to purchase property on Merritt Island, just north of the Air Force's existing Atlantic Missile Range (AMR) at Cape Canaveral. Air Force launch pads were to be used for the Mercury and Gemini missions. Working out the relationship between NASA and the Air Force was not straightforward, because issues of relative financial responsibilities and of control over various phases of a launch were involved. What follows is only the first part of Kurt Debus's letter, and the appendices to the agreement are not included here.

Document I-40

[1]

JAN 2 1963

LO - DIR

Captain John K. Holcomb
Office of Manned Space Flight
Code MLO
National Aeronautics and Space Administration
Washington 25, D.C.

Dear Captain Holcomb:

Reference draft DOD/NASA Agreement dated 20 December 1962 regarding management of Merritt Island and AMR.

I have reviewed reference draft agreement and submit the following comments and recommendations:

    a.   GENERAL COMMENTS

(1) The agreement represents a significant improvement over the Webb-Gilpatrick Agreement of 24 August 1961 in the management, logistics, and administrative areas, but is greatly inferior in the technical and mission support areas. If approved, the draft agreement would clearly relinquish NASA management control of vital mission support functions. It would also prohibit NASA/LOC [Launch Operations Center] from continuing development activities which have significantly contributed to NASA and DOD programs during the past ten years and which could be even more important to NASA programs in the future.

(2) The agreement does not provide for sufficient latitude and independent actions on the part of either the Director, LOC, or the Commander, AMR, and will retard progress by requiring joint planning and actions where this is not necessary. This is not to say that the Director, LOC, and Commander, AMR, should not continue as in the past to make best use of the resources made available to either organization.

[2] (3) The restrictive nature of the portions agreed on by the NASA/DOD negotiating team, and the unresolved problem areas, appears to stem from a basic fear on the part of the DOD negotiators that NASA/LOC wants to "take over the Range." The agreement is in many areas more restrictive on NASA activities than our present practices and agreements with the Commander, AMR. If this be the case, the fear is completely unfounded. However, I strongly believe that NASA cannot delegate their responsibility for the fulfillment and execution of assigned programs, including assuring that the necessary supporting functions meet the program milestones and requirements in an economical and timely fashion. This is not contrary to, but rather consistent with, the concept of NASA retaining responsibility for and control of vital support, but making full use of the capabilities and experience of DOD in executing these functions.

(4) Many of the above objections could be removed by NASA retaining the funding control of all functions which are vital to NASA programs, particularly the [Manned Lunar Landing Program]. To accomplish this, I strongly recommend that the agreement be changed, as indicated below in the specific comments, so that NASA will seek appropriations for and control the appropriated funds, including reprogramming, for the development and operation of Merritt Island. Items which are for the sole use or support of DOD programs should be excepted. . . .

<hr>

Document I-41

[1]

Agreement between the Department of Defense and the National Aeronautics and Space Administration Regarding Management of the Atlantic Missile Range of DoD and the Merritt Island Launch Area of NASA

I. Purpose and Scope

A. It is the purpose of this Agreement to set forth the general concept of operations by DoD and NASA and to fix responsibility for specific functions carried out at the installations listed below.

B. This Agreement applies to the following:

1. Atlantic Missile Range (Administered by Air Force Missile Test Center—AFMTC)

The installations listed below are hereinafter referred to collectively as the Atlantic Missile Range.

a. Cape Canaveral. The tract now owned or leased by the Department of Defense. including the DoD-owned and leased facilities at Port Canaveral.

b. Patrick Air Force Base.

c. Sites other than Cape Canaveral within the Continental United States for instrumentation and equipment in support of the AFMTC mission (See IV-A-1)

d. DoD downrange instrumentation stations such as those which are presently located in the Bahamas, Puerto Rico, the West Indies, the South Atlantic Ocean and on the African Continent; the DoD air-borne and ship-borne instrumentation stations deployed in the Atlantic and the Indian Oceans; and the logistic bases in these tracts in support of the instrumentation stations.

2. Merritt Island Launch Area MILA (Administered by the NASA Launch Operations Center—LOC)

      a. The tract north and west of Cape Canaveral now being purchased by NASA, hereinafter referred to as MILA, excluding the TITAN III site, which is considered a part of AMR.

[2] II. Effect on Existing Agreements or Arrangements

    A. This Agreement supersedes the "Agreement between DoD and NASA Relating to the Launch Site for the Manned Lunar Landing Program" executed by the Deputy Secretary of Defense and the Administrator of NASA on August 24, 1961.
    B. Should the provisions of this Agreement be inconsistent with DoD and NASA regulations, or with the terms of previously executed agreements between DoD and NASA, including agreements covering the MERCURY and GEMINI programs, the provisions of this Agreement will govern.

III. General Concept

    A. The DoD will continue to be the single manager responsible for the development, operation, and management of range facilities of the Atlantic Missile Range as a national asset, providing common range services to all missile and space vehicle launch programs of the DoD and NASA. The DoD will similarly be responsible for range operation functions at MILA unless, for compelling technical or operational reasons, it is decided jointly that these should not be integrated under single manage-ment. . . .
    B. In recognition of the acquisition by NASA of MILA and its anticipated use, predominantly in support of the Manned Lunar Landing Program, and in order to provide more direct control by NASA of the MILA development and operation, the Merritt Island Launch Area is considered a NASA installation, separate and distinct from the Atlantic Missile Range. NASA will be fully responsible for master planning and the development of MILA and will be the host agency at MILA for the providing of facilities and services to DoD, as DoD is host at Cape Canaveral and elsewhere on the AMR.
    C. In order to ensure a maximum of mutual assistance, and a minimum of duplication, both DoD and NASA will inform each other of their plans and requirements and will consult fully regarding their activities. Consultation and decision-making under this agreement will normally be carried out at the local level. However, in the event that either the Director, LOC, or the Commander, AFMTC, feels in a particular situation that there is an important area of disagreement which is vital to the accomplishment of the missions assigned to this organization, the responsible local authority will refer the matter to a higher level for joint resolution.

IV. Responsibilities

    A. General
[3]   1. The Air Force Missile Test Center (AFMTC) is the DoD executive agent and single manager of the AMR and will establish policies and procedures for the operation of that installation. Its mission is to develop, operate, and manage range facilities and to provide range services to all range users. It does not have responsibility for preparation and launching of missiles or space vehicles.
        2. The Launch Operations Center (LOC) is the NASA executive agent and single manager of the MILA and will establish the policies and procedures for that installation. In addition, the LOC has certain responsibilities within NASA for preparation and launching of space vehicles at Cape Canaveral and MILA. The LOC is the focal point for all NASA relations with AFMTC, including the MERCURY and GEMINI programs.

3.   Within the terms of this Agreement, the agency designated as responsible for a given function will either perform that function or have full power to determine how and by whom that function will be performed. AFMTC and LOC will work out arrangements for the actual performance of functions in accordance with these responsibilities. These arrangements should contain clear guidelines regarding the extent of delegation intended in order that the parties can resolve at the outset the manner in which one agency is willing to undertake to perform a particular function that is the responsibility of the other.

B.   Master Planning. The DoD and NASA will be responsible for master planning of their respective installations. This will include: compiling the total requirements for facilities and equipment to be located at the installations in question; designation of areas for future use (zoning); selection of specific location for facilities (siting); and planning for area development in implementation of the above functions. Each agency will be responsible for the timely identification and resolution of problems relating to the compatibility of one master plan to the other. It is intended that, to the maximum extent possible, final master planning authority will be delegated to AFMTC for AMR and LOC for MILA in connection with facilities funded by the other agency.

C.   Development and Operations

1.   Responsibility for these functions at the AMR and MILA will be divided as indicated below and as set forth in the appendices attached to this Agreement. To the extent functions not listed in the appendices require the assignment of responsibility between AMR and MILA, such [4] assignment, consistent with the terms of this Agreement, may be made by local agreement between AFMTC and LOC. It is intended that such local agreements will lead to the management and utilization of resources so as to minimize costs and maximize efficiency.

Category 1.

Within its own installations, each agency will be responsible for those logistic and administrative functions which have no necessary interdependence or intercoupling with the similar function performed at the installation of the other. . . .

Category 2.

Regardless of location, each agency will be responsible for mission specific functions directly associated with the handling, preparation, launching, and in-flight control of its missiles or space vehicles, and with ground-support equipment for its missiles or vehicles. This does not preclude the establishment of special arrangements (e.g., the current arrangements for assembly by [the U.S. Air Force] of ATLAS boosters for NASA payloads etc., which are unaffected by this over-all agreement) in those cases where the payload of one agency is launched by the booster of the other. . . .

Category 3.

Range operation functions which are of such a nature that division of responsibility between agencies is impractical or undesirable will be the responsibility of the DoD. . . .

Category 4.

Other range operation functions are of such a nature that any division in the responsibility for their performance must be in accordance with clearly specified ground rules for the particular function, in order to avoid operational or management difficulties. . . .

It is recognized that the matter of compatibility between instrumentation at the AMR and that necessary for the NASA worldwide tracking network [5] in the areas of telemetry and electronic tracking is a matter of special concern. This arises from the fact that planning for and operation of the NASA network is the responsibility of NASA whereas DoD has responsibility for planning for and operation of the AMR. Compatibility will be achieved by joint consultation between the two agencies beginning with early planning stages and taking into account both economical and technical aspects of the problem. Issues which cannot be resolved between LOC and AFMTC will be referred to a higher level for joint resolution.

D.   Acquisition of Resources

The agency having responsibility for a particular function will be responsible for managing the acquisition or modification of facilities and equipment to perform the function. This includes the construction, development/procurement, installation, checkout, calibration, spare parts provisioning, and other services required to place the facility or equipment into operation and to maintain it. The master planning agency will, in each case, review and concur in criteria and specifications as being compatible with the master plan, with minimum construction standards, and with connecting utilities. Where AFMTC acquires or modifies facilities and equipment which are critical requirements in achieving NASA program milestones, review and comment of LOC will be obtained on specifications, criteria and implementing schedules prior to initiation of procurement. In such cases, AFMTC will keep LOC informed of progress in meeting requirements with particular reference to any problems which might result in schedule delays.

E.   Funding

1.   Each agency will budget and fund, for the acquisition of facilities and equipment necessary to perform the functions for which it is responsible. (However, for FY 1963 and 1964, current budget and funding arrangements will remain in effect.) It is contemplated that certain equipment and facilities . . . may be required for NASA's sole use or for earlier acquisition than needed to accomplish the general purpose functions of the AFMTC.

When such circumstances arise, NASA will fund for the acquisition of such equipment and facilities. It is intended that LOC and AFMTC will consult in advance in all such matters. The design, acquisition, and operation of such equipment and facilities will be the responsibility of AFMTC. Accountability for it will be transferred to AFMTC in accordance with paragraph F below.

2.   DoD and NASA will undertake jointly to study the matter of budgeting for and funding of the general administrative, management, [6] maintenance and operations cost[s] of AMR in order to determine whether NASA should provide to DoD a prorata share of such costs based on the relationship of NASA program workload to total workload. There will be no change in funding arrangements for FY 1963. After FY 1963, each agency will budget and fund for the administrative, management, maintenance and operations costs of the functions for which they are responsible, until otherwise decided as a result of the study and reflected in an amendment to this agreement. Until such times, responsibilities as delineated in this agreement will govern.

3.   When requirements for additional range resources are generated subsequent to the normal programming and budget preparation cycles (established for purposes of LOC and AFMTC planning, as one year before the beginning of the fiscal year in which work must start to meet the requirement), the following guidance will be applicable:

a.   For new, amended, incomplete, redirected or additional expression of range requirements by LOC, NASA will be responsible to arrange for the necessary resources, including fund[s], to be made available to the AFMTC in accordance with established procedures.

b.   When shortages of range resources occur within the area of responsibility of the AFMTC unless otherwise agreed between DoD and NASA, AFMTC will be responsible to arrange for resources, including funds, from within those available to DoD in order to support the requirements.

F.   Accountability

Regardless of funding responsibility, accountability for real property and equipment heretofore or hereafter acquired will rest with the agency having responsibility for the performance of the function to which the particular facilities or equipment are related. The right to modify and assign use of real property and equipment will rest with the agency holding accountability.

G. Community Relations

Community relations matters, which include the activities of DoD and NASA which have a significant impact upon such community interests as schools, housing, highways, public transportation, public utilities, community development, civic affairs, local manpower problems, local government, and related subjects, will be handled by each agency [7] for its own installation. Before dealing with the outside community or with other government agencies, with regard to such matters, the problems and requirements of both DoD and NASA will be considered jointly, using such coordinating boards or other procedures as AFMTC and LOC consider expedient.

H. Public Information

Each agency will be responsible for public information matters related to its own activities at either the AMR or Merritt Island. Coordination prior to release by either agency of information bearing upon the activities of the other will be accomplished between AFMTC and LOC, with full recognition being given in such releases to any contribution of the other agency to the particular program or event.

I. Visitor Control

Subject to applicable security regulations, DoD and NASA will be responsible for the visitor control policies and practices with regard to their own programs, both at the AMR and Merritt Island. Prior to visits by U.S. dignitaries and high foreign officials, the DoD and NASA, jointly and in conjunction with the Department of State (for foreign visitors) [,] will determine the purpose of the visit, identify the host agency at the AMR or MILA, as appropriate, and will develop sufficient details regarding the visit so as to avert misunderstanding or confusion at the time of the visit.

J. Labor Relations

DoD and NASA will each be responsible for labor relations matters relating to their respective programs. AFMTC and LOC will keep each other informed concerning the labor relations policies of each agency, and will coordinate their activities in the labor relations area.

K. Security

Each agency will be responsible for over-all security administration at its installations, except for security clearance matters involving the personnel of the other agency. In addition to establishing and enforcing restrictions and safeguards pertaining to its own operations, each agency will enforce such additional security regulations and orders established by the other agency as are necessary to safeguard the operations of the other agency.

[8] V. Implementation of this Agreement

A. The terms of this Agreement will be implemented as rapidly as is deemed practicable by mutual agreement of the Commander, AFMTC, and the Director, LOC; in no case will their implementation be delayed beyond June 30, 1963.

B. AFMTC and LOC are authorized to enter into such local level agreements as are necessary to effectuate the provisions of this Agreement. Issues which cannot be resolved at the local level will be forwarded promptly for resolution at higher level.

Robert S. McNamara
Secretary of Defense
1/17/63

James E. Webb
Administrator, NASA
1/17/63

Document I-42

Document title: "Minutes of the Management Council," Office of Manned Space Flight, May 29, 1962.

Source: NASA Historical Reference Collection, NASA History Office, NASA Headquarters, Washington, D.C.

One of the major issues facing the managers of the Apollo program was what kind of launch operations complex to construct. One option would have been to transport each stage of the Saturn V booster and the Apollo spacecraft separately to the launch pad and assemble and test them there. This was the approach that had been employed for all rocket launches to date. An alternative was to create a massive new enclosed facility where the "stack" could be assembled and tested before being taken to the launch pad. One of the advantages of this approach is that, in principle, there could be six launch campaigns going on at the same time—one on each of two launch pads and one in each of four bays within the assembly building. This also would avoid tying up a launch pad for months at a time during vehicle assembly.

At the May 1962 Management Council meeting at which the decision on which approach to take was made, Wernher von Braun argued that the high capital costs of the assembly building approach were justified only if the United States intended to maintain a high launch rate of Saturn boosters for a number of years. In the optimism of the early Apollo years, the decision was made to follow the assembly building approach. What follows are the beginning of those meeting minutes and Review Item number 9, which focused on the launch facilities.

[original marked "CONFIDENTIAL," crossed out by hand]

Minutes of the Management Council

OFFICE OF MANNED SPACE FLIGHT

May 29, 1962

The sixth meeting of the Management Council convened at 0900 on Tuesday, May 29, 1962, at Marshall Space Flight Center, Huntsville, Alabama.

All Members of the Council were present.

The next meeting will be held in the Office of Manned Space Flight, Washington, D.C., on Friday, June 22, 1962.

Review Items . . .

9. Launch facilities for Saturn C-5; to discuss impact of spacecraft servicing requirements, launch rates, etc., on the technical aspects of Complex 39; to outline the factors which weigh heavily of the requirements for Complex 39.

Dr. Debus presented the current picture on the need for Complex 39 as a vertical assembly, checkout, transport, and launch facility. He said that, under current project firing rates, we are at about the "break even" point when choosing between the mobile and fixed concepts from the standpoint of economics.

Dr. von Braun pointed out that the fundamental question is whether we believe "a space program is here to stay, and will continue to grow," in which case he believes a vertical assembly facility is vital.

Mr. Gilruth and Mr. Williams questioned the effect that a favorable decision on the mobile concept for Complex 39 would have on the accessibility for servicing of space vehicles on the pad.

Mr. Rosen said that he didn't disagree with any of the advantages claimed for the mobile concept, but suggested that there has been insufficient consideration of the disadvantages, and recommended that these should be studied further.

IT WAS DECIDED THAT:

a.   THE MOBILE LAUNCHER CONCEPT IS APPROVED.

b.   CLOSE COORDINATION BETWEEN ALL DESIGN ACTIVITIES AND [THE LAUNCH OPERATIONS CENTER] MUST TAKE PLACE TO ASSURE COMPATIBILITY OF THE FLIGHT AND GROUND EQUIPMENT WHEN USING THE MOBILE CONCEPT. FLIGHT VEHICLE EQUIPMENT WILL BE GIVEN PRIORITY IN ANY DESIGN COMPROMISES REQUIRED BETWEEN FLIGHT EQUIPMENT AND GROUND EQUIPMENT. . . .

Document I-43

Document title: James E. Webb, Administrator, NASA, Memorandum to Associate Administrator for Manned Space Flight, NASA, "Termination of the Contract for Procurement of Long Lead Time Items for Vehicles 516 and 517," August 1, 1968.

Document I-44

Document title: W.R. Lucas, Deputy Director, Technical, Marshall Space Flight Center, NASA, Memorandum to Philip E. Culbertson, NASA Headquarters, "Long Term Storage and Launch of a Saturn V Vehicle in the Mid-1980's," May 24, 1972.

Document I-45

Document title: George M. Low, Deputy Administrator, NASA, Memorandum to Associate Administrator for Manned Space Flight, NASA, "Leftover Saturn Hardware," June 2, 1972.

Document I-46

Document title: Dale D. Myers, Associate Administrator for Manned Space Flight, NASA, Memorandum to Administrator, "Saturn V Production Capability," August 3, 1972

Source: All in NASA Historical Reference Collection, NASA History Office, NASA Headquarters, Washington, D.C.

The Saturn V was a remarkable engineering achievement, but it was extremely expensive to operate and was useful primarily for very large space missions. Even before the first mission to the Moon, NASA Administrator James E. Webb sensed that the political support for a continued large-scale space effort was unlikely to be sustained, whoever won the 1968 presidential election. He proved prescient, and in 1972, when it became clear that the Nixon administration would not grant NASA the budget needed both to develop the Space Shuttle and to continue to use the Saturn V, the NASA leadership reluctantly gave up the capability to produce the vehicle. Note that the enclosures to the Lucas memorandum in Document I-43 do not appear here.

Document I-43

AUG 1 1968

MEMORANDUM to M/Associate Administrator for Manned Space Flight

SUBJECT: Termination of the Contract for Procurement of Long Lead Time Items for Vehicles 516 and 517

REFERENCE: N memorandum to the Administrator, dated June 2, 1968, same subject
              D memorandum to the Administrator, dated July 31, 1968
              AD memorandum to M, dated July 13, 1967

After reviewing the referenced documentation and in consideration of the FY 1969 budget situation, your request to expend additional funds for the procurement of long lead time items for the S-IC stages of the 516 and 517 vehicles is disapproved. This decision, in effect, limits at this time the production effort on Saturn through vehicle 515. No further work should be authorized for the development and fabrication of vehicles 516 and 517.

> James E. Webb
> Administrator

HBF:kh 7/30/68 ext. 24463

Document I-44

[1]

NATIONAL AERONAUTICS AND SPACE ADMINISTRATION
GEORGE C. MARSHALL SPACE FLIGHT CENTER
MARSHALL SPACE FLIGHT CENTER, ALABAMA 35812

REPLY TO
ATTN OF: DEP-T   May 24, 1972

TO: NASA Headquarters
            Attn: Mr. Philip E. Culbertson

FROM: DEP-T/W. R. Lucas

SUBJECT: Long Term Storage and Launch of a Saturn V Vehicle in the Mid-1980's

This is in response to your request of May 9 for information concerning the cost of maintaining present reliability of the Saturn V vehicle as a function of long time storage and the cost of storing and maintaining a capability to launch a Saturn V in the mid-1980's.

First, it is extremely difficult to estimate the cost of maintaining the current reliability of the Saturn V launch vehicle for approximately 13 additional years since this time is so far beyond our experience. For example, a June 1985 launch of one of our available Saturn V's would mean that the age of some of the critical components from start of stage assembly would exceed 18 years. We are prepared to state, based upon tests and other experience, that there is no significant degradation of some of the more sensitive components, for example engine soft goods, up to ten years, provided the storage environment is closely controlled. However, we do not understand quantitatively the effects of aging on our systems beyond the ten year period.

To gain confidence in components between ten and twenty years old, we would have to establish the requirement to do single engine static firings three years prior to launch, utilizing spare J-2 engines. The J-2 engine would be selected because it contains most of the commonly used softgoods (most likely to deteriorate) on the vehicle. In addition, selected electrical, electronic and mechanical critical components that were stored with the vehicle would be subjected to functional tests and teardown inspection. There would be no remove-and-replace activity on the vehicles unless determined necessary by this test program.

[2] In addition to the vehicle hardware reliability concern, there is another vital element to consider. The present Apollo and Skylab Programs depend on the full-time, dedicated involvement of carefully selected, highly skilled individuals within both contractor and Civil Service ranks. Many of the key individuals can trace their experience back to the beginning of the Apollo Program. Every Saturn V launch to date, particularly the Apollo 16, has required their real time decisions to convert a potential launch scrub or mission loss situation into a mission success. By the 1980's, this present capability will be practically nonexistent. It must be rebuilt with individuals possessing possibly more advanced technical knowledge of new vehicles but who would lack specific knowledge of the Saturn V systems. Therefore, these individuals must be provided the means and the time to become technically proficient with the Saturn V system. All records pertaining to design, qualification, manufacturing and assembly processes, handling, checkout, and launch preparation and launch must be preserved.

In addition to the above, there are other potential problem areas which deserve a brief comment:

- Advanced computer processing systems may not be compatible with the developed Saturn software programs.
- It is not feasible to environmentally control all critical components of the system. For example, the [Launch Umbilical Tower] and the stage transporters will be exposed to atmospheric conditions requiring possible major refurbishment.
- Certain critical spare parts would be impossible to replace if an unforeseen problem required an unusual demand for replacement parts.
- There may be an impact to the Shuttle flight program at [Kennedy Space Center] and related activities at [Marshall Space Flight Center] in order to concentrate the manpower on the Saturn launch preparation activity.

In summary, we have very little basis for extrapolating reliability of Saturn vehicles beyond the proposed six to fifteen year period of inactivity. Undoubtedly some degradation would occur. If it is intended to use a Saturn V in the mid-1980's, the earlier the requirement is identified, the better will be our confidence in maintaining a reasonable reliability at a tolerable cost.

[3] In conjunction with [Kennedy Space Center], we examined the major factors influencing the cost of a program to maintain the capability to launch one of the two unassigned Saturn V launch vehicles (SA-514 and SA-515) with confidence in the mid-1980's. Comments and cost estimates from [Kennedy] are included. The examination was conducted in accordance with the guidelines and assumptions presented in enclosure 1. The approach taken would require the present contractors to prepare the stages, spares and documentation for long term storage before their present contracts expire; store the stages and spares at [Kennedy]; maintain the documentation at [Marshall Space Flight Center]; and then identify the required post storage activities to be performed.

A summary of the cost and manpower phasing is presented in enclosures 2 and 3. You will note that the Post Storage and Launch Phase contains two options: the first option utilizes only Civil Service and support contractors; the other option utilizes a single prime contractor for this phase. This choice is left open because it is not possible to predict at this time the density of workload within the Civil Service ranks during the mid-1980's. For example, during the mid-1980's the Shuttle will be flying from [Kennedy]. The priority of this activity in relation to a Saturn V mission will determine the availability of Civil Service personnel. Depending on which option is chosen, the estimated total program cost for a Saturn V launch in mid-1985 will range from 206.0 M to 298.7 M.

W. R. Lucas
Deputy Director, Technical

3 Enclosures

cc: See page 4

Document I-45

|1|

June 2, 1972

MEMORANDUM

TO: M/Associate Administrator for Manned Space Flight

FROM: AD/Deputy Administrator

SUBJECT: Leftover Saturn Hardware

The purpose of this memorandum is to document the meeting you and I held on the way back from Houston after Apollo 16. I realize that you have already issued instructions to meet some of the decisions of that meeting, but for completeness I will document all of the decisions in this memorandum. They were as follows:

1.   It will be determined whether or not there exists a possibility of a new NASA mission in the middle 1970's that might make use of the remaining Saturn V's. You will solicit ideas from the [Office of Manned Space Flight] organization to see whether such a mission might be worthwhile, and I will work with remaining elements of the organization.

2.   You will formally ask the Department of Defense whether they foresee a need for either the leftover Saturn V hardware or, for that matter, for any future build of Saturn V's for DOD purposes.

3.   You will conduct a study to determine whether it is profitable to maintain the tooling or even the existing Saturn V hardware for possible missions in the 1980's, assuming that there will be no missions in the 1970's.

|2| 4.   You will identify the costs for storing the existing hardware as well as the costs for maintaining the tooling, etc.

5.   Assuming that no 1970 missions are identified and that it is not worthwhile to maintain the capability for the 1980's, you will prepare the document that we will staff through the OMB [Office of Management and Budget] and others in the Executive

Branch leading to a decision by NASA to terminate the Saturn V capability. I assume that this will be completed some time in the summer or early fall of 1972.

George M. Low

cc: A/Dr. Fletcher
     ADA/Mr. Shapley
     B/Mr. Lilly

bcc: AXC/Beran
      AX/Clements
      AX/Hoban

AD/GMLow:smm 6/2/72

Document I-46

[1]

REPLY TO
ATTN OF: MBB-1

MEMORANDUM

TO: Administrator

THRU: B/Assistant Administrator for Administration
             D/Assistant Administrator for Organization & Management

FROM: M/Associate Administrator for Manned Space Flight

SUBJECT: Saturn V Production Capability

    The purpose of this memorandum is to obtain your approval to cancel the two-per-year Saturn V production capability requirement.
    As you know, when the decision was made to retain Saturn V industrial assets, we took action to store, maintain and preserve tooling, equipment and facilities capable of producing up to two Saturn V Vehicles per year at the following primary locations:

> Manufacturing Sites:
> Michoud Assembly Facility, Louisiana
> Seal Beach Assembly Facility, California
> McDonnell Douglas, Huntington Beach, California
> International Business Machines, Huntsville, Alabama
> Rocketdyne, Canoga Park, California
>
> Test Sites:
> Mississippi Test Facility, Mississippi (S-IC only)
> McDonnell Douglas, Sacramento Test Site, California
> Rocketdyne, Santa Susana Test Site, California
> Rocketdyne, Edwards Air Force Base, California

The approximate acquisition value of the government-owned Saturn V tooling, equipment and facilities presently retained at these locations is $585M. The approximate annual cost of maintaining these assets after we have discontinued flight support for ongoing programs will be $6M. Lower maintenance costs in FY 1973 and 1974 are made possible by continuing current "in place" storage and by making the most efficient use of existing Saturn contractor man-power.

[2] The possibility of future Saturn V missions, the potential utilization of Saturn V industrial assets by the Shuttle Program, and the relatively low cost of maintenance made it prudent to retain Saturn V industrial assets until their utility could be confirmed. I have re-examined this requirement in view of the exceedingly stringent expenditure limitation facing us in FY 1973 and the advent of the Shuttle Program, and I have determined that:

1.   Existing Saturn IB flight hardware is adequate to conduct anticipated space missions prior to Shuttle [Initial Operational Capability].

2.   Beyond 1978 there is significant potential interference between planned Shuttle activities at [Kennedy Space Center] and Saturn launched missions. For example, [Launch Complex] 39A and B will have been modified for Shuttle use.

3.   Approximately $100M of Saturn V assets will be directly applicable to the Shuttle Program.

4.   By taking action now and with actual Saturn asset dispositioning being deferred until FY 1974 or later, it is anticipated that up to $2.9M in cost savings will accrue in FY 1973.

After careful consideration of these factors, I believe that the retention of the two-per-year Saturn V production capability is no longer prudent. Accordingly, I request your approval to cancel this requirement.

Dale D. Myers

APPROVED: Original signed by George M. Low|
For James C. Fletcher
Administrator

Approved subject to notification of OMB, and subject to "no objection" by OMB.

GML

Chapter Two

Developing the Space Shuttle[1]

by Ray A. Williamson

Early Concepts of a Reusable Launch Vehicle

Spaceflight advocates have long dreamed of building reusable launchers because they offer relative operational simplicity and the potential of significantly reduced costs compared to expendable vehicles. However, they are also technologically much more difficult to achieve. German experimenters were the first to examine seriously what developing a reusable launch vehicle (RLV) might require. During the 1920s and 1930s, they argued the advantages and disadvantages of space transportation, but were far from having the technology to realize their dreams. Austrian engineer Eugen M. Sänger, for example, envisioned a rocket-powered bomber that would be launched from a rocket sled in Germany at a staging velocity of Mach 1.5. It would burn rocket fuel to propel it to Mach 10, then skip across the upper reaches of the atmosphere and drop a bomb on New York City. The high-flying vehicle would then continue to skip across the top of the atmosphere to land again near its takeoff point. This idea was never picked up by the German air force, but Sänger revived a civilian version of it after the war. In 1963, he proposed a two-stage vehicle in which a large aircraft booster would accelerate to supersonic speeds, carrying a relatively small RLV to high altitudes, where it would be launched into low-Earth orbit (LEO).[2] Although his idea was advocated by Eurospace, the industrial consortium formed to promote the development of space activities, it was not seriously pursued until the mid-1980s, when Dornier and other German companies began to explore the concept, only to drop it later as too expensive and technically risky.[3]

As Sänger's concepts clearly illustrated, technological developments from several different disciplines must converge to make an RLV feasible. Successful launch and return depends on all systems functioning in concert during the entire mission cycle as they pass through different environmental regimes. In the launch phase, the reusable vehicle and

---

1. In addition to the discussion of the Space Shuttle in this essay and the documents associated with it, there are several other places in the *Exploring the Unknown* series in which substantial attention is paid to issues related to the Space Shuttle, with related documents included. In particular, Chapter Three of Volume I discusses the presidential decision to develop the Space Shuttle; see John M. Logsdon, gen. ed., with Linda J. Lear, Jannelle Warren-Findley, Ray A. Williamson, and Dwayne A. Day, *Exploring the Unknown: Selected Documents in the History of the U.S. Civil Space Program, Volume I, Organizing for Exploration* (Washington, DC: NASA SP-4407, 1995), 1: 386–88, 546–59. Chapter Two of Volume II discusses NASA-Department of Defense relations with respect to the Shuttle; see John M. Logsdon, gen. ed., with Dwayne A. Day and Roger D. Launius, *Exploring the Unknown: Selected Documents in the History of the U.S. Civil Space Program, Volume II: External Relationships* (Washington, DC: NASA SP-4407, 1996), 2: 263–69, 364–410. Chapter Three of this volume discusses issues associated with the use of the Shuttle to launch commercial and foreign payloads. Future volumes will contain discussion and documents related to the use of the Shuttle as an orbital research facility.

2. Irene Sänger-Bredt, "The Silver Bird Story, a Memoir," in R. Cargill Hall, ed., *Essays on the History of Rocketry and Astronautics: Proceedings of the Third Through the Sixth History Symposia of the International Academy of Astronautics*, Vol. 1 (Washington, DC: NASA, 1977), pp. 195–228. (Reprinted as Vol. 7-1, American Astronautical Society History Series, 1986.)

3. Helmut Muller, "The High-Flying Legacy of Eugen Sänger," *Air & Space*, August/September 1987, pp. 92–99.

its booster, with any associated propellant tankage, must operate as a powerful rocket, lifting hundreds of thousands of pounds into LEO. While in space, the reusable vehicle functions as a maneuverable orbiting spacecraft in which aerodynamic considerations are moot. However, when reentering the atmosphere and slowing to subsonic speeds, aerodynamics and heat management quickly become extremely important, because the reusable vehicle must fly through the atmosphere, first at hypersonic speeds (greater than Mach 5), then at supersonic and, ultimately, at subsonic speeds. Finally, the vehicle must fly or glide to a safe landing. Because RLVs must be capable of flying again and again, and because they must reenter the atmosphere, they are subject to stresses on the materials and overall structure that expendable launchers do not have to withstand. Hence, building an RLV imposes extraordinarily high demands on materials and systems.

The conceptual origins of the world's first partially reusable vehicle for launch, NASA's Space Shuttle, reach back at least to the mid-1950s, when the Department of Defense (DOD) began to explore the feasibility of an RLV in space for a variety of military applications, including piloted reconnaissance, anti-satellite interception, and weapons delivery. The Air Force considered a wide variety of concepts, ranging from gliders launched by expendable rockets to a single-stage-to-orbit Aerospaceplane that bore a remarkable resemblance to the conceptual design for the National Aerospace Plane (NASP) of the late 1980s. The X-20 Dyna-Soar (Dynamic Soaring), the Air Force's late 1950s project to develop a reusable piloted glider, would also have had a small payload capacity.[4] NASA joined the Dyna-Soar project in November 1958.[5] The Air Force and NASA envisioned a delta-winged glider that would take one pilot to orbit, carry out a mission, and glide back to a runway landing. It would have been boosted into orbit atop a Titan II or III. As planned, the Dyna-Soar program included extensive wind tunnel tests and an ambitious set of airdrops from a B-52 aircraft. The Air Force chose six Dyna-Soar pilots, who began their training in June 1961. However, Dyna-Soar always competed for funding with other programs, including NASA's Project Gemini after 1961. Rising costs and other competing priorities led to the program's cancellation in December 1963.

Nevertheless, the testing that began during the Dyna-Soar program continued in other Air Force projects, such as the Aerothermodynamic-Elastic Structural Systems Environment Tests (ASSET) and Precision Recovery Including Maneuvering Entry (PRIME) projects. ASSET began in 1960 and was designed to test heat resistant metals and high-speed reentry and glide. PRIME was a follow-on project that began in 1966 and tested unpiloted lifting bodies (so called because they have a high ratio of lift over drag) that were boosted into space atop Atlas launchers. The Air Force also tested several models of piloted lifting bodies that were generally carried to high altitudes and released to a gliding landing. Among other things, these programs demonstrated that sufficient control could be achieved with a lifting body to land safely without a powered approach. This result later proved of great importance in the design of the Space Shuttle orbiter.

In 1957, the Air Force commissioned a conceptual study that examined recoverable space boosters.[6] From this came the concept called the Recoverable Orbital Launch System, which Air Force designers hoped would be capable of taking off horizontally and reaching orbits as high as 300 miles with a small payload. In a design that preceded the NASP concept, it would have had a hydrogen-fueled propulsion system that took its source of oxygen directly from the air by compressing and liquefying it in a "scramjet" engine,

---

4. Clarence J. Geiger, "History of the X-20A Dyna-Soar," Air Force Systems Command Historical Publications Series 63-50-I, October 1963. (Report originally classified, but declassified in 1975.)

5. See Chapter Two in Logsdon, gen. ed., *Exploring the Unknown*, 2: 249–62, for a complementary account of the Dyna-Soar program.

6. See Air Force Study Requirement SR-89774 (1957), Air Force Historical Research Agency, Maxwell Air Force Base, AL.

capable of operating at hypersonic speeds.[7] Designers quickly saw that the challenge of designing a propulsion system, or systems, capable of operating through three speed regimes—subsonic, supersonic, and hypersonic—placed extreme demands on available engine and materials technology. It was clearly not possible to build a single-stage-to-orbit vehicle with the technologies of the day.[8]

In 1962, in an effort to save the reusable concept, Air Force designers turned to a two-stage design for a concept they began to call the Aerospaceplane. Seven aerospace companies received contracts for the initial design.[9] Through these and several follow-on contracts, the companies not only produced paper studies, but undertook research on ramjet and scramjet propulsion, explored new structures and materials, and made significant advances in understanding hypersonic aerodynamics. However, reality never lived up to the designers' aspirations. By October 1963, after watching the Aerospaceplane program for some time with concern, DOD's Scientific Advisory Board reached the conclusion that the program was leading the Air Force to neglect conventional problems in launch research.[10] The Aerospaceplane program was quickly shut down.

NASA also sponsored a series of studies investigating reusable concepts for a variety of crews and payload sizes. By June 1964, NASA's Ad Hoc Committee on Hypersonic Lifting Vehicles with Propulsion issued a report urging the development of a two-stage reusable launcher.[11]

During the early 1960s, under government sponsorship, all of the major aerospace companies also developed their own version of a two-stage launch vehicle employing a lifting-body reentry vehicle. In each of these studies, the industrial concerns depended to a high degree on NASA and the Air Force to furnish the initial configuration on which to base their own version. The firms were concerned about straying too far from the concepts that their government "customers" were promoting.[12] This continued the practice evident in Project Mercury, in which the government agencies not only set the design goals and laid out the technical specifications but also instructed industry how to achieve them.

Origins of the Space Shuttle Program

No single action or decision similar to President Kennedy's May 25, 1961, "we should go to the moon" speech marks the beginning of the focused NASA program to develop the Space Shuttle. Rather, the program emerged over time in increments while NASA was simultaneously completing work on the Saturn V and launching the Apollo astronauts to the Moon and back. By the time President Nixon made the 1972 decision to proceed with Space Shuttle development, most major aspects of its design had been set.[13]

---

7. A scramjet (supersonic combustion ramjet) is an engine in which air compression, fuel mixing, and combustion all occur at supersonic speed.

8. Some even advocated refueling the Recoverable Orbital Launch System in hypersonic flight, using the X-15 to validate the concept. Fortunately, this extremely risky and dangerous concept was never tried. See Richard P. Hallion and James O. Young, "Space Shuttle: Fulfillment of a Dream," in Richard P. Hallion, ed., *The Hypersonic Revolution: Eight Case Studies in the History of Hypersonic Technology*, Volume II (Dayton, OH: Wright-Patterson Air Force Base, Special Staff Office, Aeronautical Systems Division, 1987), p. 948.

9. Boeing, Douglas, General Dynamics, Goodyear, Lockheed, North American Aviation, and Republic received contracts for system design studies. General Dynamics, Douglas, and North American received funding for detailed development plans. Martin built a full-scale model that explored the concept of incorporating the wings with the fuselage.

10. Hallion and Young, "Space Shuttle: Fulfillment of a Dream," p. 951.

11. Report of the NASA Special Ad Hoc Panel on Hypersonic Lifting Vehicles with Propulsion, June 1964. See also the memorandum from Floyd L. Thompson to James Webb, June 18, 1964. Copies in the NASA Historical Reference Collection, NASA History Office, NASA Headquarters, Washington, DC.

12. "In each case, whether dealing with Air Force-inspired configurations or NASA-inspired ones, contractors generally danced to an Air Force or NASA tune as regards the overall configuration itself." Hallion and Young, "Space Shuttle: Fulfillment of a Dream," p. 957.

13. See Logsdon, gen. ed., *Exploring the Unknown*, 1: 386–88.

As early as August 24, 1965, more than two years before the first Saturn V rose from the launch pad, the Air Force and NASA established an Ad Hoc Subpanel on Reusable Launch Vehicle Technology under the joint DOD-NASA Aeronautics and Astronautics Coordinating Board. Its objective was to determine the status of the technology base needed to support the development of an RLV. The report, which was issued in September 1966, concluded that many cost and technical uncertainties needed to be resolved, but it projected a bright future for human activities in Earth orbit. [II-1, II-2] Because the panel could find no single launch concept that would satisfy both NASA and DOD, it included ideas for a variety of fully reusable and partially reusable vehicles. Interestingly, the panel projected that partially reusable vehicles would be much cheaper to develop than fully reusable ones. Even so, engineers within both NASA and the Air Force continued to focus on fully reusable launch systems for several years, in the belief that once the difficult design and development problems were solved, such systems would prove much less costly to operate.[14] Some designers favored fully reusable designs that would employ a reusable booster and a cryogenic-powered orbiting vehicle. Others felt that the surest path to success was a small lifting body mounted on top of an expendable launch vehicle, such as a Titan III. Other design concepts lay between these two extremes.

As NASA began to think in depth about its post-Apollo human spaceflight programs after 1966, its top-priority objective became gaining approval for an orbital space laboratory—a space station. NASA planners also began to recognize that there was a need to reduce the costs of transporting crews and supplies to such an orbital outpost if it was to be affordable to operate. This, in turn, led to a focus on an Earth-to-orbit transportation system—a space shuttle. The idea that such a vehicle was an essential element in whatever might follow Apollo was first publicly discussed in an August 1968 talk by NASA Associate Administrator for Manned Space Flight George Mueller. [II-3]

In December 1968, as planning for the post-Apollo space program gained momentum, NASA convened the Space Shuttle Task Group to determine the agency's needs for space transportation. [II-4] This task group set out the basic missions and characteristics of the kind of vehicle that NASA hoped to gain approval to develop. Through the Manned Spacecraft Center and Marshall Space Flight Center, the Space Shuttle Task Group in mid-1969 issued a request for proposals (RFP) for what it termed an Integral Launch and Reentry Vehicle (ILRV) system. The RFP specified an emphasis on "economy and safety rather than optimized payload performance."[15] The eight-month studies that resulted formed the beginning of the Space Shuttle Phase A study effort.[16] Four aerospace contractors won ILRV study contracts—General Dynamics, Lockheed, McDonnell Douglas, and North American Rockwell.

The Space Shuttle Task Group final report, issued in July 1969, concluded that an ILRV should be capable of:

- Space station logistical support
- Orbital launch and retrieval of satellites

---

14. In the 1980s and 1990s, the goal of achieving vastly cheaper operational costs continued to elude designers. For a discussion of the technical issues, see U.S. Congress, Office of Technology Assessment, *Reducing Launch Operations Costs: New Technologies and Practices,* OTA-TM-ISC-28 (Washington, DC: U.S. Government Printing Office, September 1988).

15. NASA Manned Spacecraft Center and Marshall Space Flight Center, "Study of Integral Launch and Reentry System," RFP MSC BG721-28-9-96C and RFP MSFC 1-7-21-00020, October 30, 1968. Copy in Johnson Space Center historical archives. Quoted in Hallion and Young, "Space Shuttle: Fulfillment of a Dream," p. 995.

16. NASA had created a four-phase project development scheme, which finally became codified in August 1968. Phase A consisted of advanced studies (or later, preliminary analysis); Phase B, project definition; Phase C, design; and Phase D, development and operations. See Hallion and Young, "Space Shuttle: Fulfillment of a Dream," pp. 995-96. See also Arnold S. Levine, *Managing NASA in the Apollo Era* (Washington, DC: NASA SP-4102, 1982), pp. 158-61.

- Launch and delivery of propulsive stages and payloads
- Orbital delivery of propellant
- Satellite servicing and maintenance
- Short-duration manned orbital missions

The report considered three classes of vehicles. Class I referred to reusable orbiting vehicles launched on expendable boosters. Class II applied to vehicles using a stage and a half. Class III meant two-stage vehicles in which both the booster and the orbiter were fully reusable.

On February 13, 1969, President Richard M. Nixon requested that a high-level study be conducted to recommend a future course of activities for the overall civilian space program.[17] The Space Task Group (STG), chaired by Vice President Spiro T. Agnew, delivered its report on September 15, 1969.[18] The STG also recommended an RLV that would:

- Provide a major improvement over the present way of doing business in terms of cost and operational capability
- Carry passengers, supplies, rocket fuel, other spacecraft, equipment, or additional rocket stages to and from LEO on a routine, aircraft-like basis
- Be directed toward supporting a broad spectrum of both DOD and NASA missions

As conceptualized in the STG report, a reusable space transportation system would have as the following components:

- A reusable chemically fueled **shuttle** operating between Earth's orbit and LEO in an airline-type mode (Figures 2–1 and 2–2)
- A chemically fueled **space tug** or vehicle for moving people and equipment to different Earth orbits and as a transfer vehicle between the lunar-orbit base and the lunar surface

Figure 2–1. This 1969 artist's rendering depicts what a fully reusable Space Shuttle would look like during takeoff. (NASA photo)

Figure 2–2. This artist's conception, also from 1969, shows a fully reusable Space Shuttle at the point of separation when the orbiter leaves the atmosphere. The larger vehicle that boosted the orbiter was then to be piloted back to Earth. (NASA photo)

17. See Logsdon, gen. ed., *Exploring the Unknown*, 1: 383–85.
18. See Document III-25 in *ibid.*, 1: 522–43.

• A reusable **nuclear stage** for transporting people, spacecraft, and supplies between Earth orbit and lunar orbit and between LEO and geosynchronous orbit and for other deep space activities[19]

Of these elements, only the Space Shuttle has been built to date.

As noted above, many aerospace engineers within both NASA and industry favored the Class III fully reusable shuttle-type vehicles because they seemed to offer the cheapest operations costs, especially at high launch rates. [II-5][20] Proponents admitted that such vehicles were much more demanding technically and also required greater development risk and costs, but they felt that if the technical issues could be overcome, such vehicles would provide the basis for an increased overall investment in space. North American Rockwell (later, Rockwell International), for example, proposed a series of Class III designs that used a large booster and orbital vehicle to carry the necessary volume of liquid oxygen/liquid hydrogen fuel. NASA's "chief designer," Maxime Faget at the Manned Spacecraft Center, advocated a two-stage concept that mounted a relatively small orbiter atop a much larger recoverable booster. [II-6] Both vehicles were powered, and both had straight wings. Faget's orbiter would carry only a small payload and had only small cross-range capability.[21] Although by January 1971 many at NASA had begun to view a partially reusable design employing an external propellant tank and a delta-wing orbiter as probably the best overall choice when weighing development costs and technical risks, NASA engineers nevertheless continued to consider the Faget concept until almost the end of 1971.[22]

The Air Force, which was also involved at senior levels in the work of the STG, was highly critical of the Faget design, arguing that reentry would put extremely high thermal and aerodynamic loads on the orbiter's straight wings. The Air Force Flight Dynamics Laboratory argued forcefully that a delta-wing design would provide a safer orbiter with much greater cross range.[23] Ultimately, the Air Force's wish for high cross range and large payload capacity, as well as reduced expectations for NASA's future budget, forced NASA to give up on the Faget concept and begin serious work on the partially reusable concept that became the final Space Shuttle design. By the time NASA had reached this decision, many other Shuttle concepts had been explored, were found wanting, and had faded from the scene. NASA awarded Phase B design study contracts for the Shuttle to McDonnell Douglas and North American Rockwell in June 1970; these studies used the Faget two-stage fully reusable concept as their baseline. NASA also awarded Lockheed and a Grumman/Boeing team additional contracts to conduct Phase A studies for systems using some expendable components, should the two-stage concepts examined in the Phase B studies prove too expensive or technically demanding. In the meantime, NASA pursued its own internal studies, in part, to improve the competence of its engineers and to give them better insight into the contractors' work.[24]

As noted earlier, logistics support for the space station was cited as one of the principal justifications for the Shuttle. However, by its September 1970 budget submission to the

---

19. Slightly paraphrased from *ibid.*, 1: 534.

20. As in the other chapters in this volume of *Exploring the Unknown*, the documents that follow this essay are not necessarily in chronological order, but rather follow in numerical sequence with the context of the essay.

21. An orbiter with high cross range is capable of altering its orbital plane significantly. The Air Force tended to favor high cross-range capability on the assumption that it might wish to fly only a single orbit and return to Earth at the same location from which it had been launched. However, during that one orbit, Earth will have rotated sufficiently to require the Shuttle to change latitude to reach the launch site, thus requiring the orbiter to have sufficient cross range. NASA had minimal need for high cross-range capability.

22. Hallion and Young, "Space Shuttle: Fulfillment of a Dream," p. 1031.

23. Eugene S. Love, "Advanced Technology and the Space Shuttle," 10th von Kármán lecture, 9th annual meeting, American Institute of Aeronautics and Astronautics, Washington, DC, January 1973 (AIAA Paper 73-31).

24. Interview of Milton Silveira by Joseph Guilmartin and John Mauer, November 14, 1984, p. 6, transcript in NASA Historical Reference Collection.

White House, NASA officials realized that the Nixon administration and Congress were unwilling to support simultaneous development of both a space station and a Space Shuttle. A complete restructuring of NASA's expectations was in order. Between September 1970 and May 1971, the focus of NASA's attention was gaining White House approval for developing a two-stage fully reusable Shuttle. By May 1971, the expectations for NASA's future budget were reduced sufficiently that having the resources needed to develop such a two-stage, fully reusable design was out of the question. NASA estimated it would need at least $10–12 billion to build a two-stage Shuttle, but with a fiscal year 1971 budget of only $3.2 billion and little hope of future funding increases, the agency was forced to examine concepts with several expendable components as a means of lowering development costs. [II-7]

An important technical issue also led to the abandonment of the fly-back reusable booster. As designs began to mature, it became clear that for this concept to be feasible, the orbiter staging velocity (that is, the velocity at which the booster would release the orbiter) had to be 12,000 to 14,000 feet per second. Achieving this velocity would require an extremely large booster incorporating enormous fuel tanks. Upon returning through the atmosphere at these velocities, the booster would have to sustain extremely high heat loads. NASA engineers became increasingly uncomfortable about their ability to build such a booster, given the technology then available and generally poor knowledge about atmospheric reentry of large structures.

The ultimate design of the Shuttle orbiter and other system components depended on decisions about five key orbiter characteristics:

- Payload bay load capacity and size
- Extent of cross-range maneuverability
- Propulsion system
- Glide or power-assisted landing
- Primary structural material[25]

The first two were of greatest concern to the Air Force. Because NASA needed Air Force support in the White House and congressional debates over the Shuttle, in January 1971, the space agency agreed to the following design criteria:

- Fifteen-foot by sixty-foot payload bay
- A total of 65,000 pounds of easterly payload lift capacity (40,000 pounds for polar orbits from Vandenberg Air Force Base)
- A cross range of 1,100 nautical miles[26]

With these decisions made, NASA was then able to focus on what combination of orbiter design, propellant tank, and booster best fit the required characteristics. Throughout 1971, Manned Spacecraft Center and Marshall Space Flight Center designers analyzed a remarkable twenty-nine different Shuttle designs, incorporating a wide variety of orbiter capacity, hydrogen and oxygen fuel tank, and boosters (see Table 2–1).[27]

While still evaluating two-stage Shuttle designs, NASA engineers had found that the existing F-1 and J-2 engines, both of which were by then out of production, were inadequate to meet the safety and weight requirements of the Shuttle without significant

25. Scott Pace, "Engineering Design and Political Choice: The Space Shuttle, 1969–1972," M.S. thesis, Massachusetts Institute of Technology, May 1982.

26. For more details on the design criteria, see Document II-32 in Logsdon, gen. ed., *Exploring the Unknown*, 2: 369–77.

27. None of these designs, however, were sized to carry 65,000 pounds to orbit (100-nautical-mile circular orbit), although several had a fifteen-foot by sixty-foot payload bay and could reach the 1,100-nautical-mile cross range.

Table 2–1. Shuttle Configurations Evaluated by the Manned Spacecraft Center (1969–1971)

Vehicle	Landing Weight (thousands)	Wing	Wing Area (ft²)	Payload Size (ft)	Payload Weight (thousands)	Body Length (in.)	Features
020	130	St (AR7)	1,275	15 by 30	20	1,272	Ext H₂, Int O₂, 4 Eng, Orbiter, SRM Booster
021	85	St (AR7)	785	15 by 40	20	1,080	Ext H₂, Int O₂, SRM Booster
122	95	St (AR5)	792	15 by 40	20	1,064	Ext H₂, Int O₂, SRM Booster
022A	—	45° LE SW	1,120	15 by 40	20	1,064	Ext H₂, Int O₂
022B	—	Delta	2,100	15 by 40	20	1,064	Ext H₂, Int O₂
023	135	Delta	2,700	15 by 40	40	1,325	Ext H₂, Int O₂, Reusable Booster
024	125	St (AR5)	1,000	15 by 60	40	1,315	(Stretched 022) Ext H₂, Int O₂
025	—	45° LE SW	1,414	15 by 60	40	1,315	(Stretched 022A) Ext H₂, Int O₂
026	125	Delta	2,500	12 by 40	40	1,200	Ext H₂, Int O₂, Reusable Booster
027	95	Delta	1,500	12 by 40	40	1,120	Ext H₂, Int O₂, Main Engine, OWB Tanks in rear, SRM Booster
028	128	Delta	2,360	15 by 40	40	1,080	(Shortened 023) Ext H₂, Int O₂, Reusable Booster
029	—	Delta	1,900	12 by 40	40	1,080	OWB Tanks Amidships, Ext Main Engine
030	105	St (AR5)	860	15 by 40	20	1,140	3 J-28 Engines
031	153	St (AR5)	1,110	15 by 60	40	—	3 J-28 Engines
032	130	Delta	2,600	15 by 40	40	1,140	3 J-28 Engines, SRM Booster
033	100	Delta	2,000	12 by 40	20	1,200	(Modified 026) SRM Booster
034	95	Delta	1,500	15 by 30	20	960	(Shortened 025 & 028), SRM Booster
035	135	45° SW	1,200	12 by 40	40	1,440	(Modified 035)
035A	135	45° SW	1,700	12 by 60	40	1,440	(Stretched 035)
036	110	Delta	2,200	15 by 40	20	1,110	3 J28/SRM Booster
036A	110	Delta	2,200	15 by 40	20	1,180	3 J28/SRM Booster
036B	110	Delta	2,200	15 by 40	20	1,110	3 J28/SRM Booster
036C	114	Delta	2,500	15 by 40	20	1,060	3 J25/Pressure-Fed Booster
037	145	Delta	2,900	15 by 60	40	1,400	3 Uprate J28/Recoverable Booster
037A	145	Delta	2,900	15 by 60	40	1,140	3 Super Uprate J25/(036) SRM Booster
038	100	Delta	2,000	15 by 40	20	1,070	550K M1Pc/Solid Booster
039	115	30° SW	1,290	15 by 40	20	1,110	3 J28/Pressure-Fed Booster
040	140	Delta	3,100	15 by 60	25	1,315	4 J28/Pressure-Fed Booster
040A	140	Delta	3,180	15 by 60	25	1,315	4 J25/Pressure-Fed Booster
040B	140	Delta	3,180	15 by 60	25	1,315	4 J28 Retractable/Pressure-Fed Booster
040C	190	60° Delta	2,900	15 by 60	40	1,315	3 MiPc, SRM Boosters
040C-1	190	50° Delta	3,200	15 by 60	40	1,315	3 MiPc, 150-ft² Canard, Twin Tail

Table 2–1 continued

Vehicle	Landing Weight (thousands)	Wing	Wing Area (ft^2)	Payload Size (ft)	Payload Weight (thousands)	Body Length (in.)	Features
040C-2	190	35°/-19° Delta	3,000	15 by 60	40	1,315	3 MiPc, 300-ft^2 Wing Clove, Twin Tail, SRM Boosters
040C-3	190	50° Delta	4,150	15 by 60	40	1,315	3 MiPc
040C-4	190	60° Delta	4,440	15 by 60	40	1,315	3 MiPc
040C-5	150	50° Delta	3,200	15 by 60	40	1,315	3 MiPc, 100-ft^2 Canard, Twin Tail
040C-6	150	55°/-19° Delta	2,800	15 by 60	40	1,315	3 MiPc, 150 ft^2 Canard, Twin Tail
041	114	30° SSW	1,290	15 by 60	15	1,300	3 J28/Pressure-Fed Booster
041A	114	30° SW	1,290	15 by 60	15	1,365	3 J25/Pressure-Fed Booster
042A	110	Delta	2,500	15 by 60	25	1,260	Glider, TIII L6 Booster
042B	105	30° SW	1,255	15 by 60	25	1,260	Glider, TIII L6 Booster
043	83	30° SW	900	10 by 30	27	770	Glider, 2 MiPc oo Ext Tank, PF Booster
044	100	60° Delta	2,000	10 by 30	25	880	2 MiPc, PF Booster
045	—	—	—	—	—	—	—
046	165	—	3,450	14 by 45	25	1,315	3 MiPc, Twin Tail, SRM Boosters
047	185	49°/-5° Delta	3,240	15 by 60	40	1,315	2 MiPc, Twin Tail, SRM Boosters
048	205	35°/-19° Delta	3,080	15 by 60	40	1,315	4 400K, 308-ft^2 Wing Clove, Twin Tail, SRM Boosters
048A	195	35°/-19° Delta	1,150/	15 by 60	40	1,315	3 Hi Pc, 350-ft^2 Wing Clove, Twin Tail, 156" SRM, 62" ASRM
049	205	75°/55° DBL Delta	3,420	15 by 60	40	1,315	3 Hi Pc, 425-ft^2 Wing Clove, Twin Tail, 178" SRM, 62" ASRM
049A	215.3	75°/55° DBL Delta	1,250/ 3,600	15 by 60	40	1,315	
050	—	—	—	—	—	—	—
051	165	35° Delta	2,000	15 by 60 up	25	1,050	3 Hi Pc Swing Engines, 156" SRM, 180-ft^2 Canard
052	175	35° Delta	2,120	15 by 60 up	25	1,250	3 Hi Pc Swing Engines, 149" SRM, 75' Bay with OWB in rear, 180-ft^2 Canard
053	185	35° Delta	2,240	15 by 60 up	25	1,250	4 CC Swing Engines, 120" SRM, 75' Bay, 190-ft^2 Canard
054	185	35° Delta	2,240	15 by 60 up	25	1,250	4 Hi Pc Swing Engines, 140" SRM, 75' Bay, 190-ft^2 Canard

Source: Richard P. Hallion and James O. Young, "Space Shuttle: Fulfillment of a Dream," in Richard P. Hallion, ed., *The Hypersonic Revolution: Eight Case Studies in the History of Hypersonic Technology*, Volume II (Dayton, Ohio: Wright-Patterson Air Force Base, Special Staff Office, Aeronautical Systems Division, 1987), pp. 1049–50.

redesign. NASA favored an engine having higher specific impulse than either of these, which would require the use of only three, rather than four, engines in the orbiter. The agency decided to build a completely new engine; in July 1971, it awarded the development contract to Rocketdyne for its staged combustion design, which became known as the Space Shuttle Main Engine (SSME).[28]

Although NASA continued to explore a wide variety of payload bay sizes and overall payload capacity during its exploration of the optimum Shuttle design, throughout 1970 and 1971, it favored a fifteen-foot by sixty-foot payload bay. After the decision to defer an attempt to gain approval for developing a Saturn V–launched space station, among the reasons for favoring a payload bay of this size was that it was compatible with the growing desire to use the Shuttle like a truck, routinely using it to place large payloads in orbit. The Air Force was also interested in the larger cargo bay for hauling some of its national security payloads. In addition, the larger bay made balancing the orbiter for launch easier and therefore carried less flight risk than a shorter payload bay. In the fall of 1971, the White House Office of Management and Budget (OMB) asked NASA to examine the benefits and drawbacks of a smaller Shuttle, having a shorter, narrower payload bay. NASA analyses showed, however, that developing a smaller orbiter would have relatively small effect on the overall inert or gross launch weight of the Shuttle system, and thus its development costs. [II-8] NASA engineers also pointed out that a larger payload bay made the handling of multiple payloads more efficient.

By late 1971, designers both within NASA and industry had begun to realize that the most cost-effective design for the Shuttle system was a vertically launched delta-winged orbiter mounted to an external tank carrying liquid oxygen and liquid hydrogen, flanked by booster rockets. [II-9, II-10] Putting all of the launch fuel and oxidizer in an external tank allowed designers to reduce the size of the orbiter. It also made the design and construction of the propellant tanks simpler and therefore cheaper. The design allowed the Shuttle to carry a greater payload as a fraction of total vehicle inert weight compared to a two-stage, fully reusable Shuttle system.[29]

Throughout the final months of 1971, OMB persisted in its pressure to lower Shuttle development costs (see Document II-7). On December 29, 1971, NASA Administrator James C. Fletcher sent OMB Deputy Director Caspar W. Weinberger a letter summarizing the results of NASA's most recent analyses, which showed that a Shuttle with a fifteen-foot by sixty-foot payload bay was still the "best buy." However, yielding to OMB pressure, NASA recommended that President Nixon approve a design with a smaller bay.[30] [II-11] Five days later, on January 3, 1972, much to NASA's surprise, President Nixon authorized the space agency to proceed with developing a Space Shuttle with the larger payload bay. There were many factors involved in the decision to authorize NASA to proceed with the Shuttle program it preferred.[31] Among them was the desire on the part of Nixon and his political advisors to initiate during the 1972 presidential election year a large aerospace program with significant employment impacts in key electoral states. [II-12] Nixon met with

---

28. Staged combustion involves partially burning the propellants before burning them completely in a second phase of combustion. NASA chose this design from among three: an "Aerospike" or plug-nozzle design that did away completely with the expansion bell and two expansion bell designs. See J.P Loftus, S.M. Andrich, M.G. Goodhart, and R.C. Kennedy, "The Evolution of the Space Shuttle Design," unpublished manuscript, Johnson Space Center, Houston, TX, 1986, pp. 15–24.

29. See Document III-30 in Logsdon, gen. ed., *Exploring the Unknown*, 1: 549–55.

30. For a fuller discussion of the process leading to Space Shuttle approval, see John M. Logsdon, "The Space Shuttle Program: A Policy Failure?," *Science*, May 30, 1986, pp. 1099–1105; Thomas Heppenheimer, *The Space Shuttle Decision: NASA's Quest for a Reusable Space Vehicle* (Washington, DC: NASA SP-4221, 1999). See also the discussion of the Shuttle decision in *ibid.*, 1: 386–88, 549–59.

31. See Document III-28 in *ibid.*, 1: 546–47.

Fletcher and NASA Deputy Administrator George M. Low on January 5, 1972; afterwards, the White House issued a statement announcing Nixon's approval of the Space Shuttle.[32]

The January 3 decision left open several issues, including whether the Shuttle's strap-on boosters would use solid or liquid fuel. [II-13] In Shuttle system configuration 040C (see Table 2–1), the external tank was flanked by two large, "strap-on" solid rocket boosters (SRBs). This design ultimately became the foundation of the Space Shuttle's configuration. Nevertheless, until March 1972, other possible designs were still on the table, and each had their supporters. For example, in preparation for choosing the booster rockets, NASA studied three general types: large solid-fuel boosters; liquid, pressure-fed boosters; and liquid, pump-fed boosters. To reduce operations costs, NASA decided to make the boosters reusable. After separation from the Shuttle at about forty kilometers altitude, they would fall back to the ocean on large parachutes and be recovered from the sea soon after launch (Figure 2–3).

Technical discussions over the relative merits of these designs centered on which type of booster was safest, most easily refurbished, and cheapest to develop and manufacture. Proponents of liquid motors pointed out that NASA and the Air Force had extensive experience with liquid motors and that they offered greater safety. Liquid engines had the distinct advantage that if system malfunctions were detected in the startup prior to launch, they could be shut down immediately and the launch safely aborted. If an engine failed after launch, it could be shut down and the launch aborted to an overseas airstrip after

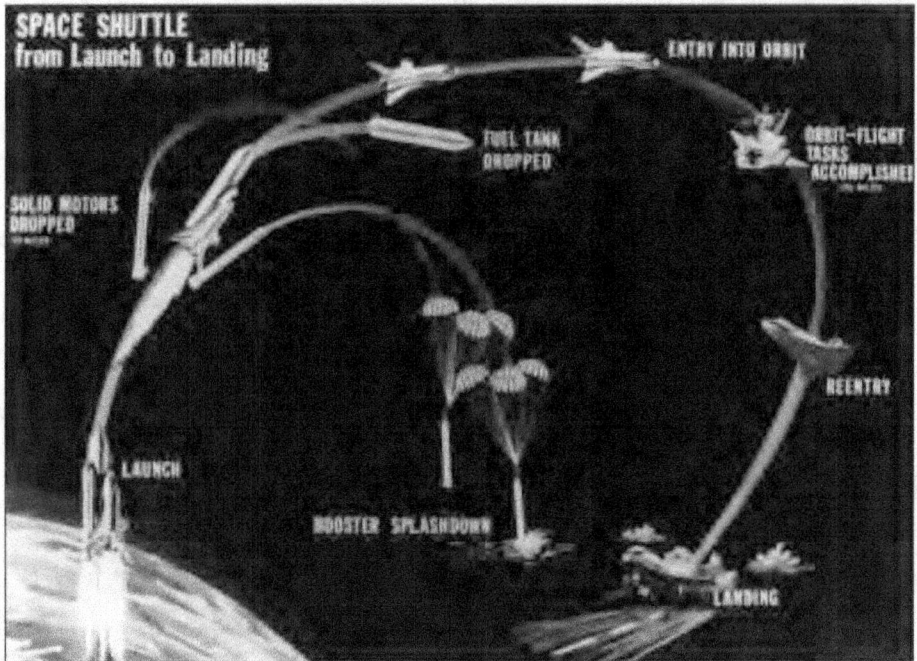

Figure 2–3. This is the standard mission profile for the partially reusable Space Shuttle that actually emerged from the political approval process. (NASA photo)

32. See Document III-32 in *ibid.*, 1: 558–59.

the boosters and the external tank were dropped off. By contrast, once the SRBs were ignited, they could not be shut down (although it was possible to terminate their thrust by blowing off the top of the booster), and the abort potential was decreased. In addition, solid rocket motors of the size NASA was considering (156-inch diameter) had never been used, although the Air Force had tested such large engines and felt they would be sufficiently reliable. Advocates of the big dumb booster designs of the 1960s felt that the pressure-fed design offered greater overall simplicity, which would contribute both to lower costs and to safety.[33] Supporters of solid rocket motors cited the high reliability of solids, as well as their lighter weight and greater simplicity compared to liquid designs.[34] Also, NASA had strong concerns about its ability to refurbish liquid rocket motors after they had been subjected to the corrosive action of an ocean bath. By March 1972, driven primarily by cost considerations, the pendulum of apparent advantages swung in favor of large solid rocket engines, and NASA officials decided to proceed with solid rocket motor development, judging that such motors offered sufficient reliability and ease of handling to be used for human spaceflight.[35] [II-14] NASA announced its choice of solid boosters on March 15, 1972, as it defended the Shuttle program before Congress. [II-15]

The prime contractor for the Shuttle orbiter still had to be decided. Grumman, Lockheed, McDonnell Douglas, and North American Rockwell had all submitted competitive designs for a Shuttle based on the Marshall Space Flight Center 040C design. A NASA-Air Force Source Evaluation Board rated North American Rockwell the highest, based on an evaluation of contractor strengths in:

- Manufacturing, test, and flight-test support
- System engineering and integration
- Subsystem engineering
- Maintainability and ground operations
- Key personnel and organizational experience
- Management approaches and techniques
- Procurement approaches and techniques

On July 26, 1972, NASA Administrator James Fletcher met with Deputy Administrator George Low and Associate Administrator for Organization and Management Richard C. McCurdy to make the final Shuttle contractor decision. This choice was essentially between North American Rockwell and Grumman, the two companies that had received the highest ratings from the Source Evaluation Board. After considerable discussion, the three adopted the board's recommendation. [II-16] In August 1972, North American Rockwell received the contract to design and develop the Shuttle orbiter. Later, Morton Thiokol was selected to produce the SRBs.[36] [II-17] NASA also selected Martin Marietta to develop the external tank. The Manned Spacecraft Center assumed responsibility for supervising overall orbiter development. Marshall Space Flight Center was to supervise the development and manufacturing of the SRB, the SSME, and the external tank, and

---

33. Arthur Schnitt and F. Kniss, "Proposed Minimum Cost Space Launch Vehicle System," Report no. TOR 0158(3415)-1, Aerospace Corporation, Los Angeles, CA, July 18, 1966. For a general discussion of the big dumb booster concept, see U.S. Congress, Office of Technology Assessment, *Big Dumb Boosters: A Low-Cost Space Transportation Option?* (Washington, DC: Office of Technology Assessment, February 1989).

34. For example, the Minuteman and Polaris, both of which use solid propellants, had proved highly reliable.

35. Eagle Engineering, Inc., "Technology Influence on the Space Shuttle Development," Report No. 86-125C, NASA Johnson Space Center, Houston, TX, June 8, 1986, pp. 5–20, 21.

36. As noted above, NASA had awarded the contract for the SSME to Rocketdyne in 1971.

Figure 2–4. Ames Research Center scientists tested the aerodynamic properties of a Space Shuttle wind tunnel model in 1973. (NASA photo)

Kennedy Space Center was to develop methods for Shuttle assembly, checkout, and launch operations.

Even after the development contracts were let, determining the best design was still a major task that required close cooperation among the design teams (Figure 2–4). During liftoff and throughout the short passage through the atmosphere, the shape and placement of each of the major Shuttle components would affect flight success. [II-18] Changes in any one of the elements—wing shape, the diameter and length of the SRBs, and the diameter of the external tank—would alter the performance of the others. Thus, the configuration of the Shuttle system and precise shapes of each component passed through several steps to reach the final overall shape and structure.[37]

North American Rockwell began fabricating Orbiter Vehicle (OV)-101 on June 4, 1974; the company rolled out the orbiter from its Palmdale, California, plant on September 17, 1976. The OV-101 lacked many subsystems needed to function in space. It was thus capable of serving only as a full-scale mockup capable of atmospheric flight; this flying testbed proved invaluable in testing the orbiter's ability to maneuver in the atmosphere and to glide to a safe landing. Flight-testing began in February 1977 at Edwards Air Force Base.

Earlier, NASA had purchased a used Boeing 747-100 to ferry the orbiters from landing sites in California and potentially other parts of the world to Kennedy Space Center for refurbishment and launch.[38] This airplane was also used to conduct flight tests with *Enterprise*, as OV-101 came to be called. A NASA committee typically chose the orbiter, but fans of the Star Trek television series had lobbied NASA and Congress to name OV-101 the title of the starship of that series. [II-19]

Enterprise underwent three major types of tests: (1) captive flight, in which NASA tested whether it could take off, fly, and land the 747 with the orbiter attached; (2) captive-active flight, in which an astronaut crew rode in *Enterprise* during captive flight; and

---

37. See Hallion and Young, "Space Shuttle: Fulfillment of a Dream," pp. 1125–42, for a summary discussion of these points.

38. Once the Shuttle began flying, NASA established backup landing sites in several other countries, should a launch failure allow an abort landing elsewhere or extraordinary conditions at both Edwards Air Force Base and Kennedy Space Center prevent landing at those two primary locations.

(3) free flight, in which *Enterprise* was released to glide back to Earth on its own. By August 1977, NASA had successfully completed the first two test phases and was ready to test the orbiter in free flight. On August 12, 1977, the 747 carried *Enterprise* to 24,100 feet, where it was released for a five-minute glide to a successful landing at Edwards.[39] After four additional test glides, NASA wound up its atmospheric flight testing program and turned to vibration and other ground tests of *Enterprise*.

Two major technical problems kept Shuttle development from proceeding smoothly: (1) a series of test failures and other problems with the SSME and (2) difficulties achieving a safe, lightweight, robust thermal protection system. SSME development proved challenging on several grounds: NASA needed a reusable, throttleable staged-combustion engine that would achieve much higher combustion chamber pressures than any previous engine. The United States had not yet built a rocket engine that was both reusable and capable of being throttled. Such an engine required high-pressure turbopumps capable of higher speeds and internal pressures than any developed to date. Reusability and the fact that the SSME would be used on a vehicle rated to carry people imposed special demands on the engine. Despite a nine-month delay in starting SSME development, caused by a Pratt & Whitney challenge to the Rocketdyne contract, as well as difficulty in procuring the necessary materials for the engine, Rocketdyne completed the first development engine in March 1975, one month ahead of schedule.

Engine tests were performed at NASA's Mississippi National Space Technology Laboratories (later named Stennis Space Center) and at the Air Force's Rocket Propulsion Laboratory at Santa Susana, California. Although the first test firing was successful, problems began to surface as the tests became more demanding. The turbopumps were particularly troublesome because their turbine blades tended to crack under the severe mechanical stresses they experienced. The engines also experienced a variety of nozzle failures during tests.[40] These problems caused significant delays in the testing program. This prompted the Senate Subcommittee on Science, Technology, and Space of the Committee on Commerce, Science, and Transportation in December 1977 to request an independent review of SSME development by the National Research Council. The report, presented in a March 31, 1978, Senate Subcommittee hearing, noted that the problems NASA was experiencing in the test program were typical of such development efforts, but also recommended a number of possible SSME modifications and a delay in the timetable for the first Shuttle flight.[41] The National Research Council committee, generally called the Covert Committee after its chair, Eugene Covert, a professor at the Massachusetts Institute of Technology, also recommended that NASA relax its goal of launching the Shuttle with the SSMEs operating at 109 percent of full power level, to reduce stress on the turbopump components.

Because NASA was then behind schedule, it decided to save SSME development time by conducting some tests using all three engines in their flight configuration. They were attached to an orbiter simulator using identical components to those on the flight article. NASA also used an external tank to supply propellant to the engines and attached it to the

39. Astronauts Fred W. Haise and Gordon G. Fullerton were the pilot and co-pilot for the first free flight of *Enterprise*.

40. Hallion and Young, "Space Shuttle: Fulfillment of a Dream," pp. 1158–59.

41. Eugene Covert, "Technical Status of the Space Shuttle Main Engine," report of the Ad Hoc Committee for Review of the Space Shuttle Main Engine Development Program, Assembly of Engineering, National Research Council. Printed in U.S. Congress, Senate Committee on Commerce, Science, and Transportation, Subcommittee on Science, Technology and Space. *Space Shuttle Main Engine Development Program. Hearing.* March 31, 1978, 95th Cong., 2d sess. (Washington, DC: U.S. Government Printing Office, 1978), pp. 16–57.

test stand in a manner identical to its connection to the SRBs on the launch pad. NASA began its main propulsion testing in April 1978, but continued to experience test delays and failures. Despite the delays and problems, the basic SSME design was considered sound. Rocketdyne proceeded with the manufacturing of the three engines needed for *Columbia* (OV-102). In May 1978, Rocketdyne finally received approval to start manufacturing the nine additional production SSMEs needed for OV-099 (*Challenger*), OV-103, and OV-104.

Nevertheless, development problems continued. One of the largest setbacks was a fire that destroyed an engine on December 27, 1978. The Covert Committee, which had been preparing a second report on the SSME program, reviewed this and an additional fire. [II-20] Once again, the committee report recommended changes in procedures and further tests, noting: "It appears unlikely that the first manned orbital flight will occur before April or May 1980."[42] The test program continued, "and by 1980 the SSME was no longer perceived to be a pacing factor for the first launch . . . the thermal protection system was considered the pacing item."[43]

Thermal protection for the Shuttle's reentry was a major issue from the earliest design concepts through the first several flights of the Shuttle. NASA engineers had solved the reentry problem for the Mercury, Gemini, and Apollo capsules by using ablative materials that heated up and burned off as the capsule encountered the upper atmosphere upon reentry. However, these capsules were not designed to suffer the rigors of multiple flights and reentries and were thus retired after use. Each Shuttle orbiter was designed to experience up to 100 launches and returns. Its thermal protection system had to be robust enough to stand repeated heating loads and the structural rigors of reentry. The system had to be relatively light to keep the orbiter's overall weight acceptably low. In addition, it had to be relatively cheap to refurbish between flights.

Between 1970 and 1973, NASA studied a wide variety of technologies to protect the orbiters' bottom and side surfaces. It investigated:

- "Hot structures," in which the entire structure took the heat load
- Heat shields separated from a lightweight orbiter structure by insulation
- Ablative heat shields over a lightweight structure
- Low-density ceramic heat shields (tiles) bonded to a lightweight structure

The "hot structures" would have required developing exotic and expensive titanium or other alloys that could dissipate reentry heating and simultaneously withstand the mechanical loads from aerodynamic pressure. The heat-resistant panels separated by insulation would transfer the mechanical load while shielding the underlying structure from atmospheric heating. This concept suffered from excessive weight and difficulties in designing the shielding to avoid buckling or excessive deflection. NASA's estimates showed that the ablative heat shields would require costly refurbishment.

Therefore, NASA chose the fourth option after extensive testing, in part because the agency decided that using tiles would lead to the lowest overall cost. A ceramic heat shield also allowed NASA engineers to use aluminum for the Shuttle orbiter's structure—

42. National Academy of Sciences, National Research Council, Assembly of Engineering, *Second Review—Technical Status of the Space Shuttle Main Engine: Report of the Ad Hoc Committee for Review of the Space Shuttle Main Engine Development Program* (Washington, DC: National Academy of Sciences, February 1979), p. 21.

43. U.S. Congress, Congressional Research Service, *United States Civilian Space Programs 1958–1978*, report prepared for the Subcommittee on Space Science and Applications of the Committee on Science and Technology, U.S. House of Representatives, 97th Cong., 1st sess. (Washington, DC: U.S. Government Printing Office, January 1981), p. 473.

a material with which they had considerable experience. The particular ceramic material chosen was foamed silica coated with borosilicate glass. The shield was divided into thousands of small tiles to enable the stiff material to conform to the shape of the orbiter skin. (The tiles are what give the orbiters' lower surfaces the look of being constructed of blocks.)

No one had ever used such materials over an aluminum structure, and many experts expressed concerns about NASA's ability to develop an appropriate means to bond the brittle, nonpliable ceramic tiles to an aluminum structure that would deform slightly under aerodynamic loads. Fitting and attaching the tiles became a major effort, one that was highly labor intensive. Each tile is approximately fifteen centimeters square and is individually cut and fitted to match its neighbor. Because every tile is slightly different in size and shape, it carries its own number and has its own documentation.[44] The orbiter nose cap and its wing leading edges, which experience heating of above 1,500 degrees Kelvin during reentry, are protected by a high-temperature, high-cost, carbon-carbon material. Other temperature-resistant materials are used on the upper parts of the orbiter.

Problems with installing the tiles caused NASA to deliver the first flight-qualified orbiter, *Columbia*, to Kennedy Space Center in early 1979 before NASA technicians had completed installation. Attaching the tiles then became the critical element in scheduling the first Shuttle launch. Originally planned for 1978, by March 1979, the schedule had slipped at least two years.[45] Work on the tiles went on twenty-four hours a day for six days a week, as technicians struggled to install more than 30,000 individual tiles. While NASA worked on methods to speed up the process, it also continued to explore better materials to develop a method that would make the tile stronger without adding weight.

In the meantime, as Rockwell and NASA engineers began to understand the extent of the aerodynamic loads the orbiter's surface would experience during the launch phase, they developed concerns that some tiles might loosen, or even fall off. Upon reentry, they feared, weakened tiles might peel away, causing the underlying aluminum structure to overheat. Thus NASA also explored various means to examine the Shuttle while in orbit to check on the tiles, and the agency began to develop a tile repair kit.[46] [II-21]

Shuttle development problems were so severe during the late 1970s that some within the Carter administration's OMB proposed that the program be cancelled. This led to a series of external reviews of the program during 1979. [II-22] Even before this recommendation, OMB had been resisting NASA's attempt to gain approval for building a fifth Shuttle orbiter. NASA believed that a five-orbiter fleet would be needed to provide adequate capability to meet anticipated launch demand. [II-23, II-24] While not authorizing the construction of a fifth orbiter (an issue NASA continued to press until the 1986 *Challenger* accident), President Jimmy Carter was persuaded that ending the program was not a good move. [II-25, II-26, II-27] After extraordinary efforts, by early 1980, NASA felt it was bringing its tile problems under control and was able to project a launch date of March 1981. [II-28]

Before NASA could launch *Columbia*, however, it had to attend to thousands of details, both large and small. In addition to the tiles, the agency had to install and test many other Shuttle orbiter subsystems. For this work, *Columbia* was rolled into the Orbiter Processing Facility at Kennedy Space Center. Because virtually everything about the Shuttle system was different from the Saturn V, launch operations crews had to learn new methods for handling the vehicle, its SRBs, and the external tank. NASA altered the Vehicle Assembly Building

    44. Paul A. Cooper and Paul F. Holloway, "The Shuttle Tile Story," *Astronautics and Aeronautics*, January 1981, pp. 24–34.

    45. U.S. Congress, House Committee on Science and Technology, 1980 NASA Authorization Hearings before a subcommittee on H.R. 1756, 96th Cong., 1st sess., February and March 1979, pt. 4, p. 1664.

    46. NASA, "On-Orbit Tile Repair Kit Being Produced," Press Release 80-10, January 23, 1980.

(VAB) and the Mobile Launch Platform that had been developed for Apollo to accommodate the Shuttle.[47] NASA also made substantial alterations to launch pads 39A and 39B.

For each Shuttle launch, the first elements of the launch system to be erected are the two large SRBs. Each is about twelve feet in diameter, 149 feet long, and composed of nine major elements—a nose cap, a frustrum, a forward skirt, four individually cast solid rocket motor segments, a nozzle, and an aft skirt. NASA technicians begin assembly of the Shuttle by attaching the aft skirt of each of the two SRBs to support posts on the Mobile Launch Platform. Then, piece by piece, technicians hoist each SRB element atop the next one and bolt it down. The motor segments are joined to their neighbors by tang-and-clevis joints and secured by steel pins located along the circumference of each joint.[48] For safety reasons, all nonessential personnel must evacuate the VAB while the SRBs are being assembled. After the two SRBs are safely bolted to the Mobile Launch Platform, a crane hoists the external tank to a vertical position and mates it with the twin SRBs. Then the orbiter is transferred from the Orbiter Processing Facility to the VAB, lifted by its nose more than 100 feet off the floor, and lowered into place and mated with the external tank.

Although NASA could have made the first launch, reentry, and touchdown in an automated mode, NASA engineers felt confident enough in the safety and reliability of the Space Shuttle system to believe that such a procedure was unnecessary.[49] [II-29] In this they were strongly supported by the astronaut corps, which was anxious to return to space. (The last crewed flight was the Apollo-Soyuz Test Project in July 1975.) Besides, preparing the orbiter for automated landing would have entailed additional expense and weight for the avionics and would have injected additional uncertainty in the interpretation of the flight results.

The first launch of the Space Shuttle *Columbia* was scheduled for April 10; it was to be piloted by astronauts John Young and Robert Crippen. After a delay caused by computer problems, the launch actually took place at 7:00 a.m. on April 12, 1981 (Figure 2–5). When the countdown clock reached T–3.8 seconds, NASA started up the SSMEs, allowing the launch directors to determine that they were firing properly. At about T+3.0 seconds, they fired up the SRBs, irrevocably committing NASA to the launch. At an altitude of 400 feet, eight seconds after lifting off the pad on a column of flame and smoke, computer instructions caused *Columbia* to roll over on its back and continue its upward climb over the Atlantic Ocean. About two minutes later, at an altitude of twenty-seven nautical miles, the SRBs, which had completed their part of the launch sequence, separated from the orbiter and fell to the ocean on orange and white parachutes. Eight minutes and fifty-two seconds after liftoff, *Columbia* reached orbit and jettisoned the nearly empty external tank, which fell back through the atmosphere into the Indian Ocean. A short burn of *Columbia*'s orbital maneuvering system rockets circularized the orbit at 130 nautical miles.

Young and Crippen orbited Earth thirty-seven times while testing the various Shuttle components, such as the large cargo bay doors, which they opened and closed. One of NASA's major concerns was the condition of the tiles. Upon opening the payload doors, the astronauts discovered that several tiles on the fairings for the orbital maneuvering and reaction control engines had separated during launch. Although the loss of these tiles, which were on the upper side of the orbiter, would not have prevented a safe reentry, Mission Control in Houston remained unsure about the condition of *Columbia*'s underside, which could not be seen from the cockpit. As the orbiter circled Earth, NASA

---

47. For example, because the Shuttle does not make use of the tower and gantry required by the Saturn V, these were removed.

48. The tang-and-clevis joints are called "field joints" because they are assembled at the launch site ("in the field") rather than at the factory.

49. The Soviet Union flew its *Buran* shuttle orbiter in an automated mode in its first and only flight in November 1988.

Figure 2–5. The Space Shuttle is finally realized with the launch of Columbia from Launch Complex 39A on April 12, 1981, on its first orbital mission. (NASA photo)

arranged for Air Force cameras to photograph *Columbia*'s underside to confirm tile integrity. Finding the tiles in apparently good order, NASA Mission Control notified the two astronauts to prepare for return.

Fifty-four hours after takeoff, *Columbia* glided to a successful landing at Edwards Air Force Base. Although *Columbia* landed at a faster speed than planned and rolled nearly 3,100 feet beyond its planned stopping point, the flight proved the feasibility of the Shuttle's design. [II-30] NASA made three more test launches with *Columbia*—on

November 12, 1981, March 22, 1982, and June 27, 1982. Each time, *Columbia* experienced some anomaly that had to be resolved.[50]

In the aftermath of the first Shuttle flight, the Reagan administration considered the longer term future of the program. A variety of uses and management approaches were evaluated; ultimately, President Ronald Reagan decided to keep NASA in the lead role in managing the Space Transportation System (STS). He reiterated the policy that once the Shuttle became operational, it would be used to launch all U.S. government missions. [II-31, II-32]

The last test flight of *Columbia* ended symbolically on July 4, 1982, when the orbiter glided to a landing before President and Mrs. Reagan and a crowd of about 750,000 visitors at Edwards Air Force Base.[51] To enhance public attention to the July 4th event, NASA had arranged to fly *Challenger*, the second of four planned orbiters, to Kennedy Space Center shortly after *Columbia* rolled to a stop. *Challenger* took off atop NASA's Boeing 747 carrier plane as Reagan was giving his speech, circled the field, and dipped its wings to the crowd. [II-33]

Space Shuttle Operational Flights—Phase I

Columbia's four successful test flights led NASA to declare that the Shuttle fleet was operational—meaning, in theory, that further development of Shuttle systems would be minimal. With *Challenger* in preparation for its first flight, and *Discovery* and *Atlantis* in production, NASA officials were now ready to push up the flight rate and extend the use of the STS to a wide variety of payloads and customers (Figure 2–6). [II-34] (Chapter 3 discusses the use of the Space Shuttle to launch commercial payloads.)

When NASA began the Shuttle's development, the agency expected the vehicle to assume the entire burden of lifting U.S. satellites and other payloads to orbit soon after reaching full operational status. [II-35] NASA also expected other nations to use the Shuttle for access to space, and the agency projected a flight rate of forty-eight per year beginning in 1980. Such a high rate would, in NASA's estimation, have led to a low cost per flight and even allowed NASA to recoup much of its investment in the Space Shuttle system.[52] By the mid- to late 1980s, NASA hoped, reduced costs for operating the Shuttle system would allow the agency to fund other projects, such as a future space station. This so-called "Shuttle funding wedge" became a tenant of NASA policy and the agency's expectations for major future projects.

The number of future projected flights allowed NASA to set its first pricing policy in 1975 to garner as many Space Shuttle flights as possible. This policy was intended in part "to effect early transition from expendable launch vehicles."[53] NASA had arrived at a price of $18 million (1975 dollars) by averaging projected development and operational costs over a total of 572 flights from 1980 through 1991.

In the early 1980s, expectations for such a high flight rate had decreased, but were still relatively high. In July 1983, for example, Rockwell International forecast that by 1988, overall U.S. demand for space transportation services for civilian and military uses would require a yearly flight rate of twenty-four launches.[54] Based on an expectation of increasingly shorter "turnaround time" for processing each orbiter, NASA expected to meet that

50. For example, during the second flight (STS-2), one of the orbiter's three fuel cells failed, causing NASA to bring *Columbia* back after only two and a half days, rather than the planned five.

51. NASA extended *Columbia*'s time in space by one orbit to accommodate the presidential visit.

52. U.S. Congress, Congressional Budget Office, *Pricing Options for the Shuttle* (Washington, DC: Congressional Budget Office, March 1985).

53. C.M. Lee and B. Stone, "STS Pricing Policy," presented at the AIAA Space Systems Conference, Washington, DC, October 18-20, 1982, p. 1.

54. Rockwell International, "Projection of Future Space Shuttle Traffic Demand," July 1983, Rockwell Corporation, Downey, CA.

Figure 2-6. The STS-8 mission on Challenger *was the first nighttime launch of the Shuttle era on August 30, 1983. (NASA photo)*

rate by 1988. Such forecasts assumed that the Shuttle would fly commercial, as well as government, payloads. It also anticipated that a fifth Shuttle orbiter would be built. [II-36]

The orbiter turned out to be much more difficult and time consuming to refurbish and prepare for launch than NASA had expected. This resulted in part from the need to correct system design deficiencies throughout the orbiter, which in turn kept the system in a state of continual development.[55] Orbiter "turnaround" time became the pacing item in efforts to improve the Shuttle launch rate. From 1983 through 1985, NASA steadily increased the flight rate until, in 1985, it was able to launch nine flights. NASA accomplished this feat in part by significantly reducing the damage to the protective tiles after liftoff and by making small improvements in the SSMEs to reduce the amount of inspection time needed.[56] Nevertheless, many observers remained skeptical that NASA would

---

55. Charles R. Gunn, "Space Shuttle Operations Experience," paper presented at the 38th Congress of the International Astronautical Federation, Brighton, England, October 1987.
    56. *Ibid.*

ever be able to reach and maintain a rate close to twenty-four flights per year, given the complications of preparing the Shuttle orbiter and other subsystems for launch.

In the early 1980s, the Reagan administration, strongly encouraged by NASA, had established the policy that all government payloads would be launched on the Shuttle and that the Delta, Atlas, and Titan expendable launch vehicles (ELVs) would be phased out. NASA ordered no more Delta or Atlas ELVs after 1982. Their manufacturers moved to shut down production lines. Because this action removed these launch vehicles from use by commercial interests, commercial communications satellite owners and a few other private payload customers were forced to use either the Shuttle or the European-built Ariane rocket. (See Chapter 3 for a discussion of the competition between the Shuttle and Ariane.)

The Shuttle was maintained under NASA control, although several groups urged policies that would put the Shuttle under the operational control of private industry (or even the Air Force). They argued that the private sector would reduce operational costs faster and more effectively than NASA. Although some officials of the Reagan administration flirted briefly with the concept, they finally concluded that, in the words of the congressional Office of Technology Assessment, the "Shuttle is an important instrument of national policy and is needed primarily for government civilian and military payloads."[57]

As noted, the operational Space Shuttle turned out to be much more complicated to operate than had been expected, took longer to refurbish, and cost much more to operate than NASA had estimated.[58] Nevertheless, between its first flight in 1981 and January 1986, it served to carry a variety of life science and engineering experiments into orbit, launched communications satellites and scientific payloads, and launched DOD payloads.[59]

From the beginning, Shuttle planners expected to launch high-inclination payloads, especially polar-orbiting payloads, from Vandenberg Air Force Base in California, because only at Vandenberg is there an available high-inclination launch path (to the south) that would not jeopardize populated areas. DOD and the National Reconnaissance Office were especially interested in using this capability to launch several reconnaissance satellites, which require polar orbit for effectiveness. DOD funded the development of launch preparation facilities and a launch pad at the site of the Space Launch Complex-6 (SLC-6, pronounced "Slick-6") to launch from Vandenberg.[60] However, the Space Shuttle proved unable to meet its payload weight goal of 65,000 pounds to LEO (twenty-eight-degree inclination), which was necessary to launch about 40,000 pounds into polar orbit. That problem, combined with the loss of *Challenger* in 1986 and the development of the Titan IV, led to the abandonment of SLC-6 as a Shuttle launch site, but only after DOD had poured several billion dollars into upgrading the launch pad and constructing appropriate supporting facilities.

The Complementary Expendable Launch Vehicle

Not everyone in the government agreed with the move toward total government dependence on the Space Shuttle. Some influential officers within the Air Force, which had the responsibility for launching all national security payloads, especially the critical

57. U.S. Congress, Office of Technology Assessment, *International Cooperation and Competition in Civilian Space Activities*, OTA-ISC-239 (Washington, DC: U.S. Government Printing Office, 1985), p. 10.
58. Roger A. Pielke, Jr., "A Reappraisal of the Space Shuttle Programme," *Space Policy*, May 1993, pp. 133–57.
59. See Logsdon, gen. ed., *Exploring the Unknown*, 2: 263–69, for a discussion of DOD disenchantment with the Space Shuttle.
60. SLC-6 was originally meant for the launch site of Dyna-Soar; it was then refurbished for the Manned Orbital Laboratory. Both programs, of course, were cancelled, so the site remained unused.

reconnaissance satellites, worried about the frequent delays in Shuttle launches and the length of time between manifesting a payload on the Shuttle and the actual flight (about twenty-four months).[61] They reasoned that any major problems encountered in a Shuttle subsystem could delay the flight of a critical payload. No matter how successful the Shuttle fleet was, there were likely to be times when it would be grounded for safety purposes, just as entire aircraft fleets may be grounded while investigators examine the causes of major subsystem failures and determine appropriate repairs. Privately, some analysts worried that the Shuttle might fail catastrophically at some point, leaving the fleet grounded for an extended period. In addition, some argued that even if NASA were able to sustain an average Shuttle flight rate of twenty-four per year, that rate would not accommodate the needs of the Air Force, along with the projected demand from civilian public- and private-sector uses.

Hence in 1983, with the strong endorsement of Secretary of the Air Force Pete Aldridge, the Air Force began to examine the benefits and costs of developing a new vehicle that it called the Complementary Expendable Launch Vehicle (CELV) to provide "assured access to space." On January 7, 1984, Secretary of Defense Caspar Weinberger approved a defense space launch strategy that included the development of a CELV with sufficient capacity to launch payloads of up to 40,000 pounds.[62]

Air Force officials chose the adjective "complementary" to avoid the appearance of competition with the Shuttle and to emphasize that the CELV would be expected to service DOD launch demand should the Shuttle be unable to meet it for any reason. Aldridge was also interested in improving Air Force launch flexibility and maintaining the technology base and production capability that might otherwise be lost.

Congressional reaction was mixed. DOD's authorization and appropriations committees generally supported the move. However, supporters of NASA's Space Shuttle expressed concern that CELV development would divert DOD attention away from the Shuttle and undercut the funding supporting Shuttle operations. The Shuttle was developed in part to serve DOD needs, which led to higher operations costs than NASA had anticipated. Continued DOD use of the Shuttle was needed to help pay for Shuttle upgrades and keep the costs of operations as low as possible.

Despite the concerns of some members of Congress, especially those of the House Committee on Science and Technology, DOD's plans nevertheless carried the day. DOD issued a request for proposals (RFP) on August 20, 1984, for the development of a launcher capable of lifting 10,000 pounds to a geostationary transfer orbit from DOD's Eastern Test Range. The initial RFP called for a total buy of ten launchers. In 1984, the Air Force had no official plans to launch the CELV from the Western Test Range at Vandenberg Air Force Base, but intended instead to rely on the Shuttle to lift payloads of up to 32,000 pounds into low-Earth polar orbit from Vandenberg.[63]

Martin Marietta won the contract to build an upgraded version of its Titan 34D in February 1985, over competing designs from General Dynamics and from NASA, which had proffered a launch vehicle based on Shuttle technology. This vehicle, which became known as the Titan IV, is capable of lifting 40,000 pounds to LEO or 10,000 pounds to geostationary transfer orbit. Martin Marietta achieved the improved payload capacity by stretching the liquid propellant tanks and by upgrading the Titan's solid rocket motors to

61. Ironically, the vehicle that resulted from the Air Force need to launch national security payloads, the Titan IV, has proved nearly as difficult to make operational and almost as costly as the Shuttle.

62. See Documents II-40 through II-44 in Logsdon, gen. ed., *Exploring the Unknown*, 2: 390–410.

63. Discussion between Congressman George Brown and Secretary of the Air Force Pete Aldridge, "Space Shuttle Requirements, Operations, and Future Plans," hearings before the Subcommittee on Space Science and Applications of the Committee on Science and Technology, U.S. House of Representatives, 98th Cong., 2d. sess., July 31–August 2, 1984, p. 86.

seven segments rather than the five and a half segments used by the Titan 34D.[64] Fairings of up to 86 feet long would accommodate Shuttle-size payloads. The Titan IV was designed with the capability to carry no upper stage, a Centaur upper stage, or an inertial upper stage (IUS).[65] The first Titan IV was launched on June 14, 1986, with an IUS upper stage. In October 1987, Martin Marietta contracted with Hercules to develop and manufacture SRBs with graphite-epoxy casings, capable of adding 8,000 pounds capacity to LEO. After the failure of the Shuttle *Challenger*, the Air Force's plans to develop the CELV seemed almost prescient.

Losing *Challenger*

Although every knowledgeable observer recognized that there was some potential for a major Shuttle failure, the press and the broader public in the early 1980s paid little attention to the risks of human spaceflight. Even those close to the Shuttle system let down their guard. As one successful launch followed another, some engineers and flight directors began to submerge their concerns about troublesome items that lay on the critical path to a safe launch. Hence, the nation was dealt an extremely rude shock when, on January 28, 1986, the orbiter *Challenger*, carrying seven crew members, seemed to disappear behind a huge fireball just over a minute after liftoff and disintegrated before the eyes of thousands of observers at the launch site and millions more watching the launch on live television coverage. It was a numbing sight, played over and over again on television, as people all over the world attempted to come to grips with what had happened.[66]

Launch vehicle reliability has always been a concern; most launch vehicles have demonstrated launch success rates of between 90 and 98 percent. Launch officials worry especially about the safety of vehicles that carry human crews. As long ago as 1977, former NASA Administrator James Fletcher had expressed his concerns to then NASA Deputy Administrator Alan M. Lovelace about the overall Space Shuttle system and whether NASA had the right people working the problem of launch reliability and safety. [II-37]

Engineers and other observers familiar with the Shuttle's many systems and points of potential weakness had their theories about the cause of the catastrophic failure, yet because of the complexity of the Shuttle system, it took careful analysis by a large team of experts to determine the exact cause. NASA began to work on the problem immediately by pulling together all of the available film footage, launch operations documents, and other materials that might be relevant to the investigation. NASA even employed a deep sea diving company to locate and retrieve parts of the failed launcher from the ocean floor. Although senior NASA officials would have preferred to carry out their own analysis outside the glare of publicity, as had been the case following the Apollo 1 fire, the highly public and dramatic loss of life that had occurred on January 28 made an independent external review almost inevitable. On February 3, President Reagan signed Executive Order 12546, which directed the establishment of a high-level commission, chaired by former Secretary of State William P. Rogers, to examine the evidence and determine not only what had happened, but also why it had. [II-38] The Presidential Commission on the Space Shuttle *Challenger* Accident, supported by NASA and other federal agencies, gathered evidence, investigated the chain of events, and held public hearings.

---

64. The first stage was stretched by almost eight feet to increase propellant volume by 10 percent, and the second stage was stretched almost two feet, resulting in increased propellant volume of 5 percent. The solid rocket motors are manufactured by the Chemical Systems Division of United Technologies.

65. With the IUS and a fifty-six-foot fairing, the Titan IV stands 174 feet tall.

66. The incident was especially numbing because NASA had worked particularly hard to generate public interest in the flight, which carried teacher Christa McAuliffe, who would have been the first teacher in space.

As the investigation revealed, the joint between the first and second motor segment was breached about fifty-nine seconds into the flight. Flames from the open joint struck the external tank and caused its liquid hydrogen and liquid oxygen tanks to rupture. At seventy-six seconds, fragments of *Challenger* could be seen against the backdrop of a large fireball, caused by the ignition of thousands of pounds of hydrogen from the external tank. The orbiter was torn apart by the enormous aerodynamic forces, which greatly exceeded the orbiter's design limits. Large parts of *Challenger* began to tumble through the atmosphere and fall back toward the Atlantic Ocean. The forward fuselage and the crew module, both of which remained largely intact, plunged into the waves a few seconds later, killing all seven astronauts on board.[67]

This description of the sequence of events during the failure of the vehicle was gained only through a meticulous examination of the photographs and the recovery and detailed inspection of many *Challenger* parts from the ocean floor. It also required a methodical analysis of the sequence of events during launch. This analysis also contributed to a more precise understanding of the O-ring failure that caused the loss of *Challenger*. Knowledge of the structural details of the SRBs became widespread as newspapers printed detailed drawings of the Shuttle system and the joint that held the motor segments together. The "tang-and-clevis" joint, which was supposed to hold the segments together with seventeen bolts and a rubber O-ring seal, received special attention from the media as well from experts, because it was this critical part of the Shuttle system that had failed. During engine firing, the joint was subject to enormous pressure. NASA and Morton Thiokol had intended to design the joint so that the O-ring would deform under pressure and fill in any small openings between the tang and clevis, preventing a "blow-by" of the hot ignition gases during motor firing. However, as NASA's own tests during SRB development had shown, the O-rings would occasionally suffer damage during firing.[68] During the second Shuttle flight (STS-2) and on several subsequent flights, the O-rings sustained both erosion and blow-by, indicating problems that could become worse. Of particular concern, as the temperature of the joint fell, the O-ring material would stiffen up and prevent it from properly squeezing into any voids, even when under pressure. Although several NASA officials and Morton Thiokol engineers were aware of the problem and the catastrophic failure it could cause, the two organizations failed to act to redesign the joint. Instead, they tried a number of other fixes, including tightening the joint and adding putty to the joint to assist the O-ring in sealing the joint.

The open hearings of the Rogers Commission, which NASA officials opposed, gave the public extraordinary insight into the almost overwhelming complexities of preparing and operating the Shuttle. In one particularly dramatic moment during the hearings, commission member Richard Feynman placed a short section of the O-ring in ice water, demonstrating on live television how inflexible the material becomes with cold. His simple demonstration dramatized a major problem that NASA officials had virtually ignored. As noted in the commission's report, "Prior to the accident, neither NASA nor Thiokol fully understood the mechanism by which the joint sealing action took place."[69]

The hearings and the report that resulted from it also exposed publicly a number of crucial management deficiencies within NASA, among which was the difficulty contractor personnel and mid-level NASA engineers had in conveying the seriousness of known technical problems to senior-level managers. [II-39] The hearings also made it clear that senior NASA officials had subtly but inexorably shifted their attitude regarding the launch

---

67. *Report of the Presidential Commission on the Space Shuttle Challenger Accident* (Washington, DC: U.S. Government Printing Office, June 6, 1986), pp. 19–39.
    68. *Ibid.*, p. 120.
    69. *Ibid.*, p. 148.

of the Shuttle. At first, the engineers had to demonstrate that the Shuttle was safe to launch. The shift was that by the time of the ill-fated *Challenger* launch (STS 51-L), they had to demonstrate that it was not safe to launch. At one point in the hearings, for example, Roger M. Boisjoly, a Morton Thiokol engineer, noted that "we were being put in a position to prove that we should not launch rather than being put in the position and prove that we had enough data to launch."[70] Decision-making regarding the Shuttle had become "a kind of Russian roulette . . . [the Shuttle] flies [with O-ring erosion] and nothing happens. Then it is suggested, therefore, that the risk is no longer so high for the next flights. We can lower our standards a little bit because we got away with it last time. . . . You got away with it, but it shouldn't be done over and over again like that."[71]

Return to Flight

Returning the Space Shuttle to space after the loss of *Challenger* was a challenging task. While the Rogers Commission investigated the technical and managerial causes of the failure, NASA had the difficult chore not only of redesigning the faulty SRBs, but also of increasing public confidence in its procedures. On March 24, 1986, well before the detailed causes of the Shuttle's failure were definitively established, the new Associate Administrator for Space Flight, former astronaut Richard H. Truly, announced a strategy for returning the Shuttle to flight status. [II-40] Among other things, his memorandum called for reassessing the entire program management structure and operation, and it laid out a plan for a "conservative return to operations."

Three weeks before Truly's memo, veteran astronaut John W. Young wrote a highly critical memorandum critiquing the management of the Shuttle program and outlining many of the steps needed to assure safety of flight. His views were representative of many who had been aware of the increasing acceptance of risk in Shuttle operations. [II-41] During the hiatus in flight, NASA examined every vulnerable element of Shuttle design and rethought Shuttle launch preparation and operations. NASA instituted many new safety procedures and replaced system components. For example, when first witnessing the huge fireball and destruction of *Challenger*, many engineers immediately concluded that one of the SSME turbopumps, which were highly susceptible to breakdown, might have failed. NASA used the "standdown" to go over the SSME piece by piece to improve its safety and reliability. NASA also increased its contractor staff at Kennedy Space Center to handle the load of new procedures for safety and quality assurance and documentation paperwork. The amount of time NASA technicians took to refurbish the orbiters after flight, to prepare the entire Shuttle system for launch, and to follow new safety and quality procedures more than doubled. The procedures were not only lengthened but became more complicated and intensive, making it increasingly doubtful that NASA could ever achieve its planned yearly launch rate of twenty-four flights, even if sufficient funding for Shuttle payloads and launch services became available to support such a rate.[72] Most important, however, NASA redesigned and tested the Shuttle's solid rocket motors so they would be much less likely to fail again, especially at the joints between motor segments.

---

70. *Ibid.*, p. 93.

71. Richard Feynman, quoted in *ibid.*, p. 148.

72. Generally missing in most NASA Space Shuttle briefings of the 1980s was a sense of the connection between launch rate and the overall costs for both payloads and Shuttle launch services. This was a case of radical optimism. Payload costs (on the launch vehicle) hovered between $40,000 and an astounding $650,000 per pound, depending on the amount of inexpensive elements in the payload (such as fuel) and the technical difficulties encountered in designing and building the spacecraft. See U.S. Congress, Office of Technology Assessment, *Affordable Spacecraft: Design and Launch Alternatives*, OTA-BP-ISC-60 (Washington, DC: U.S. Government Printing Office, January 1990).

The shock of losing *Challenger* and its crew also forced officials within the Reagan administration to reconsider what types of payloads the Shuttle would carry. For example, well before the failure, some observers had complained that using the Shuttle to launch commercial communications satellites, which could routinely be launched by ELVs, was a waste of federal resources and competed with possible commercial ELV efforts (see Chapter 3). In August 1986, the administration issued a statement on Shuttle use, followed by a formal policy statement in December. [II-42, II-43] That policy restricted Shuttle payloads to those requiring the unique capabilities of the Shuttle or needing the Shuttle for national security purposes. In particular, the Shuttle would no longer be used to launch commercial communications satellites.

The costs of losing *Challenger* were high, not only to the crew members and their families, but also in economic terms. NASA's Office of Space Flight estimated that the nation lost about seventy equivalent Shuttle flights over a period of ten years as a result of the loss of *Challenger*, as well as the loss of two Titan 34Ds and the Atlas-Centaur within a few months.[73] Europe's Ariane launched many of these lost payloads. Others were launched much later on ELVs or were never launched.[74]

The Reagan administration and Congress moved relatively quickly to replace the lost orbiter. NASA was able to proceed promptly with construction because, in April 1983, the agency had awarded Rockwell International a contract to construct long-lead-time structural spares, which were to have been completed by 1987. In part, the 1983 decision was prompted by the concern that eventually a fifth orbiter would be needed to handle the expected demand for Space Shuttle launch services. NASA officials also wanted to have crucial replacement parts on hand in case of a major failure of the Shuttle system. The administration requested funding to build a replacement orbiter in mid-1986. In an unusual move, Congress approved the entire package of funding of $2.1 million as part of a supplemental appropriations for fiscal year 1987.[75] The new vehicle (OV-105) was delivered to Kennedy Space Center in May 1991 and made its first flight in May 1992.[76] Congress had directed NASA to establish a contest to name the orbiter, involving elementary and secondary school students. In May 1989, President George Bush announced that the vehicle would be named *Endeavour*, after Captain Cook's famous ship.

On September 29, 1988, the Shuttle *Discovery* lifted off Pad 39B at the Kennedy Space Center, conveying a crew of five into orbit (STS-26). [II-44] *Discovery* also carried the replacement for NASA's Tracking and Data Relay Satellite (TDRS), one of the payloads lost when *Challenger* exploded in January 1986. The successful flight of *Discovery* and launch of TDRS held special significance because it marked the return of the Space Shuttle program to flight status and the end of a painful reevaluation of U.S. access to space. As an editorial in *Aviation Week & Space Technology* opined, "The launch, witnessed by the largest gathering of spectators and press since the Apollo 11 launch to the Moon in 1969, was balm to the wounds remaining from the *Challenger* accident. It was a long time coming. . . . It was a moment worth waiting for. . . . The *Discovery* mission should be savored as a triumph for NASA, the U.S. space program and the nation." The article also quoted

---

73. Cited in U.S. Congress, Office of Technology Assessment, *Launch Options for the Future: A Buyer's Guide*, OTA-ISC-383 (Washington, DC: U.S. Government Printing Office, July 1988), p. 23.

74. Most of the payloads eventually launch on ELVs had to be reconfigured, as the support points had been configured for horizontal integration into the Shuttle, rather than the vertical configuration required for ELV launch. This shift sometimes imposed substantial additional costs.

75. Normally, Congress is reluctant to fund an entire project in one appropriation because of the impact on the budget of any one year. However, proponents of the fifth orbiter successfully argued that full funding would result in lower overall costs. In fact, the funds for the fifth orbiter were taken from excess Air Force appropriations.

76. This first flight of OV-105 was used to rescue the Intelsat VI satellite, which had been left stranded in LEO.

Kennedy Space Center Director Forrest McCartney, who observed that "[it] was a great day for America . . . today we stand tall."[77]

The second and last Shuttle flight of 1988 (STS-27) took place nine weeks later on December 2, during which the orbiter *Atlantis* carried a classified DOD satellite into high-inclination orbit.[78] The success of this flight added to NASA's (and DOD's) confidence in the revised launch procedures.

The loss of *Challenger* had forced NASA to reexamine the risks of human spaceflight, to examine more closely the methods used to evaluate and reduce such risks, and to be more forthcoming with the American public about them. Some NASA officials had inadvertently slipped into thinking that the Space Shuttle was nearly as reliable as a commercial aircraft. However, aircraft typically have empirically derived reliabilities (successful flights divided by attempts) approaching 99.9999 percent, based on many thousands of flights of essentially identical vehicles. Prior to the Shuttle's first launch, NASA had faced the difficulty of estimating flight risks based on detailed estimates of previous experience with subsystems, extensive testing of new subsystems, and the amount of redundancy built into critical systems. Based on such considerations, NASA designed each orbiter to have a 97-percent probability of lasting 100 flights, which leads to a requirement that each individual Shuttle flight have a reliability of at least 99.97 percent. Actual Shuttle reliability was uncertain, but one NASA-funded study estimated that it lies between 97 and 99 percent.[79]

After operations begin, estimations of reliability can also be based on statistical analysis of observed successes and failures, although with most launch vehicles such analysis involves the statistics of small numbers.[80] For example, using a simple statistical analysis, the congressional Office of Technology Assessment estimated for illustrative purposes that if STS reliability were assumed to be 98 percent, NASA would face a fifty-fifty chance of losing an orbiter within thirty-four flights.[81] [II-45] Whatever the actual reliability, this analysis led to the conclusion that reducing, rather than enhancing, the flight rate would be a prudent way to reduce Shuttle losses over time. The 1986 policy that encouraged federal agencies to launch on commercial ELVs when possible helped reduce the pressure on Space Shuttle launches. It also increased the resilience of the launch fleet because it made it possible to recover from a launch failure of a single vehicle more quickly than was true prior to January 1986—a concern of great importance to military planners who must have the greatest possible access to space.[82] However, too few Shuttle flights might increase flight risks, because the skill level of Shuttle launch crews might degrade between launches. Since 1988, NASA has kept the rate of Shuttle flights relatively low (five to seven per

77. "Back to Space!," *Aviation Week & Space Technology*, October 3, 1988, p. 7.

78. According to news sources, *Atlantis* carried a Lacrosse imaging radar satellite, a supposition that is strengthened by the fact that *Atlantis* entered a 57-degree orbit. See Craig Covault, "Atlantis' Radar Satellite Payload Opens New Reconnaissance Era," *Aviation Week & Space Technology*, December 12, 1988, pp. 26–28.

79. L-Systems, Inc., *Shuttle/Shuttle—Cooperations, Risks, and Cost Analyses*, LSYS-88-008 (El Segundo, CA: L-Systems, Inc., 1988).

80. Most statistical analyses of launch system reliability are further hampered by the changes that are made in the system after a failure to improve it; this introduces new unknowns into the analysis. Furthermore, for the STS, each of the four orbiters are somewhat different, and many upgrades and other changes are made in the subsystems between flights. Therefore, each launch can in many respects be considered as nearly the first of its kind. Nevertheless, one can obtain a rough statistical estimate of reliability by assuming that all Shuttle launches are roughly identical.

81. U.S. Congress, Office of Technology Assessment, *Round Trip to Orbit: Human Spaceflight Alternatives* (Washington, DC: U.S. Government Printing Office, August 1989), pp. 6, 25.

82. Resilience is a measure of the ability to recover from a launch failure. High resilience can be accomplished by repairing the failure quickly and employing a launch surge strategy to catch up on waiting launches or by using other launch vehicles (assuming launch vehicles are relatively interchangeable). Before the development of the Titan IV, heavy payloads could only be launched on the Shuttle.

year) and improved its on-time launch performance, suggesting that such a rate provides a good balance between safety and costs.

The Soviet Shuttle

Before NASA officials were able to savor fully the return of the Space Shuttle to flight status, the Soviet Union demonstrated its capacity to build and launch its own shuttle. In a move that mirrored the increasing openness of Soviet society during the regime of General Secretary Mikhail Gorbachev, early in 1988, Soviet officials released drawings and descriptions of their space shuttle.[83] Later in the year, on November 15, rocket engineers successfully launched the shuttle *Buran* (meaning "snowstorm") into orbit, attached to the all-liquid Energiya heavy-lift launch vehicle.[84] The flight was automated; no crew members were aboard (Figure 2-7). After two orbits, flight controllers landed *Buran* on a runway about ten kilometers from the Baikonur Cosmodrome launch pad.

Although *Buran* superficially resembled the U.S. Shuttle orbiter, in detail its concept was rather different. For one thing, in keeping with the Russian approach to new human spaceflight undertakings, the first flight was fully automatic—no cosmonauts were aboard, although the orbiter was reportedly capable of carrying ten crew members. Second, *Buran* carried no rocket engines. Finally, unlike the integrated SRBs, external tank, and SSMEs of the U.S. Shuttle, Energia was a stand-alone vehicle capable of launching up to 220,000 pounds to LEO, including the *Buran* orbiter. Although it lasted only two orbits, the flight was an impressive achievement, but one that was not followed up either with additional flights or the crafting of other orbiters. While the weakness of the Soviet space program had not yet become fully apparent in the United States, the program was past its zenith. By 1991, the Soviet Union and its economy had collapsed, taking with them the will to continue to invest large sums in space achievements. In a few years, *Buran* became an exhibit in a Moscow park, and the Energiya launcher was never used again to lift payloads into orbit.

Figure 2-7. The former Soviet Union's unmanned shuttle, Buran, stood ready on the launch pad with the Energiya launcher in late 1988. It would make only one flight.

---

83. "Soviet Union Developing Range of Manned, Unmanned Launchers," *Aviation Week & Space Technology*, March 28, 1988, pp. 52, 53, 58.

84. Craig Covault, "Soviet Shuttle Launched on Energia Booster," *Aviation Week & Space Technology*, November 21, 1988, pp. 18–21.

Variations on the Shuttle Theme

Beginning well before the Space Shuttle actually flew, engineers considered a wide variety of technical options for improving or extending the Shuttle's basic capabilities. These included adding to its lift capacity, carrying civilian passengers, and extending the stay time on orbit. The impetus for such studies derived from the firm belief among some observers that once the Shuttle became operational, the demand for launch services would grow quickly, making it attractive to add significantly to overall launch capacity. Among the ideas driving such thinking was the photovoltaic solar power satellite, which if built would have required lofting millions of kilograms of materials into geosynchronous orbit and space workers into LEO.[85] Concepts developed during the mid-1970s ranged from simply adding additional smaller solid rockets to the SRBs, to substituting large liquid rocket boosters for the SRBs, to building a fly-back booster.[86] Concepts also included ideas as diverse as a passenger-carrying orbiter capable of taking several tens of passengers to and from orbit and a strictly-cargo vehicle based on using the SRBs, the external tank, the SSMEs, and a cargo canister to substitute for the orbiter.

In general, these ideas never got beyond the concept stage. Yet, by the late 1980s, as space station planners struggled with the realities of lofting a station into orbit and resupplying it, some experts began to revive such concepts. Among other options, they considered building a heavy-lift launch vehicle that would be capable of launching large station payloads to orbit. The specter of losing an orbiter in the course of station construction, and the large number of Shuttle flights (more than twenty) required for the station then under consideration, led to studies of an alternative, larger cargo vehicle to reduce the number of orbiter flights. The Advanced Launch System (ALS) then under consideration (see Chapter 4) might have served such a purpose, but some NASA engineers argued for a cargo vehicle based on the Space Shuttle.

Initially, this was called the Shuttle-Derived Vehicle; later, the concept became known as the Shuttle-C, for cargo.[87] [II-46] Because the design of the Shuttle puts the SSMEs necessary for part of the propulsion on the orbiter itself, the Shuttle-C cargo carrier would also need to carry liquid engines to reach orbit. NASA considered the option of using the reusable SSMEs in a boat-tail configuration and dropping them off to be recovered in the ocean, but the agency found recovery and refurbishment too costly.[88] NASA engineers decided instead to employ SSMEs that had flown enough times that they were no longer sufficiently reliable for human flight, then letting them burn up in the atmosphere after use. As the concept was developed, the Shuttle-C would have been capable of lofting about 178,000 pounds to orbit from Kennedy Space Center. Ultimately, after nearly four years of study, NASA dropped its Shuttle-C efforts, in large part because OMB deemed the vehicle too costly. Furthermore, the move away from using the Shuttle launch for science payloads that could fly on ELVs removed most of the non–space station launch pressure on the Shuttle.

---

85. Peter E. Glaser, "Power from the Sun: Its Future," *Science* 162 (November 22, 1968): 857–86. For a description and assessment of solar power satellite concepts of the late 1970s, see U.S. Congress, Office of Technology Assessment, *Solar Power Satellites*, OTA-E-144 (Washington, DC: U.S. Government Printing Office, August 1981).

86. M.W. Jack Bell, "Space Shuttle Vehicle Growth Options," paper presented at the American Institute of Aeronautics Conference on Large Space Platforms: Future Needs and Capabilities, Los Angeles, CA, September 27–29, 1978.

87. Theresa M. Foley, "NASA May Seek Proposals for Shuttle-Derived Booster," *Aviation Week & Space Technology*, June 29, 1987, pp. 24–25.

88. Craig Covault, "Shuttle-C Unmanned Heavy Booster Could Simplify Space Station Launch," *Aviation Week & Space Technology*, August 15, 1988, pp. 87–88.

The Advanced Solid Rocket Motor

The failure of the Space Shuttle's solid rocket motor had repercussions for NASA's Shuttle program that extended far beyond the redesign of the motor. Proponents of both liquid boosters (pump-fed and pressure-fed) and more advanced solid rocket designs argued within NASA and before Congress that a major overhaul was needed. In addition to providing additional safety, the proposed designs would have improved the payload capacity of the Shuttle, which fell far short of the expected 65,000 pounds placed in the standard twenty-eight-degree LEO 110 nautical miles above Earth's surface. As a result of weight growth during manufacture and early operations, the Shuttle was capable of carrying a maximum payload to this orbit of only 48,000 pounds. However, some payloads, particularly space station components, were expected to weigh more.

During the period after the Shuttle returned to flight, NASA engineers explored two new solid rocket designs—the Advanced Solid Rocket Motor (ASRM) and an improved Redesigned Solid Rocket Motor (RSRM). The ASRM was a totally new design that would use a new manufacturing process, allowing the entire motor to be poured at one time. It would therefore not have joints that might fail. Proponents argued that the ASRM would provide greater safety than segmented boosters. After conducting detailed engineering studies of both liquid- and solid-fuel designs and comparing costs and safety, NASA decided in early 1989 to proceed with the ASRM on the basis that it would result in lower overall costs with comparable flight safety.[89] In March 1989, NASA's Aerospace Safety Advisory Panel noted that "on the basis of safety and reliability alone it is questionable whether the ASRM would be superior to the RSRM . . . until the ASRM has a similar background of testing and flight experience."[90]

Yet, NASA's own analysis disagreed with these findings, and in late April 1989, the agency awarded two contracts for the ASRM to a partnership formed between Aerojet and Lockheed. One contract supported the design and development of the ASRM; the second contract was for the design, construction, and operation of an automated solid rocket motor production facility. NASA designated Yellow Creek, Mississippi, as its preferred government-owned/contractor-operated ASRM production site and the Stennis Space Center in Mississippi as the motor test location. NASA estimated that ASRMs could be ready for a first launch in 1994 or 1995. Agency officials also expected that the ASRM program would help promote a competitive solid rocket motor industry.[91]

The ASRM was never built. After NASA built the plant in Yellow Creek, Mississippi, and began to outfit it, Congress began to have second thoughts about the increasing costs of the ASRM program. In October 1993, Congress voted to shut down the ASRM program as a cost-saving move. NASA then decided to put greater emphasis on improving the RSRM.

Space Shuttle in the 1990s

Once NASA was assured that the redesigned solid rocket motors worked safely, that the operation of the SSME improved, and that other safety-related issues were addressed, the space agency began to operate the Space Shuttle on a more regular basis, and launches had fewer delays. In fact, by the late 1990s, NASA felt that it could hand over the day-to-day

---

89. Proponents of solid rocket motors argued that such motors, if properly designed, are nearly as safe as liquid rocket motors that are by their very nature much more complicated and suffer from a greater number of possible failure modes.

90. Aerospace Safety Advisory Panel, *Annual Report for 1988* (Washington, DC: NASA Headquarters, Code Q-1, March 1989), p. 3.

91. NASA, "Space Shuttle Advanced Solid Rocket Motor—Acquisition Plan," March 31, 1988, p. 3, NASA Historical Reference Collection.

operations of the Shuttle to a private contractor, United Space Alliance. [II-47] The reusability of the orbiter also made it possible for NASA to demonstrate the Shuttle's ability to return payloads from orbit. For example, in 1990, STS-32 returned from the Long Duration Exposure Facility, which had been in orbit since 1984, when it was deployed by STS 41-C. After the communications satellite Intelsat VI was placed in an unusable orbit by a Titan III rocket in March 1990, NASA astronauts aboard STS-49 in May 1992 captured the satellite and redeployed it after attaching a new perigee kick motor to place it in geosynchronous orbit. In December 1993, the Shuttle rendezvoused with the Hubble Space Telescope, which had been launched with a misshapen primary mirror; the Shuttle crew was able to install equipment on the telescope to correct this mistake and perform other servicing tasks. Such feats, while demonstrating the utility and flexibility of the Space Shuttle, were generally overshadowed by the Shuttle's high operating costs, and NASA began gradually to focus more on the use of the Space Shuttle for use in constructing and operating the International Space Station.

The 1993 agreement between the Russian Federation and the United States to include Russia as a partner in the International Space Station had a major effect on Space Shuttle's operation during the 1990s.[92] On one hand, Russia agreed to launch part of the station and to assist in resupply, reducing the burden on the Shuttle. On the other hand, the United States agreed to place the station in a 51.6-degree orbit, which reduces the payload the Shuttle can carry to an orbit with that high of an inclination. Furthermore, Russia and the United States agreed to a combined Shuttle-*Mir* program as a precursor to International Space Station's construction. As NASA argued before Congress, this program would not only give NASA and the Russian Space Agency valuable experience in working together before the launch and assembly of the International Space Station, it would also test the Shuttle system's ability to reach a high orbit reliably with a tightly constrained launch window.

The first Shuttle launch to the Russian space station *Mir* took place during June 1995 on STS-71 (Figures 2–8, 2–9, and 2–10). On June 29, the Shuttle *Atlantis* docked with *Mir* to deliver two Russian cosmonauts and to return NASA astronaut Norman Thagard to Earth after 115 days aboard the Russian station. The Shuttle-*Mir* program was completed with STS-91 in June 1998 after nine successful dockings with *Mir.* On December 4, 1999, the Shuttle *Endeavour* (STS-88) launched the first component of the International Space Station into orbit, marking at long last the start of the Shuttle's use for which it was primarily designed—transport to and from a permanently inhabited orbital space station.

Conclusion

As the documents following this essay illustrate, the design of the Space Shuttle was a compromise among many technical and political considerations. During its conception, right on through to its development and use, virtually every element of the Shuttle's design and use was criticized by someone—sometimes for technical reasons, sometimes for its high costs, and sometimes for questionable NASA decisions. In retrospect, perhaps the most serious of the criticisms was that leveled at the set of policies that led to the attempt to require the use of the Space Shuttle for all U.S. space transportation needs.[93] Nevertheless, this compromise design, while expensive and complicated to operate, is today the world's most advanced and versatile launch system. Although NASA and its contractors have explored numerous alternatives to launching human crews to and from space (see Chapter 4), none are likely to replace the Space Shuttle for at least another decade.

---

92. U.S. Congress, Office of Technology Assessment, *U.S.-Russian Cooperation in Space,* OTA-ISS-618 (Washington, DC: U.S. Government Printing Office, April 1995).
93. Logsdon, "The Space Shuttle Program: A Policy Failure?"

Figure 2–8. A member of the crew on the Russian space station Mir *took this photo of the orbiter* Atlantis *over the southern Aral Sea prior to rendezvous. With the payload doors open, the Spacelab science module and the docking mechanism can be seen on June 28, 1995. (NASA photo)*

Figure 2–9. Taken the same day, this photo shows Atlantis *approaching the docking node on the Kristall module of the* Mir *space station. (NASA photo)*

Figure 2–10. This shows the historic docking of the two spacecraft on June 29, 1995, with the two crews meeting in the insets. (NASA litho)

Document II-1

Document title: Ad Hoc Subpanel on Reusable Launch Vehicle Technology, "Report for presentation to the Supporting Space Research and Technology Panel," September 14, 1966, pp. 1–8.

Document II-2

Document title: Supporting Space Research and Technology Panel, "Final Report, Ad Hoc Subpanel on Reusable Launch Vehicle Technology," submitted to the Aeronautics and Astronautics Coordinating Board, September 22, 1966, pp. 7–10.

Source: Both in NASA Historical Reference Collection, NASA History Office, NASA Headquarters, Washington, D.C.

During the 1960s, the Aeronautics and Astronautics Coordinating Board (AACB) was the primary coordinating body between NASA and the Department of Defense (DOD) on aeronautics and space issues. Both agencies had begun to think through their future space transportation needs by 1965. The AACB Supporting Space Research and Technology (SSRT) Panel established an "ad hoc subpanel" to examine the technology needs if a reusable launch vehicle (RLV) concept were to be pursued. Although there had been some prior thinking within government and industry on such vehicles, this group's work was among the first to give focused attention to the technological and economic foundations for an RLV development effort. Only the summary section of the subpanel report to the SSRT Panel appears here, as well as only the memo and comments on economic aspects of reusability from the final report submitted to the AACB.

Document II-1

[original stamped "CONFIDENTIAL," "OFFICIAL USE ONLY," and "UNCLASSIFIED"]

Report
of the
AD HOC SUBPANEL ON REUSABLE LAUNCH VEHICLE TECHNOLOGY
Supporting Space Research and Technology Panel
Aeronautics and Astronautics Coordinating Board

**for presentation
to the
Supporting Space Research and Technology Panel**

September 14, 1966 . . .

[1]   SECTION I
SUMMARY

The Aeronautics and Astronautics Coordinating Board (AACB) established the Ad Hoc Subpanel on Reusable Launch Vehicle Technology (SSRT) to review and assess the adequacy of the technologies which directly support reusable launch vehicle systems. As defined in the Terms of Reference (Appendix A), "This supporting technology includes aerodynamics, structures and materials associated with such vehicles, as well as lifting reentry, recovery devices, and supersonic combustion engines and rocket propulsion."

Due to the large number of technologies involved, the Subpanel has been selective in its reviews both as to subject matter and detail.

It is important to note that no single, most desirable vehicle concept could be identified by the Subpanel for satisfying future DOD and NASA objectives. Consequently, a number of reusable launch vehicle configurations were selected by the Subpanel and operating modes of greatest potential interest to the DOD and NASA were defined to provide a realistic means for the identification and assessment of the critical supporting technologies. The selected vehicle concepts included both fully recoverable and partially recoverable reusable vehicles. These advanced concepts were specifically chosen to be typical and representative of future development possibilities, and to reflect a time-phased evolutionary pattern of growth capability consistent with potential needs beyond the early 1970's. Figure 21 [not reprinted here] summarizes the technology status for the selected vehicle concepts; technologies considered critical are highlighted. This report is basically concerned with these critical technology areas.

In deriving these representative configurations a review was made of current launch vehicle and recoverable spacecraft capabilities, extensive planning studies conducted on future vehicle configurations, and current projections of future capability goals. On the basis of this review, it appeared that the present stable of launch vehicles provides a substantial spectrum of payload delivery capability and that the present vehicles either in use or under development could fulfill the requirements of both agencies in terms of payload capability for the next seven to ten years.

While it appears technically feasible to recover selected ballistic stages and components of the present launch vehicle systems (i.e., S-IC), it is not clear that ballistic stage recovery and reuse would be economically justifiable or operationally advantageous even for the case of Saturn V stages (modest launch frequency). [2] Basic questions, concerning system design and operations which critically affect the estimated economic impact, remain for the ballistic mode of recovery. In view of the possible economic gain, an experimental program could aid in reducing uncertainties relating to ballistic flight and terminal recovery, refurbishment operations, and stage or major subsystem reuse, as applicable to both existing stages, i.e., Titan, Saturn, and future ballistic launch vehicle systems.

The most likely area for a new or substantially uprated launch vehicle system in the future appears to be in the 60,000 to 100,000 lb. payload delivery category. This potential need is predicated on the basis of higher energy orbit requirements and a consistent historical trend toward heavier payloads for manned space flight systems, rather than specific planned missions. In this regard, it is also noted that the manned spacecraft system will impose additional weight on the launch vehicle, particularly if substantial on-orbit spacecraft propulsion and reentry aerodynamic maneuvering capability are required. There is also a possible need for a very large vehicle to provide a payload delivery capability, considerably beyond the Saturn V or uprated Saturn V capabilities, for NASA deep space missions.

When requirements dictate the development of a substantially new launch vehicle, partially and fully reusable concepts must compete with advanced expendable concepts in the selection of the most economical and operationally desirable approach. Research and development costs of reusable launch vehicles result in significant amortization penalties at the projected launch rates. On the other hand, a vehicle capable of autonomous, reliable operation can be made less dependent on world-wide support activities during launch, on-orbit, and recovery, and may thereby permit a significant reduction in surface support operations, the economic value of which has not been adequately assessed. In any event, both the expendable and reusable avenues to future vehicle development should remain open.

In the area of spacecraft, it appears highly probable that an advanced unmanned or manned spacecraft capable of land recovery and reuse may be required in the mid-1970 time period. Current spacecraft systems are well suited to today's programs but are limited in terms of their applicability to more ambitious operational programs. Air snatch of data capsules from space has been demonstrated. Remarkable success has been demonstrated in manned space operations. The Gemini and Apollo spacecraft can meet the current manned spacecraft requirements. However, the basic characteristics of these spacecraft are fixed in terms of size, shape and operational modes. These spacecraft have limited cross-range capability (approximately 40 n.mi. [nautical miles] cross-track during return from low earth orbit), are constrained during launch and for return by [3] sea state, atmospheric and daylight conditions, and are exposed to water recovery which can increase the costs associated with recovery and refurbishment operations. These systems are supported by extensive deployment of surface forces during launch and reentry, and by extensive ground station support during orbital operations.

It seems probably that future desired spacecraft capabilities will include unmanned and manned reusable vehicles having capabilities of autonomous operation on orbit and the ability to touch down at selected land sites under unfavorable weather conditions. The current and planned programs of both agencies appear to be well directed toward this goal. The critical technical areas associated with such a spacecraft are also shown on Figure 21. Pursuit of these critical areas is considered by this Subpanel as a technology goal of major importance. A further technology goal of equal importance and somewhat longer term significance is the development of technology associated with an integral upper-stage spacecraft which could offer improved operational capabilities. This goal includes virtually all of the technological problem areas related to reusable vehicles and, consequently, offers a convenient framework for organizing the technology activities recommended for the coming years. Such an integral upper-stage spacecraft is included in the selected vehicle concepts, and the critical or limiting technologies associated with it are shown on Figure 21.

The technologies assessed by this Subpanel are limited to aerodynamics, structures and materials, rocket propulsion, and air-breathing propulsion. No attempt has been made to assess the technologies associated with guidance, space power, command and control, and other functions which will be required of future space systems.

The most serious deficiency in the aerodynamics of reusable launch vehicles is the small amount of wind-tunnel data on realistic vehicle configurations incorporating necessary stability, control, propulsion, heat protection, terminal descent, and landing features. The limited configuration analysis and testing possible at current levels of effort are insufficient to assess impact of technology uncertainties on system capabilities for design optimization, system evaluation, or development decisions. Early development of an operational system would require excessive design conservatism with weight and performance penalties.

Aerodynamics technology is sufficiently well advanced to support the development of reusable ballistic spacecraft, except for land-landing systems, and is advancing at a reasonable rate on moderate L/D [lift-to-drag ratio] lifting-body configurations. The technology of higher L/D spacecraft and of integral upper-stage/spacecraft combinations is less developed.

[4] Present aerodynamic test facilities do not adequately simulate the high-speed flight environments of reusable launch vehicles and lifting reentry spacecraft. The most critical need is for hypersonic facilities which can achieve high Reynolds numbers and adequately simulate turbulent flow on large detailed models of complete configurations. In addition, high-enthalpy facilities are needed to determine real gas effects at high hypersonic speeds.

There are a number of pressing structures and materials problems associated with reusable launch vehicles and advanced maneuvering spacecraft which will pace the availability of efficient operational designs. Vertical take-off and horizontal landing launch vehicles pose problems in thermal protection systems which will have long life and can be reliably inspected and reused, or can be refurbished and reused. Tank configurations and arrangements compatible with good aerodynamic designs pose significant structural problems. More advanced vertical and horizontal take-off launch vehicles will pose additional problems in fabrication of lightweight structures employing cryogenic tankage having long life and capable of many reuses. Second-generation maneuvering reentry spacecraft capable of reuse are similarly paced by long-life thermal protection systems capable of refurbishment at low cost. Reusable spacecraft integral with the upper stage of the launch vehicle combine the most severe structural design problems. Definition of realistic configurations would greatly assist structures and materials programs in attacking these problems.

It is not generally realized that demonstrated durabilities of existing large rocket engines offer promise of up to 50 reuses before major overhaul. However, routine inspection, maintenance and refurbishment would be difficult and costly for reusable applications and engine modifications to enhance reusability may be very costly for these cases. Thus, use of existing rocket engines in future reusable systems, while feasible, may not yield the desired economies of operation, and should only be considered in conjunction with Near Term, partially reusable vehicle concepts. The advanced high-performance O_2H_2 [liquid oxygen/hydrogen] engine demonstration program of DOD and NASA will provide a basis for future engine development specifically for reusable vehicles. Reuse and low maintenance cost is a design objective of this engine technology demonstration program. This program is a forerunner of future engines applicable to first and second reusable O_2H_2 stages and high-performance expendable stages. For expendable first-stage applications, this liquid rocket concept must compete with demonstrated large solid motor technology. The accumulative large solid motor technology capability is expected to receive consideration in any new large launch vehicle definition and development.

Advanced spacecraft are expected to utilize existing storable propellant technology in initial operational phases. While multi-start [5] space propulsion systems have been successfully flown, these engines were not designed with low-cost maintenance criterion. High-energy propellant technology is of interest for reusable spacecraft requiring high orbital maneuvering capability. An advanced development program having applicability to such spacecraft is presently planned by DOD.

Air-breathing propulsion systems offer promise for horizontal take-off horizontal landing first-stage use in the Mid Term period. For this application, a hydrogen fueled turboramjet utilizing subsonic combustion could be developed by the mid-1970's. However, the required capability has not been fully demonstrated to date. Of primary importance is high installed thrust-to-weight turbomachinery. A hypersonic air-breathing system would present substantial vehicle integration problems; effective coordination with future aerodynamics and structures/materials efforts related to these applications is required. More advanced air-breathing propulsion systems involving supersonic combustion are too indistinct at this time to permit anything more than a preliminary assessment in terms of applicability to reusable first-stage launch vehicles. Further applied research is needed to establish performance and fully define the interrelated aero-thermo-structural problems of supersonic combustion propulsion systems. A major problem in developing an air-breathing propulsion system is ground test facilities. While current facilities are adequate for large full-scale turbomachinery development to Mach 3.5, these facilities are inadequate for large ramjet development to substantially higher Mach numbers. Small-scale ramjet research can be conducted adequately to about Mach 7.

At this point it is concluded that system design, integration, and evaluation studies of promising reusable launch vehicle and spacecraft concepts are needed to provide specific and continuing guidance to technology programs. Such studies would provide realistic configurations of sufficient interest to warrant point designs and wind-tunnel testing, and would assure necessary consideration of the more promising structures and thermal protection systems, propulsion system integration, control, terminal descent, and landing features. These studies should be highly selective and provide a basis for effective coordination and balance between the various technology disciplines.

The Subpanel has found a substantial amount of research and advanced technology effort being performed in aerodynamics, materials, structures, and propulsion that is applicable to reusable launch vehicles and spacecraft. These activities are summarized in tabular form in Section V. However, much of this effort is directed primarily toward advanced manned spacecraft that are recovered but not necessarily reused, manned hypersonic-cruise vehicles, and expendable launch vehicles.

[6] There is no assurance that these activities alone will provide the balanced, integrated technology base needed to support a reusable vehicle or spacecraft development decision in the future.

The Subpanel has not been entirely successful in sharply defining boundary conditions within which the various technologies should be advanced. The difficulties experienced by the Subpanel, however, are in part a reflection of the disciplinary rather than systems approach employed in this area in recent years. The approach recommended for future activities, consisting of technology programs integrated and guided by means of selective system studies, should contribute substantially in defining more precisely and solving the problem areas limiting the evaluation and future design of effective reusable configurations. General recommendations are included for each technology area within this report. However, the Subpanel has identified the major areas which should receive priority as discussed in preceding paragraphs.

The Subpanel has found a strong mutual interest in a[n] uninhibited and effective two-way flow of information between DOD and NASA on essentially all aspects of the research and development activities discussed herein. Present DOD/NASA coordination procedures are adequate in the area of technologies associated with reusable launch vehicles and spacecraft. Continuation of this Ad Hoc Subpanel is considered unnecessary.

The technology goals and recommendations of this Subpanel should be of value to the field organizations of both agencies in planning their future technology programs in the areas discussed in this report. The Subpanel recommends that the Supporting Space Research and Technology Panel review the area of reusable launch vehicles and reusable spacecraft in the future to assure that the following principal recommendations of this Ad Hoc Subpanel on Reusable Launch Vehicle Technology are pursued. These principal recommendations are:

1.   Selective systems design, integration and evaluation studies should be initiated to provide a definitive basis for establishing suitable technology goals, for guiding the direction of technology programs, and to assure effective coordination and balance between interrelated efforts in the various technological disciplines involved.

2.   Aerodynamics configuration research on reusable launch vehicles should be increased in conjunction with the above system analyses to permit quantitative assessment of limiting technologies and evaluation of promising concepts in terms of their technical feasibility, sensitivity to aerodynamic parameters, operational capabilities, and costs.

[7] 3.   Where systems studies and configuration research identify areas of sufficient technological uncertainty on reusable configurations of interest, the required technological programs should be undertaken to assure that valid comparisons of such reusable configurations can be made with advanced expendable launch vehicle concepts, and to provide an adequate technological basis for future development decisions.

4.   Greater effort should be applied to investigation of the deployment and performance characteristics of maneuverable terminal descent systems for soft earth landing of either ballistic or decoupled lifting reentry spacecraft.

5.   Configuration research in wind tunnels on advanced maneuvering spacecraft and integral upper-stage combination configurations should be increased to determine their aerodynamic characteristics and performance capabilities.

6.   New hypersonic facilities and modifications to existing facilities should be provided to enable testing large models at high Reynolds numbers and high enthalpy in order to more adequately simulate turbulent flow and real gas effects.

7.   Additional structures and materials effort is required specifically supporting the long-life low-cost refurbishable thermal protection systems required for reusable launch vehicles and maneuvering reentry spacecraft. This effort should be carefully directed and guided by the systems studies.

8.   Analytical studies should be conducted using advanced air-breathing propulsion systems for reusable launch vehicles in the Mid and Far Term time periods. These studies should be incorporated with advanced vehicle configurations and should be closely coupled with the configuration and wind-tunnel studies recommended under Aerodynamics in this report.

9.   Turboaccelerator engine component and demonstrator technology programs should be sustained to assure the turboaccelerator-type engine can be available for Mid Term applications if required.

10. Supersonic combustion component research and demonstrator technology programs should be supported to insure acquisition of technology for future broad application, including possibly an advanced launch vehicle stage.

11. If provisions are made for ground-based test facilities in which full-scale research and development of air-breathing component systems and engines can be conducted (Mach 0–8), reusable launch vehicle propulsion requirements should be considered in defining such a facility.

[8] 12. Studies are needed to define an experimental program which could aid in reducing uncertainties relating to ballistic flight and terminal recovery operations, refurbishment operations, and subsequent vehicle stage or subsystem reuse; experience gained from a flight test program of a current vehicle stage could provide preliminary feasibility demonstration of recovery and the first significant data on ballistic stage recovery and reuse operations.

Document II-2

[original stamped "OFFICIAL USE ONLY"]

[7] TO: Co-Chairmen
            Aeronautics and Astronautics Coordinating Board

SUBJECT: Final Report, Ad Hoc Subpanel on Reusable Launch Vehicle Technology

On 24 August 1965 the AACB established the Ad Hoc Subpanel on Reusable Launch Vehicle Technology under the Supporting Space Research and Technology Panel. The work of this Subpanel is now complete. The Subpanel's findings and recommendations were presented on 14 September 1966 to a joint meeting of the Supporting Space Research and Technology and Launch Vehicle Panels.

The SSRT Panel feels that the attached final report is responsive to the Terms of Reference set down by the AACB and that the Subpanel is to be commended. The document provides valuable guidance to both DOD and NASA for future technology programs relating to reusable launch vehicles and maneuvering reentry spacecraft. Although not included in the Terms of Reference, the SSRT Panel also requested the Subpanel on Reusable Launch Vehicle Technology to prepare a brief assessment of the economic aspects of reusable vehicles, including its views on the relative order of "payoff" in recovery and reuse of spacecraft and launch vehicle stages. The Subpanel has responded with the attached statement.

The SSRT Panel agrees with the summary conclusions and general recommendations of the Subpanel as presented in this report. However, we feel that an economic study in depth is required to provide more specific guidelines for developing the most meaning-ful technology to yield the greatest payoff. We recommend, therefore, that an additional study, focused on the economic aspects of spacecraft and launch vehicle stages, be con-ducted by an appropriate group.

We consider the findings of the Subpanel of sufficient interest to warrant a one-hour presentation at the AACB meeting on 22 September 1966 and request the necessary time be so scheduled.

Mac C. Adams   Donald M. MacArthur
Chairman, SSRT Panel   Vice Chairman, SSRT Panel
Date: 9/22/66   Date: 22 Sept. '66

Attachments (As stated)

[8]   AD HOC SUBPANEL ON REUSABLE LAUNCH VEHICLE TECHNOLOGY

Comments on Economic Aspects of Reusability
Requested by SSRT Panel

The Subpanel concentrated its efforts on the objectives in the Terms of Reference—i.e., to examine the technologies related to reusable vehicles. The Subpanel was not asked to justify reusable launch vehicles nor to determine the conditions under which a reusable launch vehicle might be economically introduced into the inventory. The Subpanel found the issue of vehicle costs to be illusory and recognizes the significance of not being able to penetrate this area since the motivation to pursue reusable vehicles inevitably will involve economic as well as operational considerations.

The difficulties experienced in the cost area were associated primarily with both development and operational cost uncertainties and the impact of future space programs and objectives on vehicle characteristics. Many past studies have made comparative cost studies of advanced reusable vehicle systems, but none were found that offered credible methods for estimating absolute costs which could be compared with confidence against the costs of the existing vehicle inventory and supporting facilities. Some of the cost uncertainties arise from assessment of the technical risks and development difficulties as well as predictions of system size and performance. Other cost estimating deficiencies are related to the economics of overall operational characteristics—such as recovery and refurbishment, intact abort capabilities, and relative independence of ground support during launch, on orbit, and reentry—for which virtually no applicable data could be found. Consequently, included among the various recommendations of the Subpanel are system studies and experimental programs specifically oriented towards acquiring mean-ingful cost and operational data in these areas for concepts of potential interest. It is believed that the conduct of these studies and experimental programs will significantly enhance the validity of future evaluations of the benefits of reusable vehicles.

Nevertheless, the Subpanel found ample reasons to be encouraged by the prospects for reusable vehicles. First, it was noted that one characteristic of the space program in the 1970's will be an increase in manned flight activity in near-earth orbits. The unquestioned requirement for spacecraft recovery in these applications, coupled with the historically demonstrated high cost of such man-rated spacecraft, makes them natural candidates for reusability. The report notes that spacecraft cost several times that of the launch vehicle on a per-pound basis. Costs per pound of spacecraft have ranged as high as $3,000 to $10,000 per pound for manned and unmanned missions with some small and special pay-load components running to $200,000 per pound. Consequently, it is felt that the princi-pal motivation for reusability will develop first in the area of land-landable recoverable spacecraft, and the experience derived from these applications coupled with continued technological advancement will stimulate greater interest in reusable launch vehicles.

[9] The current large launch vehicles such as TITAN IIIC and SATURN IB are now capable of delivering payloads to low earth orbit for $700 to $1000 per pound. This figure could be reduced to $500 per pound in the future. Past studies of advanced reusable launch vehicles have estimated transportation costs at $100 to $200 per pound of payload. Such optimistic assumptions could, however, be achieved only by an investment in a new reusable vehicle development estimated to range from three to seven billion dollars, depending on the system selected and the respective degree of reuse. Such a large development cost and the estimated high unit production costs would of necessity require system and facilities amortization over extended periods of possibly ten, fifteen or twenty years at projected launch rates.

The Subpanel notes that a partially reusable launch vehicle involving recovery and reuse of a stage or certain major components would cost less to develop and might be amortized in a shorter period with fewer flights. For this reason the Subpanel has emphasized that partially reusable concepts could be competitive with uprated existing systems and advanced expendable vehicles in the 1975 period.

The following perspective on relative order of payoff in reusable space vehicle systems has been developed from a consideration of both technical and economic factors:

1. Recoverable manned spacecraft of demonstrated high costs as well as future unmanned spacecraft with expensive payloads operating in low earth orbits are the first natural candidates for land recovery and reusability.

2. The decision to develop a new launch vehicle will be based on a major new requirement which cannot be met effectively by an existing uprated vehicle rather than on an economic basis alone. At such a time in the future the most likely choice will be between a competitive partially reusable launch vehicle and an advanced expendable system.

3. The integral upper-stage/spacecraft combination is next in relative payoff. Extremely difficult technological problems are encountered due to the severe reentry environment, the probable use of all-cryogenic propellants, and the attendant large surface areas and structural weight penalties. These technical goals are of major importance in our program planning because they also combine the most difficult technical problems of fully reusable launch vehicle systems.

4. Reusable launch vehicles propelled by advanced air-breathing propulsion systems (ABPS) will probably not become operationally attractive until the late 1970's because of the technical difficulties and development time required for such complex systems. Some of the technology required will be developed by the hypersonic aircraft program.

All of these factors and the uncertainties in development and operating costs surrounding reusable launch vehicle concepts and the need for additional studies and technological efforts to resolve these uncertainties are considered by the Subpanel to provide cause for sustained interest in reusable [10] vehicles and to justify the recommendations in this report to establish the technology base associated with such vehicles. Recoverable land-landable spacecraft should also receive early consideration for reusability since these vehicles will afford an excellent opportunity for reducing space operations costs.

M.B. Ames, Jr.
Chairman
Date: SEP 22 1966

Howard P. Barfield
Vice Chairman
Date: SEP 22 1966

Document II-3

Document title: Dr. George E. Mueller, Associate Administrator for Manned Space Flight, NASA, "Honorary Fellowship Acceptance," address delivered to the British Interplanetary Society, University College, London, England, August 10, 1968, pp. 1–10, 16–17.

Source: NASA Historical Reference Collection, NASA History Office, NASA Headquarters, Washington, DC.

This 1968 speech to the British Interplanetary Society by NASA's Associate Administrator for Manned Space Flight, George Mueller, was one of the first attempts to set out a comprehensive vision for the future of the U.S. human spaceflight program after Apollo. Central to making his vision feasible, said Mueller, was a reusable Earth-to-orbit launch system—a "space shuttle." This was one of the first public uses of the term by a senior NASA official. The twelve figures referred to in this speech are omitted here.

[1] I am greatly honored by your action to extend to me the privilege of Honorary Fellowship in the British Interplanetary Society. In bestowing this distinction, you are recognizing the magnificent effort of so many of our people who are taking the initial steps in space exploration. On their behalf and my own, I thank you.

There has indeed been great progress in the seven years since man first ventured out of Earth's atmosphere. In this short span of time, minute in terms of the history of mankind, man's ability to live and work in space has been validated. When two Astronauts step through the hatch of the Lunar Module onto the surface of the moon, man will have come through the threshold of the present into the future. We hope to achieve this goal—the dream of man since time began—within the next year.

With Apollo and the earlier programs, strides have been taken toward the control of a new region of our environment. The learning and testing which were the primary purposes of the Mercury and Gemini programs produced significant accomplishments.

[2] The data accumulated provided a sufficient sample for all to conclude that man can live and work in space for at least 14 days. None of the flight results indicated that there was a physiological or psychological limit to the time he might yet stay in space. Future programs will have to determine these limits if they exist.

The Saturn V launch vehicle is now the foundation of the U.S. manned space program. It is being qualified to make the journey to the moon and back and to carry out the forward programs now planned. It is, however, only the forerunner of other transportation systems which will be needed to extend our knowledge and initiate our utilization of the space environment.

I believe that the exploitation of space is limited in concept and extent by the very high cost of putting payload into orbit, and the inaccessibility of objects after they have been launched. Therefore, I would forecast that the next major thrust in space will be the development of an economical launch vehicle for shuttling between Earth and the installations, such as the orbiting space stations which will soon be operating in space.

The Orbital Workshop shown in the first figure (Figure 1), now under development, is a space station utilizing for its components and its logistics support, stages, modules and spacecraft which were developed in the Apollo Program. It will provide accommodation for 3 people and their equipment for up to a year in orbit.

[3] The Orbital Workshop is the progenitor of space stations that should be used for the conduct of the many scientific, technological and commercial experiments and processes which planners are now describing.

These space stations will be used as laboratories in orbit and will provide the facilities to study and understand the nature of space. They will provide observatories to view the sun, the planets and the stars beyond the atmospheric veil of earth. Stations in orbit will provide bases for continuous observation of the earth and its atmosphere on an operational basis—for meteorological and oceanographic uses, for earth resource data gathering and evaluation, for communications and broadcasting and for ground traffic control. As these stations evolve, other uses will include the manufacture of specialized items utilizing the unique characteristics of the space environments. The basic nature of space offers some natural conditions and circumstances that are not achievable here on earth.

One of the applications of these stations that has intrigued planners for many years has been their use as fuel and supply bases, and as transfer points enroute to high or distant orbits, to lunar distance, or toward the planets.

The orbit of such a transfer station will normally be of low inclination and low altitude for reasons of economy, safety, convenience and flexibility. Many of the missions [4] that require orbit changes could use such a space station with specialized spacecraft which could maneuver to place payloads in desired orbits, either higher or lower in altitude and/or inclination, or to rendezvous with established satellites for inspection, maintenance or retrieval.

Another possibility are operations between a close earth orbit and synchronous orbit as illustrated in the next figure (Figure 2). In these activities, for example, a continuous broadcast satellite could be installed, checked-out or, at a later time, maintained. The service crews could then return to the space station in low-orbit. Or, as shown in the next figure (Figure 3), a spacecraft, fitted for lunar operations, could take on fuel and other supplies from the low-earth orbiting space station.

The performance of a Lunar Module as an example of a transfer vehicle could shift about 225,000 pounds from a 100 nautical mile orbit to a 300 nautical mile orbit and return to the space station. If we use a nuclear powered stage we could transfer 38,000 pounds of payload to synchronous orbit and return, or a payload of 45,000 pounds to lunar orbit and return (Figure 4).

Essential to the continuous operation of the space station will be the capability to resupply expendables as well as to change and/or augment crews and laboratory equipment. A basic consideration is the relationship between the original cost of the space station and the costs accumulated by resupply support operations. Our [5] studies show that using today's hardware, the resupply for a single three-man orbital space station for a year equals the original cost of the space station. This type of cost analysis has led us to carefully evaluate concepts for more efficient resupply systems.

Manufacturing in space, fuel and supply storage for deep space operations, life support for crews on board space stations, require not tons, but thousands of tons of material, to be shuttled in and out of space.

Therefore, there is a real requirement for an efficient earth to orbit transportation system—an economical space shuttle. This need has been under study by long range aerospace planners for over a decade. The objective of these investigations is to find a design that will yield an order of magnitude reduction in operating costs. The elements to which we must look for cost reductions are aircraft manufacturing techniques, aircraft development test procedures, maximum flexibility for multiple use and volume production, long life components for repetitive reuse, and airline maintenance and handling procedures for economy of operation.

The desirable operating characteristics of a space shuttle which would satisfy the needs which have been described are listed on this chart (Figure 5). The shuttle ideally would be able to operate in a mode similar to that of large commercial air transports and be compatible with the environment of major airports.

[6] It would take off vertically, as shown in this concept (Figure 6), from a small pad at an airbase or major airport.

Crews similar in size to those required for intercontinental jet dispatch would service the craft for launch.

The space shuttle, upon its return from orbit, would reenter the atmosphere and glide to a runway landing, with practically no noise. The landing would be completely automated with prime dependence upon the spacecraft guidance system but with ground control backup.

Cryogenic tank trucks containing liquid oxygen and liquid hydrogen would refuel the craft on its pad. Seven years of accident-free experience in handling cryogenic fuels have advanced this technology to practical safety. These non-toxic fuels are 10 times more powerful than gasoline and have demonstrated their efficiency.

The cockpit of the space shuttle would be similar to that of the large intercontinental jet aircraft, containing all instrumentation essential to complete on-board checkout, as shown in this illustration (Figure 7).

Programmable automatic equipment would perform the systems and subsystems tests necessary for take-off and flight support. Malfunction detection would be automatic.

I assume that continental and intercontinental air traffic control centers will have been established so that the space shuttle could take its place in the air traffic and space traffic patterns under these controls.

[7] Interestingly enough, the basic design described above [for] an economical space shuttle from earth to orbit could also be applied to terrestrial point-to-point transport.

If the space shuttle were used as a global transport for point-to-point traffic in military, commercial or cargo service, its safety and comfort standards could be comparable to those of large transport jets.

The economics of the space shuttle must be evaluated in comparison with today's means of accomplishing similar missions.

Until now it has been essential to optimize space transportation systems on the basis of performance. Only a decade ago, technology was pushed to its limits in order to barely achieve orbital flight. Our first Vanguards and Explorers cost in the order of $1,000,000 per pound of payload to fly into space. The next chart (Figure 8) illustrates the economy achieved by the Saturn V, which delivers payload at a cost roughly 3 orders of magnitude less than Explorer I. Extrapolating, we could reasonably expect a cost reduction of at least another order of magnitude, given the will to accomplish it, with present techniques.

If, however, the development of a space shuttle such as I have described were implemented, it seems that a reduction in cost by two orders of magnitude is achievable.

[8] Any significant technological breakthrough in such areas as propulsion and structures would accelerate this process.

The use of a space shuttle for point-to-point global transportation would depend upon its cost equivalence to the then operational supersonic or hypersonic equipment in commercial use.

Current aerospace contractor studies show that, if the cost of rocket engine replacement parts can be reduced to the current level of those of jet engines, the total operating cost of a space shuttle flying a nominal route (New York to Tokyo or 5,850 nautical miles) would be 10.6 cents per passenger nautical mile. Comparison cost rates and times for cruise aircraft and space shuttle are shown in the next table (Figure 9). Although more than supersonic transport, it is less than hypersonic transport even now.

Turning now to the basic elements on which such a cost reduction depends, I believe that a pattern exists in aviation practice for decreasing both development and operating costs of space vehicles. Reliability and hardware maturity are achieved in aircraft flight testing by incrementally expanding the test regime until the full operational envelope is covered, with full recovery of the article for analysis and correction of deficiences [sic] after each flight.

The next chart (Figure 10) displays the contrasting patterns of man hours required for checkout for delivery [9] of spacecraft as against aircraft, as a function of numbers of vehicles.

A second important factor is the cost savings resulting from repetitive use of the same equipment. However, since some components of a space vehicle cost considerably more than others, cost effectiveness evaluations were applied to the various systems and elements of a space shuttle, along with their relation to recovery costs.

The next figure (Figure 11) shows that electronics, engines, power supply, environmental control system and airframe costs exceed the cost per unit volume criterion for recovery, based on our present experience. Therefore, the sub-systems which can be considered for disposal are the adapters and the large tanks for propellants.

This analysis leads to a promising design, the "Drop Tank" configuration shown in the next illustration (Figure 12). It consists of a core vehicle which contains all of the required functional elements for boost and subsequent reentry plus external propellant tanks. The core vehicle is designed for vertical take-off and horizontal landing, and contains all of the high cost equipment including the high chamber-pressure lox/hydrogen engines.

Attached to the sides of the core are large inexpensively manufactured expendable propellant tanks which carry the major part of the fuel required for boost. When the propellant in the external tanks is depleted, the tanks are jettisoned. The [10] remainder of the boost velocity increment required to attain orbital velocity, orbital maneuvers and retrograde is supplied from propellant tanks located inside the core vehicle.

This concept for a space shuttle, extrapolated from a number of proposals, is technologically within the present state of the art.

One problem is, of course, the germination period of from 7 to 15 years for new designs. Jet power, available in 1946, came in to commercial use on the Boeing 707 in 1958. Driving against traditional time lags, the Saturn V system has been developed and used within 9 years of its conception.

It is reasonable to conclude, then, that a space shuttle development program, initiated now, could not be brought to fruition before the end of the 1970's. . . .

[16] No really meaningful estimate of the number of space shuttle vehicles which will be required can be given at this time, for that number is a function, not only of the [17] various missions which the space shuttle will be called upon [to] perform, but it is also a function of the existence of the machine itself. It is interesting to note that in 1954, Business Week, [an] authoritative U.S. magazine, stated that 50 large computers would be required by U.S. industry "in the foreseeable future." Today over 100,000 are in service, all larger and more complex than the original.

In 1945, the then President of one of the world's leading airlines said that he thought 30 large aircraft (D.C.4. vintage) would carry all traffic he could anticipate across the North Atlantic. In the first few years of its existence, nobody needed the telephone. So we see that the space shuttle, by its very existence, may generate the traffic it requires to make it economical.

Arthur Clarke, in THE PROMISE OF SPACE, wrote that ". . . the exploitation of the foreseeable techniques to their limit could result in truly commercial space transport being in sight by the end of this century."

The space shuttle is another step toward our destiny, another hand-hold on our future. We will go where we choose—on our earth—throughout our solar system and through our galaxy—eventually to live on other worlds of our universe. Man will never be satisfied with less than that.

Document II-4

Document title: NASA, Space Shuttle Task Group Report, "Volume II, Desired System Characteristics," revised, June 12, 1969.

Source: Space Policy Institute, George Washington University, Washington, D.C.

As NASA began to investigate the desirability and feasibility of developing a reusable space trans-portation system as part of its post-Apollo activities, the agency created an internal task force to exam-ine the Shuttle concept prior to requesting industry studies. This task force was chaired by Leroy E. Day. Its work represented the first comprehensive NASA examination of a Space Shuttle. There were five volumes in the task group study. In Volume II, an initial listing of the desired characteristics and capabilities of a Space Shuttle were identified; only the summary section appears here.

NASA SPACE SHUTTLE
TASK GROUP REPORT

Volume II

Desired Systems Characteristics

Prepared by:
NASA SPACE SHUTTLE TASK GROUP

JUNE 12, 1969
(REVISED)

RESTRICTED TO GOVERNMENT AGENCY USE ONLY

[no pagination] I. SUMMARY

A. Discussion

The purpose of this volume is to desribe [sic] the basic operational concepts and desirable systems characteristics required of a space shuttle vehicle designed for econom-ic and functionally efficient fulfillment of NASA missions. Total system economics are achievable through the application of operational and system design concepts currently used in air cargo carrier and commercial airlines. A total listing of the desired system char-acteristics may be found in Part B of this section. Ground rules appear in Section II and vehicle basic design precepts are listed below in Section III General.

The desirable system characteristics related to mission functions appear in Section IV thru VIII which consist of pre-flight, launch, on-orbit, return and post flight phases.

Pre-Flight Phase

Large potential cost reductions can be realized by abandoning present day approach-es to launch site vehicle integration, vehicle to payload integration and complete vehicle preflight checkout. An onboard vehicle checkout, system test, and functional analysis sys-tem eliminates extensive and costly ground based equipment. To minimize cost even fur-ther, an integrated launch, loading, and refurbishment facility should be provided to serve logistics and servicing functions. Crew and passenger safety dictates that the ready-to-launch vehicle include provisions to safe the vehicle and perform quick egress. Cost sav-ings will not be implemented at the expense of reduced crew and passenger safety. Major emphasis is placed upon design concepts that return the entire function after liftoff. The vehicle will have design conservatism and all major system redundancies such that single point failures having potential abort implications are minimized. Simplified vehicle ground handling, payload integration, propellant loading and launch pad erection procedures are desirable to provide system flexibility.

Launch Phase

The flight crew and onboard systems should have the capability of performing all tasks during launch. The vehicle should be capable of an all azimuth capability. The vehicle should be designed to lift off within a 60 sec. launch window.

On-Orbit Phase

The onboard autonomous checkout provisions needed for pre-flight lends itself to mission period onboard decision making and will preclude extensive ground based support in the form of real time telemetry and tracking. Present day capabilities have already proven the feasibility of conducting guidance and navigation functions onboard for the entire mission. System operation is to be implemented such that a two-man crew can readily perform all the task[s] associated with launch, orbital flight, rendezvous, docking, reentry, and landing. It is necessary that one man operation be feasible where passenger safety so dictates.

Crew and passenger safety requires a "return to base" mission termination capability for all flight phases starting at lift-off. Major emphasis is placed upon vehicle design concepts which provide crew and passenger safe return. The vehicle will have design conservatism and system redundancies to eliminate failures having potential mission abort implications.

A shirtsleeve environment is desired and this characteristic applies to all mission phases including passenger transfer.

Cargo transfer should be automated as much as possible and require little if any EVA. Cargo handling provisions should be located on the space station.

Docking procedures should be simplified by automatic onboard approach and docking systems ending with a "hard" docked configuration. The vehicle should be capable of rendezvous and docking with passive satellites.

The cargo delivery phase will include a variety of cargos [sic] and cargo/passenger mixes for a variety of missions that have been stipulated. In addition, consideration must be given to special purpose cargo modules to support scientific and commercial satellite placement, maintenance, servicing, retrieval and return. Replacement equipment, liquid propellants, and other expendables have to be handled appropriately and these provisions must be available without modification to the basic vehicle. Inherent cargo adaptability and flexibility are essential for a low cost system that is to be useful for the forecast missions.

Return Phase

Consistent with the autonomous philosophy the vehicle should be self sustaining for the entire (7-day) mission period and capable of all onboard checkout prior to a return.

A once per day return to a landing site selected before deorbit is deemed adequate, and should assist in reducing weather problems developing at the landing field after the deorbit maneuver.

The vehicle design will be commensurate with a reentry cross range of 250 nautical miles to 400 nautical miles. Additional range capability would provide mission flexibility.

Horizontal landings normally will be made at standard jet airfields and should require runways of approximately 10,000 feet. In view of the return to base mission abort concept, thrust augmentation during landing approach and a resulting capability to "go around" will be provided. If an alternate site is used for landing, the ability to ferry the shuttle vehicle back to the primary base for maintenance and prelaunch checkout would be very desirable.

Vehicle landing visibility, handling qualities and landing characteristics should not be more demanding on the pilot than on operational high performance, commercial land based aircraft. Day, night, all weather and automatic landing capability should be provided for all reusable stages.

Post Flight Phase

All reusable stages should have self-ferry flight capability for transport between airports and on-board provisions to quickly place the vehicle in a safe condition following landing. Onboard check-out and ease of module replacement should result in a design goal turn around time (from landing to launch) of less than two weeks. All sub-systems should be designed for minimum maintenance, modular replacement and make maximum usage of standard aircraft type maintenance.

Additional cost advantages will accrue if troubleshooting, repair, replacement, and refurbishment are considered in the design. There is an obvious need to do extensive inspection of the shuttle vehicle heat shield elements and basic structure which will be made less difficult by proper design provisions. Present developments also indicate a need for easy engine replacement even though a number of flights on a single engine are anticipated.

The specific desired systems characteristics are presented in the remaining sections along with rationale substantiations.

B. Listing of Desired Characteristics

Ground Rules

1. All criteria and characteristics deal with the vehicle after it reaches operational status.
2. The vehicle launch site will be located at [the Eastern Test Range].
3. Vehicles should nominally be operated to orbit with a full payload.

General

1. The vehicle should have the following typical capabilities:
   a. up to 50 000 lb up/down cargo
   b. seven days on-orbit life
   c. 2000 ft/sec on-orbit delta velocity for circularization, transfer, rendezvous, docking, launch dispersions, de-orbit and contingencies
2. The vehicle configuration should provide for safe mission termination for major malfunctions occurring during the prelaunch preparations and subsequent to lift-off. The desired safe mission termination capabilities should allow for crew passenger egress prior to lift-off and for intact abort following lift-off.
3. Vehicle preflight and inflight checkout systems should be on-board.
4. The vehicle should have a two man flight crew and should be flyable by a single crewman.
5. The vehicle trajectory design load factors should be 3g to accommodate passengers. The vehicle may be flown on a 4g trajectory when not carrying passengers.
6. The launch site, the primary landing site and the servicing facility should be at the same general location to minimize costs.
7. The vehicle should be designed for maximum on board autonomy such that ground mission operations can be minimized to reduce cost.
8. The vehicle systems should be developed to provide redundant full mission capability and should avoid minimum requirement, minimum performance backup systems concepts.

9. Multiple redundancy system techniques should be adopted that minimize or eliminate system transients caused by system component failures.
10. Subsystems should be designed to fail operational after the failure of the most critical component and fail safe after the second failure. Electronic systems should be designed to fail operational after failure of the two most critical components and fail safe after the third failure.
11. Crew station displays should be designed to eliminate toggle switches and electromechanical gauges and meters, and replace these components with all electronic displays.
12. The crew and passenger environment should be "shirtsleeve."
13. Space-to-ground communications should be available via [a] satellite communications system.
14. The vehicle communications system should provide for the two-way self-validating data transmission.
15. Cargo elements containing hazardous material should have self-contained protective devices or provisions.
16. The vehicle and its systems shall be capable of use for 200 mission cycles with a minimum of maintenance. Capability for a large number of mission cycles is desired.
17. Flexibility will allow technology growth to be incorporated in the vehicle.
18. Standardized electronic interface systems should be developed that interface with a standardized redundant multiplex data bus system.
19. For missions other than logistics, EVA capability should be provided at the expense of the allocated payload weight. The design of the vehicle should not preclude EVA capability.
20. Design of the deployment hatch and deployment mechanism should be compatible with dimensions of the payload bay.

Pre-Flight Phase
1. Systems sensitivity to weather conditions during assembly, checkout and launch should be minimized.
2. Systems sensitivity to fluid consumables loading should be minimized.
3. Contamination control (clean room) operations should be minimized.
4. Payload integration features should include accommodating a variety of payload types which are self-sustaining. Prelaunch payload integration procedures similar to current air-cargo carrier operations are desired.
5. The vehicle should have minimal assembly and checkout requirements at the launch site.

Launch Phase
1. An all azimuth launch capability is desired.
2. Reusable boost stages should be designed for manned operations. The vehicle should be capable of operating in an unmanned mode by using the capability of the automatic landing system.
3. For rendezvous missions, the vehicle should be designed to liftoff within a 60 second launch window.
4. The vehicle should be capable of rendezvous with any low altitude manned satellite in less than 48 hours.

On-Orbit Phase
1. All guidance and navigation functions should be performed on board. The guidance and navigation system should be simple to operate and should not restrict vehicle attitude.

2. A three axis translational system and a three axis attitude control system is required. These systems should be designed to minimum coupling of motions with an attitude and/or translational thruster inoperative.
3. The vehicle should be equipped with an automatic approach and docking capability.
4. The vehicle should be "hard" docked to the space station/base and docking to accommodate personnel and cargo transfer should nominally be accomplished in a single operation.
5. To eliminate interface complications when the vehicles are docked, the vehicle atmosphere and total pressure should be the same as the space station/base.
6. Personnel/cargo transfer should nominally be IVA.
7. Limited transfer of cargo should be possible through the personnel transfer hatch.
8. Total vehicle self-sustaining lifetime should be seven days.
9. Provisions for deployment and retrival [sic] of maximum cylindrical payloads is desired. Normal operation should not include EVA.
10. The vehicle should be capable of rendezvous, station keeping and docking with a passive satellite.

Return Phase

1. Opportunity to return should be available at least once per 24 hours to a single landing site selected prior to lift off. More frequent emergency returns are possible using alternate sites. Consideration should be given to shorter times for specific missions.
2. Return guidance and navigation capability should be onboard.
3. The vehicle should have design characteristics (i.e., planform [sic] loading and trimmable attitude) and reentry flight parameters that will provide low heating rate profiles necessary for maximum utilization of refurbishable thermal protection materials.
4. The vehicle should be capable of making more than one landing attempt at the selected landing site.
5. Landing visibility should be comparable to high performance aircraft standards.
6. Landing characteristics and handling qualities should not require skills more demanding than those required for operational, land-based aircraft.
7. The vehicle should have the capability to land horizontally on runways of approximately 10 000 feet.
8. The vehicle should utilize a landing safety criteria as a guideline for vehicle design.
9. An automatic landing capability should be provided for zero-zero visibility conditions. A manual landing capability should be provided. When the automatic landing system information is not available, the manual landing capability will be capable of meeting the minimum [Federal Aviation Administration] certified requirements.

Post-Flight Phase

1. All reusable stages should be capable of self-ferry flights between airports.
2. The vehicle design should include proper on- board provisions to quickly and easily place the vehicle in a safe condition following landing.
3. Total vehicle turnaround time from landing to launch readiness should be less than two weeks. The removal and replacement time should be minimized with on-board checkout and module accessibility.
4. Subsystems should be designed for minimum maintenance with modular design for removal and replacement making maximum use of aircraft practice. . . .

Document II-5

Document title: Charles J. Donlan, Acting Director, Space Shuttle Program, NASA, to Distribution, "Transmittal of NASA paper 'Space Shuttle Systems Definition Evolution,' " July 11, 1972.

Source: NASA Historical Reference Collection, NASA History Office, NASA Headquarters, Washington, D.C.

Between 1969, when NASA began to seriously study Space Shuttle concepts, and the selection of a final Shuttle configuration in March 1972, many versions were examined. This paper provides an overview of the Space Shuttle configuration studies. The long distribution list is omitted here.

July 11, 1972

TO: Distribution

FROM: MH/Acting Director, Space Shuttle Program

SUBJECT: Transmittal of NASA paper "Space Shuttle Systems Definition Evolution"

   Attached is a paper which documents the evolution of the Space Shuttle configuration.
   I believe this evolution to be a remarkable example of what is generally meant by the term "Systems Engineering." I hope you will find this [en]capsulated history of the shuttle useful to you in discussions of NASA programs.
   Dr. Fletcher has sent this paper to Mr. William Anders, Dr. David and others at the White House, and Mr. Casper [sic; should be "Caspar"] Weinberger at OMB.

Charles J. Donlan . . .

[1]   Space Shuttle
                System Definition Evolution

<u>INTRODUCTION</u>

   In March 1970, President Nixon established six specific objectives for the Nation's Space Program. One of these objectives was to <u>reduce substantially the cost of space operations</u>. The reusable Space Shuttle was identified as one way of achieving that cost objective while providing a new capability suitable for a wide range of scientific, defense and commercial uses. Since that time NASA has conducted extensive in-depth system engineering studies, technology efforts and economic studies to evolve a reusable Space Shuttle system definition that would provide an optimum new space capability within projected budget constraints. This two year systems definition effort culminated on January 5, 1972, when the President announced his decision to proceed with the development of the reusable Space Shuttle. The following chronology summarizes the system definition evolution of the Space Shuttle that led to the President's decision.
   A large number of system concepts have been examined in the search for a configuration that would afford the best relationship between development costs and operational costs. In addition to the technical work, comprehensive economic studies have been completed which scrutinized a substantial number of combinations of traffic models and shuttle systems to help determine the proper compromise between the recurring operational costs and the non-recurring development costs. Figure MH 71-7518B shows the evolution

to the present solid rocket booster with an external tank orbiter and some of the other configurations studied.

[2] REUSABLE FLYBACK SYSTEMS

The initial studies[,] begun in 1969–70, addressed a fully reusable shuttle system which emphasized minimum refurbishment, autonomous on-board checkout, minimum turnaround time, and had the lowest operational cost of any system studied. The operational cost, about $4.1M per flight, is about the same as for the Thor Delta launch vehicle—the most widely used launch vehicle in the United States. The development costs of the fully reusable system, however, approach $10B and reflect the extensive research and development activity associated with developing two large piloted vehicles that possess both the features of a rocket launch vehicle and a hypersonic aircraft.

Further studies yielded a system with a smaller more efficient orbiter by the use of expendable hydrogen tanks, rather than propellant tanks located in the orbiter. The booster staging velocity was lowered from 11,000 feet per second for the fully reusable system to 7,000 feet per second. This allowed use of a heat sink booster so that the development costs were lowered to $8.1B. The expendable tankage, of course, meant somewhat higher operational costs of $4.5M per flight. The high risk and high peak annual funding associated with developing two piloted vehicles still existed and studies for lower cost systems continued.

Eventually, by removing both the liquid oxygen and liquid hydrogen from within the orbiter, NASA was able to devise a much smaller, lower cost orbiter with a single expendable combined propellant tank. The size of the orbiter and its development costs were dramatically reduced while retaining equal performance capability by utilizing this expendable tank for both liquid propellants. The selected orbiter is a delta wing aircraft powered by high pressure hydrogen-oxygen engines.

[3] Time phasing some of the orbiter subsystems received considerable study effort. This was known as the Mark I/Mark II shuttle system. The Mark I orbiter was to use available ablative thermal protection, a J-2S engine developed as an extension of the existing Saturn J-2 engine, and other state-of-the-art components such as existing avionics. Improved subsystems such as fully reusable thermal protection and the new high pressure engine would be phased into later orbiters to achieve the operational system (Mark II). This time phasing reduced expenditures early in the development cycle but the Mark I system had reduced payload and crossrange capability as well as an increased turnaround time of one month. This represented a severe loss in operational capability. Furthermore, the total development costs to achieve the full Mark II system actually increased.

Additional studies indicated that further reductions in orbiter development costs could only be achieved at the expense of compromising the objectives of providing the required flexible orbital capability at low operational costs. The possibility of reducing total systems costs through reducing the size of the payload bay in the orbiter from 4.6 x 18 meters (15 x 60 feet) to 4.3 x 14 meters (14 x 45 feet) and reducing the payload capabity [sic] for a due east launch from 29,500 kilograms (65,000 pounds) to 20,400 kilograms (45,000 pounds) was considered. The additional cost savings were estimated to be only about $70 million in the development program. Furthermore, the orbiter with the smaller payload compartment was unable to accommodate about 10 percent of the projected civil missions and about 37 percent of the projected military missions for a typical mission model for the period 1979–1990. Therefore, the smaller shuttle would have required retention of large expendable boosters in the U.S. launch vehicle inventory to handle the larger payloads[,] thus incurring higher costs than were achievable with the base-line shuttle system.

[4] The Mark I/Mark II Concept which was studied would have used Saturn F-1 engines but nevertheless would have been a costly and relatively high risk undertaking since again,

two manned returnable vehicles were required to be developed. Its development cost is estimated at between $6B and $7B with a cost per flight of approximately $7M. In a further attempt to reduce the development cost, studies were initiated to examine a shuttle configuration utilizing an unmanned ballistic booster.

EVOLUTION TO THE CURRENT SHUTTLE CONFIGURATION

The introduction of the external tank orbiter had a major impact on the booster element of the shuttle system. Since the orbiter became much more efficient, it became possible to let it take even more of the burden of propelling the shuttle into orbit. Staging could therefore occur at about 5,000 feet per second. An important advantage from the use of the external tank orbiter was the opportunity to utilize ballistic liquid boosters or solid rocket motor boosters that are efficient at the lower staging velocities. Their use promised the greatest reduction in development costs.

The ballistic unmanned boosters studied included both pressure-fed and pump-fed liquid propellant boosters and solid propellant boosters. The two liquids compared as follows:

- In the pressure-fed system, the engine would have been a major new development. In the pump-fed system, it would have been a modified F-1 engine (the engines used in the Saturn V booster).
- New manufacturing techniques would be required for the pressure-fed booster; conventional techniques developed for Saturn would be used for the pump-fed.
[5] • Major modification of facilities would be required for the pressure-fed booster; to a large extent, existing facilities could be used for the pump-fed booster with minor modifications.
- The stiff, thick walls of the pressure-fed booster could withstand a moderately high impact velocity, and thus it lent itself to booster recovery. Recovery of the thin-walled pump-fed booster appeared to be of much higher risk.

It was concluded that the pump-fed system had cost advantages and lower technical risk in all aspects except the recovery risk, which appeared large. Of the two liquids, the pump-fed concept was deemed more advantageous in spite of the need to develop complex recovery systems.

Having examined the liquid booster class, a comparison was then made against solid rocket motor configuration. Conventional expendable pump-fed systems currently exist in the series burn configuration where the orbiter engines are ignited after booster shutdown and separation. However, a parallel burn configuration where booster and orbiter engines are both ignited at lift-off takes maximum advantage of the high performance orbiter engines. This parallel burn configuration is particularly attractive for the solids where it is desirable to stage at a low velocity and to minimize the size of solids for operational cost reasons. The pump-fed liquid booster in the series configuration was therefore compared with the parallel burn solid rocket motor booster.

Due to the high cost for each pump-fed booster, recovery refurbishment and reusability are essential[,] while for the [solid rocket motor] this is not so critical. Essentially, the net cost of losing a liquid [6] booster would be much greater than losing a solid, jeopardizing the ability of the shuttle to attain the low costs of recurrent operations. In addition, providing recovery would entail major developmental risks for the liquid but would be simpler for the solids.

Development costs of the solid booster are estimated to be about $700 million lower than those of the liquid booster. Environmental effects for both liquid and solid systems were about the same with one exception—propellants and their exhaust products. The liquid booster would use RP, a kerosene-like rocket propellant, and liquid oxygen, and its exhaust products would be chiefly carbon monoxide, water vapor, and carbon dioxide, along with smaller quantities of hydrocarbons and ammonia. The chief emissions from the solid rocket motors are hydrogen chloride, carbon monoxide, water vapor, and aluminum oxide.

It was finally determined that, of the unmanned ballistic boosters, the solid booster recoverable system with parallel orbiter burn would give the lowest development cost ($5.15B), least capital risk per flight, and lowest technical risk of development. In addition, economic studies have shown that this system will provide the highest rate of return on investment. Environmental effects would be minor, although it would be necessary to impose additional but acceptable constraints on launch associated with the likelihood of rain.

SUMMARY

Preliminary design studies of the initial two-stage fully reusable concept showed that the size of the system and its development cost could be greatly reduced through the use of an external expendable liquid-hydrogen tank for the orbiter, [7] with a small increase in operating costs per launch. Further study showed that additional cost savings and technical advantages in the development program would accrue if both the liquid-oxygen and liquid-hydrogen for the orbiter were carried in an external tank jettisoned from orbit. This change permitted the orbiter vehicle to be significantly smaller and more efficient[,] thereby simplifying the booster development and reducing substantially the development and procurement costs at the expense of some additional increase in the recurring cost per flight. Consideration of all factors led to the selection of the solid rocket motor booster, parallel burn system for the Space Shuttle. All configuration comparative issues have been studied in great detail both in and outside of NASA, to evolve this most cost-effective space transportation system.

[no page number]

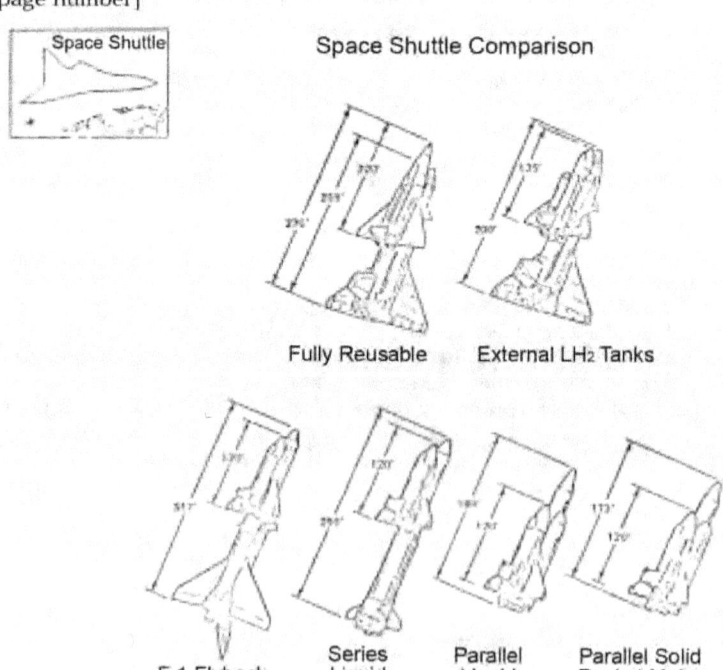

Space Shuttle

Space Shuttle Comparison

Fully Reusable   External LH₂ Tanks

F-1 Flyback   Series   Parallel   Parallel Solid
               Liquid   Liquid   Rocket Motor

Document II-6

Document title: Maxime A. Faget and Milton A. Silveira, NASA Manned Spacecraft Center, "Fundamental Design Considerations for an Earth-Surface-to-Orbit Shuttle," presented at the XXIst International Congress of the International Astronautical Federation, Constance, German Federal Republic, October 4–10, 1970.

Source: NASA Historical Reference Collection, NASA History Office, NASA Headquarters, Washington, D.C.

In 1970, Maxime Faget was the head of engineering at NASA's Manned Spacecraft Center, and Milton Silveira was one of his associates. Faget had played a key role in the design of the Mercury, Gemini, and Apollo spacecraft. His concept for a two-stage fully reusable Space Shuttle was the NASA baseline for the program until the combination of Department of Defense requirements for cross-range capability and White House budget constraints forced NASA to investigate alternative Shuttle configurations. Note that only the first five of the seventeen figures appear here.

[1]   Fundamental Design Considerations for an
               Earth-Surface-to-Orbit Shuttle

By Maxime A. Faget and Milton A. Silveira
NASA Manned Spacecraft Center
Houston, Texas 77058

The design of a reusable earth-surface-to-orbit shuttle is receiving an ever-increasing amount of study. A complete discussion of only the most significant design considerations would be more than sufficient to occupy the entire time available at this conference. Therefore, I plan only to discuss those aspects that should greatly affect the cost or performance of the vehicle and to limit this discussion to fundamentals and basic trade-offs. Although this approach may not be very rewarding to those who are already deeply involved in the shuttle program, I believe it may provide others with some understanding of the more interesting design considerations.

Although single-stage and stage-and-a-half arrangements are also being studied, the most promising configuration appears to be a fully reusable vehicle with two stages—a booster and an orbiter. Such a vehicle not only has the advantage of complete reusability, but would also perform quite well. Several arrangements that may be used to join the two stages during launch are shown in figure 1. Although the tandem arrangement is the most conventional, it is undesirable because the interstage structure must be jettisoned. More importantly, the tandem [2] arrangement suffers a penalty in structural weight to counteract the effect of increased bending moments between stages. In the other two arrangements, "belly to belly" and "back to back," the weights are approximately the same. The choice between these two systems depends upon factors such as aerodynamics, control, detailed mechanical-interface design, and separation dynamics (including orbiter-plume effects).

During a mission, both stages will undergo three distinct flight phases that will significantly affect their design. These flight phases are launch, entry, and landing. During launch, the vehicle is the most heavily loaded and undergoes the greatest dynamic pressure and noise levels. During entry, the heating rates and total heat load are the primary considerations; while, during the landing phase, good subsonic flying characteristics are the most important considerations. The task of the designer is to define a vehicle that can suitably accommodate these flight phases and that will at the same time be of reasonable size and cost. That this is no simple task is illustrated in figure 2, which shows a typical distribution between inert weight and propellant weight for the booster and orbiter. The gross weight of the booster is approximately five times that of the orbiter. The payload is also shown to be a very small portion of the orbiter weight. In fact, for most designs, the payload usually varies between 0.5 and 1 percent of the gross lift-off weight.

A better understanding of weight apportionment may be obtained from figure 3, which shows a breakdown of the inert weight for a typical [3] booster and orbiter. It can readily be seen that the heaviest items are the structure, the propellant tanks, the thermal-protection system, the cruise-capability and the propulsion system. Thus, significant improvements in performance must be obtained by lowering the weight of one or more of these major weight items. For instance, the propulsion-system weight might be reduced by using lighter weight engines or by reducing the requirements for gimbal actuation. A major reduction in weight could be obtained by completely eliminating the gimbals. In this case, steering might be accomplished by differentially throttling opposing engines and by taking advantage of the aerodynamic control surfaces.

The requirement for cruise capability of the orbiter could be eliminated completely if its subsonic flying characteristics were adequate for an unpowered landing. Numerous flight tests, including some with aircraft of the same landing weight as the orbiter, have been conducted using this technique at the NASA Flight Research Center. These tests indicate that this technique should be completely acceptable. In the case of the booster, substantial savings in the cruise-fuel weight can be achieved if landings are made down range.

A basic consideration in the structural design of both the booster and the orbiter is the load-carrying ability of the propellant tanks. Historically, launch-vehicle tanks have been used to carry the acceleration and bending loads. In fact, it is quite clear that the inert weight would [4] otherwise have been substantially greater. It should not be surprising, therefore, to find that the tank structure can be advantageously used to carry loads during entry and landing maneuvers as well as during launch. The direct application of the tank structure to primary fuselage loads in the booster is shown in figure 4.

The payload compartment on the orbiter becomes a major consideration in the arrangement of tanks. Three of the most straightforward arrangements that might be considered are shown in figure 5. If the payload is of sufficiently low fineness ratio, it can be located immediately ahead of the propellant tanks, which would be arranged in a conventional tandem manner. This arrangement would not only result in maximum volumetric efficiency in fuselage packaging but also in benefits from the ideal use of the tank walls for carrying fuselage loads. This arrangement is best suited for low-fineness-ratio payload compartments for which many potential payloads are too long. This arrangement also brings about a very large variation in center of mass with payload weight, which hampers aerodynamic balance.

For very long payloads, a high fineness-ratio payload compartment can be located above a twin-lobe tank. With this arrangement, the payload can be carried directly above the vehicle center of mass, and any special aerodynamic balance considerations can thereby be avoided. The shortcomings of this arrangement would be the large cross-sectional area and skin area of the fuselage brought about by any attempt to accommodate large-diameter payloads.

[5] A third arrangement that is well suited to intermediate-fineness-ratio payload compartments is also shown in figure 5. In this arrangement, the liquid oxygen is carried in two tanks directly under the payload and the hydrogen is carried in a single tank at the rear. The vehicle center of mass would vary slightly with payload weight. In this case, it is not clear whether there would be an advantage in using the liquid-oxygen tanks as a load path. During the launch phase, the liquid oxygen in these tanks accounts for 60 percent of the weight of the orbiter. Therefore, the heavy structural paths that must be provided to support the tanks might also contribute to the transmission of other loads, such as fuselage bending.

The thermal-protection system accounts for an appreciable portion of the weight of the booster and orbiter. It is also usually the most expensive part of the spacecraft structure to build. One way to reduce the requirements for the thermal-protection system is to reduce the thermal load. This reduction can be accomplished by using a lower lift-to-drag ratio (L/D) for the entry trajectory. As shown in figure 6, the total heat load is significantly lower for a trajectory with an L/D of 0.5. The L/D of 0.5 provides sufficient cross range for the majority of the missions yet does not exceed acceptable passenger and crew deceleration load factors.

An L/D of 0.5 can be obtained by entering the atmosphere at an angle of attack near 60°. At this angle of attack, the flight of the vehicle is governed by essentially the same consideration as are semi-ballistic [6] entry vehicles such as the Apollo command module. This concept is illustrated in figure 7. The vehicle is not only easy to stabilize in this attitude, but it is easily controlled using reaction control jets, as has been done in the past. Computer-driven flight simulations using wind-tunnel-derived aerodynamic-stability coefficients have shown that such entries are well within the reaction control system capability and, in fact, require very little propellant.

A benefit almost equal to the thermal advantage of this type of entry lies in the fact that the vehicle need only be designed to fly subsonically. Thus, the cost of numerous hours of wind-tunnel testing and various aerodynamic and stability augmentation system "fixes" can be avoided because vehicles of the type shown remain stable in the high-angle-of-attack attitude through entry and descent over the entire speed range down to low subsonic speeds.

Once subsonic speeds and a sufficiently low altitude for conventional flight have been achieved, a transitional maneuver must be made. This maneuver would be accomplished by depressing the elevator, diving until sufficient aerodynamic pressure is obtained, and then pulling out of the dive. A computer simulation of such a maneuver is shown in figure 8. In addition to computer confirmation, the feasibility of this maneuver has been proven in tests using a 0.1-scale radio-controlled model dropped from a helicopter. To obtain the most effective subsonic aerodynamic vehicle after transition, a straight-wing configuration has a considerable [7] advantage over a delta-wing configuration, as shown in figure 9. Not only will the straight-wing vehicle produce a higher L/D, but it will also produce a higher lift coefficient. Furthermore, the lift for a straight-wing vehicle can be increased by the use of flaps. On the other hand, the straight-wing vehicle must be equipped with a tail to provide trim and control moments. However, for the type of vehicle shown, a delta wing would have to have approximately four times as much area as a straight wing to achieve the same landing speed as vehicles of comparable size and weight. The L/D operating range during approach and landing for typical space shuttles using straight and delta wings is shown in figure 10. It should be noted that, during the terminal phase of the landing when the lift coefficient is increased as velocity is decreased, the delta-wing vehicle would experience a decreasing L/D—a highly undesirable flight characteristic, if unpowered vehicles are to be seriously considered.

Although both the orbiter and booster would undergo aerodynamic heating during entry, the primary concern is the thermal environment of the orbiter. The heating rates predicted for one orbiter design are illustrated in figure 11, which shows the heating-rate history for the stagnation point of a reference sphere of a radius of 30.48 centimeters (1 foot). It should be noted that the heating rate is reasonably low and that duration of the significant portion of the heat pulse is slightly longer than 10 minutes. The equilibrium-temperature distribution on the lower surface of the orbiter at the time of peak heating rate is shown on figure 12. [8] The temperatures shown are those that would be obtained if the heat were being reradiated from a skin with an emissivity value of 0.85.

Although an entry strategy can be adopted that will minimize the heating rate and load, the cost and weight of the thermal-protection system will still be major considerations in the program. There are several ways to design for hot surfaces. The use of hot structure may be feasible for certain places; however, if the temperature exceeds the working range of titanium, the structure may become quite heavy. Thus, it would be advantageous to look for ways of insulating the structure from the hot skin. The method given the most attention to date is the use of shingles made of an appropriate refractory material. In such a scheme, insulation and structural standoffs would be required to support the hot skin as shown in figure 13.

A scheme that shows promise of reducing both weight and cost is the use of external insulation. In the simplest application, this insulation would be bonded in sufficient thickness to a "cold" structural skin. This material, of relatively recent development, exhibits the capability of withstanding repeated temperature cycling up to 1400° C (2500° F). Coatings to prevent material abrasion and water absorption have also been tested on two different external-insulation materials. Samples of the materials were fastened under the fuselage of a transport airplane behind the nose wheel; these materials showed no adverse effects from numerous landings and other flight conditions. The application of [9] external insulation is shown schematically in figure 14.

Perhaps one of the best methods for dealing with entry heating in certain areas is the use of replaceable ablative panels. Regions such as wing and tail leading edges (fig. 14), which would require expensive and complex treatment, can be protected quite easily with ablators.

Although the booster will encounter a far less severe thermal environment during entry than the orbiter, its thermal-protection system may represent a significant portion of the program cost, because extensive surface areas will be exposed to entry heating. It may be possible to avoid much of the thermal-protection-system cost by relying on the heat capacity of the skin as a thermal sink. The heating histories of the upper and lower surfaces of a piggy-back booster fuselage are shown in figure 15. During launch, the upper surface receives higher heating rates as a result of orbiter interference with the flow in this region. During entry, however, the lower surface receives appreciably higher heating. The required aluminum-skin thicknesses for structural loads and thermal capacity about the fuselage cross section are indicated in figure 16. The maximum temperature of the skin is limited to 300° F to avoid changing the material properties. In the diagrams on the left-hand side of figure 16, the flat-bottomed fuselage cross section is left unmodified. It can be seen that the exposed skin of the aluminum tank is more than sufficiently thick to absorb the flight heat load without modification. This skin thickness of the fairing on the lower surface is determined by the thermal load, however. [10] In this case, the booster was found to weigh approximately 6800 kilograms (15,000 pounds) more than one with a thin refractory metal skin and under-surface insulation. However, if the aerodynamic fairing were removed, leaving the tank skin exposed around the entire section, as shown in the right-hand side of figure 16, the weight would be approximately the same as with the high-temperature skin. In this case, the lower-temperature skin must be made considerably thicker than necessary for structural loads. Although no weight advantage would result, a considerable cost savings might be realized as a result of design and manufacturing simplification. In a similar manner, both aluminum and magnesium wings and tails using load-carrying skins might greatly reduce the cost with little or no weight penalty.

Numerous important design considerations have been discussed. However, the cost and performance of the shuttle are more likely functions of the various operating requirements than of the skill of the designers. The effect of some of the more important operational requirements on the gross lift-off weight or payload is illustrated on figure 17 for one shuttle design. The basic vehicle would carry a 11[,]340-kilogram (25,000-pound) payload at a gross lift-off weight of 1,590,000 kilograms (3,500,000 pounds). It would have a 370-kilometer (200-nautical-mile) cross range capability with a payload compartment 4.6 meters (15 feet) in diameter by 18.3 meters (60 feet) in length. The orbiter landing would be made with air-breathing engines with sufficient thrust and fuel for a "wave-off, go around" [11] maneuver. Sufficient fuel would be carried in the booster to cruise back to the launch site after entry. Also shown in figure 17 are the amount the size of the vehicle could be decreased or the amount the payload could be increased if each of the above operational requirements were deleted and the savings that could be accomplished by halving the volume of the payload compartment. Also shown is the weight penalty associated with increasing the cross-range capability to 2780 kilometers (1500 nautical miles).

[12]

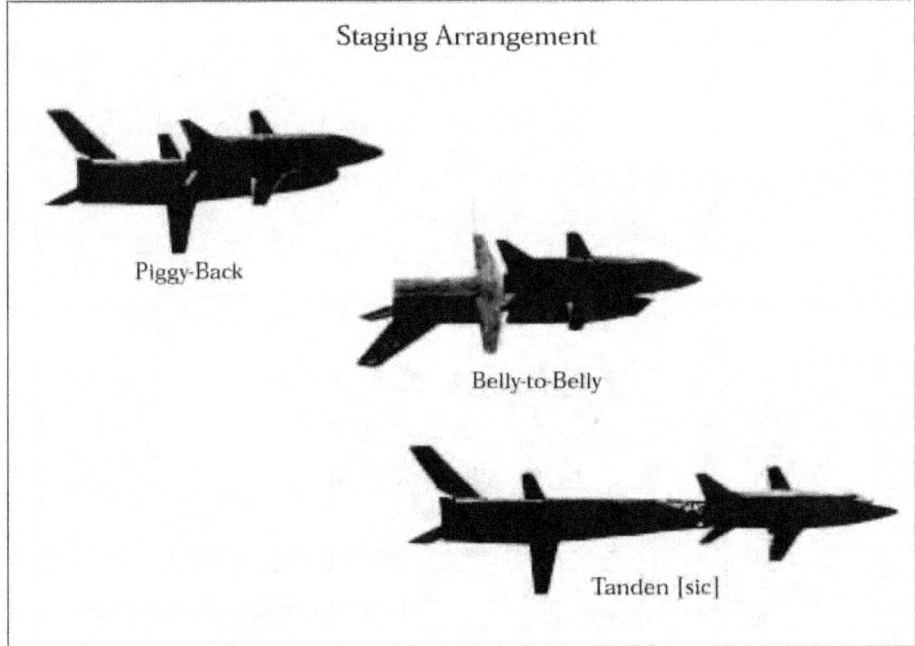

Figure 1

[13]

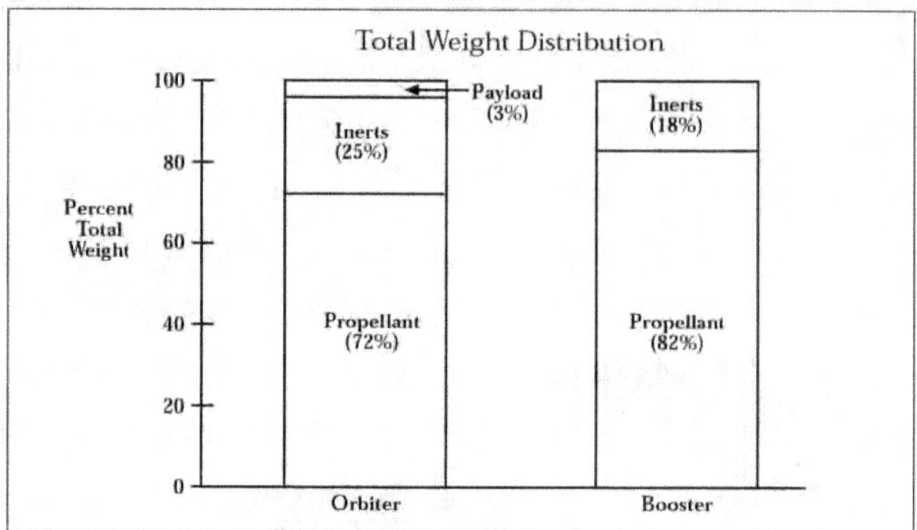

Figure 2

[14]

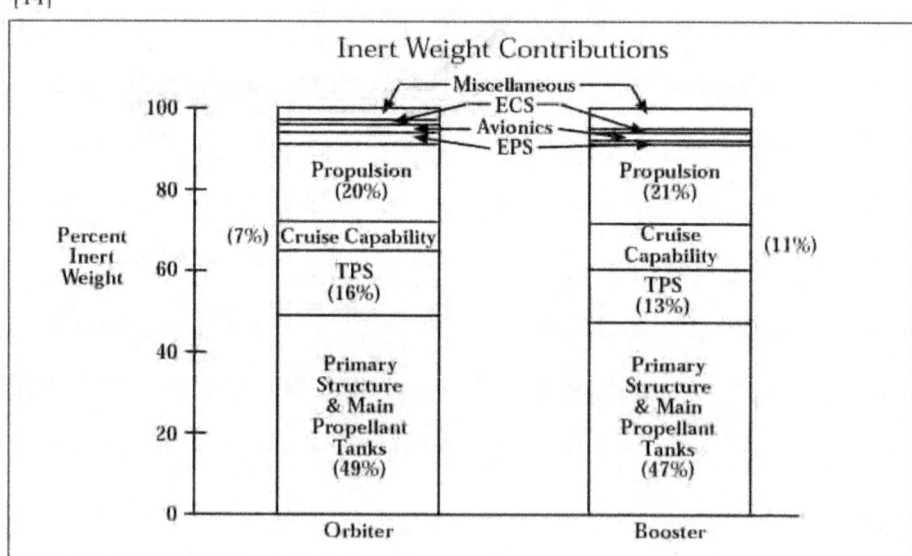

Figure 3

[15]

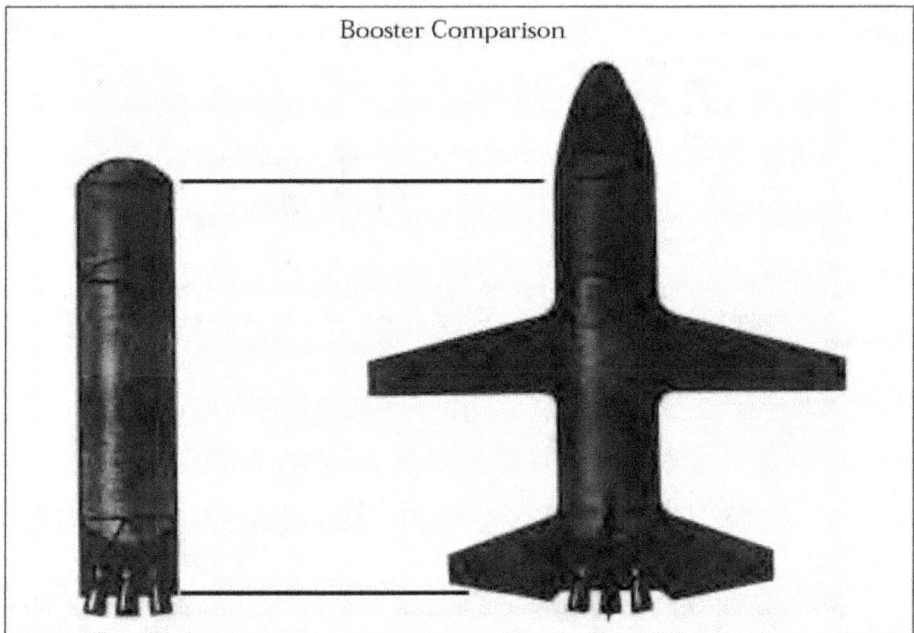

Figure 4

[16]

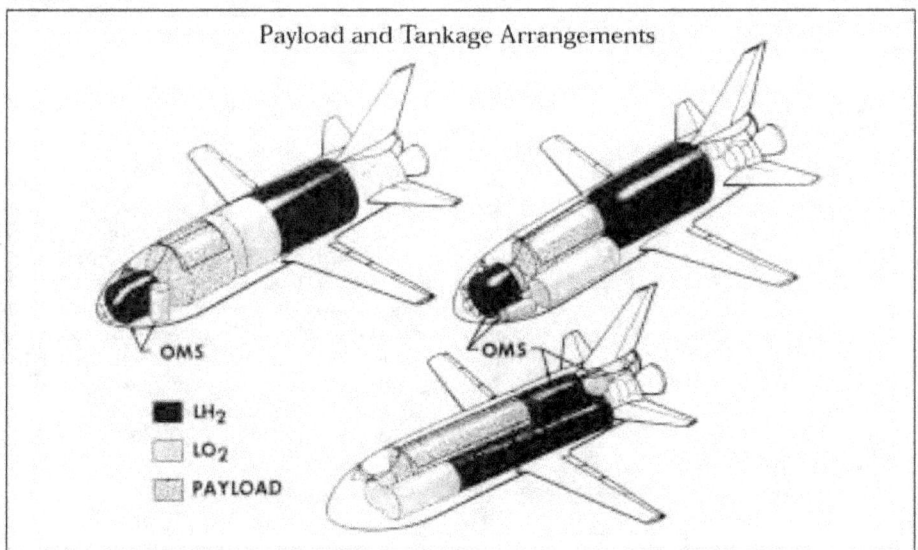

Figure 5

Document II-7

Document title: Office of Management and Budget, "Documentation of the Space Shuttle Decision Process," February 4, 1972.

Source: NASA Historical Reference Collection, NASA History Office, NASA Headquarters, Washington, D.C.

The decision process leading to approval of Space Shuttle development was extremely complex. It involved intense, often conflict-filled interactions among NASA, the Bureau of the Budget (BOB), which became the Office of Management and Budget (OMB) in 1970, the Office of Science and Technology (OST) and its President's Science Advisory Committee (PSAC), and other White House staff. This chronology of the Shuttle decision process was prepared by OMB staff in the Economics, Science, and Technology Program Division (ESTPD) a month after the positive presidential decision to proceed with the Shuttle.

---

[1]   2/4/72

Documentation of the Space Shuttle Decision Process

Reference	Date	Description
1. Telephone call BOB	1/7/70	BOB staff alterted [sic] NASA staff to forthcoming (Earl Rhode) to NASA request for economic analysis of shuttle compared with alternatives; analysis was to include life cycle costs of meeting specific NASA/DOD mission requirements.
2. Letter, Director Mayo to Dr. Paine	1/20/70	BOB identified space shuttle as a major policy issue for FY 1972.
3. Memo, Tom Newman (NASA) to Earl Rhode	2/17/70	NASA proposed to analyze one alternative to the fully reusable shuttle, i.e. the current expendable.
4. Letter, Director Mayo to Dr. Paine	3/18/70	BOB requested Major Program Issue study, "Analysis of Alternative Systems for Reducing the Cost of Payload in Orbit." Requested use of 10% discount rate with sensitivity tests. Enclosure referred to NASA in-house studies and suggested they be integrated into a systems study which would include total non-recurring and recurring costs of launch vehicles (fully reusable shuttle, partially reusable shuttle, current expendables, and new low-cost expendable) and payloads. Due dates: Interim Report - 5/1/70; Final - 7/1/70.

Reference	Date	Description
5. NASA Interim Report	4/24/70	NASA submitted Interim Report, "Alternative Systems for Reducing the Cost of Payloads in Orbit, an Economic Analysis," to OMB. Report concludes that internal rate of return analysis ranks alternatives as follows: • fully reusable shuttle • new low-cost expendable • partially reusable shuttle • current expendable Economics of space tug not addressed
[2] 6. Memo, Robert Lindlay [sic] (NASA) to Earl Rhode	6/18/70	NASA suggested that comparing present values is more meaningful than comparing internal rates-of-return of alternatives.
7. Letter, John Young to William Lilly (NASA)	6/29/70	BOB commented on in-house NASA interim report (4/24/70) and requested final report by August 15, 1970. Attachment requested that final report hold [Office of Space Science and Applications] annual budget to $750 M and examine sensitivity of space station IOC.
8. NASA study contracts	June 1970	NASA issued contracts to Mathematica (economic analysis), Aerospace Corp. (cost estimating), and Lockheed (payload effects) for 11 month studies (7/70 to 6/71) of space shuttle. Robert Lindley of NASA designated to be project monitor.
9. Memo, Earl Rhode to John Young	7/23/70	Mathematica meeting (7/9/70)—described initial meeting of OMB, NASA, and Mathematica representatives. Pending further study, Mathematica's analysis agreed with those of NASA interim report of 4/24/70.
10. NASA Second Report to OMB	8/15/70	NASA submitted second report, "Economic Analysis, Alternative Systems for Reducing the Cost of Payloads in Orbit" to OMB. Relative ranking of alternatives unchanged. Report stated that ultimate goal is fully reusable and therefore the "hybrid (partially reusable shuttle) has been dropped from contention. . . ." Payload effects more important than launch cost effects. Space tug economics not addressed.
11. Letter, Dr. Low to Director	9/30/70	Recommended $180 M for proceeding with detailed design and development in FY 1972.

Reference	Date	Description
[3] 12. Letter, Mr. Rice to Dr. Low	12/17/70	Contained language describing shuttle decision—develop engine; design airframe (FY 1972 budget).
13. NASA briefing to Taft) on	12/15/70	Robert Lindly [sic] briefed new OMB staff (Dan OMB results of economic studies.
14. Letter, Director to Dr. Low	2/19/71	Allowance of $100 M (BA) reiterating shuttle decision and requesting opportunity to review shuttle studies.
15. NASA baseline design	1/25/71	NASA defined baseline requirements (65,000 pounds payload; 1100 [nautical mile] cross range; 550,000 pounds main engine thrust).
16. Memo, Dan Taft to John Young	1/22/71	Suggests three-tier approach to evaluation of shuttle economic studies, including that OMB encourage OST to convene a PSAC space shuttle panel.
17. Director Shultz letter to Dr. Low	2/27/71	Reiterates need for final economic analysis of shuttle.
18. Robert Lindley briefing to Dan Taft	3/10/71	NASA explained rationale for baseline design requirements.
19. Letter, Robert Lindley to Dan Taft	3/29/71	Mathematica interim report "Benefit Cost Analysis of New Space Transportation Systems"—3/15/71 submitted to OMB.
20. John Sullivan on-board	5/5/71	Economist hired by OMB to review analysis of shuttle economics.
21. NASA briefing to OMB [4]	5/7/71	Subject: Current status of space shuttle. NASA planned to release vehicle RFP for fully reusable shuttle in Aug. 1971. Stage 1 1/2 shuttle discarded because: • not technically feasible • potential drop-tank solution • didn't meet requirement of all-azimuth capability
22. Meeting, Dr. Low and Dan Taft	5/14/71	Arranged at Dr. Low's request prior to FY 1973 Preview. Dr. Low expressed belief that annual NASA funding levels of $4.5–5.0 B were reasonable to expect. Fully reusable system desired. Some concern about peak shuttle funding.
23. Letter, Mr. Rice to Dr. Fletcher	5/17/71	OMB suggested 5-year NASA plan with Base Plan peak of $3.2 B per year.

Reference	Date	Description
24. 1973 Preview, Science and Space Program	5/17/71	ESTPD analysis indicated that fully reusable Shuttle not cost-effective when compared with Titans. Guidance was to continue study of alter-native configurations including stage 1 1/2.
25. Memo, EST[PD] staff to NASA staff	5/24/71	Commented (primarily directed at Mathematica Report) on NASA briefing of 5/7/71. Questioned whether the then postulated due dates of Mathematica final report (June 1971) and Aerospace final report (August 1971) weren't reversed (Aerospace provided input to Mathematica). Asked whether Mathematica final report would include: • partially reusable (stage and one half) • reusable tug IOC in 1985 rather than 1979
26. Letter, John Sullivan to Dr. Klaus Heiss (Mathematica)	5/27/71	Sent with NASA concurrence. Commented on Mathematica Interim Report (March 1971) on page-by-page basis. Suggested more sensitivity analysis of the mission model. Enumerated weaknesses in the input data from Aerospace (cost estimates) and Lockheed (payload study).
[5] 27. Material. Provided to Robert Lindley by John Sullivan	6/71	Informal OMB comments on Aerospace interim report 4/12/71 (e.g., no dispersions presented for cost estimates), and Lockheed interim report—12/22/70 (costs of payload refurbishment and maintenance were assumed rather than estimated) sent to NASA.
28. Letter, Dr. Fletcher to Mr. Rice	6/1/71	Informed OMB that NASA was examining phased approach (orbiter first) with interim expendable booster. NASA preferred 2 1/2 stage system.
29. Meeting, William Lilly (NASA) and Dan Taft	6/7/71	Discussed schedule for shuttle decisions and alternatives being examined.
30. Memo, John Sullivan to John Young	6/9/71	Mathematica meeting (6/2/7)—OMB, NASA, and Mathematica to discuss inadequacies of Mathematica Interim Report, "Benefit Cost Analysis of New Space Transportation Systems"—3/71). Specific OMB criticisms (e.g., lack of alternatives, unrealistic IOC dates for space tug and space station, and additional sensitivity analysis required) provided in advance of meeting.
31. Memo, John Sullivan to John Young	6/23/71	Presented proposed game plan for staff analysis of shuttle studies, e.g., a staff paper to be completed September 1971.

Reference	Date	Description
32. Letter, Mr. Rice to Dr. David	7/14/71	Prepared by ESTPD—detailed specific questions regarding alternatives to the 2 1/2 stage shuttle which the PSAC shuttle panel might address.
33. Letter, Klaus Heiss (Mathematica) to John Sullivan	7/15/71	Detailed replies to OMB written comments (5/27/71) on Mathematica Interim Report.
[6] 34. Letter, Mr. Rice to Dr. Fletcher	7/20/71	Prepared by ESTPD—stated emphasis should be placed on substantially reducing overall investment cost; requested additional information on economics of alternative lower-cost systems. Referred to follow-on letter at staff level (see meeting 7/21/71 below—Reference 35).
35. Meeting with NASA Budget Office	7/21/71	Discussed draft of staff letter requesting (substantial) additional analysis be submitted by 8/16/70 including: • alternative configurations: 1 1/2 stage, 2 1/2 stage. • brief report on feasibility of designing recoverable satellites with an expendable launch system. • analysis of a shuttle (35,000 lb. payload capability, 12 x 40' payload bay, low cross range) in context of specific mission model (smaller than NASA baseline model). (Results: EST[PD] staff worked with NASA staff on economics of alternative configurations of full sized shuttle. Budget Office organized several meetings between EST[PD] staff and staff from Shuttle Program Office. NASA Budget Office felt that workload was too heavy to allow analyses of 12 x 40' shuttle.)
36. Mathematica Follow-up Report	7/23/71	OMB received Mathematica follow-up report—5/31/71. Report further refined analysis of fully reusable shuttle.
37. Meeting with Shuttle Program Manager	7/26/71	Discussed the 29 shuttle performance and technical requirements (as detailed in NASA document—2/12/71) including their interactions, tradeoffs, and alternatives. (NASA comments: No alternatives to any requirement)
[7] 38. Meeting with Advanced Missions—OMSF	8/2/71	Discussed the reusable space tug. Learned that there were many versions of the tug and that analysis of the tug was assigned low priority by NASA (e.g., Phase A studies hadn't started; tug economic studies just underway (more in Ref. 60)).

Reference	Date	Description
39. Letter, Mr. Weinberger to Dr. Fletcher	8/2/71	OMB informed NASA of FY 1973 Planning Ceiling of $2,835 M BA and $2,975 M outlays.
40. First meeting PSAC Shuttle Panel	August 13–15, 1971	Presentations by NASA, Airframe Contractors, Aerospace Corp., Mathematica, Inc., Lockheed, and Air Force. Contractors concentrated on fully reusable and included limited discussion of the 2 1/2 stage and the thrust augmented 1 1/2 stage. NASA pushed 2 1/2 stage, but indicated serious peak funding problem (more in Ref. 42).
41. FY 1972 Apportionment action	9/20/71	$25 M held in reserve pending decisions in context of FY 1973 budget.

42. Subsequent meetings of PSAC	9/23– 11/18/71	Presentations (Selective list)
		 • Several by Air Force, both projecting lower launch rates than that used in Mathematica Reports.
		 • Sept. 24, 1971 NASA (Dale Myers) presentation
		   – Mark I/II approach outlined, but mentioned would study several booster options including flyback (S-I-C), T III-L, solids, and pressure-fed.
		   – Revised economic analysis (by Lindley of NASA): if feasible, 1 1/2 stage is preferred to 2 1/2 stage and Mark I/II.
		 • October 15, 1971—Panel Chairmen's analysis of gliders and 3 stage vehicles (reusable 1st stage, expendable 2nd, powered orbiter).
		 • November 17, 1971—NASA presented report of studies including first definition of pressure-fed booster.
		 • November 18, 1971—Dr. Low emphasized latest NASA thinking (pressure-fed booster) would result in loss [with] future peaking problem than design on which FY 73 budget based. Runout costs of NASA budget placed at about $3.6 B with new starts.

43. Letter, Dr. Fletcher to Director Shultz	9/30/71	Transmittal letter for FY 1973 budget. NASA recommended Mark I/II phased technology approach with flyback booster as baseline but reference to ballistic booster study.
44. OMB Staff Paper	10/4/71	Final draft of "The Future Space Transportation System—An Economic Analysis of the Options" which concluded that the new expendable (Titans plus Big Gemini) was more cost-effective than shuttle. Also concluded that shuttle with non-flyback booster (current configuration) was more cost-effective than one with baseline flyback booster (see Ref. 43).

[8]

Reference	Date	Description
45. 1973 Budget Hearing	10/7/71	Manned Space Flight hearing for FY 1973 budget.
46. OMB Staff Paper	10/14/71	Final draft of "The U.S. Civilian Space Program—A Look at the Options" which discussed post-Apollo/Skylab plan. Included an analysis of the shuttle (see Ref. 44).
[9] 47. Memo, John Sullivan to Mr. Rice	10/19/71	Discussed PSAC Shuttle Planel [sic] meeting of 10/15/71. Majority of members concluded that large shuttle not cost-effective but that alternatives must preserve option for manned space flight.
48. 1973 Director's Review—Session on Space and General Research	10/22/71	ESTP[D] recommended that shuttle program be cancelled or if this not feasible that decision be defined to FY 1974. Various options identified by PSAC Panel Chairmen were discussed (including small glider). Guidance was that lower cost alternative to NASA shuttle (large orbiter with flyback booster) be developed by NASA.
49. Memo, Dan Taft to Mr. Rice	11/3/71	Discussed NASA FY 1973 Budget decisions in light of Director's Review. Suggested that rather than define a particular shuttle design, OMB provide NASA with program criteria. Criteria for initiating reduced–cost shuttle definition were attached.
50. Meeting with NASA	11/23/71	OMB staff reviewed with NASA project staff the latest data on all design options.
51. Memo, NASA unit to Mr. Rice	11/29/71	Analyzed effect of reducing orbiter size on shuttle payload benefits. Conclusion based on available data: large shuttle not cost-effective; 10 x 20' or 20 x 40' shuttle would provide intangible benefits such as national prestige.
52. Meeting with NASA	11/30/71	Dr. Low presented interim comparison of costs of booster options.
[10] 53. Memorandum for the President	12/2/71	Presented options for future manned space program. Recommended that OMB and OST work with NASA on the reorientation of shuttle effort to define a reduced-cost shuttle (investment $4–5 B).
54. Aerospace Report	12/2/71	OMB receives 4 volumes (of 5) of Aerospace Final Report, dated August 1971, but apparently printed in early November (one volume was still in draft).
55. Talking Paper	12/7/71	Presented to NASA 12/10/71. Discussed guidance on 15 items. Stated that no decision had been made on whether to develop the shuttle.

Reference	Date	Description
56. Draft Memo from John Sullivan	12/9/71	Delivered to Mr. Rice. Described latest NASA mission model and concluded that manned missions accounted for 50% of NASA's shuttle benefits.
57. Meeting with NASA	12/11/71	OMB (Mr. Rice) presented to NASA a series of general concepts, specific assumptions, and guidelines for the Shuttle program including 10 x 30' orbiter and $4 B [research, development, test, and evaluation]. NASA agreed to study various sized shuttle options.
58. Memo, Dan Taft for Mr. Weinberger	12/16/71	Discussed FY 1973 NASA appeal and attached draft Memo for President on space shuttle decision. Suggested understanding be reached about closure of a manned space flight center.
59. Draft Memorandum for the President	12/16/71	Discussed capabilities, size, and cost of the space shuttle as a Presidential issue remaining in the NASA FY 1973 budget. Recommended that NASA be directed to define a shuttle system subject to certain constraints, including $5 B for R&D plus investment.
[11] 60. Meeting with Advanced Missions—OMSF	12/16/71	Discussed status of economics of reusable tug studies. No progress had been made since contract issued in August 1971 (see Reference 38).
61. Memo, John Sullivan to Mr. Rice	12/17/71	Presented partial analysis (e.g., didn't discount dollars) of economics of reducing orbiter size. Concluded that DOD was primary loser if payload-bay length were reduced to 40' and that roughly 60% of shuttle savings accrued to NASA. (This memo superseded that of 11/29/71—reference in light of recently acquired data.)
62. Memo, John Sullivan to Mr. Rice	12/28/71	Updates staff economic analysis of large shuttle. Concluded that neither configuration (pressure-fed, solid motor) was cost-effective when compared with Titan plus Big [Gemini].
63. Talking paper for Mr. Rice	12/29/71	Prepared by EST[PD] staff. Included breakdown of investment cost for large orbiter with pressure-fed or solid-rocket booster.
64. Letter, Dr. Fletcher to Mr. Weinberger	12/29/71	Reported results of study of options and concluded that 15 x 60' orbiter a "best buy" and 14 x 45' the minimum acceptable size.
65. OMB Meeting with NASA	12/29/71	Meeting with Dr. Fletcher and Low on NASA's study of options.

Reference	Date	Description
66. Memo, Dr. David (OST) to Director	12/30/71	Strongly recommended that smaller shuttle (12 x 40' or 10 x 20') be selected to preserve a balanced space program.
[12] 67. Memo, NASA unit to Mr. Rice	12/30/71	Analyzed NASA's position on shuttle as stated in letter of 12/29/71 (see Ref. 62). Suggested NASA cost estimates were very uncertain. Noted that investment costs should be kept in mind and inclusion would bring estimated cost of large orbiter plus pressure-fed booster to $9 B. Presented brief analysis of smaller shuttle (e.g., lower launch costs of substantial payload capture).
68. OMB Questions for NASA	12/31/71	List of questions provided to Dr. Low concerning overall fiscal constraints, payload requirements, and cost estimates.
69. Letter, Dr. Fletcher to Mr. Weinberger	1/3/72	Reiterated previous conclusion that 15 x 60' orbiter a "best buy." Answers to OMB questions (see Ref. 68) were vague, e.g., smaller orbiter would lose many missions.
70. Memo, Dr. David (OST) to Director OMB	1/3/72	Urged that decision on specific characteristics of space shuttle be delayed for several months pending review by NASA of lower cost alternatives. Noted that shuttle decision will commit R&D funds until early 1980's.
71. Memo, Mr. Rice to Director	1/3/72	Prepared by ESTP[D]. Recommended that proposals in Dr. David's memo of 1/3/72 be adopted. Recommended that NASA be directed to design a shuttle within total investment cost (including facilities[,] vehicles, and contingency of $5 B and peak annual funding of $3.2 B).
72. Meeting with NASA	1/3/72	Decided to develop shuttle (up to 15 x 60'); study 14 x 45'; and decide booster later (pressure-fed vs. solid rocket motors).
[13] 73. Letter, Dr. Fletcher to Mr. Weinberger	1/4/72	Documented decision on shuttle.
74. Statement by the President	1/5/72	President announced decision to develop Shuttle.
75. Letter, Dr. Low to Mr. Rice	1/11/72	Described NASA further study of orbiter size (15 x 60' vs. 14 x 45') and booster (pressure-fed vs. solid-rocket) in order to make decisions by March 1, 1972.

Reference	Date	Description
76. Memo, Dan Taft to Mr. Rice	1/27/72	Summarized draft letter, Director to Dr. Fletcher prepared by EST[PD] staff. Attached was an OMB staff analysis of shuttle options, which recommended that OMB concerns be transmitted informally to NASA management. Stressed the risks associated with particular choices.

Document II-8

Document title: George M. Low, Deputy Administrator, NASA, to Donald B. Rice, Assistant Director, Office of Management and Budget, November 22, 1971, with attached: "Space Shuttle Configurations."

Source: NASA Historical Reference Collection, NASA History Office, NASA Headquarters, Washington, D.C.

During the final months of the White House-NASA debate over whether to develop a reusable space transportation system, as well as what kind of system to develop, the White House, supported by a panel of the President's Science Advisory Committee, suggested that NASA consider an unpowered glider launched on top of an expendable launch vehicle. This concept would have been rather similar to the Air Force Dyna-Soar program, which had been cancelled in 1963. NASA resisted this suggestion, arguing that the development savings of such a concept would be outweighed by its operating costs at the flight rate NASA was anticipating. NASA Deputy Administrator George Low used the tradeoff curve contained in this document to argue that both development costs and cost per flight needed to be taken into consideration in the decision over what system to develop.

NOV 22 1971

Mr. Donald Rice
Assistant Director
Office of Management and Budget
Executive Office of the President
Washington, D.C. 20503

Dear Don:

In accordance with your request, I am sending you a reconstruction of the diagram that I sketched on your blackboard the other day, together with a discussion of the material represented on the diagram.

We also discussed comparative information for small and large shuttles and small and large gliders. The earliest we will be able to provide this information is Monday, November 29. I was unable to push this to an even earlier date, although I would have liked to have done so.

Please let me know if I can provide any additional information.

Sincerely yours,

George M. Low
Deputy Administrator

Enclosures
cc: Jack Young, OMB
bcc: A/Fletcher   AAD/von Braun
      ADA/Shapley   D/McCurdy
      M/Myers
      B/Lilly

AD/GML:rej:11-19-71

                              ****

[1]   Space Shuttle Configurations

For the past 18 months, seven aerospace companies and NASA have studied and
evolved various designs of the space shuttle.

As a result of these design efforts, and as a result of tradeoffs between development
costs and operating costs, the shuttle system efficiency has been greatly improved. The
result is a class of configurations that costs much less to develop than earlier configurations, is much smaller but can carry the required payload, and is still "productive" in terms
of operating costs.

Definitions

The following configurations have been considered.

1.   Two-stage Fully Reusable. This was the preferred configuration at the beginning of the
"Phase B" design effort. The "orbiter" carried all of its propellant (hydrogen and oxygen) internally, and was very large (larger than a 707). The "booster" was huge (like a 747), also used
hydrogen and oxygen propellants, and used the same high pressure engines as the orbiter.

[2] 2.   Two-stage Reusable with External Hydrogen Tanks. (sometimes called "baseline")
Midway through the "Phase B" studies, it became apparent that by carrying the hydrogen
in tanks external to the orbiter, the size of the orbiter could be reduced, and the development cost could be reduced somewhat as well. A secondary effect also resulted: since
the orbiter became more efficient, it became possible to let it take more of the burden of
propelling the shuttle into orbit (lower staging velocity). The booster requirements were
thereby lessened, resulting in further savings in complexity and cost.

3.   Mark I/Mark II (MkI/II). In this step, further advantage was taken of the evolution started in the previous step. For the orbiter, oxygen[,] as well as hydrogen, would be
carried in external tanks, leading to an even smaller orbiter (smaller than a DC 9). Some
of the subsystems would be phased, starting out in the Mark I model with more nearly
existing technology in areas such as the heat shield and avionics, and phasing in more
advanced versions later in Mark II.

Four different booster configurations are being considered in conjunction with the same
Mark I/Mark II orbiter. They are the Flyback Booster, the Pressure-fed Ballistic Booster, the
[3] Parallel-staged Pressure-fed Booster, and the Parallel-staged Solid Rocket Booster.

3a.   MkI/II - Flyback Booster. This booster is evolved from the first stage of the Saturn
V. It uses conventional propellants, and the Saturn F-1 engines, but has wings so that it can
fly back to the launch site.

3b.   MkI/II - Pressure-fed Ballistic Booster. With the compact, efficient MkI/II orbiter,
it became possible to take another step in reducing booster complexity: take off the wings,
make it unmanned, let it fly ballistically, and recover it with parachutes. At the same time,
simplify the propulsion system by using gas pressure to force the propellants through the
engines, instead of pumps and turbines. in this configuration, the booster still propels the
orbiter to a velocity of 5,000–6,000 feet per second, at which time it is jettisoned and the
orbiter takes over.

3c. MkI/II - Parallel-staged Pressure-fed Booster. In this configuration, the booster and orbiter are mounted side-by-side—for example, twin boosters mounted under the orbiter's wings. All booster and orbiter engines are ignited before takeoff. The boosters are jettisoned after their propellant [4] is depleted, and then recovered. The orbiter continues to burn into orbit. This further simplifies the booster propulsion system, since the booster engines now no longer need to be steerable: all of the steering is done with the orbiter engines.

3d. MkI/II - Parallel-staged Solid Rocket Booster. In this configuration, the booster described in the previous paragraph is replaced with solid rocket motors. These, however, would not be recoverable.

4. Glider. This vehicle requires two propulsive stages to put a winged recoverable payload carrier into orbit. It could make use of a recoverable or a non-recoverable first stage. The second stage would be non-recoverable. It differs from the Mark I/Mark II orbiter in one significant way: the engines and the electronics to go with the propulsion system are placed into the external tank, thus making it into a stage, and then thrown away during each flight; in the orbiter the engines and electronics are recovered and reused.

Comparison of Configurations

The various shuttle configurations are best compared on a plot of Cost Per Flight versus Development Cost (see attached [5] figure). For the purpose of this comparison, the Development Cost is defined to include all Design, Development, Test and Evaluation (DDT&E) costs. It does not include costs for operational hardware, facilities, or flight operations.

On this plot, the two-stage fully reusable configuration shows a development cost of nearly $10 billion, at a cost per flight of less than $5 million. The "baseline" configuration has a development cost of $8 billion, with about the same cost per flight as the first configuration.

The Mark I/Mark II orbiter with all four booster configurations falls within a range of development costs between $4.5 and $6.5 billion, with operating costs ranging from $6 to $12 million per flight. The parallel-staged solid rocket configuration is the cheapest to develop, and the most expensive to operate within that range. The flyback-booster version is at the opposite end of the range. The two pressure-fed booster configurations fall in between, with the parallel staged one being closer to the left of the box, and the series staged one closer to the right. (Mark II will cost somewhat more to develop, but less to operate than Mark I. Development costs shown in the figure are for the full Mark II capability.)

[6] All of these configurations carry the same payload: 65,000 pounds due east, or 40,000 pounds into polar orbit, in a 15 ft. by 60 ft. bay. The only glider for which information is now available is smaller: it carries a payload of less than half that weight and volume. It was designed to fit on a Titan III L (a new booster 4 times as heavy as the largest existing Titan III) and a new second stage. Very preliminary estimates give a development cost of around $3 billion, and a cost per flight of $30 million.

A glider with the same payload capacity as the orbiter, together with its booster stages, would probably cost as much to develop as the low-cost configurations of the shuttle, since there is little difference in complexity. However, operating costs would remain high as long as one or more stages are thrown away. Conversely, a smaller shuttle (with a payload of the size now considered for the glider) would cost less to develop and operate than the present shuttle configurations.

Conclusions

NASA has not yet made a final configuration selection. However, for practical purposes, the two-stage fully reusable and the baseline configurations can be discarded because of [7] their high development cost. At the other extreme, the glider, as presently proposed, also does not appear to be promising. When compared on the basis of the same payload, it will probably not offer a significant saving in development cost, but will be expensive to operate. (Definitive numbers on this tentative conclusion are not yet available.)

This leaves the Mark I/Mark II configurations with four booster options: flyback, pressure-fed, parallel-staged pressure-fed, and parallel-staged solid rocket boosters. The MkI/II orbiter has been studied extensively and is well defined. Booster studies are not yet as complete, but the pressure-fed options look very promising.

The most promising candidate configuration today is the Mark I/Mark II orbiter with the parallel-staged pressure-fed booster.

[no page number]

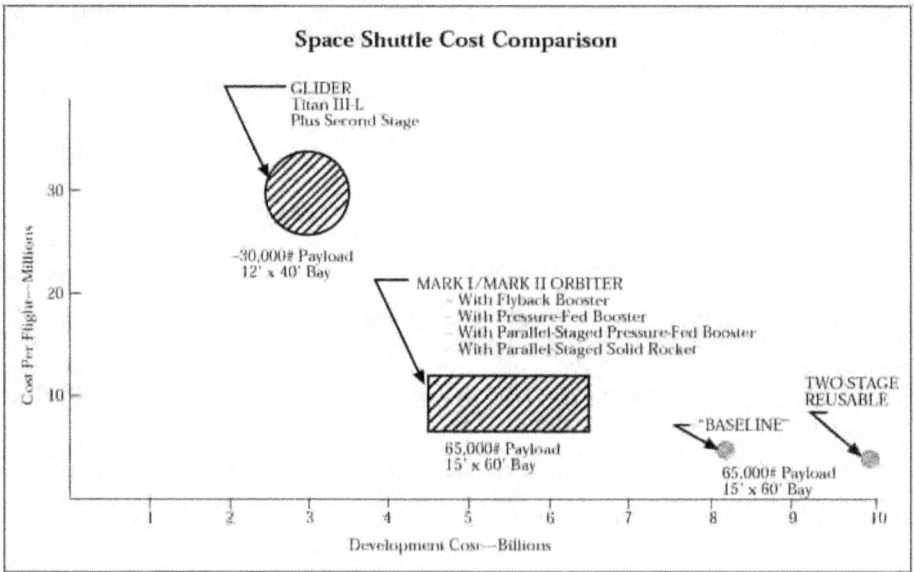

Document II-9

Document title: Charles J. Donlan, Acting Director, Space Shuttle Program, to Deputy Administrator, "Additional Space Shuttle Information," December 5, 1971.

Source: George Low Papers, Rensselaer Polytechnic Institute Library, Troy, New York.

As NASA struggled to gain White House approval for Space Shuttle development in December 1971, there was a constant need for information to support the particular Shuttle concept NASA was promoting. Charles Donlan, a career NACA/NASA engineer, was Acting Director of the Space Shuttle program at NASA Headquarters. He provided this memorandum on Shuttle design choices and costs to NASA Deputy Administrator George M. Low, the NASA "point man" in dealing with the White House on the Shuttle decision. A particular item of controversy between NASA and the White House was whether there were significant cost savings associated if a smaller Space Shuttle orbiter, rather than the one with the fifteen-foot by sixty-foot payload bay that NASA was advocating, were to be approved. Donlan's memo suggests why NASA thought that such savings would not be substantial.

[no page number]

December 5, 1971

TO: AD/Deputy Administrator

FROM: MH/Acting Director, Space Shuttle Program

SUBJECT: Additional Space Shuttle Information

The attached information is in response to your request for additional rationale and analysis in support of the shuttle selection. Should you need any additional information over the weekend, please feel free to call me at my home on 765-4625.

Charles J. Donlan

None of this information has been transmitted to OMB.

CJD

12/7
This is a rewritten version less the Incremental Cost-Benefit Analysis Section.

[1]   Selection of the Delta Wing Configuration

The delta wing orbiter configuration was selected on two accounts: (1) to obtain the operational benefits of cross range and (2) in recognition of the fundamental aerodynamic superiority of the delta configuration in the supersonic/hypersonic flight regime.

1. Cross Range Consideration
    The cross range requirement for the shuttle has been subjected to many critical reviews. Whereas the initial request for a 1500n.m. [nautical mile] cross range capability originated as an Air Force requirement, it became evident with increased depth of study that a substantial degree of aerodynamic maneuvering capability at hypersonic and supersonic speeds is fundamental to the operation of the orbiter. It is a requisite to safe abort being required to turn hypersonically for the immediate return to base for selected abort modes. It is also required to fly the cross range to the launch site from once-around abort or to an off-track landing site for a down-range abort. It affords frequent normal opportunities to return to base from orbit on a due east mission from [the Eastern Test Range]. The ground tracks for these returns vary greatly[,] enabling selection of reentry routes over sparsely populated land mass or water in the event that sonic boom over pressures are judged to be of objectionable levels for densely populated areas. The minimum cross range performance compatible with these operational needs, as determined by once-around abort, is 1100n.m. This requirement also serves to satisfy an important Air Force mission requiring one orbit return. The 1100n.m. cross range capability can most effectively be supplied with a highly swept or delta wing configuration. A straight wing configuration cannot satisfy this cross range requirement. Technical rationale in support of this thesis are contained in sections 2 and 3.

2. Aerothermodynamic Considerations
    Apart from operational requirements for cross range, the selection of a delta wing configuration in preference to a straight wing is strongly influenced by basic aerodynamic considerations. The critical periods during reentry of the shuttle are the hypersonic/supersonic flight regime and the accomplishment of transition.

[2] The delta wing configuration is stable throughout a wide range of angles of attack in this regime and by modulating angle of attack and bank angle can take full advantage of trajectory shaping, cross range and high altitude, supersonic transition options. The flow field over the delta vehicle tends to be relatively smooth[,] producing uniform, predictable, aerodynamic heating gradients. Also the delta vehicle experiences relatively low and uniform temperatures, 600–800°F, over the sides and upper surfaces of the vehicle. The flow over the sides of the fuselage is smoothly blended; there are no shock interactions and few, if any, hot spots. These conditions are favorable to straightforward, accurate heating prediction and confidence in the design of the thermal protection system.

The straight wing configuration suffers from unsteady flow and buffeting in the transonic regime. In the hypersonic regime the flow fields are complex with strong bow and wing shock interactions with the vehicle. The strong interference flow field results in high local temperatures and severe temperature gradients on the wing, body and tail. Vortex flows in wing-body and tail-body junctures tend to result in local hot spots on the fuselage. Fuselage side temperatures range from 900–1300°F making the analysis and design of the [thermal protection system] a complex problem. For these reasons, the delta configuration lends itself more readily to solution of the critical aerodynamic problems of the shuttle.

3. Growth Potential

The problem of growth from a cargo bay of 12'X40' to 15'X60', or even 12'X60', is not a straightforward change for either a straight wing or delta orbiter configuration. Fuselage stretch of subsonic transport aircraft are not indicative of the problem of a hypersonic/supersonic orbiter booster configuration. The stretch of an orbiter will significantly alter the hypersonic flow field[,] resulting in greatly different stability and control and thermal characteristics for the orbiter as well as the complex launch configuration of orbiter and booster combined. Extensive aerodynamic, static and dynamic ground tests, plus additional flight test development would be required, approaching that required for another orbiter configuration.

[3] The question of reductions of orbiter weight and dimensions have been examined for bay size of 15'X60' down to 12'X40'. These reductions are limited to less than about 15%–20% of the vehicle dry weight for practical design reasons. For example, reduction in payload bay diameter from 15' to 12' cannot be fully realized in orbiter weight saving because of the necessity to provide a boat tail of approximately 15' diameter to accommodate the rocket engines.

In summary, the stretch of an orbiter accommodating 12'X40' payloads to 15'X60', or 12'X60', payloads is not considered practical. The most cost effective system is one sized properly at the outset for its intended use.

[no page number]   PROGRAM COST DIFFERENCES FOR LARGE
                          AND SMALL ORBITER SYSTEMS

The following is an explanation as to why the cost differential for the program using the large orbiter with the pressure fed booster [PFB] ($5.7 billion) and the small orbiter with the pressure fed booster ($5.1 billion) is not greater than $600 million. In other words, what elements in these two programs remain fixed and what elements are scaled with size?

The 65K twin RAO [rocket assisted orbiter] and the 30K twin RAO system dry weights estimated by GAEC [Greenbelt Aerospace Engineering Corporation] and MSFC [Marshall Space Flight Center] are listed in the attached Table I. Although the system payload capability decreases 54% (65K to 30K) the percentage decrease in system dry weight is only 26% to 30%.

Utilizing cost estimating relationships (CER's) and changes in orbiter dry weight, GAEC estimates the delta DDT&E [design, development, test, and evaluation] cost and [sic "at" meant] $600M and MSFC at $400M. Attached Table II details the MSFC estimates and delta differences of the large and small orbiter DDT&E costs. The major changes in the MSFC analysis (23%) is in the structures and thermal protection system. There would be, realistically, little or no impact on avionics, environmental control and life support systems, and relatively minor changes in the electrical power system. Program functions such as management, systems engineering and integration, and installation, assembly and checkout are not directly related to orbiter size and therefore will not change appreciably. Subsystem develop-ment testing and program support which includes such items as crew equipment, simulators, and development propellants varies as a percentage of the DDT&E effort. Basic tooling, jigs, test stands, handling equipment, dollys [sic], and some [ground support equipment] would be less expensive due to sealing effects.

In summary, a cost savings of $500M in orbiter DDT&E would appear to be reasonably attainable in scaling a 65K orbiter system to a 30K orbiter system. In arriving at the total delta of $0.6B between the configurations approximately $100M can be attributed to the reduced weight of the twin pressure fed booster. There is no change in [Space Shuttle main engine] HiPc cost estimates. The engine thrust size used by GAEC for both orbiters was 350K (sea level thrust) and MSFC used 415K (sea level thrust) for both orbiters.

[no page number]

TABLE I
DRY WEIGHT ESTIMATES

	ORBITER		HO TANK		TWIN PFB		SYSTEM		_	%
	65K	30K	65K	30K	65K	30K	65K	30K		
MSFC	140K	116K	79K	67K	352K	216K	571K	399K	172K	30
GAEC	159K	118K	89K	63K	675K	500K	923K	681K	242K	26

[no page number]

TABLE II
MSFC ORBITER DDT&E (TWIN PFB)
FY 1971 DOLLARS INCLUDING FEE

	65K ORB.	30K ORB.	_	% OF TOTAL
STRUCTURE	641	539	102	23
PROPULSION	437	357	80	18
ME	(116)	(94)		
OMS	(43)	(36)		
ACPS	(216)	(176)		
ABES	(62)	(51)		
AVIONICS	511	499	12	3
POWER	348	299	49	11
ECLSS	174	171	3	1
INSTALL., ASSY., & C/O	58	54	4	1
SUBSYS. DEV. TESTING	310	249	61	14
SYSTEM ENG. &				
INTEGRATION	236	214	22	5
PROGRAM SUPPORT	517	444	73	16
MANAGEMENT	123	106	17	4
TOOLING	150	130	20	4
TOTAL	$3505M	$3062M	$443M	100%

[no page number]   Comparative Analysis of Cost Per Flight

The following discussion explains why it costs an incremental $4.1 million per flight when solids are used instead of the pressure fed booster with a larger orbiter; and only $800,000 per flight when solids are substituted for the pressure fed booster on the smaller orbiter.

The difference is due to the differing cost of solid rocket motors (SRM's). The 65K orbiter utilizes 156" SRM's at $6.2M per set while the small orbiter utilizes 120" SRM's at $2.9M per set. The following data indicates the reasons for this increase in SRM cost from 120" to 156".

ITEM	156"	120"
Number of segments	3	7
Motor weight	1490K	705K
Propellant weight	1367K	644K
Burn-out weight	123K	61K
Prop. cost of $0.60/lb.	0.8M	0.386M
Case Cost	2.3M	1.061M

The attached table compares the cost per flight for both systems using pressure fed boosters and SRM's.

[no page number] Comparative Analysis of Cost Per Flight (Dollars in millions)

	Large System 65K East 13' X 60'		Small System 30K East 12' X 40'	
	Twin PFB (156")	Twin SRM (120")	Twin PFB	Twin SRM
Orbiter + HO tank	5.2	5.2	4.6	4.6
Booster	2.1*	6.2**	2.1*	2.9**
Cost per Flight	7.3	11.4	6.7	7.5

	COST PER FLIGHT			COST OF SRM SET	
	Twin PFB	Twin SRM	—		
65K Orbiter	7.3	11.4 (156" solids)	4.1	2 156" solids @ 3.1	6.2
30K	6.7	7.5 (120" solids)	.8	2 120" solids @ 1.45	2.9
			3.3		3.3

* No differential is shown for refurbishment/turnaround cost for the 65K twin PFB vs. the 30K twin PFB. The 30K twin PFB will probably be less expensive to refurbish and turn-around; e.g., smaller recovery chutes, less surface area to clean and process. However, the current definition and understanding of these costs precludes identifying quantitative differentials between the two booster systems at this time.

** Expended hardware.

Document II-10

Document title: Mathematica, "Economic Analysis of the Space Shuttle System,"
Executive Summary, prepared for NASA, January 31, 1972.

Source: Space Policy Institute, George Washington University, Washington, D.C.

*In 1970, at the urging of the Office of Management and Budget, NASA, as it tried to gain approval
to begin Space Shuttle development, contracted with an independent economic analysis group,
Mathematica, Inc., to carry out an analysis of the economic benefits of such development. This was
the first time that NASA attempted in advance to project the economic benefits of a proposed development effort. Mathematica was headed by the prestigious economist Oskar Morgenstern; in charge of
the NASA effort was his associate Klauss Heiss. Mathematica's initial analysis was submitted to
NASA in May 1971; it compared a generic Space Shuttle concept with the use of existing expendable
launch vehicles as a means of providing space transportation over the 1978–90 period. This document summarizes the results of a second round of analysis, which compared various Space Shuttle concepts with the use of expendable vehicles (the four figures and table mentioned in this executive
summary do not appear here. Because this formal report would not have been completed by the time
that Space Shuttle decisions were anticipated in the November–December 1971 time period—while the
report is dated January 31, 1972, it actually was not submitted until May 1972—in October 1971,
Morgenstern and Heiss submitted to NASA Administrator James Fletcher a memorandum (see
Document III-30 of* Exploring the Unknown, Volume I*) summarizing their results.*

Economic Analysis of the
Space Shuttle System

EXECUTIVE SUMMARY

[0-1]   0.1 CONCLUSIONS

The major conclusions of the Economic Analysis of the Space Shuttle System are:
* *The development of a Space Shuttle System is economically feasible assuming a level of space
  activity equal to the average of the United States unmanned program of the last eight years.*
* *A Thrust Assisted Orbiter Shuttle (TAOS) with external hydrogen/oxygen tanks is the economically preferred choice among the many Space Shuttle configurations so far investigated.
  Early examples of such concepts are RATO of McDonnell Douglas, TAHO of Grumman-
  Boeing, and similar concepts studied by North American Rockwell and [Lockheed Missile
  and Space Company]-Lockheed; these concepts are now commonly known as rocket assisted
  orbiters (RAO).*
* *The choice of thrust assist for the orbiter Shuttle is still open. The main economic alternatives are pressure fed boosters and solid rocket motors, either using parallel burn. A third economic alternative to these versions is to use series burn boosters.* [italics added for
  emphasis; original was all capital letters]
These conclusions are based on the following results of the economic analysis:

[0-2]   0.2 THE ECONOMIC WORTH OF A SPACE SHUTTLE SYSTEM

0.2.1   Results of the May 31, 1971 Analysis

The major findings of the economic analysis of new Space Transportation Systems
reported on May 31, 1971, which were prepared for the National Aeronautics and Space

Administration, are concerned with the analysis of the economic value of a reusable Space Transportation System without any particular concern as to which, among the many alternative Space Shuttle Systems[,] would, in the end, be identified as the most economic system.

Figure 0.1 shows the summary of the major results of the May 31, 1971 analysis. In this analysis we report only the results of the "Equal Capability" analyses, the most conservative approach to evaluate new technologies. "Equal Budget" analyses were also performed and those calculations give even more favorable economic results (see also May 31, 1971 analysis). On the horizontal axis the numbers of Space Shuttle flights between 1978 and 1990 are shown as ranging between 450 and 900 flights for that period. On the vertical axis the allowable non-recurring cost for the development of the launch vehicle—that is, the Space Shuttle as well as the Space Tug and the required launch sites—are shown in billions of undiscounted 1970 dollars. The benefit lines shown in this figure show how the allowable non-recurring costs—that is, the benefits to be associated with a fully reusable Space Transportation System—increase as the flight level expected for the 1980's increases between 450 and 900 flights. Overall, this is very much a function of the particular rate of discount (or social rate of interest) chosen and applied to the analysis. Three summaries are shown in Figure 0.1: the results of 5%, 10% and 15% social rates of discount respectively. We may wish to use them interchangeably. Since all the costs as well as the calculated cost savings were expressed in constant dollars, the interest rates applied are real interest rates which do not include elements of inflation. As shown at a 10% rate of interest, the allowable non-recurring cost would vary from about $12.8 billion (about 500 Space Shuttle flights in the 1980's), up to $20 billion at a flight level of about 850 flights for the same period. The shaded vertical lines in Figure 0.1 show, first, the average U.S. flight level in terms of Shuttle flights between 1964 and 1969 (61 flights per year) and reflect also the funding average between the years 1963 and 1971. Also shown are the average USSR flights for the period 1965 to 1970 (65 flights per year). Furthermore, the baseline mission model of 736 flights, at that time, is shown on the right side of the darkly shaded area where the left boundary of that area is defined by a reduced mission model of around 600 flights for Space Program 3 in that analysis. Since then, we have used in our present analysis a reduced baseline mission model of 514 flights with a potential overall level of 624 space flights. Thus, in the last six months, the analysis of the Space Shuttle System has been extended downwards to cover substantially the region between 450 and 600 flights. Also shown in Figure 0.1 are the then estimated non-recurring costs of $12.8 billion for a two-stage fully reusable Space Shuttle System* as well as the Space Tug and the required installations. We show the estimated economic potential of a reusable Space Transportation System in terms of allowable non-recurring costs as a function of several economic variables, among them the expected space activity level, the social rate of discount, and the type of cost-effectiveness analysis. The major findings of that effort are:

The major economic potential identified for Space Transportation Systems in the 1980's is the lowering of space program costs due to the reuse, refurbishment, and updating of satellite payloads. The fully reusable, two-stage Shuttle is the major system considered in the May 31, 1971 report, but not the only system to achieve reuse, refurbishment and updating of payloads. Payloads were assumed to be refurbished on the ground, with refurbishment costs varying between 30% and 40%. The launch costs of the Space Shuttle and Space Tug needed to recover and place the refurbished payloads are also allowed for. We strongly recommended in May that other systems be studied to determine the extent and the cost at which they can achieve reuse, refurbishment, and updating of payloads.

---

*   The selected Space Shuttle System is no longer a two-stage fully reusable system and has substantially reduced non-recurring costs [see section 0.2].

[pages 0-3 and 0-4, Figure 0.1, omitted]
[0-5] The cost reductions identified originate in three distinct areas:

(a) The research, development, test and evaluation (RDT&E) phase of new payloads (satellites);
(b) The construction and operating costs of payloads (satellites) for different space missions;
(c) The cost of launching payloads into orbit.

The projected non-recurring cost associated with developing the Space Shuttle and Tug as configured in May, 1971, (a two-stage system) is shown by the economic analysis to be covered by the identified benefits provided the United States intends to operate a space program with the number of flights equal to the unmanned space program activities of the United States in the 1960's. The direct costs (payload and transportation.) of space activity carried out by a Space Shuttle System are expected to be about one-half of the direct costs of the current expendable transportation system.

Manned space flight options—for example, a manned lunar option—are also analyzed. They show that a Space Shuttle System offers economic advantages also in terms of transportation costs for some large lunar and planetary (or defense) space flight options for the 1980's. These advantages were not considered when formulating the basic conclusions of the economic study, due to the great uncertainty of these options being adopted by the United States.

The choice of the social discount rate has a major influence on the economics of a new Space Transportation System. Differences in the rate applied to the analysis outweigh many other important issues usually raised—and analyzed—in the context of large scale RDT&E projects, including uncertainties in the cost data. As shown in this report, the social rate of discount influences not only the overall worth of a new Space Transportation System, but also the choice of specific technical configurations in deciding among alternative technical approaches to bring about a reusable Space Transportation System.

The May 31, 1971 report concludes that the economic justification of a reusable Space Transportation System is not tied to the question of [0-6] manned versus unmanned space flight. Space programs used and analyzed are in line with the activity and funding levels of the unmanned United States space program of the 1960's (NASA, DoD, and commercial users included). If a substantial number of manned space flights were to be undertaken in the 1980's, a Space Shuttle System would also contribute significantly to lowering the costs of such missions and activities.

The May 31, 1971 report analyzes the economically allowable nonrecurring cost of a reusable Space Transportation System. It is the task of the present report to identify the economically best reusable Space Transportation System among all the possible required alternatives.

A major point of the May 31st report is: any investment can only be justified by its goals. This applies to business as well as to government, hence also to NASA. A new, reusable Space Transportation System should only be introduced if it can be shown, conclusively, what it is to be used for and that the intended uses are meaningful to those who have to appropriate the funds, and to those from whom the funds are raised, as well as to the various government agencies that undertake space activities. The space goals can be political (rivalry with the space programs of other countries), military (to meet military space efforts of other countries who use the potential of space to meet needs of national security), scientific (for example, astronomy), or commercial (for example, earth resources applications). All these goals will, of course, be mixed into one national space program, representing to various degrees a joint demand for space transportation with a varying mix of payloads.

0.2.2 Updated Economic Results on the Economic Worth of a Space Shuttle System

Since May 31, 1971 our efforts concentrated on two major questions: first, to what extent is the overall economic worth of a Space Shuttle System modified by new inputs given to our study; and, second, which of the many alternative Space Shuttle configurations is the most economical.

The new inputs reflect a substantially modified NASA and DoD Baseline Mission Model for the 1980's, and make a new assessment of payload [0-7] effects for different missions; very importantly, new alternative Space Shuttle Systems that still promised the achievement of most of the objectives of the Space Shuttle program[,] but at considerably reduced non-recurring costs in the 1970's, were considered.

Table 0.1 shows the estimated complete direct life-cycle costs for a NASA and DoD U.S. space program from 1979 to 1990 (twelve years) of 514 Space Shuttle flights, or an average of 43 Space Shuttle flights per year, in this period. This space program is based on the NASA Baseline Mission Model, including scientific and application missions as well as some manned space flight activity, and a modified DoD mission model.

As can be seen from Table 0.1, the same facts hold for the basis of the economic analysis of the Space Shuttle System as in the May 31, 1971 report:

(1) The Space Shuttle System has substantially higher research, development and investment costs (non-recurring costs) associated with it than any of the current expendable or new expendable systems. This remains true, although the non-recurring costs of the Thrust Assisted Orbiter Shuttle (TAOS) System are substantially lower than the corresponding fully reusable two-stage Shuttle System costs of May, 1971.

(2) The TAOS Space Shuttle System promises reductions in the recurring launch costs of Space Transportation.

(3) The Space Shuttle System promises a reduction in the costs of satellite payloads through reuse, refurbishment, in-orbit checkout of payloads, and possible updating and maintenance of payloads in orbit or on the ground.

It is the combined reduction in launch costs and payload costs that underly [sic] the economic justifications of the TAOS Space Shuttle System. These life-cycle costs are the starting point and the basis of our economic analysis. A wide variety of alternative Space Shuttle Systems was investigated by us with a wide variety of technical changes when compared with the May, 1971 Space Shuttle configuration.

[page 0-8, Table 0.1, omitted]

[0-9] On each of these changes a substantial set of alternative calculations was made, in keeping with the analyses and methodology already developed.

The results of the updated economic analysis are shown in the next three figures. In Figure 0.2 the estimated non-recurring costs of alternative Space Shuttle Systems are shown on the horizontal axis. These nonrecurring costs include the full non-recurring costs of the Space Shuttle System with at least the same capabilities as those given by the expendable Space Transportation System. Where the economic analysis of a space program indicated the continued use of expendable rockets—e.g., Scout Rockets—then these system costs have been included as Space Shuttle System costs. Similarly, in the time of the Space Shuttle System phase-in—to replace expendable Space Transportation Systems—the cost of expendable systems, as required, is also included as a Space Shuttle cost. Most important, the non-recurring costs of the Space Tug, which gives the Space Shuttle System the capability to deploy and bring back payloads from all earth orbits when economically justified, are fully included. Finally, the non-recurring costs, as used in our analysis, also include the costs of two launch sites ([Eastern Test Range] and [Western Test Range]). It is on the basis of these non-recurring costs that the economic evaluation of the Space Shuttle System has been carried out.

The estimated non-recurring costs also include fleet investment. An estimated five Space Shuttles will be required to fulfill the NASA and DoD Baseline Mission Models for the 1980's. Fleet investment includes the orbiter procurement cost for all configurations considered, but reusable booster costs have been amortized as a recurring cost except for the manned flyback booster case.

Not shown in Figure 0.2 are the RDT&E and investment costs to the First Manned Orbited Flight (FMOF) of the Thrust Assisted Orbiter Shuttle (TAOS), estimated now by NASA at $5.5 billion. The estimates of alternative Space Shuttle Systems in Figure 0.2 are grouped into two classes: first, the modified two-stage reusable Space Shuttle Systems that were investigated in the past months as alternatives to the two-stage fully reusable Space Shuttle System of May 31, 1971. These systems all have associated [page 0-10, Figure 0.2, omitted] [0-11] with them lower non-recurring costs than the estimate for the original fully reusable Space Shuttle System. Considerable variation existed with regard to the non-recurring costs of these modified two-stage (manned booster) systems. In addition, therefore, we show the mean of these estimates as well as the standard deviation (σ) of the non-recurring cost estimates of these systems. As shown in Figure 0.2, the mean of the non-recurring costs of such modified two-stage Space Shuttle, Systems is $11.5 billion, the standard deviation is $1.44 billion.

Similarly, also shown in Figure 0.2 are estimated total non-recurring costs of Thrust Assisted Orbiter Space Shuttle Systems (TAOS) that include a wide variety of technical choices, all having in common that only the orbiter is manned, with external hydrogen/oxygen tanks[,] and all are assisted at takeoff by either solid rocket motors or pressure fed rocket systems. The mean of the non-recurring cost estimates of such systems is $7.5 billion. These include about $1.6 billion for the non-recurring costs of the Space Tug and the additional required launch site. They also include a fleet of 5 Space Shuttles, each estimated at about $300 million. When Space Tug and [Western Test Range] costs are excluded ($1.6 billion), as well as 3 Space Shuttle vehicles (about $900 million), then the estimated non-recurring costs in the 1970's (comparable, roughly, to FMOF costs) are estimated to be $5.0 billion (1970 dollars). The standard deviation of this estimate is $900 million, again in 1970 dollars.

Using these alternative Space Shuttle Systems, a comprehensive set of economic analyses was performed along the lines of the May 31, 1971 report to determine the economic benefits of a Space Shuttle System. In Figure 0.3 the results of the equal capability cost-effectiveness analysis are shown, at a 10 percent social rate of discount, directly comparable to the results of May 31, 1971 as shown in Figure 0.1. The benefits are expressed in Allowable Non-Recurring Costs, thus making the benefits shown directly comparable to the estimated non-recurring costs of Figure 0.2.

Major variations were introduced in the space program activities of the 1980s, concentrating on the lower role of expected space activities of the 1980's and beyond. While in the May 31st analysis the area of interest—based on historical, unmanned activities of the United States (and the Soviet [page 0-12, Figure 0.3, omitted] [0-13] Union)—was confined to between 500 and 900 Space Shuttle flights in the 1978 to 1990 period, the present analysis was confined to look at the range of Space Shuttle flights between 400 and 650 Space Shuttle flights, with major variations in the analysis at 514 and 624 flights.

Two separate benefit lines were arrived at and are shown in Figure 0.3: first, the analysis concentrating around 514 Space Shuttle flights shows the economic results with the exclusion of some DoD missions that are particularly suited for Space Shuttle operations; second, the analysis concentrating at around 624 Space Shuttle flights takes the same NASA mission model, now, however, including on the DoD side the missions omitted in the first analysis.

With regard to the lower benefit line, we conclude that at 514 flights in the 1979–1990 period, the estimated benefits of a Space Shuttle System are $10.2 billion in 1970 dollars with a variance of $940 million expressed in allowable non-recurring costs. The economic "break even" point is reached at an annual space activity level of about 30 Space Shuttle flights, carrying satellite payloads. This annual level of NASA and DoD space activity in the 1980's and beyond will justify the development of the TAOS Space Shuttle at a social rate of discount of 10 percent.

When, on the other side, Space Shuttle related DoD missions are included, the economic analysis shows, at 624 Space Shuttle flights in the 1979 to 1990 period, an estimated benefit of $13.9 billion of allowable non-recurring costs, with a standard deviation of ±$1.45 billion. As activity levels are increased or decreased around these space programs, the expected benefits of a Space Shuttle System increase or decrease as shown by the two benefit lines in Figure 0.3. The TAOS Space Shuttle System will "break even" at an annual activity level of about 25 Space Shuttle flights, carrying satellite payloads, when the "624" mission model is taken as representative of U.S. space activities in DoD and NASA for the 1980's.

Again, we want to emphasize that these results reflect the benefits of a Space Shuttle System when applying a 10 percent real social rate of discount to the complete economic analysis.

[0-14] By combining Figures 0.2 and 0.3 we can directly judge the results of the economic analysis of a Space Shuttle System.

In Figure 0.4, we show on the vertical axis the estimated nonrecurring costs—as developed in Figure 0.2—and also the benefits of a Space Shuttle System in terms of "allowable non-recurring costs" as developed in Figure 0.3. The estimated non-recurring costs of the TAOS Space Shuttle Systems are emphasized and the expected standard deviation of these costs is shown by the shaded area around the non-recurring cost estimate of TAOS. Similarly, the benefit lines as developed in Figure 0.3 are shown; the standard deviation around these estimates is indicated again by the shaded areas.

From the results as shown in Figure 0.4, *we conclude that the development of a TAOS Space Shuttle System is economically justified,* [italics added for emphasis; original was all capital letters] within a level of space activities between 300 and 360 Shuttle flights in the 1979–1990 period, or about 25 to 30 Space Shuttle flights per year, well within the U.S. Space Program including NASA and DoD. If the NASA and DoD mission models are taken at face value (624 Space Shuttle flights in the 1979–1990 period), the estimated benefits of a Space Shuttle are 13.9 billion with a standard deviation of ±$1.45 billion expressed in 1970 dollars (at a 10% social rate of discount). If parts of the expected U.S. Space Program are substantially modified (514 Space Shuttle flight level in the 1979–1990 period), the estimated benefits of a Space Shuttle System are $10.2 billion, with a standard deviation of $940 million (at a 10% social rate of discount).

The estimated non-recurring costs directly comparable to the benefits expressed in "allowable" non-recurring costs of a TAOS Space Shuttle System are $7.5 billion with a standard deviation of $960 million.

Since the complete economic evaluation of the Space Shuttle System as summarized here *reflects the results when using a 10 percent real social rate of discount, the economic results in support of the TAOS Space Shuttle development have to be regarded as very strong in the context of United States national priorities.* [italics added for emphasis; original was all capital letters] [page 0-15, Figure 0.4, omitted]

Document II-11

Document title: James C. Fletcher, Administrator, NASA, to Caspar W. Weinberger, Deputy Director, Office of Management and Budget, December 29, 1971.

Source: NASA Historical Reference Collection, NASA History Office, NASA Headquarters, Washington, D.C.

Under continuing pressure from the White House Office of Management and Budget to lower the development costs of a Space Shuttle, NASA in late December 1971 reluctantly changed its recommended Shuttle configuration to one with a smaller payload capacity. In this letter, NASA Administrator James Fletcher made what he believed to be NASA's final arguments for Shuttle approval by the White House. The debate over which Shuttle configuration to approve continued over the New Year's weekend. On January 3, NASA learned that it had received presidential approval to develop its "best buy" Shuttle, rather than the smaller system recommended in this letter. While the development costs projected for the Shuttle in Fletcher's letter were not greatly off the system's final costs, the cost-per-flight estimates proved to be much lower than the actual expense.

[1]   NATIONAL AERONAUTICS AND SPACE ADMINISTRATION
                          Washington D.C. 20546

OFFICE OF THE ADMINISTRATOR   December 29, 1971

Honorable Caspar W. Weinberger
Deputy Director
Office of Management and Budget
Executive Office of the President
Washington, D.C. 20503

Dear Cap:

The purpose of this letter is to report the results of recent studies of several space shuttle options, and to recommended a course of action to be taken in the FY 1973 budget.

SUMMARY

We have concluded that the full capability 15 x 60' - 65,000# payload shuttle still represents a "best buy," and in ordinary times should be developed. However, in recognition of the extremely severe near-term budgetary problems, we are recommending a somewhat smaller vehicle—one with a 14 x 45' – 45,000# payload capability, at a somewhat reduced overall cost.

This is the smallest vehicle that we can still consider to be useful for manned flight as well as a variety of unmanned payloads. However, it will not accommodate many DOD payloads and some planetary payloads. [2] Also, it will not accommodate a space tug together with a payload, and will therefore not provide an effective capability to return payloads or propulsive stages from high "synchronous" orbits, where most applications payloads are placed.

BACKGROUND

Early in 1971, after completion of feasibility studies, NASA focused on a shuttle configuration that would replace all of the existing launch vehicles (except the very small Scout, and the very large Saturn V); would provide for a continuation of manned space flight; and would have the lowest possible cost per flight. This configuration has a 15 x 60' – 65,000# payload bay; a very large orbiter; and a huge fly-back booster. It would cost $10 billion to develop, and $4.1 million per flight.

We then set out to optimize the configuration for the best balance between development cost and operating cost, while retaining the full 15 x 60' – 65,000# capability [3] that is required to accommodate all NASA and DOD payloads. The result: a much smaller orbiter with external jettisonable tanks; and a ballistic reusable booster. The development cost was cut nearly in half, to $5.5 billion, while the cost per flight increased to $7.7 million. Although the cost per pound of payload in orbit increased from $63 to $118, we felt this to be worth the huge savings in development cost.

During the course of our studies as well as at the request of the "Flax Committee" we also looked at smaller payload compartments. More recently in a meeting with Don Rice, we were asked to examine shuttle costs with an even smaller performance capability. Specifically, we were asked 2 1/2 weeks ago to look at a 10 x 30' – 30,000# payload capability, with the added guideline that the development cost should be less than $4 billion, and the cost per flight less than $5 million. (We have not been able to meet these cost objectives.) We have now compared costs and payload capabilities of five different shuttle options, and have reached certain conclusions.

[4]   RESULTS OF RECENT STUDIES

Payload Capabilities: We analyzed five different shuttle options, with different payload bay sizes and payload weight carrying capabilities. There are:

	Bay Size	Payload Weight*
Case 1	10 x 30	30,000
Case 2	12 x 40	30,000
Case 2A	14 x 45	45,000
Case 3	14 x 50	65,000
Case 4	15 x 60	65,000

[* in equivalent "due east" orbits]

Case 4 is the basic shuttle configuration, and will accommodate all NASA and DOD payloads. None of the other configurations will do this.

As the payload bay is decreased in length, many of the DOD payloads are eliminated at the 50-foot length, as are some NASA planetary payloads. At the 50-foot length we also lose the capability to fly a space-tug/payload combination for synchronous orbit applications payloads.

A 45-foot length appears to be the minimum practical size for many manned space flight modules, as well as many [5] space science payloads, and applications payloads, with a one-way delivery capability. The 30-foot length eliminates nearly all DOD payloads, some important space science payloads, most applications payloads, all planetary payloads, and useful manned nodules.

A similar analysis shows that the space shuttle bay diameter should be 14'. This requirement stems primarily from manned flight considerations. The proposed 10-foot diameter would lead to an outside module diameter of 9 feet (1-foot clearance requirement), and an inside diameter of 8 feet. By the time this is "squared off," cabling and plumbing are added, as well as consoles, cabinets, and other accommodations, this size is unacceptable. Note also that Skylab is 22 feet in diameter, and the Apollo Command Module is 13 feet. Some science, applications and planetary payloads are also better accommodated in a 14-foot diameter.

The payload weight requirement of 60,000 to 65,000 pounds was set by the space tug as well as by DOD payloads. Without the tug, the manned modules establish a requirement [6] of 45,000 pounds. (Actually these modules will only weigh 15,000 to 20,000 pounds. However, they must be boosted to an orbit of 270 miles at a 55-degree inclination; this requires an equivalent "due east" payload capability of 45,000 pounds).

In summary then, if a decision is made to develop a shuttle with less than full payload capability, the 14 x 45' – 45,000# option appears to be the minimum useful configuration. It will not handle many DOD payloads; it will not handle some planetary payloads; and it will not handle the space tug in combination with a payload. However, it will accommodate manned spaceflight modules, a one-way capability to synchronous orbit for civilian applications payloads, most other NASA payloads, and some DOD payloads.

Cost Comparison: The results of the studies, in terms of costs, are shown in the attached table. (The definitions of "development" and "operating" costs are the same as used in previous studies and discussions. Amounts are in 1971 dollars.) The cost trends shown were established [7] independently by NASA and by two contractors. The main conclusion is that development costs do not vary sharply from one option to the next— cost differences between adjacent options are about $200 million.

In other words, the most important cost reductions were achieved through the basic configuration changes (with the same payload capability) undertaken by NASA during the past year. A variation in payload size and weight has only smaller effects on development cost. For this reason, the full capability shuttle must still be considered to be a "best buy."

Development cost, for any given shuttle size, can be further reduced by using solid rocket motors instead of the pressure-fed liquid reusable booster. For the 14 x 45' – 45,000# shuttle we estimate that the development cost could be reduced from $5 billion to $4.3 billion. However, this would be at the expense of increased operating costs: from $7.5 million per flight to $10–$13 million per flight.

|8| RECOMMENDED SHUTTLE

On the basis of the studies just completed, NASA would ordinarily recommend proceeding with the full capability 15 x 60' payload shuttle. However, in recognition of severe budgetary pressures we have concluded that a lesser capability still provides a useful vehicle, and therefore recommend proceeding with the 14 x 45' – 45,000# shuttle. With a pressure-fed liquid booster, this shuttle is estimated to cost $5 billion to develop and $7.5 million per flight.

BOOSTER OPTIONS

The question of a liquid as opposed to a solid booster is not yet completely settled. There are some open technical questions concerning noise, interference effects, thrust-vector-control requirements, and quality control requirements for manned flights. Also, differences in operating costs have not yet been determined with accuracy. For these reasons, we recommend that two booster options should be considered for the next two months in conjunction with the recommended orbiter.

They are:

[9] Option 1 – Pressure-fed liquid booster
              – Shuttle development cost $5 billion
              – Shuttle operating cost $7.5 million/flight
                 (There remain some uncertainties that might drive this as high as
                 $9 million/flight.)
    Option 2 – Solid rocket motor booster
              – Shuttle development cost $4.3 billion
              – Shuttle operating cost $10–$13 million/flight

We would then select the appropriate booster on or about March 1, 1972, based on technical considerations, as well as the best balance between minimum development and minimum operating costs.

FUNDING CONTINGENCY

The cost figures mentioned so far represent NASA's best estimate of the actual costs expected during the course of the shuttle development. They are based on actual experience in NASA and DOD aircraft and space programs, and in addition contain a 15% factor for research and development changes. It is our intention to manage the program to bring it in at those costs.

[10] Nevertheless, we believe that we should include a contingency against future cost growths due to technical problems, in recognition of the very advanced nature of this development. We believe a 20% contingency would be appropriate. Approval of a $5 billion program would thus constitute a commitment by NASA to make every effort to produce the desired system for under $5 billion, but in no case more than $6 billion.

DECISION TO PROCEED

The various shuttle studies have progressed to the point where a decision to proceed with full shuttle development should now be made.

Further delays would not produce significant new results. The orbiter is fully defined. Although a question of solid versus liquid boosters remains open, the range of variables involved in the booster decision is not large, and a decision can be made at an early date. No substantial cost savings can be realized by further studies. (All of the most recent cost refinements for a given payload size have been less than the overall cost uncertainties inherent in a large R&D undertaking.)

[11] On the other hand, additional delays would have many unsettling effects. In the aerospace industry, the existing shuttle teams will soon be dissipated, unless fully funded by the government. Last year's strong Congressional support for the shuttle may be lost this year if the Administration cannot present equally strong support. And within NASA, many of the best people will be lost, with a resulting loss in overall morale.

In other words, there is a great deal to be gained, and nothing to be lost, by making a decision to proceed now.

Elements of the Decision: The decision would entail the following elements:
1. A statement that shuttle development will proceed.
2. That the orbiter payload bay size should be 14 x 45' – 45,000 pounds.
3. That NASA will commit to do the job for a development cost of $5 billion (plus a maximum contingency of 1 billion) for the liquid booster option (less for the solid booster option); and that NASA will select the proper booster on the [12] basis of technical considerations as well as the best balance between minimum development and operational costs.

Required Actions: To implement this decision, the following actions are required:
1. By OMB: inclusion of $200 million R&D funds plus $28 million [Construction of Facilities] funds, together with appropriate narrative, in the FY 1973 budget.
2. By NASA:
    (a) Notification of contractor of intent to issue RFP in March, 1972.
    (b) Selection of one of two booster options by March, 1972.
    (c) Contractor selection in June or July, 1972.

I look forward to our meeting this afternoon, and will then be able to answer any questions you may have.

Sincerely,

James C. Fletcher
Administrator

[no page number]   RESULTS OF STUDIES

CASE	1	2	2A	3	4
PAYLOAD BAY (FT.)	10 X 30	12 X 40	14 X 45	14 X 50	15 X 60
PAYLOAD WEIGHT (LBS.)	30,000	30,000	45,000	65,000	65,000
DEVELOPMENT COST (BILLIONS)	4.7	4.9	5.0	5.2	5.5
OPERATING COST ($MILLION/FLT.)	6.6	7.0	7.5	7.6	7.7
PAYLOAD COSTS ($/POUND)	220	223	167	115	118

Document II-12

Document title: Arnold R. Weber, Associate Director, Office of Management and Budget, Memorandum for Peter Flanigan, "Space Shuttle Program," June 10, 1971, with attached: "NASA's Internal Organization for the Space Shuttle Project" and "NASA's Space Shuttle Program."

Source: NASA Historical Reference Collection, NASA History Office, NASA Headquarters, Washington, D.C.

An issue of political interest to the White House during the debate over whether to approve Space Shuttle development was the potential employment impact of the program. This was particularly the case because, if approved, the program would begin during the 1972 presidential election year. Peter Flanigan was President Nixon's assistant with oversight responsibility for NASA.

[no page number]   EXECUTIVE OFFICE OF THE PRESIDENT
                      OFFICE OF MANAGEMENT AND BUDGET
                      WASHINGTON, D.C. 20503

MEMORANDUM FOR PETER FLANIGAN

Subject: Space Shuttle Program

Attached are the two papers on the impact of the space shuttle on the aerospace industry which you requested.

You should be aware that these employment estimates are preliminary. As the paper indicates no decision on development has been made. The critical contractor selections will not be made until the Administration has approved the project. NASA expects approval in August, but it may be delayed until late 1971 when the 1973 Budget is decided. If the decision is delayed the employment impacts will also be delayed by approximately 6 months.

                                    Arnold R. Weber
                                    Associate Director

Attachments

                         ****

[no page number]   June 10, 1971

NASA's Internal Organization for the Space Shuttle Project

NASA has decided that the responsibility for program manage-ment of the space shuttle (including systems engineering and coordination of field center activities) will be assigned to the Manned Spacecraft Center (MSC), Houston, Texas. The major consideration in this decision was the determination by NASA's top management that a field center should have this responsibility and that a large project management organization (like that of the Apollo program) should not be established at NASA headquarters in Washington. The NASA decision reflects a conclusion on the part of NASA management that the responsibilities for integration and coordination of the shuttle program should be directly carried out by a field center which has sufficient technical competence to run the program.

Because of their technical capabilities and unique experience in the manned space program, the only centers seriously considered for assignment of the shuttle management responsibility were MSC, Houston, and Marshall Space Flight Center (MSFC), Huntsville, Alabama. The decision to award the program management and coordination responsibility to MSC, Houston, was made on the basis that MSC had the most experience applicable to the particular portion of the space shuttle program which is likely to be the most difficult to accomplish, namely the orbiter. NASA also felt that with the assignment of responsibility for the development of the shuttle engine and booster to MSFC, Huntsville, together with continuing Skylab responsibilities at MSFC, that the workload balance would be better if MSC received the overall program management assignment.

The employment impact of this decision is minimal because the responsibility will be fulfilled by reassignments of personnel currently at MSC. Of course, this affects only the NASA organization and management responsibilities. No decision had yet been made on which contractors would be utilized for the shuttle, assuming the shuttle program is approved.

NASA has two unmanned space flight centers in California. The Jet Propulsion Lab in Pasadena and the Ames Research Center in the Bay area. These centers do not have the capability to manage a manned space flight program of the magnitude of the space shuttle.

****

[1]   NASA's Space Shuttle Program

The space shuttle would be a reusable space transportation system, consisting of an orbiter and a booster, which would carry NASA and DOD payloads to and from earth orbit beginning in 1979. The shuttle would replace all but the very smallest and very largest (Saturn V) expendable rockets. The investment costs (research and development, facilities, and initial fleet) of the shuttle would be about $14 billion through FY 1979 when the shuttle would, under NASA's schedule, become operational.

Thus far, the Administration has not approved NASA's plan for the fully reusable shuttle. The 1972 budget provides $100 million for initial development of the engine (the longest lead-time item) and continuing design of the shuttle airframe. However, the initiation of development of the airframe is contingent upon favorable assessment of technical and economic studies and a positive decision by the Administration that NASA can proceed with fullscale development. NASA is now completing the various studies required including an economic analysis.

1. Engines

NASA intends to announce a contractor selection on the engine near the end of June. This is a firm date based on presently budgeted funds. There are three contractors currently competing for the engine contract:

a. Aerojet General . . . Sacramento, California
b. Rocketdyne (North American Rockwell) . . . Canoga Park, California
c. Pratt and Whitney . . . West Palm Beach, Florida

Anticipated Employment:

6/71	12/71	6/72	12/72
500	1500	2500	3500

2. Airframe

NASA's current schedule calls for an Administration decision on the shuttle airframe in August, followed by issuance of a Request for Proposals (RFP) in early September, and contractor selection in December. However, in order to look at alternative phasing plans, NASA is [2] seriously considering stretching out this schedule by several months. There are currently two contractor teams competing for the major shuttle contract (airframe):

1. McDonnell Douglas . . . Los Angeles, California and St. Louis, Mo.
2. North American Rockwell . . . Los Angeles, California

If NASA received the go-ahead decision on the airframe in September, the following contractor employment pattern would be likely:

Anticipated Employment:

Decision Time	6/71	12/71	6/72	12/72
August 1971	500	1500	4000	7000
January 1971	500	500	1500	4000

Thus, although a peak of 70,000 jobs might ultimately result from the shuttle in the mid-1970's, the number of actual jobs by the end of CY 1972 would be relatively small.

3.   Launch Site

A NASA evaluation group is reviewing alternative launch sites including Cape Kennedy, Fla[.]; Edwards Air Force Base, Claifornia [sic]; White Sands, N.M.; and Wendover Air Force Base, Utah. From a cost standpoint, Cape Kennedy has the advantage (investment cost of $3–400 million vs. $800 million–$1 billion required elsewhere). A recommendation is expected in September.

There would be no employment impact at the launch site during 1972. Employment would peak at about 6,000 in 1980.

Alternatives to NASA's current plan which would decrease near-term costs and employment include a phased development of the shuttle (orbiter first), a partially reusable shuttle with expendable drop tanks, and improved fully expendable rockets. The FY 1973 budget will be a key decision point for the shuttle alternatives.

Document II-13

Document title: James C. Fletcher, Administrator, NASA, to Caspar W. Weinberger, Deputy Director, Office of Management and Budget, January 4, 1972.

Source: NASA Historical Reference Collection, NASA History Office, NASA Headquarters, Washington, D.C.

On December 29, 1971, NASA provided the Office of Management and Budget (OMB) its recommendation that a Space Shuttle with a smaller 14-foot-by-45-foot payload bay be developed. At a January 3, 1972, meeting in the office of OMB Director George Shultz, NASA learned that the White House, and perhaps President Richard Nixon himself, had decided to give NASA approval to develop the "full-size" Shuttle that the space agency had been advocating prior to its December 29 recommendation. There were a variety of programmatic reasons for this decision. In addition, there was a desire among Nixon's political advisors to begin a major aerospace project during the 1972 presidential election year. (Congress had canceled the Supersonic Transport program in 1971.) Such a project would have important employment impacts in key electoral states, such as Texas, California, and Washington.

[1]   NATIONAL AERONAUTICS AND SPACE ADMINISTRATION
WASHINGTON, D.C. 20546

OFFICE OF THE ADMINISTRATOR   January 4, 1972

Honorable Caspar W. Weinberger
Deputy Director
Office of Management and Budget
Washington, D. C. 20503

Dear Cap:

The purpose of this letter is to document the decision reached yesterday concerning the space shuttle.

NASA will proceed with the development of the space shuttle. The shuttle orbiter will have a 15x60-foot payload bay, and a 65,000-pound payload capability. It will be boosted either by a pressure-fed liquid recoverable booster or by solid rocket motors. NASA will make a decision between these two booster options before requests for proposals are issued in the spring of 1972.

NASA and industry will also continue to study, for the next several weeks, a somewhat smaller version of the orbiter, with a 14x45-foot, 45,000-pound payload capability, with the pressure-fed liquid and solid rocket motor booster options. The main purpose of studying this smaller shuttle is to determine whether or not significant savings in operational costs can be realized, with solid rocket motors, at this smaller size. The decision between the larger (15x60 – 65,000#) and smaller (14x45 – 45,000#) shuttle will also be reached by NASA before requests for proposals are issued in the spring.

The basic decision to proceed with the shuttle development will be announced by the White House. Following that announce-ment, NASA will inform the aerospace industry of the details of the decision, as stated in this letter.

[2] Thank you for your support in bringing about the decision to go ahead with the space shuttle.

Sincerely,

James C. Fletcher
Administrator

Document II-14

Document title: James C. Fletcher, Administrator, NASA, to Caspar W. Weinberger, Deputy Director, Office of Management and Budget, March 6, 1972.

Source: NASA Historical Reference Collection, NASA History Office, NASA Headquarters, Washington D.C.

With this letter, NASA informed the Office of Management and Budget (OMB) about its final choice of a Space Shuttle configuration, as well as the reasoning behind that choice. The key was the desire to hold down development costs, even if that meant higher per-flight costs for the Shuttle because of the choice of solid-fueled rather than liquid-fueled boosters. The letter also reflected the continuing tension between NASA and OMB with respect to the budget committed to the Shuttle program—an issue that was to continue to constrain the program during its development.

[1]   NATIONAL AERONAUTICS AND SPACE ADMINISTRATION
                WASHINGTON, D.C. 20546

OFFICE OF THE ADMINISTRATOR   MAR 6 1972

Honorable Caspar W. Weinberger
Deputy Director
Office of Management and Budget
Executive Office of the President
Washington, D.C. 20503

Dear Cap:

With regard to the space shuttle, we decided in George Shultz' office on January 3 that we would develop a shuttle with a 15x60' – 65,000# payload capability. At that time I urged that we look further at what kind of a booster to use—liquid or solid—and decide that issue in the spring. In addition, I proposed at that time that we would continue to look at a somewhat smaller size shuttle (14x45' – 45,000# payload) for the sole purpose of determining whether or not, if we choose the solid booster, substantial cost savings could be obtained from the use of the smaller vehicle.

Our studies have now been completed, and we have reached the following conclusions:

1. The use of solid boosters in the parallel staged configuration represents the optimum choice from combined technical and budgetary points of view.

2. Our prior decision to incorporate the larger payload capability is confirmed by our subsequent analysis from an overall program point of view, notwithstanding our choice of the solid rocket booster.

We plan to announce these conclusions shortly before or at a hearing before the Manned Space Flight Subcommittee of the House Committee on Science and Astronautics, scheduled for March 16, 1972. Issuance of the RFP will come as soon as [2] possible thereafter. As I told you earlier, the Committee has demanded a firm decision by the time of our appearance regarding shuttle configuration and choice of booster. In order to assure timely passage of the President's shuttle program by the Congress, our legislative experts believe it essential that the Committee's firm deadline be met. Since we met last Friday, a scheduling problem with our Senate Authorization Committee has also developed. This may require an announcement of the decision on March 15, one day earlier.

The decision concerning liquid or solid boosters was a difficult one. It involves a trade-off between future benefits (at the time the shuttle becomes operational) and earlier savings in the immediate years ahead: liquid boosters have lower potential operating costs, while solid boosters have lower development costs. The decision concerns development risk which is lower for the solids because the technical unknowns are less, and also risks in operational costs which favor the solids because the economic exposure of failing to recover a booster is much less.

Another approach in reaching this decision involved adding all costs together—development, investment and operating. However, the conclusions here are heavily dependent on the mission model, with the liquid booster favored if we assume a large number of flights per year, and the solids if the number of flights per year is less.

Based on the results of our contractor studies and our inhouse estimates, and with our great concern about holding down development costs in these years of tight fiscal constraint, our decision must be in favor of the solid booster. We feel quite confident of being able to develop the solid-boosted shuttle for less than the $5.5 billion committed to you last January and, hopefully, when we have developed the data more firmly we may be able to commit to a smaller overrun amount that the 20 percent mentioned in my January 23 letter. [underlined by hand]

[3] From the budgetary point of view, perhaps the most important consideration is that we have selected the configuration which, for a given payload size and weight, entails the lowest development cost. Thus there would seem to be no budgetary interest in further delay.

Our reaffirmation of the payload size is based on the facts that the differences in development and operational costs between the larger and smaller versions have been verified to be very small; that these savings would nowhere near compensate for the future savings that would be lost because of the many important payloads which cannot be accommodated in the smaller shuttle; and that the President's expressed desire to make the shuttle a useful vehicle for military space operations could not be fulfilled with the smaller shuttle.

George Shultz' letter of February 16 transmitted a number of detailed questions on matters relating to the booster decision and payload size reaffirmation. We intend to provide answers to as many of these as possible before March 15 but, because of the short timetable under which recent studies have been made, the bulk of the material needed for proper response will not be finalized for submission to your office until March 13. George Low has arranged to meet with Don Rice on March 7 to present and discuss the material then available and to identify on a timely basis any matters of special concern.

We will present our plans, along with supporting data, to members of your staff, to other members of the White House who have been involved with the shuttle, and to a staff committee of outside experts which will convene after March 10 to review in depth our conclusions and considerations which support them.

During our meeting of March 3, 1972, we also discussed another matter: that of an expenditure ceiling of $3.2 billion of outlays during the time of shuttle development stated as a "previous understanding" in George Shultz' letter of [4] February 16. I told you that this had not been my understanding; instead I had planned on our new obligational authority to remain essentially constant at the FY 1973 level—$3,379 billion—over the next several years. You and I did not settle this matter, but you agreed that the issue is separate from the shuttle decision and should be considered later in the context of the FY 1974 budget, and not now.

In summary:

1. We plan to develop a shuttle making use of solid boosters in the parallel-staged configuration. From the budgetary point of view, this is the lowest development cost option.

2. Our analysis has reaffirmed the previous conclusion reached in January that the shuttle should have a 15x60, 65,000# payload capability.

3. We need to iron out our differences concerning NASA's constant budget—whether this is based on FY 1973 outlays or [new obligational authority]. However, you have agreed that this is not an issue involved in this immediate decision—it will be discussed in terms of the FY 1974 budget preparations at a later date.

We look forward to working with you in the future as we have in the past toward the success of this most important program.

Sincerely,

James C. Fletcher
Administrator

Document II-15

Document title: James C. Fletcher, Administrator, NASA, to Senator Walter F. Mondale, April 25, 1972.

Source: NASA Historical Reference Collection, NASA History Office, NASA Headquarters, Washington, D.C.

Democratic Senator Walter Mondale of Minnesota was a skeptic with respect to the wisdom of developing the Space Shuttle from the time the program was first proposed in 1970. In this letter to Mondale, NASA Administrator Fletcher provides a top-level overview of the expectations for the Shuttle program shortly after it was approved by President Nixon, as well as a final configuration selected. The March 15 Appendix to the Space Shuttle Fact Sheet to which Fletcher refers as an enclosure does not appear here.

[1]   NATIONAL AERONAUTICS AND SPACE ADMINISTRATION
                          Washington, D.C. 20546

                                                                    April 25, 1972

OFFICE OF THE ADMINISTRATOR

Honorable Walter F. Mondale
United States Senate
Washington, D.C. 20510

Dear Senator Mondale:

This is in further response to your letter of February 23, 1972, on the space shuttle. The answers to your 22 questions are numbered as in your letter; to save space I have given a brief indication of the subject of each question in lieu of repeating the question in its entirety. All cost estimates are stated in current dollars.

1. Projected Costs of the Space Shuttle

The estimated costs of the space shuttle program are as given below. These estimates correspond to those in the Appendix to the Space Shuttle Fact Sheet, as revised March 15, 1972 (copy enclosed).
   a. Development and initial investment costs:
       (1) Development cost, based on the use of the recoverable parallel-burn solid rocket motor booster configuration now selected, and with prudent provision for potential cost increases as development proceeds . . . $5.15 billion
       (2) Facilities costs for development and initial operations, including launch and landing facilities to be provided at the Kennedy Space Center . . . $.3 billion
[2]   (3) Investment for initial operating inventory. This is subject to future decisions based on requirements in the late 1970's and early 1980's. On the reasonable assumption that 3 production and 2 refurbished orbiters will be needed, we have allowed in our projections a total of . . . $1.0 billion
       Total development and initial investment $6.45 billion
   b. The later additional investment costs required at and after the end of this decade to fly a reasonable mission model all through the 1980's are estimated at $1.6 billion. This amount includes the $500 million estimated to be required for the sec-

ond operational launch and landing site, to be located at Vandenberg AFB, California, and provision for the development and investment costs of the reusable space tug required for more economical operations at synchronous orbit.

c. Shuttle operating costs are estimated at $10.5 million per flight. These are not costs of the space shuttle program but will be part of the cost of the space missions to be flown, just as the cost of the Titan III C launch vehicles which will be used in the Viking program is considered a part of the cost of the Viking program. For each mission, the shuttle operating costs will replace the costs of the expendable launch vehicles that would otherwise be used.

d. All of the costs of the space shuttle program, plus all of the development and operating costs of a balanced total space program using the space shuttle, can be accommodated within a total space budget (NASA, DOD, and other users) which does not exceed the current total annual level (in current dollars). Approval of the space shuttle program does not represent a "built-in" commitment to higher space budgets in future years.

2. Future Budget Requests for Space Shuttle

The annual budget requests for the space shuttle for the next six years will rise from the $228 million in the FY 1973 budget to a peak of about $1.2 billion in [3] FY 1976 and FY 1977 and then decline. As I have testified to the responsible Congressional Committees in their review of NASA's FY 1973 budget, all expenses of the space shuttle program and the other elements of a balanced total NASA program can be accommodated within a total annual NASA budget at the $3.4 bullion level recommended by the President for FY 1973 (in current dollars). Again, approval of the space shuttle does not represent a "built-in" commitment to higher NASA budgets in future years.

3. Costs of "Old" Shuttle and Letter to N.Y. Times

As I have testified on a number of occasions, our studies during the past year showed that the development envisaged a year ago (fully reusable with fly-back booster) would have been about $10 billion, almost twice the $5.15 billion cost of the configuration we are now proceeding with. However, the figures of "$10 to $14 billion" mentioned in my letter to the New York Times of January 28, 1972, were not NASA's figures but were figures which had been used erroneously in an article in the Times. The purpose of my letter was to correct the misunderstanding evidenced by the use of these figures by the Times.

4. Space Shuttle Booster

The decisions on the shuttle booster configurations which were noted as open in the statement issued January 5, 1972, have now been made and were announced on March 15, 1972. A parallel burn configuration using two solid rocket motors designed for water recovery, refurbishment, and subsequent reuse has been selected. In our cost estimating we have assumed 20 reuses.

5. Cost of Shuttle Flights

The major elements of the cost-per-shuttle flight are the refurbishment of the orbiter after each flight, the replacement of the orbiter's hydrogen-oxygen tank that is expended on each flight, and the recovery and refurbishment of the solid rocket booster. The $10.5 million cost-per-shuttle flight is based on a careful assessment of NASA and contractor studies of each of the principle elements of cost. Since some of these [4] elements are related to the industry competition for the shuttle development contract now underway, we are following the policy of not making the details of our estimates public at this time.

6. Lower Payload Capabilities

NASA and industry have made exhaustive studies of the effects of lowering the payload capability of the space shuttle. The results established that savings in development cost were relatively small and that the reduction in payload capability would result in a substantial net increase in the cost of the overall space program through 1990.

7. Requirement for 65,000 Pound Payload Capability

The requirement for a 65,000 pound payload capability for the shuttle results from consideration of (1) the maximum single mission weight requirements that can reasonably be expected in the 1980's and beyond, and (2) the shuttle performance required to place lighter payloads in higher altitude and higher inclination orbits.

Shuttle capability equivalent to 65,000 pounds launched due east in a 100 nautical mile orbit will be required for a number of other missions. Examples are earth resources types of satellites expected to be in use in the early 1980's. These satellites are expected to weigh about 6,000 pounds, but because of the high inclination and high altitude of the orbits required, and the weight of the propulsion stage required to reach these orbits, the shuttle capabilities required correspond approximately to the 65,000 pound, 100 nautical mile due east capability that has been specified for the space shuttle.

8. Cost per Pound in Orbit

Cost per pound in orbit when fully loaded is simply an index of the efficiency of a launch vehicle. [5] This amount is computed by dividing the average cost per launch into the total weight the launch vehicle can place in a standard reference 100 nautical mile due east orbit. This index is one of many indicators which show the relative efficiency of the space shuttle compared to current expendable launch vehicles. The index of $160 per pound for the space shuttle compares to an updated estimate of $900 per pound for the Titan III C, the most efficient current launch vehicle by this standard.

The shuttle does not have to be fully loaded to achieve economies, any more than a 230 horsepower car has to be operated at full power to be efficient, or a 150-passenger airplane has to be fully loaded to show a profit. With the loadings required to carry out the specific mission models studied, it was found that savings averaging one billion per year would result from use of the shuttle.

9. Savings in Space Transportation Expenses

The savings the shuttle will make possible are not related only to the cost of launch vehicle procurement but to total transportation expense in the broadest sense of all the costs necessary to accomplish useful missions in space. For the mission model discussed in the enclosed Fact Sheet Appendix, savings through the use of the shuttle over the 12-year period 1979–1990 are estimated at $5.1 billion in launch and launch-related costs and another $8.3 billion in payload development and procurement costs, for a total savings over the period of $13.4 billion and an average savings of over $1 billion per year.

10. Space Station

NASA's space station studies have been completed and there are no present plans for development, production, or specific missions. The mission model study referred to above assumes that a 6-man space station might be operational in the mid-1980's. The non-recurring costs of development and investment for a space station of this type has been estimated at about $3 billion. An amount of this magnitude is compatible with my earlier statement [6] that all costs of the space shuttle program and the other elements of a balanced total space program can be accommodated in an overall space budget at about the present annual levels. The decision to proceed with the space shuttle does not commit the Nation to proceed with a space station.

11. Mathematica Study

The Mathematica, Inc. study concludes that the space shuttle can be justified on economic grounds for a wide range of possible mission models. Mathematica studied in detail a range of discrete mission models calling for from 681 to 403 shuttle flights over a 12-year period (1979–1990). When these results were extended to even lower numbers of flights, Mathematica found that even with a 10% discount rate the break-even point occurred at 360 flights over the 12-year period. Thus the shuttle would represent a good investment even if the total number of flights did not exceed an average of 30 per year, or even less if a period longer than 12 years had been assumed for the useful life of the shuttle. (It should be noted that both Atlas and Thor boosters have been in use for over 13 years.) The Mathematica conclusions do not depend on the weight of the payloads associated with the program. A copy of the final Mathematica, Inc. report and related reports by Aerospace Corporation and Lockheed were sent to you some time ago.

12. Assessment of Mathematica Study

The Mathematica, Inc. study has been subjected to review by NASA management and within the Office of Management and Budget, and has been presented by Dr. Oskar Morgenstern of Mathematica to a number of other professional economists and to the Senate Committee on Aeronautical and Space Sciences. I am not aware of any professionally competent adverse criticisms of either the methodology or the findings. On the other hand, many of us, including myself, believe that the constraints placed by Dr. Morgenstern and his people on the scope of the study, whereby they excluded the benefits of any missions which would be beyond today's state-of-the-art, or which would not be possible of performance using expendable vehicles, represented an extremely conservative approach which has resulted in an understatement of the real advantages that will result from the introduction of the space shuttle.

[7] 13. Military Use of the Shuttle

The space shuttle can be used for both civil and military missions; in both cases the number and nature of the missions to be flown are matters for future decision. In the mission model referred to in the enclosed Fast Sheet Appendix, military missions represent substantially less than one-half of the total.

14 and 15. Cost per Pound of Payloads

The cost per pound of scientific and technical payloads is not particularly useful as a general measure. First, it can vary greatly depending on the design and use of the payloads. Second, it cannot be directly related to launch vehicle capabilities because the weights to be place[d] in orbit must also include the propulsion stages, fuel, and the other equipment required for placing and deploying the scientific and technical payloads in the proper orbits. Third, the unit cost of a given payload type varies substantially because the initial development cost is generally high compared to the cost of producing additional payloads of the same type. Finally, the utility of the shuttle does not depend on its being fully loaded.

For these reasons the cost per pound of payload cannot be used to estimate the cost of a space program that would be required to utilize the space shuttle and our cost studies have been based on the total estimated costs of specific missions in a variety of specific mission models.

To validate the general studies which indicated that the shuttle will make possible substantial savings in payload development and production costs, the Lockheed Company made an engineering analysis in depth of the Orbiting Astronomical Observatory (OAO) satellites. This showed that the relaxation of the size and weight constraints imposed by expendable launch vehicles would permit a reduction in development cost from the actual cost of $168 million to about $85 million, and in the unit production cost from $33 million to about $18 million. In this redesign the "dry weight" of the satellite was increased from 4,800 pounds to about 7,700 pounds. Thus, it can be calculated that in the case of OAO the payload cost per pound when designed for the shuttle wold be less than one-third of what it was for the expendable launch vehicle.

[8] 16. Impact of Defense Requirements on Shuttle Costs

The basic design and performance characteristics of the space shuttle system are essentially the same for both civil and military requirements. For example, the "cross range" requirement, which permits the shuttle to land after one orbit at the same site from which it was launched. is required by NASA and all users for safety reasons to make it possible to abort a flight during the first orbit. No part of the development cost of the basic space shuttle configuration is attributable solely to requirements of the Department of Defense.

17. Estimated Launch Costs for 400 Missions

As indicated in the enclosed Fact Sheet Appendix, the total launch and launch related costs for a 580 mission, 12-year mission module would be about $13.2 billion with conventional launch vehicles and about $8.1 with the shuttle. For comparison, a similar model with about 400 flights, also over a 12-year period, the corresponding costs would be about $8.4 billion without the space shuttle and $6.0 billion with the space shuttle. Of course, there is no reason to assume that space flights will stop after 12 years or after any particular number of flights. Regardless of the number of flights, savings will continue to be generated as long as the shuttle is used.

18. Use of Titan III C

In studying the comprehensive costs of future space programs with and without the space shuttle, the Aerospace Corporation and Mathematica, Inc. worked out the lowest possible cost program using expendable launch vehicles to compare with the cost of a program accomplishing the same missions with the space shuttle. As indicated above in the answer to question 11, the results were that savings of over a billion dollars a year could be expected from the use of the space shuttle with a realistic mission model, and that the shuttle would still be a good investment at a 10% discount rate with as few as 30 shuttle flights per year. The expendable launch vehicle program used in these studies made optimum use of the capabilities of the Titan III C as a "workhorse" for NASA missions for which it would be appropriate.

19. Technological Unknowns

As stated above, the $5.15 billion estimate for development of the space shuttle includes prudent provisions for unforeseen requirements requiring special attention in research and development.

[9] 20. In-orbit Repair of Satellites

Repair and maintenance of satellites in orbit is technically and practically feasible when the satellites have been designed with this in mind.

21. Retrieval of Satellites

While it is conceivable that in some cases the cost advantage of retrieval from orbit and reuse of satellites might be offset by technological obsolescence, the trade-off studies of this point by the Aerospace Corporation have clearly shown advantages of satellite recovery and refurbishment as an operating mode in most cases. Actual decisions on retrieval of particular satellites can be made on the basis of specific technical and economic analyses on a case-by-case basis.

22. Space Tug

The space tug is an essential future element of the total space transportation system of which the space shuttle is the cornerstone. It will be a reusable vehicle to place and, if desired, retrieve satellites requiring synchronous or other high orbits. However, until the tug is available, the shuttle can place satellites into these orbits by using expendable energy stages like the present Agena or Centaur. The space tug is currently in the study phase. The mission model referred to in the enclosed Fact Sheet Appendix assumes that the tug will become available in 1985. Development and investment costs for the space tug are estimated at about $800 million. This amount is included as a later investment cost in the economic analyses in the Fact Sheet Appendix and in the answer to Question 1 above.

I trust that the foregoing answers are responsive to your questions.

Sincerely,

James C. Fletcher
Administrator

Enclosure

Document II-16

Document title: James C. Fletcher, Administrator, George M. Low, Deputy Administrator, and Richard McCurdy, Associate Administrator for Organization and Management, NASA, Memorandum for the Record, "Selection of Contractor for Space Shuttle Program," September 18, 1972.

Source: NASA Historical Reference Collection, NASA History Office, NASA Headquarters, Washington, D.C.

Once it received White House and congressional approval to initiate Space Shuttle development, NASA moved quickly to select the prime contractor for the program. This document, signed by the NASA officials responsible for that selection, was initially prepared to explain the reasoning behind the choice to the General Accounting Office and the losing industrial bidders. When NASA discovered that the Wall Street Journal was about to run a story based on a leaked copy of the document, it released the paper to the press on October 4, 1972.

[1]   Selection of Contractor for
                         Space Shuttle Program

On July 19, 20, and 21, 1972, Dr. Low and I, along with other senior officials from Headquarters, Manned Spacecraft Center, Marshall Space Flight Center, Kennedy Space Center, and the U.S. Air Force, met with the Source Evaluation Board appointed to evaluate proposals for the design, development, and production of the Space Shuttle orbiter vehicle and for integration of all elements and support of the Space Shuttle system. Mr. McCurdy returned to NASA Headquarters on July 22, and received a full briefing from the Board on July 22 and 23.

The Space Shuttle program will provide the United States a new space transportation capability that will reduce substantially the cost of space operations and support a wide range of scientific, defense, and commercial uses. The Space Shuttle system will consist of a reusable orbiter vehicle capable of entry maneuvering and aerodynamic flight, reusable solid rocket motors (SRM's) which will burn during launch in parallel with the orbiter main engines, and an expendable main propellant external tank. The Government will procure the SRM's, main engines, air breathing engines, and tanks separately and furnish them to the contractor selected in this competition. Following a competitive solicitation, NASA earlier this year awarded a contract to the Rocketdyne Division of North American Rockwell Corporation for design, development, and production of the Shuttle main engines.

The Space Shuttle orbiter vehicle program as presently planned will consist of four increments. The first is for initial design. The second is for completion of design, development, test, and evaluation (DDT&E) including the delivery of two orbiter vehicles. The third increment is for production of three orbiters and the upgrading and retrofit of the two orbiters previously used for DDT&E.

[2] Increment four is the operational phase of the shuttle system. The proposed contract will be for the initial design work compromising increment one, including preliminary design review, covering a performance period of approximately two years. The proposed contract will also contain an option provision which will provide to the Government the right to require the contractor to perform through the completion of DDT&E, which will constitute increment two of the contract. The contractor selected, upon completion of increments one and two, will be expected to perform increments three and four. Horizontal flight testing is expected to begin in 1976 and manned orbital flights in 1978.

The Shuttle is to be operational by 1980. The contract will be awarded on a cost-plus-fixed-fee basis with an award fee feature.

For several years preceding this procurement, NASA has conducted extensive studies of the feasibility of a Space Shuttle system and the needs it would serve; the configuration to be adopted; and the technology of components and materials to be used. All the companies proposing for this procurement participated in such studies under NASA contracts. The results of the studies were published and made available to all proposers.

The Source Evaluation Board solicited 8 firms for this procurement. Twenty-nine others requested and received copies of the request for proposals. The following 4 companies submitted proposals:

Grumman Aerospace Corporation
Lockheed Missiles and Space Company, Inc., Space Systems Division
McDonnell Douglas Corporation
North American Rockwell Corporation, Space Division

Prior to the issuance of the RFP, the Board established mission suitability evaluation criteria consisting of technical criteria in the areas of manufacturing, test, and [3] flight test support; system engineering and integration; and subsystem engineering; criteria in the areas of organization, key personnel, and related experience; management approaches and techniques; and procurement approaches and techniques. The Board assigned weights to these criteria and established a scoring system. A statement of the criteria and a general indication of their relative importance were included in the RFP.

To assist it in the evaluation, the Board established technical, maintainability and ground operations, management, and cost teams. Each team was supported by panels and expert advisors. In all, 416 people representing seven NASA centers, NASA Headquarters, and the Air Force participated in the evaluation.

With the assistance of the teams and panels, the Board conducted an initial evaluation of the proposals prior to any written or oral discussion, and rated the proposals in the following order of suitability to meet the Government's requirement:

1. North American Rockwell
2. Grumman
3. McDonnell Douglas
4. Lockheed

The Board determined that all four proposals were within the competitive range. It sent written questions to the competing firms and invited them to participate in oral discussions at the Manned Spacecraft Center concerning their proposals. Following the discussions, the Board received further responses and final revisions to the proposals. The Board conducted its final evaluation and ranked the proposals in the following order of suitability to meet the Government's requirement:

1. North American Rockwell
2. Grumman
3. McDonnell Douglas
4. Lockheed

[4] North American received the highest score in mission suitability and an overall rating in the good to very good range. The North American design provided the lightest dry weight of any of the designs submitted. For guidance, navigation, and control, North American used a triple-redundant single-string approach which the Board considered to

be very good as a simple design with minimum interfaces. Its good understanding of all electrical power subsystems reflected the very thorough studies that North American made following the Apollo 13 accident, which had its origin in an electrical subsystem. However, the Board considered North American's choice of a male-female concept for docking to be less advantageous than the androgynous method proposed by the other companies.

North American presented an excellent analysis of maintainability from the standpoint of design criteria and goals to achieve optimum turnaround conditions and timing between flights. It designed its orbiter vehicle with very good overall accessibility for maintenance. North American's requirement for a trolley to support the orbiter on the ground was regarded as a weakness, since the trolley complicates the ground systems and causes operational constraints.

North American's greatest advantages over the other offerors, within the mission suitability area, were in management. Its proposal showed efficient centralized control of the program, with a readily identifiable chief engineer and deputy. While all the offerors had well qualified key personnel, the Board reported that North American's top project management team was the best overall, the individuals having very good experience and demonstrated competence relevant to their assigned positions. As a company, North American has strong experience in manned space flight, including especially the Apollo command and service modules and the Saturn II second stage vehicle. A minor weakness is the company's lack of recent experience with large operational airframes.

Grumman received the second highest score, very close behind that of North American, with an overall rating in the good to very good range. Grumman's greatest strengths were of [5] a technical nature. In general, Grumman's design went to a greater depth of detail than those of the other companies. Its detailed weight estimates were substantiated by the design details. It was rated very good in its design of primary structure based on simple, straightforward, and reliable load paths, showing a thorough understanding of potential problems and positive solutions. On the other hand, it presented complex designs for the guidance, navigation, and control system, and for data processing.

Grumman did a very good job in proposing design features to enhance maintainability. The provisions it made for access throughout the vehicle were outstanding. The design approach to the external tank also was strong; the tank does not require pressure stabilization on the ground and can support the orbiter. The Board was concerned about Grumman's proposal to place the liquid oxygen and liquid hydrogen fill and drain couplings in the same umbilical plate.

In the management area, Grumman presented a strong organization with well-integrated assistance from its principal subcontractor, Martin-Marietta Corporation. The Board reported that the team of key personnel was strong, but had limited experience in large cryogenic systems. As a company, Grumman has good experience in manned space flight, particularly the lunar module, and management of large programs involving spacecraft and aircraft. The Board reported that Grumman's proposal showed evidence of indepth comprehensive planning of its overall management approach; but concluded that the program plan presented lacked balance. Grumman proposes to incorporate detailed specifications and plans as baselines in the contract early in the program and to build up its work force rapidly to an early manpower peak. This poses the risk of premature hardening of the specifications and premature commitment of resource[s] during the course of the program.

McDonnell Douglas received the third highest score, with an overall rating of good. It ranked third in most of the areas of the evaluation. Distinctive design features of the McDonnell configuration included an underslung internal [6] air-breathing engine system . . . package, which retains a full payload bay capability with the [air-breathing engine system] installed; and the largest fuselage volume. McDonnell proposed a very good reaction control subsystem, with a plug nozzle to minimize re-entry heating effect. Its radiator design for the payload by doors was the best presented. However, McDonnell's external tank design with a common bulkhead between the liquid oxygen and liquid hydrogen tanks had undesirable operational and manufacturing characteristics and would require insulation inside the hydrogen tank. McDonnell proposed horizontal flight testing at Edwards Air Force Base, California, with an early shift to Kennedy Space Center, which would require both sites to be equipped with full data handling capability.

In maintainability, the Board stated that the McDonnell proposal did not reflect adequate application of the company's experience in the design of the DC-10 for maintainability. Furthermore, the ground operations portion of the proposal did not reflect adequately the recent launch vehicle experience of McDonnell Douglas in the Apollo program. McDonnell planned to vent its liquid hydrogen tank to the atmosphere during ground operations, creating a risk of fire or explosion. On the positive side, it provided a good recovery technique for the expended solid rocket motors.

McDonnell's organization of the eastern and western segments of the company was relatively complex. It proposed to carry out engineering functions at both locations according to the category of work involved, thereby complicating the assignment of one overall engineering responsibility. McDonnell presented a strong management team; however, some of the managers were proposed in project assignments differing from the areas of their main experience. As a corporation, McDonnell was considered to have superior related experience, including manned space programs, a wide range of major Government projects, and experience with large commercial airframes. Its principal subcontractor, TRW, also had good experience in its assigned avionics area. The McDonnell management approach was not specific in many areas and failed [7] to show integration of computerized systems. Furthermore, different management systems in St. Louis and California caused a loss of visibility and a likelihood of serial information flow from one to the other.

Lockheed received the lowest score with an overall rating of fair. It designed a configuration that was distinctive in adopting thrust vector control for the solid rocket boosters for better ascent control, and in extending the solid rocket nozzles well aft of the orbiter and external tank, so as to reduce nozzle cant angle, reduce thermal effects, and reduce acoustic levels. Its design was the heaviest proposed. In general, the reason for the relatively low Lockheed evaluation was its lack of consistent technical depth. Lockheed's proposal for aborting a mission leaves a 65 second gap during which there is no provision for abort. Its proposed vehicle required a landing speed slightly higher than that specified in the RFP. Lockheed introduced unnecessary complexity through the use of a wide variety of structural materials and advanced processes, and through the use of complex subsystems for mechanical power, environmental control, and avionics. Lockheed did a good job in communications and tracking, and planned to phase in its automatic landing system early in the program. It also produces a silica material which is considered to be very good for the thermal protection system.

Lockheed enhanced the maintainability of its reaction control and orbit maneuvering systems by proposing modular systems, but obstructed accessibility by burying the OMS module in the main engine compartment. The vented honeycomb structure of Lockheed's vehicle was susceptible to moisture; the proposal did not discuss interstructural purging of it. Lockheed presented strengths in proposing two tail service masts to fill and drain liquid oxygen and liquid hydrogen separately, and in providing a liquid hydrogen vent through the tail service mast.

Lockheed proposed to subcontract all the major components of the orbiter. Under this arrangement, the major subcontractors would do the greater share of their own design [8] work, with Lockheed doing the overall design and systems integration. The Board expressed concern that this plan would generate complex organization interfaces, which Lockheed did not sufficiently address in its proposal. Within its own organization, Lockheed placed the system engineering and orbiter vehicle engineering groups in separate organizations, both reporting to the program manager. The key personnel proposed for this job were rated as good, but lacking the overall strength and balance of the teams proposed by their competitors. In general, the experience of the key personnel group is in missile development and space design studies. The individuals lacked experience in manned space flight; and relatively few of them reflect the broad aircraft experience of the Lockheed organization. As a company, Lockheed similarly has relatively little manned space flight experience, although it has wide experience in major Government programs, commercial airframes, and space payloads.

All the proposals contained estimates of the costs to be incurred under the proposed contract, as well as broader estimates of the cost to be incurred by the Government in carrying out the development program; estimates of production costs; and estimates of operational costs per flight. The Board conducted detailed analyses of the cost proposals and of the supporting information furnished by the offerors to gain insight into the probable cost of the program and into probable cost differences among the offerors. The costs as estimated in the proposals differed widely, with North American the lowest, followed closely by Lockheed, and with Grumman and McDonnell substantially higher.

The Board studied the cost implication of the designs proposed and concluded that the design differences among the companies would not account for significant differences in cost. The exception to this was Lockheed, whose design was heavier and more complex than those of the other companies, so that its vehicle should cost more to build and operate. There were differences in salary and indirect rates among the companies, causing differences [9] in the cost of a man-year's work from one company to another; but such differences were not large.

The wide differences in the cost estimates were due essentially to widely differing estimates of the number of man-years required for the job. In turn, the widely varying manpower estimates reflected different treatments of unknowns and contingencies for program growth.

The Board made adjustments to the proposed costs of all the companies, reflecting its view of the cost of correcting identified weaknesses, and its view of proposers' estimates thought to be in error for various portions of the work. However, these adjustments were relatively small; the Board did not attempt to normalize the remaining large differentials in manpower that the competitors had proposed. That is, the Board did not estimate the different number of man-years required for the different companies to do the job because the actual work will depend, to a considerable extent, on the management approaches applied by each company.

The Board looked at management approaches and planning for the program to gauge the effect of such approaches on the confidence that could be placed in the cost estimates. This evaluation favored North American. The management techniques proposed [by] North American should provide earlier identification of cost problems. Its program planning lent conference to its ability to control costs, by planting a constrained buildup of resources in the beginning of the program, so as to avoid the commitment of large resources of manpower and other resources to the job during the early period when problems were emerging and changes being made.

Grumman, and to a lesser extent McDonnell, proposed to build up their forces considerably more rapidly at the beginning of the program than previous NASA experience with large programs would indicate to be desirable. This approach increases the likelihood of significant cost growth resulting from development problems, which typically occur in early program phases. Also, Grumman's approach did not appear to be supported by the milestones it designated for program performance.

[10] The Lockheed approach inspired less confidence than the others in its cost for a number of reasons. Its design was more complex than the others, giving rise to a probable cost differential. Furthermore, its estimating techniques, its management plans, and its technical approaches all were set forth in its proposal with a lack of depth which contributed to an impression that many unforeseen problems might arise to jeopardize the company's control over its costs.

In answer to our questions, the Board said it was not able to assign dollar values to its judgments of cost risk inherent in the program approaches of the competitors; but unanimously concluded that the North American proposal would result in the lowest final cost to the Government. It believed that Grumman would probably be the second lowest in cost, but that its rapid manpower buildup and its general emphasis on schedule over cost involved greater risk of cost growth than North American's slower buildup and more cost-centered emphasis. McDonnell was believed to be the next in line, with higher cost resulting from its higher rate of cost per man-year and resulting from the risk of cost growth in its program plan for a rapid buildup of its forces, though not so rapid as that of Grumman. Lockheed was evaluated as having the highest probable cost because of its design and because the uneven quality of its proposal impaired confidence in its ability to avoid costly problems during performance.

On July 24, we met separately with the chief executive officers of the four competitors, together with their Shuttle program managers and other senior corporate representatives. We scheduled these meeting[s] because of the unusual importance of this procurement, in order to meet with the top management of each competing corporation and ascertain its views on management of the Space Shuttle program and the extent of top level corporate interest in the program. These meetings were held in addition to our established source selection procedures, and were held with the agreement of all four competing companies.

[11] On July 25, we met with a small group of key NASA personnel who had heard the presentation of the Source Evaluation Board and who carry responsibilities related to the procurement. Their views on the presentation and findings were solicited and given. They then withdrew.

Dr. Low, Mr. McCurdy, and I met again on July 26 and care-fully considered the presentation and the comments of the key personnel involved. It was apparent to us that the competition had been keen and that the four companies involved were worthy competitors offering impressive experience and capabilities for this major program. We noted at the outset that McDonnell and Lockheed ranked significantly lower than the other two companies in most areas of the Board's technical and management evaluation. Since these companies offered no probabl[e] cost savings in relation to the higher-ranked firms, our deliberations tended to focus on North American and Grumman.

The mission suitability competition between these two companies was close, as reflected by a narrow differential in their point scores. Each company had its own areas of strength in which it was superior to the other. On the basis of our careful review of the Board report and its presentation, and the comments of the key personnel involved, we concluded that the overall advantage did indeed lie with North American as indicated by the final mission suitability scores.

In our view, the cost considerations led to the same result. North American's cost proposal was substantially below that of Grumman, based largely on a smaller number of man-years. We kept in mind that estimates for cost reimbursement contracts do not carry as much assurance as fixed price proposals. But, the lower North American proposal, which was considered reasonable, will enable the Government to negotiate a lower dollar fee. It also enhances the possibility that NASA and the contractor will give earlier and closer attention to cost-generating problems and changes as they arise. More fundamentally, we were impressed with [12] the orderly approach to the work planned by North American, with its special attention to cost control. This latter is indicated by its relatively restrained buildup of forces during the early period of the program when problems can be expected to be encountered and changes made.

Because North American Rockwell attained the highest score from a mission suitability standpoint, because its cost proposal was lowest and credible, and because its approaches to program performance gave high confidence to us, to the Board, and to the Manned Space Flight Center Directors, that it will indeed produce the Shuttle at the lowest cost, we selected North American Rockwell Corporation, Space Division, for the award.

[signature]   9/18/72
James C. Fletcher   Date
Administrator

CONCUR:

[signature]   9-15-72
George M. Low   Date
Deputy Administrator

[signature]   Sept. 14, 1972
Richard C. McCurdy   Date
Associate Administrator for
Organization and Management

Document II-17

Document title: The Comptroller General of the United States, Decision in the Matter of Protest by Lockheed Propulsion Company, File B-173677, June 24, 1974, pp. 1, 18–23.

Source: NASA Historical Reference Collection, NASA History Office, NASA Headquarters, Washington, D.C.

In contrast to the 1972 selection of North American Rockwell as the prime contractor for the Space Shuttle orbiter, which went relatively smoothly, the selection of Thiokol as the provider of Shuttle Solid Rocket Motors (SRM) was more controversial. After NASA announced that it had selected Thiokol as the SRM contractor in December 1973, Lockheed Propulsion filed a formal protest with the U.S. government. One of the responsibilities of the Comptroller General, who is appointed for a 15-year term as head of the Congressional General Accounting Office (GAO) to ensure his independence, is to rule on such protests. Although Comptroller General Elmer B. Staats recommended in this decision that NASA should reconsider its selection of Thiokol, the space agency did not accept this nonbinding recommendation. The following are two excerpts from the decision.

[1] DECISION

THE COMPTROLLER GENERAL
OF THE UNITED STATES
Washington, D.C. 20549

FILE: B-173677

DATE: June 24, 1974

MATTER OF: Lockheed Propulsion Company
Thiokol Corporation

DIGEST: 1. On basis of GAO review of NASA evaluation of cost-plus-award-fee proposals for Solid Rocket Motor Project of Space Shuttle Program covering 15-year period in estimated price range of $800 million, it is recommended that NASA determine whether, in view of substantial net decrease in probable cost between two lowest proposers, selection decision should be reconsidered. . . .

[18] Chronology of Procurement and Selection

The RFP was issued on July 16, 1973, to four prospective sources—Thiokol, Lockheed, UTC, and Aerojet. Technical and cost proposals were [19] submitted on August 27 and 30, 1973, respectively, by the four firms. From the latter date until October 20, 1973, the SEB [Source Evaluation Board], according to the Source Evaluation Plan, evaluated and scored the proposals and established preliminary rankings for the offerors. During the period from September 24 through October 10, 1973, oral and written discussions were conducted with all of the offerors. All offerors filed timely best and final offers by the cut-off date of October 15, 1973. After the cutoff date, final reports of the SEB's evaluation teams were submitted to the SEB.

The four proposers were ranked and scored in mission suitability as follows:

	Score	Overall Adjective Rating
Lockheed	714	Very Good
Thiokol	710	Very Good
UTC	710	Very Good
Aerojet	655	Good

The SEB was of the opinion that all proposers had the requisite capability and experience to accomplish the SRM project. Furthermore, the SEB evaluated Thiokol as the lowest most probable cost performer by $122 million ($RY) with Lockheed evaluated second lowest. Both proposers estimated total program cost to be in the $800 million ($RY) range. The SEB compiled a report of its findings which was presented to the SSO [Space Shuttle Office] and was the basis of its oral presentation to the SSO on November 19, 1973. The SSO, after selecting Thiokol for final negotiations, issued a selection statement on December 12, 1973, which states, in pertinent part, as follows:

"In considering the results of the Board's evaluation, we first noted that in Mission Suitability scoring the summation resulted essentially in a stand-off amongst the top three scorers (Lockheed, Thiokol and UTC) though with a varying mix of advantages and disadvantages contributing to the total. Within this group, Lockheed's main strengths were in the technical categories of scoring, while they trailed in the management areas. Thiokol led in the management areas but trailed in the technical areas, and UTC fell generally between these two.

We noted that Aerojet ranked significantly lower than the other three competitors in the Mission Suitability evaluation, and the proposal offered no cost advantages in relation to the higher ranked firms. Accordingly, we agreed that Aerojet should no longer be considered in contention for selection.

[20] "We noted that the Board's analysis of cost factors indicated that Thiokol could do a more economical job than any of the other proposers in both the development and the production phases of the program; and that, accordingly, the cost per flight to be expected from a Thiokol-built motor would be the lowest. We agreed with the Board's conclusion that this would be the case. We noted also that a choice of Thiokol would give the agency the lowest level of funding requirements for SRM work not only in an overall sense but also in the first few years of the program. We, therefore, concluded that any selection other than Thiokol would give rise to an additional cost of appreciable size.

"We noted that within the project logic and the cost proposals, there was a substantial difference in basic approach caused by the varying amount of new facilities needed by the several proposers. Their situations ranged from Thiokol, who needed little new facilities investment to do the job, to Lockheed, who proposed creation of a new facility complex on the Gulf Coast to handle the program, commencing at an early date and building up to full size by the production phase. The prospect of such a major new facility raises a question regarding the basic operational economics involved, and also a question of what other important benefits or drawbacks there might be to such a plan. In regard to the economics proper, the Board's evaluation made it clear that such an investment could not at this time, under any reasonable view of the forecasted economic factors, be considered likely to pay its way as against Thiokol's existing facility. As regards other considerations, we recognized that it may well be advantageous, when the major production phase arrives, to plan to have two or more suppliers in the country capable of competing for the manufacture of SRM's in quantity; however, there is no need to embark upon the construction of a new major facility at this time in order to secure these benefits in a timely manner.

"We found no other factors bearing upon the selection that ranked in weight with the foregoing.

"We reviewed the Mission Suitability factors in the light of our judgment that cost favored Thiokol. We concluded that the main criticisms of the Thiokol proposal in the Mission Suitability evaluation were technical in nature, were readily correctable, and the cost to correct did not negate the sizeable Thiokol cost advantage. Accordingly, we selected Thiokol for final negotiations."

Award of the contract has been withheld pending resolution of this protest.

[21]   CHRONOLOGY OF PROTEST

Lockheed filed notices of protest by letters dated December 5, 6, and 14, 1973. On January 9 and 21, 1974, Lockheed furnished protest details which were forwarded promptly to NASA requesting a complete report responsive to the protest. By this time, Thiokol, UTC, and Aerojet had expressed active interest in the protest. On or about February 15, NASA awarded a 90-day interim contract to Thiokol for studies, analysis, planning and design in support of the integration of the SRM into the Space Shuttle

System. Lockheed protested the award of the interim contract shortly thereafter. NASA filed a report, through the Assistant Administrator for Procurement, on March 11, 1974. The report was distributed to all interested parties for comment.

The report revealed to the protester and interested parties previously unknown significant cost information and other evaluation details upon which the selection of Thiokol was based. Prior to this, Lockheed had been unsuccessful in obtaining such information from NASA. Lockheed filed extensive comments on the NASA report on April 9, 1974, wherein, for the first time, specific contentions based on the previously unavailable significant cost information and other details were made. On April 23, a bid protest conference was held at GAO [General Accounting Office] attended by all interested parties and NASA. The formal record was then closed except for possible questions GAO might have to ask of Lockheed, Thiokol, and NASA. On May 8, questions were posed to Lockheed, Thiokol and NASA, all of whom responded to GAO by the May 15 deadline. About that time, Lockheed protested any possible extension by NASA of the interim contract to Thiokol. NASA extended the interim contract for 45 days or until approximately July 1. On May 20, further questions were raised with NASA by GAO. A response was received on May 24 and Lockheed filed comments thereon on May 30, 1974.

DECISION

This decision was reached after a thorough and comprehensive review of the voluminous documentation submitted by Lockheed, Thiokol and NASA, as well as presentations made at the bid protest conference. To assist in the resolution of the many issues raised by the protest, GAO assembled an audit team at the Marshall Space Flight Center where the procurement file is located. NASA's workpapers and other material were reviewed by the GAO team. From shortly after the protest was filed, the GAO review was performed at the Center simultaneously with the procedural steps in the bid protest process. Site visits were [22] made to Lockheed and Thiokol. While, in the interest of clarity of presentation, this decision does not respond specifically to each matter brought to our attention, we thoroughly considered all available information and documentation.

The Lockheed protest charges that the entire NASA evaluation was marred by plain mistakes, inconsistency, arbitrary judgments, and improper procedures. Lockheed states an adequate and proper cost evaluation would have resulted in its proposal being evaluated low by an amount significantly in excess of $100 million and conceivably in excess of $200 million. Furthermore, Lockheed argues that it was prejudiced by improper correction in Thiokol's design, improper crediting of Thiokol proposal features not conforming to the RFP, improper reliance on uncertain cost estimates, and improper disregard of future competition as a factor. The effect of these alleged prejudicial occurrences in combination with the alleged improprieties in the evaluation of cost made the selection of Thiokol improper, and is said to have wrongfully denied Lockheed the award of the SRM contract.

On the other hand, NASA vigorously defends the selection of Thiokol as the lowest cost proposer citing a most probable cost difference of $122 million ($RY) [real year dollars] which "must be regarded by NASA as the potential savings attainable by contracting with Thiokol." NASA maintains that the SEB evaluation as adopted by the SSO properly concluded that both Thiokol and Lockheed were essentially equal in the mission suitability scoring and "other factors" evaluation.

GAO's examination and review revealed no reasonable basis to question the SSO's decision based on scored mission suitability and unscored "other factors" evaluations. Nor did the review find that the reliance on cost represented an unreasonable exercise of discretion. However, as set forth in more detail below, we recommend that the SSO determine whether, in light of the GAO findings that the most probable cost differences between Lockheed and Thiokol were significantly less than those reported by the SEB and relied upon by the SSO, the selection decision should be reconsidered.

Before proceeding with a discussion of the issues, it is noted that a substantial amount of information and documents furnished GAO with the NASA report of March 11 and in its answers to GAO questions of May 8 were withheld from the protester and interested parties at the request of NASA. According to NASA, that material contains business confidential material and descriptions of confidential proprietary manufacturing processes, the disclosure of which would be in violation of law. Also not released to the protester and interested parties were SEB analyses of probable cost based on the proposals submitted to be further used by NASA in the negotiation of the SRM [23] contract and material generated prior to final negotiations. In addition, while NASA has publicly released the significant evaluated cost differences where the SEB made adjustments to proposed costs between Thiokol and Lockheed, the specific amounts of the adjustments have not been released except in rare instances.

The discussions of the protest issues that follow are presented in a context which safeguards the confidential or proprietary aspects of the data. . . .

Document II-18

Document title: Gerald J. Mossinghoff, Assistant General Counsel for General Law, Memorandum for the Record, "Classification of the Space Shuttle as a 'Space Vehicle' and not an 'Aircraft,'" September 25, 1975.

Source: NASA Historical Reference Collection, NASA History Office, NASA Headquarters, Washington, D.C.

One of the many unique features of the Space Shuttle was that it would glide to a landing after reentering the atmosphere from orbit. This raised the question in some minds of whether the Shuttle had to be treated as an aircraft during its atmospheric reentry; doing so would have made it subject to the regulations of the Federal Aviation Administration (FAA). NASA's position, spelled out in this memorandum, was that the Space Shuttle should not be so categorized.

[1]   NATIONAL AERONAUTICS AND SPACE ADMINISTRATION
                               Washington, D.C. 20546

REPLY TO
ATTN OF: GG(75-18103)   September 25, 1975

MEMORANDUM FOR THE RECORD

FROM: GG/Assistant General Counsel for General Law

SUBJECT: Classification of the Space Shuttle as a "Space Vehicle" and not an "Aircraft"

This memorandum records my response to a question asked by Mr. Seth Taylor, a GAO official reviewing the Space Shuttle program at the Johnson Space Center, as to whether the Space Shuttle is an "aircraft" within the meaning of the Federal Aviation Act of 1958, as amended, and FAA's implementing regulations. My response to this question was based on discussions with Messrs. Hosenball, Griffin and Doyle and with Mr. Charles Anderson, Deputy Chief Counsel of FAA.

The issue of whether the Shuttle could be classified as an aircraft is significant. If it were to be so classified, it would be subject to a number of provisions of the Federal Aviation Act of 1958, as amended, and FAA regulations, including possibly, Section 611 of the act concerning aircraft noise and sonic boom (if it were also determined that the Shuttle was "engaged in carrying persons or property for commercial purposes"). Classification of the Shuttle as an aircraft would also have international implications.

I informed Mr. Taylor in a telephone conversation this morning that it is NASA's firm position that the Space Shuttle is a space vehicle and not an aircraft within the meaning of the Federal Aviation Act. I informed him that this position was based on the following:

(1) The National Aeronautics and Space Act of 1958 authorizes NASA to develop, test and operate both "aeronautical and space vehicles." The legislative history makes clear that aeronautical vehicles are those designed for operation "within the atmosphere" whereas space or astronautical vehicles are designed for operation "primarily outside the atmosphere, although often passing through the atmosphere on the way to outer space." Based on this history, although there is [2] legally no precise dividing line between the atmosphere and outer space, it is clear that the Space Shuttle is a space vehicle under our act, and not an aeronautical vehicle.

(2) Although the definition of aircraft in the Federal Aviation Act is quite broad ("any contrivance now known or hereafter invented, used, or designed for navigation of or flight in the air"), the fact that something falls within this literal definition does not mean that it legally will be considered an aircraft even by the FAA, which recognizes "non-aircraft" airborne objects, for example, surface-effects (air-cushion) vehicles.

(3) NASA's authorizing committees, when describing the Space Shuttle in reports accompanying our annual authorization acts, have consistently characterized it as a "reuseable [sic] manned space vehicle."

(4) Under our interpretation of the Convention on International Liability for Damage Caused by Space Objects, the Space Shuttle would clearly be a "space object" so as to impose absolute liability upon the United States for "damage caused [by it] on the surface of the earth or to aircraft in flight."

(5) Similarly, Space Shuttle flights would be registrable [sic] under the Convention on Registration of Objects Launched into Outer Space, which will be transmitted to the Senate for ratification later this year. We understand that a staff study of the Senate Committee on Aeronautical and Space Sciences regarding that treaty will specifically point out that the Space Shuttle is a "space vehicle having characteristics of a launch vehicle or rocket and a recoverable spacecraft [that] would be registrable [sic] under this convention as an object launched into outer space." If the Shuttle were determined to be an "aircraft" it would be registrable [sic] under the Federal Aviation Act, and that would be logically inconsistent with our (and the Senate staff's) views of the treaty.

Mr. Anderson of FAA was not in a position formally to concur in our interpretation, since the matter has not been raised within the FAA. He did indicate that in his view our position was reasonable and consistent with the intent of the Federal Aviation Act and FAA's regulations, particularly since Shuttle operations would still be subject to FAA coordination to the extent that the Shuttle would operate in the "navigable airspace of the United States." For example, FAA is specifically authorized by Section 307 of the Federal Aviation Act [3] to prescribe rules and regulations "for the prevention of collision . . . between aircraft and [other] airborne objects."

Mr. Taylor indicated that he plans to quote NASA in his report as saying that under the National Aeronautics and Space Act, the Shuttle is not an aircraft, and that, therefore, FAA's regulations regarding aircraft noise and sonic boom would not apply to Shuttle operations.

[signed Gerald J. Mossinghoff]

Document II-19

Document title: John F. Yardley, Associate Administrator for Space Transportation Systems, NASA, to Director, Public Affairs, NASA, Memorandum, "Recommended Orbiter Names," May 26, 1978, with attached: "Recommended List of Orbiter Names."

Source: NASA Historical Reference Collection, NASA History Office, NASA Headquarters, Washington, D.C.

As Space Shuttle development reached its final stages in 1978, NASA needed to select names for the four Space Shuttle orbiters that had been approved. Reacting to pressure from fans of the television series Star Trek, NASA had already named the Shuttle test vehicle used for atmospheric flight experiments Enterprise. After this memorandum was prepared, NASA decided to name the Shuttle orbiters after sailing ships of earlier exploratory expeditions.

[no page number]   MAY 26 1978

M-1
MEMORANDUM

TO: LF-6/Director, Public Affairs

FROM: M-1/Associate Administrator for Space Transportation Systems

SUBJECT: Recommended Orbiter Names

In accordance with paragraph 4 of [NASA Management Instruction] 7620.1A I convened and chaired a meeting of an ad hoc committee to recommend names for Space Shuttle Orbiters. The committee consisted of Mike Malkin, Roy Day, Chet Lee, Dave Garrett and Dan Nebrig.

We elected to recommend names having significant relationship to the heritage of the United Sates or to the Shuttle's mission of exploration.

The attached list of names is recommended in descending order of preference. The committee further recommends that the name Enterprise be reserved for Orbiter Five, assuming that there is a fifth orbiter, to carry on the name assigned to Orbiter 101 during the [Approach and Landing Test] Program.

[signed John F. Yardley]

Enclosure

****

[no page number]   RECOMMENDED LIST OF ORBITER NAMES
(In descending order of preference)

1. Constitution
2. Independence
3. America
4. Constellation
5. Enterprise
6. Discoverer
7. Endeavour
8. Liberty
9. Freedom
10. Eagle
11. Kitty Hawk
12. Pathfinder
13. Adventurer
14. Prospector
15. Peace

Document II-20

Document title: Eugene E. Covert, Chairman, Ad Hoc Committee for Review of the Space Shuttle Main Engine Development Program, National Research Council, Statement before the Subcommittee on Science, Technology, and Space, Committee on Commerce, Space, and Transportation, U.S. Senate, February 22, 1979.

Source: NASA Historical Reference Collection, NASA History Office, NASA Headquarters, Washington, D.C.

One of the major technical challenges during Space Shuttle development was the Shuttle's main engine. The engine was fueled by cryogenic (extremely cold) liquid hydrogen and liquid oxygen, and its turbopumps operated at high speed and pressure. To assess progress in solving the Shuttle's main engine development problems and the readiness of the Shuttle for launch, the Congress asked NASA to convene a panel of the independent National Research Council. That panel was chaired by Massachusetts Institute of Technology Professor Eugene Covert. This congressional testimony summarizes the panel's efforts and recommendations.

STATEMENT
OF

Prof. Eugene E. Covert
Chairman, ad hoc Committee for Review
of the Space Shuttle Main Engine Development Program
National Research Council

Before the
Subcommittee on Science, Technology, and Space
Committee on Commerce, Science, and Transportation
U.S. Senate

February 22, 1979

[1] Mr. Chairman and members of the Subcommitee [sic]:

Once again it is my privilege to present to you a summary of the findings of the National Research Council's ad hoc Committee for the Review of the Space Shuttle Main Engine. The ad hoc Committee has published a report entitled "Second Review—Technical Status of the Space Shuttle Main Engine" and dated February 1979, which, with your concurrence, I will submit for the record. I am here, today, as a result of your request of a year ago that the ad hoc Committee review the progress made in the program since its first review in early 1978.

When I presented the findings of the ad hoc Committee's first review in March 1978, a number of modifications had been made to the engine to correct problems that had been encountered up to that time and, in fact, prompted your original request in December 1977 for the National Research Council to review the status of the Space Shuttle Main Engine development. Most of the modifications had been made but were untested[,] making it impossible to assess their adequacy.

Our committee met for two days on October 30-31, 1978 at the National Space Technology Laboratories, Bay St. Louis, Mississippi and, while there, the committee members were able to witness a main engine test run of 823 seconds at about 87 percent rated power which simulated a case in which a flight would be aborted and the orbiter would return to land at the launch site.

In December 1978 as the committee's second report was nearing completion, two fires were encountered in the engine development program. [2] On December 5, 1978 a fire occurred as a result of a leak in the heat exchanger of a test engine—a component the review committee had singled out for concern in its first report. On December 27, a fire originating in the main oxidizer valve virtually destroyed a second engine.

As a consequence of these incidents, you requested that we review the causes and impact of these problems and reexamine our findings and conclusions before submitting our second report. Thus, our committee met again February 1-2, 1979, at the National Academy of Sciences in Washington, D.C., to receive the accounts of the origins and consequences of the incidents and to deliberate on their implications for the development of the shuttle's main engine.

Our report, therefore, contains the findings and recommendations as a result of meetings in October 1978 and February 1979.

As mentioned last year, the concept of Space Shuttle Main Engine is simple. Hydrogen and oxygen in liquid form are pumped from a tank into a combustion chamber where the hydrogen is burned and the products of the combustion, primarily superheated steam, is ejected at very high velocity through a nozzle.

While the basic concept is simple—cold liquid hydrogen and oxygen coming in and superheated steam propelled out—the various flow paths need meticulous control.
[3] The space shuttle engine has severe requirements of light weight, compactness, high absolute thrust, and high thrust per pound of fuel burned. To meet these design requirements the engine components must operate at very high power density levels. The various components are closely coupled and interactive, i.e., the output of one component, of a pump for example, affects the performance of all the elements of the engine between the rocket nozzle and the pump. Thus it must operate within relatively close tolerances with regard to pressure and temperature.

Major developments of new flight vehicles have traditionally proceeded by stages, with provision made for alternative approaches along the way in the design and construction of components. Customarily, the overall system is separated into clusters and sub-clusters representing different components and functioning assemblies. These are designed and tested separately under simulated operational conditions. If necessary, a redesign is initiated to correct any problems or malfunctions. Ultimately the component or assembly is qualified first for peak performance and then long-term service. When major innovations are required, an alternative design may be initiated in parallel with the original. The alternative design may be constructed and tested separately. Later, the best design, possibly with modifications, is chosen and the others are discarded. When component performances have been validated, entire assemblies are then tested for coordinated functions. Finally, the complete engine is put through a series of tests under full power to assess proof-of-flight capability. Such a step-by-step approach provides opportunities to test each piece under conditions that are intended to exceed the demands of operational [4] performance. In this way, it becomes possible to uncover unexpected weaknesses and to plan for contingencies that may arise in flight testing. However, in the case of the shuttle engine, a "success–dependent" strategy for developing the engine is being used. This strategy is a departure from the traditional procedure of development stages in that component-development testing was foreshortened and the quantity of spare parts was severely limited. The success-dependent procedure was intended to offer potential savings in cost and time, by eliminating parallel and possibly redundant development and test hardware. However, as the committee noted in its earlier report, when malfunctions occur during the testing of the prototype of the operational engine, new hardware may need to be designed, constructed, and retrofitted, resulting in delays.

NASA plans to conduct the first six manned orbital flights of the space shuttle at 100 percent power, the "rated power level" of the main engines. Later flights will require a thrust level that is 9 percent greater than rated power in order to launch the full shuttle payloads into orbit. Therefore, the later development of the operational engine includes increasing its thrust rating to at least 109 percent of rated power, which is sometimes called full power level, and maintaining an engine for safe and reliable operation over a life of 55 missions, which means about 7 1/2 hours or 27,000 seconds of engine running time with essentially no repairs or refurbishment.

The development of such a life span for 55 missions is not necessary for the orbital flight test program. In fact, to require such a lifetime capability by 1979 or 1980 would be unrealistic and would result in inordinate increases in the risk of failure and of delays in the overall space shuttle program. Since the attainment of high reliability, [5] long engine life, and performance (in terms of thrust level) cannot be attained simultaneously within the schedule, the current approach is to emphasize reliability at the rated power level (100% thrust) at the expense of engine life and full power thrust level (109% rated power level). The committee considers this order of priority appropriate. Thus, the successful completion of the first six orbital flight tests does not signify the end of the main engine development. Further development to elevate the thrust level to 109% of rated power with high reliability and 7 1/2 hours life must continue. In fact, even after the space shuttle becomes an operational earth to near-earth-orbit transportation system, a sustaining engineering program will be needed just as it is for all new transportation systems.

The committee's assessment described here takes account of two sets of problems. One set was considered in its earlier review and continue[s] to cause concern. The other set is those that have appeared during the tests since March 1978. In its first report, the committee had concentrated on the engine for the first manned flight. In the subsequent review it gave more emphasis both to immediate issues and to longer range issues related to the main engine system.

Rather than base my discussion on a chronological sequence of events, let me first discuss the two fires that occurred in December. After NASA and Rocketdyne had examined the engines and evaluated the failures, they reported their findings to the committee.

[6] 1. Engine 0007

During the first checkout test for the preflight certification of engine 0007 on December 5, 1978, an explosion occurred in the heat exchanger at 3.5 seconds of the planned run of 50 seconds. The source of the failure was attributed to a leak in the coil tubing of the heat exchanger, which was caused, according to the explanation, by a weakness in the tubing that occurred during arc welding while an adjacent bracket was reworked with the heat exchanger still in the engine. The weakness went undetected because existing procedures did not call for a detailed inspection or proof-test of the reworked part. As a result, the heat exchanger and high-pressure oxygen turbopump, both integral to the engine, were damaged[,] although these and other major components of the engine are considered reusable. Inspection procedures and pressure testing have now been established for similar repairs.

The committee recognizes the description of the cause of the explosion as a possible order of events but points out that there are two other ways the failure in the tubing could have occurred—i.e., very high internal pressure caused by a restriction, such as debris, in the tubing, or slow growth of a flaw in the tubing. In any event, the committee recommends that NASA and Rocketdyne establish inspections and proof testing or rebalance procedures as appropriate for all reworked parts. The paucity of development hardware in the program, coupled with the ambitious test schedule, makes the use of refurbished parts a certainty. While the practice of using reworked parts and subcomponents provides valuable experience in the development of an engine for a 7 1/2-hour life cycle, it increases the chance for flaws or malfunctions, with the consequent risk of failures. Therefore, [7] the committee considers it necessary for NASA and Rocketdyne to develop appropriate inspection procedures with a sense of urgency. Because the committee considers a failure in any part of the oxygen system to be potentially catastrophic, the accident in Engine 0007 reinforces the committee's concern, expressed in its first report, about a single-point failure in the heat exchanger.

2. Engine 2001

Engine 2001 had passed the acceptance test in January 1978 and completed four Main Propulsion Test Article runs between April and July 1978—accumulating a total of 287 seconds of test time. After this series of tests, the engine was returned to Rocketdyne for a turbopump retrofit. Then, during the third of a new series of acceptance tests, at 255.6 seconds of its test run, fire broke out in the main oxidizer valve, leading to extensive damage to the engine and the A-1 test stand. The failure was caused by a sequence of events: pressure oscillation in the oxygen flow led to vibrations in the main oxidizer valve inlet sleeve, which were sufficient to loosen one of eight retainer screws and allow fretting between metal parts; this resulted in enough friction to heat the metal to its ignition point in pure oxygen.

Actions to avoid fretting in the future include replacing the thin metal shims with ground shims, coating the surfaces with an oxygen-compatible dry lubricant (During a discussion of dry lubricants, the committee concluded that more study is needed to make a convincing case that lubricants can be used safely.), and replacing the cap screws [8] with screws with a conical shoulder, and providing conical seats incorporating a locking device. The committee supports the need for remedial changes.

In the design goals of compactness and lightness in the closely coupled main shuttle engine, vibrations of fluid-mechanical origin may occur. This provides considerable potential for rubbing or fretting. The committee recommends, therefore, that all fasteners should be examined for loosening and wherever feasible all means of eliminating such loosening should be incorporated.

Rocketdyne has initiated an investigation into the source of the vibrations in the main oxidizer valve and potential remedies. The committee considers further investigation into this problem to be important to pursue.

The Committee is not gravely concerned by these two incidents. Rather these incidents are considered to be a normal part of a development program. In a sense these incidents constitute the price one pays when undertaking to develop any hardware whose performance is beyond the state-of-the-art.

Significant progress has been made in the program in the past year. The rate of testing had proceeded toward NASA's goal of 80,000 seconds of test time before the first manned flight.

Since the Committee's first report, accumulated test time as of December 27, 1978, has more than doubled to 34,810 seconds in 394 firings. Of this total, a little more than 10,000 seconds has been at 100 percent thrust, or the main engine's rated power level, and seven tests have been run at 100 percent power for the full 520 seconds that the main engine operates during the launch. Furthermore, a test run at 102 percent has been completed as well as the abort and return-to-launch-site run that I mentioned previously.

[9] The fact most of this time was accumulated on a single engine implies the Committee's conclusion that a successful engine can be built was correct. Further[,] the teardown inspection has been most instructive to NASA and Rocketdyne. As the lessons learned from these engines are incorporated in the design and their value proved by tests, the rate of accumulation of time at high power levels will increase rapidly. This will herald the successful manned orbital flight.

The Committee, in its March 1978 report, had recommended that NASA and Rocketdyne explore means to acquire and operate a component-development test rig for the rotating machinery of the main engines. Instead, NASA and Rocketdyne have chosen to use a rocket engine itself for this purpose. To this end, test stand A-3 at Santa Suzanna [sic], California, has been reactivated.

The Committee considers this approach to be far from ideal and could lead to long delays in the event of major failure such as a failure in a high-pressure turbopump. Such a failure could result during tests to explore the functional limits of components including the red-line limits for operational safety because the engine is used as a test stand.

An additional consideration is the possibility of an unexpected failure of a component during the operational life of the shuttle. The sustaining engineering program needed to support the shuttle may be more economically and more effectively carried out through the use of a component test stand. The Committee is concerned that replacement hardware will not be available when components fail and, worse, spare engines will not be available for use as new test stands to replace those lost due to component failures.

The Committee considers that it is not too late to develop a component test rig. A component test rig would be valuable not only in the testing process to extend the life of the engine to 7 1/2 hours, but also for the [10] sustaining engineering that is likely to prove necessary over the useful life of the shuttle. While recognizing that engines will have to be used for some time to test components, the Committee urges that appropriate actions be taken to require a component testing.

The main oxidizer valve incident underscores the earlier finding by the Committee that parts and components need to be tested individually before they are assembled and tested as an engine system. If the main oxidizer valve had been mounted in a test stand so that its compliance could have been the same as in the engine assembly and tested, the vibration and fretting might have been identified early in the test program.

One of the effects of both incidents is to highlight the shortage of spare parts and components—not only to ensure that the test and development program can be completed on a reasonably early schedule but to provide enough hardware for the manned flight tests and later operational missions. The situation appears more critical now than at the time of the Committee's last review.

In the report, the Committee has noted that replacement parts and additional engines to be used as test stands will be needed to advance the progress of the program in the event of malfunctions and accidents. To ignore this in a development program as highly complex as this[,] one is to take inescapable risks.

The high probability that failures will occur in the component development process, whether conducted on a separate test stand or as part of the engine test program[,] is a further consideration. The Committee is concerned that when components fail, replacement hardware is not included in the program and, worse, spare engines will not be available for use as new test stands. More test hardware is needed in the program and the Committee considers it to be urgent that a plan to acquire more test hardware be implemented soon if a costly delay in the program is to be avoided.

[11] Another area of continuing concern to the Committee is the oxygen heat exchanger. You will recall that the Committee was concerned that failure in the oxygen heat exchanger, which is located in a hot, hydrogen-rich environment, could be a catastrophe.

NASA and Rocketdyne have complied with the Committee's recommendation to explore alternative designs to relocate or reconfigure the oxidizer heat exchanger. Short of relocating it in a less perilous location, there is no practical way to eliminate the threat to the total system in the event of a failure. Since March 1978, NASA and Rocketdyne have made a design study of a "line-replaceable" heat exchanger that can be more readily and thoroughly inspected in the field—one that could be replaced and would be less subject to damage during fabrication. As designed, the new heat exchanger would be about 170 pounds heavier than the present 20-pound heat exchanger. The accident investigation team has determined that the cause of the fire in the December 5, 1978 incident was a leak, located in a relatively inaccessible region of the heat exchanger. It is of the utmost importance that the heat exchanger be readily accessible for routine inspections in the field. The Committee, therefore, recommends that NASA and Rocketdyne move ahead with the construction and testing of the line-replaceable heat exchanger for installation in the shuttle main engine as early as practicable.

The engine development program includes a Preliminary Flight Certification that consists of a set of ground tests on a single engine. The purpose of these tests is to certify the engine configuration for use in the first six orbital flights.

In its initial report, the Committee made certain recommendations with respect to the minimum test requirements for Preliminary Flight Certification [12] that should be fulfilled in tests on a single "flight-configured engine" before the first manned orbital flight. NASA and Rocketdyne now propose Preliminary Flight Certification test requirements that are essentially in agreement with and, in some aspects, more stringent than the Committee's recommendation. The proposed requirements call for an accumulation of 5,000 seconds of engine test time, including at least 3,000 seconds at rated power level and 425 seconds at 102 percent rated power level, as well as one aborted-flight simulation involving either abort-to-orbit (665 seconds at rated power level) or abort with return-to-launch-site (823 seconds at rated power level). The Committee endorses this set of Preliminary Flight Certification requirements as an adequate demonstration of the engine's performance and reliability for the first manned orbital flight.

However, the Committee is concerned that because of design changes, the engines to be used in the orbital flight tests are not to be of the same configuration as the engine to be tested in the Prelimary [sic] Flight Certification. The differences are significant. While the Committee continues to recommend the use of a "flight-configured engine" for the Preliminary Flight Certification tests, it concludes that the Certification, as presently scheduled, is premature. The Committee recommends that currently planned testing continue but that the formal Preliminary Flight Certification be delayed until the configuration of the engine to be certified is the same as the actual flight engines in all respects affecting safety.

The Preliminary Flight Certification should be viewed as a formal event. If there are any configuration differences, NASA, Rocketdyne, and in particular the Material Review

Board and the NASA Aerospace Safety Advisory Panel should agree in advance on the acceptability of the configuration to be certified. Similarly, if any changes are made during Preliminary Flight Certification testing, the acceptability should be redetermined by the same groups.

[13] In addition to the conclusions and recommendations that I have discussed[,] the Committee in its report also makes a number of recommendations as follows:

A. With respect to the first manned orbital flight, the Committee recommends that:
   - NASA and Rocketdyne should prepare a detailed technical case for the method for determining platform crack growth rates, the intervals and procedures [sic] of inspection, and the criteria for the replacement of turbine blades in the high-pressure fuel turbopump. The case should explain the rationale and demonstrate that engine operation with some platform cracks is not harmful.
   - One engine should be removed from the shuttle orbiter following the first flight and a complete tear down and inspection performed for signs of wear or stress.

B. For later in the manned orbital flight program, the committee recommends that:
   - An agreed upon list of engine components should be tested to the point where individual components have each accumulated 10,000 seconds of test time before the sixth orbital flight. The test time is to be accumulated on a schedule that maintains about 3:1 ratio between total time in ground tests and total time in flight on any single component or assembly.

C. For the longer term in the shuttle flight program, the committee recommends that:
   - A plan to acquire additional engine test and development hardware should be prepared and implemented in the program as soon as possible.
   - The turbine bearing retention system on the low-pressure fuel turbopump should be redesigned to reduce any relative motion or fretting between the bearing and its journal or housing, eliminating the need for dry film lubrication.
   - Tests of the high-pressure fuel turbopump should take place with uncoated turbine blades, and if test results indicate that a coating is not warranted, its use should be discontinued.

[14]
   - A program should be established to gain an understanding of the source of platform cracks in the high-pressure fuel turbopump turbine blades—a program designed to lead to crack prevention. An aggressive program should be undertaken to gain an understanding of the cracking of the high-pressure fuel turbopump turbine blades in order to prevent its occurrence.
   - A study should be undertaken to define the primary cause of the oxidizer injector post failure to provide a "fix" without the need for shields, thus eliminating the source of increased turbine inlet temperature in the high-pressure fuel turbopump.
   - The design, development, construction, and testing of an alternative high-pressure oxygen turbopump should continue in order to be ready to be installed in the engines by 1983 or 1984.

In conclusion, I will state that the National Research Council's review committee continues to have confidence that the space shuttle main engines will perform safely for the first flight once the recommended tests are successfully completed. However, the committee considers that more test hardware, components and engines, are needed to accomplish the required testing. For the longer term a component-development test rig will be needed.

With regard to schedule, the committee offers the following comments: The first flight could occur in November 1979 as suggested by NASA officials only if the test program to be accomplished encounters minimal or no difficulties. In other words, this launch date depends upon a completely successful test program.

This is highly improbable. Some components and parts of the engine to be tested for Preliminary Flight Certification are new relative to the configurations used in the research and development engines previously tested. This procedure reduces the probability of success of the Preliminary Flight Certification. The existence of any components with very little test time, such as the P-6 engine controller and the new impellor for the high-pressure fuel turbopump, leads the committee to the conclusion that an early failure-free Preliminary Flight Certification is unlikely. Any failure (not necessarily catastrophic) will lead to program delays. This is particularly true because of the existing shortage of development engines, spare parts, and test stands. These shortages undermine the expectation for an early manned orbital flight.

From the standpoint of the engines, the committee feels that a first manned orbital flight is not likely to occur before April or May 1980, and even this date is contingent upon adequate hardware for the engine test program.

Document II-21

Document title: John F. Yardley, Associate Administrator for Space Transportation Systems, NASA, to Director, Space Shuttle Program, NASA, "Study of TPS Inspection and Repair On-Orbit," June 14, 1979.

Source: NASA Historical Reference Collection, NASA History Office, NASA Headquarters, Washington, D.C.

One of the major concerns with respect to Space Shuttle safety was the possibility of pieces ("tiles") of the orbiter's thermal protection system (TPS) becoming loose or falling off during launch. If tiles were missing at critical points on the orbiter airframe, there was a danger that the heat of reentry could burn through the orbiter's surface and cause a potentially catastrophic accident. NASA was so concerned about this possibility that various means of in-orbit inspection and repair, if necessary, were considered.

NASA
National Aeronautics and
Space Administration

Washington, D.C. 20546   June 14, 1979
Attn of:

TO: MH-7/Director, Space Shuttle Program

FROM: M/Associate Administrator for Space Transportation Systems

SUBJECT: Study of TPS Inspection and Repair on-Orbit

I talked to Aaron Cohen regarding the subject study and suggested he look at a "Piton" approach. This would be an EVA using light plastic handles, bonded as the Astronaut goes to the tile surface with stickey-back [sic] tape. He would then retrieve these by peeling them off as he returns. They would be designed so that if one was missed it would still burn off harmlessly before it would create critical shocks from a heat point of view. I asked him to review the lost tile test data and try to define the critical areas which would have to be inspected so as to cut down the necessary EVA time and labor. I also asked that he intensify efforts for easy on-orbit repairs for the various types of damage that one could conceive. He agreed to initiate such a study immediately.

You are requested to follow up on this.

[signed John F. Yardley]

Document II-22

Document Title: William A. Anders, Consultant, to Dr. Robert Frosch, Administrator, NASA, September 19, 1979.

Source: NASA Historical Reference Collection, NASA History Office, NASA Headquarters, Washington, D.C.

The status of the Space Shuttle program was of significant concern to the Carter administration. In a July 12, 1979, letter to NASA Administrator Robert Frosch, President Jimmy Carter suggested that NASA seek independent outside judgments of the program's status. In response, Frosch appointed three external consultants to make a top-level assessment of the program. One was William A. Anders, a former astronaut, a member of the Nuclear Regulatory Commission, and an executive with the General Electric Company; he had been executive secretary of the National Aeronautics and Space Council in the White House at the time of the decision to develop the Space Shuttle. The second was retired Vice Admiral Levering Smith, former director of the Navy's Strategic Systems Project. The third was Robert Charpie, president of the Cabot Corporation. Each consultant conducted his own review and reported separately to Frosch. This letter was the report of Anders.

[1]

Baker Hall 4-4
Advanced Management Program
Harvard Business School
Soldiers Field
Boston, MA 02163
September 19, 1979

Dr. Robert Frosch
Administrator
NASA Headquarters
600 Independence Avenue
Washington, DC 20500

Dear Dr. Frosch:

I was pleased to accept the invitation to look, as an outsider, at the current status of the Space Transportation System program and to report my observations and recommendations. I have not been directly involved with the Space Shuttle for some years, but since I was a "midwife" to its birth I feel I am in a relatively good position—seven years later—to measure how the program is meeting its objectives.

Operating as an individual, I could only examine the broader questions and problem areas. But, rather than a disadvantage, I believe this has helped give me a perspective that has served the main purpose for the request for a fresh and independent general assessment. My ability to obtain the necessary information for my evaluation has been due to the fine cooperation I have received not only from you and your headquarters staff, but also from the NASA Centers, Aerospace Safety Advisory Panel, contractors, the Air Firce [sic], OMB, NSC, OSTP, and many others. I was surprised not only by their degree of assistance but also by the consistent pattern of their stories which I have factored into my own thinking in order to provide you with my observations and recommendations. The following summarizes my verbal preliminary report to you.

Observations

1. <u>Need</u>. The concepts underlying the original national commitment appear even more valid today. Plans are proceeding to develop a vehicle that will be the base of a family tree of reusable launch vehicles—cost effective trucks hauling freight to and from orbit. Already, this concept is effecting payload design and operations plans in a beneficial way. I also sense that the originally reserved attitude of the DOD has rather recently begun to swing around to one of support and increasing vision of expanded use of the system.

[2] 2. <u>Funding and Management</u>. Many problems in the management of the program have been cited by a host of reviewers. In my view most of these have really been symptoms of the basic problem—<u>underbudgeting</u> by successive administrations coupled to a progressively <u>overoptimistic</u> view of what work should be attempted on reduced resources. NASA, flush from their outstanding achievement of putting men on the moon and convinced that a shuttle program was vital to our nation, probably had tended to underestimate the degree of some of the technical challenges of the STS and, as problems became more obvious, probably has buckled too easily to budget pressure. The Nixon Administration did not live up to agreements of initial funding and subsequent budget levels nor was the contingency recommended by NASA allowed. Support by subsequent administrations has not been strong. While permitting the program to continue, the emphasis has been to pressure NASA to reduce its annual costs below those required to maintain program schedule and management efficiency. The impact of this approach, inevitably, has been to push NASA towards a higher risk and less efficient program where qualification testing is done concurrently with vehicle manufacture and work performance shortfalls are pushed into succeeding years—in essence, schedule slip was substituted for adequate funding levels and contingency. This, in turn, has led to a need for continual reprogramming of work (very inefficient) and a stretch in the completion date and overall cost. NASA managers have had to become so caught up in the budget battle each year that their program focus tended to shift toward that of achievement of an annual level rather than the completion of a difficult technical project.

As the RTD&E program draws to a close and with schedule now a re-emphasized ingredient, these chickens are coming home to roost. The program still faces technical challenges, and increased costs and schedule delays must be faced up to. Though I am not able to develop a credible estimate of the funds required to complete the program through delivery of the presently scheduled operational vehicles, the number is finite and very likely the magnitude of the contingency requested but denied at the program's birth.

3. <u>Technical</u>. The status of development and testing does not appear to be unusual for a program of this nature. Though real technical challenges remain (especially with regard to the thermal protection system–tile launch survivability, main engine performance and reliability, and the hydraulic power unit of the orbiter) and concern is high, there are no obvious "show stoppers" at this time. Programs addressing critical technical areas are underway but program schedule and hardware performance margins appear worrisomely thin.

[3] 4. <u>Program Management</u> has evolved to exploit individual and organizational strengths and styles. Though management has been adequate in the technical/development area, more attention to program control (cost and schedule status projection and reporting) and operational considerations are now required. This will require increased staffing and some reallocation of duties to insure [sic] that key managers are not stretched too thin. Also, improved reporting and communication in both the program management line and at the policy levels in and out of NASA is required.

5. <u>NASA Credibility</u>. Though NASA might have had an optimistic approach to the STS program for too long and thus helped get themselves into the cost/schedule/performance box we now find the agency, the overall performance of the program—considering its size and challenge—has been quite good. If there is a credibility problem, it appears to me to be more due to inadequate communication at the top level (Congress/OMB/OSTP/NSC/DOD) than to some major programatic [sic] or organic weakness in NASA. All those involved (in and out of NASA) have been, or should have been, reasonably aware of budget problems and what has transpired over the past several years. If NASA has a credibility problem, I believe it is more due to a tendency to be overly accommodating to budget pressure for the sake of preserving a national commitment to a[n] STS rather than to a lack of candor. Backbiting and finger-pointing will serve no useful purpose at this juncture of an important national effort. If there is a "problem," enough blame can be developed to spread around (maybe even to midwives!). Now is the time of all involved to resist carping and kabitzing [sic; "kibitzing" meant] and get behind the program.

With these observations in mind, I would make the following general and specific recommendations:

1. First, and by far the most important, you should prepare a concise statement of the major technical and operational problems to be solved, a realistic schedule for shuttle availability around which others can plan with reasonable certainty, and the cost for following such a schedule. Though the program was probably helped initially by "management-by-schedule-contingency" and work "roll over," this approach appears to have become counter-productive a couple of years ago. Care should be taken to insure [sic] that excessive optimism is weeded out and that adequate contingency reserves (cost and schedule) are now provided. This should be reviewed with the Secretary of the Air Force and other major users for adequacy and then presented to the President as a NASA (Frosch) commitment.

[4] 2. The associate administrator, John Yardly [sic; Yardley], has become the STS program director and generally has done a remarkable job. Nonetheless, he is now being stretched too thin and should be relieved of his other duties to concentrate on managing the RTD&E program through first manned orbital flight (FMOF)—but still at the associate administrator/policy level. As mentioned earlier, he needs more staff to accomplish the required upgrading in program and cost control. Additionally, the communications link between you and the STS program needs strengthening.

3. Organizational steps should be taken to obtain increased attention to and priority for the operational aspects of the STS. I believe there would be multiple benefits to assigning this area (presently part of Yardly's [sic]) to someone from DOD and current on DOD space priorities. This more operationally focused individual might take over the non-RTD&E/ FMOF responsibilities of the present Associate Administrator and should also be at the policy level.

4. Though the safety margins may be adequate under an aircraft testing philosophy (tuning the shuttle to airline-like operation has been a key program guide star), the shuttle is still the preeminent U.S. spacecraft, and much like Apollo, bears the burden of being a significant part of the image of U.S. technical capability. Though I would test fly the shuttle on FMOF (if problems are addressed as expected), I would worry more about it than I did for Apollo Eight due to narrower safety margins (e.g. fallout from reduced hardware qualifications and unmanned flight testing). I believe that this narrower-than-Apollo-margins situation should be brought to the attention of the President for his review of any national and international political/policy implications along with your revised program estimate.

5. Improve external communications by periodic (at least once per month) meetings with the Secretary of the Air Force and the Director of OSTP (and probably OMB). These should be only with principals in attendance. Obviously, improved communication is also necessary with NASA's Congressional leadership. Candor and cooperation are key ingredients to success here.

I hope you and others will find these views useful and that the recommended readjustments and additional commitments are made. These, plus a commitment of support by the President and the Congress[,] will not only help overcome questions of NASA's credibility but will provide reasonable assurance that you and your team will be able to deliver a new and vital capability [5] to our nation. But, the pressures to rationalize and cut corners will likely be great. The time has come for NASA to be fully candid with itself about the remaining challenges and for you to help our national leadership pull together on this important program.

I would be pleased to discuss this further with you if you wish.

Sincerely,

William A. Anders
Consultant

Document II-23

Document title: James C. Fletcher, Administrator, NASA, to James T. Lynn, Director, Office of Management and Budget, October 22, 1976

Source: NASA Historical Reference Collection, NASA History Office, NASA Headquarters, Washington, D.C.

A persistent issue from the time the Space Shuttle was first approved until the Challenger accident was the number of Shuttle orbiters needed to meet U.S. space transportation needs. The 1972 decision on Space Shuttle development authorized NASA to build three orbiters; NASA consistently argued that two additional arguments were needed. A secondary but important issue was which organization should pay for the additional orbiters. Some suggested that because launching Department of Defense (DOD) payloads would be a significant part of Shuttle use, DOD should pay for at least one of the additional orbiters. The enclosure (a NASA-Air Force joint study executive summary) does not appear here.

[1] National Aeronautics and
Space Administration

Washington, D.C. 20546

Office of the Administrator   October 22, 1976

Honorable James T. Lynn
Director
Office of Management and Budget
Washington, DC 20503

Dear Jim:

This letter and the enclosed executive summary of the Joint NASA/USAF Study on Space Shuttle Orbiter Procurement and Related Issues respond to your letters of June 8, 1976, which requested that NASA and DOD undertake such a study. Following the September 10, 1976, briefing, and the draft report of September 15, 1976, provided to your staff, we initiated the final NASA/DOD update and are now providing the final comprehensive written joint study report to you.

The Space Shuttle is being developed as the major component of the nation's first line operational Space Transportation System. The Space Shuttle will reestablish our nation's preeminence in manned space flight, enhance our national prestige, and provide new military and civil space capabilities. Designed to meet the growing needs of space transportation, its use will be open to all nations of the world under appropriate safeguards for the national interest. Already, the development is being shared with other nations through cooperative agreements with the European Space Agency and Canada. It is the only meaningful new manned space program currently under development in the Western world. The Space Shuttle offers a wide range of applications for space exploitation in areas such as weather, earth resources, space science, communications, and space industrialization. The manned reusable vehicle and its associated technology will permit routine space operations that will contribute significantly to national strength and to improving the way of life for all mankind.

In order to achieve the full economic and operational benefits of the Space Shuttle, there must be enough orbiters to provide for the full space transportation requirements of the nation. The approved NASA program will provide for the first three; those Shuttle orbiters required beyond the initial three are not included at present in either the NASA or DOD approved program. An FY 1978 start on the procurement of additional Shuttle orbiters will be required to maintain reasonable schedules and to avoid the severe cost penalties of a break in production.

[2] The latest national traffic model described in the study postulates an eventual steady-state rate of 60 space flights per year. The fleet size analysis shows that, allowing for appropriate maintenance periods, turn-around times, good scheduling performance, and potential attrition of an orbiter, a fleet of five orbiters is the minimum fleet size which should be acquired to support the national requirements projected during the 1980–1991 period. This five orbiter fleet is more cost effective in supporting the national traffic model than the Space Shuttle/expendable launch vehicle mix alternatives required with a three or four orbiter fleet.

The future space capability of the nation is a dominant consideration in establishing the fleet size for the national Space Transportation System. Therefore, we must assess the fleet size and orbiter procurement decisions in relation to their impact on this future space capability. A decision now not to procure the additional orbiters would impose a tight operational ceiling on our future space capability which could adversely impact this nation's leadership in space technology and the attainment of the significant benefits to mankind we are certain will evolve through new and innovative uses of the Space Shuttle fleet. Attrition of an orbiter from the three orbiter fleet would significantly worsen this posture. The establishment of a five orbiter fleet capability would provide the impetus for all classes of users—both civil and defense—to transition as early as possible from expendable vehicles to the Space Shuttle and to make serious plans and investments in developing new and unique uses of the Space Shuttle to enhance the benefits from future exploitation of space. Since the initiation of the national Space Shuttle program in 1972, NASA and DOD have funded or developed budget plans for over $11 billion in FY 1978 budget dollars toward development and support of a viable national Space Transportation System. We believe it is prudent to add the approximate additional ten percent to this significant investment to practically double our space flight capability and to provide the fleet size we believe is the minimum essential to move forward in the exploitation of space and to enhance our national strength and prestige.

The projected cost for two additional orbiters is $1.177 billion in FY 1978 budget dollars. To require funding of this amount from within the tightly constrained currently projected budgets of either DOD or NASA would have a severe impact on either agency's capability to accomplish its planned national objectives. It is agreed with DOD that funding responsibility for the additional orbiters should be placed where the responsibility for management and performance now rests: with NASA. DOD believes that the funds already in their budget for facilities, payload transition, [3] and upper stage development constitute a "fair share" investment in the Space Transportation System as related to their planned utilization.

It is recommended, therefore, that the U.S. commit itself to a five-orbiter fleet and that the funding for the two additional orbiters be provided to NASA as an add-on to the currently projected NASA budget. The required cost of $1.177 billion in FY 1978 budget dollars would be spread over a period of seven years, with $41 million in accrued costs required in FY 1978 and $289 million in accrued costs required in the peak year, FY 1982. Even with these additions, the NASA budget would continue to be well below the levels projected and publicized at the time the Space Shuttle was first approved.

Sincerely,

James C. Fletcher
Administrator
National Aeronautics and Space Administration

Enclosure

Document II-24

Document title: James T. McIntyre, Jr., Acting Director, Office of Management and Budget, to Robert A. Frosch, Administrator, NASA, December 23, 1977.

Source: Jimmy Carter Presidential Library, Atlanta, Georgia.

A continuing issue throughout the Carter administration (1977–81) and indeed up to the time of the Challenger accident was how many Space Shuttle orbiters to build. NASA argued that a five-orbiter fleet was needed to meet the anticipated demand for Shuttle launches. This argument was first made to the Carter White House in late 1977, as Jimmy Carter formulated his first budget since entering office. Both the Carter and Reagan administrations resisted NASA's arguments; ultimately, only "structural spares" for an additional orbiter were authorized. Those spare pieces were the basis for developing a replacement orbiter for Challenger.

---

[1]   DEC 23 1977

Honorable Robert A. Frosch
Administrator
National Aeronautics and Space
    Administration
Washington, D.C. 20546

Dear Bob:

The interpretation in your December 21 letter that "The President decided that an option for a fifth orbiter should be negotiated now . . .," is not a correct reading of the President's decision. The decision was clearly to support a four orbiter option, with NASA authorized to negotiate an option <u>for</u> an option to proceed with a fifth orbiter. Thus, two decisions would have to be made in the outyears: 1) a decision in the context of the FY 1981 budget on whether to provide additional funds for the option; and 2) a decision then or later to exercise that option.

The President stated his explicit concern that no action be taken that might be interpreted as a possible commitment <u>now</u> by the Government to build a fifth orbiter. The option for a fifth orbiter should be kept open for <u>future</u> Presidential consideration and it is NASA's obligation to assure that no actions, contractual or otherwise, are taken that might tend to pre-empt the President's future decision on a fifth orbiter.

The President's decision on space shuttle orbiters can be summarized as follows: The Administration has reviewed the projected uses of the space shuttle in the 1980's and has concluded that:

- Early transition from expendable launch vehicles to use of the space shuttle for civilian and military purposes should be encouraged with operations from launch sites on both coasts by 1984.
- A total fleet of four operational orbiters will meet civilian and military shuttle flight requirements and funds to proceed with production of a four-orbiter fleet are provided in the NASA budget for FY 1979.
- Additional orbiters can be considered for funding in future years in the event that projected flight rates (or the loss of an orbiter) warrant augmentation of the operational orbiter fleet.

[2] A brief summary of the President's shuttle decision, along the lines outlined above, will be included in the President's budget document.

Finally, the President's 1979 budget and the run-out projections of the NASA program that support the President's budget should not include any future-year allowance for the cost of maintaining orbiter production capability through FY 1983. A decision on whether to maintain such a capability or to go ahead with production of a fifth orbiter can be raised for Presidential consideration and budget decision in FY 1981, or later, as circumstances warrant.

                                        Sincerely,

                                        James T. McIntyre, Jr.
                                        Acting Director

Document II-25

Document title: Robert A. Frosch, Administrator, NASA, to President Jimmy Carter, November 9, 1979.

Source: NASA Historical Reference Collection, NASA History Office, NASA Headquarters, Washington, D.C.

At the end of 1979, NASA still believed that the initial Space Shuttle launch would take place some time during 1980. In this letter, prepared in anticipation of a presidential meeting on the Space Shuttle (see Document II-27), NASA Administrator Robert Frosch outlined for President Carter the various reviews that had already taken place and those scheduled before a final decision to commit to a Shuttle launch attempt. Alan Lovelace, mentioned in the letter, was NASA's Deputy Administrator.

---

[1] **NASA**
National Aeronautics and
Space Administration
Washington, D.C.
20546
Office of the Administrator

November 9, 1979

The President
The White House
Washington, DC 20500

Dear Mr. President:

The first launch of the Space Shuttle will, of course, have a high level of domestic and international interest and visibility, and I want to outline for you the pre-launch steps I will take. I can review these briefly at our meeting on Wednesday.

The basic philosophy underlying the Shuttle launch decision, as with other launch decisions by NASA over the years, is that we will launch when ready, and not before. This means that we will launch when we have accomplished each of the pre-launch tasks we set ourselves, and are thus satisfied that every practical effort has been made to reduce risk—to assure crew safety and mission success. And it means that once ready, we avoid any risks inherent in further delay.

I would not want to give the impression that the reviews and activities listed below are all-inclusive. NASA has in place a formal, structured set of procedures leading through a hierarchial [sic] structure of tests and reviews far too numerous to recount here. The list below excludes, for example, the monthly Shuttle program reviews chaired by Deputy Administrator Lovelace, numerous contractor, field center and program office activities and my periodic discussions with the astronauts. I am highlighting here the major, top-level requirements in two categories: in-line program efforts and independent internal and external reviews and analyses.

In-line Efforts

– Design Certification Review. Conducted last April, this review led to acceptance for first flight of the basic Shuttle design. Several open questions (e.g., effects of tank icing) were identified, and work on them is continuing. All such matters will be closed well in advance of launch.

[2] – Mission Rules Review. Extensive, precise mission rules are established before flight, laying out the limits of action by all parties—flight crew and ground control. The final review will be completed by February, and the rules adopted after study by Dr. Lovelace.

– Flight Certification Program. Each major system and subsystem will undergo a thorough, vigorously preplanned series of tests designed, conducted and documented to provide maximum confidence in successful performance in the flight environment. These tests have been underway for much of the past year, and will continue into the Spring of next year.

– Flight Readiness Firing. We will conduct a number of full Shuttle system tests of the flight vehicle on the launch pad. Several of these critical tests will involve proceeding through a full countdown to the point of ignition. One will continue on to actual ignition of the liquid full rocket engines for 20 seconds. Thus all launch systems will be exercised except for solid rocket ignition and lift-off, which will be simulated.

– Flight Readiness Reviews. A comprehensive series of reviews to determine the flight readiness of each of the elements of the Shuttle—for example, reviewing the entire development history and certification firing experience of the liquid fuel engines—will culminate in a final two-day review attended by Dr. Lovelace. The results of this review will be presented to me, and I will then make my decision in light of these reviews and the additional steps outlined below.

Independent Reviews

As referenced in my report to you on Shuttle Management, the statutory Aerospace Safety Advisory Panel, Professor Covert's Committee on the Shuttle main engine, and Dr. Ashley's Committee on the thermal protection system have been at work for some time—the safety panel for nearly a decade—on Shuttle issues. Ashley and Covert have reported their findings to me and their groups have been disbanded. They will continue to stay abreast of our work, and I will consult personally with them before approving the first flight.

– Aerospace Safety Advisory Panel. In addition to its continuing reports, the panel will attend the flight readiness review and will report its assessment separately to Dr. Lovelace and me.

[3] – Chief Engineer. The NASA Chief Engineer is organizationally independent of our program line elements. His independent assessment of the flight certification test program will be available during the readiness reviews and to Dr. Lovelace and me.

– Prime Contractor Management. I will discuss with the corporate managers of the major Shuttle prime contractors—Rockwell, Martin-Marietta, and Thiokol—their assessments of our readiness prior to making my decision.

– Flight Crew. I will talk with the flight crew immediately prior to my launch decision, to be sure that they are satisfied with all that has been accomplished.

When, in light of these steps, I have made the decision to launch, I will notify you immediately in writing that these steps have been taken, and of the scheduled launch date, which will then be about one week away. As a final step, Dr. Lovelace and I will participate in a review of all pertinent factors one day before the scheduled launch, to determine if any anomaly during final preparations warrants rescheduling.

It must be recognized that there will always be some risk in any space mission. It is my task to understand and minimize that risk. The process I have outlined will enable me to exercise my responsibility for approving the first Shuttle launch with the greatest possible confidence.

Respectfully,

Robert A. Frosch
Administrator

Document II-26

Document title: Brigadier General Robert Rosenberg, National Security Council, "Why Shuttle Is Needed," undated but November 1979.

Source: NASA Historical Reference Collection, NASA History Office, NASA Headquarters, Washington, D.C.

The staff of the Office of Management and Budget (OMB) in the fall of 1979 raised the possibility of terminating the Space Shuttle program, given its technical problems and schedule delays and the desire of the Carter administration to reduce the federal budget. Air Force Brigadier General Robert Rosenberg, on detail to the staff of the National Security Council in the White House, prepared this brief paper as a counter to OMB's position. The paper was one of the inputs to a November 14, 1979, presidential meeting on the Space Shuttle (see Document II-27).

[1]   WHY SHUTTLE IS NEEDED

Maintenance of world leadership in space and associated technologies is essential to the long-term political and strategic position of the United States in world affairs and in the pursuit of our national goals and policy. Shuttle will be the world's first reusable space vehicle and because of its reduced operational costs, increased operational capability and flexibility the Shuttle will propel the U.S. space program a generation ahead of foreign capabilities and technologies. Without Shuttle plus the inherent ability of man to operate in space, the capability to exploit space effectively to maintain world leadership will be impossible.

Foreign focus on space is evidenced by their intense interest in developing competitive expendable boosters should the U.S. falter or retreat in its space leadership. The U.S. has no current manned space operating capability. The <u>currently operational military</u> and civil Soviet manned space program could provide them with significant scientific, technical, political, and strategic advantages which cannot be overcome with an expendable launch vehicle-based U.S. space program. If we do not expend the thought, the effort, and the money required, then another and more progressive nation will. <u>It will dominate space, and it will dominate the world.</u>

Loss of space leadership by termination of deferral of Shuttle operations will be comparable to losing U.S. lead in the airline industry. Significant loss of jobs would occur, sizable dollar outflow would result as U.S. industries, especially communications, move to foreign launch capabilities. Foreign interests would use foreign boosters rather than Shuttle, thus increasing trade deficits. American industry use of foreign boosters would be encouraged if Shuttle is not available because of foreign government subsidies for low cost launch services. Foreign governments would seek, as part of bargaining strategies[,] to obtain sensitive American technology from the American customers. In the long term, American customers, because of profit incentives, would become captive of the selected foreign booster as a result of additional costs required to redesign operational satellites for compatibility on subsequent U.S. boosters.

Shuttle will encourage a more vigorous space program which is one of the most powerful ways of stimulating U.S. economic growth through R&D. This would increase productivity, improve our international competitive posture, support U.S. private capital formation and provide anti-inflationary effects.

From a national security aspect, shuttle capability allows large structures to be built in space for enhanced monitoring of arms control agreements over present day capabilities.

Shuttle will allow adequate advances in our military space capabilities as required to counter the growing Soviet space threat to deny others the use of space.

Shuttle represents a U.S. commitment to itself and the world. American and foreign users have made significant financial investments based on its availability in the early 1980's. Revocation of this commitment [2] would seriously erode the U.S. posture as a world leader that honors its pledges.

Ten European nations, with the encouragement and agreement of the U.S., are completing development at their own expense (over $500 million) of the manned Spacelab for use solely with the Shuttle in the 1980's. The Spacelab provides unique capabilities for manned scientific and technological advances which could provide significant benefits in new developments to the people on earth (e.g. new medical advances, new lightweight materials). Termination or extensive delay of the Shuttle not only would thwart such advances but would be regarded as an act of bad faith and seriously undermine the confidence of our allies and unaligned nations in other U.S. commitments.

Shuttle will dramatically change the economics of space. Our largest expendable booster is Titan III. Shuttle will be able to reach low-earth orbit with roughly twice the payload at less than half the cost. And Shuttle payload costs will also be lower because of the relaxation of design requirements made possible by the Shuttle's large cargo bay, moderate launch and flight environment, and on-orbit maintenance capability.

Current boosters have but one purpose—to launch payloads. Shuttle has many purposes. It has been designed to service and refurbish satellites, retrieve and return to earth payloads weighing up to 32,000 pounds, perform dedicated experimentation and technology development missions, carry passengers in relative comfort, and, with suitable upper stage propulsion, launch from orbit satellites and spacecraft whose missions require the attainment of super-orbital velocities.

It will enable us to assemble large structures in space—an essential capability if we are to fully use the space environment to help solve earth-based problems.

Satellites taken into space can be carefully checked out in earth orbit before being orbitally inserted. Thus, loss of expensive payloads due to launch induced malfunctions will be eliminated.

Payload bay volume permits pooling of payloads, reducing flight costs.

Finally, Shuttle will allow us to return space-produced products to earth.

Shuttle will stimulate advancing technology in virtually every field in which the U.S. excels. The direct economic contributions of the Shuttle will grow to major proportions as we expand the industrialization of space. One recent study estimates that some 2,000,000 direct jobs will be created by the year 2010 through an active space industrialization program made possible by Shuttle.

Strong national support and prestige is focused on Shuttle as a means for maintaining space dominance as evidenced by broad user interest and recent space policy statements. Significant delay or abandonment of the Shuttle and manned space capabilities at this time would be viewed as a loss of national pride and direction. The notion that we are forced for short term economic reasons to abandon a major area of endeavor in which we have achieved world leadership at great cost is simply not credible.

Document II-27

Document title: Office of Management and Budget, Background Paper, "Meeting on the Space Shuttle," November 14, 1979.

Source: Jimmy Carter Presidential Library, Atlanta, Georgia.

This paper was prepared as background for a White House meeting with President Jimmy Carter to discuss the status and future of the Space Shuttle program. During 1978 and 1979, the Shuttle program had experienced a series of technical problems leading to schedule delays and the need for additional budget resources. Some of the staff within the Office of Management and Budget had even recommended that Carter cancel the program. This meeting was called to inform Carter of the results of several external reviews of the Shuttle program and of NASA's actions to overcome various Shuttle development problems. The outcome of the meeting was a presidential decision to proceed with the Shuttle program as planned. This paper included four "tabs" as attachments; only Tab A appears here. The classified Tab B is not reprinted here; however, Tab C appears as Document II-25, and Tab D appears as Document II-26.

[no page number]   EXECUTIVE OFFICE OF THE PRESIDENT
                       OFFICE OF MANAGEMENT AND BUDGET
                               WASHINGTON, D.C. 20503

Meeting on the Space Shuttle . . .

Wednesday, November 14, 1979
10:30 (1 hour)
The Oval Office (15 minutes
with Dr. Frosch)
Cabinet Room (45 minutes
with others)

From: Zbigniew Brzezinski
James T. McIntyre, Jr.
Frank Press

I.   PURPOSE

To discuss with Dr. Frosch and your advisors the status of the Space Shuttle Program and actions being taken to deal with current problems.

II.   BACKGROUND, AGENDA, PARTICIPANTS, AND PRESS PLAN

Background: On July 11 you wrote to Dr. Robert Frosch, Administrator of NASA, requesting that he appoint a few highly competent and independent individuals to assist him in making a comprehensive review of the Space Shuttle Program. Dr. Frosch wrote to you on October 10 on the management actions he is taking to deal with problems in the program, including the views of the independent advisors. At that time we agreed to meet with you, Dr. Frosch, and Harold Brown to report on the shuttle's technical, schedule, and budget status, as part of the FY 1981 Budget process.

Agenda:

10:30:   The President meets with Dr. Frosch in the Oval Office.

10:45:   The President and Dr. Frosch meet with Dr. Hans Mark (Secretary of the Air Force, representing Harold Brown), [Executive Office of the President] senior staff, and others in the Cabinet Room.

10:50:   Dr. Frosch makes a viewgraph presentation on the Space Shuttle Program.

11:10:   Question and Discussion Period.

11:30:   Meeting adjourns.

[2] Participants: The President, Bob Frosch (NASA Administrator), Hans Mark (Secretary of the Air Force), Alan Lovelace (NASA Deputy Director), John Yardley (Shuttle Program Manager), Bill Lilly (NASA Comptroller), Zbigniew Brzezinski, David Aaron, Jim McIntyre, John White, Bo Cutter, Randy Jayne, Curt Hessler, Frank Press, and Ben Huberman.

Press Plan: No press coverage planned.

III. FURTHER BACKGROUND AND TALKING POINTS

Additional background materials are attached:

Tab A—a "shuttle program assessment" which summarizes a larger paper developed by OMB working with NASA and with NSC and OSTP.

Tab B—a classified assessment of the backup options, decision dates, and costs for national security launches.

Tab C—a letter from Dr. Frosch on the steps he is taking to assure safety of the shuttle flight crews.

Tab D—a brief NSC staff paper on "why we need the shuttle" which addresses shuttle capabilities and the program's implications for the United States.

Your advisors remain convinced, despite recent technical and cost problems, that the shuttle program should be continued on its present schedule. At this late date, it would not be economic or prudent to cut the program short, to slow it down, or to redirect it in some radical fashion.

Although covered in Dr. Frosch's presentation and in some detail in the attached program assessment (at Tab A), we would highlight for you several problems where the personal judgments of Bob Frosch and Hans Mark are especially important for you to hear:

(1) Management: Dr. Frosch reported to you last month on changes he is making to improve management of the shuttle program. The key changes involve increasing audit/cost oversight of the program at all levels and creating a new NASA Associate Administrator position to plan and run the shuttle system once it becomes operational.
   –   Are these reforms progressing satisfactorily?

[3]   (2) <u>Schedule risks:</u> The main technical risk to launching the First Manned Orbital Flight (FMOF) by September 1980 is the development and testing program for the main engines.

- What is the probability that this[,] or other problems, will push FMOF beyond September 1981?
- What is the likelihood of a slippage of 6 months or more beyond September?

    (3) <u>Program costs:</u> NASA presently believes that shuttle costs (including program reserves) will exceed the projections in the 1980 budget by at least $520 million in FY 1980, and at least $720 million in FY 1981.

- In light of ongoing cost reviews at NASA and the recent problems with main engine development, are larger add-ons likely? How much larger?
- How firm can the cost numbers be in the FY 1981 Budget?

    (4) <u>Contingency plans:</u> Slippage [of] FMOF beyond September 1980 would require some of the scheduled commercial customers to use systems other than the shuttle. An extended slippage might require some DOD flights to use other launch vehicles.

- What is the status of contingency planning at NASA to provide alternative launch capabilities for civilian payloads? When must such contingency decisions be made? What are the alternatives in the civil flight program if shuttle slips 12 to 18 months? What will be the cost of providing backup systems (to the U.S. Government and to the private users)?
- What is the status of contingency planning at DOD? When would we know if shuttle cannot support critical SALT[Strategic Arms Limitation Talks]-related missions?
- To what extent do firm civilian and military payloads depend upon attainment of the high-performance (109 percent) shuttle engines?

****

[1]   Tab A
11/12/79

National Aeronautics and Space Administration
Space Shuttle Program
Assessment Summary

<u>OVERVIEW</u>

This assessment:

- <u>Sets forth the current Shuttle development schedule.</u> Key dates are August/September 1980 for First Manned Orbital Flight (FMOF) and late 1981 for First Operational Flight (FOF) at Kennedy Space Center.

- <u>Identifies key problems</u> in achieving the schedule. Major findings:
  - NASA is having difficulty with main engine development, testing, and certification and the main engine problem is the current major threat to schedule and budget. However, if engine problems can be overcome, the program appears to have adequate schedule margins to meet FMOF by September 1980 and all other launch dates thereafter.
  - Schedule slippage for FMOF beyond September 1980 would impact some commercial and national security missions. Although further slips are not now fore-

seen, our ability to forecast further problems is quite limited. Therefore, funding decisions are required soon to continue to protect back-up launch options for some national security and commercial payloads.

– Describes the safety problem. One of the three senior outside consultants appointed to review the program on your behalf expressed concern about "narrower-than-Apollo" safety margins. Major findings:
  • While Apollo conducted more beyond-design-limits testing and launched some early flights unmanned, there is little difference in design safety factors between Apollo and Shuttle and the Apollo moon missions had single point failure vulnerability and fewer abort options than Shuttle.
  • NASA is undertaking a continuing, detailed flight readiness review to culminate on launch minus-one day with a decision by the Administrator. Also, NASA plans a first flight with uniquely wide safety margins.

– Analyzes costs and cost overruns. NASA completed a major cost and schedule review this summer, after the April budget amendment, and developed more conservative and precise estimates for their 1981 budget plan that provide coverage for FMOF through September 1980. The April budget amendment and the re-estimates raised Shuttle funding, compared to the January FY 1980 budget, by $520 million in FY 1980 and $727 in FY 1981. However, because of the recent engine problems, it is now only 50% probable that FMOF will occur by September 1980 and NASA estimates that an additional $50–100 million in FY 1981 may be required above the previous FY 1981 budget request.

|2| – Identifies the options, decision dates, and costs for providing backup options for national security and commercial launches. Major findings:
  • Most national security missions currently manifested on Shuttle have back-up options secured; long-lead protection for national security launches with a continued full commitment to the Shuttle would require five-year costs of approximately $250 million.
  • Many commercial users currently manifested on the Shuttle have back-up options secured, but some would require a minor uprating of standard launch vehicles (which NASA has proposed to cost-share with industry) and some can use only the Shuttle. Several commercial users have already had to revert to expendable launch vehicles as a result of Shuttle schedule slips and others will do so if further slips occur, thereby losing their potential savings from flying on the Shuttle. A few users are currently negotiating with foreigners for launch services.
  • Some commercial and national security payloads have been Shuttle-optimized to the point that a major redesign would be required to adapt to expendable boosters.

– Identifies the management improvements underway to meet schedule and cost goals, especially the establishment of a new Shuttle Operations unit within NASA and the search for appropriate staff to create a responsive service organization.

SHUTTLE SCHEDULE AND RISK OF FURTHER DELAY

The current major schedule milestones are summarized below.

	Current Estimate
First Manned Orbital Flight (FMOF) with Orbiter OV-102 "Columbia"	June/July 1980 at 10% probability and August/September 1980 at 50% probability
Initial Operational Capability (IOC) at KSC and First Operational Flight (FOF)	September/December 1981 (depends on 4 successful test flights after FMOF)
Initial Operational Capability at Vandenberg AFB (VAFB)	December 1983 (depends on completion of VAFB facilities and timely delivery of production orbiters)
Production Orbiter Deliveries OV-099 "Challenger" OV-103 "Discovery" OV-104 "Atlantis"	June 1982 September 1983 December 1984

[3] The most critical program milestones are First Manned Orbital Flight (FMOF), First Operational Flight (FOF), and delivery of orbiters OV-099 and OV-103. For the FMOF, the current schedule supports a late June/early July launch with only 10% probablility [sic] and late August/early September launch with 50% probability. NASA's 1981 budget request assumed a July launch with 50% probability, but provided coverage for a September launch.

Major milestones in the achievement of First Manned Orbital Flight to meet the August/September 1980 current estimate are shown below:

Event	Current Estimate Completion Date	Significance
Resumption of Main Propulsion Testing	December 1979	Would indicate that previous problems have been cleared.
Orbiter Rollout from Orbiter Processing Facility	Feb./March 1980	Would signal completion of all orbiter manufacturing, tile pull tests, and most orbiter systems tests.
Orbiter Transfer to Pad	March 1980	Would signal successful mating of major systems.
Certification of Remaining Orbiter Systems and Software	March/April 1980	Would signal completion of most detailed systems testing, including interaction of ground and flight systems.
Thermal Protection System Certification	April 1980	Would mean key safety concerns [are] resolved.

Event	Current Estimate Completion Date	Significance
Flight Readiness Firing	May 1980	Would indicated [sic] successful group operation of main engines and all other Shuttle systems except the solid rocket boosters.
Certification of Main Engines for First Flight	June 1980	Would mean engines are ready.
FMOF	July/Sept. 1980	Full system demonstration.

This schedule involves a high degree of concurrency between actual hardware deliveries/flight preparations and the completion of certification testing. The high degree of concurrency causes significant risk of further delay from adverse test results.

Key risks in meeting the FMOF date (September 1980) are that:

—   The main engine certification presents the major threat. The last main propulsion test on November 4 experienced two engine problems which, although not catastrophic, will require additional analysis, component testing and possibly engine modifications before flight. Once main [4] propulsion testing can begin again (now estimated for mid-December 1979), at least 7 additional successful tests will be required before FMOF, with a minimum turn-around time of 3 weeks between tests. Delays from any further engine problems would depend on how flight-related the problem is—a well understood fatigue problem might not affect first flight.

—   Further schedule slippage could also occur if:
   •   Actual test experience or review of expected loads on the thermal protection tiles mandate a replacement of a high percentage of tiles. Completion of tile installation could be delayed until March if planned re-installation rates are not achieved or current reject rates increase sharply. Current progress rates (October data) do not raise serious concerns here.
   •   Major problems are discovered during upcoming integration tests of whether all the software and subtle and profound difficulties are found in making all software programs and associated hardware play together.
   •   A large number of small problems develop, delaying orbiter rollout (e.g., orbiter certification, hardware/software certification).

—   A minimum 4 week to 6 weeks' delay could occur if any problem is encountered when the Shuttle is on the Pad that requires it to return to the Vehicle Assembly Building or to the Orbiter Processing Facility. If such a problem occurs, it is most likely to be identified at the Flight Readiness Firing 6 weeks before flight, when for the first time the main engines, all electronics, and the auxillary [sic] power units will all be tested together.

However, 75% of all Shuttle testing has been completed satisfactorily and most major final hardware systems are already in place for the complete Space Transportation System.

Key risks in meeting the First Operational Flight (FOF) date (September 1981):

- any delay in FMOF;
- problems encountered in the four orbital test flights; and
- major problems in performing "turnaround" of the orbiter in preparation for next launch.

Reducing turnaround time from several months in the orbital flight tests to several weeks in mature operations is critical to operational cost-effectiveness—and is currently highly uncertain.

Most key milestones after FMOF are keyed to achieving planned performance improvements (e.g., higher engine thrust levels).

Achievement of the full 109% power level for abort conditions alone would be sufficient to support planned launches in the 1982–1983 period. Achievement [5] of the planned weight savings in the follow-on orbiters, the external tank, and the booseters [sic], and some fractional (about 102%) improvement in sustained engine performance will provide adequate margins for even the most demanding mission now planned. NASA is also considering development of extra strap-on rockets which could provide even greater performance capabilities in later years (post 1984).

For the production and delivery of later orbiters (i.e., OV-099, OV-103, OV-104) current schedule margins appear achievable.

FLIGHT SAFETY

One of the three senior outside consultants appointed to review the Shuttle program on behalf of the President, former Astronaut William Anders, expressed the view that the Shuttle system had narrower-than-Apollo safety margins because of reduced hardware qualification testing and lack of unmanned flight testing for the Shuttle program. NASA program management believes that the Shuttle compares more favorably to the Apollo program when examined in detail.

The Space Shuttle has been designed with factors of safety basically comparable to those of the Apollo program. For example, the Shuttle has been designed to a 1.4 structural factor of safety as was Apollo. The Apollo structural ground testing was conducted to 1.4 of limit loads; the orbiter has been tested to 1.2 of limit loads; and the external tank and solid rocket booster have been tested to 1.4 of limit loads.

While there is little difference in design safety factors between Apollo and Shuttle, Apollo did conduct more testing beyond design limits and did launch some early flights unmanned. However, the Apollo moon missions had single point failure vulnerability and fewer abort options than Shuttle. In the judgment of NASA program management, some Apollo flights were considerably more risky than Shuttle FMOF.

As a conservative approach, NASA plans to fly a benign mission for the first flight that will restrict the limit loads to 80% of design, which will increase the factor of safety on FMOF from 1.5 to 1.8 of design limit loads.

Launch delays for FMOF will not adversely affect flight safety margins. For flight safety, required testing must be complete before a commitment to flight can be made.

Criteria/procedures for determining flight readiness consist of an organized series of detailed technical reviews with top level management overviews, commencing with the Design Certification Review last April, and concluding with a launch-minus-one day review. The Administrator will make the final decision on flight readiness.

Outside advisory groups provide an independent assessment of problem areas and provide action recommendations. Examples include:

- The Aerospace Safety Advisory Panel which has reviewed the program from the outset and reports their findings at least annually to the Administrator.
[6] - Ad Hoc Groups, such as the Covert Committee on the main engines, the Ashley Committee on the thermal protection tiles, and the Wilkerson Committee on the hydraulics system.

In addition, NASA has had internal reviews utilizing organizations from NASA Centers not normally involved with day-to-day operations of the Shuttle Program, and NASA has used DOD expertise where applicable to trouble spots (e.g., safe handling of solid rocket propellants).

SHUTTLE PROGRAM COSTS

Cost estimates for total Shuttle development increased substantially in the past two years because:

- Annual program costs were tightly constrained and estimates of work that could be accomplished within annual cost limits were overly optimistic, causing much work to be postponed and overall schedules to slip.
- The future cost impact of postponed work was difficult to estimate. Management and program control resources were insufficient to develop precise estimates.
- The firmness of critical launch dates beginning in 1982 and the extent to which those missions were dependent on the Shuttle was not established until about a year ago (although the Shuttle was from the beginning designed to meet the space transportation needs of national security programs beginning in the mid-1980s).
- A series of technical problems developed late in the Shuttle development program. For instance:
  • In December 1978 and July 1979, the engine test program was interrupted by major component failures.
  • Parts of the first flight orbiter were shipped to the assembly facility without completing manufacturing and the assembled orbiter was shipped to Kennedy Space Center with substantial open work remaining.
  • The installation of thermal protection tiles at the launch site proved far more difficult than anticipated and interferred [sic] with other required manufacturing and check-out work on the orbiter.

This summer, in preparing its FY 1981 budget plan, NASA:

- Moved the expected FMOF launch date to September 1980 (versus late 1979 in the FY 1980 budget plan);
- Undertook a major cost review of all work remaining; and
- Developed unusually careful estimates for program reserves keyed specifically to potential problems.

[7] The funding increases requested in NASA's FY 1981 budget plan are displayed in Figure 1. Points worth noting:

- Shuttle budget costs have increased over the January 1980 budget by approximately $520 million (requiring an FY 1980 supplemental of $300 million) and by about $730 million in FY 1981.
- The estimate of total cost at completion has increased by nearly $2.9 billion (1981 dollars) since the January FY 1980 budget plan. This implies a total cost increase for the Shuttle program of about 20% in constant 1971 dollars over the estimate made in 1971 ($5.15 billion, 1971 dollars) at the program's inception. Other large high technology projects have experienced similar cost overruns.
- The NASA FY 1981 estimates will probably have to be increased by another $50–100 million to reflect the November 4 engine test failure. Further engine problems, causing a delay in FMOF beyond September 1980, could entail further 1981 budget increases.
- NASA is presently conducting another baseline cost review to be completed in late November, and Dr. Frosch's management changes will also yield new and better cost estimates well into 1980. If new technical problems do not arise, we do not not [sic] expect these reviews to alter the budget request significantly, but we cannot be sure.
- None of these estimates include the budget cost of providing back-up capability for national security missions to cover the contingency of shuttle failure.

CONTINGENCY PLANNING

Current Flight Assignments

At the present time, NASA has firm commitments from 15 different users who plan to fly 60 payloads during the first 3 years of STS Operations. NASA has manifested these payloads on 38 flights with the first flight scheduled for September 1981.

Except national security and NASA-critical missions, payloads are generally accommodated on a first-come-first-served basis, and given flight assignments which assure compatibility of shared payloads.

Current NASA planning provides for launching DOD payloads on their requested launch dates in accordance with their top priority as provided by [Policy Directive]-37.

Back-Up Options, Decision Dates and Costs for Commercial/Foreign Users

There are 27 non-national security payloads currently manifested that are configured only for the Shuttle. Most of these are NASA missions, but several commercial payloads in this category would experience large costs from a schedule slippage because of their commercial commitments to provide services on fixed timetables.

[8]   11/12/79

Figure 1

Space Shuttle Investment Funding
(Budget Authority, $ in Millions)

	Prior Funding	FY 1981 Request	Projected Total Funding in Constant FY 1981 Dollars	
			To Complete	At Completion
NASA				
Shuttle DDT&E and Orbiter Production	7,577	1,733	4,434	13,744 [1]
Construction of Facilities	390	10	45	445
Operations Capability Development	192	131	352	675
NASA Total	8,159	1,874	4,831	14,864

	Prior Funding	FY 1981 Request	Projected Total Funding in Constant FY 1981 Dollars	
			To Complete	At Completion
DOD				
R&D and Procurement	903	336	638	1,887
Military Construction	192	105	50	347
DOD Total	1,095 [2]	441	698	2,234

[1] History of recent changes for NASA Shuttle Development and Production before November 1978 engine problems:

	Estimate at Completion ($ M BA)
FY 1980 January Budget Projection (FY 1980 $)	10,856
Runout of FY 1980 Budget Amendment	+811
Inflation Adjustment of Projections to FY 1981 $	+156
Revised Estimates in FY 1981 Request	+1,921
Estimate at Completion, FY 1981 Budget Request (September)	13,744

[2] FY 1980 congressional appropriations action incomplete.

Except for the above and 3 payloads for which booster compatibility is undetermined and two users who already have special booster commitments, all remaining non-national security users would be compatible with either standard or uprated versions of the Delta vehicle. The uprating for the Delta has not yet been initiated, but NASA has proposed to share the $6–8 million cost of the upgrade with the six commercial users in this category. Generally, commercial users pay the full cost of protecting back-up launch options for their payloads. Federal participation in the Delta uprating is being considered as it would be necessary for some NASA payloads. Once developed, commercial users would pay for the cost of the uprated hardware they use.

Users of standard versions of the Delta each forego $17–24 million in potential savings if they have to launch on a Delta instead of Shuttle. These users must choose and commit to booster or Shuttle at one month after FMOF or nine months before their needed Delta launch, whichever comes earlier. Three users have already commited [sic] to Delta because of earlier Shuttle schedule slippage, and the recent Shuttle engine problem will likely cause three more to commit to Delta.

[9] Except for the firm commitment to Delta uprating, which must be made in December 1979, there are currently no other pending Federal funding decisions associated with commercial/foreign back-up boosters.

None of the key national security missions would use the Delta vehicle.

The European Space Agency (ESA) is currently developing a booster called Ariane which, if its first flight now scheduled for December 1979 is successful, would be a potential back-up for Delta class payloads that require uprating and for the heavier Atlas-Centaur class payloads. Some U.S. commercial users are negotiating with the French for use of Ariane.

However, NASA believes it is unlikely that payloads compatible with an upgraded Delta would shift to Ariane if a commitment to upgrading is made in December because the Delta is a proven launch vehicle.

Back-Up Options, Decision Dates, and Costs for National Security Launches

See Tab B for a classified discussion of planned flights and options.

[10] STATUS OF SHUTTLE MANAGEMENT ACTIONS

The management changes proposed earlier by Dr. Frosch are now underway. Status of the major actions is summarized below:

Proposed Action	Current Status
Establish a new and responsive service organization for Shuttle Operations headed by an Associate Administrator.	Search for a qualified candidate is underway but will probably take another 2–3 months. A request for 52 additional positions to support the reorganization is currently being reviewed by OMB.

Proposed Action	Current Status
Develop a revised financial operating baseline as part of the FY 1981 budget process.	NASA is continuing to gather new subcontractor data through the month of November. This review is not expected to change overall requests as now identified.
Provide additional financial, schedule, and program analytical manpower at each level of program structure.	Gradual change underway as appropriate new people are selected; total change will probably take 2–3 months. OMB is reviewing a NASA request for 89 additional positions in FY 1980 and 1981. . . .

Document II-28

Document title: Robert A. Frosch, Administrator, NASA, Special Announcement, "Examination of the Shuttle Program," August 18, 1980.

Source: NASA Historical Reference Collection, NASA History Office, NASA Headquarters, Washington, D.C.

With this announcement, NASA Administrator Robert Frosch indicated that NASA's top managers had reached agreement that Space Shuttle development was far enough along—and remaining problems well enough understood—to set a date for an initial Space Shuttle launch.

[1] Special
Announcement

NASA
National
Aeronautics and
Space
Administration

Date: August 18, 1980
Subject: Examination of the Shuttle Program

During the past year we have carried out a very detailed examination of the Shuttle Program using experts and specialists from outside of NASA as well as many of our own people, and have conferred repeatedly with the prospective "users" of the Shuttle. This examination has greatly improved our understanding of the program and of the capabilities of the NASA/contractor organization to solve those problems which remain ahead as we prepare for the first flight.

Based on this broader understanding, Dr. Lovelace and I have arrived at a number of conclusions—conclusions which have the full concurrence of the Shuttle management and the Directors of the Centers with principal responsibility in the program.

1. The extraordinary attention which has been given to the Thermal Protection System over the past 18 months has greatly enhanced our knowledge of the system's requirements and our confidence in its capabilities. This has allowed us to define and schedule the remaining tile effort and to plan the first flight without including the Manned Maneuvering Unit and the Tile Repair Kit, although their development will be continued to allow them to be incorporated into the flight if later tests indicate that it is desirable to do so.

2.　The formal certification process of the Shuttle main engine is about 70% complete. Despite recent problems, it is clear that the basic design has been proven. Our confidence in the engine is thus much greater than it was a year ago.

3.　The Flight Certification Assessment is essentially complete. Overall, it has provided a strong endorsement to the manner in which the program is being carried out.

[2] During the last week of July, Dr. Lovelace and I met with agency and contractor Shuttle Program management and, based on these conclusions, reached four key decisions.

1.　We have baselined the remaining TPS and other necessary work on the orbiter and concluded that it can be completed in time for Columbia's rollout from the Orbiter Processing Facility at KSC on or before November 23, 1980.

2.　We have adopted a 15-week work schedule for activities necessary from [Orbiter Processing Facility] rollout to launch.

3.　We intend to launch by the end of March 1981, although we recognize that this is a tight schedule.

4.　Although we have not reassessed the requirements for the total flight test program, we expect that the planned 18-month [Orbital Flight Test] program will lead to an initial operational capability in September 1982.

There is a time in a major national program to join together in a concerted drive for the finish. This time has come. Not one of us believes that we have set an easy course. But not one of us can identify any single aspect of the Shuttle which, from what we know today, is "not achievable." Meeting these milestones will require exceptional dedication to the task to be done; exceptional judgment in refraining from doing those things which are not required to be done; and exceptional leadership to provide the opportunity for each of us to contribute his or her full share to that task.

Chris Kraft, Bill Lucas. Dick Smith, John Yardley, Al Lovelace and I pledge our best efforts to this task. We urge you to join us in moving forward toward this worthy goal.

Robert A Frosch
Administrator

Document II-29

Document title: NASA, Lyndon B. Johnson Space Center, "Major Safety Concerns: Space Shuttle Program," JSC 09990C, November 8, 1976, Preface and pp. 1-1–2-3, 5-1–A-6.

Source: NASA Historical Reference Collection, NASA History Office, NASA Headquarters, Washington, D.C.

Crew and vehicle safety was a constant concern throughout Space Shuttle development. NASA's lead center for the Shuttle, Johnson Space Center, conducted periodic assessments of safety issues. Safety risks were divided into three categories: (1) those that could be addressed through remedial actions; (2) those that might be addressed, but at a high cost; and (3) those that were inherent in the particular design chosen for the Shuttle and could not be ameliorated without changing the design. This document reviews the status of each of these risk categories as of late 1976, mid-way in Shuttle development. Sections 3.0 and 4.0 do not appear here.

Major Safety Concerns
Space Shuttle Program

National Aeronautics and Space Administration
LYNDON B. JOHNSON SPACE CENTER
Houston, Texas

November 8, 1976

PREFACE

This document provides risk management data for management overview purposes and facilitates a periodic independent assessment of the cumulative residual risks. The document also provides technical information and status on open concerns, and rationale for concern closures and accepted risks.

This document is updated quarterly to reflect changes in status of major safety concerns and to add newly selected major safety concerns. This issue is a complete revision of JSC 09990B, dated June 28, 1976. . . .

Jerome B. Hammack
Chief, Safety Division

M.L. Raines
Director
Safety, Reliability, and Quality Assurance . . .

[1-1] 1.0 <u>SUMMARY</u>. This document provides a summary of the major Space Shuttle Program safety concerns selected by the Johnson Space Center [JSC] in conjunction with other NASA Centers and the integration contractor. The document provides program management visibility of open safety concerns, closed safety concerns, and accepted risks and will be updated on a quarterly basis. The number and status of safety concerns included in the previous issue and this issue are as follows:

	June 28, 1976 issue	November 8, 1976 issue
Open Safety Concerns	21	24
Closed Safety Concerns	16	18
Accepted Risks	8	9

This issue contains six new open safety concerns. The new safety concerns are (1) SRB (Solid Rocket Booster) and SSME (Space Shuttle Main Engine) thermal effects on ejected crewmen and the escape systems during ascent; (2) APU (Auxiliary Power Unit) exhaust combustion damage to TPS (Thermal Protection Subsystem) for orbiter 101; (3) inability to control the orbiter with two adjacent blown tires; (4) nonredundancy of static hydraulic fluid seals; (5) the R/SB (Rudder/Speed Broke) actuation system has several failure points which could cause loss of vehicle and crew; and (6) inability to accurately calibrate the air data inputs in the supersonic/transonic regions for first OFT (Orbital Flight Test) may affect orbiter approach and landing capability. A safety concern on the nonredundancy of SRB static seals is being prepared. Three previously open safety concerns, (1) fire detection and suppression provisions in the orbiter aft fuselage;

(2) shuttle potential collision with the tower on lift-off; and (3) SCA (Shuttle Carrier Aircraft) empennage/aft fuselage buffet with orbiter tailcone off, have been closed. The first concern was closed as an accepted risk; the other two were closed as "hazard controlled." Rationale for closures are contained in the section 4.0 and 5.0 summary writeups.

Action is being taken to resolve the 24 open safety concerns. The Johnson Space Center Safety Division is participating in the resolution of these concerns and will track them to satisfactory resolution. Safety assessments of the 18 closed safety concerns have been performed, resulting in the Johnson Space Center Safety Division's concurrence in closing the concerns. Rationale for the nine accepted risks has been assessed and is considered satisfactory.

Safety assessments of the aggregate risk are iterative and culminate in the release of a safety assessment document for each approach and landing test and orbital flight test mission. The capability to assess risks in aggregate at any particular time in a program is dependent upon program maturity. At this phase of the orbital flight test program, the design, operational analysis, and planning is not complete. After the critical design review, when the design is approved and operational data have been developed, a complete assessment will be accomplished, and open concerns identified. The Approach and Landing Test Project has completed the critical design review and is scheduled for the design certification review in December of 1976. The initial release of the safety assessment document for the approach and landing test project will be updated to support this review. The Johnson Space Center Safety Division considers the shuttle program aggregate risk for Orbital Flight Test and Approach and Landing Test acceptable considering the accepted risks and the action being taken for resolution of the open concerns.

[2-1] 2.0 <u>INTRODUCTION</u>

2.1 GENERAL

2.1.1 Safety Concern Definition. A safety concern is a potentially hazardous condition associated with a design or operation that has the potential of injury to personnel and/or damage to hardware. Each safety concern will require resolution by elimination, control, or acceptance of the risk. Safety concerns are identified at the shuttle element and system levels as a result of safety analyses, hazard analyses, special studies, failure mode and effects analyses, trade studies, etc.

2.1.2 Element Level Safety Concerns. Concerns identified at the element level are evaluated by the element contractors and NASA element project offices to determine design or procedural changes required or if changes are not feasible, to develop rationale for acceptance of the identified risks. These concerns are also provided to the Space Shuttle Program Office and the integration contractor, Rockwell/Space Division, for evaluation of system impact.

2.1.3 System Level Safety Concerns. System level safety concerns are identified through performance of system level safety analyses and evaluation of element analyses and assessments. The JSC Safety Division, in conjunction with the NASA project offices and the integration contractor, provide[s] recommendations to the Space Shuttle Program Office on design or procedural changes required or if changes are not feasible, the rationale for accepting the identified risks.

2.1.4 Selection Criteria for Major Safety Concerns. The JSC Safety Division, in conjunction with the NASA project offices and the integration contractor, recommends candidate safety concerns for this document. The candidate safety concerns are presented to the

SR&QA [Safety, Reliability, and Quality Assurance] Major Safety Concerns Screening Board, chaired by the JSC SR&QA Director, and is composed of Safety representatives from JSC, MSFC [Marshall Space Flight Center], Headquarters, and Rockwell/Space Division. This board selects the safety concerns to be contained in this document. A concern will be included in the document if any board member considers it appropriate. The concerns in this document, which represent only a small number of the concerns identified and being processed by the system and element contractors, were selected after considering factors such as:

    a.   Whether or not sufficient analysis and/or testing has been completed to determine the magnitude and probable occurrence of the potential risk.
    b.   Can the hazard be eliminated or the control verified?
    c.   Will the hazard probably not be eliminated or controlled because of other programmatic considerations?
    d.   Will the hazard resolution decision timing result in program impact?

2.1.5 Assessment of Aggregate Risk. In addition to the activity of identifying and resolving safety concerns, there is an ongoing safety assessment activity which considers the aggregate risk associated with each mission. This mission assessment will culminate in the documentation of the results of safety analyses approximately a year before both the Approach and Landing Test and the Orbital Flight Test missions and will be revised to support the DCR (Design Certification Review) and the Flight Readiness Review. The first issue of the ALT assessment is contained in JSC 10888, ALT (Approach and Landing Test Project Safety Assessment), dated June 7, 1976.
    Appendix A contains summary discussions of space shuttle design features that represent inherent risks and that were considered to be justified on the basis of past space program maturity and established technology. These features are considered acceptable risks by program management and constitute a baseline risk posture.

[2-2] 2.2 PURPOSE. This document provides risk management data for management overview purposes and facilitates a periodic independent assessment of the cumulative residual risks. The document also provides technical information and status on open concerns, and rationale for concern closures and accepted risks.

2.3 SCOPE. This document contains summaries of selected safety concerns, affecting the space shuttle system, space shuttle elements, shuttle carrier aircraft, and approach and landing test. Safety concerns associated with both ground operations and flight operations are included. Concerns associated with payloads, ground support equipment, major ground tests, and Government furnished equipment will be added in future issues of the document as identified. This document is published and presented to the program manager quarterly. Individual data packages for each concern are maintained current in the form of chronological records of actions taken and supporting documentation. The data in this issue are current, as of October 22, 1976.

2.4 ORGANIZATION. Section 3.0 provides summaries of open safety concerns. These summaries will remain in this section until the concern is closed or the risk is accepted by program management.
    A safety concern will be closed in one of two ways:
    a.   The hazard is eliminated.
    b.   The hazard is controlled.
    A concern is closed if the hazard is eliminated by design and the design has been approved by program management and documented. A concern is also closed if the haz-

ard has been reduced to an "acceptable level," (controlled hazard) and the design has been approved by program management and documented. The criteria for "acceptable level" are that the hazard is not catastrophic (time or means are available for corrective action) and the hazard is not critical (emergency action in a timely manner is not required).

A concern is identified as an accepted risk if the decision, supported by technical rationale, has been made and documented by program management.

Concerns are numbered sequentially and are categorized as shown below:

<u>Concern Designators</u>

<u>Designator</u>	<u>Category</u>
INTG--	Shuttle level concern
O--	Orbiter concern
ET-	External tank concern
SSME--	Space shuttle main engine concern
SRB--	Solid rocket booster concern
ALT-	Approach and landing test concern
P/L-	Payload concern
GSE-	Ground support equipment concern
MGT-	Major ground test concern
GFE--	Government-furnished equipment concern

[2-3] Section 4.0 contains summaries of safety concerns closed subsequent to the previous issue of this document. Titles of selected safety concerns that have been closed by the JSC Safety Division based on hazard elimination or control and have been reported as closed in a previous issue of this document are contained in this section.

Section 5.0 contains summaries of accepted risks and rationale for acceptance of the identified risks.

Appendix A contains summaries of the space shuttle design features. . . .

The JSC Safety Division, mail code NS, is responsible for the preparation and maintenance of this document. The name and FTS (Federal telecommunications system) phone number for the organization with prime responsibility for working each concern is provided after each concern title in sections 3.0, 4.0, and 5.0. The listed organization may be contacted for additional information. . . .

[5-1] 5.0 ACCEPTED RISKS. This section contains summaries of safety concerns and rationale for acceptance of the identified program risks. These accepted risks have resulted from program decisions made relative to space shuttle system concept tradeoffs, and detail design selections. These summaries will remain as a permanent part of the document.

A listing of the accepted risks recorded in this section are as follows. The organization with prime responsibility for the concern is also identified.

INTG-1 SSME (Space Shuttle Main Engine) Heat Exchanger Leakage (JSC Safety Division/525-3126)

INTG-4 On-Orbit Rescue During Early Orbital Flights (JSC Safety Division/525 3126)

INTG-7 Manual Guidance Capability During Ascent (JSC Safety Division/525-3126)

INTG-9 Emergency Drain System Provisions for ET (External Tank) (JSC Safety Division/525-3126)

INTG-14 Fire Detection and Suppression Provisions in the Orbiter Aft Fuselage on the Launch Pad (Rockwell/Space Division/985-1416)

0-2   Smoke Sensor Provisions in the Orbiter Crew Cabin for Orbiter 101 (JSC Safety Division/525-3126)

0-3   Lack of Redundant Elevon Hydraulic Actuators (JSC Safety Division/5253126)

0-7   Bird Impact with the Orbiter Windshield (JSC Safety Division/525-3126)

0-9   Thermal Windshield Panes (JSC Safety Division/525-3126)

Accepted risk safety concerns added and revised in this revision are discussed below. INTG-14 Fire Detection and Suppression Provisions in the Orbiter Aft Fuselage on the Launch Pad, was closed as an accepted risk and transferred from section 3.0, Open Safety Concerns, to this section.

[5-2]   ACCEPTED RISK
SAFETY CONCERN

IDENTIFICATION NO. INTG-1   DATE December 10, 1975
REVISED October 22, 1976

TITLE
SSME (Space Shuttle Main Engine) Heat Exchanger Leakage

SAFETY CONCERN

A leaking heat exchanger coil could result in the flow of hydrogen through the LO_2 (liquid oxygen) tank pressurization line to the pogo suppressor and the ET (external tank) LO_2 tank. Ignition of the resulting oxygen/hydrogen mixture could result in explosion and loss of the crew/vehicle.

DISCUSSION

Heat exchanger coil failure was identified in the SSME FMECA (failure modes, effects, and criticality analysis), RSS-8553-2, as a criticality category I failure which could result in explosion of the ET and loss of the crew and vehicle. Addition of the pogo suppressor aggravated the potential problem by lowering the heat exchanger outlet pressure.

MSFC recommended a modification to the LO_2 pressurant system, PCIN (program change identification number) S00927, which manifolded the heat exchanger outlets as a means of eliminating the effect on the ET. The change was disapproved by level II PRCB (Program Requirements Control Board). Rockwell/Space Division has performed a system level hazard analysis, MCR (master change record) 922, which concluded the baseline heat exchanger design is adequate and that the risk of failure is acceptably low. Heat exchanger leak checks will be performed periodically during the development phase.

JSC Safety Division evaluated the addition of shutoff valves to the heat exchanger inlet and outlet to permit isolation of a failed heat exchanger and recommended their incorporation to the PRCB on September 5, 1975. The PRCB did not concur with the recommendation.

Following the disapproval of PCIN S00927, the PRCB assigned actions to JSC/LA2 to (1) assess the Rocketdyne report resulting from the SSME Margin Review of the heat exchanger design; and (2) assess the Rocketdyne heat exchanger test program. JSC/NA was assigned an action to develop, in conjunction with MSFC, an SSME heat exchanger Product Assurance Control Plan.

In October 1975 MSFC issued a special task assignment to Rocketdyne for the performance of a design evaluation of a single tube heat exchanger coil. This action was apparently in response to the concern expressed by JSC/EA about the use of the bifurcated tube heat exchanger coil design.

JSC/EP presented, at the April 30, 1976, PRCB, an assessment of the Rocketdyne single tube heat exchanger coil study which indicated that although the design appears to be technically feasible, its incorporation would be too late to be an in-line block change. The design was recommended as a backup concept in case of technical problems with the baseline design.

[5-2a] STATUS/DISPOSITION
Accepted risk. Following the presentations on the outstanding SSME heat exchanger issues on April 30, 1976, the decision of the PRCB was to retain the existing baseline (bifurcated tube) heat exchanger coil design.
The rationale for acceptance includes the following:

a. Analysis shows that the heat exchanger coil design is adequate for 24 service lives.
b. Testing of 12 units is planned to substantiate design strength and service life capability.
c.   A heat exchanger product assurance control plan has been prepared to ensure that the necessary steps are taken during the design, development, manufacture and testing of the heat exchanger. An individual "Pedigree Report" will be provided with each heat exchanger.
d. Heat exchanger leak tests will be performed on a decreasing frequency until a frequency of once every 12 flights is achieved.

The following continuing actions are noted for information.
MSFC/Main Engine Project Office was assigned an action to investigate the feasibility of flowing LN_2 (liquid nitrogen) through the heat exchanger in place of LO_2 during main propulsion test. Cost and schedule impacts will be reported to the shuttle program manager.
On October 6, 1976, the SR&QA director delegated ND/J. A. Jones the responsibility for seeing that a viable plan is prepared and implemented which will assure added emphasis is placed on this critical item.

[5-3]   ACCEPTED RISK
                             SAFETY CONCERN

IDENTIFICATION NO. INTG-4   DATE December 19, 1975
                                             REVISED June 28, 1976
TITLE
On-Orbit Rescue During Early Orbital Flights

SAFETY CONCERN
In June 1973, Rockwell/Space Division, through MCR (master change record) 210, identified rescue capabilities and deficiencies. The first vertical flight vehicle (Orbiter 102) will fly all six of the orbital flight test missions and the early operational missions before a rescue orbiter will be on-dock at KSC. Thus, the level I requirement (attachment A to JSC 07700, Volume 1, Space Shuttle Program Requirements Document) for rescue cannot be met, and any failure precluding orbiter return from orbit will result in loss of crew. In August 1973, JSC Safety Division identified various shuttle conditions which would require rescue. In November 1973, Rockwell/Space Division presented a summary of the issues to date and the first proposal to fly a command module as a rescue vehicle. In March 1974, KSC summarized the impacts of the following options available to meet the rescue requirement.

a. Retain the capability for a CSM (command and service module) launch.
b. Install a CSM in the orbiter payload bay.
c. Delay the Orbiter 102 launch until a rescue orbiter is ready.
d. Compress the schedule to bring a rescue orbiter on-dock simultaneously with Orbiter 102.

The Space Shuttle Program Office subsequently performed a study to evaluate the pros and cons of the various rescue options.

STATUS/DISPOSITION
Accepted risk. The JSC Space Shuttle Program Office presented the various rescue options in a briefing to NASA Headquarters in May 1974. A recommendation was accepted that rescue capability not be provided for the early orbital flights. The recommendation was based on the rationale that the probability of a failure that would preclude a safe return from orbit is sufficiently low to allow the risk to be accepted.

[5-4]   ACCEPTED RISK
                                    SAFETY CONCERN

IDENTIFICATION NO. INTG-7   DATE December 10, 1975
                                             REVISED October 22, 1976
TITLE
Manual Guidance Capability During Ascent

SAFETY CONCERN
Lack of manual guidance capability during ascent could result in loss of vehicle and crew in the event of a malfunction of the primary automatic guidance control system.

DISCUSSION
A requirement to provide digital processed manual control inputs to the computers for all flight phases was contained in paragraph 3.3.1.2.3.3.2 of JSC 07700, Volume X; however, the paragraph lacked definitive requirements.
In the absence of definitive requirements for manual guidance control during ascent, a Level II change request (SO1575, dated April 1975) was initiated to delete manual guidance control during that mission phase. This was intended to simplify the software and display requirements, but was withdrawn prior to PRCB (Program Requirements Control Board) disposition.
The Level II change request was amended to delete manual guidance only during the nominal ascent mode. The Safety Division reviewed and approved this change based on the limited usefulness of this particular ability. This change request was subsequently approved by the PRCB on August 21, 1975. As a result, the crew had only the capability to manually throttle the SSME's (space shuttle main engines) during mated ascent. Manual guidance was provided during abort situations.
In February 1976, the CCB (Configuration Control Board) established that the software will be designed to a single fault tolerant baseline for the primary system during the ascent phase. Manual guidance and throttling capabilities were deleted as a result of this decision.

STATUS/DISPOSITION
Accepted risk. The deletion of manual guidance and throttling was reevaluated in May 1976 by the OFT (orbital flight test) Baseline Review Board. Reinstatement was discussed at that time, but final disposition was deferred, pending the results of further

studies by the Spacecraft Software Division and the Avionics System Engineering Division. On June 15, 1976, the PRCB reversed the CCB decision and gave specific direction to retain manual throttling during ascent and manual guidance and throttling during aborts as Level II requirements. The JSC Safety Division concurred in the decision to retain these contingency capabilities. Manual guidance will be provided only during abort situations. The software will not be mechanized to return to automatic guidance following manual guidance selection. The imposed requirements are acceptable to define manual control capability during ascent.

[5-5]   ACCEPTED RISK
                               SAFETY CONCERN

IDENTIFICATION NO. INTG-9   DATE December 10, 1975
                                               REVISED October 22, 1976

TITLE
Emergency Drain System Provisions for ET (External Tank)

SAFETY CONCERN
    There is no provision for draining the LO_2 (liquid oxygen) and LH_2 (liquid hydrogen) from the ET except through the orbiter feedlines and the propellant lines in the aft fuselage. The concern is that detanking during an emergency must be accomplished through a system which may be involved in the emergency.

DISCUSSION
    This concern was identified in the Space Shuttle External Tank Preliminary Hazards Analysis Report, MMC-ET-RA01-0, Hazard No. 2.003A, dated August 30, 1974.
    The baseline orbiter/ET configuraiton [sic] requires that propellant transfer operations (fill and drain) be accomplished through the orbiter feedlines. A significant leak in either the ET or orbiter plumbing would require detanking through the leaking component, thereby increasing the potential for fire/explosion. In the event of a fire, it would be necessary to drain propellants through lines which pass through the fire affected area.
    The propellant drain rates are known to be slower than the fill rates. The time required to drain the tanks through the orbiter during an emergency backout operation may be excessive.
    The baseline design of the ET propellant fill and drain system was accepted at the Orbiter 102 PDR (preliminary design review) in January 1975. The issue was discussed at the ET critical design review held November 10 through 14, 1975. The shuttle system contractor submitted a RID (review item dispostion [sic]) which recommended that the hazard be submitted for closure as an acceptable residual hazard. The RID (P-27) was disapproved with the recommendation that the issue be submitted to the System Safety Subpanel for disposition. The concern was discussed at the subpanel meeting on January 29, 1976. A decision was made to reflect this concern as an accepted risk, based on the December 1975 discussion between the Space Shuttle Program Manager and the JSC Safety Division Chief.

STATUS/DISPOSITION
    Accepted risk. The Space Shuttle Program Manager made a decision at the December 1975 review of JSC 09990 to reflect this concern as an accepted risk. The JSC Safety Division reviewed the following rationale for acceptance of the risk and considers the rationale satisfactory:

a.   Relocating fill/drain system to ET would require the same number of operating components so that the reliability would not be improved by relocation.

[5-5a] b. Relocating fill/drain system to ET would move the control valves, etc., approximately 20 feet on the vehicle. Any major incident within the orbiter that might require offloading the ET would be transmitted this 20 feet in less time (approximately 10 minutes) than the ET could be drained—regardless of the location of the fill/drain system (orbiter or ET).

c.   Relocating fill/drain valves and components to ET would cause these parts to be thrown away with the ET.

Analyses of ET, orbiter main propulsion system, and space shuttle main engine hazards which potentially affect this concern are continuing by MSFC, JSC, and Rockwell/Space Division as part of the on-going hazard analysis activity. Component tests, single engine tests[, and] main propulsion system tests will also be analyzed to provide confidence in this system. KSC emergency procedures will be reviewed when they become available.

[5-6]   ACCEPTED RISK
                         SAFETY CONCERN

IDENTIFICATION NO. INTG-14   DATE March 8, 1976
                                          REVISED October 22, 1976

TITLE
Fire Detection and Suppression Provisions in the Orbiter Aft Fuselage on the Launch Pad

SAFETY CONCERN
No fire detection or suppression is provided in the aft fuselage. Leakage of flammable fluids and/or oxidizers in excess of specification allowable leakage may create a fire/explosion potential in the orbiter aft fuselage on the launch pad. The primary concern is during main engine operation when an LO_2 (liquid oxygen) system failure may result in an uncontrolled LO_2 supported fire.

DISCUSSION
The baseline orbiter design does not contain any provision for fire detection/suppression within the aft fuselage. In the event of a major LO_2 fed fire in the main engine compartment, the existing fixed facility fire detection/suppression capability may not provide adequate response to prevent flight vehicle loss and damage to the mobile launch platform/launch pad. A GN_2 (gaseous nitrogen) purge is provided in the aft fuselage to produce an inert atmosphere and prevent the accumulation of hazardous gases due to leakage within specification. It would not provide protection from the results of large propellant leaks of the type that could result from a major mechanical failure of an engine. A hazardous gas detection system samples the compartment atmosphere during prelaunch to detect the presence of N_2H_2 (hydrazine), MMH (monomethylhydrazine), N_2O_2 (nitrogen tetroxide), H_2 (hydrogen), [and] O_2 (oxygen). Detection of out-of-limit conditions indicating more than allowable leakage could result in a launch scrub and detanking of propellants.

PCIN (Program Change Identification Number) S01581 proposed the addition of a level II requirement for KSC ground support of fire detection and water deluge systems for the orbiter aft fuselage. MCR (Master Change Record) 1994 was issued to provide design concepts and impacts for the addition of fire detection and deluge systems. The study recommended against incorporation of any of the systems evaluated, concluding that reliable detection and response between SSME (Space Shuttle Main Engine) ignition

($\cong$ T–3.5 seconds) and launch commit was doubtful and that none of the deluge concepts added further assurance of safety for the facility, vehicle, or crew. In an effort to augment launch pad fire suppression capability, the PRCB (Program Requirements Control Board) directed continued investigation of a manually actuated, fuselage penetrating, water deluge system. KSC objected to the use of such a system without additional fire detection capability in the aft fuselage. The effort was terminated due to cost and weight impacts.

JSC Safety Division recommended further consideration of a fire suppression method using Halon 1301 and was directed to perform a feasibility study. The resulting PCIN S01581A proposed substitution of Halon 1301 for the aft fuselage GN_2 purge late in the countdown to produce a 40 percent (volumetric) concentration of Halon 1301 at liftoff. The Halon 1301/nitrogen mixture was shown to provide approximately 4.5 times the inerting capability of nitrogen alone for propellant leakage within specification. Negative factors included cost, weight, excessive delta pressure, and potential environmental impact. The PRCB disapproved PCIN S01581A.

[5-6a] The JSC Safety Division position is that the availability of a water deluge system during and subsequent to propellant loading would provide an additional launch pad safety margin. The Halon 1301, carried in the aft fuselage during the initial ascent phase[,] would also provide protection against small fires which could lead to larger fires.

STATUS/DISPOSITION

Accepted risk. The Space Shuttle Program Manager accepted this risk at the August 20, 1976, PRCB when a decision was made not to provide a fire detection or suppression capability in the orbiter aft fuselage.

The rationale for his accepting this risk included the following: The GN_2 purge will prevent accumulation of hazardous gases within specification. The hazardous gas detection system will provide detection of out-of-limit hazardous conditions in the aft fuselage, permitting termination of the countdown and detanking of propellants. In the event of a fire, the facility water deluge system will provide some fire suppression capability. The use of Halon 1301 does not appear to offer a sufficient improvement over the present GN_2 system to warrant the attendant orbiter and facility design changes and the additional costs. The addition of an aft fuselage water deluge sytem [sic] was disapproved for these same reasons.

[5-7]   ACCEPTED RISK
                               SAFETY CONCERN

IDENTIFICATION NO. 0-2   DATE December 10, 1975
                                               REVISED October 22, 1976
TITLE
Smoke Sensor Provisions in the Orbiter Crew Cabin for Orbiter 101

SAFETY CONCERN

The baseline orbiter 101 design consisted of a single smoke sensor installed on the orbiter flight deck, and a single sensor installed in the environmental control life support system equipment bay located beneath the middeck floor of orbiter 101. A single sensor failure could allow significant damage to occur in either area before a fire is detected.

DISCUSSION

Orbiter 101 Delta PDR (preliminary design review) RID (review item disposition) No. 05.04.27 (R-1), "Smoke Detector Redundancy," identified lack of redundancy for smoke detection for orbiter 101. It is the position of the JSC Safety Division that sensors that detect emergency conditions such as fire should be redundant. The single sensors are not redundant because they do not sample the same volume of air simultaneously.

STATUS/DISPOSITION

Accepted risk. The Orbiter Project Office accepted the risk for orbiter 101 and the JSC Safety Division concurred based on the following rationale:

a.   The short length of approach and landing test flights.
b.   Emergency breathing apparatus available to the flight crew.
c.   The lesser amount of cabin avionics on orbiter 101.
d.   The major wiring and cost impact of relocation of detectors.

This decision was made at the August 28, 1975, techincal [sic] status review.

[5-8]   ACCEPTED RISK
                                        SAFETY CONCERN

IDENTIFICATION NO. 0-3   DATE December 10, 1975
                                                REVISED October 22, 1976
TITLE
Lack of Redundant Elevon Hydraulic Actuators

SAFETY CONCERN

Normal operation of the four orbiter elevons is required for flight control. Each elevon is powered by a single linear hydraulic actuator, the failure of which would result in the loss of the orbiter and onboard personnel.

DISCUSSION

The original orbiter baseline design employed dual tandem elevon actuators. This concept provided flight control redundancy consistent with normal commercial and military aircraft designs. Weight reduction requirements led to a JSC decision in 1973 to develop and use a single actuator having a number of single failure modes that could cause loss of an orbiter. The resultant actuator, while based on existing electro-hydraulic actuator designs, is a unique design with no previous operational experience. A January 8, 1974, proposal to further reduce weight by changing to an electromechanical actuator system was rejected by the Orbiter Configuration Control Board.

The JSC Safety Division investigated civil and military requirements for control surface operational redundancy. It was found that single actuators are often allowable, but not in criticality I applications. Research of 289 military aircraft accident/incident reports for the period 1965 through 1971 related to hydraulic actuators revealed that 51.6 percent of all actuator failures were structural failures. This increases to 77.1 percent if seal failures, which are considered as structural failures during Rockwell/Space Division failure mode and effects analyses, are included. (It should be noted that most structural failures resulted from metal fatigue.)

The Safety Division also reviewed the structural stress analysis approach being used in orbiter actuator design with no specific weaknesses noted. Several F-4 aircraft, however, which were analyzed and tested prior to flight were lost due to structural failure of a single hydraulic system component.

The conclusion and recommendation of the Safety Division was that the dual tandem actuator concept should be retained. This position was rejected by the project manager and the single actuator concept was implemented. After this decision, SR&QA recommended that special controls and actions be imposed through a product control plan to provide better assurance that the actuators will perform properly.

[5-8a] STATUS/DISPOSITION

Accepted risk. The technical direction to baseline the single elevon actuators was contained in JSC letter BC42-73/87. The JSC Director concurred with this decision. The rationale for acceptance of the single elevon actuators are:

a. Establishing adequate design margins
b. Inclusion of the actuators in the orbiter structure fracture control plan
c. Implementation of a special product control plan for the actuators
d. Extensive qualification testing
e. Low incidence of failure history for similar designs
f. Minimal life cycle requirements
g. Proven supplier
h. Redundant external dynamic seals.

The JSC Safety Division concurs with the closure of this concern as an accepted risk. Due to problems noted during developmental, certification, and vehicle testing, the actuator design, production controls, and test programs are being reevaluated by Rockwell/Space Division, E&D, and SR&QA to provide the highest possible confidence in the single actuator design.

The NASA chief engineer is also conducting an independent assessment of the adequacy of the present design. Results of these activities and implementation of any corrective actions deemed necessary are expected to be complete by January 1, 1977.

[5-9]   ACCEPTED RISK
                              SAFETY CONCERN

IDENTIFICATION NO. 0-7   DATE December 10, 1975

TITLE
Bird Impact with the Orbiter Windshield

SAFETY CONCERN

There is a possibility of a bird strike at low altitudes resulting in the penetration of the orbiter windshields.

DISCUSSION

The windshield glass system specification does not contain a requirement for the orbiter windshield to be designed for bird impact. Rockwell/Space Division completed an analysis on August 24, 1973, which concluded that "the orbiter windshield could withstand the impact of a four pound bird at 230 knots." The analysis also showed that the probability of no catastrophic window failure because of bird impact was 0.0005 for the life of the orbiter.

STATUS/DISPOSITION

Accepted risk. Based on the low probability of a bird impact and the inherent capability of the orbiter window system to withstand bird impact, the Orbiter Project Office directed Rockwell/Space Division in letter ES2, dated November 29, 1973, to proceed with the window design within the previously established specification.

[5-10]   ACCEPTED RISK
                                    SAFETY CONCERN

IDENTIFICATION NO. 0-9   DATE December 10, 1975

TITLE
Thermal Windshield Panes

SAFETY CONCERN
    During the entry phase of the orbiter, a loss of the thermal (exterior) window pane
could cause loss of crew and orbiter. The redundant (middle) pane does not have the total
heat capacity of the thermal (exterior) pane. Loss of the thermal (external) pane could
result in a loss of the redundant (middle) pane and subsequent loss of the pressure
(inner) pane and orbiter cabin atmosphere.

DISCUSSION
    Rockwell/Space Division reported at the TSR (technical status review) on November
8, 1974, that the 100 in2 hole requirement in the orbiter CEI (contract end item) was
unrealistic. Tests at Corning Glass in Canton, New York, showed that the panes would have
to crack in at least three directions to produce a hole. It was also shown that crack geometry was such that crack segments would be held by window pane retainers and remain in
place. In this case, thermal integrity would not be completely lost. It was concluded that
under these conditions the redundant (middle) pane is capable of sustaining a crack failure of the thermal (exterior) pane and that the requirement of being able to withstand
the thermal load from a 100 in^2 hole in the thermal (exterior) pane is not needed.

STATUS/DISPOSITION
    Accepted risk. Analysis performed by Rockwell/Space Division and presented at the
November 20, 1974, TSR showed that the redundant (middle) pane is capable of sustaining a crack failure of the thermal (exterior) pane without losing the thermal entry capability. The Orbiter Level III Configuration Control Board on November 11, 1974,
approved Rockwell/Space Division's recommendation to "eliminate consideration of a
hole appearing in the thermal (exterior) pane at fracture during entry." The change was
approved by specification change notice 01-0124 to the Orbiter CEI MJ070-0001-1A.

[A-1]   APPENDIX A
                           SPACE SHUTTLE DESIGN FEATURES

    This appendix contains summary discussions of space shuttle design features that rep-
resent inherent risks and that are considered to be justified on the basis of past space program maturity and established technology. The discussions are as follows:

    1. Intact Abort
    2. Unpowered Landing
    3. Outward Opening Hatch
    4. SRB (Solid Rocket Booster) Thrust Termination
    5. Manned First Vertical Flight

[A-2]   DESIGN FEATURE

IDENTIFICATION NO. 1   DATE December 10, 197
                                         REVISED March 8, 1976

TITLE
Intact Abort

DESIGN FEATURE
    Return to launch site, abort once around, and abort to orbit intact abort modes are
provided.

DISCUSSION
    System, operational, and payload interface requirements have been established to
provide the necessary features for overall safe launch and landing capability with established intact aborts. Critical flight vehicle subsystems (except primary structure, thermal
protection system, and pressure vessels) are required to be at least fail safe. The design
adequacy of the structure and thermal protection systems is enhanced by the use of adequate design margins and testing. During the launch countdown sequence, the main
engines are ignited approximately 3.5 seconds before launch commit. This provides time
to assure proper main engine operation and allows the engines to reach 90 percent thrust
level before launch commit. Pogo suppression devices have been designed for the space
shuttle main engines. Intact abort capability is based on the capability of the combined
vehicle to continue flight through separation of the SRB's (solid rocket boosters). The
SRM's (solid rocket motors) have been designed with the same fail safe redundancy
required for the rest of the vehicle. The failure histories of other large solid propellant
motors were reviewed in the design process. Where the specific failures were found to be
applicable to the shuttle SRM design, increased factors of safety were used. The SRM's will
be tested to verify that their performance characteristics meet the required specifications.
    The following landing operational capabilities during intact abort modes enhance
orbiter and crew safety. The orbiter has an automatic and manual landing system. The
aerodynamic cross-range capability to return the orbiter to the launch site after one revolution is provided. The orbiter is capable of landing with the full 65,000 lb. payload under
specified landing constraints. In the event of a launch abort, orbiter landing safety is
enhanced by dumping orbiter and payload propellants.
    This feature provides the capability for orbiter center-of-gravity adjustment, reducing
landing weight, and reducing the quantity of potentially hazardous fluids on board during
landing. The program requirements for payloads to have self-contained provisions against
payload generated hazards enhances safety during both the launch and landing phases.
    During the orbital flight test phase, crew ejection seats have been provided. The ejection
seats will provide an additional means of crew escape during launch and landing. The seats will
provide escape capability at altitudes below 75,000 feet and velocities below Mach 2.7. The seats
will also provide crew escape on the ground in the event of an emergency after touchdown.

[A-3]   DESIGN FEATURE

IDENTIFICATION NO. 2   DATE December 10, 1975

TITLE
Unpowered Landing

DESIGN FEATURE
    The energy controlled landing design feature allows the orbiter to glide to a safe land-
ing at primary and contingency landing sites.

DISCUSSION

The energy controlled landing design feature requires the orbiter to be capable of aerodynamic flight which is controlled by elevons, rudder, speed broke and body flaps. Initial attitude control during entry is provided by RCS (reaction control subsystem) jets until dynamic pressure begins to build on the aerodynamic surfaces which then control in combination with the RCS. The RCS is then deactivated and aerodynamic surfaces provide control to landing. At 70,000 feet altitude, the TAEM (terminal area energy management) phase begins where the orbiter maneuvers to dissipate energy using the speed brake until the final approach begins at 10,000 feet. During final approach, the glide slope and altitude are controlled by modulation of the speed brake.

Energy controlled landing has always been a baseline feature of the shuttle program. ABE's (airbreathing engines) were also a requirement for design mission 2. ABE's were not required for design missions 1 and 3 if maximum payloads were required. ABE's were to provide 15 minutes loiter time at 10,000 feet altitude to allow operational assessment of conditions at completion of reentry. ABE's were deleted from the shuttle program in 1972 to simplify orbiter design and more than double payload for design mission 2. Design simplicity results from structural considerations and not having to space rate or isolate an air-breathing propulsion system.

RCS, avionics and hydraulic power redundancy, the large landing foot print, and similar landing techniques verified by other aircraft enhance the energy controlled landing feature of the orbiter. The aft RCS modules which provide orbiter attitude control during the entry phase of landing are redundant. Each module has multiple thrusters in each control axis, and the RCS is connected to the [Orbital Maneuvering System] propellant supply for supplemental propellant. The GN&C (guidance, navigation and control) system has both automatic and manual control capability. The GN&C system has redundant inertial measurement units (3), computers (4), and TACAN (tactical air navigation) receivers (3). The 3-sigma GN&C error is well within the energy management and landing foot print capability. The possible Earth landing area is 1085 nautical miles crossrange by 5000 nautical miles downrange. Three independent auxiliary power units, hydraulic pumps and supply lines provide aeroflight control pressure to the elevon, rudder, speed brakes, and body flap actuators. These aerodynamic control surfaces on the orbiter are standard aircraft type control surfaces used by all high performance commercial and military aircraft. The energy controlled landing is not a new concept as several experimental aircraft have used this landing technique, with the X-15 and X-24B being two examples.

[A-4]   DESIGN FEATURE

IDENTIFICATION NO. 3   DATE December 10, 1975
                                      REVISED March 8, 1976

TITLE
Outward Opening Hatch

DESIGN FEATURE

The outward opening side hatch allows rapid egress of onboard personnel during both prelaunch and postrollout phases of a mission.

DISCUSSION

The 40-inch diameter hatch is located on the left side of the orbiter at the middeck level. The hatch can be opened either from the inside or the outside. Opening from the inside is accomplished by unlocking the actuator and rotating an inplace handle. The handle is protected from inadvertent operation during flight by an actuator lock. Opening from the outside is accomplished by removing a thermal protection system plug and inserting a GSE (ground support equipment) tool that provides for unlocking and unlatching. Hatch latching is achieved by 18 overcenter latches located around the circumference of the hatch. The latches are interconnected by two rigid linkages, each of which includes 9 of the 18 latches. The linkages are moved by a bell crank and achieve latching/unlatching. The hatch actuation, latching, and locking mechanisms are similar to the proven Apollo spacecraft design. The hatch includes dual pressure seals.

The outward opening design feature of the orbiter ingress/egress hatch was primarily influenced by previous spacecraft experience that demonstrated the need for rapid crew egress during ground contingencies. The orbiter will be exposed to many of the same prelaunch hazards as on previous spacecraft. With the orbiter passenger-carrying capability, more personnel will be involved, making a rapid egress capability even more important.

Opening of the hatch in-flight is precluded by an actuator lock and fail safe latch design. Any 9 of the 18 hatch latches are adequate to carry the delta-pressure load on orbit and to provide sufficient pressure integrity to allow a safe orbiter return. During the orbital flight test phase, proper latch position will be verified by performing two electrical continuity tests (one for each linkage) from the outside by GSE. During the orbiter operational phase, proper latch position after premission closeout will be verified visually from the inside. The hatch dual pressure seals will be verified after premission closeout by a pressure integrity test.

[A-5]   DESIGN FEATURE

IDENTIFICATION NO. 4   DATE December 10, 1975
                                              REVISED March 8, 1976
TITLE
SRB (Solid Rocket Booster) Thrust Termination

DESIGN FEATURE
    The SRB is designed to provide 120 seconds of thrust for the shuttle without thrust termination capability.

DISCUSSION
    This design feature requires the SRB to provide thrust until propellant depletion. The approximately 1,110,000 pounds of propellant per booster provides 120 seconds of thrust at varying thrust levels to maintain acceptable shuttle loads. Thrust termination capability is not required for nominal and intact abort missions and was only considered for contingency abort situations. Thrust termination would have neutralized SRB thrust by initiating SRB thrust through the forward dome.

    The requirement for SRB thrust termination was removed from the shuttle program in April 1973. SRM's (solid rocket motors) are inherently reliable, and the SRB's have been designed with increased design margins. The failure histories of other large solid propellant motors were reviewed in the design process. Where the specific failures were found to be applicable to the shuttle SRM design, increased factors of safety were used. The SRM's will be tested to verify that their performance characteristics meet the required specifications. Contingency aborts, where SRB thrust termination could be used, are not required as stated in JSC 07700, Volume X, unless

    a.   Thrust from two or more SSME's (space shuttle main engines) is lost.
    b.   TVC (thrust vector control) from two or three SSME's is lost.
    c.   SRB TVC on two or more axes is lost.

[A-6]   DESIGN FEATURE

IDENTIFICATION NO. 5   DATE December 10, 1975

TITLE
Manned First Vertical Flight

DESIGN FEATURE
    The space shuttle is designed to be manned on the first vertical flight.

DISCUSSION
    The shuttle design and operational planning is based on a manned first vertical flight.
Accommodations for two crewmen, as well as the capability for manual system control, will
be included in the shuttle design. Previous manned spacecraft programs have flown
unmanned developmental missions before committing to a manned flight.
    Manning of the shuttle first vertical flight increases the probability of mission success
and decreases the probability of vehicle loss. Man in the loop provides significant backup
capability in evaluation, checkout, and operation of shuttle subsystems. Manual control
during ascent or entry could prevent loss of vehicle in the event of a failure in the primary automatic systems. Reconfiguration of essential systems or overriding of automatic systems could be accomplished if conditions are not consistent with premission planning.
Mission completion could be accomplished in the event of a total loss of communications.
    Mission profiles for the first vertical flight have been selected for maximum safety.
Ascent and entry trajectories will feature low dynamic pressure, low structural and thermal loads and control system gains set to enhance control capability. Launch and landing
will not be performed under adverse environmental conditions. The optimum center of
gravity will be established with low payload weights. The mission will terminate at Edwards
Air Force Base which has maximum usable landing space. Additional consumables will be
allocated for contingencies. Ejection seats and pressure suits will be provided to allow ejection at altitudes up to 75,000 feet.

Document II-30

Document title: Associate Administrator for Space Transportation Systems, NASA, to
Administrator, NASA, "STS-1 Mission Assessment," May 12, 1981, with attached:
"Postflight Mission Operation Report," No. M-989-81-01, Foreword and pp. 1–3, 6–7, 9,
11, 13–14.

Source: NASA Historical Reference Collection, NASA History Office, NASA
Headquarters, Washington, D.C.

*More than two years behind its original scheduled, and after a two-day delay because of computer prob-
lems, the first Space Shuttle launch took place on April 12, 1981. (By coincidence, this was twenty
years to the day after the first human spaceflight by Soviet cosmonaut Yuri Gagarin.) This report is a
top-level summary of the successful mission's results; only the first part of the report appears here, without the five figures and a table.*

Postflight
Mission Operation Report
No. M-989-81-01
May 12, 1981

TO: A/Administrator

FROM: M/Associate Administrator for Space
Transportation Systems

SUBJECT: STS-1 Mission Assessment

The first flight of the four-flight Orbital Flight Test phase of the Space Shuttle Program has been accomplished. The STS-1 mission was launched from Pad A of Launch Complex 39, Kennedy Space Center, Florida, on April 12, 1981. The landing was on dry lake bed Runway 23 at Edwards Air Force Base, California, on April 14, 1981.

This letter formally submits the STS-1 Postflight Mission Operation Report (MOR). It includes my previously submitted formal statement of the mission's objective (see STS-1 Prelaunch MOR No. M-989-81-01, dated April 3, 1981), on which I have annotated my formal assessment of the mission's success.

The mission's objective to demonstrate a safe ascent and return of the Orbiter and crew was accomplished, and I judge the mission to have been a success.

John F. Yardley

****

Postflight Mission Operation Report

[no page number]   <u>Foreword</u>

MISSION OPERATION REPORTS are published expressly for the use of NASA Senior Management, as required by the Administrator in NASA Management Instruction HQMI 8610.1A, effective October 1, 1974. The purpose of these reports is to provide NASA Senior Management with timely, complete, and definitive information on flight mission plans, and to establish official Mission Objectives which provide the basis for assessment of mission accomplishment.

Prelaunch reports are prepared and issued for each flight project just prior to launch. Following launch, updating reports for each mission are issued to keep General Management currently informed of definitive mission results as provided in NASA Management Instruction HQMI 8610.1A.

Primary distribution of these reports is intended for personnel having program/project management responsibilities which sometimes results in a highly technical orientation. The Office of Public Affairs publishes a comprehensive series of reports on NASA flight missions which are available for dissemination to the Press.

MISSION OPERATIONS REPORTS for the Space Shuttle Program Orbital Flight Tests are comprised of a PRELAUNCH REPORT and a POSTFLIGHT REPORT for each test flight. In addition, the ORBITAL FLIGHT TEST REFERENCE DOCUMENT, issued on a one-time basis, describes equipment, facility, and management systems common to all of the test flights. . . .

[1]   INTRODUCTION

STS-1 mission, the first of four manned orbital flights planned for the Orbital Flight Test (OFT) phase of the Space Shuttle Program, was completed on April 14, 1981.

The Space Shuttle is the prime element of the U.S. Space Transportation System (STS) for space research and applications in future decades. The primary goal of the OFT phase of the Space Shuttle program is to demonstrate a capability for routine prelaunch, launch orbital, entry, approach, landing, and turnaround operations. The Space Shuttle flight system for OFT consists of the Orbiter Columbia with its three main engines, an external tank [ET] and two solid rocket boosters. The Orbiter, its main engines and the retrievable booster components are reusable elements; the tank is expended on each launch.

The first OFT flight was designed to maximize crew and vehicle safety by reducing ascent and entry aerodynamic loads on the vehicle as much as possible. Each successive flight will be planned to expand the operating envelope including increased ascent and entry loads and varied launch and entry payload weight and center-of-gravity locations.

The first flight of the OFT program was designated STS-1. This STS-1 POSTFLIGHT REPORT assesses the achievement of the STS-1 mission objective and provides a detailed description of the STS-1 flight. Descriptions of Space Shuttle Flight vehicles, systems, facilities, mission support, and mission management, which are common to all of the OFT missions, are contained in the ORBITAL FLIGHT TEST REFERENCE DOCUMENT (MOR M-989-811), dated March 2, 1981.

[2]   NASA MISSION OBJECTIVE FOR STS-1

The NASA mission objective for the STS-1 mission is to demonstrate a safe ascent and return of the Orbiter and crew.

John F. Yardley
Associate Administrator for Space Transportation Systems
Date: April 3, 1981

                     ASSESSMENT OF THE STS-1 MISSION

The STS-1 mission is judged to have been a success.

John F. Yardley
Associate Administrator for Space Transportation Systems
Date: 5/12/81

[3]   STS-1 MISSION DESCRIPTION GENERAL

General

The first flight of the Space Shuttle was completed at 10: 20: 58 a.m., PST, on April 14, 1981, with touchdown of the Orbiter Columbia. The landing was at Edwards Air Force Base (EAFB), California, 54 hours, 20 minutes, and 54.1 seconds after launch from the Kennedy Space Center (KSC), Florida. A profile of the STS-1 mission, with event times shown in actual ground elapsed time (GET) in hours: minutes: seconds, is illustrated in Figure 1.

The commander of the mission was John W. Young, and the pilot was Robert L. Crippen.

The data presented in this document are based on quick-look reports. Detailed analysis of all data is continuing, and a final evaluation report prepared by the Integrated Systems Evaluation Team will be issued by the Johnson Space Center (JSC) Space Shuttle Program Office prior to the flight readiness review for the STS-2 mission.

Final Countdown (Figure 2)

The STS-1 mission was launched at 7: 00: 03.9 a.m., EST, on April 12, 1981, following a scrubbed attempt on April 10. The countdown on April 10 proceeded normally until T–20 minutes (20 minutes prior to launch in the countdown sequence) when the Orbiter general purpose computers (GPC's) were scheduled for transition from the vehicle checkout mode to the vehicle flight configuration mode. The launch was held for the maximum time and scrubbed when the four primary GPC's would not provide the correct timing for the backup flight system GPC. Analysis and testing indicated the primary set of GPC's provided incorrect timing to the backup flight system at initialization and caused the launch scrub.

The problem resulted from a Primary Ascent Software System (PASS) skew during initialization. The PASS GPC's were reinitialized and dumped to verify that the timing skew problem had cleared. During the second final countdown attempt on April 12, transition of the primary set of Orbiter GPC's and the backup flight system GPC occurred normally at T–20 minutes.

The launch pad damage from the STS-1 launch was less than predicted. All launch facilities, systems, and support equipment performed as designed.

|pages 4 and 5 were Figures 1 and 2| [6] Ascent

The STS-1 mission was launched from Pad A of Launch Complex 39 at KSC on an azimuth of 66.96 degrees. Lift-off was achieved with both solid rocket boosters (SRB's) igniting and the Space Shuttle main engines (SSME's) operating at rated power level (100%). The SSME's were throttled down to 65% thrust level for maximum dynamic pressure control and back up to 100% thrust level at the predicted times. The maximum dynamic pressure of ascent was encountered at a GET 56 seconds. (The 00: 00: 00 point (hr: min: sec) of GET is SRB ignition.) The SRB separation command was initiated at 00: 02: 10.4 GET following SRB burnout.

Second stage flight utilized the SSME's at 100% thrust level until 3 g's were reached, and 3 g's were maintained by SSME throttling until approximately 6 seconds before main engine cutoff (MECO) when the engines were throttled to 65% where they remained until MECO. MECO occurred at 00: 08: 34.4 GET, and separation from the ET occurred 23.7 seconds later. After separation of the ET, the Orbiter was inserted into an orbit of 133.7-n. mi. [nautical mile] apogee and 132.7-n. mi. perigee, with a 40.3-degree inclination. This orbit was achieved by two orbital maneuvering system firings (OMS-1 and OMS-2). The 86.3-second OMS-1 manuever [sic] was initiated at 00: 10: 34.1 GET with the 75.0-second OMS-2 maneuver occurring at 00: 44: 02.1 GET at the apogee of the orbit resulting from the OMS-1 burn.

The ascent trajectory was as planned with all events up through payload bay door opening and radiator deployment occurring normally. Prior to the initial OMS burn, the chamber pressure measurements for both engines were reading off-scale high on the ground. The crew indicated proper operation of the onboard indicators. This was traced subsequently to a ground calibration problem.

Some real-time data were lost, as expected, during SRB operations because of signal attenuation due to the SRB plume. Other communications losses during orbit number one were encountered at the IOS (Indian Ocean Station) where two-way S-band lockup was not obtained, at the Yarragadee Tracking Station in western Australia where UHF voice was intermittent, and at the Orroral Valley Tracking Station in eastern Australia where no S-band downlink voice was received. Communications subsequent to orbit one were excellent.

The main propulsion system performed normally with two apparent transducer failures and an unexpected rise in the pogo precharge pressure. Data indicate the precharge pressure exceeded 1425 psia and this situation is being analyzed. This is not believed to be a problem since engine operation was satisfactory for the remainder of the ascent burn. [7] The auxiliary power units (APU's) operated as expected with no apparent problems. The hydraulic systems also operated normally, although all three water spray boiler and vent temperatures were off-scale low. Additionally, lubrication oil temperatures were lighter than expected. These conditions were caused by freezing of preload water in the spray boilers. The icing in the boilers quickly thawed when the APU heat output increased.

The fuel cells, cryogenics, and electrical power distribution systems all performed satisfactorily with no anomolies [sic]. The lift-off electrical loads were about 23 kw [kilowatts], some 5 to 7 kw lower than predicted.

The structural, mechanical, and thermal systems all performed well.

Solid Rocket Booster and External Tank Disposal

SRB Disposal—The SRB's were jettisoned after burnout (ignition + 130.4 seconds) on tumbling free-fall trajectories. Both SRB's fell within the predicted impact footprint in the Atlantic Ocean approximately 140 n. mi. northeast of KSC. Splashdown occurred approximately 7 minutes, 10 seconds after lift-off. The SRB's were recovered by retrieval ships. The boosters were found floating high in the buoy position indicating good water entry (Figure 3). One was dewatered with the nozzle plug and the other with the "barb" back-up fixture. The solid rocket motor cases, frustums, and remaining items, except for 2 (of 6) SRB parachutes which were not retrieved, have been returned to KSC for inspection and processing.

ET Disposal—ET separation from the orbiter was nominal at 0: 08: 58.1 GET. After separation, the ET followed a ballistic trajectory, and upon its return into the atmosphere, it began to break up about 100,000 feet above the planned breakup altitude of 180,000 feet. Photography of the ET separation taken from the Orbiter indicates that the ET tumble system failed to activate. The tumble system, which is activated before separation by signals from the Orbiter to a pyrotechnic valve inside the liquid oxygen tank nose cap, is designed to prevent aerodynamic skip during reentry to ensure that tank debris will fall within a preplanned disposal area. Verbal reports from the ET tracking ship, [the U.S. Navy ship] Arnold, positioned in the Indian Ocean were that the debris foot print was larger than expected. Tracking data are being returned from the ship on an expedited basis for evaluation.

Onorbit

The STS-1 orbital operations phase was initiated at the completion of the OMS-2 maneuver. Day 1 of the STS-1 flight was concerned primarily with configuring the Orbiter for onorbit operation (i.e., opening payload bay doors, reconfiguring software, IMU alinements [sic]). After opening the payload bay doors, the crew [page 8 was Figure 3] [9] directed the onboard TV camera at the OMS pods, showing some thermal protection system (TPS) damage on both pods (Figure 4). An assessment of the thermal and structural loads for the area of the TPS damage on the OMS pods was conducted. The assessment of the structural loads on the TPS, assuming worst case descent conditions, indicated sufficient margin existed to insure [sic] that additional damage would not occur due to the entry environment.

The DFI PCM recorder was noted to be in the continuous record mode about 1 hour into the flight. Attempts to place the recorder in the high sample mode were unsuccessful with the recorder apparently not responding to mode switch changes. Because of this condition, the recorder was stopped by removing power. Data review indicated that the mode switch was placed in the high sample mode at the planned time. An in-flight test showed that the recorder was not responding to mode switch changes. A procedure was developed for the crew to further troubleshoot the recorder and determine its status for entry and landing. All DFI PCM data continued to be transmitted and recorded over tracking stations.

An attempt was made to replace the DFI PCM recorder with the ascent wideband recorder; however, the crew could not remove all of the fasteners holding the panel covering the recorder and the replacement was not made. Postflight troubleshooting of the DFI PCM recorder revealed that a loose shim had jammed the tape mechanism.

Orbiter temperatures remained within acceptable limits. The flight control systems checks using one APU went as planned.

At 6: 20: 46.5 GET, after verification of critical vehicle systems, the first of 2 OMS maneuvers was initiated to transfer the Orbiter to a higher orbit. The OMS-3 firing was completed as planned, and at 7: 05: 32.5 GET, the OMS-4 manuever [sic] was initiated raising the orbit to a 148-n. mi. apogee by a 147.9-n. mi. perigee.

The firing time for OMS-3 was 28.8 seconds and for OMS-4 was 33.1 seconds. The propellant remaining after the manuevers [sic] was at the predicted levels, indicating satisfactory system performance.

The right OMS pitch gimbal primary channel exhibited degraded performance during gimbal checks for the maneuvers, and a fault summary message of right-OMS-pitch-gimbal-fail was noted. The data were reviewed, and the analysis concluded that the gimbal drive actuator rate did not meet specification performance requirements, and the primary was used as a backup for the deorbit maneuver.

Four reaction control system (RCS) maneuvers were performed to verify that all thrusters were operating properly.

[page 10 was Figure 4] [11] At 40: 02: 39 GET, the APU-2 gas generator injector bed temperature dropped to 236° F (normal range: 350° F to 410° F), indicating the loss of gas generator heater B. The heater was switched from the B to the A system and the temperatures began increasing. Approximately 4.5 hours later, the gas generator injector bed temperatures were again decreasing. The heater was switched to the B system, but no increase was noted. It was then returned to system A, but no increase in temperature was realized, indicating loss of both heaters. It was determined through a real-time ground test that APU 2 would start satisfactorily at a bed temperature as low as 70° F. The temperature was predicted to be higher than 70° F for APU start for deorbit[;] however, a start override was required and was accomplished successfully.

During the flight control system checkout, the horizontal situation indicator (HSI) compass card did not respond properly. The indicator was off 5 degrees during the "low" test and did not drive at all during the repeated "high" test. A test procedure was performed by the crew and the indicator again failed to respond, with the card appearing stuck. Later, during checkout, the crew reported normal HSI function.

The Y-star tracker experienced an anomaly. Bright object protection was being provided by an interim backup circuit which senses light in the field of view and was latching the shutter closed. The crew opened the shutter via an override command for subsequent alinements [sic].

The onorbit electrical loads were about 15 to 25kw, some 2kw lower than predicted.

Descent and Landing

Entry preparation was accomplished according to the crew activity plan and without problems. The deorbit maneuver using both OMS engines was initiated at approximately 53: 21: 31.1 GET during the 36th orbit, and 31 minutes later, the Orbiter entered the communications blackout period of approximately 6-minutes duration. A nominal reentry was flown and touchdown (Figure 5) was made at 180 knots at 10: 20: 58 a.m., PST, on dry lake bed Runway 23 at Edwards Air Force Base, California. Total runway rollout distance from the touchdown point was 8993 feet. Postrollout operations were accomplished without incident, and ground cooling was connected about 16 minutes after landing. The flight crew egressed the Orbiter 1 hour and 8 minutes after landing. This occurred after a delay for the ground crew to clear hazardous vapors detected in the vicinity of the Orbiter side hatch.

Structural, power, and heat rejection entry loads were generally lower than predicted as were the APU, RCS, and active thermal control subsystem consumables usage. Orbiter structure backface temperatures were also lower than expected.

[page 12 was Figure 5] [13] Ground Operations/Turnaround

Following flight crew egress and Orbiter safing, the Columbia was towed to the Mate/Demate Device for weight and balance checks, purge of the main propulsion system, and propellant detanking. An inspection of the Orbiter was performed. The most significant discrepancies were a delamination of a section of graphite epoxy structure on the right OMS pod due to overheating and lesser overheating damage on the left OMS pod, a 1.25-inch cut through 5 of 17 plies on the inboard tire of the left main landing gear, and the loss of sleeve and bearing pieces from the up-lock roller of the right main landing gear which were found on the approach path 4-miles short of the touchdown point. A detailed inspection of the thermal protection tiles revealed minor damage to approximately 400 tiles. About 200 tiles will require replacement—100 as a result of flight damage and 100 identified prior to STS-1 as suitable for one flight. After completion of inspection activities, the tailcone assembly was installed. The tailcone is an aerodynamic fairing that attaches to the aft end of the Orbiter for ferry flight. The ferry flight departed Edwards at 10:16 a.m., PDT, on April 27, 1981. Following a stop at Tinker AFB, Oklahoma, for refueling, the 747 and Orbiter remained overnight[,] then proceeded to the Kennedy Space Center Shuttle Landing Facility, landing at 11: 25 a.m., EDT, on April 28. After demating from the 747 aircraft, the Columbia was towed to the Orbiter Processing Facility to begin processing for reuse in the STS-2 mission.

[14]   STS-1 TEST AND SUPPLEMENTARY OBJECTIVES

All of the crew activity objectives assigned to STS-1 were accomplished based on early data available from the flight. The functional test objectives (FTO's) and functional supplementary objectives (FSO's) accomplished are listed in Table 1. The FTO's describe functions that were required to be performed to satisfy STS-1 and/or OFT objectives. The FSO's describe functions which were necessary to satisfy supplementary objectives. Some FTO/FSO results may be partially incomplete, dependent on the amount of data lost due to the DFI PCM recorder problem described on page 9 of this report. . . .

Document II-31

Document title: Allen J. Lenz, Staff Director, National Security Council, Memorandum for Martin Anderson, Assistant to the President for Policy Development, *et al.*, "Space Shuttle Policy," July 17, 1981, with attached: "Presidential Directive: Space Transportation Policy" and "Space Policy Review: Terms of Reference."

Source: NASA Historical Reference Collection, NASA History Office, NASA Headquarters, Washington, D.C.

The administration of President Ronald Reagan entered the White House in January 1981. With respect to the space sector, one of the first issues addressed was policy for the Space Shuttle, which had its initial launch on April 12, 1981. The terms of reference for the Shuttle policy review suggest the wide variety of ideas under discussion in mid-1981 about the future use and management of the Shuttle.

[each page stamped "UNCLASSIFIED" and "Declassified/Released on 1-25-96 under provisions of E.O. 12958 by D. van Tassel, National Security Council"]

[1]   NATIONAL SECURITY COUNCIL
                         WASHINGTON, DC 20506

                             July 17, 1981

MEMORANDUM FOR

          MARTIN ANDERSON
          Assistant to the President for Policy Development

          EDWIN HARPER
          Assistant to the President and Deputy Director, Office of Management and
          Budget

          THE HONORABLE VERNE ORR
          The Secretary of the Air Force

          RICHARD DARMAN
          Deputy Assistant to the President and Deputy to the Chief of Staff

          GEORGE A. KEYWORTH, II
          Director, Office of Science and Technology–Designate

          JAMES BEGGS
          Administrator, National Aeronautics and Space Administration–Designate

          HANS MARK
          Deputy Administrator, National Aeronautics and Space
          Administration–Designate

          WILLIAM SCHNEIDER
          Associate Director for National Security and International Affairs, Office of
          Management and Budget

SUBJECT: Space Shuttle Policy

    Attached for your review are drafts of the proposed Policy Statement and the Terms of Reference for the Space Policy Review which were discussed at the meeting held on June 10. An abbreviated version of the Terms of Reference was distributed to the National Security Council members for their review before the meeting to be held on July 23.

[2] It is requested that any comments be submitted to the [National Security Council] Staff by [close of business] July 24. After all comments are received and reviewed, a meeting will be shceduled [sic] for discussion on these items.

Allen J. Lenz
Staff Director

Attachments

****

[no page number]   # Presidential Directive
Space Transportation Policy

This directive establishes national policy that shall guide the activities related to the Space Transportation System (STS). The United States will continue to develop the STS through the National Aeronautics and Space Administration in cooperation with the Department of Defense to service all authorized space users. The STS will be the primary space launch system for both U.S. military and civil government missions. The transition to the Shuttle should occur as expeditiously as practical.

The STS is a national program requiring sustained commitments by all departments and agencies. NASA will assure the Shuttle's utility to the civil government and non-government users. The Department of Defense, through the Air Force, will assure the Shuttle's utility to defense and integrate all national security missions into the Shuttle system. Launch priority will be provided to national security missions, and such missions may use the Shuttle orbiters as dedicated mission vehicles. Any changes to key STS program milestones or cancellation of key milestones will require my approval.

The Shuttle affects broad policy considerations. Accordingly, the Director of the Office of Science and Technology Policy will examine, in cooperation with appropriate agencies, whether new national directions in space policy are warranted in order to ensure U.S. leadership, to assure that the STS is managed in the most effective manner, to meet the future needs of space users, and to protect U.S. national security needs. Any goals or initiatives considered as new directions in space policy should be analyzed to determine if they are consistent with fiscal and economic priorities.

****

[1]   # Space Policy Review
Terms of Reference

SECTION I.   Future Launch Vehicle Needs

A. Identify, and assess the future needs for defense, commercial and scientific launch vehicles or space platforms that may be required in the late 1980's and beyond.
B. Review existing studies that have been undertaken by various agencies on launch vehicle requirements and evaluate the implications of meeting future launch requirements by the Shuttle alone. Evaluate the national security, arms control, political and economic implications of the Shuttle as the only launch capability.
C. Assess the current defense and civil launch vehicle backup strategy and rate of transition to the Shuttle. Assess the budgetary, national security and political implications of this strategy and possible alternatives.

D. Determine what launch capabilities will be required for future [anti-satellite] options that we may pursue. Specifically, address the launch requirements for possible space based [anti-satellite] systems.

E. Assess the implications of using the Shuttle in active military operations. Assess the feasibility of using the Shuttle as an integral part of a weapon system. What advantages do unmanned launch vehicles provide over the Shuttle? Moreover, to what ends would we be willing to use the Shuttle in an [anti-satellite] mode. What steps are necessary to protect against Shuttle disruption?

F. Assess how the Shuttle will be used in times of crises or war to launch satellites. If the conflict is protracted, do we run the risk of losing all our space launch assets through attrition? Will the nation accept a conscientious decision to expose the Shuttle flight crews to anti-satellite attact [sic]?

G. Determine what launch vehicle requirements are needed for survivability and enduring space borne systems.

H. Assess whether we should embark on a completely new unmanned (expendable or recoverable) launch capability [2] to meet future non-Shuttle needs, if any. Consider development schemes that build on existing technologies or systems—i.e., Delta, Titan, Shuttle main engine, new ballistic missile systems.

I. In the near term, development and enhancement of the Shuttle's operations and performance may be required for both civilian and national defense applications. Examine what the long term future requirements and opportunities will be for both civilian and defense applications. Review and make recommendations on Shuttle system improvements to include, but not limited to, upper stages.

SECTION II.   Shuttle Organizational Responsibilities and Capabilities

A. Review the present policy that NASA should continue to be responsible for overall management and operations of the Shuttle.

B. Evaluate the implications—political, budgetary, foreign policy, national security, legislative—for NASA to continue not only as the Shuttle developmental agent, but the operational manager of the Shuttle in the 1980's and beyond.

　　1. Examine the implications (same as in B above) for two parallel operational Shuttle systems: a civil effort under NASA and a defense effort under the Air Force.

[3] 2. Examine the implications (same as in B above) for the Air Force to take over as operational manager of the Shuttle for all users—including foreign—in the 1980's and beyond.

　　3. Evaluate the implications (same as in B above) of establishing a separate space transportation agency that would provide launch service to all users.

　　4. Examine the implications (same as in B above) of turning the Shuttle over to the US private sector to meet all US Government launch service requirements short of what launch vehicles, if any, would be required by Defense in crisis and conflict situations.

C. For each alternative examined under 1–4 above, state the most appropriate phasing of each alternative.

SECTION III.   Update Existing Policy

A. Review Presidential Directive-37—National Space Policy. Evaluate the existing relationships among the civil, defense, and intelligence space sectors. Determine if the guidance set forth in the National Space Act 1958 remains valid. Assess the implications of: (1) Provision for emergency use of civil systems by Defense. (2) Survivability of space systems. (3) Wider dissemination of classified space products.

B. Review Presidential Directive-42—Civil and Further National Space Policy. Determine whether the framework for space science activities remains valid. Assess the decisions that directed NASA to reenter long term communciations [sic] R&D and charged [the National Telecommunications and Information Administration] to aggregate the public service communications market. Determine whether technology sharing among space sectors should continue to be pursued.

C. Review Presidential Directive-54—Civil Operation Remote Sensing. Assess any outstanding issues concerning operational land remote sensing. Evaluate the decision that separate weather satellite systems should be pursued in the future. Consider if an[y] oceans satellites are initiated, whether joint NASA, Defense, and Commerce management should continue as national policy.

[4]D. Critique the above directives and place in context with the Reagan Administration policies and direction. Keep in mind the long term commercial and national security interests of the nation; assess the process by which decisions that affect these interests are coordinated within the US Government, i.e., preparations for UN Outer Space Committee, Moon Treaty, etc.

SECTION IV.   Legislation

A. Examine what legislative initiatives, if any, will be necessary to implement any changes in US space policy.

Document II-32

Document title: The White House, National Security Decision Directive 8, "Space Transportation System," November 13, 1981.

Source: NASA Historical Reference Collection, NASA History Office, NASA Headquarters, Washington, D.C.

In its first space policy statement after taking office in January 1981, the Reagan administration set out its policy for the Space Transportation System (Space Shuttle). This policy, assigning NASA the continuing lead role in the Shuttle program and declaring the Shuttle to be the primary launch system for all government missions, essentially endorsed existing plans developed during the Carter administration. Some had argued during the preceding months for different management arrangements for the Shuttle.

[stamped "Declassified on 6/14/90"]

THE WHITE HOUSE
WASHINGTON

November 13, 1981

National Security Decision
Directive Number 8

Space Transportation System

Recognizing the importance of space programs in the broad commercial, civil, and national security needs, the United States is committed to a vigorous effort that will ensure leadership in these areas. The Space Transportation System (STS) is a vital element in fulfilling these needs.

This decision establishes national policy that shall guide the activities related to the STS. The United States will continue to develop the STS through the National Aeronautics and Space Administration in cooperation with the Department of Defense to service all authorized space users. The STS will be the primary space launch system for both United States military and civil government missions. The transition to the Shuttle should occur as expeditiously as practical.

The STS is a national program requiring sustained commitments by all departments and agencies. NASA will assure the Shuttle's utility to the civil government and non-government users. In coordination with NASA, the Department of Defense will assure the Shuttle's utility to defense and integrate national security missions into the Shuttle system. Launch priority will be provided to national security missions, and such missions may use the Shuttle orbiters as dedicated mission vehicles. Major changes to STS program capabilities will require my approval.

Document II-33

Document title: President Ronald Reagan, "Remarks on the Completion of the Fourth Mission of the Space Shuttle *Columbia*," July 4, 1982, pp. 869–872.

Source: *Public Papers of the Presidents of the United States: Ronald Reagan* (Washington, DC: U.S. Government Printing Office, 1982–1991).

In a ceremony replete with patriotism, on July 4, 1982, President Ronald Reagan witnessed the landing of the fourth mission of the Space Shuttle Columbia *at Edwards Air Force Base in the California desert. He used the occasion to issue a new statement of National Space Policy [see* Exploring the Unknown, *Volume I, Document III-38], to make a tentative initial commitment to the development of a space station, and to declare the Space Transportation System operational. The* Columbia *mission was commanded by veteran astronaut Thomas K. (Ken) Mattingly, referred to by Reagan here as "T.K."; the pilot was Henry (Hank) Hartsfield. As Reagan spoke, the NASA carrier aircraft with the second Shuttle orbiter,* Challenger, *on top of its fuselage took off from Edwards.*

[869] **United States Space Policy**

Remarks on the Completion of the Fourth Mission of the Space Shuttle Columbia.
July 4, 1982

The President. T.K. and Hank—as you can see, we've gotten well acquainted already—you've just given the American people a Fourth of July present to remember. I think all of us, all of us who've just witnessed the magnificent sight of the *Columbia* touching down in the California desert, feel a real swelling of pride in our chests.

In the early days of our Republic, Americans watched Yankee Clippers glide across the many oceans of the world, manned by proud and energetic individuals breaking [870] records for time and distance, showing our flag, and opening up new vistas of commerce and communications. Well, today, I think you have helped recreate the anticipation and excitement felt in those homeports as those gallant ships were spotted on the horizon heading in after a long voyage.

Today we celebrate the 206th anniversary of our independence. Through our history, we've never shrunk before a challenge. The conquest of new frontiers for the betterment of our homes and families is a crucial part of our national character, something which you so ably represent today. The space program in general and the shuttle program in particular have gone a long way to help our country recapture its spirit of vitality and confidence. The pioneer spirit still flourishes in America. In the future, as in the past, our freedom, independence, and national well-being will be tied to new achievements, new discoveries, and pushing back new frontiers.

The fourth landing of the *Columbia* is the historical equivalent to the driving of the golden spike which completed the first transcontinental railroad. It marks our entrance into a new era. The test flights are over. The groundwork has been laid. And now we will move forward to capitalize on the tremendous potential offered by the ultimate frontier of space. Beginning with the next flight, the *Columbia* and her sister ships will be fully operational, ready to provide economical and routine access to space for scientific exploration, commercial ventures, and for tasks related to the national security.

Simultaneously, we must look aggressively to the future by demonstrating the potential of the shuttle and establishing a more permanent presence in space.

We've only peered over the edge of our accomplishment, yet already the space program has improved the lives of every American. The aerospace industry provides meaningful employment to over a million of our citizens, many working directly on the space program, others using the knowledge developed in space programs to keep us the world leader in aviation. In fact, technological innovations traced directly to the space program boost our standard of living and provide employment for our people in such diverse fields as communications, computers, health care, energy efficiency, consumer products, and environmental protection. It's been estimated, for example, that information from satellites has saved hundreds of millions of dollars per year in agriculture, shipping, and fishing.

The space shuttle will open up even more impressive possibilities, permitting us to use the near weightlessness and near-perfect vacuum of space to produce special alloys, metals, glasses, crystals, and biological materials impossible to manufacture on Earth. Similarly, in the area of national security, our space systems have opened unique opportunities for peace by providing advanced methods of verifying strategic arms control agreements. The shuttle we just saw land carried two kinds of payloads, one funded entirely by private industry, and the other, related to our national security, sponsored by the Air Force.

This versatility of the *Columbia* and her sister ships will serve the American people well, yet we must never forget that the benefits we receive are due to our country's commitment made a decade ago to remain the world leader in space technology.

To ensure that the American people keep reaping the benefits of space and to provide a general direction for our future efforts, I recently approved a national space policy statement which is being released today. Our goals for space are ambitious, yet achievable. They include continued space activity for economic and scientific benefits, expanding private investment and involvement in space-related activities, promoting international uses of space, cooperating with other nations to maintain the freedom of space for all activities that enhance the security and welfare of mankind, strengthening our own security by exploring new methods of using space as a means of maintaining the peace.

There are those who thought the closing of the western frontier marked an end to America's greatest period of vitality. Yet we're crossing new frontiers every day. The high technology now being developed, much of it by byproduct of the space effort, offers us and future generations of Americans opportunities never dreamed of a few years ago. Today we celebrate American [871] independence confident that the limits of our freedom and prosperity have again been expanded by meeting the challenge of the frontier.

We also honor two pathfinders. They reaffirm to all of us that as long as there are frontiers to be explored and conquered, Americans will lead the way. They and the other astronauts have shown the world that Americans still have the know-how and Americans still have the true grit that tackled a savage wilderness.

Charles Lindbergh once said that "Short-term survival may depend on the knowledge of nuclear physicists and the performance of supersonic aircraft, but long-term survival depends alone on the character of man." That, too, is our challenge.

Hank and T.K., we're proud of you. We need not fear for the future of our nation as long as we've got men like you to serve as our inspiration. Thank you both, and God bless you for what you're doing.

Before I introduce you, if you'll all just look—well, I'm sure down in front maybe you can't see—but way out there on the end of the runway, the space shuttle *Challenger*, affixed atop a 747, is about to start on the first leg of a journey that will eventually put it into space in November. It's headed for Florida now, and I believe they're ready to take off.

Challenger, you are free to take off now.

And now it's my pleasure to introduce to you two sons of Auburn, Captain T.K. Mattingly and Colonel Hank Hartsfield.

Captain Mattingly. Thank you.

Mr. President, you mentioned something about people having a desire to maintain a presence in space. Not very many hours ago I know two guys who really wanted to maintain that presence in space a while longer. That is, you never get tired of it. The most remarkable thing, besides the machine and the team that put it together, is that it's a new discovery every minute and every day.

The machine we built is a first stepping stone. Here comes the second one. We're standing in front of its pathfinder, and there's more to come. Where we're going to go in the future is something that depends on you. [At this point, the Space Shuttle *Challenger* and its transporting aircraft passed overhead, en route to Kennedy Space Center at Cape Canaveral, Florida.] And maybe that's our second stage.

I'd like to thank you for being here today. It's really a privilege for us to be part of this celebration. I don't feel like—it isn't our celebration at all. We were just lucky enough to be here.

The people that make all this work are the thousands of designers and engineers that made it work. And as the President pointed out, all the technology in the world is just a tool. And the only thing that makes the difference between our technology and the trip that we've just had and the sights that we've seen and the things that we've thought and the ideas that that's spurred—all the difference between that and just plain old technology is the people that made it happen. And the country is blessed with having a team that's dedicated to the United States and to the exploration and exploitation of space. And I am just as proud as I can be to be a part of that NASA team.

There's one other thing that I'd like to say, and I'll let Hank talk to you. Hank's had to endure me for a long time now, and he probably thinks that this last year has been the longest year of his life. And it's certainly had more hours packed into it than most. But throughout it all, this guy has maintained a sense of humor and an industry that's second to none. And this is the finest pilot that ever flew in a spacecraft.

Hank.

Colonel Hartsfield. It's kind of tough to follow that. I can only echo the words of the President and T.K. I am very proud to be here and be a part of the shuttle program.

I think back to 206 years ago when our forefathers ushered in a new era of true democracy for the world. And here today I think we have ushered in a new era also—a fully operational space transportation system. We've got a real fine vehicle there. That vehicle performed far beyond my expectations, and I think T.K. and I brought all you folks about the best spacecraft that's ever been built. It will be tough for *Challenger* and the ones coming down the line to top it.

[872] But as Ken said, the people that put all this together are the important part. T.K. and I are only just a little tip of the pyramid, and we're standing on the top of a huge number of people who have dedicated their lives and their efforts to making it all work. It can't

be done without you folks. And I'm convinced, as T.K. is, that American technology is the greatest in the world, because we have the best people in the world, people who are willing to work.

I think that the future is going to hold something for us that at this point we cannot even imagine. In the short time that I was there in space, I thought of some things that could only be done there. And when we start sending people up routinely, as the President pointed out, we just opened a railroad. T.K. referred to it once as "opening up the freeway." Once they're built, we know no bounds to what we can do. And I am very, very proud to be a part of this initial effort.

Thank you.

The President. Come on up, both of you. I just want to, again, tell you how proud we are of you. And today, as we celebrate our 206th anniversary of our independence, let us remember we're a prosperous people and a strong people because we're a free people.

Well, God bless you all and a happy Fourth of July.

Now, here they come. [At this point, a band played "God Bless America."]

Happy Fourth of July. And, you know, this has got to beat firecrackers.

Document II-34

Document title: NASA, "National Space Transportation System: Analysis of Policy Issues," August 1982, pp. 5–12.

Source: NASA Historical Reference Collection, NASA History Office, NASA Headquarters, Washington, D.C.

This report contains NASA's views on the future of U.S. space transportation as of 1982, after the Space Shuttle was declared operational. NASA repeated its long-running arguments that at least five Shuttle orbiters were necessary, that the Shuttle could launch all U.S. payloads without being complemented by expendable launch vehicles, and that NASA should remain the manager of Space Shuttle operations, with the Department of Defense as a cooperating partner. What follows is the executive summary from the report.

National Space Transportation System
Analysis of Policy Issues

August, 1982

Prepared by

National Aeronautics and Space Administration
Washington, D.C. 20546 . . .

[5] EXECUTIVE SUMMARY

At the direction of the President, the Office of Science and Technology Policy (OSTP) of the Executive Office of the President was commissioned in the Fall of 1981 to develop a space policy that sets national priorities and goals. Under OSTP auspices, an interagency Space Policy Study Steering Group with a supporting Working Group was formed to develop issues and recommend attendant policy statements. NASA was assigned the study lead on three key policy issues:

a. The best institutional placement for the organization to manage and operate the
    Space Shuttle
b. Mixed fleet of expendable launch vehicles and the reusable Space Shuttle versus
    Space Shuttle only
c. Future launch vehicle needs for defense, commercial, and scientific purposes

In January 1982, a Space Policy Task Team was established within NASA, composed of
key people from Headquarters and its field centers. The Task Team met periodically over
the course of the study to develop and analyze the issues, to debate the strengths and
weaknesses of alternate organization placements, mixed fleet and new launch vehicles,
and to interview key executives from government and industry to insure [sic] that a broad
perspective on all space policy issues emerged.

The original intent was to conduct a joint NASA/Department of Defense (DOD)
study. This mutual effort was not possible, however, because of incompatibilities between
the NASA and Air Force Task Team schedules. Although the conclusions and recommendations of this report primarily represent the NASA Task Team's evaluation of the key
issues, there has been limited involvement with other agencies including DOD. Part I of
this report presents an Executive Summary, Part II an Introduction, and the three key
issues are addressed in Parts III, IV, and V.

As a result of NASA's study, the following policy statement was developed for submission to the interagency Space Policy Study Steering Group.

RECOMMENDED SPACE TRANSPORTATION POLICY STATEMENT

The National Space Transportation System (NSTS) is composed of the Space
Transportation System (Space Shuttles and Upper Stages) and Expendable Launch
Vehicles (ELV's). The Space Shuttle is the primary U.S. space launch system. The U.S.
Government is fully committed to maintaining the U.S. as the world leader in space transportation and to a continuing Space Transportation System (STS) capability with a capacity to meet the needs of all users. All new U.S. Government spacecraft shall be designed to
take advantage of the unique capabilities of the STS. ELV operations shall be continued
until the capability of the STS is sufficient to meet the needs of all users. National Security
considerations may dictate maintaining ELV's or developing other launch capabilities.

Commercialization of part or all of the STS or ELV's shall be encouraged when it is
determined to be in the best national interest.

[6] The planned STS performance to low earth orbit is sufficient for the foreseeable
future. Optimization of the STS shall be continued to include an economical upper stage
delivery capacity, an efficient local orbital maneuvering capability, and other overall STS
enhancements as necessary to increase the effectiveness and utilization of the STS for
national and international users.

The STS shall continue to be managed and operated in the near-term under an institutional arrangement consistent with existing NASA/DOD agreements; however, the flexibility to evolve toward a different institutional arrangement shall be maintained.

The STS organization, consistent with NASA/DOD responsibilities, shall:

*   Maintain U.S. role of world leader in space transportation
*   Provide priority operations for national security missions
*   Provide economical STS operations
*   Expand the STS capabilities to fully exploit the opportunities of space
*   Provide access to space for all customers
*   Promote civil and U.S. military space operations
*   Develop new markets for the STS

- Meet the challenge of international competition
- Insure [sic] smooth transition of U.S. Government payloads to the Space Shuttle

PART A — THE NATIONAL SPACE TRANSPORTATION SYSTEM OPERATIONS PLACEMENT STUDY

This part of NASA's Analysis of Policy Issues identifies the best institutional setting for the NSTS within a spectrum of placement options ranging from within the Government:

a. NASA
b. DOD
c. A partnership or joint arrangement between NASA and DOD
d. Another existing Federal agency, e.g., Transportation or Commerce
e. A new Federal agency, or
f. Some form of government corporation

to a setting outside the Government in a private corporation. The NSTS is a national resource composed of the entire fleet of U.S. space launch vehicles. NASA and the DOD currently share responsibility for managing and operating this fleet. NASA is responsible for the management and operation of the Space Shuttle, Delta, Atlas-Centaur, and Scout space launch vehicles, while the Air Force as the executive agent for the DOD is responsible for the Atlas E/F and Titan vehicles.

The NASA Task Team conducted this study in two phases. In Phase One, the focus was on envisioning the national priorities and scope of related space activities over the next decade and beyond (the far term) and identifying the best institutional setting for that time frame. In Phase Two, the focus was on the realities of organizations, plans, and events now in place and in motion to identify the best near-term organizational setting supportive of the far-term.

[7] The NASA Task Team concludes that, in the far term, the civil and military sector will continue to expand. Furthermore, the unique military demands for assured access to space in times of crisis and conflict, secure operations, and possible special vehicle configurations unsuited for general purpose civil usage may eventually lead to separate civil and military space operations. The cost of separate operations need not be large[,] provided the two sectors share common production, logistics, and other support services. Today, national defense ELV operations are separate from civil operations. This arrangement may eventually reemerge in the Space Shuttle era. With the expectation that the eventual separation of national defense operations from civil operations is likely, the NASA Task Team turned its attention to determining the organizational setting best suited to operate the civil segment in the far term.

The NASA Task Team concludes that NASA, a new government corporation, or a private corporation are all suitable for managing civil operations. NASA's strengths include its in-place manned space flight organization, management structure, experience base with a proven record of success, well-established linkages to the domestic and international civil user communities, and its recognized advocacy for visible, peaceful, and open use of space. Many perceive Shuttle operations as diluting NASA's principal focus of advancing the leading edge of science and technology; hence, managing an operational organization would be an unsuitable assignment in the far term. The NASA Task Team does not accept this view and believes that the operation of the NSTS complements and supports NASA's role in science and manned space flight technology development. A government corporation and a private corporation in the far term are also deemed suitable organizational settings, because both offer flexibility and innovation (i.e., marketing incentives, unfettered personnel, procurement and pricing policies), and access to debt and/or equity markets versus sole dependence on appropriations. The primary advantage of a government-controlled corporation over a private corporation is the assurance that the interests of all users are protected and, in particular, the needs of the national defense sector are adequately coordinated with DOD operations.

The establishment of either a government or private corporation is not practical in the near term, as executive and congressional assent is required, and it is questionable whether the STS could be a self-sustaining enterprise with sufficient return on investment at this stage of the Shuttle Program. Transferring the NSTS in the far term to either an existing federal agency or a new federal agency offers no significant advantage over continued management within NASA, other than to symbolize renewed national commitment to an expanding space transportation system or to permit NASA to focus more sharply on research and development. Further, creation of a new agency is inconsistent with the Reagan Administration's goal to curb the size and scope of government and reduce the proliferation of agencies. In summary, the NASA Task Team concludes that it is too early to determine the best civil sector far-term organizational placement (NASA, Government Corporation, or Private Corporation). However, the choice of a near-term interim organization placement must not impede or preclude the eventual transition to one of these institutional settings. The timing of this transition will be dictated by the maturity of the Space Shuttle Program, the growth of space traffic, and the economic viability of [the] Space Shuttle in an evolving and uncertain internationally competitive space transportation marketplace.

[8] Phase Two of the study focused on the near-term placement options. The near term extends to the point when the DOD brings on-line their launch and landing operations capability at Vandenberg Air Force Base (VAFB) and their Consolidated Satellite Operations Center (CSOC); that is, in the 1986–1987 time frame. The NASA Task Team concludes that only three organizational placement options are viable: a NASA/DOD partnership, a NASA/DOD joint operation, or management exclusively by NASA. These options are viable principally because of the necessity for continuity in the Space Shuttle program during the emerging early operations and because NASA's experience base cannot be readily uprooted and transplanted to another organization without possible harm both to the Nation's space transportation leadership and to NASA itself. The alternative of placing the NSTS solely in DOD was also considered but rejected since, in the NASA Task Team's judgment, DOD is not prepared to conduct STS operations, nor is this placement consistent with the Nation's emphasis on open and peaceful use of space. Consequently, the most attractive near-term organizational placement options are described as follows:

a.   NASA/DOD Partnership—This organizational design is the currently evolving arrangement consistent with a 1979 memorandum of understanding between NASA and DOD. NASA is responsible for the overall Space Shuttle management and operations until DOD's VAFB and CSOC become operational. Until that time, NASA is the executive agent for launching and operating military missions with DOD as the mission manager. With the activation of VAFB, DOD will launch both military and civil missions from the West Coast as NASA now does from the East Coast. NASA is the mission manager for civil missions, and DOD is the mission manager for military missions. Responsibility for DOD Shuttle mission control shifts from NASA to DOD with the activation of CSOC.

The NASA/DOD partnership is in place and successfully evolving; has a proven record of success; enjoys a unity of control, responsibility, and accountability; and assures continued linkages of the early phase of Shuttle operations to the development experience base within NASA. It also provides the DOD with adequate means for control of military operations through the mission manager function. The partnership and division of resonsibilities [sic] are suitable both for the near and far term. However, as the utilization of space expands, the national security requirements will probably dictate the need for separation of military and civil operations. In the interim, to improve and better balance the partnership, the organization should be fortified with greater DOD participation in the functional line management and policy development of the NSTS. This balance can be encouraged by assignment of DOD detailees to key management roles within NASA and, in turn, assignment of NASA personnel to DOD functional management roles. The intent is to assure DOD participatory management control and visibility and to prepare DOD for eventual quasi-autonomous Shuttle operations in the far term. The partnership organization also preserves the separate NASA and DOD ELV programs undisturbed within their respective agency settings.

   b.   NASA/DOD Joint Operation—This organizational design is a variant of the partnership wherein, in the extreme form, a totally new management structure is developed and constituted with personnel from NASA and DOD to manage and operate the NSTS. Policy, program direction, allocation of resources and other responsibilities are jointly determined and approved.

[9] The definition and implementation of this option is difficult and complex. The linkages between the new organization and DOD, NASA and Congress, its relationships with the NASA field centers and DOD installations, and the hierarchical levels in the organization where responsibility is shared results in a complex organization that would be difficult to implement. While such an organizational setting may provide greater DOD assurance of control and responsiveness to national defense needs, the injection of additional civil involvement into the DOD operations at all organizational levels may complicate security measures and impede the open and growing use of space by the civil sector. Furthermore, once a joint operation is created in the near term, it would be difficult to sever and disjoint in the far term. While such an organization may be established in the near term and function into the far term, its drawbacks were judged to outweigh its potential advantages.

   c.   NASA only—This organizational concept places all responsibility for the Space Shuttle management and operations within NASA. Accordingly, the partnership with DOD is dissolved and NASA oversees the activation of VAFB and the development of the Inertial Upper Stage (IUS) and other related DOD activities. The principal advantage of this option is that it consolidates all space transportation; however, it symbolizes primarily civil use of space and relegates DOD to a position totally dependent upon NASA for space access and manned operations. With NASA as sole operator of the Space Shuttle, the DOD does not have control and direction of national security operations. Consequently, the Task Team rejects this option as neither politically nor practically desirable.

   In summary, the evolving NASA/DOD partnership, strengthened by infusion of DOD personnel into the functional line management of [the] Space Shuttle, is the organizational setting best suited to direct near-term NSTS operations. This partnership provides the flexibility for a far-term transition to a new institutional setting if required.

   In support of the National Space Policy, the NASA Task Team recommends that the NASA/DOD partnership, consistent with the current interagency memorandum of understanding, serve as the organization for managing the NSTS. Flexibility should be maintained to evolve toward a different institutional placement for the far-term if necessary.

PART B — MIXED FLEET ASSESSMENT

   This portion of the NASA Space Policy Study assesses the mixed fleet issue. Three specific subissues are addressed:

Subissue 1 — ELV's requirements for national defense.
Subissue 2 — ELV's required to augment STS.
Subissue 3 — Cost of Shuttle versus mixed fleet.

Subissue 1 is primarily a DOD issue but was addressed by the NASA Task Team. The concern is primarily one of confidence. Until the Shuttle operations mature, Titans must be available if needed to launch critical national defense missions. The issue, then, is how long should the Titans be available at ETR [Eastern Test Range] and WTR [Western Test Range]. The consensus of the NASA group was that by the end of 1984, adequate confidence in the STS launch capability from ETR [10] should exist since 15 to 20 STS flights from ETR will have been flown, and there will be 3 or 4 orbiters in the inventory. WTR is scheduled to be ready for STS launches in October 1985. At least one flight must be successfully accomplished to provide adequate confidence that the WTR launch facility can support critical launch schedules; therefore, Titans will be needed at WTR until late 1985, at least.

A detailed analysis was conducted to answer subissues 2 and 3 relating to the total needs for ELV's and Shuttles. The most current NASA and DOD mission models were used and coordinated with DOD and other government agencies. Cases were run with an all-STS fleet compared with various combinations of ELV's. Then sensitivity analyses were run against each of the cases for the following parameters: the number of flights per year per orbiter, the amount of orbiter downtime, and the total number of flights.

To augment the STS (subissue 2), ELV's are manifested currently into 1986. How long past 1986 they are needed is primarily a function of when the fifth orbiter is delivered. The conclusion from this analysis is that some or all of the ELV's will be needed until the fifth orbiter is delivered. Exactly which ELV's will be needed depends on which assumptions turn out to be valid, but it is possible that all current ELV's will be needed.

As for the cost of an all STS fleet versus a mixed fleet (subissue 3), this analysis conclusively demonstrates that with more STS flights and correspondingly less ELV flights, the total space transportation cost to the U.S. Government decreases, even when STS investment costs include more orbiters, facilities, production capability, and other related costs. In addition, the unique capabilities of the Shuttle provide additional significant economic benefits to STS users.

A parametric analysis was conducted to determine how many orbiters were needed and their need dates. Figure I.1 assumes a nominal mission model and summarizes the effects of number of flights per year per orbiter and orbiter downtime.
[original placement of Figure I.1] [11] Combining all variables in a worst case resulted in a need date for the fifth orbiter in 1986 and the sixth orbiter in 1990. Combining all variables into a best case resulted in no need for a fifth or a sixth orbiter.

After determining that these sensitivity analyses resulted in such a broad range of need dates, a detailed evaluation was conducted to determine the most likely set of parameters with which to define the conclusions. The Task Team concluded that 8 flights per year per orbiter, 50% equivalent downtime of one orbiter, and a 295 flight mission model was the most likely set of conditions. Using that set of conditions, the fifth orbiter is needed in 1990 and the need for the sixth orbiter is marginal. However, for a slightly higher orbiter downtime and/or slightly less flights per year per orbiter, the fifth orbiter is needed in 1986. Considering the possibilities of a catastrophic loss of one of the orbiters, the fifth orbiter is needed as soon as possible, and the sixth orbiter in 1990.

For cases in which one orbiter is down 50% of the time or more, the economic assessment concludes that it is less expensive to purchase another orbiter to fly the shortfall in the mission model than to purchase more ELV's. That is, it is more economical to purchase another orbiter rather than ELV's any time the demand is equivalent to 50% or greater of one orbiter's capability.

Sensitivity Assessment Results 295 Mission Model					
		Percent Downtime			Flts/Yr/Orbiter
		100%	50%	0%	
Orbiter Requirements	5th Orbiter	Worst Case 1986	1986	1990	6.5
	6th Orbiter	1990	1990	Never	
	5th Orbiter	1986	1990	1990	8.0
	6th Orbiter	1991	Marginal 1991	Never	
	5th Orbiter	1990	1990	Best Case Never	10.0
	6th Orbiter	Never	Never	Never	
Darklined box nominal case: 50% downtime, 8 flights per year per orbiter					

Figure I.1

In summary, for all cases and conditions evaluated, an all-STS fleet is more cost effective than a combined STS and ELV fleet. More STS flights and correspondingly less ELV flights result in the least total launch costs. Considering all factors, the fifth orbiter should be available as soon as possible after 1986, and the sixth orbiter should be delivered no later than the 1990/1991 time frame. Those availabilities will meet the present launch rates with the most current mission model and still provide an adequate margin for orbiter downtime. Although the requirements are not as firm as those for a fifth and sixth orbiter, it is likely that a seventh orbiter will be needed in 1991.

Peripheral to assessing the mixed fleet, the desirability and practicality of commercializing some or all of the ELV fleet was considered. There is such a vast array of issues that a recommendation could not be made in the alloted [sic] time of this study. However, the consensus of the Task Team was that the U.S. Government should encourage and promote commercialization of ELV's and actively seek to resolve questions and issues posed by interested commercial organizations.

PART C — FUTURE LAUNCH VEHICLE NEEDS ASSESSMENT

The purpose of this part of the study is to identify new launch vehicle needs over and above the capability of the current NSTS.

Within the assessment guidelines and groundrules, the chief of which is a projection of minimal growth of the NASA budget, a Space Shuttle fleet can support all the needs for earth to orbit transportation for the foreseeable future. There is no clear identified requirement for a heavy lift new or shuttle-derived large capacity launch vehicle through the late [12] 1990's. This conclusion could change if there is a rapid escalation of the launch demand and/or the NASA budget. Needs for redundancy would be driven by National Security considerations and were not addressed by NASA.

There is a need for a new high energy upper stage, compatible with the Space Shuttle and future space-based operational concepts, capable of placing about 15,000 pounds of mission payload into geostationary orbit from Cape Canaveral. To satisfy the needs of the commercial users and to compete with the Ariane, the new high energy upper stage must provide a low cost, economical delivery system and accommodate multiple payloads.

There can be considerable improvement in cost effectiveness and performance from space-based operations of high energy upper stages by making up the Space Shuttle orbiter load factor with propellant and transferring the propellant to on-orbit storage. This improvement could raise the effective payload orbited per year by 30–40%. It is important that this high energy upper stage have the capability for upgrading to a manned, reusable vehicle.

Finally, the NASA Task Team determined that a need exists for a small space transportation system to provide local transportation operations within the general vicinity of the orbiter. This local space transportation vehicle would be used for satellite placement and retrieval operations associated with satellite services.

CONCLUSIONS

The NASA Space Policy Task Team has analyzed three key issues on the future of the Space Transportation System. The issues were: (1) the National Space Transportation System institutional placement assessment; (2) mixed fleet assessment; and (3) future launch vehicle needs assessment.

The Team concludes that the current NASA/DOD partnership is the most suitable institutional arrangement for managing and operating the National Space Transportation System. Secondly, the NSTS fleet should consist of both expendable launch vehicles and reusable Space Shuttles until the capability of the STS is sufficient to permit the phaseout of ELV's. Finally, future launch vehicle needs should include development of a high energy upper stage and a small, specialized vehicle to provide transportation within the general vicinity of the orbiter.

Document II-35

Document title: Chester Lee, Director, STS Operations, to Manager, Space Shuttle Payload Integration and Development Program Office, Johnson Space Center, and Manager, STS Projects Office, Kennedy Space Center, "Guidelines for Development of the Flight Assignment Baseline," November 20, 1978

Source: Space Policy Institute, George Washington University, Washington, D.C.

As the first manned orbital flight (FMOF) of the Space Shuttle approached, NASA began to develop a schedule (called a "manifest") for when various payloads would fly aboard early Shuttle missions. This memorandum from Director of STS Operations Chester Lee set out the guidelines for establishing that schedule; the two enclosures do not appear here. The memo also stated NASA policy for the priority to be assigned if there were conflicts among candidate payloads for a particular mission. There were several changes from the Shuttle schedule set out by Lee. The FMOF was delayed from September 1979 to April 1981. The Orbital Flight Test (OFT) program was shortened from six to four flights, with the first operational flight occurring on November 11, 1982.

The Inertial Upper Stage was a solid-fueled system for transferring payloads from the Shuttle orbit to other orbits. OV-099 was the second Space Shuttle orbiter, named Challenger. The Tracking and Data Relay Satellite System (TDRSS) was a major early Shuttle payload. Galileo was a mission to orbit Jupiter that was repeatedly delayed, first by problems with the spacecraft, then by the Challenger accident in January 1986. Galileo was finally launched in October 1989.

[1] NASA
National Aeronautics and
Space Administration

Washington, D.C.
20546   [stamped NOV 20 1978]

Reply to Attn of MOB-6

TO: Johnson Space Center
           Attn: PA/Manager, Shuttle Payload Integration
                 and Development Program Office

           Kennedy Space Center
           Attn: SP/Manager, STS Projects Office

FROM: MO-6/Director, STS Operations

SUBJECT: Guidelines for Development of the Flight Assignment Baseline

REF: Memo from JSC/Lunney to MO-6/Lee dated 10/25/78,
           Subj: STS Flight Assignment Baseline

In response to the JSC reference memo, I agree that it is extremely important that an updated Flight Assignment Baseline be issued as soon as possible; however, some adjustments will be necessary to the proposed guidelines and recommended manifest accompanying the referenced memo. The following guidelines should be followed during the development of the next issue of the Flight Assignment Baseline.

1. FMOF—September 28, 1979.
2. Six OFT flights.
3. First Operational Flight (OV-102)—February 27, 1981.
4. Minimum orbiter configuration change between flights; retain energy kits once installed.
5. IUS ground turnaround capability—60 calendar days.
6. OV-099 delivery to KSC—September 1981.
7. TDRSS turnaround—launch centers to be separated by at least 90 calendar days with at least 60 calendar days allowed between launch of TDRS-B and Spacelab 1.
[2] 8. Plan Galileo on OV-099 but retain capability to accommodate on OV-102. For Galileo performance assessment, see Code M memo entitled, "Reassessment of Galileo Mission Performance[,]" dated November 17,1978.
9. SSME Performance:
    a. 100/109 prior to September 1981
    b. 109/109 September 1981 and subsequent flights.
10. Lighter weight ET available for Galileo mission and subsequent flights.
    a. Nominal ET weight reduction—4,000 lb.
    b. Special tailoring available for critical performance missions (i.e., Galileo to 6,000 lbs.)
11. Shuttle performance calculations to include variable I_s targeting where applicable and an OMS loading for a two sigma ET abort reserve on [abort-once-around] for applicable missions.

12. 3,000 lb. STS reserve based on the September 1978 Shuttle Systems Weight and Performance Report is to be used until further notification.
13. Shuttle capability based on latest official Shuttle program performance improvements and weight reduction data (i.e., 700 lb. SRB weight reduction).
14. Select SRB configuration for maximum thrust with seasonal change.

The Level I payload launch requirements to be used for the baseline are indicated in Enclosure 1. The general policy in regards to mission priorities is to accommodate user requirements as close as possible on a first come, first served basis. In the event of conflicting requirements, the priorities to be followed are: (1) space programs requiring urgent STS support to maintain national security mission capabilities; (2) significant Science and Technology missions and/or missions with critical launch window constraints; (3) committed reimbursable missions; (4) routine Science and Technology and late request missions; (5) space available requests.

[3] Within the above guidelines, the current mission priorities are as follows:

1. TDRSS initiation (TDRS A and B)—An operational TDRSS is required by the STS in order to initiate the NASA Spacelab program.
2. Spacelabs 1 and 2—These are verification flights that should be accommodated as soon as possible in order to establish an operational Spacelab program. Spacelab 1 should be flown as soon as feasible after TDRS A and B are operational, and Spacelab 2 as soon as practical after the Galileo mission (to avoid the weight penalty of the 5th Cryo set if OV-102 is required for Galileo).
3. Galileo—This is an important planetary mission with a critical launch window constraint.
4. After the above payloads are satisfied, the remaining payloads listed in Enclosure (1) should be accommodated as indicated by the accompanying booking date.
5. Payloads listed without booking dates should then be accommodated in any remaining cargo space.

In view of a number of related decisions expected by the end of this year or early next year, I believe that we should target the revised flight assignment baseline for early next year (January or February 1979). This issue should address the flights from STS-1 to at least through FY 1982, utilizing the enclosed mission sequence as a starting point. JSC and KSC should accomplish the analyses, i.e., performance, weight, [control and guidance], Shuttle configuration, launch window, turnaround times, mission compatibility, etc., to verify that the flight assignments are feasible. I would also like the next issue to indicate the accommodation of Small Self-Contained Payloads (SSCP's) in the most efficient manner possible, assuming we have available a structure that can be mounted across the bay and carry an average of 12 SSCP's. Your analyses should also provide adequate information to Headquarters to permit late [4] changes before publication. To this end, your recommended flight assignments are to be submitted to Headquarters for final review and approval before promulgation.

Chester M. Lee

2 Enclosures
1. STS Operations Payload Manifest
2. Preliminary Flight Assignment Baseline

Document II-36

Document title: National Security Council, Senior Interagency Group (Space), "Issue Paper on the Space Transportation System's (STS) Fifth Orbiter," late 1982.

Source: Ronald Reagan Presidential Library, Simi Valley, California.

The appropriate number of Space Shuttle orbiters to build was a persistent issue during the first decade of the program. The Reagan administration used the Senior Interagency Group (Space, known as) SIG (Space) of the National Security Council as its top policy-making body for space issues. SIG (Space) discussed the issue of continuing Shuttle production in the context of preparing the fiscal year 1984 budget. The recommended option of continuing to maintain Shuttle production capability by building "structural spares" was accepted; the existence of those spares made it possible to replace Challenger *more quickly than otherwise would have been the case.*

---

[1]   NATIONAL SECURITY COUNCIL
                      SENIOR INTERAGENCY GROUP (SPACE)

Issue Paper on the
Space Transportation System's (STS) Fifth Orbiter

ISSUE: Should Orbiter production capability, in the form of the initiation of a fifth
        Orbiter, be supported in the NASA FY 84 budget?

BACKGROUND:

In 1969, NASA adopted a program plan to develop a manned Space Transportation System (STS) based largely on reuseable [sic] components; this system was conceived to provide cost-effective, routine, manned access to space. Economics and politics, as well as technology, were all critical factors in the decision process that led to President Nixon's approval of the STS development in 1972.

The number of Orbiters in the STS fleet required for responsive and dependable operations has been the subject of intense scrutiny by NASA, Congress, and various Administrations for most of the 1970's. Original planning envisioned a five-Orbiter fleet. The estimates of STS demand and the numbers of Orbiters to fulfill demand have fluctuated through the life of the program. These fluctuations have led to a series of examinations of fleet size.

In preparation for the FY 1977 budget, the Office of Management and Budget (OMB) undertook a review of the STS mission model and examined the need for more Orbiters than the two then on order. To support the OMB review, NASA and the DOD jointly reviewed the requirements and issued a statement that five Orbiters were essential to meet National requirements. The Ford Administration concurred with this assessment.

In 1977, the Carter Administration, again reviewed the question of Orbiter fleet size. The result was that funding for a restricted four-Orbiter fleet would be requested by NASA. During subsequent consideration of the FY 78 budget, the Senate authorization stated that: (1) a five-Orbiter fleet was an option which should be kept open; and that (2) interrupting production of Orbiters between the fourth and fifth Orbiter would have cost penalties. In February 1980, NASA testified before Congress that, due to slips in all parts of the STS program, a delay in fifth-Orbiter funding until FY 1982 would probably not cause substantial penalties. Subsequently, NASA's testimony in both the FY 82 and FY 83 budget hearings has underscored the need to maintain Orbiter production and the need to commit to the fifth Orbiter as soon as possible.

[2] National Space Policy (NSDD-42, July 4, 1982) states:
- The United States is fully committed to maintaining world leadership in space transportation with an STS capacity sufficient to meet appropriate National needs.
- The STS is the primary U.S. Government space launch system.
- The STS shall be available to authorized users—domestic and foreign, commercial and government.
- The STS program requires sustained commitments by all affected departments and agencies.
- Major changes to the STS program capabilities will require Presidential approval.

Space Assistance and Cooperation Policy (NSDD-50, August 6, 1982) states:
- U.S. space launch assistance will be available to interested countries, international organizations, or foreign business entities for those spacecraft projects which are for peaceful purposes.

The fourth Orbiter is currently scheduled for December 1984 delivery. Unless a decision is made to continue Orbiter production in FY 84, the production line base of facilities, personnel, and major subcontractors will be closed down.

DISCUSSION:

The decision for or against the fifth Orbiter is in reality a decision whether or not to truncate the production program and, therefore, the system capability at a point which will assure that, under the most favorable conditions, the maximum flight rate will be limited to approximately 26 flights per year. The second equally important consideration is the question of the loss of an Orbiter and, without a production base, can an operationally viable and responsive system, capable of absorbing problems and contingencies, be assured to meet U.S. launch requirements?

[3] Demand & Capacity

Many projections of both launch demand and STS capacity have been made during the past 10 years. Current projections of demand and capacity indicate that it is probable that by 1988 launch demand will approximate the capacity of the four-Orbiter fleet, providing there are no major operational problems reducing that capacity. The fact is, however, that 5 to 15-year projections of demand in an environment as dynamic as space are not reliable enough to use as a major argument on this issue. Similarly, NASA's first five launches have not provided an adequate data base to use for sound projections of capacity over the manifest period. The fundamental question which must be decided is: is it wise today to dismantle the capability to produce and repair the key element of the primary U.S. Government launch system?

This issue must be clearly faced by those government users who are beginning to develop spacecraft compatible only with the Shuttle and to plan program operations dependent upon the unique capabilities provided by the Shuttle. It must also be faced by commercial and foreign users who have committed to use of this capability.

Other Issues

Prematurely constraining the U.S. to a four-Orbiter fleet could erode confidence in the STS as a viable, dependable approach to space transportation. Foreign nations' perception of the U.S. as a questionable source of launch services may be reinforced; we must offer them a service that is available and reliable to their needs as they, not we, perceive them. Both U.S. and foreign commercial customers could also view the STS capacity as inadequate to assure firm launch dates. The business community is primarily concerned with schedules; significant launch delays rapidly translate into large economic penalties. Perception that the U.S. is turning away from its commitment to a fully exploitable STS could accelerate the transition of foreign and commercial customers from STS planning to other options.

Abandoning the reimbursable market would constitute an abrupt change in policy and would place the entire burden of STS operations on the U.S. Government. This will seriously undermine the entire concept of a viable Shuttle based Space Transportation System.

CONCLUSIONS:

National Space Policy commits the U.S. to maintain U.S. world leadership in space transportation.

World leadership requires a strong, responsive, dependable and cost effective STS operation.

National Space Policy mandates the STS be made available to authorized commercial and foreign users.

[4] The U.S. Government should not turn only to its own needs at this time and abandon the commercial and foreign market, as this would constitute an abrupt change in policy and be counter to U.S. interests as it would increase the costs to the U.S. Government users. Any actions that result in increased STS operation costs to the U.S. Government should be avoided.

Projected STS four Orbiter fleet capacity only marginally satisfies launch service demand starting in about 1988 with little or no reserve. These 5-year projections of capability and demand, however, are uncertain.

A STS three-Orbiter fleet (loss of one Orbiter) does not satisfy launch service demand in 1988.

The U.S. Government should not abandon STS Orbiter and ELV production bases concurrently as planned. Prudent and sound management requires maintaining Orbiter production capability until STS fleet capacity and demand are better known.

A balanced, low-risk option should be selected that preserves basic production capabilities, assures maximum insensitivity to errors in projecting system capabilities, as well as demand, and yet retains the flexibility to adopt new options when firm data is available.

OPTIONS:

The SIG considered three options: (1) close-out Orbiter production capability after the delivery of the fourth Orbiter (December 1984); (2) maintain Orbiter production capability for an additional 1 to 2 years by manufacturing selected structural parts and major structural assemblies; and (3) continue full Orbiter production leading to delivery of the fifth Orbiter in late 1988.

I.   Close-Out Orbiter Production Capability

Shuts down the production line and tooling for fabricating all major structural assemblies (i.e., wings, fuselage, tail, etc.).

The cost (in millions of FY 84 dollars) is:

FY 84	FY 85	FY 86	FY 87	Total
$ 65	$ 85	$ 40	$ 40	$230

PRO—Avoids costs of $1.6B through FY 89.

CON—Limits the Space Shuttle fleet to a maximum capacity of about 26 flights per year through at least 1992. (If one Orbiter is lost, capability is reduced 25 percent.) Increases lead time to delivery for a replacement Orbiter to 6 to 8 years from startup. Costs to restart production line and deliver a fifth unit will be substantially higher (perhaps 20–30%). U.S. jobs for approximately 6,000 direct production people will be lost in FY 84 and a total of about 45,000 jobs over 6 years.

[5] II.   Maintain Orbiter Production Capability

This option maintains the production capacity of selected structural parts and major structural assemblies for an additional 1 to 2 years (depending on the element) beyond the normal close-down of the fourth Orbiter. The selected structural parts are those most likely to be damaged in handling incidents or landing accidents (rudder, elevons, speed brake, landing gear, landing gear doors). The major structural assemblies are the wings, engine compartment, crew module, including the nose and cockpit, the mid and aft fuselage sections, payload doors, vertical tail, etc. The cost (in millions of FY 84 dollars) is:

FY 84	FY 85	FY 86	FY 87	TOTAL
$90–110	$100–120	$90–115	$60–90	$350–435

PRO
Capacity to fabricate and repair major structural assemblies is maintained. The decision to assemble and complete the fifth Orbiter can be postponed 1 to 2 years until uncertainties concerning the "inherent" capacity of a four-Orbiter fleet and the far term demand for launch services are better understood. Should an additional Orbiter not be required, the cost savings can be significant (about $1.0B) depending upon the time of the decision. The delivered major structural assemblies, if not assembled into an Orbiter, are valuable insurance assets for major repairs. If a fifth Orbiter is required, the lead time from start to delivery is reduced about 1 to 2 years.

CON

Does not increase STS reserve capacity and instill commercial and foreign confidence in the resilience of the system. Critical skills to integrate, install, assemble and test an Orbiter will be lost. Total cost of the fifth Orbiter, if ordered, will be higher and could not be delivered before about 1990.

III. Continue Full Orbiter Production

All elements of production are continued through the delivery of the fifth Orbiter in late 1988.

The cost (in millions of FY 84 dollars) is:

FY 84	FY 85	FY 86	FY 87	FY 88	FY 89	TOTAL
$200	$325	$350	$350	$320	$ 50	$1,595

[6] PRO

Hedges against program uncertainties by bringing on-line reserve capacity at the earliest time and the lowest cost. The production base is available to respond to major repairs and structural maintenance needs. If experience shows the Orbiter is not required, unassembled components are valuable spares.

CON

Uncertainty in STS capacity and demand prevent conclusively establishing the need for a fifth Orbiter. These uncertainties should first be resolved before committing to another Orbiter.

POSITIONS OF THE SIG (SPACE) MEMBERSHIP

The Senior Interagency Group (Space) met on December 3 to consider NASA's request to continue Orbiter production with the start of the fifth Orbiter in FY 84. Provided below are the stated positions of the voting members and observers.

Agency	Option Selected
DOD	2 or 3
[Joint Chiefs of Staff]	2 or 3
NASA	2 or 3
[Central Intelligence Agency]	2
Commerce	3
State	2
[Arms Control and Disarmament Agency]	3
OMB	2
OSTP	2

In summary, there were no votes for Option I. All members felt that, as a minimum, the production line should be maintained (Option II) if only as a hedge against the uncertainties in STS capacity and demand. The Department of Commerce urged NASA to continue evaluation of possibilities for commercial funding of the fifth Orbiter.

RECOMMENDATION

Option II (maintain Orbiter production capability) is recommended for approval. This option is judged to be in the best overall interest of the United States. It satisfies the main concerns of the majority of the SIG (Space) members most affected by the decision (i.e., terminating total U.S. space launch production capacity) and preserves the option for full production of a fifth Orbiter at a future date should optimistic estimates of demand actually materialize. It also preserves the capability [7] to repair and replace major components of a Shuttle Orbiter that might be damaged in an accident or other unfavorable event. Furthermore, the marginal difference in cost between Options I and II (approximately $230 Million versus $400 Million over the FY 84–87 period) is judged to be worth the significant gain to the Nation.

Document II-37

Document title: James C. Fletcher, Consultant, Memorandum to Al Lovelace, "Personal Concern about the Launch Phase of Space Shuttle," July 7, 1977.

Source: NASA Historical Reference Collection, NASA History Office, NASA Headquarters, Washington, D.C.

Even after he left his position as NASA administrator after President Jimmy was inaugurated, James Fletcher served as an occasional consultant to top NASA management. In this memorandum to NASA Deputy Administrator Al Lovelace, Fletcher raises some issues troubling him about whether there was enough attention being paid by experienced individuals to issues related to the Space Shuttle as a launch vehicle, as contrasted to an orbiting spacecraft. The "Bob" mentioned at the end of the memorandum was new NASA Administrator Robert A. Frosch; Walt Williams mentioned in the first line was at Johnson Space Center.

[1]   EYES ONLY

July 7, 1977

MEMORANDUM TO AL LOVELACE

SUBJECT: Personal Concern about the Launch Phase of Space Shuttle

As I have mentioned to you (and Walt Williams) several times previously, I am still concerned about the reliability of the Space Shuttle during the launch phase, i.e., from just before lift-off until the shuttle is in orbit. To illustrate the concern, perhaps a simple reminder of the statistics would be helpful. Thor-Deltas have a reliability of about 95%, Atlas Centaurs probably less, Titan III's about the same, and there isn't enough data on Titan Centaurs. I don't have the recent data on Minuteman and Polaris but they are much simpler launch vehicles than Thor, Atlas and Titan. Even Nike Hercules isn't much better than 97%. We are aiming at 99% or better on the shuttle.

The Saturn's had a remarkable success record although admittedly the sample was small. Huntsville had the experience of three decades working as a team before Saturn's were launched. During the Thor/Jupiter competition, it was my impression that Jupiter had a better success record even though the Thor team (Douglas, Rocketdyne, AC Spark Plug and TRW Systems (STL) [Space Technology Laboratories]) was usually behind and learned from Jupiter's experience. At TRW we knew Huntsville had the expertise but felt that because of the Army's "arsenal" concept, they would be slow in getting Jupiter into production. This turned out to be the case and because of this and other reasons, only Thor survived.

The Saturn launch vehicle was more complicated than any of its predecessors and at least the leadership at Huntsville knew it (Wernher von Braun, Ernst Stuhlinger,

Eberhardt Rees and, I believe, Rocco Petrone). Somehow [2] all their past experience and a great deal of team effort (each member trusting the other, no one afraid to admit a mistake, no real gambles taken and, above all, an enormous amount of hard work) made the Saturn a superb launch vehicle. Some would call it "luck" but those of us familiar with the Huntsville team knew better.

My very great concern is that no one at NASA that I know of is looking at the Space Shuttle as a launch vehicle! That is perhaps an exaggeration because someone on Bob Thompson's staff (loaned by Huntsville, I believe) is probably now designated as "Mr. Launch Phase" but I haven't met him and I don't know how he interfaces with Bob and with Huntsville. Let's hope, in my ignorance, that I've misjudged the difficulty of the problem. Regardless of this, I'm certain that it is still true that Huntsville from Bill Lucas on down, does not feel it has the same responsibility for the Shuttle (during the launch phase) that it had for Saturn. In fact, Bob Thompson and Chris Kraft feel this is Houston's responsibility just as the orbit phase and the re-entry phase are also their responsibility.

Let's, then, take a look at the people involved in Shuttle development. John Yardley, a brilliant engineer, has had vast experience wit-h Space Vehicles starting with Mercury and especially Gemini. Before that it was with airplanes. Mike Malkin is a physicist with electronics background but no rocket experience. I was unable to isolate anyone on John's staff who had had rocket (i.e., launch vehicle) experience except those transferred over from [the Office of Space Science] (Joe Mahon's group) and these latter had had oversight (i.e., staff) background, not design background.

Bob Thompson, Chris Kraft, Aaron Cohen and most of the others at Houston (of course, I don't know all of them) have had backgrounds similar to John's. Dale Myers, John's predecessor, and George Low also had similar backgrounds. In fact George often remarked that when it came to launch vehicles—especially Saturn—he pretty well left that up to Huntsville. (George was perhaps overly modest [3] because when we had problems with stress corrosion on the Saturn 1B's used with Skylab and Apollo-Soyuz, George was able to locate the appropriate people to deal with the problem and solve it. Most of them were from Huntsville, however, or outside of NASA.)

Who, then, are the "experts" on launch vehicles nowadays? Besides Huntsville, you could probably include Walt Williams and perhaps some at Goddard and Lewis, but I doubt if any of these are design, development or test engineers. I'm not really sure—Walt would know. As far as I know, there are none at Rockwell. There are some at Martin-Denver, General Dynamics and [McDonnell Douglas Astronautics Company] (West Coast) but I'm under the impression that these had intensive oversight initially from TRW Systems (STL) and later from Aerospace Corporation. There are, of course, systems engineers at TRW and Aerospace who have had launch vehicle backgrounds.

None of these have built anything as complex as the Saturn; and the Shuttle, because of its unusual configuration, is probably much more complicated than the Saturn! Why doesn't space vehicle experience or aircraft experience help during the launch phase? It may but the design problems are entirely different—you can't expect workarounds during the launch phase—there simply isn't enough time. Yet that's how unreliability problems are dealt with in aircraft and space vehicles ("manned" especially). You simulate them on the ground, then you flight test and when there are problems you fix them on the next test. You can't do this with launch vehicles—at least not with men aboard. All fixes must be automatic, in real time, in a matter of seconds which pretty well excludes any help from the ground. There was much discussion about this in the Air Force in the early 50's and that's why STL was set up—to convert "aircraft" engineers into "rocket" engineers by putting an outside group of "systems engineers" in charge of what they called "Systems Engineering and Technical Direction." Gradually, aircraft companies learned how to build rockets. In my opinion the real strength of STL came from JPL [Jet Propulsion Laboratory] (Louis Dunn, Frank Lehan and others) and Huntsville (Dolph Thiel-Thor) and some others with guided missile background (myself and others from Hughes Aircraft, Bell Labs, etc.). [4] JPL, at that time, had the principal "American-born" ballistic missile experts and they were pioneers in early smaller weapons (V2 testing, WAC Corporal, Corporal and Sergeant). JPL and Huntsville put up the first U.S. satellite, Explorer I.

What does all this mean, other than an "old-timer" saying how things used to be? In simple terms, if the state-of-the-art has developed rapidly since early Saturn days so that launch vehicles can be designed and tested "scientifically," i.e., without the intuitive background that a team learns only by trial-and-error, my concerns should be ignored.

If, in fact, the state-of-the-art has not developed so much (and I'm afraid I'm not a good judge of this having been away from it so long), NASA may have a problem. If I were alone in this concern, I'd probably not even bring up the subject. However, Rocco Petrone brought it up several times and our unresponsiveness may have contributed to his leaving. Wernher was concerned when we first announced Houston as the lead center. George Mueller (Dale Myers' predecessor) was furious with me over that decision and still believes we made a mistake. Bill Lucas and Dick Smith have individually expressed themselves to George Low and to me but, of course, they can't be regarded as unbiased. Lately, both have been "good soldiers" and I believe have tried very hard to cooperate with both John Yardley and with Houston. Sam Phillips, who ran both Apollo and Minuteman, was neutral; he said "any organization could be made to work."

When you put all this long-winded background together, what does it add up to? I'm afraid I don't have a good answer. I simply had to unload it on you partly to clear my conscience and partly to help make you especially cautious as you approach the [First Manned Orbital Flight] date. I have suggested a quiet review which you did make, but only Walt Williams in that review group had launch vehicle background. Eberhardt, Kurt Debus, Ernst Stuhlinger and perhaps some TRW Systems types might be helpful. I'm afraid Lockheed or Boeing [5] wouldn't be much help since Polaris and Minuteman are so much simpler in every respect than the Shuttle.

Anyway, the problem is yours and Bob's now. I intend to say no more about it.

James C. Fletcher
Consultant

Document II-38

Document title: President Ronald Reagan, Executive Order 12546, "Presidential Commission on the Space Shuttle *Challenger* Accident," February 3, 1986.

Source: NASA Historical Reference Collection, NASA History Office, NASA Headquarters, Washington, D.C.

In the aftermath of the Apollo 1 accident in January 1967, NASA organized its own investigation of the causes of the accident and the steps needed to correct them. After the Challenger accident, however, with NASA headed by an acting administrator with close White House ties, President Ronald Reagan created a Presidential Commission on the Space Shuttle Challenger Accident on February 6, 1986. This commission was chaired by former Secretary of State William Rogers and became known as the Rogers Commission.

[1]   EXECUTIVE ORDER

Presidential Commission on the Space Shuttle *Challenger* Accident

By the authority vested in me as President by the Constitution and statutes of the United States of America, including the Federal Advisory Committee Act, as amended (5 U.S.C. App. I), and in order to establish a commission of distinguished Americans to investigate the accident to the Space Shuttle *Challenger*, it is hereby ordered as follows:

Section 1. Establishment. (a) There is established the Presidential Commission on the Space Shuttle *Challenger* Accident. The Commission shall be composed of not more than 20 members appointed or designated by the President. The members shall be drawn among distinguished leaders of the government, and the scientific, technical, and management communities.

(b) The President shall designate a Chairman and a Vice Chairman from among the members of the Commission.

Sec. 2. Functions. (a) The Commission shall investigate the accident to the Space Shuttle *Challenger*, which occurred on January 28, 1986.

(b) The Commission shall:

(1) Review the circumstances surrounding the accident to establish the probable cause or causes of the accident; and

(2) Develop recommendations for corrective or other action based upon the Commission findings and determinations.

(c) The Commission shall submit its final report to the President and the Administrator of the National Aeronautics and Space Administration within one hundred and twenty days of the date of this Order.

Sec. 3. Administration. (a) The heads of Executive departments and agencies shall, to the extent permitted by law, provide the Commission with such information as it may require for purposes of carrying out its functions.

(b) Members of the Commission shall serve without compensation for their worry on the Commission. However, members appointed from among private citizens of the United States may be allowed travel expenses, including per diem in lieu of subsistence, to the extent permitted by law for persons serving intermittently in the government service (5 U.S.C. 5701–5707).

(c) To the extent permitted by law, and subject to the availability of appropriations, the Administrator of the National Aeronautics and Space Administration shall provide the Commission with such administrative services, funds, facilities, staff, and other support services as may be necessary for the performance of its functions.

Sec. 4. General Provisions. (a) Notwithstanding the provisions of any other Executive Order, the functions of the President under the Federal Advisory Committee Act which are applicable to the Commission, except that of reporting annually to the Congress, shall be performed by the Administrator of the National Aeronautics and Space Administration in accordance with guidelines and procedures established by the Administrator of General Services.

(b) The Commission shall terminate 60 days after submitting its final report.

RONALD REAGAN

THE WHITE HOUSE
    February 3, 1986.

Document II-39

Document title: Presidential Commission on the Space Shuttle *Challenger* Accident, "Report at a Glance," June 6, 1986.

Source: NASA Historical Reference Collection, NASA History Office, NASA Headquarters, Washington, D.C.

After four months of intense investigation, the Rogers Commission issued its final report, identifying schedule pressure and failures in communication as major nontechnical factors contributing to the accident. It also recommending a number of changes in NASA organization and procedures to reme- dy problems made visible by the Challenger *accident. The commission included former well-known astronauts Sally K. Ride and Neil A. Armstrong (who was vice chairman), as well as legendary test pilot Chuck Yeager. This "Report at a Glance" is a synopsis of excerpts from the final report.*

Report to the President

By The
PRESIDENTIAL
COMMISSION
on the Space Shuttle
Challenger Accident

Report at a Glance

[no page number]   Presidential Commission
                                              on the
                              Space Shuttle Challenger Accident

June 6, 1986

Dear Mr. President:

On behalf of the Commission, it is my privilege to present the report of the Presidential Commission on the Space Shuttle Challenger Accident.

Since being sworn in on February 6, 1986, the Commission has been able to conduct a comprehensive investigation of the Challenger accident. This report documents our findings and makes recommendations for your consideration.

Our objective has been not only to prevent any recurrence of the failure related to this accident, but to the extent possible to reduce other risks in future flights. However, the Commission did not construe its mandate to require a detailed evaluation of the entire Shuttle system. It fully recognizes that the risk associated with space flight cannot be totally eliminated.

Each member of the Commission shared the pain and anguish the nation felt at the loss of seven brave Americans in the Challenger accident on January 28, 1986.

The nation's task now is to move ahead to return to safe space flight and to its recognized position of leadership in space. There could be no more fitting tribute to the Challenger crew than to do so.

Sincerely,

William P. Rogers
Chairman

The President of the United States
The White House
Washington, D.C. 20500 . . .

[no page number] **Preface**

The accident of Space Shuttle Challenger, mission 51-L, interrupting for a time one of the most productive engineering, scientific and exploratory programs in history, evoked a wide range of deeply felt public responses. There was grief and sadness for the loss of seven brave members of the crew; firm national resolve that those men and women be forever enshrined in the annals of American heroes, and a determination, based on that resolve and in their memory, to strengthen the Space Shuttle program so that this tragic event will become a milestone on the way to achieving the full potential that space offers to mankind.

The President, who was moved and troubled by this accident in a very personal way, appointed an independent Commission made up of persons not connected with the mission to investigate it. The mandate of the Commission was to:

1. Review the circumstances surrounding the accident to establish the probable cause or causes of the accident; and

2. Develop recommendations for corrective or other action based upon the Commission's findings and determinations.

Immediately after being appointed, the Commission moved forward with its investigation and, with the full support of the White House, held public hearings dealing with the facts leading up to the accident. In a closed society other options are available; in an open society—unless classified matters are involved—other options are not, either as matter of law or as a practical matter.

In this case a vigorous investigation and full disclosure of the facts were necessary. The way to deal with a failure of this magnitude is to disclose all the facts fully and openly; to take immediate steps to correct mistakes that led to the failure; and to continue the program with renewed confidence and determination.

The Commission construed its mandate somewhat broadly to include recommendations on safety matters not necessarily involved in this accident but which require attention to make future flights safer. Careful attention was given to concerns expressed by astronauts because the Space Shuttle program will only succeed if the highly qualified men and women who fly the Shuttle have confidence in the system.

However, the Commission did not construe its mandate to require a detailed investigation of all aspects of the Space Shuttle program; to review budgetary matters; or to interfere with or supersede Congress in any way in the performance of its duties. Rather, the Commission focused its attention on the safety aspects of future flights based on the lessons learned from the investigation with the objective being to return to safe flight.

Congress recognized the desirability, in the first instance, of having a single investigation of this national tragedy. It very responsibly agreed to await the Commission's findings
before deciding what further action might be necessary to carry out its responsibilities.

For the first several days after the accident—possibly because of the trauma resulting
from the accident—NASA appeared to be withholding information about the accident
from the public. After the Commission began its work, and at its suggestion, NASA began
releasing a great deal of information that helped to reassure the public that all aspects of
the accident were being investigated and that the full story was being told in an orderly
and thorough manner.

Following the suggestion of the Commission, NASA established several teams of persons not involved in the mission 51-L launch process to support the Commission and its
panels. These NASA teams have cooperated with the Commission in every aspect of its
work. The result has been a comprehensive and complete investigation.

The Commission believes that its investigation and report have been responsive to the
request of the President and hopes that they will serve the best interests of the nation in
restoring the United States space program to its preeminent position in the world.

[no page number] *Chapter III*

The Accident

Just after liftoff at .678 seconds into the flight, photographic data show a strong puff
of gray smoke was spurting from the vicinity of the aft field joint on the right Solid Rocket
Booster. The two pad 39B cameras that would have recorded the precise location of the
puff were inoperative. Computer graphic analysis of film from other cameras indicated
the initial smoke came from the 270 to 310-degree sector of the circumference of the aft
field joint of the right Solid Rocket Booster. This area of the solid booster faces the
External Tank. The vaporized material streaming from the joint indicated there was not
complete sealing action within the joint.

Eight more distinctive puffs of increasingly blacker smoke were recorded between
.836 and 2.500 seconds. The smoke appeared to puff upwards from the joint. While each
smoke puff was being left behind by the upward flight of the Shuttle, the next fresh puff
could be seen near the level of the joint. The multiple smoke puffs in this sequence
occurred at about four times per second, approximating the frequency of the structural
load dynamics and resultant joint flexing. Computer graphics applied to NASA photos
from a variety of cameras in this sequence again placed the smoke puffs' origin in the
270- to 310-degree sector of the original smoke spurt.

As the Shuttle increased its upward velocity, it flew past the emerging and expanding
smoke puffs. The last smoke was seen above the field joint at 2.733 seconds.

The black color and dense composition of the smoke puffs suggest that the grease,
joint insulation and rubber O-rings in the joint seal were being burned and eroded by the
hot propellant gases.

At approximately 37 seconds, Challenger encountered the first of several high-
altitude wind shear conditions, which lasted until about 64 seconds. The wind shear created forces on the vehicle with relatively large fluctuations. These were immediately
sensed and countered by the guidance, navigation and control system.

The steering system (thrust vector control) of the Solid Rocket Booster responded to
all commands and wind shear effects. The wind shear caused the steering system to be
more active than on any previous flight.

Both the Shuttle main engines and the solid rockets operated at reduced thrust
approaching and passing through the area of maximum dynamic pressure of 720 pounds
per square foot. Main engines had been throttled up to 104 percent thrust and the Solid

Rocket Boosters were increasing their thrust when the first flickering flame appeared on the right Solid Rocket Booster in the area of the aft field joint. This first very small flame was detected on image enhanced film at 58.788 seconds into the flight. It appeared to originate at about 305 degrees around the booster circumference at or near the aft field joint.

One film frame later from the same camera, the flame was visible without image enhancement. It grew into a continuous, well-defined plume at 59.262 seconds. At about the same time (60 seconds), telemetry showed a pressure differential between the chamber pressures in the right and left boosters. The right booster chamber pressure was lower, confirming the growing leak in the area of the field joint.

As the flame plume increased in size, it was deflected rearward by the aerodynamic slipstream and circumferentially by the protruding structure of the upper ring attaching the booster to the External Tank. These deflections directed the flame plume onto the surface of the External Tank. This sequence of flame spreading is confirmed by analysis of the recovered wreckage. The growing flame also impinged on the strut attaching the Solid Rocket Booster to the External Tank.

The first visual indication that swirling flame from the right Solid Rocket Booster breached the External Tank was at 64.660 seconds when there was an abrupt change in the shape and color of the plume. This indicated that it was mixing with leaking hydrogen from the External Tank. Telemetered changes in the hydrogen tank pressurization [no page number] confirmed the leak. Within 45 milliseconds of the breach of the External Tank, a bright sustained glow developed on the black-tiled underside of the Challenger between it and the External Tank.

Beginning at about 72 seconds, a series of events occurred extremely rapidly that terminated the flight. Telemetered data indicate a wide variety of flight system actions that support the visual evidence of the photos as the Shuttle struggled futilely against the forces that were destroying it.

At about 72.20 seconds the lower strut linking the Solid Rocket Booster and the External Tank was severed or pulled away from the weakened hydrogen tank permitting the right Solid Rocket Booster to rotate around the upper attachment strut. This rotation is indicated by divergent yaw and pitch rates between the left and right Solid Rocket Boosters.

At 73.124 seconds, a circumferential white vapor pattern was observed blooming from the side of the External Tank bottom dome. This was the beginning of the structural failure of the hydrogen tank that culminated in the entire aft dome dropping away. This released massive amounts of liquid hydrogen from the tank and created a sudden forward thrust of about 2.8 million pounds, pushing the hydrogen tank upward into the intertank structure. At about the same time, the rotating right Solid Rocket Booster impacted the intertank structure and the lower part of the liquid oxygen tank. These structures failed at 73.137 seconds as evidenced by the white vapors appearing in the intertank region.

Within milliseconds there was massive, almost explosive, burning of the hydrogen streaming from the failed tank bottom and the liquid oxygen breach in the area of the intertank.

At this point in its trajectory, while traveling at a Mach number of 1.92 at an altitude of 46,000 feet, the Challenger was totally enveloped in the explosive burn. The Challenger's reaction control system ruptured and a hypergolic burn of its propellants occurred as it exited the oxygen-hydrogen flames. The reddish brown colors of the hypergolic fuel burn are visible on the edge of the main fireball. The Orbiter, under severe aerodynamic loads, broke into several large sections which emerged from the fireball. Separate sections that can be identified on film include the main engine/tail section with the engines still burning, one wing of the Orbiter, and the forward fuselage trailing a mass of umbilical lines pulled loose from the payload bay.

[no page number] **STS 51-L Sequence of Major Events**

Mission Time (GMT, in hr:min:sec)	Event	Elapsed Time (secs.)	Source
16:37:53.444	ME-3 Ignition Command	−6.566	GPC
37:53.564	ME-2 Ignition Command	−6.446	GPC
37:53.684	ME-1 Ignition Command	−6.326	GPC
38:00.010	SRM Ignition Command (T=O)	0.000	GPC
38:00.018	Holddown Post 2 PIC firing	0.008	E8 Camera
38:00.260	First Continuous Vertical Motion	0.250	E9 Camera
38:00.688	Confirmed smoke above field joint on RH SRM	0.678	E60 Camera
38:00.846	Eight puffs of smoke (from 0.836 thru 2.500 sec MET)	0.836	E63 Camera
38:02.743	Last positive evidence of smoke above right aft SRB/ET attach ring	2.733	CZR-1 Camera
38:03.385	Last positive visual indication of smoke	3.375	E60 Camera
38:04.349	SSME 104% Command	4.339	E41M2076D
38:05.684	RH SRM pressure 11.8 psi above nominal	5.674	B47P2302C
38:07.734	Roll maneuver initiated	7.724	V90R5301C
38:19.869	SSME 94% Command	19.859	E41M2076D
38:21.134	Roll maneuver completed	21.124	V90R5301C
38:35.389	SSME 65% Command	35.379	E41M2076D
38:37.000	Roll and Yaw Attitude Response to Wind (36.990 to 62.990 sec)	36.990	V95H352nC
38:51.870	SSME 104% Command	51.860	E41M2076D
38:58.798	First evidence of flame on RH SRM	58.788	E207 Camera
38:59.010	Reconstructed Max Q (720 psf)	59.000	BET
38:59.272	Continuous well defined plume on RH SRM	59.262	E207 Camera
38:59.763	Flame from RH SRM in +Z direction (seen from south side of vehicle)	59.753	E204 Camera
39:00.014	RM pressure divergence (RH vs. LH)	60.004	B47P2302
39:00.248	First evidence of plume deflection, intermittent	60.238	E207 Camera
39:00.258	First evidence of SRB plume attaching to ET ring frame	60.248	E203 Camera
39:00.998	First evidence of plume deflection, continuous	60.988	E207 Camera
39:01.734	Peak roll rate response to wind	61.724	V90R5301C
39:02.094	Peak TVC response to wind	62.084	B58H1150C
39:02.414	Peak yaw rate response to wind	62.404	V90R5341C
39:02.494	RH outboard elevon actuator hinge moment spike	62.484	V58P0966C
39:03.934	RH outboard elevon actuator delta pressure change	63.924	V58P0966C
39:03.974	Start of planned pitch rate maneuver	63.964	V90R5321C
39:04.670	Change in anomalous plume shape (LH$_2$ tank leak near 2058 ring frame)	64.660	E204 Camera
39:04.715	Bright sustained glow on sides of ET	64.705	E204 Camera
39:04.947	Start SSME gimbal angle large pitch variations	64.937	V58H1100A
39:05.174	Beginning of transient motion due to changes in aero forces due to plume	65.164	V90R5321C

Mission Time (GMT, in hr:min:sec)	Event	Elapsed Time (secs.)	Source
39:05.534	LH outboard elevon actuator delta pressure change	65.524	V58P0866C
39:06.774	Start ET LH$_2$ ullage pressure deviations	66.764	T41P1700C
39:12.214	Start divergent yaw rates (RH vs. LH SRB)	72.204	V90R2528C

[no page number]

Mission Time (GMT, in hr:min:sec)	Event	Elapsed Time (secs.)	Source
39:12.294	Start divergent pitch rates (RH vs. LH SRB)	72.284	V90R2525C
39:12.488	SRB major high-rate actuator command	72.478	V79H2111A
39:12.507	SSME roll gimbal rates 5 deg/sec	72.497	V58HI100A
39:12.535	Vehicle max +Y lateral acceleration (+.227 g)	72.525	V98AI581C
39:12.574	SRB major high-rate actuator motion	72.564	B58HI151C
39:12.574	Start of H$_2$ tank pressure decrease with control valves open	2 flow 72.564	T41PI700C
39:12.634	Last state vector downlinked	72.624	Data reduction
39:12.974	Start of sharp MPS LOX inlet pressure drop	72.964	V41PI330C
39:13.020	Last full computer frame of TORS data	73.010	Data reduction
39:13.054	Start of sharp MPS LH$_2$ inlet pressure drop	73.044	V41PI100C
39:13.055	Vehicle max −Y lateral acceleration (−.254 g)	73.045	V98AI581C
39:13.134	Circumferential white pattern on ET aft dome (LH$_2$ tank failure)	73.124	E204 Camera
39:13.134	RH SRM pressure 19 psi lower than LH SRM	73.124	B47P2302C
39:13.147	First hint of vapor at intertank	73.137	E207 Camera
39:13.153	All engine systems start responding to loss of fuel and LOX inlet pressure	73.143	SSME team
39:13.172	Sudden cloud along ET between intertank and aft dome	73.162	E207 Camera
39:13.201	Flash between Orbiter and LH$_2$ tank	73.191	E204 Camera
39:13.221	SSME telemetry data interference from 73.211 to 73.303	73.211	
39:13.223	Flash near SRB fwd attach and brightening of flash between Orbiter and ET	73.213	E204 Camera
39:13.292	First indication intense white flash at SRB fwd attach point	73.282	E204 Camera
39:13.337	Greatly increased intensity of white flash	73.327	E204 Camera
39:13.387	Start RCS jet chamber pressure fluctuations	73.377	V42P1552A
39:13.393	All engines approaching HPFT discharge temp redline limits	73.383	E41Tn010D
39:13.492	ME-2 HPFT disch. temp Chan. A vote for shutdown; 2 strikes on Chan. B	73.482	MEC data
39:13.492	ME-2 controller last time word update	73.482	MEC data
39:13.513	ME-3 in shutdown due to HPFT discharge temperature redline exceedance	73.503	MEC data
39:13.513	ME-3 controller last time word update	73.503	MEC data
39:13.533	ME-1 in shutdown due to HPFT discharge temperature redline exceedance	73.523	Calculation
39:13.553	ME-1 last telemetered data point	73.543	Calculation
39:13.628	Last validated Orbiter telemetry measurement	73.618	V46P0I20A

Mission Time (GMT, in hr:min:sec)	Event	Elapsed Time (secs.)	Source
39:13.641	End of last reconstructed data frame with valid synchronization and frame count	73.631	Data reduction
39:14.140	Last radio frequency signal from Orbiter	74.130	Data reduction
39:14.597	Bright flash in vicinity of Orbiter nose	74.587	E204 Camera
39:16.447	RH SRB nose cap sep/chute deployment	76.437	E207 Camera
39:50.260	RH SRB RSS destruct	110.250	E202 Camera
39:50.262	LH SRB RSS destruct	110.252	E230 Camera

[no page number]

ACT POS	–	Actuator Position
APU	–	Auxilliary [sic] Power Unit
BET	–	Best Estimated Trajectory
CH	–	Channel
DISC	–	Discharge
ET	–	External Tank
GG	–	Gas Generator
GPC	–	General Purpose Computer
GMT	–	Greenwich Mean Time
HPFT	–	High Pressure Fuel Turbopump
LH	–	Lefthand
LH$_2$	–	Liquid Hydrogen
LO$_2$	–	Liquid Oxygen (same as LOX)
MAX Q	–	Maximum Dynamic Pressure
ME	–	Main Engine (same as SSME)
MEC	–	Main Engine Controller
MET	–	Mission Elapsed Time
MPS	–	Main Propulsion System
PC	–	Chamber Pressure
PIC	–	Pyrotechnics Intitiator [sic] Controller
psf	–	Pounds per square foot
RCS	–	Reaction Control System
RGA	–	Rate Gyro Assembly
RH	–	Righthand
RSS	–	Range Safety System
SRM	–	Solid Rocket Motor
SSME	–	Space Shuttle Main Engine
TEMP	–	Temperature
TVC	–	Thrust Vector Control

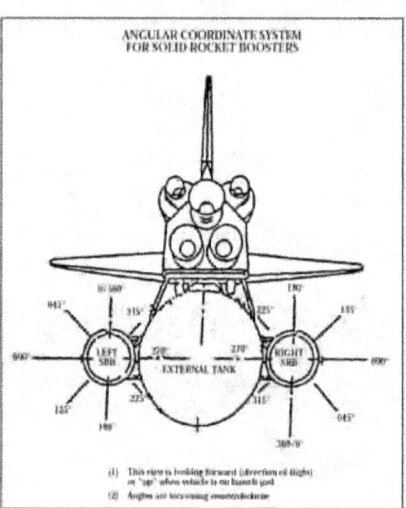

ANGULAR COORDINATE SYSTEM FOR SOLID ROCKET BOOSTERS

NOTE: The Shuttle coordinate system used in Chapter 3 is relative to the Orbiter, as follows:

+X direction = forward (tail to nose)
–X direction = rearward (nose to tail)
+Y direction = right (toward the right wing tip)
–Y direction = left (toward the left wing tip)
+Z direction = down
–Z direction = up

[no page number] *Chapter IV*

The Cause of the Accident

The consensus of the Commission and participating investigative agencies is that the loss of the Space Shuttle Challenger was caused by a failure in the joint between the two lower segments of the right Solid Rocket Motor. The specific failure was the destruction of the seals that are intended to prevent hot gases from leaking through the joint during the propellant burn of the rocket motor. The evidence assembled by the Commission indicates that no other element of the Space Shuttle system contributed to this failure.

In arriving at this conclusion, the Commission reviewed in detail all available data, reports and records; directed and supervised numerous tests, analyses, and experiments by NASA, civilian contractors and various government agencies; and then developed specific failure scenarios and the range of most probable causative factors.

Findings

1. A combustion gas leak through the right Solid Rocket Motor aft field joint initiated at or shortly after ignition eventually weakened and/or penetrated the External Tank initiating vehicle structural breakup and loss of the Space Shuttle Challenger during STS Mission 51-L.
2. The evidence shows that no other STS 51-L Shuttle element or the payload contributed to the causes of the right Solid Rocket Motor aft field joint combustion gas leak. Sabotage was not a factor.
3. Evidence examined in the review of Space Shuttle material, manufacturing, assembly, quality control, and processing of nonconformance reports found no flight hardware shipped to the launch site that fell outside the limits of Shuttle design specifications.
4. Launch site activities, including assembly and preparation, from receipt of the flight hardware to launch were generally in accord with established procedures and were not considered a factor in the accident.
5. Launch site records show that the right Solid Rocket Motor segments were assembled using approved procedures. However, significant out-of-round conditions existed between the two segments joined at the right Solid Rocket Motor aft field joint (the joint that failed).
   a. While the assembly conditions had the potential of generating debris or damage that could cause O-ring seal failure, these were not considered factors in this accident.
   b. The diameters of the two Solid Rocket Motor segments had grown as a result of prior use.
   c. The growth resulted in a condition at time of launch wherein the maximum gap between the tang and clevis in the region of the joint's O-rings was no more than .008 inches and the average gap would have been .004 inches.
   d. With a tang-to-clevis gap of .004 inches, the O-ring in the joint would be compressed to the extent that it pressed against all three walls of the O-ring retaining channel.
   e. The lack of roundness of the segments was such that the smallest tang-to-clevis clearance occurred at the initiation of the assembly operation at positions of 120 degrees and 300 degrees around the circumference of the aft field joint. It is uncertain if this tight condition and the resultant greater compression of the O-rings at these points persisted to the time of launch.

[no page number]
6. The ambient temperature at time of launch was 36 degrees Fahrenheit, or 15 degrees lower than the next coldest previous launch.

    a.   The temperature at the 300 degree position on the right aft field joint circum-ference was estimated to be 28 degrees ± 5 degrees Fahrenheit. This was the cold-est point on the joint.

    b.   Temperature on the opposite side of the right Solid Rocket Booster facing the sun was estimated to be about 50 degrees Fahrenheit.

7.   Other joints on the left and right Solid Rocket Boosters experienced similar combi-nations of tang-to-clevis gap clearance and temperature. It is not known whether these joints experienced distress during the flight of 51-L.

8.   Experimental evidence indicates that due to several effects associated with the Solid Rocket Booster's ignition and combustion pressures and associated vehicle motions, the gap between the tang and the clevis will open as much as .017 and .029 inches at the secondary and primary O-rings, respectively.

    a.   This opening begins upon ignition, reaches its maximum rate of opening at about 200–300 milliseconds, and is essentially complete at 600 milliseconds when the Solid Rocket Booster reaches its operating pressure.

    b.   The External Tank and right Solid Rocket Booster are connected by several struts, including one at 310 degrees near the aft field joint that failed. This strut's effect on the joint dynamics is to enhance the opening of the gap between the tang and clevis by about 10–20 percent in the region of 300–320 degrees.

9.   O-ring resiliency is directly related to its temperature.

    a.   A warm O-ring that has been compressed will return to its original shape much quicker than will a cold O-ring when compression is relieved. Thus, a warm O-ring will follow the opening of the tang-to-clevis gap. A cold O-ring may not.

    b.   A compressed O-ring at 75 degrees Fahrenheit is five times more responsive in returning to its uncompressed shape than a cold O-ring at 30 degrees Fahrenheit.

    c.   As a result it is probable that the O-rings in the right solid booster aft field joint were not following the opening of the gap between the tang and clevis at time of ignition.

10. Experiments indicate that the primary mechanism that actuates O-ring sealing is the application of gas pressure to the upstream (high-pressure) side of the O-ring as it sits in its groove or channel.

    a.   For this pressure actuation to work most effectively, a space between the O-ring and its upstream channel wall should exist during pressurization.

    b.   A tang-to-clevis gap of .004 inches, as probably existed in the failed joint, would have initially compressed the O-ring to the degree that no clearance existed between the O-ring and its upstream channel wall and the other two surfaces of the channel.

    c.   At the cold launch temperature experienced, the O-ring would be very slow in returning to its normal rounded shape. It would not follow the opening of the tang-to-clevis gap. It would remain in its compressed position in the O-ring chan-nel and not provide a space between itself and the upstream channel wall. Thus, it is probable the O-ring would not be pressure actuated to seal the gap in time to preclude joint failure due to blow-by and erosion from hot combustion gases.

11. The sealing characteristics of the Solid Rocket Booster O-rings are enhanced by time-ly application of motor pressure.

    a.   Ideally, motor pressure should be applied to actuate the O-ring and seal the joint prior to significant opening of the tang-to-clevis gap (100 to 200 milliseconds after motor ignition).

    b.   Experimental evidence indicates that temperature, humidity and other variables in the putty compound used to seal the joint can delay pressure application to the joint by 500 milliseconds or more.

    c.   This delay in pressure could be a factor in initial joint failure.

[no page number]
12. Of 21 launches with ambient temperatures of 61 degrees Fahrenheit or greater, only four showed signs of O-ring thermal distress; i.e., erosion or blow-by and soot. Each of the launches below 61 degrees Fahrenheit resulted in one or more O-rings showing signs of thermal distress.
   a.   Of these improper joint sealing actions, one-half occurred in the aft field joints, 20 percent in the center field joints, and 30 percent in the upper field joints. The division between left and right Solid Rockter [sic] Boosters was roughly equal.
   b.   Each instance of thermal O-ring distress was accompanied by a leak path in the insulating putty. The leak path connects the rocket's combustion chamber with the O-ring region of the tang and clevis. Joints that actuated without incident may also have had these leak paths.
13. There is a possibility that there was water in the clevis of the STS 51-L joints since water was found in the STS-9 joints during a destack operation after exposure to less rainfall than STS 51-L. At time of launch, it was cold enough that water present in the joint would freeze. Tests show that ice in the joint can inhibit proper secondary seal performance.
14. A series of puffs of smoke were observed emanating from the 51-L aft field joint area of the right Solid Rocket Booster between 0.678 and 2.500 seconds after ignition of the Shuttle Solid Rocket Motors.
   a.   The puffs appeared at a frequency of about three puffs per second. This roughly matches the natural structural frequency of the solids at lift off and is reflected in slight cyclic changes of the tang-to-clevis gap opening.
   b.   The puffs were seen to be moving upward along the surface of the booster above the aft field joint.
   c.   The smoke was estimated to originate at a circumferential position of between 270 degrees and 315 degrees on the booster aft field joint, emerging from the top of the joint.
15. This smoke from the aft field joint at Shuttle lift off was the first sign of the failure of the Solid Rocket Booster O-ring seals on STS 51-L.
16. The leak was again clearly evident as a flame at approximately 58 seconds into the flight. It is possible that the leak was continuous but unobservable or non-existent in portions of the intervening period. It is possible in either case that thrust vectoring and normal vehicle response to wind shear as well as planned maneuvers reinitiated or magnified the leakage from a degraded seal in the period preceding the observed flames. The estimated position of the flame, centered at a point 307 degrees around the circumference of the aft field joint, was confirmed by the recovery of two fragments of the right Solid Rocket Booster.
   a.   A small leak could have been present that may have grown to breach the joint in flame at a time on the order of 58 to 60 seconds after lift off.
   b.   Alternatively, the O-ring gap could have been resealed by deposition of a fragile buildup of aluminum oxide and other combustion debris. This resealed section of the joint could have been disturbed by thrust vectoring, Space Shuttle motion and flight loads induced by changing winds aloft.
   c.   The winds aloft caused control actions in the time interval of 32 seconds to 62 seconds into the flight that were typical of the largest values experienced on previous missions.

Conclusion

In view of the findings, the Commission concluded that the cause of the Challenger accident was the failure of the pressure seal in the aft field joint of the right Solid Rocket Motor. The failure was due to a faulty design unacceptably sensitive to a number of factors. These factors were the effects of temperature, physical dimensions, the character of materials, the effects of reusability, processing, and the reaction of the joint to dynamic loading.

[no page number] *Chapter V*

The Contributing Cause of the Accident

The decision to launch the Challenger was flawed. Those who made that decision were unaware of the recent history of problems concerning the O-rings and the joint and were unaware of the initial written recommendation of the contractor advising against the launch at temperatures below 53 degrees Fahrenheit and the continuing opposition of the engineers at Thiokol after the management reversed its position. They did not have a clear understanding of Rockwell's concern that it was not safe to launch because of ice on the pad. If the decisionmakers had known all of the facts, it is highly unlikely that they would have decided to launch 51-L on January 28, 1986.

Findings

1.   The Commission concluded that there was a serious flaw in the decision making process leading up to the launch of flight 51-L. A well structured and managed system emphasizing safety would have flagged the rising doubts about the Solid Rocket Booster joint seal. Had these matters been clearly stated and emphasized in the flight readiness process in terms reflecting the views of most of the Thiokol engineers and at least some of the Marshall engineers, it seems likely that the launch of 51-L might not have occurred when it did.
2.   The waiving of launch constraints appears to have been at the expense of flight safety. There was no system which made it imperative that launch constraints and waivers of launch constraints be considered by all levels of management.
3.   The Commission is troubled by what appears to be a propensity of management at Marshall to contain potentially serious problems and to attempt to resolve them internally rather than communicate them forward. This tendency is altogether at odds with the need for Marshall to function as part of a system working toward successful flight missions, interfacing and communicating with the other parts of the system that work to the same end.
4.   The Commission concluded that the Thiokol Management reversed its position and recommended the launch of 51-L, at the urging of Marshall and contrary to the views of its engineers in order to accommodate a major customer.

Findings

The Commission is concerned about three aspects of the ice-on-the-pad issue.
1.   An analysis of all of the testimony and interviews establishes that Rockwell's recommendation on launch was ambiguous. The Commission finds it difficult, as did Mr. [Arnold] Aldrich, to conclude that there was a no-launch recommendation. Moreover, all parties were asked specifically to contact Aldrich or Moore about launch objections due to weather. Rockwell made no phone calls or further objections to Aldrich or other NASA officials after the 9:00 Mission Management Team meeting and subsequent to the resumption of the countdown.
2.   The Commission is also concerned about the NASA response to the Rockwell position at the 9:00 a.m. meeting. While it is understood that decisions have to be made in launching a Shuttle, the Commission is not convinced Levels I and II appropriately considered Rockwell's concern about the ice. However ambiguous Rockwell's position was, it

is clear that they did tell NASA that the ice was an unknown condition. Given the extent of the ice on the pad . . ., the admitted unknown effect of the Solid Rocket Motor and Space Shuttle Main Engines ignition on the ice, as well as the fact that debris striking the Orbiter was a potential flight [no page number] safety hazard, the Commission finds the decision to launch questionable under those circumstances. In this situation, NASA appeared to be requiring a contractor to prove that it was not safe to launch, rather than proving it was safe. Nevertheless, the Commission has determined that the ice was not a cause of the 51-L accident and does not conclude that NASA's decision to launch specifically overrode a no-launch recommendation by an element contractor.

3. The Commission concluded that the freeze protection plan for launch pad 39B was inadequate. The Commission believes that the severe cold and presence of so much ice on the fixed service structure made it inadvisable to launch on the morning of January 28, and that margins of safety were whittled down too far.

Additionally, access to the crew emergency slide wire baskets was hazardous due to ice conditions. Had the crew been required to evacuate the Orbiter on the launch pad, they would have been running on an icy surface. The Commission believes the crew should have been made aware of the situation, and based on the seriousness of the condition, greater consideration should have been given to delaying the launch.

Chapter VI

An Accident Rooted in History

Early Design

The Space Shuttle's Solid Rocket Booster problem began with the faulty design of its joint and increased as both NASA and contractor management first failed to recognize it as a problem, then failed to fix it and finally treated it as an acceptable flight risk.

Morton Thiokol, Inc., the contractor, did not accept the implication of tests early in the program that the design had a serious and unanticipated flaw. NASA did not accept the judgment of its engineers that the design was unacceptable, and as the joint problems grew in number and severity NASA minimized them in management briefings and reports. Thiokol's stated position was that "the condition is not desirable but is acceptable."

Neither Thiokol nor NASA expected the rubber O-rings sealing the joints to be touched by hot gases of motor ignition, much less to be partially burned. However, as tests and then flights confirmed damage to the sealing rings, the reaction by both NASA and Thiokol was to increase the amount of damage considered "acceptable." At no time did management either recommend a redesign of the joint or call for the Shuttle's grounding until the problem was solved.

Findings

The genesis of the Challenger accident—the failure of the joint of the right Solid Rocket Motor—began with decisions made in the design of the joint and in the failure by both Thiokol and NASA's Solid Rocket Booster project office to understand and respond to facts obtained during testing.

The Commission has concluded that neither Thiokol nor NASA responded adequately to internal warnings about the faulty seal design. Furthermore, Thiokol and NASA did not make a timely attempt to develop and verify a new seal after the initial design was shown to be deficient. Neither organization developed a solution to the unexpected occurrences of O-ring erosion and blow-by even though this problem was experienced frequently during the Shuttle flight history. Instead, Thiokol and NASA management came to accept erosion and blow-by as unavoidable and an acceptable flight risk. Specifically, the Commission has found that:

1.   The joint test and certification program was inadequate. There was no require-
     ment to configure the qualifications test motor as it would be in flight, and the
     motors were static tested in a horizontal position, not in the vertical flight position.
[no page number]
2.   Prior to the accident, neither NASA nor Thiokol fully understood the mechanism
     by which the joint sealing action took place.
3.   NASA and Thiokol accepted escalating risk apparently because they "got away
     with it last time." As Commissioner Feynman observed, the decision making was:
          "a kind of Russian roulette. . . . [The Shuttle] flies [with O-ring erosion] and
          nothing happens. Then it is suggested, therefore, that the risk is no longer so
          high for the next flights. We can lower our standards a little bit because we
          got away with it last time. . . . You got away with it, but it shouldn't be done
          over and over again like that."
4.   NASA's system for tracking anomalies for Flight Readiness Reviews failed in that,
     despite a history of persistent O-ring erosion and blow-by, flight was still permitted.
     It failed again in the strange sequence of six consecutive launch constraint waivers
     prior to 51-L, permitting it to fly without any record of a waiver, or even of an explic-
     it constraint. Tracking and continuing only anomalies that are "outside the data
     base" of prior flight allowed major problems to be removed from, and lost by, the
     reporting system.
5.   The O-ring erosion history presented to Level I at NASA Headquarters in August
     1985 was sufficiently detailed to require corrective action prior to the next flight.
6.   A careful analysis of the flight history of O-ring performance would have revealed
     the correlation of O-ring damage and low temperature. Neither NASA nor
     Thiokol carried out such an analysis; consequently, they were unprepared to
     properly evaluate the risks of launching the 51-L mission in conditions more
     extreme than they had encountered before.

Chapter VII

The Silent Safety Program

The Commission was surprised to realize after many hours of testimony that NASA's safety staff was never mentioned. No witness related the approval or disapproval of the reliability engineers, and none expressed the satisfaction or dissatisfaction of the quality assurance staff. No one thought to invite a safety representative or a reliability and quali-ty assurance engineer to the January 27, 1986, teleconference between Marshall and Thiokol. Similarly, there was no representative of safety on the Mission Management Team that made key decisions during the countdown on January 28, 1986. The Commission is concerned about the symptoms that it sees.

The unrelenting pressure to meet the demands of an accelerating flight schedule might have been adequately handled by NASA if it had insisted upon the exactingly thor-ough procedures that were its hallmark during the Apollo program. An extensive and redundant safety program comprising interdependent safety, reliability and quality assur-ance functions existed during and after the lunar program to discover any potential safe-ty problems. Between that period and 1986, however, the program became ineffective. This loss of effectiveness seriously degraded the checks and balances essential for main-taining flight safety.

[no page number] On April 3, 1986, Arnold Aldrich, the Space Shuttle program manager, appeared before the Commission at a public hearing in Washington, D.C. He described five different communication or organization failures that affected the launch decision on January 28, 1986. Four of those failures relate directly to faults within the safety program. These faults include a lack of problem reporting requirements, inadequate trend analysis, misrepresentation of criticality and lack of involvement in critical discussions. A properly staffed, supported, and robust safety organization might well have avoided these faults and thus eliminated the communication failures.

NASA has a safety program to ensure that the communication failures to which Mr. Aldrich referred do not occur. In the case of mission 51-L, that program fell short.

Findings

1. Reductions in the safety, reliability and quality assurance work force at Marshall and NASA Headquarters have seriously limited capability in those vital functions.
2. Organizational structures at Kennedy and Marshall have placed safety, reliability and quality assurance offices under the supervision of the very organizations and activities whose efforts they are to check.
3. Problem reporting requirements are not concise and fail to get critical information to the proper levels of management.
4. Little or no trend analysis was performed on O-ring erosion and blow-by problems. As the flight rate increased, the Marshall safety, reliability and quality assurance work force was decreasing, which adversely affected mission safety.
5. Five weeks after the 51-L accident, the criticality of the Solid Rocket Motor field joint was still not properly documented in the problem reporting system at Marshall.

Chapter VIII

Pressures on the System

With the 1982 completion of the orbital flight test series, NASA began a planned acceleration of the Space Shuttle launch schedule. One early plan contemplated an eventual rate of a mission a week, but realism forced several downward revisions. In 1985, NASA published a projection calling for an annual rate of 24 flights by 1990. Long before the Challenger accident, however, it was becoming obvious that even the modified goal of two flights a month was overambitious.

In establishing the schedule, NASA had not provided adequate resources for its attainment. As a result, the capabilities of the system were strained by the modest nine-mission rate of 1985, and the evidence suggests that NASA would not have been able to accomplish the 15 flights scheduled for 1986. These are the major conclusions of a Commission examination of the pressures and problems attendant upon the accelerated launch schedule.

[no page number] **Findings**

1. The capabilities of the system were stretched to the limit to support the flight rate in winter 1985/1986. Projections into the spring and summer of 1986 showed a clear trend; the system, as it existed, would have been unable to deliver crew training software for scheduled flights by the designated dates. The result would have been an unacceptable compression of the time available for the crews to accomplish their required training.
2. Spare parts are in critically short supply. The Shuttle program made a conscious decision to postpone spare parts procurements in favor of budget items of perceived high-

er priority. Lack of spare parts would likely have limited flight operations in 1986.

3. Stated manifesting policies are not enforced. Numerous late manifest changes (after the cargo integration review) have been made to both major payloads and minor payloads throughout the Shuttle program.

- Late changes to major payloads or program requirements can require extensive resources (money, manpower, facilities) to implement.
- If many late changes to "minor" payloads occur, resources are quickly absorbed.
- Payload specialists frequently were added to a flight well after announced deadlines.
- Late changes to a mission adversely affect the training and development of procedures for subsequent missions.

4. The scheduled flight rate did not accurately reflect the capabilities and resources.

- The flight rate was not reduced to accommodate periods of adjustment in the capacity of the work force. There was no margin in the system to accommodate unforeseen hardware problems.
- Resources were primarily directed toward supporting the flights and thus not enough were available to improve and expand facilities needed to support a higher flight rate.

5. Training simulators may be the limiting factor on the flight rate: the two current simulators cannot train crews for more than 12–15 flights per year.

6. When flights come in rapid succession, current requirements do not ensure that critical anomalies occurring during one flight are identified and addressed appropriately before the next flight.

Chapter IX

Other Safety Considerations

In the course of its investigation, the Commission became aware of a number of matters that played no part in the mission 51-L accident but nonetheless hold a potential for safety problems in the future.

Some of these matters, those involving operational concerns, were brought directly to the Commission's attention by the NASA astronaut office. They were the subject of a special hearing.

Other areas of concern came to light as the Commission pursued various lines of investigation in its attempt to isolate the cause of the accident. These inquiries examined such aspects as the development and operation of each of the elements of the Space Shuttle—the Orbiter, its main engines and the External Tank; the procedures employed in the processing and assembly of 51-L, and launch damage.

[no page number] This chapter examines potential risks in two general areas. The first embraces critical aspects of a Shuttle flight; for example, considerations related to a possible premature mission termination during the ascent phase and the risk factors connected with the demanding approach and landing phase. The other focuses on testing, processing and assembling the various elements of the Shuttle.

Ascent: A Critical Phase

The events of flight 51-L dramatically illustrated the dangers of the first stage of a Space Shuttle ascent. The accident also focused attention on the issues of Orbiter abort capabilities and crew escape. Of particular concern to the Commission are the current abort capabilities, options to improve those capabilities, options for crew escape and the performance of the range safety system.

It is not the Commission's intent to second-guess the Space Shuttle design or try to depict escape provisions that might have saved the 51-L crew. In fact, the events that led to destruction of the Challenger progressed very rapidly and without warning. Under those circumstances, the Commission believes it is highly unlikely that any of the systems discussed below, or any combination of those systems, would have saved the flight 51-L crew.

Findings

1. The Space Shuttle System was not designed to survive a failure of the Solid Rocket Boosters. There are no corrective actions that can be taken if the boosters do not operate properly after ignition, i.e., there is no ability to separate an Orbiter safely from thrusting boosters and no ability for the crew to escape the vehicle during first-stage ascent.
- Neither the Mission Control Team nor the 51-L crew had any warning of impending disaster.
- Even if there had been warning, there were no actions available to the crew or the Mission Control Team to avert the disaster.

Landing: Another Critical Phase

The consequences of faulty performance in any dynamic and demanding flight environment can be catastrophic. The Commission was concerned that an insufficient safety margin may have existed in areas other than Shuttle ascent. Entry and landing of the Shuttle are dynamic and demanding with all the risks and complications inherent in flying a heavyweight glider with a very steep glide path. Since the Shuttle crew cannot divert to any alternate landing site after entry, the landing decision must be both timely and accurate. In addition, the landing gear, which includes wheels, tires and brakes, must function properly.

In summary, although there are valid programmatic reasons to land routinely at Kennedy, there are concerns that suggest that this is not wise under the present circumstances. While planned landings at Edwards carry a cost in dollars and days, the realities of weather cannot be ignored. Shuttle program officials must recognize that Edwards is a permanent, essential part of the program. The cost associated with regular, scheduled landing and turnaround operations at Edwards is thus a necessary program cost.

Decisions governing Space Shuttle operations must be consistent with the philosophy that unnecessary risks have to be eliminated. Such decisions cannot be made without a clear understanding of margins of safety in each part of the system.

Unfortunately, margins of safety cannot be assured if performance characteristics are not thoroughly understood, nor can they be deduced from a previous flight's "success."

The Shuttle Program cannot afford to operate outside its experience in the areas of tires, brakes, and weather, with the capabilities of the system today. Pending a clear understanding of all landing and deceleration systems, and a resolution of the problems encountered to date in Shuttle landings, the most conservative course must be followed in order to minimize risk during this dynamic phase of flight.

[no page number] Shuttle Elements

The Space Shuttle Main Engine teams at Marshall and Rocketdyne have developed engines that have achieved their performance goals and have performed extremely well. Nevertheless the main engines continue to be highly complex and critical components of the Shuttle that involve an element of risk principally because important components of the engines degrade more rapidly with flight use than anticipated. Both NASA and Rocketdyne have taken steps to contain that risk. An important aspect of the main engine program has been the extensive "hot fire" ground tests. Unfortunately, the vitality of the test program has been reduced because of budgetary constraints.

The number of engine test firings per month has decreased over the past two years. Yet this test program has not yet demonstrated the limits of engine operation parameters or included tests over the full operating envelope to show full engine capability. In addition, tests have not yet been deliberately conducted to the point of failure to determine actual engine operating margins.

[no page number] **Recommendations**

The Commission has conducted an extensive investigation of the Challenger accident to determine the probable cause and necessary corrective actions. Based on the findings and determinations of its investigation, the Commission has unanimously adopted recommendations to help assure the return to safe flight.

The Commission urges that the Administrator of NASA submit, one year from now, a report to the President on the progress that NASA has made in effecting the Commission's recommendations set forth below:

— I —

Design. The faulty Solid Rocket Motor joint and seal must be changed. This could be a new design eliminating the joint or a redesign of the current joint and seal. No design options should be prematurely precluded because of schedule, cost or reliance on existing hardware. All Solid Rocket Motor joints should satisfy the following requirements:

- The joints should be fully understood, tested and verified.
- The integrity of the structure and of the seals of all joints should be not less than that of the case walls throughout the design envelope.
- The integrity of the joints should be insensitive to:
    - Dimensional tolerances.
    - Transportation and handling.
    - Assembly procedures.
    - Inspection and test procedures.
    - Environmental effects.
    - Internal case operating pressure.
    - Recovery and reuse effects.
    - Flight and water impact loads.
- The certification of the new design should include:
    - Tests which duplicate the actual launch configuration as closely as possible.
    - Tests over the full range of operating conditions, including temperature.
- Full consideration should be given to conducting static firings of the exact flight configuration in a vertical attitude.

Independent Oversight. The Administrator of NASA should request the National Research Council to form an independent Solid Rocket Motor design oversight committee to implement the Commission's design recommendations and oversee the design effort. This committee should:

- Review and evaluate certification requirements.
- Provide technical oversight of the design, test program and certification.
- Report to the Administrator of NASA on the adequacy of the design and make appropriate recommendations.

[no page number]   — II —

Shuttle Management Structure. The Shuttle Program Structure should be reviewed. The project managers for the various elements of the Shuttle program felt more accountable to their center management than to the Shuttle program organization. Shuttle element funding, work package definition, and vital program information frequently bypass the National STS (Shuttle) Program Manager.

A redefinition of the Program Manager's responsibility is essential. This redefinition should give the Program Manager the requisite authority for all ongoing STS operations. Program funding and all Shuttle Program work at the centers should be placed clearly under the Program Manager's authority.

Astronauts in Management. The Commission observes that there appears to be a departure from the philosophy of the 1960s and 1970s relating to the use of astronauts in management positions. These individuals brought to their positions flight experience and a keen appreciation of operations and flight safety.
- NASA should encourage the transition of qualified astronauts into agency management positions.
- The function of the Flight Crew Operations director should be elevated in the NASA organization structure.

Shuttle Safety Panel. NASA should establish an STS Safety Advisory Panel reporting to the STS Program Manager. The charter of this panel should include Shuttle operational issues, launch commit criteria, flight rules, flight readiness and risk management. The panel should include representation from the safety organization, mission operations, and the astronaut office.

— III —

Criticality Review and Hazard Analysis. NASA and the primary Shuttle contractors should review all Criticality 1, 1R, 2, and 2R items and hazard analyses. This review should identify those items that must be improved prior to flight to ensure mission success and flight safety. An Audit Panel, appointed by the National Research Council, should verify the adequacy of the effort and report directly to the Administrator of NASA.

— IV —

Safety Organization. NASA should establish an Office of Safety, Reliability and Quality Assurance to be headed by an Associate Administrator, reporting directly to the NASA Administrator. It would have direct authority for safety, reliability, and quality assurance throughout the agency. The office should be assigned the work force to ensure adequate oversight of its functions and should be independent of other NASA functional and program responsibilities.

The responsibilities of this office should include:
- The safety, reliability and quality assurance functions as they relate to all NASA activities and programs.
- Direction of reporting and documentation of problems, problem resolution and trends associated with flight safety.

[no page number]   — V —

Improved Communications. The Commission found that Marshall Space Flight Center project managers, because of a tendency at Marshall to management isolation, failed to provide full and timely information bearing on the safety of flight 51-L to other vital elements of Shuttle program management.

* NASA should take energetic steps to eliminate this tendency at Marshall Space Flight Center, whether by changes of personnel, organization, indoctrination or all three.
* A policy should be developed which governs the imposition and removal of Shuttle launch constraints.
* Flight Readiness Reviews and Mission Management Team meetings should be recorded.
* The flight crew commander, or a designated representative, should attend the Flight Readiness Review, participate in acceptance of the vehicle for flight, and certify that the crew is properly prepared for flight.

— VI —

Landing Safety. NASA must take actions to improve landing safety.

* The tire, brake and nosewheel steering systems must be improved. These systems do not have sufficient safety margin, particularly at abort landing sites.
* The specific conditions under which planned landings at Kennedy would be acceptable should be determined. Criteria must be established for tires, brakes and nosewheel steering. Until the systems meet those criteria in high fidelity testing that is verified at Edwards, landing at Kennedy should not be planned.
* Committing to a specific landing site requires that landing area weather be forecast more than an hour in advance. During unpredictable weather periods at Kennedy, program officials should plan on Edwards landings. Increased landings at Edwards may necessitate a dual ferry capability.

— VII —

Launch Abort and Crew Escape. The Shuttle program management considered first-stage abort options and crew escape options several times during the history of the program, but because of limited utility, technical infeasibility, or program cost and schedule, no systems were implemented. The Commission recommends that NASA:

* Make all efforts to provide a crew escape system for use during controlled gliding flight.
* Make every effort to increase the range of flight conditions under which an emergency runway landing can be successfully conducted in the event that two or three main engines fail early in ascent.

[no page number]   — VIII —

Flight Rate. The nation's reliance on the Shuttle as its principal space launch capability created a relentless pressure on NASA to increase the flight rate. Such reliance on a single launch capability should be avoided in the future.

NASA must establish a flight rate that is consistent with its resources. A firm payload assignment policy should be established. The policy should include rigorous controls on cargo manifest changes to limit the pressures such changes exert on schedules and crew training.

— IX —

Maintenance Safeguards. Installation, test, and maintenance procedures must be especially rigorous for Space Shuttle items designated Criticality 1. NASA should establish a system of analyzing and reporting performance trends of such items.

Maintenance procedures for such items should be specified in the Critical Items List, especially for those such as the liquid-fueled main engines, which require unstinting maintenance and overhaul.

With regard to the Orbiters, NASA should:
- Develop and execute a comprehensive maintenance inspection plan.
- Perform periodic structural inspections when scheduled and not permit them to be waived.
- Restore and support the maintenance and spare parts programs, and stop the practice of removing parts from one Orbiter to supply another.

Concluding Thought

The Commission urges that NASA continue to receive the support of the Administration and the nation. The agency constitutes a national resource that plays a critical role in space exploration and development. It also provides a symbol of national pride and technological leadership.

The Commission applauds NASA's spectacular achievements of the past and anticipates impressive achievements to come. The findings and recommendations presented in this report are intended to contribute to the future NASA successes that the nation both expects and requires as the 21st century approaches. . . .

Document II-40

Document title: Richard H. Truly, Associate Administrator for Space Flight, NASA, to Distribution, "Strategy for Safely Returning the Space Shuttle to Flight Status," March 24, 1986.

Source: NASA Historical Reference Collection, NASA History Office, NASA Headquarters, Washington, D.C.

In the aftermath of the Challenger *accident, former astronaut Richard Truly returned to NASA as Associate Administrator for Space Flight, with lead responsibility for returning the Space Shuttle safely to flight. Within a few weeks after taking office, Truly set forth a strategy for achieving this objective; this strategy preceded the June 1986 recommendations of the Rogers Commission and provided the framework within which NASA operated over the next thirty months until the September 29, 1988, return-to-flight Shuttle launch (STS-26).*

[1] National Aeronautics and   MAR 24 1986
Space Administration
Washington, D.C.
20546

TO: Distribution

FROM: M/Associate Administrator for Space Flight

SUBJECT: Strategy for Safely Returning the Space Shuttle to Flight Status

This memorandum defines the comprehensive strategy and major actions that, when completed, will allow resumption of the NSTS flight schedule. NASA headquarters (particularly the Office of Space Flight), the [Office of Space Flight] centers, the National Space Transportation System (NSTS) program organization and its various contractors will use this guidance to proceed with the realistic, practical actions necessary to return to the

NSTS flight schedule with emphasis on flight safety. This guidance is intended to direct planning for the first year of flight while putting into motion those activities required to establish a realistic and an achievable launch rate that will be safely sustainable. We intend to move as quickly as practicable to complete these actions and return to safe and effective operation of the National Space Transportation System.

Guidance for the following subjects is included:

- ACTIONS REQUIRED PRIOR TO THE NEXT FLIGHT
- FIRST FLIGHT/FIRST YEAR OPERATIONS
- DEVELOPMENT OF SUSTAINABLE SAFE FLIGHT RATE

ACTIONS REQUIRED PRIOR TO THE NEXT FLIGHT

Reassess Entire Program Management Structure and Operation

The NSTS program management philosophy, structure, reporting channels and decision-making process will be thoroughly reviewed and those changes implemented which are required to assure confidence and safety in the overall program, including the commit to launch process. Additionally, the Level I/II/III budget and management relationships will be reviewed to insure [sic] that they do not adversely affect the NSTS decision process.

[2] Solid Rocket Motor (SRM) Joint Redesign

A dedicated SRM joint design group will be established at [Marshall Space Flight Center], with selective participation from other NASA centers and external organizations, to recommend a program plan to quantify the SRM joints problem and to accomplish the SRM joints redesign. The design must be reviewed in detail by the program to include [Preliminary Design Review, Critical Design Review, Design Certification Review], independent analysis, DM-QM testing, and any other factors necessary to assure that the overall SRM is safe to commit to launch. The type and content of post-flight inspections for the redesigned joints and other flight components will be developed in detail, with criteria developed for commitment to the next launch as well as reusability of the specific flight hardware components.

Design Requirements Reverification

A review of the NSTS Design Requirements (Vol. 07700) will be conducted to insure [sic] that all systems design requirements are properly defined. This review will be followed by a delta [Design Certification Review] for all program elements to assure the individual projects are in compliance with the requirements.

Complete CIL/OMI Review

All Category 1 and 1R critical items will be subjected to a total review with a complete reapproval process implemented. Those items which are not revalidated by this review must be redesigned, certified, and qualified for flight. The review process will include a review of the OMI's [Operational Maintenance Inspections], OMRSD's [Operational Maintenance Readiness Support Documents], and other supporting documentation which is pertinent to the test, checkout, or assembly process of the Category 1 and 1R flight hardware. KSC [Kennedy Space Center] will continue to be responsible for all OMI's with design center concurrence required for those which affect Category 1 and 1R items. Category 2 and 3 CIL's [Critical Item Lists] will be reviewed for reacceptance and to verify their proper categorization.

Complete OMRSD Review

The OMRSD will be reviewed to insure [sic] that the requirements defined in it are complete and that the required testing is consistent with the results of the CIL review. Inspection/retest requirements will be modified as necessary to assure flight safety.

Launch/Abort Reassessment

The launch and launch abort rules and philosophy will be assessed to assure that the launch and flight rules, range safety systems/operational procedures, landing aids, runway configuration and length, performance vs. [takeoff and landing] exposure, abort weights, runway surface, and other landing related capabilities provide an acceptable margin of safety to [3] the vehicle and crew. Additionally, the weather forecasting capability will be reviewed and improved where possible to allow for the most accurate reporting.

FIRST FLIGHT/FIRST YEAR OPERATIONS

First Flight

The subject of first flight mission design will require extensive review to assure that we are proceeding in an orderly, conservative, safe manner. To permit the process to begin, the following specific planning guidance applies to the first planned mission:

- daylight KSC launch
- conservative flight design to minimize [takeoff and landing] exposure
- repeat payload (not a new payload class)
- no waiver on landing weight
- conservative launch/launch abort/landing weather
- NASA-only flight crew
- engine thrust within the experience base
- no active ascent/entry [Detailed Test Objectives]
- conservative mission rules
- early, stable flight plan with supporting flight software and training load
- daylight [Edwards] landing (lakebed or runway 22)

First Year

The planning for the flight schedule for the first year of operation will reflect a launch rate consistent with this conservative approach. The specific number of flights to be planned for the first year will be developed as soon as possible and will consider KSC and VAFB [Vandenberg Air Force Base] work flow, software development, controller/crew training, etc. Changes to flight plans, ascent trajectories, manifest, etc., will be minimized in the interest of program stability. Decisions on each launch will be made after thorough review of the previous mission's SRM joint performance, all other specified critical systems performance and resolution of anomalies.

In general, the first year of operation will be maintained within the current flight experience base, and any expansion of the base, including new classes of payloads, will be approved only after very thorough safety review. Specifically, 109 percent thrust levels will not be flown until satisfactory completion of the MPT testing currently being planned, and the first use of the Filament Wound Case will not occur with the first use of 109 percent SSME thrust level. Every effort will be made to conduct the first VAFB flight on an expeditious and safe schedule which supports national security requirements.

[4] DEVELOPMENT OF SUSTAINABLE SAFE FLIGHT RATE

The ultimate safe, sustainable flight rate, and the buildup to that rate, will be developed utilizing a "bottoms-up" approach in which all required work for the standard flow as defined in the OMRSD is identified and that work is optimized in relation to the available work force. Factors such as the manifest, nonscheduled work, in-flight anomaly resolution, mods, processing team workloads, work balancing across shifts, etc., will be considered, as well as timely mission planning, flight product development and achievable software delivery capability to support flight controllers and crew training. This development will consider the availability of the third orbiter facility, the availability of spares, as well as the effects of supporting VAFB launch site operations.

THE BOTTOM LINE

The Associate Adminstrator [sic] for Space Flight will take the action for reassessment of the NSTS program management structure. The NSTS Program Manager at Johnson Space Center is directed to initiate and coordinate all other actions required to implement this strategy for return to safe Shuttle flight.

I know that the business of space flight can never be made to be totally risk-free, but this conservative return to operations will continue our strong NASA/Industry team effort to recover from the Challenger accident. Many of these items have already been initiated at some level in our organizations, and I am fully aware of the tremendous amount of dedicated work which must be accomplished. I do know that our nation's future in space is dependent on the individuals who must carry this strategy out safely and successfully. Please give this the widest possible distribution to your people. It is they who must understand it, and they who must do it.

Richard H. Truly

Document II-41

Document title: John W. Young, Chief, Astronaut Office, to Director, Flight Crew Operations, "One Part of the 51-L Accident—Space Shuttle Program Flight Safety," March 4, 1986, with attached: "Examples of Uncertain Operational and Engineering Conditions or Events Which We 'Routinely' Accept Now in the Space Shuttle Program."

Source: NASA Historical Reference Collection, NASA History Office, NASA Headquarters, Washington, D.C.

Even before the January 28, 1986, accident that destroyed the Challenger *orbiter and killed its seven-person crew, those close to the Space Shuttle program realized that, although the Space Transportation System had been declared operational on July 4, 1982, there were continuing developmental problems with the vehicle. In fact, the O-ring failure that led to the* Challenger *accident was much less feared as a source of a catastrophic failure than many other system problems. In this memorandum, written six weeks after the* Challenger *accident, John Young, an experienced astronaut from the Gemini and Apollo programs and the commander of the first Space Shuttle flight, reflects his concern with various Shuttle issues. In 1986 at the time of this memo, Young was head of the astronaut office at NASA's Johnson Space Center.*

[1]

US Government MEMORANDUM		Lyndon B. Johnson Space Center NASA	
REFER TO CB-86-075	DATE March 4, 1986	INITIATOR CB/JW Young/cgh:3/3/86:3897	ENCL 1
TO: CA/Director, Flight Crew Operations FROM: CB/Chief, Astronaut Office		CC CA2/R.C. Zwieg DA3/D.R. Puddy CB/All Astronauts TA/J.F. Honeycutt CC/J.S. Algranti NASA Hqs/M/R.H. Truly	
		SIGNATURE John W. Young	
SUB: One Part of the 51-L Accident—Space Shuttle Program Flight Safety			

Background. The enclosure lists some conditions and events that are present in the Space Shuttle Program at this time. These are not all the conditions of serious concern. These situations increase, to some <u>unquantifiable</u> and <u>unassessable</u> extent, the Space Shuttle Program, the Space Shuttle, and the flightcrew risk. I have talked to individuals—working level and mid-level engineers, operators, and managers—who are seriously concerned about each of these accepted situations in their areas of specialty. These accepted conditions could have been or are now <u>potentially</u> as catastrophic to the Space Shuttle Program as the 51-L accident.

From watching the Presidential Commission open session interviews on televison [sic], it is clear that none of the direct participants have the faintest doubt that they did anything but absolutely the correct thing in launching 51-L at every step of the way. While it is difficult to believe that any humans can have such complete and total confidence, it is even more difficult to understand a management system that allows us to fly a solid rocket booster single-seal design that explosively, [and] dynamically verifies its criticality 1 performance in its application. This is because the prelaunch leak check pressurized that criticality 1 primary seal <u>away</u> from its proper <u>sealing position</u>. Sealing then relied on the single dynamic action of solid rocket motor ignition to properly seal the primary seal. The proper sealing has to be accomplished within <u>milliseconds</u>. If proper sealing did not occur, in Morton Thiokol Inc.'s own words, "subscale testing verified seal resiliency unable to follow gap opening in metal parts—<u>no</u> secondary seal activation if primary seal <u>fails</u>." There is only one driving reason that such a potentially dangerous system would ever be allowed to fly—launch schedule pressure.

The enclosure lists several other potentially dangerous examples that you can be sure were accepted for the very same reason. Unlike the secret seal, which no one that we know, knew about, everyone knows about these items. These examples are the way we do business.

Future Considerations. The preliminary launch schedule of future flights has launches at 9 the first year, 14 the second, and 18 the third. Our Space Shuttle machinery is not airline machinery. As the launch rate [2] increases, we will start having directly increasing numbers of various conditions and events like the enclosure where things are not working normally and management will still want to go fly. We have already, as the enclosure shows in part, launched with less than certain full reliability and full redundancy of the systems, including the flightcrews, that we operate. We are under continuing pressure to launch without full-up avionics from computers to other sensors.

The Space Shuttle is it; it is the state-of-the-art as the space reusable machine. For examples: When will we start seeing the effects of our short-term tile waterproofing? When will we start seeing more effects of the true long duration, true environmental conditions testing that we are running in our vehicles right now? These are the tests we have been performing for years now on such things at [sic; "as" meant] aluminum covered by tiles, and storable, reusable propellants—hydrazine and hypergolics? When will our less-than-infant mortality rate checkout on the vehicles' mechanical systems catch up with us? When does main engine systems' hydrogen embrittlement start worrying us? What will be the effects of these systems failing when we experience them on the vehicles after launch? What about our already accepted risks such as arming the communication satellite booster motors in the payload bay before we launch them? When, in the next 100 communication satellite launches, will that risk take a Shuttle down with it due to some obscure test deletion, tired personnel performance, or a waiver plus the incredible failure?

The Space Shuttle is, by its very pioneering nature in reusability and its state-of-the-art systems, an inherently risky machine to operate. We must be very careful with it just to launch it successfully and get it back everytime.

An Urgent Request. By whatever management method it takes, we must make Flight Safety First. People being responsible for making Flight Safety First when the launch schedule is First cannot possibly make Flight Safety First no matter what they say. The enclosure shows that these goals have always been opposite ones. It also shows overall Flight Safety does not win in these cases. Flight Safety, to be safe, has to have real teeth in it. It will not be free. For starters, we should not allow any increase in the inherent risk of operating the Space Shuttle just to increase the launch rate, or reduce operating costs, or fly unsafe payloads.

If we have to put tough risk assessment or hazards analysis on all of the real-time operational management decision-making process for the life of this Program, then we need to do it. If we do not consider Flight Safety First all the time at all levels of NASA, this machinery and this Program will NOT make it. If the management system is not continuously self-assessing with respect to Flight Safety of the inherently hazardous business that we are in, it will NOT last. If the management system is not big enough to STOP the Space Shuttle Program whenever necessary to make Flight Safety corrections, it will NOT survive and neither will our three Space Shuttles or their flightcrews.

****

[no page number]

EXAMPLES OF UNCERTAIN OPERATIONAL AND ENGINEERING CONDITIONS
OR EVENTS WHICH WE "ROUTINELY" ACCEPT NOW
IN THE SPACE SHUTTLE PROGRAM

Time	Condition/Event	Results and Potential Effect
Oct–Dec 1984	Orbiter/external tank quick disconnect fittings flapper valves extremely sensitive to rigging angle and low tip loads (55–70 lbs.). Valve open capability could be compromised by small changes in flow such as a partial external tank line liner failure.	If any of four flapper valves close, the result is loss of vehicle and crew. Designs to aleviate [sic] the valve sensitivity to these loads and to allow precise rigging were turned down.
Aug–Sep 1985	51-I (OV-103) launched through two cloud decks. It was raining from the lower cloud deck. Moderate turbulence and rain on Shuttle landing abort runway 33. Light rain on Runway 15 after liftoff. Rain severity prediction is not possible in the dynamic weather around the Cape.	New Mission Rule: "Consideration will be given to light to moderate rain" for return to landing site abort. Tile damage could be severe. Assessed by an Engineering WAG as not more than 65 drag counts. If the tile damage assessment was realistic, winds in storms plus tile damage drag might lose the vehicle and crew in an abort.
Oct 1985	61-A, OV-099, aft left [Reaction Control System] Regulator A locked up incorrectly. The cause of lockup was not known. Decision made to fly with redundancy in Regulator B. Regulator B indicated failed after orbital insertion. Two days and seven hours were needed to get the Orbiter to "blowdown" capability that would handle a partial entry.	For 2 days, discussions in Mission Control were to use the improperly gained no-yaw-jet flight control system. This is a get-me-down catastrophy [sic] system in the primary guidance control and navigation system. The no-yaw-jet system is not in the back-up flight system. If the no-yaw-jet system were used and the switch to the BFS occurred, loss of vehicle and crew would result.
Dec 1985	61-C, OV-102, launched with leaking coolant in fuel cell 1. Leak isolated to fuel cell externally by inspection, but exact effect of leak in orbit was not known for certain. Recently, fuel cell 2 in OV-102 was discovered to be leaking coolant internally. 61-E would have launched with fuel cell 1 leaking to meet turnaround (and maybe fuel cell 2?).	Potential loss of fuel cell 1 was discussed before launch. This would result in a full duration priority flight. Loss of the next fuel cell in that timeframe is high risk. It results in complex crew procedures juggling systems to keep low loads on last fuel cell and loss of avionics redundancy for entry at the next planned landing site.

Time	Condition/Event	Results and Potential Effect
Dec 1985	61-C, OV-102, scheduled to land at KSC [Kennedy Space Center] to speed turnaround had used tires on Orbiter. Due to KSC runway surface conditions, Langley [Research Center] recommended the use of new tires. New nosewheel steering system on OV-102 not fail ops. At least nine single-point failures result in loss of nosewheel steering. All new nosewheel steering system failures not analyzed or completely understood.	All work to make nosewheel steering fail operational was deleted. If nosewheel steering required by single leaking tire or crosswinds and not available, heavy damage will result when Orbiter leaves the runway for unprepared runway shoulders at the Shuttle Landing Facility.
Jan 1985	61-C, OV102, launch scrub revealed criticality of [liquid oxygen] prevalve failure to close at Main Engine cutoff [MECO]. Failure to close at MECO is an uncertain event. [Johnson Space Center] stated pump damage. [Marshall Space Flight Center] stated catastrophy [sic].	Potential loss of vehicle and crew if each of three valves does not operate reliably at Main Engine cutoff. No test in Program to assess criticality of these three valves. Valves have single-point failures that should be fixed if the valves are criticality 1.

Document II-42

Document title: President Ronald Reagan, "Statement by the President," August 15, 1986.

Source: The White House, Office of the Press Secretary, Washington, D.C.

Document II-43

Document title: The White House, Fact Sheet, NSDD-254, "United States Space Launch Strategy," December 27, 1986.

Source: NASA Historical Reference Collection, NASA History Office, NASA Headquarters, Washington, D.C.

In the aftermath of the Challenger *accident, there was intense debate inside the Reagan administration over both whether to build an additional Space Shuttle orbiter and what policy should govern Shuttle use once the vehicle returned to flight. Advocates of creating a U.S. space launch industry based on the commercialization of expendable launch vehicles (ELVs) were successful, over NASA's opposition, in their argument that the Shuttle should no longer be used to launch commercial payloads, thereby creating a potential market for commercial launch providers. (See Chapter 3 of this volume.)*

The presidential directive in December 1986 formalized the August 15 decision by President Reagan. The Space Shuttle would no longer be used to launch commercial or foreign payloads unless its unique capabilities were required, or there were overriding national security or foreign policy reasons for doing so. It also codified U.S. policy that government access to space would be provided by a "mixed fleet" of Space Shuttle orbiters and ELVs, rather than solely by the Shuttle.

Document II-42

[no page number]   THE WHITE HOUSE
Office of the Press Secretary

For Immediate Release   August 15, 1986

Statement by the President

I am announcing today two steps that will ensure America's leadership in space exploration and utilization. First, the United States will, in FY 1987, start building a fourth Space shuttle to take the place of Challenger which was destroyed on January 28th. This decision will bring our shuttle fleet up to strength and enable the United States to safely and energetically project a manned presence in space.

Without the fourth orbiter, NASA's capabilities would be severely limited and long-term projects for the development of space would have to be either postponed, or even canceled. A fourth orbiter will enable our shuttles to accomplish the mission for which they were originally intended and permit the United State[s] to move forward with new exciting endeavors like the building of a permanently manned space station.

My second announcement concerns the fundamental direction of the space program. NASA and our shuttles will continue to lead the way, breaking new ground, pioneering new technology, and pushing back the frontiers. It has been determined, however, that NASA will no longer be in the business of launching private satellites.

The private sector, with its ingenuity and cost effectiveness, will be playing an increasingly important role in the American space effort. Free enterprise corporations will become a highly competitive method of launching commercial satellites and doing those things which do not require a manned presence in space. These private firms are essential in clearing away the backlog that has built up during this time when our shuttles are being modified.

We must always set our sights on tomorrow. NASA and our shuttle can't be committing their scarce resources to things which can be done better and cheaper by the private sector. Instead, NASA and the four shuttles should be dedicated to payloads important to national security and foreign policy, and, even more, on exploration, pioneering, and developing new technologies and uses of space. NASA will keep America on the leading edge of change; the private sector will take over from there. Together, they will ensure that our country has a robust, balanced, and safe space program.

It has been over 6 months since the tragic loss of the Challenger and her gallant crew. We have done everything humanly possible to discover the organizational and technical causes of the disaster and to correct the situation. The greatest tribute we can pay to those brave pathfinders who gave their lives on the Challenger is to move forward and rededicate ourselves to America's leadership in space.

Document II-43

[1]   <u>FACT SHEET</u>

United States Space Launch Strategy

<u>Introduction</u>

On December 27, 1986, the President signed a directive which establishes U.S. nation-al policy to launch satellites and missions into space to support U.S. national security, civil, and commercial goals using space. It is essential that U.S. space launch operations be as efficient as possible consistent with available funding and safety concerns; and that U.S. space launch assets provide a balanced, robust, flexible space launch capability which can function independently of failures in any single launch vehicle system, allow a return to regularly scheduled launch operations, meet continuing requirements, help make up for lost launch opportunities and reassert global space leadership.

This directive supersedes the National Security Launch Strategy policy directive of February 25, 1985. Other previous space policy directives remain valid but are modified accordingly.

<u>National Space Launch Capability</u>

The U.S. national space launch capability will be based on a balanced mix of launch-ers, consisting of the Space Transportation System (STS) and expendable launch vehicles (ELVs). The elements of this mix will be defined to best support the mission needs of the national security, civil government and commercial sectors of U.S. space activities. Critical mission needs will be supported, whenever necessary, by both the STS and ELVs so as to provide added assurance that payloads can be launched regardless of specific launch vehi-cle availabilities.

a.   <u>National Security Space Transportation</u>. The national security space sector will use both the STS and ELVs. Selected critical payloads will be designed for dual-compati-bility, i.e., capable of being launched by either the STS or the ELVs.

<u>Implementation</u>: The Department of Defense (DOD) will procure additional ELVs to maintain a balanced launch capability and to provide access to space. The DOD will implement procedures to assure payload/launch vehicle compatibility and schedul-ing, and maintain launch capability for ELVs at both the East and West Coast launch sites.

b.   <u>Civil Government Space Transportation</u>. The unique STS (Shuttle) capability to provide manned access to space will be exploited in those areas that offer the greatest national return. [2] The STS fleet will maintain the Nation's capability to support critical programs requiring manned presence and other unique STS capabilities. NASA will use the Shuttle where the unique capabilities of the STS are required to support civil research and development programs.

<u>Implementation</u>: NASA will procure STS structural spares and other necessary equipment needed to sustain the existing three-orbiter fleet and will do so in an expedi-tious and cost-effective manner. Funding for procurement of a replacement fourth orbiter will begin in FY 1987 based on an [Office of Management and Budget]-approved pro-gram. NASA will establish sustainable STS flight rates to provide for planning and bud-geting of Government space programs. The recommendations of the President's Commission on the Space Shuttle Challenger Accident will be considered and incorpo-rated as appropriate. The STS will be phased out from providing launch services for com-mercial and foreign payloads that do not require a manned presence or the unique

capabilities of the STS. NASA will not maintain an ELV adjunct to the STS. If there is a need for additional NASA capacity for government launches, then NASA is authorized to contract for necessary ELV launch services.

　　c.　Commercial Space Transportation. The principles and policy of domestic exploitation of space for commercial purposes are enunciated in a policy directive entitled Commercialization of Expendable Launch Vehicles, dated May 16, 1983. Those principles and policies remain valid.

　　　　Implementation: NASA shall no longer provide launch services for commercial and foreign payloads unless those spacecraft have unique, specific reasons to be launched aboard the Shuttle. Those reasons are: the spacecraft must be man-tended or the spacecraft is important for national security or foreign policy purposes. Satellite manufacturers whose spacecraft do not meet those criteria will be provided as realistic an appraisal as possible by NASA of when they could be scheduled on the Shuttle launch manifest prior to the 1995 commercial contract mandatory termination date.

Document II-44

Document title: H. Guyford Stever, Chair, Panel on Redesign of Space Shuttle Solid Rocket Booster, Committee on NASA Scientific and Technological Program Reviews, National Research Council, to James C. Fletcher, Administrator, NASA, Seventh Interim Report, September 9, 1986.

Source: NASA Historical Reference Collection, NASA History Office, NASA Headquarters, Washington, D.C.

Responding to a suggestion of the Rogers Commission on the Challenger *accident, NASA asked the National Research Council to form a panel of independent technical experts to evaluate NASA's efforts to redesign, test, and certify the Space Shuttle solid rocket booster (SRBs) and to oversee the manufacture of the two specific boosters to be used in the first Shuttle flight after the accident, which took place on September 29, 1988. In this letter report to NASA Administrator James Fletcher, panel chair H. Guyford Stever, a former science advisor to President Gerald Ford, communicated the panel's views on various aspects of the booster effort. Most importantly, he indicated in the last sentence of the letter that the panel had no basis for objecting to NASA's plans for launching the Shuttle in late September, thereby clearing the way for the Shuttle's return to flight.*

[1]　　　　　　　　　　NATIONAL RESEARCH COUNCIL
　　　　　COMMISSION ON ENGINEERING AND TECHNICAL SYSTEMS
　　　　　　　2101 Constitution Avenue　　Washington, D.C. 20418

　　　　　　　　　　COMMITTEE ON NASA SCIENTIFIC
　　　　　　　AND TECHNOLOGICAL PROGRAM REVIEWS
　　　　　　　　　Panel on Redesign of Space Shuttle
　　　　　　　　　　　Solid Rocket Booster

　　　　　　　　　　　　　　　　　　　　September 9, 1988

The Honorable James C. Fletcher
Administrator
National Aeronautics and Space Administration
400 Maryland Avenue, S.W., Room 7137
Washington, D.C. 20546

Dear Jim:

I am pleased to submit herewith the seventh interim report of the National Research Council's Panel for the Technical Evaluation of NASA's Redesign of the Space Shuttle Solid Rocket Booster.

The preflight program for testing the redesigned solid rocket booster has been completed and the Shuttle is expected to be returned to service soon. This report provides our assessment of the new design and its certification program, including production and quality control issues, and our findings on the status of the program at this time. Our conclusions are based on engineering judgment and the results of tests, the number of which has been necessarily small.

Since our last report, the Panel has conducted four formal meetings and members of the Panel have attended a number of test readiness reviews; the QM-6, QM-7, and PVM-1 static tests; technical interchange meetings on the outer boot ring, aft skirt, and insulation debond problem; the design certification review; and an inspection of the stacked boosters to be used in the next flight (STS-26). I also presented testimony on the progress of the redesign effort to the Senate Appropriations Subcommittee on HUD and Independent Agencies on June 8th. Since June 1986 the Panel has participated in or observed more than 90 meetings, reviews, site visits, and tests and I have testified before the Congress on four occasions.

Assessment of the Redesigned SRB

Approach of the Redesign. The redesign program was organized to concentrate its resources on a "baseline" design, thereby avoiding a dilution of effort in both the design and testing [2] phases. The redesign was also constrained to make maximum use of existing or previously ordered hardware, owing to the long time it takes to acquire new cases.

The consequences of these restraints were: (1) With few exceptions, no alternative design of a major component was carried to full-scale, full-duration testing. The principal exception, an alternative design for the nozzle outer boot ring, turned out to be needed and, because it was available, many months of delay were avoided when the original baseline design failed in the DM-9 static test. (2) The development effort aimed at solving problems with the baseline design rather than providing the technological basis for selecting the best design. (3) The time and cost to return to flight were minimized.

Early in the recovery effort, we urged NASA to give more thorough consideration to alternative designs, of which many were and are potentially promising. We recognized, however, the advantages of the baseline approach for returning to flight as soon as possible and believe that it has proved effective in this case.

Results from the Redesign. The redesign program has been aimed at improving the design features that may have contributed to the Challenger accident as well as other components that were performing less than satisfactorily or that were identified as having inadequate factors of safety. It also included an extensive analytical effort, a subscale test program, and full-scale, short-duration testing which provided much improved understanding of the design and its limitations. The most important results of the program are outlined below.

Case field joint. Five features of the design of the original case field joint are thought to have contributed to the accident. (1) The sealing surfaces of the original design opened excessively during the ignition transient; this motion has been greatly reduced by the addition of the capture feature with its interference fit and extra O-ring. (2) The O-ring material used in the original design has poor low temperature resilience; since no suitable alternative is currently available, the redesign employs the same material, but seals are heated to maintain proper resilience. Also, greater care is taken to assure the quality of O-ring materials and manufacture. (3) The O-ring grooves were too narrow to take full advantage of the effects of pressurization for making the seal; the seal grooves have been widened in the redesign. (4) The original system for verifying the seals [3] pushed the primary O-ring in the wrong direction to be an effective seal upon pressurization; the new vent port and leak check procedures assure proper seating of both the primary and secondary seals. (5) The O-ring seals could be exposed to jets of combustion gases through blowholes in the putty of the original design; this exposure has been reduced or eliminated by replacing the putty with a thermal barrier of bonded insulation (the so-called J-seal) The interference fit also helps to protect the seals from exposure to hot gas jets in the event of a defect in the bonded insulation. In our opinion, NASA has a reasonable basis for concluding that these changes have corrected the previous design deficiencies.

Case-to-nozzle joint. The same redesign principles were applied where possible to the case-to-nozzle joint which, while not involved in the accident, had previously shown problems similar to those observed in the case field joint. Joint motion has now been restricted by radial bolts added to the design. The preloading of these and other bolts is more carefully controlled than previously. The O-rings are in a heated environment, as they were in the original design, but with more careful control of temperature. The O-ring grooves have been widened, although the design selected cannot assure that the primary seal is seated in the proper direction. Bonded insulation and an extra (wiper) O-ring are provided to protect the seals from combustion gases; in this case, however, the assembly process, dictated by the geometry, tends to allow voids and blowholes to form in the adhesive that forms the insulation bond.

The potential for gas flow through blowholes in the adhesive and the potential leak paths around the additional bolts create less certainty about the reliability of this joint than the case field joint. However, the joint has performed well in tests. When realistic blowholes were deliberately introduced into the joint during tests, the volumes of gas that flowed through them were less than the amount needed to jeopardize the seals. Therefore, while uncertainties remain, the new design appears to represent a significant improvement over the original. Additional work is required to develop and demonstrate assembly techniques that yield a more reproducible product. Until this is accomplished, very careful attention should be paid to quality control in assembly. In addition, the performance of the case-to-nozzle joint in flight must be monitored to verify that the additional stress of occasional blowholes in the adhesive does not threaten to compromise its function.

Igniter joint. Some of the redesign principles were also applied to the igniter joint. For example, the inner sealing [4] surfaces of the igniter in the original design opened too quickly on ignition for the gasket seal to respond as required by the specifications; the redesign employs a more substantial preload on the igniter bolts to restrict this motion. Beyond this, the addition of a heater to improve the resilience of the seal material for cold weather launches is planned; however, the heater has not yet been demonstrated or qualified. Test results indicate that the new design of this joint represents an improvement over the original.

Case factory joint. The vulcanized insulation that completely covers the inside of this joint acts as its only qualified pressure seal. The layup and thickness of the insulation have been modified to enhance safety. No structural changes were made in the case factory joint to reduce the relative motion of the metal parts and the O-rings are not heated, so the two O-rings do not meet the formal requirements for seals. This joint, therefore, does not have redundant, verifiable seals that will operate independently throughout motor burn. The insulation over the factory joints has performed satisfactorily in both flight and ground tests, which have demonstrated that the insulation forms a highly reliable seal. Furthermore, the O-rings may well provide redundant sealing action if called upon well after the ignition transient.

Nozzle ablative parts. Flight experience before the accident suggested the need to improve the thermal performance of carbon cloth phenolic parts in the nozzle. The results of a nozzle technology development program initiated before the accident led to improved control of the materials used to make the parts as well as to changes in the cloth layup patterns. A limited number of ground tests suggest that the thermal performance of these parts has been substantially improved. However, the performance of these components can be sensitive to manufacturing variables so operational flight data should be monitored very carefully.

The redesign of one nozzle ablative part, the outer boot ring, proved in test to be structurally deficient and it was necessary to turn to an alternative design. Based on test results to date, the current baseline ("structural support") design of the outer boot ring appears to be substantially better than either the original design or the first redesign ("involute"). Some degree of uncertainty exists, however, regarding the structural loads on this component when vent holes, which were designed to assure the equalization of pressure across the part, become plugged. NASA's analysis of the "worst case" pressure differential due to plugging indicates that the situation is unlikely to threaten the safety of flight.

[5] Booster components. During the redesign activity, two booster components were found to have structural safety factors that did not satisfy specifications. The aft attachment to the external tank and the aft skirt were both redesigned. Only the new design of the former appears to be satisfactory; the modified aft skirt has failed to meet the ultimate design load condition required in the specifications. We support NASA's decision to grant a waiver of the requirement for the aft skirt for the first flight since the safety of flight is not in question. The current skirt is heavier than the original design by several hundred pounds without apparent improvement in its strength. We conclude that further work on the aft skirt is needed to meet the design requirements.

Very recently, we learned that during an evaluation of booster parts in storage for future use, a crack was found in a strut. We understand that the struts installed in STS-26 had been inspected in accordance with procedures and proof tested and that the implications of the occurrence of this defect for STS-26 are being evaluated.

Assessment of the Certification Program

The certification program is aimed at verifying that the design meets the contract specifications and at determining if it is qualified for manned flight. The program includes analytical studies as well as development and qualification tests.

Concurrency. In the case of the redesigned SRB, the certification program was conducted in parallel with the manufacture and assembly of the first several pairs of flight motors. The objective, as with the baseline design approach, was to return to flight as quickly as possible. The assumption inherent in the parallel approach was that the certification program would successfully demonstrate that the baseline design meets requirements.

In practice, several changes in the first flight set were made after the two boosters were constructed when certification activities identified deficiencies. For example, the alternative outer boot ring design replaced the original redesign; new igniter bolts were installed to restrict the gap opening; and insulation debonds were repaired. Other changes were identified but were judged by NASA not to be sufficiently important to incur the delays that would have been required to incorporate them on the first flight set. These changes, [6] including stronger bolts in all internal nozzle joints, adjustable vent port plugs, improvements in the aft skirt to meet design requirements, and improvements in case-liner bonding processes[,] presumably will be introduced in future flights.

We conclude that the concurrent approach to certification and manufacture of flight sets was an appropriate strategy. The resumption of flight will clearly occur much earlier than otherwise would have been the case and program management has demonstrated diligence in making changes when tests or analyses indicated priority needs.

The Test Program. The test program comprises work to validate the mechanical and thermal integrity of the design and to confirm that it operates as intended. For example, assuring mechanical integrity is the primary focus of hydroproofing, structural, and assembly tests of various kinds that are intended to determine structural margins of safety or practicality of assembly. Also in this category are tests to determine the aging characteristics of nonmetallic parts, such as compression set in O-rings and insulation. Aging tests for components other than the propellant had been quite limited in the SRB program.

Mechanical and thermal integrity as well as operational characteristics were examined in a series of experiments in which propellant was burned. A design feature was often first tested in subscale motors, then in full-scale but short-duration test beds, and finally in full-scale, full-duration static motor firings. In addition to the usual testing of articles under nominal conditions, the redesign program included tests of four types that had not been conducted previously in the shuttle program. (1) The motor will have been test fired while conditioned to the highest and lowest operating temperature specified in the design requirements, with the low temperature test (QM-8) coming after the resumption of flight but before a cold temperature launch. (2) Some test articles, including one full-duration motor, were subjected to external dynamic forces to simulate the loads experienced at launch and during flight. (3) Both short-duration and full-duration firings were conducted to determine the tolerance of the design to flaws that might be introduced during manufacturing or assembly but not be detected by inspection. (4) The performance of seals was tested in both short-duration and full-duration firings by breaching the upstream barriers that normally would protect the O-rings from combustion gases.

[7] While not every test that the Panel and others might have desired was conducted before the return to flight, it is clear that the current test program has been considerably more extensive and thorough than the test program that preceded the first Shuttle launch. As a consequence, we conclude that NASA can have commensurately more confidence in the redesigned SRM [solid rocket motor] than it had in the original design.

Much was accomplished in the testing program that will also be valuable to NASA in developing future generations of solid rocket motors. Unfortunately, both because the focus was on the baseline design and because NASA did not have an ongoing program for developing advanced technology with applications to the motor, except for nozzle phenolics, the redesign program has not taken full advantage of its subscale test capability. We believe deeply in the value of technology programs as the basis for future design, development, and operations, building as they do the understanding needed to approach the future.

The Analytic Program. In addition to testing, NASA also relies on analytical studies to help verify that the design meets requirements, especially in those circumstances where tests cannot be conducted for practical reasons. For example, the factor of safety for thermal loads, i.e., the ability of the design to withstand thermal loads in excess of those experienced under "worst case" operating conditions, cannot be demonstrated by test. The requirements, nonetheless, specify the factor of safety to be achieved and demonstrated.

While NASA may have no other choice but to rely on analysis in these circumstances, it nonetheless appears to us as if the program has in some cases placed undue confidence in the results of analytical studies, particularly regarding structural integrity. Analyses incorporate a variety of assumptions and too seldom are estimates made of the effects of assumptions on the accuracy or precision of the results.

Modeling the behavior of complex structures subject to three-dimensional loads is a challenging task; the efficacy of analytical models must be verified by appropriate experiments. The nozzle ablative parts, for example, are complex inhomogeneous, anisotropic structures and their physics and chemistry may not be adequately captured by existing analytical models. The analysis of the outer boot ring, for example, did not account for torsion, used incorrect loads, and had an inappropriate failure criterion, yet the results of analysis were originally used to select the baseline design. A similar caution is warranted for the application of current models to plastic deformation of metal parts and to complex structures, such as the aft skirt.

[8] Future Verification Activities. As indicated earlier, the low temperature certification static motor test is scheduled to be conducted before the first cold temperature launch. After mission STS-26, a full-scale case joint is to be subjected to multiple cycles of pressure loading and then burst to identify effects of multiple uses and validate structural analyses.

It is also our understanding that the first six flights were intended to provide data as part of the verification program. We have been disappointed to learn that the instrumentation required for this purpose will not be flown after the third flight, apparently for budgetary reasons. No amount of ground testing can simulate with complete fidelity the conditions of flight, which is the environment that counts; there is no substitute for flight data for identifying anomalous behavior or verifying preflight calculations. We are concerned that once flight instruments have been deleted from the program, it will be difficult to get them back on the flight articles. We recommend, therefore, that NASA set aside funds for flight instrumentation beyond the third flight; the agency should identify critical needs for operational data, based on the results of the first several flights, that can be met with instrumentation on future flights.

Production and Quality Control

Because an SRM flight article cannot be operationally tested before it is used, defining and maintaining controls on the materials, processes, and parts used in its manufacture are essential to establish confidence in the reliability of the booster. The goal is to define the most effective manufacturing processes, then to assure that each motor is as much as possible identical to all of the motors that have been tested and flown before it. Careful workmanship and diligent supervision are essential in working toward this goal.

Manufacturing Processes. Among the thousands of processes used to make parts of the booster, four have been the focus of our attention because, while progress has been made, they have not yet been completely developed, demonstrated, and controlled. These processes are: manufacturing nozzle ablative parts, which are particularly vulnerable to single point failures; manufacturing high quality O-rings; assembling the case-to-nozzle joint without forming blowholes or voids in the polysulfide adhesive; and bonding elastomers to metal surfaces, including both the insulation-to-case bond and bonds within the flexible nozzle bearing. The features of the respective processes that determine the quality of their [9] products have not yet been conclusively identified. We recommend that technology development be vigorously pursued to resolve these uncertainties and that the changes be carefully tested before being introduced into flight articles.

Uncertainties also remain in developing the specifications governing the purchase of some critical materials, particularly those whose formulas or preparations are proprietary. Considerable progress has been made in specifying O-ring materials, but that work is not yet complete. Much work still needs to be done regarding adhesives and other bonding agents.

Quality Control. Diligence in assuring the required quality in materials and processes is a demanding, never-ending task. It appears to us that NASA and its contractors appreciate the central importance of quality control and have been working hard to improve the record of achievement. For example, progress has been made, both at Kennedy Space Center and at Morton-Thiokol, Inc., toward establishing and maintaining standards of cleanliness. Progress has also been made in nondestructive inspection and evaluation of materials, parts, and assembled articles.

Considerable attention has also been paid to the problem of measuring case segments. The case field joint design requires relatively precise control of the dimensions of the capture feature and the mating clevis leg: a few thousandths of an inch on a cylinder approximately 12 feet in diameter may be critical not only for its intended operation but also for reuse. Making accurate, precise measurements in this context has not proven to be easy. We concur that NASA should continue to develop and then employ the best demonstrated technique for making the required measurements.

The analysis of failure modes and effects, which was extensive, identified a very large number of items that will be subject to mandatory inspection. The number is larger than before the accident primarily because of greater care and attention to detail in the current assessment. The number is so large, however, that the program runs the risk of getting overwhelmed with potentially insignificant details. We concur with the recommendation of the [National Research Council's] Committee on Shuttle Criticality Review and Hazard Analysis Audit that means be devised for establishing priorities so that the inspection program can focus its attention on the truly important items.

[10] Deviations in STS-26 and Subsequent Flights. While the goal is to make each booster identical, deviations from design requirements and discrepancies in materials and processes always occur in the normal course of events. These are formally reviewed to assess their potential consequences for a successful mission and only those judged not to affect safety or reliability adversely are accepted.

The STS-26 boosters have a considerable number of such deviations and waivers. The Panel has reviewed NASA's process for evaluating and approving them and finds it to be satisfactory. We have also reviewed a few of the more significant items. While we have not been able to make a thorough assessment in each case, we have found nothing which demonstrates that NASA's evaluations are in error.

Among the deviations, the first flight set contains some parts, and was manufactured using some processes, that will be changed for future missions because improvements were identified during the development and certification process but after the STS-26 boosters were assembled. As described earlier, NASA concluded that the benefits to be derived from making certain changes were not worth the associated delays. Included in this category are: O-rings in the safe and arm device that have less than the specified squeeze on the rotor shaft; fully threaded bolts in nozzle internal joint #5; case-liner edge bonds built without the benefit of the most recent process controls; nozzles that have been subjected to removal and replacement of the outer boot ring; and the so-called custom-fitted vent port plugs. In addition, putty in the igniter joints of the STS-26 boosters—and one of the STS-27 boosters—has been mechanically tamped to reduce the potential for blowholes although this process will not be followed in the future. Each of these unique features of the SRBs on STS-26 has been tested in at least one static motor ground test.

While flights beyond STS-26 will also have waivers and deviations for the reasons described in the first paragraph of this section, we are concerned that many deviations and waivers arise because the related design requirements are incorrect or impractical and will not ever be met. NASA should reduce the number of such deviations and waivers by changing the requirements where there is no practical expectation of meeting them in the future and where the resulting reliability, performance, and operating constraints, if any, are acceptable.

[11] Current Status of the Program

As noted in our first report, dated August 1, 1986, four interdependent factors influence the program: safety, schedule, cost, and performance. Improving the design for enhanced safety and reliability has been the prime consideration, but schedule and costs have also had important influences on the redesign program.

Many design changes have improved safety and reduced risks. Some design changes may have introduced new, as yet unrecognized risks. Some risks, associated with elements of the design that were not changed, remain as they were before the Challenger accident. On balance, based primarily on changes in design of the case field joint, case-to-nozzle joint, and the nozzle ablative components, we believe that the overall level of safety and reliability has been substantially improved.

More might have been accomplished if the program were unconstrained by the need to return to flight as soon as possible and by limitations in budget and other resources. But such constraints were practical necessities and our impression is that they were not unreasonable in this case.

More can still be accomplished to improve safety and reliability after flights resume, however, as a number of important issues in the design and verification program have been deferred or are still unresolved. Among these are: the adequacy of new procedures for making the case-insulation bonds for future flight articles; the adequacy of repairs to case-insulation bonds; the structural performance of nozzle parts and bonds; the occurrence and effects of blowholes in the adhesive in the case-to-nozzle joint; effects of long term storage on installed elastomeric seals and bonds; the accuracy and reliability of measuring and matching case segments; the adequacy of the aft skirt design; the potential need to prevent the establishment of differential pressure across the nozzle flexible boot; the verification of structural analysis by a burst test of a full-scale case; the potential for achieving the required number of reuses of case segments; and the removal of materials that contain asbestos. Additional issues or concerns can be expected to arise from flight experience, which is the true test of the redesign.

We have previously recommended that NASA undertake a program to continue to reduce risks, enhance reliability, and reduce costs associated with the redesigned SRB after flight resumes. Having such a program, which should address both issues unresolved in the redesign to date and concerns that [12] arise from flight experience, requires planning to assure appropriate continuity in technical efforts and of personnel and to be capable of introducing improvements into an ongoing operational program. NASA's commitment to and budget for such a program is essential. In our opinion, the prospect of an advanced solid rocket motor, which might not be available until the middle of the next decade at the earliest, does not warrant a relaxation in NASA's diligence to provide the safest practical space transportation system in the interim. We strongly reiterate our earlier recommendation.

Our focus today, however, is on the return to flight: mission STS-26. NASA and its contractors have worked diligently on the redesign and testing program and deserve to be recognized for their efforts. The redesigned solid rocket boosters have incorporated a large number of improvements that should result in considerably enhanced safety and reliability, hence reduced risk. Risks remain, however. And readiness to fly depends as much, if not more, on confidence in manufacturing and assembly as on the redesign, which our Panel has evaluated over the past 28 months. Whether the level of risk is acceptable is a matter that NASA must judge. Based on the Panel's assessments and observations regarding the redesigned solid rocket boosters, we have no basis for objection to the current launch schedule for STS-26.

Sincerely,

H. Guyford Stever
Chairman

Document II-45

Document title: Office of Technology Assessment, "Shuttle Fleet Attrition if Orbiter Recovery Reliability is 98 Percent," August 1989, p. 6.

Source: U.S. Congress, Office of Technology Assessment, Round Trip to Orbit: Human Spaceflight Alternatives—Special Report, OTA-ISC-419 (Washington, DC: U.S. Government Printing Office, August 1989).

The Office of Technology Assessment (1972–1995) was a congressional support organization providing in-depth technical analysis for the House and Senate. It prepared a number of reports on space issues during the 1980s and 1990s. In a 1989 report, it assessed the sensitive issue of the statistical likelihood of another major Space Shuttle accident.

[6]

Figure I-1—Shuttle Fleet Attrition if Orbiter Recovery Reliability is 98 Percent.

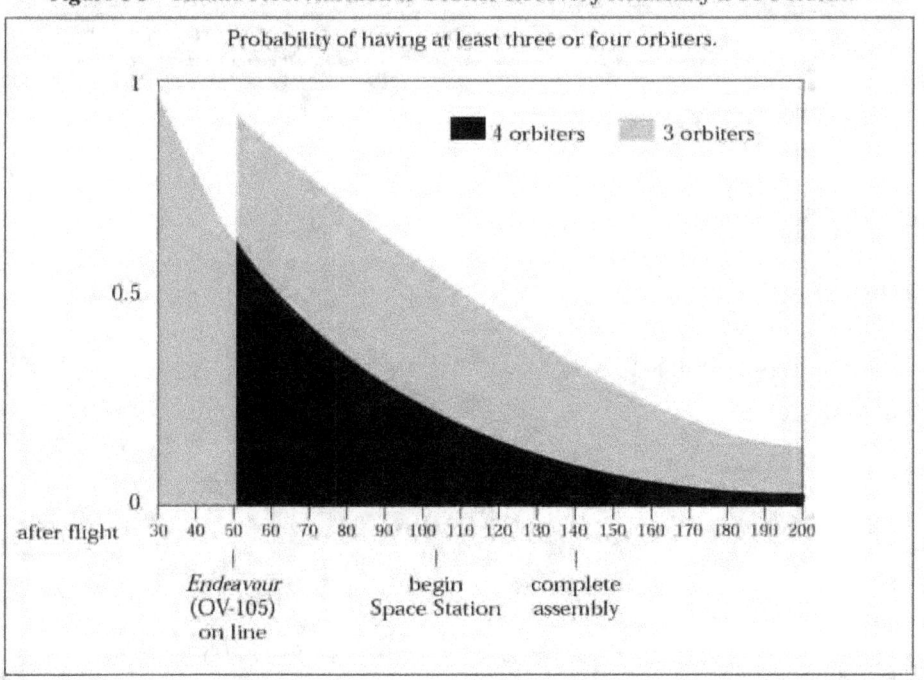

Shuttle reliability is uncertain, but has been estimated to range between 97 and 99 percent.[1] If the Shuttle reliability is 98 percent, there would be a 50–50 chance of losing an orbiter within 34 flights. At a rate of 11 flights per year, there would be a 50 percent probability of losing an orbiter in a period of just over three years. The probability of maintaining at least three orbiters in the Shuttle fleet declines to less than 50 percent after flight 113.

Although loss of an orbiter would not necessarily result in loss of life, it would severely impede the progress of the civilian space program, as it would likely lead to a long standdown of the orbiter fleet while the cause of the failure was determined and repaired. Seen in terms of Space Station construction, if the probability of recovering an orbiter were 98 percent, the probability of retaining four operational orbiters would be only 28 percent when Space Station construction begins on flight 92 and only 12 percent when the Phase I Space Station is completed 42 flights later.

Document II-46

Document title: Dale D. Myers, Deputy Administrator, NASA, to Robert K. Dawson, Associate Director for Natural Resources, Energy and Science, Office of Management and Budget, January 20, 1988, with attachment on the benefits of the Shuttle-C, December 1987.

Source: NASA Historical Reference Collection, NASA History Office, NASA Headquarters, Washington, D.C.

In the late 1980s, NASA retained its belief that the Space Shuttle was a well-designed, robust system and that there should be maximum use of Shuttle hardware in meeting future space transportation requirements. Accordingly, NASA developed the concept of a cargo-carrying vehicle, the Shuttle-C, to lift heavy payloads, particularly Space Station Freedom, into orbit. NASA never received White House or congressional approval to initiate Shuttle-C development. Figure 2 of the enclosure with this letter does not appear here.

[1] NASA
National Aeronautics and
Space Administration

Washington, D.C.
20546

Office of the Administrator   JAN 20 1988

Mr. Robert K. Dawson
Associate Director for Natural Resources, Energy and Science
Executive Office of the President
Office of Management and Budget
Washington, DC 20503

---

1. L-Systems, Inc., *Shuttle/Shuttle-C Operations, Risks, and Cost Analysis*, LSYS-88-008 (El Segundo, CA: 1988).

Dear Mr. Dawson:

As requested in your letter of August 6, 1987, our assessment of the potential benefits and the cost effectiveness of the Shuttle-C is enclosed. Our System Definition Studies will contribute further to a final answer. It is important to remember that our consideration of a Shuttle-C capability is part of a broader space transportation strategy. Returning the Space Shuttle to safe flight, supporting the flight rate buildup, and replacing the Challenger remain our highest priorities. Further, the Advanced Solid Rocket Motor (ASRM) is a critical and necessary investment to meet requirements, increase safety and reliability, and regain performance. Accordingly, we have decided that the ASRM should be given greater priority than a near-term heavy–lift launch vehicle and are proceeding with this to help meet the Nation's overall defined requirements.

In addition to our manned Shuttle capability and our existing expendable launch vehicles, we must plan a more robust national space launch capability which should include a heavy-lift capability. The Advanced Launch System (ALS) Studies and the Shuttle-C studies will provide the basis for formulation of national launch vehicle development strategy. As you are aware, the ALS studies are focused on new systems and new facilities, and the Shuttle-C studies are focused on maximum utilization of Shuttle hardware and facilities. The Shuttle-C appears to offer an affordable, limited flight rate, [and] reliable near-term capability.

Identification of Shuttle-C budget requirements, including funding and schedule, was included in NASA's Fiscal Year 1989 budget submission. We recognize that our overall national and NASA budget posture does not allow proceeding with a design and development decision at this time. However, we presently have sufficient information to decide in favor of conducting the second phase of System Definition Studies. Completing the Phase II definition study tasks will facilitate a much better cost assessment based on a preliminary design analysis. The next major cost review will be conducted in the summer of 1988 when many of the Shuttle-C systems definition tasks will have been completed.

[2] Our studies to date indicate that for a limited flight rate heavy-lift capability, Shuttle-C is more cost effective than other systems contemplated for the mid-1990's. It can be on line earlier, at lower development cost, and higher reliability than other systems since it utilizes major elements of Space Shuttle propulsion, tankage, and engines which have been qualified for the stringent requirements of manned space flight. It will benefit from the continuing production base of the overall STS program. The Shuttle-C would provide assured access to space for STS, Titan IV/Centaur-class planetary, and national security payloads. The Shuttle-C could also launch national security and civil payloads from Vandenberg. Moreover, it is an appropriate response to the Soviet Energiya launch vehicle and provides a comparable capability.

I have concluded that the Shuttle-C concept offers a potential step toward a more robust national launch posture and, with your concurrence, I plan to implement Phase II. Although Phase II does not begin until mid-March, a decision is needed by February 20 to avoid a gap in the study effort. Completing the systems definition effort is essential to establishing a preliminary design and to provide a valid cost data base to review with DOD as a part of the ALS deliberations.

I would look forward to your approval of our Phase II study. If you would be interested in a briefing of our progress in this area to date, we would be pleased to supply it.

Sincerely,

Dale D. Myers
Deputy Administrator

Enclosure

****

[1]   National Aeronautics and Space Administration
                                        Washington, DC 20546

INFORMATION REQUESTED:

An assessment of the benefits and cost-effectiveness of an SDV: (1) for the Space Station including overall funding and schedule, (2) for other approved NASA programs, including the Shuttle, and (3) whether these benefits could also be obtained with the current or improved expendable launch vehicles (ELV's) or other ALS version.

RESPONSE:

SPACE STATION BENEFITS

The use of the Shuttle-C concept could benefit the Space Station Program in several ways. The Shuttle-C concept provides the capability to launch fully integrated Space Station modules, it provides a reduction in the total number of flights needed to achieve permanently manned operational capability, and it provides a large logistics capability.

Launching fully Integrated Space Station modules with the Shuttle-C would reduce the need to integrate the modules on orbit during assembly. For example, the fully integrated Space Station lab module estimated at 69,300 pounds would require 29,800 pounds of hardware to be off-loaded prior to launch on the Shuttle. Such hardware would then be launched on additional Shuttle flights, installed, and integrated on orbit. With Shuttle-C, the fully integrated 69,300 pound lab module could be launched on one flight, thereby reducing the extravehicular/intravehicular activity time and enhancing reliability.

The Shuttle-C concept of compatible interfaces with the Shuttle provides flexibility in Space Station launch packaging by its increased volume and weight capability. The recent Space Station Transportation Studies identified how the number of STS flights could be reduced from 19 to 7 by adding five Shuttle-C flights. The assembly period timespan could be reduced, if desired, from the present 36 months to as little as 18 months. The number of launch package end items to be assembled on orbit is reduced from 45 to 34. Phase I assembly could, thus, be completed several months earlier than with the STS alone and with a net reduction of seven flights and no changes in Space Station design. Further, the Shuttle-C would provide significant increased flexibility and robustness in schedule and weight margin for Station assembly. For example, because of the inherent large payload capacity of Shuttle-C, late hardware articles could be delivered to the Station as an aggregate payload on one Shuttle-C. This resiliency could possibly permit the compression, or catch-up, of the assembly schedule that may not be feasible with the Shuttle alone. Slips in hardware manifested for Shuttle-C could be accommodated without a large remanifesting effort for subsequent STS launches.

The current baseline for Space Station resupply requires annual delivery weight of approximately 180,000 pounds, including crew rotation and logistics. With 103,000 pounds of payload capability to the Space Station, Shuttle-C could help accommodate resupply requirements.

Studies are planned to investigate the feasibility of launching the Crew [illegible word] Return Vehicle on the Shuttle-C.

[2]   BENEFITS TO OTHER NASA PROGRAMS

The Shuttle-C could benefit several proposed new initiatives and planned programs. Shuttle-C would provide design options to payloads now planned for manifesting on smaller and more constraining vehicles. The extra payload margin could be used to carry additional scientific instruments or to make cost trades.

The projected Shuttle-C capability could place 56,000 pounds in sun-synchronous orbit (445 NM [nautical miles]/98.7 degrees) or 20,000 pounds in geosynchronous orbit (22,000 NM) using a Centaur upper stage adapted for Shuttle-C. Polar platforms and other payloads not requiring crew interaction could be off-loaded to Shuttle-C. Shuttle-C would also allow the launch of co-orbiting platforms on the same launch vehicle. It would assure alternate launch capability for all Titan/Centaur-class payloads.

NASA has examined the use of the Shuttle-C for several planned planetary exploration missions, including the Comet Rendezvous Asteriod [sic] Flyby (CRAF)—the first of the planned Mariner Mark II missions, Cassini (the second planned Mariner Mark II mission), and the Mars Rover Sample Return (MRSR). CRAF is proposed as a new start for Fiscal Year 1989, and it is currently planned for launch on a Titan IV/Centaur.

The benefit of the Shuttle-C/Centaur G-Prime for any of these missions derives from the fact that the Shuttle-C can deliver the spacecraft and a fully loaded Centaur to low Earth orbit. This is a significant improvement over the current Titan IV, wherein approximately one-third of the Centaur propellants are expended in order to achieve the initial parking orbit.

Additional performance provided by the Shuttle-C allows added mission and spacecraft system flexibility and permits tradeoffs of one or more of the following to enhance the mission:

1. Extended observation time,
2. Additional flexibility in the selection of scientifically interesting targets,
3. Additional spacecraft propellant for operations and maneuvers and/or additional satellite encounters,
4. Increased payload mass to enable addition of a second penetrator/probe or other science instruments, and
5. Shorter trip time.

Shuttle-C offers a significant advantage for the MRSR mission by launching the rover orbiter, ascent and descent systems, and sample return vehicle in a single launch as opposed to the requirement for two separate launches if the Titan IV/Centaur were used.

[3]   SHUTTLE BENEFITS

In addition to serving as an alternate launch capability, the Shuttle-C would provide four major benefits to the Shuttle: (1) An unmanned flight test bed for new or enhanced Shuttle capabilities and advanced systems, (2) reduced unit costs from increased production rates, (3) cost savings by use of older Shuttle engines, and (4) increased transportation resiliency from the combination of the two systems.

Several propulsion enhancements are under study as improvements to the Shuttle, including the advanced solid rocket motor; the liquid rocket booster, which would replace the solid rocket booster; and, possibly, new liquid engine systems. Although these systems would be designed for high reliability, use of the unmanned Shuttle-C vehicle for the initial flight would give added confidence and demonstrate performance without any risk of human life.

A second benefit from Shuttle-C is that the increase in production rates of STS common components (e.g., engines, computers) will reduce unit costs. In some areas, such as avionics, there is also the potential of losing Shuttle subcontractors because of the low production rates, which may be alleviated by Shuttle-C needs.

A third benefit would be that older STS engines could be put to productive use. The Shuttle was designed for reusable engines. These engines are being qualified for 20 flights and, with our conservatism for manned flight, will be flown only ten times. That amortizes the engines and, although they are worth little to the Shuttle, they would still be highly reliable engines for an unmanned flight. If such a costing approach were taken, with Shuttle-C using amortized engines, the cost per launch for Shuttle-C could be considered to be approximately $240M per flight in 1986 dollars.

The use of a mixed Shuttle/Shuttle-C fleet is also expected to provide increased transportation resiliency. A parametric study is currently under way which will provide an analysis in terms of resiliency (the probability of satisfying flight rate requirements), availability (fraction of the time operational), mean time to failure risk, surge capability, and cost effectiveness.

BENEFITS FROM ALTERNATE USE OF EXPENDABLE LAUNCH VEHICLES (ELV'S)

Another benefit of a Shuttle-C mixed fleet derives from its overall reduction in cost per flight over alternate launch vehicles. Some of the preliminary estimated trends are discussed below.

Figure 1. Launch vehicle operations cost estimates compares the dollars per pound to 160 NM of various existing and planned launch systems. All existing and planned expendable systems exhibit higher operations cost than the projected marginal costs associated with Shuttle-C.

Figure 2. Life cycle cost comparison of Shuttle-C versus interim Advanced Launch System (ALS) takes into account both the Design, Development, Test, and Evaluation (DDT&E) and operations costs and compares the resulting life cycle costs of Shuttle-C and a representative interim ALS concept over a range of [illegible words] life cycle cost of a [4] representative interim ALS concept is shown as a band corresponding to the cost both with and without a strongback for payload support with the shroud. The Shuttle-C has lower DDT&E requirements than the representative interim ALS concept and lower operational cost for the same mission model. For three million pounds to orbit (corresponding to 27 Shuttle-C and 32 ALS flights, respectively), Shuttle-C has undiscounted life cycle costs of about two-thirds of the life cycle costs associated with the representative concept.

The projected Shuttle-C launch marginal cost per payload pound is substantially lower than any available ELV. The ALS program goal of reducing launch costs of the objective ALS by a factor of ten would make the ALS more cost effective at higher flight rates, but, until the ALS is available in the late 1990's, the Shuttle-C would be the most cost effective means of launching large unmanned payloads.

###

December 1987

[5]

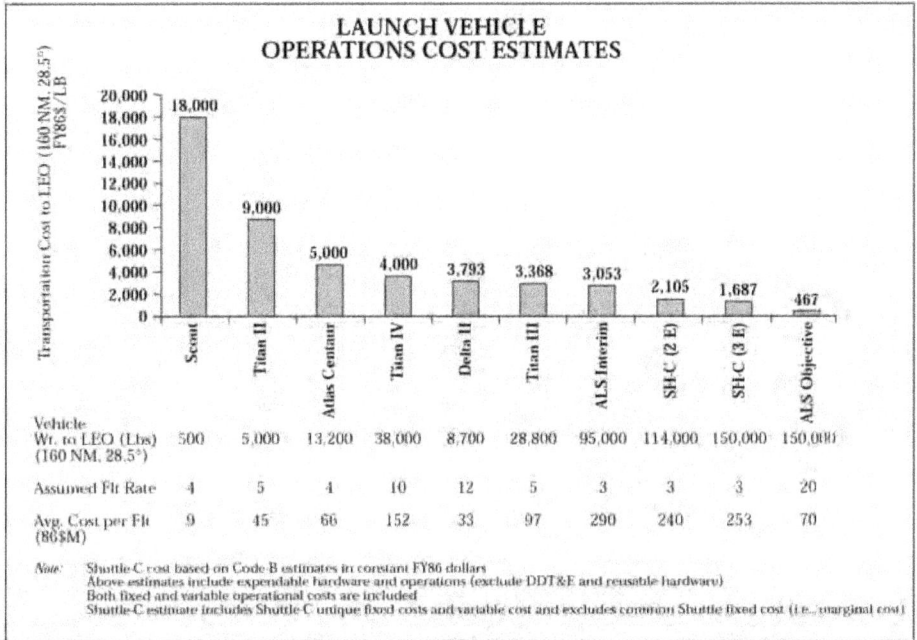

LAUNCH VEHICLE OPERATIONS COST ESTIMATES

Vehicle	Scout	Titan II	Atlas Centaur	Titan IV	Delta II	Titan III	ALS Interim	SH-C (2 E)	SH-C (3 E)	ALS Objective
Wt. to LEO (Lbs) (160 NM, 28.5°)	500	5,000	13,200	38,000	8,700	28,800	95,000	114,000	150,000	150,000
Assumed Flt Rate	4	5	4	10	12	5	3	3	3	20
Avg. Cost per Flt (86$M)	9	45	66	152	33	97	290	240	253	70

Note: Shuttle-C cost based on Code-B estimates in constant FY86 dollars
Above estimates include expendable hardware and operations (exclude DDT&E and reusable hardware)
Both fixed and variable operational costs are included
Shuttle-C estimate includes Shuttle-C unique fixed costs and variable cost and excludes common Shuttle fixed cost (i.e., marginal cost)

Document II-47

Document title: "Report of the Space Shuttle Management Independent Review Team," February 1995, pp. iii–iv, vii–x, A-1–A-2.

Source: Space Policy Institute, George Washington University, Washington, D.C.

In the years after its post-accident return to flight, the Space Shuttle became a reliable space transportation system, but it was also extremely costly to operate. NASA Administrator Daniel S. Goldin, who came to the space agency in April 1992, set as two of his priorities reducing the cost of Shuttle operations and finding a way to remove NASA from the day-to-day repetitive operations of various systems. His goal was to free up NASA financial and human resources for an increased focus on research and development activities. In November 1994, Goldin chartered a Space Shuttle Management Independent Review Team, chaired by former Johnson Space Center Director Christopher Kraft, to examine possible changes in the management of Space Shuttle operations that would lead to reduced costs while maintaining or improving the safety of the system. NASA accepted the team's recommendation to consolidate Shuttle operations under a single business entity. That entity became United Space Alliance, an equal partnership between the Lockheed Martin Corporation and the Boeing Company. What appears here are the preface, acknowledgments, executive summary, and first appendix of the review team's report.

Report of the
Space Shuttle Management
Independent Review Team

The undersigned present the report of the
Space Shuttle Management Independent Review

Dr. Christopher Kraft, Chairman

Col. Frank Borman   George Jeffs

Robert Lindstrom   Thomas Maultsby

Isom Rigell

[iii]   **PREFACE**

The space shuttle is recognized throughout the world's technical community as the consummate vehicle for space transportation. Its performance in placing humans and payloads in orbit and returning products and satellites to Earth is unmatched. Since the vehicle was declared operational in the mid-1980s, however, it has been severely criticized for the high cost of operation. In addition, many of the promises made for the shuttle have never been realized for a number of reasons. For example: 1) the number of flights per year that were forecast never materialized; 2) the Challenger accident temporarily cast doubt on shuttle reliability; 3) the number of payloads by other U.S. Government agencies (particularly the Department of Defense) was overestimated, with many transferred to other launch vehicles; 4) policy (e.g., National Space Policy) and statutory changes were made to discourage the use of the shuttle as a launch vehicle except for missions that require human presence or other unique shuttle capabilities; 5) NASA continued to operate the shuttle in a quasi-research and development mode; this was exacerbated by the Challenger accident.

The NASA Administrator has attempted by various means, and with reasonable success, to reduce the total cost of operating the shuttle. In recent years, NASA has reduced the shuttle's direct operating costs by approximately 25 percent—a valiant effort considering the scrutiny the shuttle receives by the government and the press. As more budget pressures are brought to bear and NASA searches for funds to use in pursuit of future programs, however, it became obvious to the Administrator that he should seek possible changes in the shuttle management structure. As a result of discussions with a number of advisors in the government, the aerospace industry, and former NASA leaders, the Administrator decided to form a team composed of some of these people to review the present shuttle operation management and to propose innovative approaches to significantly decrease total operating costs while maintaining systems safety.

If NASA is successful in bringing about a new approach to spaceflight operations, it will add to NASA's credibility as an agency on the forefront of reinventing government and provide a model for the management of future programs and their transition to the private sector.

[iv]   **ACKNOWLEDGMENTS**

The team chairman wishes to acknowledge the work of the official team members, the advisors, and the NASA team members. The team members all served in the best interest of the country's space program and those who read this report must recognize that the team did its utmost to provide candid and useful inputs to the future conduct of spaceflight.

The team wishes to compliment the people both in NASA and the aerospace industry for their lucid and frank presentations and discussions on the many facets of the Space

Shuttle Program. As is the usual case, many of the ideas presented herein came from these dedicated and competent people.

Jeff Bantle and Cliff Farmer provided a great deal of effort to bring the written word to paper and are typical of the fine young people that reside in NASA. They are all anxious to continue a productive and exciting space program, which will provoke new knowledge and technology.

Christopher C. Kraft . . .

[vii] **EXECUTIVE SUMMARY**

At the request of the NASA Administrator a team was formed to review the Space Shuttle Program and propose a new management system that could significantly reduce operating costs. Composed of a group of people with broad and extensive experience in spaceflight and related areas, the team received briefings from the NASA organizations and most of the supporting contractors involved in the Shuttle Program. In addition, a number of chief executives from the supporting contractors provided advice and suggestions.

The team found that the present management system has functioned reasonably well despite its diffuse structure. The team also determined that the shuttle has become a mature and reliable system, and—in terms of a manned rocket-propelled space launch system—is about as safe as today's technology will provide. In addition, NASA has reduced shuttle operating costs by about 25 percent over the past 3 years.

The program, however, remains in a quasi-development mode and yearly costs remain higher than required. Given the current NASA-contractor structure and incentives, it is difficult to establish cost reduction as a primary goal and implement changes to achieve efficiencies. As a result, the team sought to create a management structure and associated environment that enables and motivates the Program to further reduce operational costs.

Accordingly, the review team concluded that the NASA Space Shuttle Program should

(1) Establish a clear set of program goals, placing a greater emphasis on cost-efficient operations and user-friendly payload integration.
(2) Redefine the management structure, separating development and operations and disengaging NASA from the daily operation of the space shuttle.
(3) Provide the necessary environment and conditions within the program to pursue these goals.

With over 65 successful launches, operations have become quite reliable. At this stage in the Shuttle Program, cost-efficient operations and user-friendly payload integration should be pursued along with safe and successful flights. If the Program is to meet the challenge of reducing costs and streamlining payload integration, it will require a major change in how the Program operates.

Given the maturity of the vehicle, a change to a new mode of management with considerably less NASA oversight is possible at this time. In addition, the bureaucracy that has developed over the program's lifetime—and particularly since the Challenger accident—will be difficult to overcome and the optimum operational effectiveness of the system will be difficult to achieve unless a new management system is provided

[vii] The team considered a number of new management approaches. These included to

(1) Stay with the present system and continue to decrease costs in the incremental fashion used to date.
(2) Implement a multi-node system, consolidating contracts in each of the major geographical areas (i.e., the Kennedy Space Center in Florida, Marshall Space Flight Center in Alabama, and Johnson Space Center in Texas), each managed by a prime contractor with continued NASA program management.

(3) Consolidate operations under a single-business entity.

The team concluded that consolidating operations under a single-business entity was the most advantageous. This single-business approach is a change from the present one of government control with industry response to that of government direction with industry operation.

The multi-node approach possesses some of the same features that cause the present system to be cumbersome and expensive. Both options (1 and 2) do not provide the centralization of control necessary to eliminate duplication, the disengaging of NASA from day-to-day direction necessary to reduce requirements, and the incentives necessary to motivate cost reduction. One of the critical deficiencies in today's program management, and one that the multi-node approach also suffers from, is the lack of a single responsible agent among all of the contractors supporting the program. As a result, no one entity feels the total responsibility for the shuttle operation; therefore, no advocate exists for overall cost reduction. This deficiency is the major fault with both the current program structure and the multi-node concept.

Several different single-business approaches were discussed with the prime contractor option considered the most achievable and practical. Other concepts, including a business consortium, joint venture, and government owned-contractor operated (GOCO) arrangement, involve complexities that are difficult to overcome in any reasonable period of time. In addition, selecting a prime contractor from among the current contractors, as opposed to an open competition, could accomplish all of the objectives in a less disruptive and more expeditious manner, realizing potential cost reductions more quickly.

The proposed single-business management system will require a steadfast commitment from both NASA and the aerospace industry to ensure success. NASA must be willing to define clear shuttle operating requirements with limited oversight. The prime contractor must be willing to assume responsibility for safe and productive operations. This requires the assignment of competent and experienced people at all levels and the direct attention of top management. For its commitment, the contractor must be rewarded with appropriate incentive fees. The government in-turn must provide similar talent in program management and a guarantee that the contractor will not be encumbered with burdensome and unnecessary oversight.

[ix] The new management approach will require the following immediate actions:

(1) Freeze the current vehicle configuration, minimizing future modifications, with such modifications delivered in bloc updates. Future bloc updates should implement modifications required to make the vehicle more re-usable and operational.
(2) Perform a requirements review, top down, with the goal of significantly reducing checkout and other requirements based upon operations experience.
(3) Consolidate and reduce program and project elements, limiting NASA involvement in operations and minimizing NASA-contractor interfaces.
(4) Restructure and reduce the overall Safety[,] Reliability[,] and Quality Assurance (SR&QA) elements—without reducing safety.
(5) Streamline payload processing and integration, minimizing costs and reducing the length of time required to integrate a payload aboard the space shuttle.
(6) Structure operational contracts to provide real incentive to reduce costs while accomplishing safe and successful missions.
(7) Allow the hiring of NASA personnel by the prime and subcontractors to ensure proper expertise and talents exist to continue with safe and successful operations.

One of the major stipulations to achieve cost reduction is to freeze the present shuttle configuration and perform only those changes required to carry out the individual flights. Currently, change and update are continual and pervasive at all levels of the program and seize a significant amount of attention, focus, and resources. To become an operational program, the shuttle configuration must be more stable. To aid in the transition process, the present NASA management system would complete the development of presently approved changes and then be phased out.

Additionally, turnaround, launch, and mission requirements should be diminished based on operational experience. Currently, for the orbiter alone, approximately 150 hardware package changeouts are performed between each flight; yet an average of only 10 in-flight anomalies, most of which are inconsequential, occur during each mission. Maturation of the vehicle checkout requirements has, clearly, not kept pace with the vehicle hardware, and redundant subsystems are not being used to provide operational flexibility.

Once the new management structure is in place, efficiencies can be realized through the consolidation, reduction, and elimination of functions. This will be a challenging task considering the diffuse state of the current NASA-contractor structure. Duplication and overlap have developed throughout the program.

One of the most apparent examples in this regard is the area of SR&QA. As a result of the *Challenger* incident, a "safety shield" philosophy has evolved[,] creating a difficult management situation. Managers, engineers, and business people are reluctant to make decisions that involve risk because of the fear of persecution. As a result, a parallel and independent SR&QA element has grown to large proportions. This is not only significant with respect to direct costs, but has an even greater impact when supporting efforts are included. Restructuring and streamlining [x] SR&QA throughout the Shuttle Program, maintaining only the necessary checks and balances, must be accomplished to achieve significant cost reduction.

As the Shuttle Program transitions to an operational program, payload processing must be streamlined, with an associated reduction in cost and length of time required to integrate a payload. As this takes place, payload operations must change from "defensive" to more customer-oriented. Toward this end, payload operations would become an integral part of mission and launch operations with attendant streamlining of organizations, people, and procedures.

To assume greater operational responsibility and risk, it will be necessary to provide the contractor with the opportunity to realize a profit. Proper contract incentives will be needed to ensure the contractor team performs the necessary steps to reduce cost. Greater and longer term sharing of cost savings, along with appropriate penalties for marginal performance, will be required to provide the contractor with the motivation to significantly reduce costs while maintaining safe and successful operations.

Finally, ensuring the NASA-contractor team has the expertise required to operate the shuttle is of significant concern. In the present aerospace industry, it may be difficult to assemble all of the necessary talent and resources to assume the responsibility for shuttle operations. Therefore, initially, this will require private industry to hire NASA personnel and/or utilize specific government engineering organizations with critical skills until these skills can be developed from within. It is also important when constructing the contractor team to recognize current expertise that has already been developed. An example of this is in the areas of orbiter obsolescence and sustaining engineering where specific expertise and experience is necessary to continue to operate the vehicle. Building the NASA-contractor team will require special attention to these types of issues.

The transition process will entail the development of a program office by the selected prime contractor. The present NASA program and project offices would be used to aid the prime contractor through the initial development of this new operating concept. As the contractors' skills mature, they would continually assume greater responsibility. The team believes this transition should be expedited with the overall transition time dependent on the specific shuttle element, the techniques employed by NASA to rearrange the contractual responsibility, and the commitment by all parties to bring about these significant changes. . . .

[A-1]   **APPENDIX A—SUMMARY OF RECOMMENDATIONS**

Recommendation 1: Establish a more balanced set of goals for the Shuttle Program, with a greater emphasis on reducing operational costs and making payload integration more user friendly. The following goals provide a better balance between operations and safety, and address the overall NASA objective of reducing the cost of access to space.

(1) Perform safe space shuttle operations while accomplishing mission objectives.
(2) Reduce the cost of space shuttle operations.
(3) Provide user-friendly payload integration.

Recommendation 2: Modify the program's management structure, separating development from operations[,] and relinquish the majority of the operational responsibility to a prime contractor.

Recommendation 3: Minimize vehicle modifications. Freeze the current vehicle hardware and software configuration. Implement future modifications using a bloc update concept. These bloc updates should be justified and only made to improve safety, reduce operating costs, make the vehicle more reusable, or test new technologies.

Recommendation 4: Initiate a requirements review, top level down, with the goal of significantly reducing requirements based on operations experience. This type of review could significantly reduce vehicle turnaround and checkout requirements based upon hardware reliability, criticality, and redundancy.

Recommendation 5: Ensure future performance upgrades to support International Space Station Alpha (ISSA) or other payloads are established through a systems engineering process to determine the most advantageous and cost-effective approach.

Recommendation 6: Reduce NASA involvement and oversight in the operation of the space shuttle, transferring responsibility of daily operations to the contractor. Space Shuttle Program and Project elements should be consolidated and reduced with NASA-contractor interfaces minimized.

Recommendation 7: Restructure and reduce the overall SR&QA element.

Recommendation 8: Streamline payload processing and integration, minimizing costs and reducing the length of time required to integrate a payload aboard the space shuttle.

Recommendation 9: Structure operational contracts to provide real incentive to accomplish safe and successful missions.

Recommendation 10: NASA must pursue innovative approaches in assembling and supporting the prime contractor team. This could include the hiring of NASA civil servants by the contractor and initially allowing the contractor to use specific government capabilities.

[A-2] Recommendation 11: All artificial barriers which preclude the shuttle from carrying certain types of payloads should be removed. This would require policy and statutory changes which currently discourage the shuttle from carrying commercial payloads.

Recommendation 12: As the prime contractor management approach develops and matures, NASA should consider further industry involvement and progression toward the privatization of the space shuttle. . . .

Chapter Three

Commercializing Space Transportation

by John M. Logsdon and Craig Reed

The precedent that the United States, through NASA, would launch satellites for pay-ing customers was set as early as 1960. That year, American Telephone and Telegraph (AT&T) asked the space agency whether it would launch, on a reimbursable basis, an experimental communications satellite to be developed by Bell Telephone Laboratories.[1] However, the notion that launching satellites could become an economically profitable undertaking, with the potential for being the basis of a commercial business, did not emerge until two decades later. It was not until the 1980s that international organizations and private-sector firms made plans to place an increasing number of communications satellites into geosynchronous orbit. This led to a series of developments during the 1980s that created a commercial launch industry in the United States and Europe, although not without conflict among governments and between the government and the private sector in the United States. Also during the 1980s, the Soviet Union and China took the initial steps toward being competitors in the commercial launch market. In the 1990s, the com-mercial launch industry saw vigorous growth, including a number of joint ventures among firms from various countries. This chapter traces the development of this sector of space transportation activity.

The Space Shuttle as a Commercial Launch Vehicle

In 1972, and for almost the following decade, the United States had a monopoly on "free world" access to space for any payload of significant size. This monopoly (except for the Saturn boosters used in the Apollo program, which were never used for launching commercial payloads) was based on adapting rockets first developed as ballistic missiles for use as space launchers—the Thor Delta (later only known as Delta), Atlas, and Titan vehicles. Only the Soviet Union, which also had adapted its ballistic missiles as space launchers, possessed launch vehicles of similar lifting capability, and in the Cold War envi-ronment of the time, there was no question that the United States would not allow the launching of Western-manufactured satellites on Soviet launchers. At the beginning of the 1970s, Europe was debating whether to develop its own autonomous means of access to space, and Japan was developing its launch capability using licensed U.S. technology under tight restrictions regarding the launch of third-country payloads. China was in the early stages of developing its space launch capability. Thus, planning for the Space Shuttle proceeded on the assumption that it would be the launcher used by all U.S. payloads and most payloads launched by other countries, international organizations such as the International Telecommunications Satellite (INTELSAT) consortium, and any other pri-vate-sector entity desiring to put a satellite into orbit. In this light, expendable launch vehi-cles (ELVs) would become obsolete once the Shuttle became operational, and therefore

1. See F.R. Kappel, President, American Telephone and Telegraph Company, to the Honorable James E. Webb, Administrator, NASA, April 5, 1961 (with several attachments), reprinted as Document I-9 in John M. Logsdon, gen. ed., with Roger D. Launius, David H. Onkst, and Stephen J. Garber, *Exploring the Unknown: Selected Documents in the History of the U.S. Civil Space Program. Volume III: Using Space* (Washington, DC: NASA Special Publication (SP)-4407, 1998), 3: 46–57.

their production, at least in the United States, would cease. The economic justification for the Space Shuttle assumed a very high launch rate, spreading the fixed costs of operating the Shuttle over many launches and thereby keeping the cost per launch low.[2]

The Space Shuttle launch price for nongovernment users announced in 1977 was $18 million (1975 dollars) for a Shuttle launch in which the whole payload bay was used, plus an insurance charge and a user's fee. For payloads requiring only a portion of the Shuttle's 15-foot by 60-foot payload bay, the charge would be reduced in proportion to the length of the payload bay used (and also the weight of the payload). Thus, a payload using only 15 feet of the bay's 60-foot length would pay only 25 percent of the launch fee. The result of this approach was that an initial Shuttle price for launching a "Delta-class" communications satellite was approximately one-half that charged at the time for a Delta ELV launch of a similar payload (Figure 3–1).[3] Thus, in the years before it actually began operation, the Space Shuttle seemed to be a very attractive way for commercial users to get their payloads into space, and there was little prospect for a U.S. commercial space launch industry operating separately from the Shuttle.

Figure 3–1. One of the principal commercial space launch vehicles of the United States, here a Delta sits on the launch pad at Cape Canaveral. (photo courtesy of McDonnell Douglas Corporation, Neg. no. GC1270-3575)

However, by the time the Space Shuttle was first launched in April 1981, two developments had tempered the optimistic assumptions of its developers. First, NASA's estimate of the number of flights through 1991 had dropped from 572 to 487; this, combined with design changes and inaccurate cost estimates, led to an anticipated average cost per flight that was 73 percent higher than had been the basis used for the original pricing policy. This made it even more urgent that NASA spread the fixed costs of Shuttle operation over as many flights as possible.[4]

Moreover, a threat to the Shuttle's dominance of the launch market had materialized. The European ELV, Ariane, had a first successful flight in December 1979 (followed by a failure in its second flight in May 1980). Europe, with France in the leading role, had decided in 1973 to develop its own ELV in 1973 after the United States had laid down what were considered unacceptable conditions under which it would launch the French-German experimental communications satellite Symphonie.[5] Launching satellites to

2.   Congressional Budget Office, *Pricing Options for the Space Shuttle* (Washington, DC: U.S. Government Printing Office, March 1985), p. 4.

3.   Office of Technology Assessment, *International Cooperation and Competition in Civilian Space Activities* (Washington, DC: U.S. Government Printing Office, 1985), p. 131.

4.   Congressional Budget Office, *Pricing Options*, p. 4.

5.   See Department of State Telegram, "Johnson Letter to Lefevre," September 7, 1971, reprinted as Document I-22 in John M. Logsdon, gen. ed., with Dwayne A. Day and Roger D. Launius, *Exploring the Unknown: Selected Documents in the History of the U.S. Civil Space Program, Volume II: External Relationships* (Washington, DC: NASA SP-4407, 1996), 2: 59–62.

geosynchronous orbit was a primary mission for Ariane, and thus the rocket was optimized for that mission. In March 1980, the developers of Ariane, again led by the French government, formed a quasi-private organization, Arianespace, to market the launcher on a commercial basis as an alternative to the Space Shuttle for launching commercial communications satellites. Arianespace soon after announced a goal of capturing one-third of the launch market—a market that previously had been a U.S. monopoly. Also in 1980, the member states of the European Space Agency (ESA), the developer of Ariane, agreed to upgrade the launcher so that it could launch *two* small communications satellites at the same time. This would make the Ariane price for a satellite launch competitive with that being offered by NASA for Shuttle launches. In June 1981, Ariane conducted its first launch for a paying customer, and in January 1982, ESA decided on a further upgrade of Ariane to make it an even more powerful and more flexible launch vehicle.[6]

This challenge to U.S. dominance in a key area of space activity produced a strong reaction in the White House, Congress, and the new NASA management team led by NASA Administrator James M. Beggs that came into office with the administration of President Ronald Reagan in 1981. In its first statement on space policy, issued November 13, 1981, the Reagan administration announced that "the United States is committed to a vigorous effort that will ensure [space] leadership" and that the Space Shuttle would be "a vital element" in providing such leadership.[7] This was followed by a July 4,1982, statement of National Space Policy, which declared in its opening paragraphs that "the Space Shuttle is to be a major factor in the future evolution of United States space programs."[8]

It was clear by the end of 1981 that a new Shuttle pricing policy was needed, both to reflect initial experience and to better compete with Ariane. In 1977, NASA had committed itself to recovering both Space Shuttle development and operating costs through launch fees; as operating costs rose, it became obvious that both they and development costs could not be recovered by the original Shuttle pricing policy. On the other hand, using the same cost recovery basis as had been used in 1977 to set a Shuttle price would have produced a launch fee so high that it would not be competitive with Ariane.

Ariane's competitiveness became vividly apparent in 1982 as GTE Spacenet became the first U.S. firm to sign a contract with Arianespace to launch its communications satellites aboard the European rocket. A 1982 NASA report suggested that "the present projection of capital lost to Ariane is estimated to be $3 billion through 1984, if every compatible U.S. customer used Ariane."[9] While there was little chance of such a shift to Ariane launches actually happening, NASA and the White House were clearly disturbed by the possibility of significant Arianespace penetration into the U.S. market (Figure 3-2).

Thus in 1982, NASA changed its definition of which costs of the Shuttle were to be recovered, and the space agency devised a new pricing policy based on this change. Any pretense of recouping the more than $5 billion cost of Shuttle development was abandoned. The price was $71 million (1982 dollars), and it was to be in effect for launches from October 1, 1985, through September 30, 1988. The earlier price ($18 million in 1975 dollars, or $38 million in 1982 dollars) would remain in effect until then.[10]

---

6. See Steven J. Isakowitz, ed., *International Reference Guide to Space Launch Systems*, 2d ed. (Washington, DC: American Institute of Aeronautics and Astronautics, 1995), for information on Ariane's development.

7. The White House, National Security Decision Directive (NSDD) 4, "Space Transportation System," November 13, 1981.

8. See NSDD 42, "National Space Policy," July 4, 1982, p. 1., reprinted as Document III-38 in John M. Logsdon, gen. ed., with Linda J. Lear, Jannelle Warren-Findley, Ray A. Williamson, and Dwayne A. Day, *Exploring the Unknown: Selected Documents in the History of the U.S. Civil Space Program, Volume I: Organizing for Exploration* (Washington, DC: NASA SP-4407, 1995), 1: 590-93.

9. *Ibid.*, p. 135.

10. Congressional Budget Office, *Pricing Options*, p. 5.

Although the competition between the Space Shuttle and Ariane hinged primarily on their comparative price, other factors were also involved. As a quasi-private organization, Arianespace was able to employ private-sector marketing techniques, flexible financing arrangements, and other methods of attracting customers. In response, NASA, while operating within the limits of its governmental character, began to actively market the Shuttle to most of the same potential customers being courted by Arianespace. A NASA advisory group in 1983 noted:

[A]n intensive high level marketing effort on behalf of Shuttle utilization is warranted. In this context, marketing means to develop and implement a broad scale and long range plan to involve increasing numbers of users in the exploration of STS [Space Transportation System, another way of designating the Shuttle] capabilities. It thus involves market analysis, planning, advertising, customer service, financing, and insurance, to name a few areas. It must be a high level, strongly led effort, with the active participation of NASA top management to the Administrator level.[11]

Figure 3–2. The Ariane launch vehicle, developed by the European Space Agency in the 1970s and first entering operational service in the early 1980s, has been one of the most important competitors for U.S. launch capability. Here an Ariane is at its launch site at the ESA launch facility in Kourou, French Guiana. (NASA photo)

Although it was highly unusual for a Federal agency to undertake such a marketing effort, NASA set about the task. In promoting the Space Shuttle, NASA's marketing people produced a glossy, colorful marketing brochure titled *We Deliver.* This document stressed the Shuttle's "remarkable suitability for delivering communications satellites to earth orbit" and its "reliability assets" that "set it apart from its expendable counterparts." The brochure also emphasized the Shuttle's "flexibility and expanded capabilities that provide the opportunity to significantly improve satellite designs by taking advantage of the new features that only the Shuttle provides." [III-1]

From 1983 to January 1986, NASA and Arianespace engaged in a vigorous global competition for available commercial launch contracts, with both viewing their success as linked to the relative standing of their supporting nations with respect to space leadership. The result was a "buyers' market" for commercial communications satellite builders and operators desiring access to space.

Origins of the Commercial Launch Industry

The July 4, 1982, statement of National Space Policy made final the decision that NASA would no longer order more Delta and Atlas launch vehicles. In addition, and with significant misgivings, the Air Force began the process of shutting down the production

---

11. Quoted in Office of Technology Assessment, *International Cooperation,* p. 341.

lines for the Titan vehicles used to launch the highest priority national security payloads. During the 1970s and early 1980s, there had been no significant government investment in upgrading ELV capabilities and facilities, once a tentative decision had been made to launch all government payloads on the Shuttle when it became operational. The aerospace industry, used to having the government fund all launcher-related research and development, did not replace government funding with industrial investment. This meant that in the early 1980s, U.S. ELVs at best embodied early 1970s technologies, as their production lines slowed to a halt.

Even so, some of the manufacturers of these proven boosters and others interested in commercial opportunities in space saw an opportunity to compete with the Shuttle and Ariane for commercial launch contracts, if only on what they considered a fair basis. They noted that the actual cost of operating the Shuttle was much greater than the price being charged by NASA to commercial users and that the Shuttle was having trouble meeting its launch schedule commitments because of its complexity and rapid transition to an operational status. There seemed to be a market opportunity for private U.S. providers of launch services, if only the U.S. government support that made it possible for NASA to keep the Shuttle price low (and the European government support that allowed Arianespace to keep the Ariane price low) could somehow be lessened.

In addition to those interested in commercializing existing ELVs, in the early 1980s, several entrepreneurs proposed developing new, privately financed space launch vehicles, particularly for the small satellite market niche that the Space Shuttle did not serve economically.[12] While they were mostly on the periphery of the space policy scene, these space commercialization entrepreneurs became increasingly visible and active during this period. One of the earliest of these ventures, for a launch vehicle called Percheron, resulted in a launch pad explosion in August 1981 (Figure 3–3). The Conestoga I, a successor to the failed Percheron developed by Space Services Inc., was successfully launched in a suborbital flight in September 1982, marking the first successful test of a privately funded U.S. launch vehicle.

The Reagan administration included many individuals interested in promoting the commercialization of space overall; they were sympathetic to those interested in commercializing launch services. Their support led to a series of moves during the

Figure 3–3. The Percheron, a privately developed rocket, sits at Matagorda Island, Texas, on August 5, 1981. (NASA photo)

---

12. Several of these attempted private ventures are discussed in Michael A.G. Michaud, *Reaching for the High Frontier: The American Pro-Space Movement, 1972–1984* (New York: Praeger, 1986), 252–70.

1983–85 period intended to create, independent of NASA, a commercial space transportation industry in the United States. A White House-mandated review concluded in April 1983 that "a U.S. commercial ELV capability would benefit both the USG [U.S. government] and the private sector and is consistent with the goals and objectives of the U.S. National [July 4, 1982] Space Policy." [III-2] Based on this assessment, the White House, on May 16, 1983, issued National Security Decision Directive (NSDD) 94, "Commercialization of Expendable Launch Vehicles." [III-3] This directive stated that "the U.S. Government fully endorses and will facilitate the commercialization of U.S. Expendable Launch Vehicles." The statement went on to say that:

> The U.S. Government will license, supervise, and/or regulate U.S. commercial expendable launch vehicle operations only to the extent required to meet its national and international obligations and to ensure public safety.
> The U.S. Government encourages the use of its national ranges for U.S. commercial expendable launch vehicle operations. . . . [T]he U.S. Government will identify and make available, on a reimbursable basis, facilities, equipment, tooling and services that are required to support the production and operation of U.S. commercial expendable launch vehicles.
> The U.S. Government will not subsidize the commercialization of expendable launch vehicles but will price the use of its facilities, equipment, and services consistent with the goal of encouraging viable commercial expendable launch vehicle launch activities.
> The U.S. Government will encourage free market competition among the various systems and concepts within the U.S. private sector . . . [and] . . . will provide equitable treatment for all commercial launch operators for the sale or lease of government equipment and facilities consistent with its . . . interests.

Under pressure to demonstrate its support for the White House policy on behalf of the commercial launch industry, a reluctant NASA issued a solicitation to prospective contractors interested in commercializing NASA's Atlas and Delta launch programs. The only responses to this formal NASA solicitation were a very tentative one from General Dynamics for the Atlas and a more positive one from Transpace Carriers, a newly formed marketing organization interested in commercializing the Delta (Figure 3–1). Notably, McDonnell Douglas, the manufacturer of the Delta, did not submit a response; its executives acknowledged that the Delta was both too limited in its lifting capability to meet future communications satellite requirements and unable to compete against the Space Shuttle.[13]

While it was willing to continue to build Delta launch vehicles for Transpace Carriers, McDonnell Douglas was unwilling to undertake the financial and business practices initiatives necessary to market and provide commercial launch services at that time. In addition, McDonnell Douglas and other industry decision makers in firms were motivated by a desire not to risk angering NASA, an important customer, by going into competition against NASA's Space Shuttle program, as well as recognizing their future business opportunities tied to the success of the Shuttle program. For McDonnell Douglas, this included a substantial company investment in a commercial upper stage for the Space Shuttle, known as the Payload Assist Module D, as well as substantial investments in commercial Shuttle materials processing facility payloads. For other major launch systems contractors, such as Martin Marietta, this included the Shuttle's external tank. Most of the other major aerospace contractors also had some vested interest in the success of the Shuttle program.

Despite the hesitation of the larger launch vehicle firms, the plans of smaller entrepreneurial firms, such as Space Services Inc., drew the attention of both Congress and the White House. In February 1984, President Reagan signed Executive Order 12465 on "Commercial Expendable Launch Vehicle Activities." Later that year, Congress passed a

---

13. *Space Business News*, August 18, 1985; Alan B. Kehlet, telephone interview by Craig Reed, March 14, 1997.

bill, signed into law by Reagan on October 30, 1984, known as the Commercial Space Launch Act. [III-4] Both of these actions were aimed at streamlining the regulatory processes that seemed to have a particularly adverse affect on the viability of smaller domestic commercial launch start-ups.

Executive Order 12465 designated the Department of Transportation as "the lead agency within the Federal government for encouraging and facilitating commercial ELV activities by the United States private sector," and it detailed a number of responsibilities the agency would have as lead agency. This designation came after a protracted executive branch competition, which pitted the Department of Transportation against the Department of Commerce for the commercial launch market oversight responsibility. The Department of Transportation argued that space launch was just one more mode of transportation and the functions and responsibilities associated with regulation and promotion of the launch industry were similar to those already performed by the Federal Aviation Administration for commercial air travel, which already was under departmental jurisdiction. The Department of Commerce argued vehemently that the job was fundamentally one of supporting the development of commerce for a new industry—a function similar to those already performed by the department on behalf of other domestic industries. Ultimately, the squabble was resolved at a "principals-only" meeting of the White House Cabinet Council on Commerce and Trade on November 16, 1983, with President Reagan presiding.[14] [III-5]

Upon acquiring the commercial space launch responsibility, then-Secretary of Transportation Elizabeth Dole established an Office of Commercial Space Transportation, whose chief purpose was to expedite applications for launch permits, and a Commercial Space Transportation Advisory Committee (COMSTAC) to serve as a means of getting industry input into the office's activities. [III-6] The Commercial Space Launch Act gave the Secretary of Transportation the exclusive legal authority to issue licenses for commercial space launches and launch operations, created a "one-stop" licensing process for launch firms, and established a minimum level of liability insurance to be carried by launch service providers.[15]

An initial focus of attention for the Office of Commercial Space Transportation was participating in the 1984–85 debate over the price to be charged for a commercial launch aboard the Space Shuttle. All involved recognized that a commercial ELV operator could not operate profitably and still offer a launch price competitive with the original Shuttle launch price, or indeed the price scheduled to be in effect beginning October 1, 1985. NASA recognized that the supporters of a private launch industry would oppose its desire to keep Shuttle prices low as a means of its attracting commercial customers. The 1983 commercial launch directive (NSDD 94) gave little comfort to the private sector in this respect; it stated that "notwithstanding the U.S. Government policy to encourage and facilitate private sector ELV entry into the space launch market, the U.S. Government will continue to make the Space Shuttle available for all authorized users—domestic and foreign, commercial and governmental." The directive also stated that "through FY1988, the price for STS flights will be maintained in accordance with the currently established NASA pricing policies."

The directive did state, however, that after October 1, 1988, "it is the Government's intent to establish a full cost recovery policy for commercial and foreign flight operations." This statement provided the primary focus for a two-year conflict between NASA with its Shuttle-oriented pricing and marketing efforts and the advocates of a commercial ELV industry. The latter recognized that perhaps their only hope for business viability was

---

14. *Space Business News*, November 24, 1983.
15. "Commercial Space Launch Act," *Statutes at Large* 98, sec. 15, 3061 (October 30, 1984).

convincing the White House to set a Shuttle price beginning October 1, 1988, that was high enough to give them a chance to be price competitive.

The initial round of this conflict took place in 1984 as the Reagan administration developed a "National Space Strategy" statement. The Department of Transportation, led by Secretary Elizabeth Dole, and the White House Office of Management and Budget (OMB) were the primary advocates during this process of having the strategy indicate that there would be a significantly higher Shuttle price after September 1988. The White House person in charge of the review leading to this strategy statement was National Security Adviser Robert (Bud) McFarlane. In a June 21, 1984, memorandum to Secretary Dole, McFarlane rejected the argument that Shuttle prices should be substantially increased, noting: "If NASA is arbitrarily forced to raise its Shuttle prices, it appears that Ariane, and not U.S. ELVs, will benefit through increased demand from payload customers. Such a result would obviously undercut the President's primary goal of maintaining U.S. space leadership." [III-7]

The reality was that the United States during the 1983–85 period was pursuing two policy goals that were clearly inconsistent: (1) creating a domestic space transportation industry based on the use of existing ELVs and (2) maximizing the number of commercial launches on the Space Shuttle in competition with Ariane. Both sides in this conflict recognized the issues at stake. The debate centered on what costs were actually to be included in a "full cost recovery" approach. The Department of Transportation and OMB argued for a definition that would increase the Shuttle price to well over $100 million per launch. NASA argued for a definition of full cost recovery that minimized the Shuttle price and, in September 1984, suggested a price after October 1, 1988, of $87 million (1982 dollars). [III-8]

In April 1985, NASA revised its position, arguing for a price that reflected only the costs of commercial and foreign missions, not also the more expensive government missions. On this basis, suggested NASA Administrator Beggs, a more appropriate price would be $71.4 million (1982 dollars) per launch. Beggs argued that he had "become increasingly convinced since last September that the Shuttle will not be able to compete effectively with the European Ariane launch vehicle at a price of $87M a flight." Beggs noted that the Central Intelligence Agency had done an analysis of Ariane's marketing strategy and had predicted "that Arianespace will raise its prices as Shuttle prices increase, but will keep them below Shuttle and any U.S. commercial ELV's." Beggs also noted that "Shuttle prices at levels above $110 Million per flight and as high as $129 Million per flight have, in fact, been proposed by other agencies who believe that such prices will permit U.S. ELV's to enter the market." Beggs's conclusion was that "the currently available U.S. ELV's cannot make inroads against Ariane" and that if the Shuttle price were set at a high level, "Arianespace could increase its prices so as to realize very large profits, and still underbid all U.S. competition." He added, "we have, not a sellers' market, but a buyers' market. It is a market where many factors are considered but where it has been shown that that a price advantage of about 5% will strongly affect the buyer's selection of launch vehicle."[16]

After intense debate, the NASA position prevailed. [III-9, III-10] On July 27, 1985, President Reagan approved a recommendation that in effect accepted NASA's definition of full Shuttle costs. The memorandum to the President recommending this choice, which was written by McFarlane, noted that doing so "will diminish the prospects for the commercialization of U.S. expendable launch vehicles," but that a higher Shuttle price "will benefit the French-built Ariane ELV rather than any prospective U.S. ELV." [III-11, III-12]

---

16. James M. Beggs, NASA Administrator, letter to President Ronald Reagan, April 24, 1985, NASA Historical Reference Collection, NASA History Office, NASA Headquarters, Washington, DC.

NASA's "victory" in this conflict over Shuttle pricing was not easily accepted by the committed advocates of the development of a U.S. commercial launch industry, most of whom were *not* in senior positions in the industry's leading firms. As mentioned earlier, those leaders had to balance considerations of their NASA business with commercial opportunities. When, six months later, the *Challenger* accident reopened the debate over whether it was appropriate to use the Space Shuttle as a commercial launch vehicle, the advocates of ELV commercialization were ready to seize the opportunity to reopen the argument.

Whatever the situation with respect to ELV-Shuttle competition, the Office of Commercial Space Transportation began in 1984 to work on removing the barriers to the emergence of a commercial launch industry. The Commercial Space Launch Act had directed the Department of Transportation, as lead agency for commercial space transportation, to identify and recommend changes to existing Federal statutes, regulations, and policies that had a potential adverse effect on launch vehicle commercialization. The office submitted to Congress in July 1985 a report that reviewed potential impediments in five areas: international treaties, Shuttle and Ariane pricing policy, insurance, tax and tariff consequences, and the licensing process. The general conclusion of the report was that the U.S. government was doing a good job in creating a policy and regulatory environment within which a commercial space transportation industry might emerge. [III-13]

Commercial Space Launches and International Trade

In June 1984, Transpace Carriers, frustrated by its inability to compete with the government-assisted Ariane—and, by implication, the Space Shuttle—filed a petition with the Office of the U.S. Trade Representative alleging that Europe's Arianespace was carrying out unfair trade practices in its provision of commercial space launch services. Specifically, Transpace Carriers claimed that Arianespace was engaging in "predatory pricing"—that is, selling launch services at a lower price to its international commercial customers than it charged ESA member states, as well as being subsidized by the French space agency, Centre Nationale d'Études Spatiales (CNES), for costs associated with launch and range facilities, services and personnel, administrative and technical personnel, and mission insurance rates. As a result of the petition, the Office of the U.S. Trade Representative initiated an investigation of these allegations as well as the broader issues of government inducements, direct and indirect government assistance, and cost and pricing policies in commercial launch services. On July 17, 1985, acting on the recommendation of U.S. Trade Representative Clayton Yeutter, President Reagan signed a memorandum rejecting the claim of Transpace Carriers on the basis that the practices of Arianespace and ESA were "not unreasonable and a burden or restriction on U.S. commerce." [III-14]

While some of the allegations were substantiated through the investigation, most were not, or at least not conclusively. The Reagan memorandum declared that "ESA practices were determined to be not sufficiently different from those of the U.S. to be actionable under Section 301[of the Trade Act of 1974]," referring to the support provided by the U.S. government to the Space Shuttle program in its pursuit of international commercial launch business, not private ELV firms. The determination further noted: "While Arianespace does not operate under purely commercial conditions, this is in large measure a result of the history of the launch services industry, which is marked by almost exclusive government involvement." In addition to this recognition of the unique political economy of the launch industry, the report also noted that "there are no international standards of reasonableness for launch services," and "it may be appropriate for the United States to approach other interested nations to reach an international understanding on guidelines for commercial satellite launch services at some point in the future."

First addressed in 1984, the relationship between the commercial space launch industry and broader issues of international trade practices has persisted as the industry has matured.

Challenger Accident and Commercial Launch Policy

On January 28, 1986, the Space Shuttle *Challenger* exploded shortly after liftoff, not only killing its seven-person crew but also reopening the debate over the appropriate role of the Space Shuttle in U.S. space transportation policy. In the following seven months, the policy debate went on in several forums and considered several issues. These included the Rogers Commission, set up by the White House to investigate the causes of the accident; the National Security Council, which considered whether to replace *Challenger* with a new Shuttle orbiter and, if so, how to pay for it; and the Cabinet Council on Commerce and Trade, soon renamed the Economic Policy Council, which considered whether policy changes with respect to the Shuttle's role in launching commercial payloads were justified in the wake of the accident.

Many of those involved in both the Economic Policy Council and National Security Council discussions were strong advocates of a greater overall private-sector role in space. Some had been supporters of the 1983–85 private-sector attempts to force NASA to increase Shuttle prices so that U.S. ELVs could compete for commercial launch contracts. Others were new to the debate, but more sympathetic to the private-sector position than had been their predecessors. They found NASA in a weakened political position after the accident, under attack for management failures leading to the Shuttle accident, and with interim leadership, because Administrator James Beggs had taken a leave of absence to fight a federal indictment for conduct prior to his coming to government in 1981.[17] They took advantage of the opportunity created by the *Challenger* accident and NASA's subsequent vulnerability to convince the White House to reverse the Shuttle-centered policy first decided in November 1981 and reinforced subsequently in various pricing and policy decisions. On August 15, 1986, President Ronald Reagan announced that "NASA will no longer be in the business of launching private satellites."[18]

Over the next month (and indeed throughout the whole post-*Challenger* policy debate), NASA attempted to find a way to retain for launch on the Shuttle as many as possible of the forty-four commercial and foreign satellites that were already under contract. NASA argued that it should continue to fly commercial payloads until a viable U.S. ELV capability was established, at which point satellites could transition to commercial expendable launchers. If NASA were forced to withdraw totally from flying commercial payloads, the agency argued, the result would be that the owners of most payloads taken off the Shuttle would contract with Ariane. The White House did not concur with NASA's argument. [III-15] On October 3, 1986, NASA Administrator James C. Fletcher announced a new Shuttle manifest that excluded most commercial communications satellites then under contract for launch. Fletcher noted that "during the intergovernmental discussions on the manifest, NASA sought to accommodate all of its customers who had signed up to fly aboard the Shuttle," but that because of constraints, including "the new national policy to accelerate the development of a viable private expendable launch industry," this had not been possible.[19]

The policy change barring the Shuttle from the commercial launch market was formalized with the December 27, 1986, signing by President Reagan of NSDD 254, "United

17. Nothing came of this indictment, and the government later apologized to Beggs for its issuance.
18. The White House, Statement by the President, August 15, 1986, NASA Historical Reference Collection.
19. NASA, Statement by Dr. James C. Fletcher, NASA Administrator, October 3, 1986, NASA Historical Reference Collection.

States Space Launch Strategy." This directive stipulated that "NASA shall no longer provide launch services for commercial and foreign payloads unless those spacecraft have
unique, specific reasons to be launched aboard the Shuttle."[20] This policy said that the
national security space sector would use a balanced mix of ELVs and Shuttle launches. It
explicitly directed NASA not to maintain an ELV capability as an adjunct to the Shuttle,
saying that if NASA needed additional launch capability beyond the Shuttle, it should contract for commercial launch services. It reaffirmed the principles in the earlier policy
directives that were aimed at encouraging and facilitating the development of a domestic
commercial launch industry. By taking NASA and the Shuttle out of competition with
commercial launch providers, NSDD 254 officially and finally opened the doors of opportunity to a whole new potential market for the U.S. launch industry.

Creating a Government Policy Framework
for the Commercial Launch Industry

Critical to U.S. industry's ability to take advantage of that potential market was the
existence of a policy framework to facilitate U.S. competitiveness in the "post-Shuttle"
environment. A major step in this direction was a new statement of National Space Policy,
which was released by the White House on February 11, 1988, together with a list of fifteen "commercial space initiatives."[21] The Commercial Space Initiative reaffirmed and
reiterated a number of guidelines for the U.S. government's encouragement of commercial launch vehicles, including the use of launch and related facilities and U.S. government pricing of the use of its facilities and services. One of the directive's goals was to
encourage viable commercial launch vehicle activities.

While much of the Commercial Space Initiative was aimed at supporting the growth
of nascent on-orbit commercial space industries, particularly materials processing and
remote sensing, several key provisions were included that impacted the commercial space
launch industry, under the heading of "Assuring a Highway to Space." These included
measures to direct federal agencies to procure expendable launch services directly from
the private sector to the fullest extent possible; a proposal for capping third-party liability
for and damage to government property resulting from a commercial launch accident;
and initiatives to explore the possible development of private space ports and the provision of vouchers to research payload owners who were manifested on the Shuttle, to
enable the purchase of commercial ELV launch services. The purchase of private-sector
launch services by federal agencies was also mandated as a matter of policy. The remaining initiatives required executive branch agencies to develop a plan for the passage of
enacting legislation, specific appropriations by Congress, or (in the case of the proposal
for developing private launch facilities) a substantial investment decision by industry.
Overall, the Commercial Space Initiative demonstrated the Reagan administration's
interest in the issues concerning the burgeoning commercial space industry.[22]

Congress also acted to redefine the policy framework set out in the Commercial Space
Launch Act of 1984 in ways that would facilitate market entry by U.S. commercial launch
providers. An effective campaign by an ad hoc group representing the interests of the

20. The White House, NSDD 254, "United States Space Launch Strategy," December 27, 1986, NASA
Historical Reference Collection.
21. See Office of the Press Secretary, "Fact Sheet: Presidential Directive on National Space Policy,"
February 11, 1988, reprinted as Document III-42 in Logsdon, gen. ed., *Exploring the Unknown, Volume I*, 1: 601–10;
Office of the Press Secretary, The White House, "The President's Space Policy and Commercial Space Initiative
to Begin the Next Century," Fact Sheet, February 11, 1988, reprinted as Document III-12 in Logsdon, gen. ed.,
Exploring the Unknown, 3: 455–60.
22. *Ibid.*, 3: 458–59.

commercial space transportation industry was crucial to convincing Congress that changes were needed, particularly in terms of limiting the amount of liability insurance required. In November 1988, both houses of Congress passed a set of amendments to the 1984 act. [III-16] The purpose of the amendments was to clarify the very general terms and requirements in the original act. More precise definitions of regulatory requirements in areas such as third-party insurance were needed by the domestic commercial launch service industry to be competitive, as its efforts to enter into the commercial market intensified. The 1988 amendments established as policy that "the United States must maintain a competitive edge in international space transportation. . . ." They also included guidelines regarding government preemption of scheduled commercial launches, negotiation of free and fair trade with international competitors, and launch vehicle research and development.

The principal impact of the 1988 amendments was to limit the total amount of liability insurance required as a condition of licensing a launch. The Commercial Space Launch Act of 1984 required commercial space launch companies to attain the maximum amount of insurance commercially available at reasonable rates. [III-17] The 1988 amendments limited the insurance required by commercial launch companies to no more than $500 million for third-party claims and no more than $100 million for the loss of or damage to U.S. government property, or to the maximum amount of insurance commercially available at reasonable rates, whichever was less. The 1988 amendments provided that the U.S. government would reimburse any claims for damages that exceeded the liability coverage required of the commercial launch companies, up to $1.5 billion above the amount of coverage they were required to obtain.[23]

Soon after George Bush became President in January 1989, he established in the Executive Office of the President a National Space Council. That council and its staff were active in commercial space launch issues, among many other topics, during the four years of the Bush administration.

The initial Bush space policy statement, issued in November 1989, was basically a reaffirmation of the February 1988 Reagan policy. President Bush signed National Space Policy Directive (NSPD) 2, "Commercial Space Launch Policy," on September 5, 1990. The directive stated as a U.S. policy goal "a free and fair market in which U.S. industry can compete." It marked a departure from the goals of the policy directives of the Reagan administration, which had stopped at the encouragement and facilitation of a "viable" U.S. commercial launch industry. The directive also established an explicit tie between U.S. commercial launch policy objectives and other formal U.S. government nonproliferation and technology transfer objectives. It proposed "a set of coordinated actions . . . for dealing with international competition in launch goods and services." The directive established as formal policy a near-term focus on trade "agreements to limit unfair competition" and a longer term focus on encouraging "technical improvements to enhance the competitiveness of U.S. launch vehicles." [III-18]

NSPD 2 stipulated that U.S. government payloads would be launched on U.S.-manufactured launch vehicles unless specifically exempted from this requirement by the President. It required U.S. government agencies to factor commercial space launch industry needs into their decisions on launch vehicle and infrastructure improvements. It also directed that "the U.S. Government will enter into ['rules of the road'] negotiations to achieve agreement with the European Space Agency (ESA), ESA member states, and others as appropriate, which defines principles of free and fair trade." It recognized the need

23. "Commercial Space Launch Amendments of 1988," *Statutes at Large* 102, sec. 9, 3906 (November 15, 1988).

for establishing a transition period for the entry of nonmarket economy launch service providers, such as China and the Soviet Union, into the market during which special competitive restraints would be imposed. Specifically, these constraints included:

* Continuing the U.S. government policy of prohibiting the export of satellites and related technologies to the Soviet Union
* Limiting the Soviets to the use of a single site for commercial launches located outside the Soviet Union and making approval for their commercial launches contingent on reaching enforceable trade and ballistic missile nonproliferation agreements
* Restricting commercial launch market entry of nonmarket economies to a framework negotiated with the United States

NSPD 2 reflected the growing tension between commercial launch policy and other national interests, as illustrated by a description of the directive in a statement by the White House Press Secretary: "It balances launch industry needs with those of other industries and with important national security interests, and establishes the long term goal of a free and fair market in which U.S. industry can compete."[24]

In July 1991, President Bush signed NSPD 4, "National Space Launch Strategy." This directive spelled out the administration's policy on the use of excess ballistic missiles for space launch and indicated plans for the development of a new national space launch system. It stated that the U.S. government would "encourage, to the maximum extent feasible, the development and growth of U.S. private sector space transportation capabilities which can compete internationally." [III-19]

While the excess ballistic missile assets issue did not significantly affect the medium and large launch vehicle builders, the U.S. government's plans to develop a new launch system had direct implications for this class of commercial launch service providers.[25] The development of the new launch system and the transition from current launch systems to the new system were to be developed, managed, and funded jointly by NASA and the Department of Defense (DOD). Despite the threats to existing launch service providers posed by the new launch system, NSPD 4 declared that the new launch system would "provide the opportunity for significant long-term benefits to the commercial space launch industry." It directed NASA and DOD to involve the U.S. private sector in the program. (See Chapter 4 for a discussion of this new launch system.)

NSPD 4 also focused directly on commercial space launch considerations, recognizing that the "improvement of space launch capabilities can facilitate the ability of the U.S. commercial space launch industry to compete." The policy encouraged U.S. government agencies to:

* Allow contractors to accommodate commercial needs when developing future launch systems and infrastructure for Government needs
* Use "best value," performance-based contracting, commercial production techniques and quality standards, and commercial products and services
* Encourage commercial and state and local government investment
* Preserve private-sector retention of technical data rights

24. Office of the Press Secretary, The White House, "Statement by the Press Secretary," September 5, 1990, NASA Historical Reference Collection.
25. The impact on medium and large launch vehicle providers was minimal because the limited lift capability (throw-weight) of those ballistic missiles becoming excess made them unable to launch most of the payloads targeted by this segment of the industry. Smaller launch vehicles, such as Orbital Science Corporation's Pegasus and Taurus, would, however, face direct competition from these excess assets.

- Remove legal or administrative impediments to forming cooperative government-industry business relationships
- Seek legislative authority for long-term commitments for launch services
- Use industry advisory groups to identify commercial sector needs and concerns

The administration of President Bill Clinton issued a new statement of National Space Transportation Policy in August 1994. (See Chapter 4.) Like the Bush administration's National Space Launch Strategy of July 1991, this policy addressed a variety of transportation-related issues, including the assignment of space transportation responsibilities between NASA and DOD. The policy also contained several elements directly relevant to the commercial space launch industry.

Turning Policy into Practice

Since its establishment in late 1983, the Department of Transportation's Office of Commercial Space Transportation has worked at defining its role and at preparing and administering the regulations and processes required to implement changing national policy and laws. [III-20, III-21] A major focus of its initial activity was developing the process for issuing the licenses required for a commercial space launch. The Department of Transportation published its final ruling on commercial space transportation licensing regulations in April 1988. In this document, the Office of Commercial Space Transportation embraced the notion that its responsibility was to ensure not just the viability of the commercial space launch industry, but its competitiveness as well. It noted that "the Secretary's mandate embraces the authority to license and otherwise regulate such activities, as well as the responsibility to encourage, facilitate and promote establishment of a competitive United States commercial space transportation industry." [III-22]

Shortly after promulgating its final ruling on licensing, the Office of Commercial Space Transportation, four years after it was established, issued its first two commercial launch licenses. The first was to Conatec, Inc., which was a small entrepreneurial firm planning to conduct suborbital launches of materials processing payloads. The second went to McDonnell Douglas for the launch of an Indian communications satellite on a Delta expendable launch vehicle in April 1989.[26]

In carrying out its mixed promotional and advocacy role, the Office of Commercial Space Transportation has had to address a wide variety of launch-related issues. These issues include, among others the following:[27]

- Defining (in a conflict with NASA) under what conditions a launch was indeed commercial and thus subject to its jurisdiction
- Determining what fees users of government launch facilities were required to pay
- Defining insurance requirements for a commercial launch
- Negotiating with NASA and the Air Force the conditions under which commercial launch providers would have access to government launch infrastructure and the priorities for various users of that infrastructure [III-23]

---

26. Office of Commercial Space Transportation, Department of Transportation, *Cleared for Launch: Annual Report to Congress: Activities Conducted Under the Commercial Space Launch Act, Fiscal Year 1988* (Washington, DC: U.S. Government Printing Office, 1989), p. 2.

27. A discussion of how these issues were handled is beyond the scope of this essay. For an in-depth analysis, see Craig R. Reed, "U.S. Commercial Space Launch Policy Implementation, 1986–1992," Ph.D diss., George Washington University, 1998.

In doing so, the Office of Commercial Space Transportation has been a partner to U.S. industry—and its advocate inside the U.S. government—in creating a new sector of commercial space activity.

A Commercial Launch Industry Emerges

The immediate reaction of a number of space users after the *Challenger* accident was to contract with Arianespace for launch services. Between the time of the accident in January 1986 and the announcement of the Space Shuttle policy change in August 1986, and despite the failure of a May 1986 launch attempt, Arianespace booked seventeen new launch customers. (There were also Delta and Titan launch failures in 1986.) Of these new contracts, eleven had been scheduled to fly on the Shuttle, and NASA had launch contract proposals outstanding for the other six.[28]

These choices were made on the assumption that the problems with Ariane would soon be fixed and the launcher would reenter operation well before alternative means of access to space might be available. Even though the August and December 1986 policy shifts had opened the window of opportunity for U.S. private-sector launch service providers to enter the commercial launch market, at that point none had made a firm decision to do so. Once they so decided, it was several years before they were ready to launch their first commercial payloads.

Given its temporary monopoly position, Arianespace not surprisingly raised its launch prices substantially over the next several years. This price increase, the absence of the Space Shuttle as a government-subsidized competitor, and the possibility of new U.S. government orders were enough to convince McDonnell Douglas, General Dynamics, and Martin Marietta to reconsider their earlier decisions not to attempt to enter the commercial launch market. Each potential U.S. supplier of commercial launch services faced a slightly different situation.

Delta

As mentioned earlier, the manufacturer of the smallest of the three U.S. launchers able to carry communications satellites into orbit, McDonnell Douglas, had decided in 1983 not to try to market the vehicle in the commercial marketplace. The company believed that the Delta did not have enough power to lift the coming generations of communications satellites and that, as long as the Shuttle was in the market, there was not likely to be enough business to justify the investments needed to increase the Delta's lifting capability.

After the 1986 policy change, McDonnell Douglas rethought its decision not to enter the commercial launch market. However, the company remained uncertain of the wisdom of such a move until it won, in January 1987, an Air Force contract for an upgraded version of the Delta, called the Delta II, to be used for launching the Global Positioning Satellite system. This contract covered most of the costs of restarting the Delta production line and of making the launcher more powerful, thereby allowing it to compete for commercial launch contracts. NASA canceled its agreement with Transpace Carriers to market the Delta, and McDonnell Douglas began to seek commercial customers for the vehicle in 1987. Although it won a few contracts for launches of lighter communications satellites, the Delta has not been a major continuing player in the commercial launch market because the weight of most communications satellites has indeed exceeded its lifting capability.

---

28. Theresa Foley, "Foreign Launches Render U.S. Vehicles Uncompetitive," *Aviation Week & Space Technology*, August 4, 1986, p. 29.

Atlas

Even before the Shuttle accident, executives of General Dynamics had been interested in trying to market the Atlas (with its Centaur upper stage) as a commercial launcher, but no decision to reopen the Atlas production line had been made. After the change in poli-cy, General Dynamics decided to resume Atlas production in 1987, even without any commercial or government contracts for additional vehicles. The company recognized that the Atlas-Centaur launcher was ideally sized to launch the communications satellites likely to be the core of the commercial market in coming years (Figure 3–4). It was willing to take the risk that it could win enough contracts for launches of these and other satellites to justify its investment in building eighteen vehicles.

In March 1988, General Dynamics won a second Air Force competition for an upgraded Atlas vehicle to launch military communications satellites and other payloads. Obtaining this government contract underwrote much of the costs of restarting Atlas production and upgrading the vehicle; thus it was critical to the firm's continuing viability in the commercial launch market. For most of the period since 1986, the Atlas-Centaur has been the primary competitor to Ariane for launching commercial communications satellites. In 1994, Martin Marietta purchased from General Dynamics the rights to produce the Atlas and the associated production capabilities.

Figure 3–4. The Atlas-Centaur two-stage rocket has been a reliable launcher for the United States since the 1960s. Here an Atlas-Centaur carries a test article for the Surveyor spacecraft program into space on December 11, 1964. (NASA photo 64-H-2808)

Titan

Well before the Shuttle accident, top Air Force and National Reconnaissance Office officials, led by Secretary of the Air Force Edward (Pete) Aldridge, had become concerned about a policy of total dependence on the Shuttle for launching intelligence satellites and other critical national security satellites. After a bitter fight with NASA, in 1985 they had succeeded in convincing the White House and Congress to approve the procurement of a limited number of what were called complementary ELVs as backups to the Shuttle. Martin Marietta won the competition in 1985 to provide this capability with a more powerful variant of its Titan 34D launcher; the variant became known as Titan IV. This contract also allowed Martin Marietta to keep the Titan production line open.

When the post-*Challenger* opportunity to enter the commercial launch market appeared, those most closely involved with Titan were able to convince the top executives of Martin Marietta to develop a commercial variant of the booster, to be called Titan III

Figure 3–5. This is an aerial view of Space Launch Complex (SLC)-4 at Vandenberg Air Force Base, a site used to launch Titan IIIB rockets. (U.S. Air Force photo)

(Figure 3–5). This was a more powerful, and more expensive, launcher than either the Delta or Atlas, and it could be commercially viable only if it could schedule two payloads on the same launch, an approach pioneered by Ariane. However, Martin Marietta was not able to find many customers willing to fly at the same time and, in June 1989, announced that it would no longer attempt to market a single launch to two customers. This had the effect of removing the Titan III from the commercial marketplace because the price of a launch was $130–150 million, which was too high for a single payload.

Not only Air Force contracts but also NASA procurements helped nurture the emerging commercial space transportation industry. On May 14, 1987, NASA announced its intent to procure from U.S. industry launch services using ELVs. NASA Administrator James Fletcher stated, "NASA's purpose in seeking expendable launch services is to lessen dependence on a single launch system, the Space Shuttle. Expendable launch vehicles will help assure access to space, add flexibility to the space program, and free the Shuttle for manned scientific, Shuttle-unique and important national security missions." NASA indicated its future intent to purchase the equivalent of three to five Delta launches per year and one to two Atlas-Centaur or Titan III launches per year.[29]

Each of the U.S. ELV providers booked their first commercial launch contract during 1987. There were six orders for Delta launches, three of which were payloads previously scheduled to fly on the Shuttle. There were four orders for Atlas-Centaur launches, three being prior Shuttle payloads. There were four Titan orders, all prior Shuttle payloads. After signing eighteen launch contracts during 1986, Arianespace added only two new orders

29. "NASA Plans Use of Expendable Launch Vehicles," NASA Press Release 87-76, May 15, 1987, NASA Historical Reference Collection.

during 1987; this slowdown in orders reflected Ariane being booked to capacity for several years, with U.S. providers being able to offer earlier launch opportunities. The Ariane launcher reentered service in September 1987 with the successful launch of two satellites.

By mid-1988, Ariane had orders for forty-four payloads, U.S. providers had eleven orders, and the owners of sixty-six payloads anticipated to fly before the end of 1994 had not chosen a launch vehicle. Thus Ariane had gone from its less than one-third market share before the *Challenger* accident to having 80 percent of the market. The Delta launched its first commercial payload in August 1989; the Titan III's first was in December 1989. The Atlas did not loft its first commercial satellite until April 1991. As the market settled down during the first half of the 1990s, Ariane was able to maintain a market share of more than 65 percent. Martin Marietta withdrew the Titan III from the commercial market after only four commercial launches. The Delta and Atlas boosters remained active competitors, with the Atlas-Centaur most closely matching Ariane's launch capability and price.

As mentioned earlier, in addition to the U.S. Delta, Atlas, and Titan launchers and the European Ariane, the Soviet Proton booster and the Chinese Long March rocket had the capability of launching commercial communications satellites. Both were marketed from the mid-1980s on as alternatives to Western launch vehicles. Both were offering launch prices significantly below their U.S. and European competitors. U.S. satellite manufacturers and operators, interested in the least expensive means of access to space, began to urge the U.S. government to facilitate their access to Soviet and Chinese launches, even if that meant less launch contracts for the U.S. space transportation industry. However, neither the Proton nor the Long March made significant market penetration. There were several reasons.

With respect to China, because the Chinese pricing policy did not meet U.S. standards for appropriate market behavior, the United States in early 1989 negotiated a quota system on Chinese launches of satellites containing U.S. components. [III-24] Such satellites were the vast majority of all commercial communications satellites. Another reason was that many satellite owners and satellite insurers were suspicious of the reliability of the Chinese Long March launcher. Also, the question of licensing satellites containing U.S. components for export to China became enmeshed in the controversies surrounding the Chinese suppression of dissent later that year. The first commercial Long March launch of a communications satellite took place in 1992.

There were no Proton launches of non-Russian communications satellites until 1996, although the Soviet Union had been attempting to market the launcher since 1983. The primary reason for this situation until the end of the Cold War was export control restrictions on the import of a Western-built communications satellite into the Soviet Union for launch. After 1992, the United States and Russia agreed on launch quotas similar to those negotiated with China. (A third quota agreement was signed between the United States and the Ukraine in 1995.) After the collapse of the Soviet Union, the Russian firms Krunichev and Energia entered into a joint venture with Lockheed to market the Proton. After Lockheed and Martin Marietta merged, this venture was enlarged and renamed International Launch Services. As mentioned above, Martin Marietta had, prior to its merger with Lockheed, purchased the portion of General Dynamics that manufactured and marketed the Atlas, and the new venture marketed the Atlas and Proton in combination.

In another joint venture called Sea Launch, the Boeing Company joined with rocket builders from the Ukraine and Russia and a Norwegian shipbuilder to develop a system for launching communications satellites from a converted off-shore oil drilling platform. Both Boeing and Lockheed Martin planned to develop a commercial variant of the evolved ELV that they were developing for DOD. Japan hoped to enter the commercial launch market in 2000 or soon after with its H-IIA vehicle. For smaller satellites, such as those comprising various mobile communications satellite constellations, a variety of smaller commercial launch vehicles were available or being developed. From its origins just over a decade earlier, the commercial launch industry had grown into a robust sector of the U.S. and global economy.

Document III-1

Document title: NASA, *We Deliver,* brochure, 1983.

Source: NASA Historical Reference Collection, NASA History Office, NASA Headquarters, Washington, D.C.

As it began to seek commercial customers for Space Shuttle launch services, once the Space Transportation System (STS) had been declared operational in July 1992, NASA prepared a glossy, colorfully illustrated marketing brochure; its text is reprinted here (graphics are not included). The brochure was produced in several languages for distribution to potential Shuttle customers around the world.

We Deliver

[1] Twenty-five years of hands-on experience [2] assures you of the most reliable, flexible, and cost-effective launch system in the world.

Space Shuttle has established a proven record as the most useful and versatile space transporter ever built. It has also demonstrated a remarkable suitability for delivering communications satellites to earth orbit. The successes of this operational space transportation system speak directly to your launch needs and concerns.

While many new and profitable business opportunities are becoming possible through the Space Shuttle, the primary focus during the 1980's will be on the delivery and operation of telecommunications satellites. Shuttle is now ready to launch these satellite payloads; now the goal of the National Aeronautics and Space Administration is to simplify its use. We know this can be done because with each increasingly smooth flight[;] there is evidence that we are substantially reducing paperwork and the time a payload must be at the launch site before liftoff. What this will mean to you is lower launch costs and less integration complexity—without compromising reliability or safety.

Supported by dedicated, "can-do" contractors, NASA has been launching telecommunications satellites for almost 25 years. In providing launch services for over 100 payloads destined for geostationary orbits, NASA has assembled a team of launch operation experts whose talent, experience, and launch record are unmatched. This team and Shuttle's extraordinary capabilities enable NASA to offer you a cost-effective launch system unequaled by any other existing or planned system on Earth.

[3] You can't beat manned reliability

In launch operations, redundancy is synonymous with reliability. We know from our experience that new or significantly modified launch vehicles normally have relatively high failure rates during early launches, mostly in nonredundant systems. This is why launch vehicles which incorporate more redundancy are less prone to such failures and are inherently more reliable.

Space Shuttle, because it is manned and reusable, has been designed with more redundancy than any other launch system ever developed, with dual or greater redundancy in all critical systems. In the limited number of areas where redundancy is not possible— [4] such as structures—the Shuttle design provides for a minimum safety factor of 1.25, and in most cases two or three times the failure point. The entire Space Shuttle vehicle is a testament to NASA's longstanding design philosophy for its manned spacecraft—maximum reliability.

This devotion to reliability has repeatedly paid off for the Shuttle. Some of the problems that occurred on its four test flights—problems that were considered only minor—would have caused launch failures in unmanned launch vehicles. So impressed was the insurance industry that, after only four flights, it lowered the insurance rates for use of the Space Shuttle. Convinced of the soundness of the Shuttle design and NASA's operational approach, insurance companies set rates for the payloads aboard the first operational Shuttle flight (STS-5) that were 20 percent lower than ever given for an expendable launch vehicle, including the most successful one, Delta, which has been flown more than 170 times.

Redundancy, however, is only one of several reliability assets that the Space Shuttle offers you, and which set it apart from its expendable counterparts. These reliability advantages derive from the unique capabilities of the Shuttle Orbiters and the flight crew. If, for example, a problem should arise during launch or after orbit insertion necessitating a mission abort (a risk estimated to be less than three percent), the Orbiter would return to Earth for a controlled landing. Your payload—instead of being lost, as would occur with an expendable launch vehicle—would be safe and ready for relaunch after only minimal further checkout. The number of payloads (designed to the same specifications as telecommunications satellite payloads) that have been planned for and successfully flown on roundtrips aboard the Shuttle have demonstrated the benign return trip environment of the Shuttle.

Moreover, Shuttle alone gives you an opportunity, once in orbit, to check out your payload and to make an unhurried decision on whether or not to proceed with deployment. Should there be significant doubt about the condition of your payload, it can be returned to Earth with the Orbiter, for relaunch on another day.

Finally, the Space Shuttle offers flexibility and expanded capabilities that provide the opportunity to significantly improve satellite designs by taking advantage of the new features that only the Shuttle provides. Such an integral design approach will offer you, the customer, the maximum benefits of the world's most versatile, operational space transportation system.

[5] [graphic only; no text]

[6] **Schedule assurance with flexibility, upon which you can depend**

Launch schedule flexibility and assurance, although seemingly contradictory, are key to the success of any space venture. With this in mind, NASA will commit to the launch schedule for your payload with no caveats linking that commitment to other payloads assigned to share your flight. In other words, NASA will launch your payload, even if your launch partner fails to show up for launch. This principle was most vividly demonstrated on the STS-8 launch, when NASA kept its commitment to launch the Indian National Satellite INSAT payload as scheduled, even after its companion payload was removed from the flight.

From the standpoint of payload availability, NASA is committed to provide replacement and reflight launch services within six to nine months of notification. This commitment assures you of the ability to maintain an operational satellite system, once it has been established. Launch scheduling flexibility can be further improved by planning an entire series of payload launches on the Shuttle, thereby providing built-in flexibility through multiple launch scheduling. This is of particular importance for the first launch of a new payload design where on-time delivery of the first payload may be a concern. By scheduling the first several launches within three to six months of each other, late delivery of the first payload would be accommodated no later than the launch date planned for the second payload.

NASA recognizes the concern of some that discovery of a generic problem in the Shuttle design could suddenly cripple the Shuttle flight schedule, but history suggests that

such a concern is unfounded. The Space Shuttle has flown a string of spectacularly successful missions. There will be three Orbiters in the operational fleet by the end of 1983
and a fourth scheduled for delivery in 1984. While there is the remote possibility that a
generic design flaw could ground the entire Shuttle fleet, such a possibility exists for any
launch vehicle. A review of our manned missions shows that such problems result in flight
delays of only a few months. But if a problem were to become a grave threat to Shuttle
operations, the involvement of the Department of Defense in the Shuttle program ensures
that a lengthy delay would not be tolerated. Any problem that could ground the Shuttle
fleet would attract the resources as well as the urgent attention of the United States government. We have no doubt that the problem would be quickly solved.

[7] [graphic only; no text]

[8] **In all the world, you won't find Shuttle's equal**

NASA recognizes the large investment required for any space venture and appreciates
the importance of accurate placement of your payload in orbit. We also understand—after
a quarter-century of experience—what it takes to launch payloads into space, time after
time after time. In short, we know how to deliver.

For the launching of your payload, we offer an unparalleled combination—the
world's most reliable space transporter and a launch team that is internationally recognized for its experience and successes. Praised by every crew that has flown it, the Space
Shuttle is launched by people who never compromise their first objective—to launch safely and successfully. This was dramatically demonstrated on the first flight of the
Challenger (STS-6). Although engine problems caused a launch delay of several weeks, we
painfully took the time necessary to understand the technical problem and the potential
for trouble in flight. As a result, STS-6 was a resounding success and the experience gained
taught us valuable lessons that are being applied to test and flight preparations for all
future flights.

NASA will have four Orbiters—Columbia, Challenger, Discovery, and Atlantis—in
operation by 1985. By the end of 1985 we will have flown more than 30 Shuttle missions,
nearly half involving deployment of payloads. We are confident that the Space Shuttle
fleet and its launch team can give you the surest, safest, and most cost-effective launch service obtainable anywhere in the world.

[9] [graphic only; no text]

[10] **You can't get a better price**

The price for launching a payload on the Shuttle is based on the share of the lifting
weight and cargo bay length required by your payload. Pricing[,] according to this continuous curve formula, assures that you will be charged only for your requirements, while
retaining the significant growth capability inherent in the Shuttle in the event those
requirements change during the period you are developing your payload. The Shuttle
price and pricing flexibility cannot be matched by any other launch system.

In evaluating the total cost associated with a launch, two other important factors must
be considered, both involving insurance. One is the effect that NASA's launch record and
experience has had on insurance rates. It has been demonstrated that this record commands for the Shuttle the lowest insurance rates in the free world. A difference of a few
percentage points in the insurance rates for a satellite or other payload program can easily mean millions of dollars saved in the insurance purchased for your launch.

The other insurance consideration involves the charges associated with postponing a
payload launch on the Shuttle. We have a commitment to all of our customers to launch
on time, and we recognize the importance of their cash-flow demands. Therefore, as an

incentive for customers to make all reasonable effort to have their payloads delivered for launch on the agreed schedule, we have established significant postponement fees. On the other hand, we appreciate the cost risk associated with postponements. Here again, the insurance industry has demonstrated its support of the Shuttle by agreeing to provide insurance, at low premium rates, to cover the postponement fees.

Considering all cost factors associated with launching your satellite or other payload into space, you can't get a better price or more for your money than the Space Shuttle.

Document III-2

Document title: Space Launch Policy Working Group, "Report on Commercialization of U.S. Expendable Launch Vehicles," April 13, 1983, pp. 1–4, 34.

Source: Ronald Reagan Presidential Library, Simi Valley, California.

Following the July 4, 1982, announcement that the Space Shuttle would be the exclusive launcher for future U.S. government missions and that the government would thus order no more expendable launch vehicles (ELVs), some in the U.S. private sector expressed interest in taking over the manufacturing, marketing, and launching of existing ELVs on a commercial basis. Others expressed interest in developing on a commercial basis launchers for markets not well served by the Shuttle. The White House established a n interagency working group to examine the issues involved in responding to this private-sector interest; this was the first examination of what might be involved within the government. What appears here are the introduction, the conclusions, and Appendix A, which lists the working group's members.

Space Launch Policy Working Group

Report on Commercialization of U.S. Expendable Launch Vehicles

April 13, 1983 . . .

[1] Introduction

The National Space Policy encourages the expansion of United States private sector investment and involvement in civil space activities. It also identifies the Space Transportation System (STS) as the primary space launch system for U.S. Government (USG) missions. Based on the projected capabilities of the STS, the USG has begun to phase out its procurement and operation of the Expendable Launch Vehicle (ELV) systems.

The U.S. private sector has expressed interest in continuing the production and operation of these ELVs as commercial ventures. Prospective commercial ELV producers/operators are seeking policy guidance in this area from the USG. The need for a prompt response from the USG has been driven by two principal factors: (a) the interested corporations must decide whether to continue ELV production before the USG orders are completed and (b) the competition for the Intelsat VI class of communication satellites. The Intelsat selection of one or more launch vehicle systems will be made in the June 1983 time frame. For these reasons, timely government action is required to provide the information the private sector needs to make business decisions.

The Space Launch Policy Working Group (Appendix A) was chartered by the Interagency Group (Space) to recommend what the US National Space Launch Policy should be with regard to (a) the increasing foreign space launch capabilities and competition, (b) US commercial launch systems and operations, and (c) maintenance and development of a capability to satisfy USG current and projected requirements.

During the course of the study, the Working Group met with many of the companies that have expressed interest in commercial ELV operations. Their commercialization plans, business concerns, production status, assessments of the potential market, and the potential benefits to the USG and the nation were all factors in the study. The Working Group also reviewed the results of a NASA study on ELV commercialization. The impact of commercial ELV operations on the USG Shuttle operations was also specifically examined. [2] This report is organized into four major sections. Section I presents the Working Group's principal conclusions. Section II contains the proposed National Security Decision Directive. Section III examines the factors pertinent to the USG decision on commercialization of existing U.S. ELVs. Finally, Section IV explores the issues that were addressed in developing a strategy to facilitate the commercialization of ELVs.

The Appendices contain supporting information. Appendix A lists the Space Launch Policy Working Group members. Appendix B provides the detailed analysis and data that supports the conclusions regarding the impacts on the USG Shuttle program resulting from the loss of commercial and foreign payloads.

[3] I. CONCLUSIONS

1. A US commercial ELV capability would benefit both the USG and the private sector and is consistent with the goals and objectives of the US National Space Policy.
2. The benefits of commercial ELV operations would offset the potential increases in total cost to the USG of the STS program which could result from the loss of commercial and foreign payloads.
3. Consistent with its needs and requirements, the USG should encourage and facilitate the commercialization of US ELVs. The USG should not subsidize the commercialization of ELVs.
4. International and national legal obligations and concerns (including those relating to public safety) require the USG to authorize, supervize [sic] and/or regulate US private sector space operations.
5. The USG should review and approve any proposed commercial launch facility and range as well as subsequent operations conducted therefrom.
6. Near-term demonstration of test flights of commercial launch vehicles will require USG review on a case-by-case basis; existing licensing authority and procedures appear to be adequate for this purpose, but should be streamlined.
7. An interagency Working Group should be established to develop and coordinate a process for the long-term licensing, supervision and/or regulation of possible routine commercial launch operations from non-national ranges.
8. The most effective means for the USG to ensure sage commercial ELV operations and compliance with US treaty obligations is to encourage the use of existing USG launch ranges. Consistent with these obligations, all commercial ELV operations conducted from a USG national range should be, at a minimum, subject to existing USG range regulations and requirements.
9. USG facilities, equipment, and services should be made available for commercial use where practical and priced in a manner that, consistent with USG needs and requirements, will facilitate and encourage commercial operations. The USG should not seek to [4] recover ELV design and development costs, or investments associated with launch facilities to which the USG retains title.
10. Any commercial launch vehicle operator should be required to provide adequate insurance to cover the loss of or damage to USG property used to support commercial operations. Additionally, the commercial operators should idemnify [sic] and

hold harmless the USG against liabilities for damage to both domestic and foreign persons and property.

11. The USG should continue to make the STS available to all authorized users—domestic and foreign, commercial and governmental. The USG must consider the effects that STS pricing for commercial and foreign flights could have on commercial launch operations. However, the price for commercial and foreign flights on the STS must be determined based on the best strategy to satisfy the economic, foreign policy, and national security interests of the United States.

* * *

[34]   APPENDIX A

WORKING GROUP MEMBERS

Charles Gunn (Co-Chairman)
National Aeronautics and Space Administration

Barton Borrasca
Office of Management and Budget

Donald Miller
Department of Commerce

James Chamberlin
Arms Control and Disarmament Agency

Jimmey Morrell
Office of Science and Technology Policy

James Harshbarger
Organization of the Joint Chiefs of Staff

George Ojalehto
Department of State

John McCarthy
Office of Science and Technology Policy

John Sharrard
Central Intelligence Agency

Thomas Maultsby (Co-Chairman)
Department of Defense

Joy Yanagida
Department of State

Document III-3

Document title: **National Security Decision Directive 94, "Commercialization of Expendable Launch Vehicles," May 16, 1983.**

Source: **Ronald Reagan Presidential Library, Simi Valley, California.**

Based on the recommendations of the Space Launch Policy Working Group as reviewed by the Senior Interagency Group (Space), President Ronald Reagan approved this policy statement on May 16, 1983.

May 16, 1983

Commercialization of Expendable Launch Vehicles

I.   INTRODUCTION

The United States Government encourages domestic commercial exploitation of space capabilities, technology, and services for U.S. national benefit. The basic goals of U.S. space launch policy are to (a) ensure a flexible and robust U.S. launch posture to maintain space transportation leadership; (b) optimize the management and operation of the STS program to achieve routine, cost-effective access to space; (c) exploit the unique attributes of the STS to enhance the capabilities of the U.S. space program; and (d) encourage the U.S. private sector development of commercial launch operations.

II. POLICY FOR COMMERCIALIZATION OF EXPENDABLE LAUNCH VEHICLES

The U.S. Government fully endorses and will facilitate the commercialization of U.S. Expendable Launch Vehicles (ELVs).

The U.S. Government will license, supervise, and/or regulate U.S. commercial ELV operations only to the extent required to meet its national and international obligations and to ensure public safety. Commercial ELV operators must comply with applicable international, national and local laws and regulations including security, safety, and environmental requirements.

The U.S. Government encourages the use of its national ranges for U.S. commercial ELV operations. Commercial launch operations conducted from a U.S. Government national range will, at a minimum, be subject to existing U.S. Government range regulations and requirements. Consistent with its needs and requirements, the U.S. Government will identify and make available, on a reimbursable basis, facilities, equipment, tooling, and services that are required to support the production and operation of U.S. commercial ELVs.

The U.S. Government will have priority use of U.S. Government facilities and support services to meet national security and critical mission requirements. The U.S. Government will make all reasonable efforts to minimize impacts on commercial operations.

[2] The U.S. Government will not subsidize the commercialization of ELVs but will price the use of its facilities, equipment, and services consistent with the goal of encouraging viable commercial ELV launch activities in accordance with the attached guidelines.

The U.S. Government will encourage free market competition among the various systems and concepts within the U.S. private sector. The U.S. Government will provide equitable treatment for all commercial launch operators for the sale or lease of government equipment and facilities consistent with its economic, foreign policy, and national security interests.

The U.S. Government will review and approve any proposed commercial launch facility and range as well as subsequent operations conducted therefrom. Near-term demonstration or test flights of commercial launch vehicles conducted from other than a U.S. Government national range will be reviewed and approved on a case-by-case basis using existing licensing authority and procedures.

III. RELATIONSHIP OF STS AND COMMERCIAL ELVS

Notwithstanding the U.S. Government policy to encourage and facilitate private sector ELV entry into the space launch market, the U.S. Government will continue to make the Space Shuttle available for all authorized users—domestic and foreign, commercial and governmental—subject to U.S. Government needs and priorities. Through FY 1988, the price for STS flights will be maintained in accordance with the currently established NASA pricing policies in order to provide market stability and assure fair competition. Beyond this period, it is the U.S. Government's intent to establish a full cost recovery policy for commercial and foreign STS flight operations.

IV. IMPLEMENTATION

An interim SIG (Space) Working Group on Commercial Launch Operations will be formed and co-chaired by the Department of State and NASA. The Working Group will be composed of members representing the SIG (Space) agencies and observers as well as other affected agencies. Additional membership, at a minimum, will include the Federal Aviation Administration and the Federal Communications Commission. This group will be used to (a) streamline the procedures used in the interim to implement existing licensing authority, (b) develop and coordinate the requirements and process for the licensing, supervision, and/or regulations applicable to routine commercial launch operations from commercial ranges, and (c) recommend the appropriate lead agency within the U.S. Government to be [3] responsible for commercial launch activities. Until a final selection of the lead agency is made, the Department of State will serve as the U.S. Government focal point for all inquiries and requests relative to commercial ELV activities.

Attachment
  Implementing Guidelines for Commercialization of Expendable
  Launch Vehicles for U.S. Government National Ranges

[4] IMPLEMENTING GUIDELINES FOR COMMERCIALIZATION OF EXPENDABLE LAUNCH VEHICLES FROM U.S. GOVERNMENT NATIONAL RANGES

A. Required U.S. Government Actions

NASA and DOD, for those functions over which they respectively have cognizance, will:

1. identify data, documentation, processes, procedures, tooling, ground support equipment and facilities that are available for commercial use;
2. identify the support services and facilities necessary for commercial launches from the U.S. Government national ranges;
3. identify the joint-use tooling, ground support equipment and facilities that the U.S. Government can make available for commercial launch operations;
4. determine the transition means, schedules, conditions, and costs for making available appropriate U.S. Government equipment, facilities and properties;
5. to the extent practical, provide, on a reasonable reimbursable basis, technical advice and assistance in operations;
6. negotiate and contract for, on a reasonable reimbursable basis, their portion of the U.S. Government services, facilities and equipment requested by the private sector for commercial launch operations;
7. as required, conduct environmental analyses necessary to ensure compliance with the National Environmental Policy Act.

B. Government Pricing Guidelines

The price for the use of U.S. Government facilities, equipment, and services will be based on the following principles:

1. price services based an those additional costs incurred by the U.S. Government;
2. the U.S. Government will not seek to recover ELV design and development costs or investments associated with facilities to which the U.S. Government retains title;
|5| 3. tooling, equipment and residual ELV hardware on hand at the completion of the U.S. Government's program will be priced on a basis that is in the best overall interest of the U.S. Government, taking into consideration that these sales will not constitute a subsidy to the private sector operator.

C. Commercial ELV Operator Requirements

The commercial ELV operator shall:

1. maintain all facilities and equipment leased from the U.S. Government to a level of readiness and repair specified by the U.S. Government;
2. provide adequate insurance to cover the loss of or damage to U.S. Government owned systems, equipment, [and] facilities used by the private sector ELV operators;
3. provide adequate insurance and agreements to indemnify and hold harmless the U.S. Government against liabilities for damage to both domestic and foreign persons and property;
4. abide by all required U.S. Government safety criteria and not hold the U.S. Government liable for damage incurred by the operator resulting from U.S. Government flight safety actions;
5. agree not to hold the U.S. Government liable for losses resulting from scheduling delays related to joint-use facilities and support services.

Document III-4

Document title: Commercial Space Launch Act of 1984, Public Law 98–575, 98 Stat. 3055, October 30, 1984.

Source: NASA Historical Reference Collection, NASA History Office, Washington, D.C.

Both Congress and the executive branch took steps during 1983 and 1984 to nurture the new commercial space transportation industry. This legislation, which originated in the House Committee on Science and Technology, provided the statutory basis for the role of the Department of Transportation to carry out its role in licensing commercial launches by the U.S. private sector and the other authorities thought needed for the industry to function. By the time the industry actually emerged in the wake of the Challenger accident, additional legal provisions were seen to be required, and this law was amended in 1988.

PUBLIC LAW 98-575—OCT. 30, 1984   [98 STAT. 3055]

Public Law 98-575
98th Congress

An Act

To facilitate commercial space launches, and for other purposes
[Oct. 30, 1984] [H.R. 3942]

Be it enacted by the Senate and House of Representatives of the United States of America in Congress assembled,
[Commercial Space Launch Act.]

SHORT TITLE

SECTION 1. This Act may be cited as the "Commercial Space Launch Act."
[49 USC app. 2601 note.]

FINDINGS

SEC. 2. The Congress finds and declares that—
[49 USC app. 2601.]

(1) the peaceful uses of outer space continue to be of great value and to offer benefits to all mankind;

(2) private applications of space technology have achieved a significant level of commercial and economic activity, and offer the potential for growth in the future, particularly in the United States;

(3) new and innovative equipment and services are being sought, created, and offered by entrepreneurs in telecommunications, information services, and remote sensing technology;

(4) the private sector in the United States has the capability of developing and providing private satellite launching and associated services that would complement the launching and associated services now available from the United States Government;

(5) the development of commercial launch vehicles and associated services would enable the United States to retain its competitive position internationally, thereby contributing to the national interest and economic well-being of the United States;

(6) provision of launch services by the private sector is consistent with the national security interests and foreign policy interests of the United States and would be facilitated by stable, minimal, and appropriate regulatory guidelines that are fairly and expeditiously applied; and

(7) the United States should encourage private sector launches and associated services and, only to the extent necessary, regulate such launches and services in order to ensure compliance with international obligations of the United States and to protect the public health and safety, safety of property, and national security interests and foreign policy interests of the United States.

PURPOSES

SEC. 3. It is therefore the purpose of this Act—
[49 USC 2602.]

(1) to promote economic growth and entrepreneurial activity through utilization of the space environment for peaceful purposes;

(2) to encourage the United States private sector to provide launch vehicles and associated launch services by simplifying and expediting the issuance and transfer of commercial launch licenses and by facilitating and encouraging the utilization of Government-developed space technology; and

(3) to designate an executive department to oversee and coordinate the conduct of commercial launch operations, to issue and transfer commercial launch licenses authorizing such activities, and to protect the public health and safety, safety of property, and national security interests and foreign policy interests of the United States.

DEFINITIONS

SEC. 4. For purposes of this Act—
[49 USC app. 2603]

(1) "agency" means an executive agency as defined by section 105 of title 5, United States Code;

(2) "launch" means to place, or attempt to place, a launch vehicle and payload, if any, in a suborbital trajectory, in Earth orbit in outer space, or otherwise in outer space;

(3) "launch property" means propellants, launch vehicles and components thereof, and other physical items constructed for or used in the launch preparation or launch of a launch vehicle;

(4) "launch services" means those activities involved in the preparation of a launch vehicle and its payload for launch and the conduct of a launch;

(5) "launch site" means the location on Earth from which a launch takes place, as defined in any license issued or transferred by the Secretary under this Act, and includes all facilities located on a launch site which are necessary to conduct a launch;

(6) "launch vehicle" means any vehicle constructed for the purpose of operating in, or placing a payload in, outer space and any suborbital rocket;

(7) "payload" means an object which a person undertakes to place in outer space by means of a launch vehicle, and includes subcomponents of the launch vehicle specifically designed or adapted for that object;

(8) "person" means any individual and any corporation, partnership, joint venture, association, or other entity organized or existing under the laws of any State or any nation;

(9) "Secretary" means the Secretary of Transportation;

(10) "State," and "United States" when used in a geographical sense, mean the several States, the District of Columbia, the Commonwealth of Puerto Rico, American Samoa, the United States Virgin Islands, Guam, and any other commonwealth, territory or possession of the United States; and

(11) "United States citizen" means—
    (A) any individual who is a citizen of the United States;
    (B) any corporation, partnership, joint venture, association, or other entity organized or existing under the laws of the United States or any State; and
    (C) any corporation, partnership, joint venture, association, or other entity which is organized or exists under the laws of a foreign nation, if the controlling interest (as defined by the Secretary in regulations) in such entity is held by an individual or entity described in subparagraph (A) or (B).

GENERAL RESPONSIBILITIES OF THE SECRETARY AND OTHER AGENCIES

SEC. 5. (a) The Secretary shall be responsible for carrying out this Act, and in doing so shall—
[49 USC app. 2604.]

(1) encourage, facilitate, and promote commercial space launches by the private sector; and

(2) consult with other agencies to provide consistent application of licensing requirements under this Act and to ensure fair and equitable treatment for all license applicants.

(b) To the extent permitted by law, Federal agencies shall assist the Secretary, as necessary, in carrying out this Act.

REQUIREMENT OF LICENSE FOR PRIVATE SPACE LAUNCH OPERATIONS

SEC. 6. (a) (1) No person shall launch a launch vehicle or operate a launch site within the United States, unless authorized by a license issued or transferred under this Act.
[49 USC app. 2605.]

(2) No United States citizen described in subparagraph (A) or (B) of section 4(11) shall launch a launch vehicle or operate a launch site outside the United States, unless authorized by a license issued or transferred under this Act.

(3) (A) No United States citizen described in subparagraph (C) of section 4(11) shall launch a launch vehicle or operate a launch site at any place which is both outside the United States and outside the territory of any foreign nation, unless authorized by a license issued or transferred under this Act. The preceding sentence shall not apply with respect to a launch or operation of a launch site if there is an agreement in force between the United States and a foreign nation which provides that such foreign nation shall exercise jurisdiction over such launch or operation.

(B) (i) Except as provided in clause (ii) of this subparagraph, this Act shall not apply to the launch of a launch vehicle or the operation of a launch site in the territory of a foreign nation by a United States citizen described in subparagraph (C) of section 4(11).

(ii) If there is an agreement in force between the United States ["International agreements" appears in the margin] and a foreign nation which provides that the United States shall exercise jurisdiction over the launch of a launch vehicle or operation of a launch site in the territory of such nation by a United States citizen described in subparagraph (C) of section 4(11), no such United States citizen shall launch a launch vehicle or operate a launch site in the territory of such nation, unless authorized by a license issued or transferred under this Act.

(b) (1) The holder of a launch license under this Act shall not launch a payload unless that payload complies with all requirements of Federal law that relate to the launch of a payload. The Secretary shall ascertain whether any license, authorization, or other permit required by Federal law for a payload which is to be launched has been obtained.

(2) If no payload license, authorization, or permit is required by any Federal law, the Secretary may take such action under this Act as the Secretary deems necessary to prevent the launch of a payload by a holder of a launch license under this Act if the Secretary determines that the launch of such payload would jeopardize the public health and safety, safety of property, or any national security interest or foreign policy interest of the United States.

(c) (1) Except as provided in this Act, no person shall be required to obtain from any agency a license, approval, waiver, or exemption for the launch of a launch vehicle or the operation of a launch site.

(2) Nothing in this Act shall affect the authority of the Federal Communications Commission under the Communications Act of 1934 (47 U.S.C. 151 *et seq.*) or the authority of the Secretary of Commerce under the Land Remote-Sensing Commercialization Act of 1984 (15 U.S.C. 4201 *et seq.*).

AUTHORITY TO ISSUE AND TRANSFER LICENSES

[49 USC app. 2606.]
SEC. 7. The Secretary may, consistent with the public health and safety, safety of property, and national security interests and foreign policy interests of the United States, issue or
transfer a license for launching one or more launch vehicles or for operating one or more
launch sites, or both, to an applicant who meets the requirements for a license under section 8 of this Act. Any license issued or transferred under this section shall be in effect for
such period of time as the Secretary may specify, in accordance with regulations issued
under this Act.

LICENSING REQUIREMENTS

[49 USC app. 2607]
SEC. 8. (a) (1) All requirements of Federal law which apply to the launch of a launch vehicle or the operation of a launch site shall be requirements for a license under this Act for
the launch of a launch vehicle or the operation of a launch site, respectively, except to the
extent provided in paragraph (2).

    (2) If the Secretary determines, in consultation with appropriate agencies, that
any requirement of Federal law that would otherwise apply to the launch of a launch vehicle or the operation of a launch site is not necessary to protect the public health and safety, safety of property, and national security interests and foreign policy interests of the
United States, the Secretary may by regulation provide that such requirement shall not be
a requirement for a license under this Act.

    (b) The Secretary may, with respect to launches and the operation of launch sites,
prescribe such additional requirements as are necessary to protect the public health and
safety, safety of property, and national security interests and foreign policy interests of the
United States.

    (c) The Secretary may, in individual cases, waive the application of any requirement
for a license under this section if the Secretary determines that such waiver is in the public interest and will not jeopardize the public health and safety, safety of property, or any
national security interest or foreign policy interest of the United States.

LICENSE APPLICATION AND APPROVAL

[49 USC app. 2608]
SEC. 9. (a) Any person may apply to the Secretary for issuance or transfer of a license
under this Act, in such form and manner as the Secretary may prescribe. The Secretary
shall establish procedures and timetables to expedite review of applications under this section and to reduce regulatory burdens for applicants.

    (b) The Secretary shall issue or transfer a license to an applicant if the Secretary
determines in writing that the applicant complies and will continue to comply with the
requirements of this Act and any regulation issued under this Act. The Secretary shall
include in such license such conditions as may be necessary to ensure compliance with
this Act, including an effective means of on-site verification that a launch or operation of
a launch site conforms to representations made in the application for a license or transfer of a license. The Secretary shall make a determination on any application not later
than 180 days after receipt of such application. If the Secretary has not made a determination within 120 days after receipt of such application, the Secretary shall inform the
applicant of any pending issues and of actions required to resolve such issues.

    (c) The Secretary, any officer or employee of the United States, or any person with
whom the Secretary has entered into a contract under section 14(b) of this Act may not

disclose any data or information under this Act which qualifies for exemption under section 552(b)(4) of title 5, United States Code, or is designated as confidential by the person or agency furnishing such data or information, unless the Secretary determines that the withholding of such data or information is contrary to the public or national interest.

SUSPENSION, REVOCATION, AND MODIFICATION OF LICENSES

[49 USC app. 2609]
SEC. 10. (a) The Secretary may suspend or revoke any license issued or transferred under this Act if the Secretary finds that the licensee has substantially failed to comply with any requirement of this Act, the license, or any regulation issued under this Act, or that the suspension or revocation is necessary to protect the public health and safety, safety of property, or any national security interest or foreign policy interest of the United States.

(b) Upon application by the licensee or upon the Secretary's own initiative, the Secretary may modify a license issued or transferred under this Act, if the Secretary finds that the modification will comply with the requirements of this Act.

(c) Unless otherwise specified by the Secretary, any suspension, revocation, or modification by the Secretary under this section—

(1) shall take effect immediately; and

(2) shall continue in effect during any review of such action under section 12 of this Act.

(d) Whenever the Secretary takes any action under this section, the Secretary shall notify the licensee in writing of the Secretary's finding and the action which the Secretary has taken or proposes to take regarding such finding.

EMERGENCY ORDERS

[Prohibition. 49 USC app. 2610.]
SEC. 11. (a) The Secretary may terminate, prohibit, or suspend immediately the launch of a launch vehicle or the operation of a launch site which is licensed under this Act if the Secretary determines that such launch or operation is detrimental to the public health and safety, safety of property, or any national security interest or foreign policy interest of the United States.

(b) An order terminating, prohibiting, or suspending any launch or operation of a launch site licensed by the Secretary under this Act shall take effect immediately and shall continue in effect during any review of such order under section 12.

ADMINISTRATIVE AND JUDICIAL REVIEW

[49 USC app. 2611]
SEC. 12. (a)(1) An applicant for a license and a proposed transferee of a license under this Act shall be entitled to a determination on the record after an opportunity for a hearing in accordance with section 554 of title 5, United States Code, of any decision of the Secretary under section 9(b) to issue or transfer a license with conditions or to deny the issuance or transfer of such license. An owner or operator of a payload shall be entitled to a determination on the record after an opportunity for a hearing in accordance with section 554 of title 5, United States Code, of any decision of the Secretary under section 6(b)(2) to prevent the launch of such payload.

(2) A licensee under this Act shall be entitled to a determination on the record after an opportunity for a hearing in accordance with section 554 of title 5, United States Code, of any decision of the Secretary—

(A) under section 10 to suspend, revoke, or modify a license; or

(B) under section 11 to terminate, prohibit, or suspend any launch or operation of a launch site licensed by the Secretary.

(b) Any final action of the Secretary under this Act to issue, transfer, deny the issuance or transfer of, suspend, revoke, or modify a license or to terminate, prohibit, or suspend any launch or operation of a launch site licensed by the Secretary or to prevent the launch of a payload shall be subject to judicial review as provided in chapter 7 of title 5, United States Code.

REGULATIONS

[49 USC app. 2612]
SEC. 13. The Secretary may issue such regulations, after notice and comment in accordance with section 553 of title 5, United States Code, as may be necessary to carry out this Act.

MONITORING OF ACTIVITIES OF LICENSEES

[49 USC app. 2613]
SEC. 14. (a) Each license issued or transferred under this Act shall require the licensee—

(1) to allow the Secretary to place Federal officers or employees or other individuals as observers at any launch site used by the licensee, at any production facility or assembly site used by a contractor of the licensee in the production or assembly of a launch vehicle, or at any site where a payload is integrated with a launch vehicle, in order to monitor the activities of the licensee or contractor at such time and to such extent as the Secretary considers reasonable and necessary to determine compliance with the license or to carry out the responsibilities of the Secretary under section 6(b) of this Act; and

(2) to cooperate with such observers in the performance of monitoring functions.

(b) The Secretary may, to the extent provided in advance by appropriation Acts, enter into a contract with any person to carry out subsection (a)(1) of this section.

USE OF GOVERNMENT PROPERTY

[49 USC app. 2614.]
SEC. 15. (a) The Secretary shall take such actions as may be necessary to facilitate and encourage the acquisition (by lease, sale, transaction in lieu of sale, or otherwise) by the private sector of launch property of the United States which is excess or is otherwise not needed for public use and of launch services, including utilities, of the United States which are otherwise not needed for public use.

(b) (1) The amount to be paid to the United States by any person who acquires launch property or launch services, including utilities, shall be established by the agency providing the property or service, in consultation with the Secretary. In the case of acquisition of launch property by sale or transaction in lieu of sale, the amount of such payment shall be the fair market value. In the case of any other type of acquisition of launch property, the amount of such payment shall be an amount equal to the direct costs (including any specific wear and tear and damage to the property) incurred by the United States as a result of the acquisition of such launch property. In the case of any acquisition of launch services, including utilities, the amount of such payment shall be an amount equal to the direct costs (including salaries of United States civilian and contractor personnel) incurred by the United States as a result of the acquisition of such launch services.

(2) The Secretary may collect any payment for launch property or launch services, with the consent of the agency establishing such payment under paragraph (1).

(3) The amount of any payment received by the United States for launch property or launch services, including utilities, under this subsection shall be deposited in the

general fund of the Treasury, and the amount of a payment for launch property (other than launch property which is excess) and launch services (including utilities) shall be credited to the appropriation from which the cost of providing such property or services was paid.

(c) The Secretary may establish requirements for liability insurance, hold harmless agreements, proof of financial responsibility, and such other assurances as may be needed to protect the United States and its agencies and personnel from liability, loss, or injury as a result of a launch or operation of a launch site involving Government facilities or personnel.

LIABILITY INSURANCE

[49 USC app. 2615.]
SEC. 16. Each person who launches a launch vehicle or operates a launch site under a license issued or transferred under this Act shall have in effect liability insurance at least in such amount as is considered by the Secretary to be necessary for such launch or operation, considering the international obligations of the United States. The Secretary shall prescribe such amount after consultation with the Attorney General and other appropriate agencies.

ENFORCEMENT AUTHORITY

[49 USC app. 2616.]
SEC. 17. (a) The Secretary shall enforce this Act. The Secretary may delegate the exercise of any enforcement authority under this Act to any officer or employee of the Department of Transportation or, with the approval of the head of another agency, any officer or employee of such agency.

(b) In carrying out this section, the Secretary may—

(1) make investigations and inquiries, and administer to or take from any person an oath, affirmation, or affidavit, concerning any matter relating to enforcement of this Act; and

(2) pursuant to any lawful process—

(A) enter at any reasonable time any launch site, production facility, or assembly site of a launch vehicle, or any site where a payload is integrated with a launch vehicle, for the purpose of inspecting any object which is subject to this Act and any records or reports required by the Secretary to be made or kept under this Act; and

(B) seize any such object, record, or report where there is probable cause to believe that such object, record, or report was used, is being used, or is likely to be used in violation of this Act.

PROHIBITED ACTS

[49 USC app. 2617]
SEC. 18. It is unlawful for any person to violate a requirement of this Act, a regulation issued under this Act, or any term, condition or restriction of any license issued or transferred by the Secretary under this Act.

CIVIL PENALTIES

[49 USC app. 2618]
SEC. 19. (a) Any person who is found by the Secretary, after notice and opportunity to be heard on the record in accordance with section 554 of title 5, United States Code, to have committed any act prohibited by section 18 shall be liable to the United States for a civil penalty of not more than $100,000 for each violation. Each day of a continuing violation shall constitute a separate violation. The amount of such civil penalty shall be assessed by

the Secretary by written notice. The Secretary may compromise, modify, or remit, with or without conditions, any civil penalty which is subject to imposition or which has been imposed under this section.

(b) If any person fails to pay a civil penalty assessed against such person after the penalty has become final or if such person appeals an order of the Secretary and the appropriate court has entered final judgment in favor of the Secretary, the Secretary shall recover the civil penalty assessed in any appropriate district court of the United States.

(c) For purposes of conducting any hearing under this section, the Secretary may (1) issue subpoenas for the attendance and testimony of witnesses and the production of relevant papers, books, documents, and other records, (2) seek enforcement of such subpoenas in the appropriate district court of the United States, and (3) administer oaths and affirmations.

CONSULTATION

[Defense and national security. 49 USC app. 2619]
SEC. 20. (a) The Secretary shall consult with the Secretary of Defense on all matters, including the issuance or transfer of each license, under this Act affecting national security. The Secretary of Defense shall be responsible for identifying and notifying the Secretary of those national security interests of the United States which are relevant to activities under this Act.

(b) The Secretary shall consult with the Secretary of State on all matters, including the issuance or transfer of each license, under this Act affecting foreign policy. The Secretary of State shall be responsible for identifying and notifying the Secretary of those foreign policy interests or obligations of the United States which are relevant to activities under this Act.

(c) The Secretary shall consult with other agencies, as appropriate, in order to carry out the provisions of this Act.

RELATIONSHIP TO OTHER LAWS AND INTERNATIONAL OBLIGATIONS

[Prohibitions. 49 USC app. 2620.]
SEC. 21. (a) No State or political subdivision of a State may adopt or have in effect any law, rule, regulation, standard, or order which is inconsistent with the provisions of this Act. Nothing in this Act shall preclude a State or a political subdivision of a State from adopting or putting into effect any law, rule, regulation, standard, or order which is consistent with this Act and is in addition to or more stringent than any requirement of or regulation issued under this Act. The Secretary may, and is encouraged to, consult with the States to simplify and expedite the approval of space launch activities.

(b) A launch vehicle or payload shall not, by reason of the launching of such vehicle or payload, be considered an export for purposes of any law controlling exports.

(c) Nothing in this Act shall apply to—
(1) any—
(A) launch or operation of a launch vehicle,
(B) operation of a launch site, or
(C) other space activity, carried out by the United States on behalf of the United States; or
(2) any planning or policies relating to any such launch, operation, or activity.

(d) The Secretary shall carry out this Act consistent with any obligation assumed by the United States in any treaty, convention, or agreement that may be in force between the United States and any foreign nation. In carrying out this Act, the Secretary shall consider applicable laws and requirements of any foreign nation.

REPORT ON LEGISLATION

[Report. 49 USC app. 2621.]
SEC. 22. (a) Not later than the last day of each fiscal year ending after the date of enactment of this Act and before October 1, 1989, the Secretary shall submit to the Committee on Science and Technology of the House of Representatives and the Committee on Commerce, Science, and Transportation of the Senate a report describing all activities undertaken under this Act, including a description of the process for the application for and approval of licenses under this Act and recommendations for legislation that may further commercial launches.

(b) Not later than July 1, 1985, the Secretary shall submit to the Committee on Science and Technology of the House of Representatives and the Committee on Commerce, Science, and Transportation of the Senate a report which identifies Federal statutes, treaties, regulations, and policies which may have an adverse effect on commercial launches and include recommendations on appropriate changes thereto.

SEVERABILITY

[49 USC app. 2622.]
SEC. 23. If any provision of this Act, or the application of such provision to any person or circumstance, is held invalid, the remainder of this Act and the application of such provision to any other person or circumstance shall not be affected by such invalidation.

AUTHORIZED APPROPRIATIONS

[49 USC app. 2623]
SEC. 24. There are authorized to be appropriated to the Secretary $4,000,000 for fiscal year 1985.

EFFECTIVE DATE

[49 USC app. 2601 note.]
SEC. 25. (a) Except for section 15 and the authority to issue regulations, this Act shall take effect 180 days after the date of enactment of this Act.

(b) Section 15 shall take effect on the date of enactment of this Act, except that nothing in this Act shall affect any agreement, including negotiations which are substantially completed, relating to the acquisition of launch property or launch services of the United States entered into on or before the date of enactment of this Act between the United States and any private party.
[Regulations.]

(c) Regulations to implement this Act shall be promulgated not later than 180 days after the date of enactment of this Act.

Approved October 30, 1984.

Document III-5

Document title: Craig L. Fuller, Memorandum for the President, "Determining the Lead Agency for Commercializing Expendable Launch Vehicles," November 16, 1983.

Source: Ronald Reagan Presidential Library, Simi Valley, California.

Following the May 1983 decision that the government would support the commercialization of expendable launch vehicles, there was a vigorous six-month debate within the executive branch about which Cabinet agency would have responsibility for this new sector of government activity. The issue finally came to President Reagan for resolution on November 16, 1983. Following a meeting of the Cabinet Council on Competitiveness and Trade (staffed by Craig Fuller) at which the issue was discussed, Secretary of Transportation Elizabeth Dole persuaded Reagan that her agency should have jurisdiction. This memorandum records that presidential decision.

THE WHITE HOUSE
WASHINGTON

November 16, 1983

MEMORANDUM FOR THE PRESIDENT

FROM: CRAIG L. FULLER [hand initialed "CLF"]
SUBJECT: Determining the Lead Agency for Commercializing Expendable Launch Vehicles

As discussed in the meeting this morning of the Cabinet Council on Commerce and Trade there are the following three options with regard to determining the lead agency for commercializing expendable launch vehicles (ELVs):

1. Take no position at this time on the lead agency for commercialization of ELVs and await further discussion of broader space commercialization issues. Congress would be told that the Administration has no position at the present time on this matter.

2. Designate the Department of Commerce the lead agency for commercialization of ELVs. An executive order would be prepared for signature and testimony would be drafted in accordance with this decision.

3. Designate the Department of Transportation the lead agency for commercialization of ELVs. An executive order would be prepared for signature and testimony would be drafted in accordance with this decision.

Once you make a decision, we will take the appropriate steps to implement it. If you wish to make a decision today, we will advise the agencies this afternoon in order to allow them an opportunity to testify at tomorrow's hearing.

Decision

    _____   1.   No decision at the present time.
    _____   2.   Designate the Department of Commerce as the lead agency.
    [initialed "RR"]   3.   Designate the Department of Transportation as the lead agency.

Document III-6

Document title: Gerald J. Mossinghoff, Assistant Secretary and Commissioner of Patents and Trademarks, to Elizabeth Hanford Dole, Secretary of Transportation, letter regarding recommendations of the Commercial Space Transportation Advisory Committee, October 31, 1984, with attached: committee recommendations, October 23, 1984.

Source: Office of Commercial Space Transportation, Department of Transportation, Washington, D.C.

As it took over responsibility for both promoting and regulating the fledgling commercial space transportation industry, the Department of Transportation created a Commercial Space Transportation Advisory Committee, soon known as COMSTAC. This committee was a means of getting the views of relevant individuals and firms on how the department could best carry out its responsibilities. This letter transmitted the recommendations of the first COMSTAC meeting to Secretary of Transportation Elizabeth Dole.

UNITED STATES DEPARTMENT OF COMMERCE
Patent and Trademark Office
ASSISTANT SECRETARY AND COMMISSIONER
OF PATENTS AND TRADEMARKS
Washington, D.C. 20231

Honorable Elizabeth Hanford Dole   [stamped OCT 31 1984]
Secretary of Transportation
400 Seventh Street, S.W.
Washington, DC 20590

Dear Madame Secretary:

On October 22 and 23, 1984, the Commercial Space Transportation Advisory Committee held its first meeting in the Department of Transportation headquarters building. Twenty-two members, along with representatives of the public and press, attended. I was privileged to act as Chairman for this meeting.

In keeping with the committee's charter, a number of matters related to the commercialization of expendable launch vehicles (ELVs) were brought before this forum. The members discussed the health of the industry, the licensing process, international competition, STS (Shuttle) policies, the cost and use of Government facilities, and international legal issues. There was an extended discussion about the focus of the nation's Shuttle program, and it was the desire of the members to have this issue reviewed further at the next meeting. The committee strongly supports and endorses the need to maintain our commercial launch capability as a national asset.

During the course of the committee's discussions, a number of recommendations were developed and reviewed by the committee. The committee agreed to the eight recommendations which are enclosed with this letter. I would be pleased to discuss these recommendations with you further at your convenience.

I greatly appreciated the opportunity to serve as Chairman for the first session, and I hope you will agree with me that the committee got off to a running start.

Yours very truly,

Gerald J. Mossinghoff
Assistant Secretary and Commissioner
of Patents and Trademarks

Enclosure
Copy to: Members of the Commercial Space Transportation Advisory Committee

****

[1] COMMERCIAL SPACE TRANSPORTATION
ADVISORY COMMITTEE
October 23, 1984

RECOMMENDATION #1: STS Pricing Policy

A. The STS and commercial launch vehicles should complement each other to provide a national space launch capability based on their own inherent advantages.
B. The commercial launch industry cannot become viable unless a free market environment is established. STS pricing should not place commercial launch operators at an unfair disadvantage. Government project managers should be free to select launch vehicles based on their merits.
C. The STS has proven its value for a variety of missions and should not have to justify itself on the basis of its market share of commercial payloads.
D. The commercial space industry—ELV, upper-stage and spacecraft producers—urgently needs a prompt decision from the Administration on STS pricing policy. Timing is critical and, unless these issues are quickly resolved, the opportunity for the development of a viable commercial ELV industry will be lost.

RECOMMENDATION #2: Launch Insurance

A. The Department of Transportation should analyze whether the Government should provide back-up launch insurance (i.e., to be an insurer of last resort) for ventures which cannot be fully insured commercially.
B. In this capacity, the Government should not compete with private sector insurance nor should it take on a regulatory role.

RECOMMENDATION #3: Financing for Start-ups of New Ventures

The Department should analyze the feasibility of establishing new mechanisms for financing new commercial space transportation ventures.

RECOMMENDATION #4: Safety Regulation

A. The Committee recognizes that the Government's responsibility is to protect the public safety, assure that our international obligations are met and protect national security.
[2] B. The private sector should have primary responsibility for reliability and mission success. The Government should have primary responsibility for public safety and protection of property.

C. While recognizing the need for adequate information to comply with our treaty oblig-ations, the Government must make every effort to protect privately funded propri-etary information and equipment.
D. The Department should consider licensing range safety officers, either from the Government or private sector, to assure that the necessary range safety measures are taken on commercial ranges.

RECOMMENDATION #5: Investigation of Accidents

The Department should address the issue of what body will investigate commercial launch, orbital or other accidents, and under what conditions. Responsibilities and autho-rization must be clearly defined.

RECOMMENDATION #6: International Competition

A. The Committee recognizes the potential for unfair international competition in pro-viding launch services and recommends that the U.S. Government employ all avail-able tools to counter any such unfair competition.
B. The U.S. should eschew countervailing subsidies as a remedy.

RECOMMENDATION #7: Commercial Use of ELV Facilities

A. Since ELV launch and support facilities and unique equipment have no value unless they are maintained, the Government should only require that the operators keep them in an operational state at no cost to the Government. This will save the Government the cost of shutting down and/or "mothballing" the facilities. Title to the property need not be consigned to the ELV launch operators.
B. The ELV launch operators should pay for all other facilities and capabilities on an additive or direct cost basis.

RECOMMENDATION #8: U.S. Delegation to United Nations
　　　　　　　　　　　　　Committee on the Peaceful Uses of Outer Space

The Department of Transportation should be represented on the U.S. Delegation to the United Nations Committee on the Peaceful Uses of Outer Space and its legal and scien-tific/technical subcommittees.

Document III-7

Document title: Robert C. McFarlane, Memorandum for The Honorable Elizabeth H. Dole, Secretary of Transportation, "STS Pricing Issue," June 21, 1984.

Source: Ronald Reagan Presidential Library, Simi Valley, California

Once assigned the lead government role in overseeing the development of a commercial space trans-portation industry, the Department of Transportation and its Secretary, Elizabeth Dole, became force-ful advocates for a policy that would price Space Shuttle launches for commercial users at a level high enough to allow the private-sector operators of expendable launch vehicles to compete with NASA and the Shuttle for commercial contracts. As the National Security Council and its Senior Interagency Group (Space) considered the Shuttle pricing issue, National Security Advisor Robert (Bud) McFarlane, its chair, pointed out to Secretary Dole the many considerations that needed to go into the decision on Shuttle pricing policy.

[1]   **THE WHITE HOUSE**
                              WASHINGTON

                                                              June 21, 1984

MEMORANDUM FOR   THE HONORABLE ELIZABETH H. DOLE
                 The Secretary of Transportation
SUBJECT: STS Pricing Issue

Following your recent telephone call, I think it would be useful to outline the rationale behind our proposed policy recommendation to the President on this issue. I believe our reasoning is sound, but I would appreciate hearing if you have a contrary view.

As I understand it, there is concern that continued low prices for Shuttle launches represent a serious obstacle to a viable U.S. commercial ELV industry. It is argued that these low prices result from government subsidies to the Shuttle which must be removed by requiring NASA to base its prices on the full recovery of costs. Supporters of this view point out that the intent of the President's objectives for commercial ELVs as contained in [Executive Order] 12465 and NSDD 94 cannot be fully implemented until these "subsidies" are removed.

As you know, NSDD 42, National Space Policy, requires NASA to assign its highest priority to making the Shuttle fully operational and cost-effective. NSDD 94 states that, beyond FY 1988, it is the government's intent that Shuttle prices be based on full-cost recovery. At issue is whether the NSDD on National Space Strategy should go beyond this policy by establishing a specific date for full-cost recovery (rather than sometime "beyond FY 1988") and whether Shuttle prices in the international marketplace should be based, at least in part, upon prices charged by other foreign launchers.

We have evidence to suggest that the French Ariane ELV would be the primary beneficiary of an increase in Shuttle prices. As you know, the French government heavily subsidizes the Ariane launcher and currently underbids U.S ELVs for most launches. We believe the French will continue to subsidize the Ariane in the interest of capturing an even larger share of the market. If NASA is arbitrarily forced to raise its Shuttle prices, it appears that Ariane, and not U.S. ELVs, will benefit through increased demand from payload customers. Such a result would obviously undercut the President's primary goal of maintaining U.S. space leadership.

[2] Removing the Shuttle as a viable contender in the international marketplace would also have other serious implications. The Space Shuttle represents a significant and highly visible instrument of our foreign policy. The Shuttle is an effective means for promoting international cooperation, good will and technological growth among our friends and allies. The flight of foreign astronauts on the Shuttle along with their payloads is one example of how the President uses the Shuttle toward these ends.

Diminishing the Shuttle's competitiveness could also be counterproductive to our other space commercialization goals. NASA is attempting to encourage commercial users to capitalize on the unique attributes offered by the manned capabilities of the Shuttle. The potential of the Shuttle to spawn new industries in such areas as materials processing and manufacture of medicines should not be discouraged. The President strongly supports initiatives to stimulate these pioneering efforts.

Finally, a reduction in foreign and domestic commercial Shuttle launches resulting from increased prices could possibly result in increased prices charged for U.S. Government launches. To the extent that NASA's fixed costs would be spread over fewer launches, civil and DOD users might be required to share the burden through increased prices for their launches. Therefore, we have reason to question whether the taxpayer's burden would be truly reduced. In short, a reduction in Shuttle subsidies to foreign and commercial users could conceivably be offset by increased prices to government users.

The bottom line is that we must proceed prudently and cautiously in resolving this issue. We must formulate a policy which is in the overall national interest and reconciles our conflicting goals. The draft NSDD on National Space Strategy currently states, in part:

> Prices for STS services and capabilities provided to commercial and foreign users on and after October 1, 1988, will reflect the full-cost of such services and capabilities consistent with the need to maintain international competitiveness in the provision of launch services. NASA, in consultation with other agencies, will develop a time-phased plan for implementing full-cost recovery for commercial and foreign STS flight operations on October 1, 1988.

This formulation advances the current policy by requiring full-cost recovery on a specific date, but recognizes the international implications of Shuttle pricing. It continues [3] to direct NASA to drive down Shuttle costs and reduce the burden on the Federal budget. We would look to the "time-phased plan" as the means for increasing our understanding of the full implications of future Shuttle pricing and to implement a course of action that we can all agree upon.

On a related subject, the National Space Strategy will also endorse DOD's requirement for assured access to space and the need for a limited number of ELVs to back up the Shuttle. The contractual and funding mechanisms to satisfy this requirement are being worked out between DOD and OMB. However, if DOD proceeds with their plan to procure ELVs, this action should serve to further underwrite our commercial ELV objectives by maintaining the industrial base.

Again, I would appreciate your views on our proposed recommendation to the President on this issue.

[signed "Bud McFarlane"]
Robert C. McFarlane

Document III-8

Document title: James M. Beggs, Administrator, NASA, to President Ronald Reagan, September 17, 1984.

Source: Ronald Reagan Presidential Library, Simi Valley, California.

The mid-1984 White House statement of National Space Strategy called for setting a Shuttle price after October 1, 1988, that would enable "full cost recovery." The precise meaning of this term was a subject of considerable debate between mid-1984 and mid-1985. In this letter, NASA Administrator James Beggs gives the space agency's initial position on an appropriate Shuttle price to meet the policy objective. The supporting documentation in the enclosure to this letter does not appear here.

[no pagination]

NASA
National Aeronautics and
Space Administration
Washington, D.C. 20546
Office of the Administrator

September 17, 1984

The President
The White House
Washington, D.C. 20500

Dear Mr. President:

This letter, with its supporting documentation, proposes a price for commercial and foreign Space Shuttle flights in the FY 1989–91 time period. Your recent instruction to us that Shuttle services be priced so as to recover the full costs of operating the system is satisfied by this proposal. I am confident that already strong public and Congressional support for the U.S. space program will only be enhanced by this approach which will reduce costs to the U.S. taxpayer.

The Shuttle is a central element in this nation's projection of world leadership in space. Shuttle flights capture the attention and imagination of people the world over. This remarkable impact is due in part to the exciting men and women who are our astronauts. And the impact also results from the fact that every Shuttle flight is a tangible demonstration of the staggering capabilities of American technology. Obviously, to take advantage of the power of Shuttle, we must fly it. And, in particular, we must use it for commercial and foreign missions which present highly visible opportunities for private citizens and foreigners to fly with us. I am comfortable that our proposed new prices for Shuttle flights will enable us to continue using the Shuttle effectively and to maintain a leadership position in the international arena.

NASA's Shuttle pricing plan is laid out in detail in the enclosed documentation. Because we are dealing with a period four years in the future, we have had to make a number of assumptions to project important factors like learning curves and flight rates. I am confident that our projections reflect the best judgement [sic] available—based as they are on twenty-five years' experience in providing launch services for both the U.S. Government and the commercial market.

Based on an in-depth analysis of our experience to date and on our projections regarding future flight demands and the system itself, we have calculated that the average full cost recovery price over the three-year period FY 1989–91 would be $83.3 million per flight. We are proposing, however, charging a list price of $87 million per Shuttle flight, with the flexibility to adjust that list price up or down by as much as 5% to accomodate [sic] special conditions relating to individual customers' situations. This baseline price is conservative relative to our calculated costs, thus giving us a safety margin for achieving your goal of full cost recovery for commercial and foreign Shuttle operations. For point of comparison, the price for a Shuttle launched today is $38 million; the price we will be charging in the FY 1986–88 period is $71 million (all figures in FY 1982$). Thus we are proposing a $49 million or 130% increase in Shuttle prices to the private sector over the next four years. Despite this steep rise, we think that the Shuttle will continue to compete successfully against the European launcher Ariane. Ariane prices are subsidized by the European governments, and we hope that they will take advantage of our price hike by raising their price.

NASA has played and will continue to play a major role in supporting the Administration's interest in promoting the development of a U.S. private sector expendable launch vehicle industry. Higher Shuttle prices will provide private sector expendable launch vehicle companies the opportunity to compete more effectively for commercial and foreign customers. Furthermore, our higher prices, coupled with more efficient private sector operations, will provide headroom so that private operators can set their prices at levels allowing significant profit margins. With the Shuttle taking the role as price leader in the launch services marketplace, the viability of private launchers will depend on whether they can compete with Ariane which has declared its intention to win against all U.S. competition.

NASA is providing strong support to your initiatives stimulating the commercial use of space. We have, throughout our history, had an active program of making our technology available to U.S. industry. Today there are in excess of sixty companies involved in NASA commercialization activities, making NASA the focal point of this nation's space commercialization activities. Through our program of joint endeavors with U.S. industry and our technical support to space entrepreneurs, commercial upper stage manufacturers and the fledgling ELV industry, we are providing opportunities for the development of space-based businesses which should help to project this country's space leadership into the future. As in the past, one of our contributions to such government-industry partnerships will continue to be reduced rates for access to the Shuttle. This approach has effectively stimulated private sector investment and involvement in space activities, the most striking example to date being the very promising McDonnell Douglas/Johnson & Johnson research on pharmaceutical manufacturing in space with which you are familiar. As new industries mature and enter into full scale space production, they will, of course, be charged the full cost recovery price for their flights.

I hope you will approve our recommended pricing plan for the period FY 1989-91. Your prompt action will permit us to continue doing our part in the important task of maintaining U.S. leadership in space transportation.

Respectfully,

James M. Beggs
Administrator

Enclosure

Document III-9

Document title: Lawrence F. Herbolsheimer, Memorandum for Craig L. Fuller, "OMB's study on U.S. ELV competitiveness," November 13, 1984.

Source: Ronald Reagan Presidential Library, Simi Valley, California.

As part of the 1984–85 debate over the appropriate price for NASA to charge commercial users of the Space Shuttle, the Office of Management and Budget (OMB) assessed the likely competitiveness of expendable launch vehicles (ELVs) in the commercial launch market. Lawrence Herbolsheimer was a White House staff person; Craig Fuller was the staff person in the White House for the Cabinet Council on Commerce and Trade with particular involvement in space commercialization issues.

THE WHITE HOUSE
WASHINGTON

November 13, 1984

MEMORANDUM FOR CRAIG L. FULLER

FROM: LAWRENCE F. HERBOLSHEIMER [signed "LFH"]
SUBJECT: OMB's study on U.S. ELV competitiveness [handwritten underlining]

I thought you would appreciate seeing a summary of OMB's conclusions and some of its reasoning about the future of ELVs in the United States. The following are the highlights:

- The U.S. has already lost its dominant market position in the commercial market and will not be able to reclaim it.
- Ariane has already gained a significant share of the market and other nations have near-term capability to participate in the market. These systems are nationally supported systems in which pricing is not necessarily related to launch costs. U.S. commercial operators will be less likely and able to compete against these subsidized systems.
- The aggregate potential supply of launch services is far in excess of near or long term demand.
- A significant increase in Shuttle prices would not necessarily assure that U.S. ELV operators will be competitive over the long term with other national systems.
- A 30% increase in STS prices over the recent NASA pricing proposal of $87 million ($ 1982) would be required to allow U.S. purely commercial ELVs to compete initially (presuming that Ariane and other suppliers continue to track the Shuttle price).
- Communications payloads will continue to be the dominant source of civil launch demand for the next 5–10 years. However, depending upon U.S. Government demand for launch services and the achievement of the Shuttle projected flight rate, the future availability of the STS for commercial and foreign launches may range as high as half of the Free World market. Such an increase in Shuttle availability would intensify pressures for the Shuttle to compete directly with ELVs for commercial traffic.

Document III-10

Document title: Gilbert D. Rye, National Security Council, Memorandum for Robert C. McFarlane, "Corporate Letters on the Shuttle Pricing Issue," July 1, 1985.

Source: Ronald Reagan Presidential Library, Simi Valley, California.

As the debate of Space Shuttle pricing reached its climax in mid-1985, two of the aerospace industry firms with ELVs that were candidates for commercialization, Martin Marietta and McDonnell Douglas, sent letters to President Reagan's National Security Advisor, Robert McFarlane, expressing doubts regarding the benefits to ELV commercialization of a high Shuttle price. John Poindexter, mentioned in this memorandum from the National Security Council's staff person on space issues, Gil Rye, was McFarlane's deputy. The letters originally attached to this memo (Tabs I, II, III, and IV) are mentioned but not included here.

MEMORANDUM   5115
NATIONAL SECURITY COUNCIL
                                                              July 1, 1985
ACTION
MEMORANDUM FOR ROBERT C. McFARLANE
FROM: GILBERT D. RYE [handwritten "Gil"]
SUBJECT: Corporate Letters on the Shuttle Pricing Issue

You have received two letters from industry on the subject of the Shuttle Pricing Issue (Tabs III and IV). As you know, John Poindexter is currently chairing the SIG (Space) in an attempt to either resolve the issue or provide options for the President's consideration. So far we have had two meetings on this subject and will have at least one more.

In the way of background, the Department of Transportation (which has purported to represent the expendable launch vehicle industries) has advocated a higher Shuttle price for foreign and domestic flights in order to allow commercial expendable launch vehicles (ELVs) to enter the marketplace and become competitive. DOT has used the President's decision for achieving full cost recovery on Shuttle launches (as promulgated in the National Space Strategy) as the basis for arguing for higher Shuttle prices. NASA has argued that an increase in the Shuttle price will benefit foreign competitors (primarily the French-built Ariane booster which has consistently underbid the Shuttle for satellite launches) rather than U.S.-built ELVs. Further, NASA argues that each Shuttle flight over and above those provided government users should be accepted as long as the price charged for these flights covers NASA's marginal cost.

In the process of deliberating on this issue, there has been some obvious lobbying including some suggestion (probably by NASA) to those contractors which manufacture ELVs (Martin Marietta, McDonnell Douglas and General Dynamics) to express their views on this subject. While NASA could possibly be accused of applying unfair pressure upon aerospace contractors that benefit from other NASA work, the fact remains that a formal position by ELV manufacturers that a higher Shuttle price would not benefit the commercialization of ELVs appears to severely undercut the DOT argument. Regardless, the suggested replies to the two letters do not take a position on resolution of this issue.

RECOMMENDATION
That you sign the letters at Tab I and II.
        Approve _____ Disapprove _____

Document III-11

Document title: Robert C. McFarlane, Memorandum for the President, "Shuttle Pricing for Foreign and Commercial Users," July 27, 1985.

Source: Ronald Reagan Presidential Library, Simi Valley, California.

After almost a year of debate, this memorandum from National Security Advisor Robert C. McFarlane, chairman of the Senior Interagency Group (Space), transmitted the group's recommendation on Shuttle pricing to President Reagan. The memorandum was prepared by Air Force Colonel Gilbert Rye, the principal National Security Council person for space issues. The four attachments to this memo (Tabs A, B, C, and D) do not appear here.

[1] MEMORANDUM ["UNCLASSIFIED" stamped over "SECRET"]   SYSTEM II
                        THE WHITE HOUSE
                        WASHINGTON

                                    The President has seen _____ [hand-initialed "RR"]

                        July 27, 1985

ACTION
MEMORANDUM FOR THE PRESIDENT
FROM: ROBERT C. McFARLANE [hand-initialed]
SUBJECT: Shuttle Pricing for Foreign and Commercial Users

Issue
Which approach to recover the full cost of the Space Transportation System (STS) services
to commercial and foreign users after 1988 best serves the overall national interest?

Facts
NASA will charge commercial and foreign Shuttle customers a price of $71M per flight dur-
ing the FY 1986 to 1988 period. The issue requiring your resolution deals with the FY 1989
to 1991 period. Two options represented for your decision: Low Auction Pricing (Option
#1) and High Auction Pricing (Option #2). The Senior Interagency Group for Space met
three times on this issue and formulated the issue paper at Tab B for your consideration.

Discussion
In our opinion, Option #1 best satisfies the overall national interest as reflected in the
space policies which you have previously promulgated. You should recognize, however,
that adoption of Option #1 will diminish the prospects for the commercialization of U.S.
expendable launch vehicles (ELVs) since the price of the Shuttle is probably the most
important factor in fulfilling this objective. During our deliberations, one factor has
remained constant. Any significant increase in the Shuttle price will benefit the French-
built Ariane ELV rather than any prospective U.S. ELV due to the commitment by France
to underbid both the Shuttle and U.S. ELVs. A higher Shuttle price would provide Ariane
with a greater share of the international market for launch services, a greater profit margin, and thereby probably provide them with a means to expand their production capacity to capture an even larger share of the market. While it is true that option #1 probably
does not reflect all conceivable costs associated with Shuttle foreign and commercial
launches, it does represent a reasonable basis for full cost recovery and is consistent with
the accounting practices utilized by Ariane.

[2] If there is a genuine market for U.S. ELVs, we feel confident that the U.S. aerospace
industry will make the necessary investment to construct an ELV that is competitive with
foreign launchers without the need for the U.S. Government to artificially raise the
Shuttle price to the level proposed in Option #2. Furthermore, you recently authorized
the Department of Defense to procure ten ELVs which will be manufactured by the Martin
Marietta Corporation. You also chartered [a] joint DOD/NASA study aimed at determining U.S. launch requirements and technology for the period of 1995 and beyond. This
study will be used as a basis for development of any required future launch capability to
satisfy the requirements of SDI [Strategic Defense Initiative], Space Station, or possibly
others. Two of the three potential U.S. ELV manufacturers have written to me indicating
that the commercialization of ELVs is not feasible at this time (Tab C). Additionally, seven
leaders from the aerospace industry have written to you indicating that any significant

increase in the Shuttle price could seriously endanger your other space commercialization objectives (Tab D). All agencies, except for the Department of Transportation, recommend that you approve Option #1.

There is some urgency in obtaining your decision on this issue. NASA must provide bids next week on two upcoming competitions that involve launches in the FY 1989 period. Also, in the absence of an Administration position on Shuttle pricing, the Congress plans to attach a provision to the FY 1986 NASA Authorization Bill which would dictate a price.

Recommendation

OK   No That you sign the NSDD at Tab A which recommends the
[hand-initialed "RR"]   ___ implementation of the low auction pricing option.

Document III-12

Document title: The White House, National Security Decision Directive Number 181, "Shuttle Pricing for Foreign and Commercial Users," July 30, 1985.

Source: Ronald Reagan Presidential Library, Simi Valley, California.

After more than a year of debate, the position that the Space Shuttle price for commercial users should be set at a level allowing the Shuttle to compete with European and Soviet launchers prevailed, even if that meant that U.S. ELVs would be unlikely to be able to enter the commercial launch market. This decision, embodied in this National Security Decision Directive, was announced by the White House on July 30, 1985.

THE WHITE HOUSE
WASHINGTON

SYSTEM II
90798

*NATIONAL SECURITY DECISION
DIRECTIVE NUMBER 181*

July 30, 1985

SHUTTLE PRICING FOR FOREIGN AND COMMERCIAL USERS

NSDD 144, National Space Strategy, directs the development of a plan for implementing full cost recovery of foreign and commercial Shuttle flights occurring after October 1, 1988.

Beginning in FY 1989, Shuttle flight capacity will be sold at auction to foreign and commercial users. The NASA Administrator will establish auction procedures to ensure maximum return to the government and equitable treatment for all potential launch customers.

The minimum acceptable bid will be $74 million (in 1982 dollars) per Shuttle equivalent. Three Shuttle equivalents per year will be available to the foreign and commercial market until two years before the launch year, at which time NASA may offer any remaining unused capacity. NASA may accept bids for multiple payloads at the auction price, subject to the above quotas. The above quotas will not apply to flights for new and innovative uses of space.

NASA will review annually Shuttle cost experience and the anticipated future effectiveness of this pricing policy in implementing National Space Policy goals under changing market conditions. NASA will submit its annual report, together with any recommendations for changes in the auction floor price or other aspects of this pricing policy, to the Assistant to the President for National Security Affairs and the Director of the Office of Management and Budget. Any policy issues resulting from this annual report may be referred to the SIG (Space).

The price charged to the Department of Defense for Shuttle flights will be negotiated separately from this foreign and commercial pricing policy and will be based on NSDD 164 and appropriate compensation for DOD services rendered in connection with Shuttle flights.

[signed "Ronald Reagan"]

Document III-13

Document title: U.S. Department of Transportation, Office of Commercial Space Transportation, "Federal Impediments to Development of a Private Commercial Launch Industry," report submitted to Congress, July 1985, pp. 1–2.

Source: Office of Commercial Space Transportation, Department of Transportation, Washington, D.C.

In the Commercial Space Launch Act of 1984, Congress directed the new Office of Commercial Space Transportation of the Department of Transportation to prepare a report that identified government impediments to the emergence of a U.S. commercial space transportation industry. This report was the response; only the introduction appears here.

Federal Impediments to Development of a Private Commercial Launch Industry

Submitted in compliance with Section 22(b) of the
Commercial Space Launch Act of 1984

July 1985

[1]   I. Introduction

The Commercial Space Launch Act of 1984 (Public Law 98-575) calls for the Secretary of Transportation to:

"Submit to the Committee on Space and Technology of the House of Representatives and the Committee on Commerce, Science, and Transportation of the Senate a report which identifies Federal statutes, treaties, regulations, and policies which may have an adverse effect on commercial launches and include recommendations on appropriate changes thereto."

This report is submitted in response to that requirement. It represents a summary of views the Department of Transportation (DOT) has formed to date as a result of its experience with the private commercial launch industry. It also reflects views on the issues specified in the above directive of private launch firms and of Government agencies with major regulatory or policy development roles affecting the industry.

In evaluating the effect of existing requirements and policies, it is important to bear in mind that the basic approach Congress took in developing a framework for commercial launch regulation was to retain all existing requirements of Federal law applicable to launches or launch site operations, as requirements for any license the Secretary of Transportation issues under the Act. In the eight months that have elapsed since enactment of the Act, and given the considerable lead time required for development of complex commercial launch marketing strategies as well as the uncertainty of the nature and scope of this developing industry, it is difficult to identify with certainty all of the requirements and policies that may present problems in the future. This report, then, should be viewed more as an initial effort to highlight existing or potential problems rather than a comprehensive inventory of them.

[2] The Administration has initiated review of several of the issues identified in the report, and is in the process of developing actions and recommendations to ameliorate them. Further, the Department is in the process of developing and promulgating regulations governing commercial space launch activities, and in certain cases these will provide the medium for dealing with concerns identified here. The provisions of the Act authorizing the Secretary to render unnecessary Federal requirements inapplicable to launch licensing—or to waive the applicability of certain requirements in individual cases—afford additional means for redressing problems as they are encountered.

Congress was far-sighted in enacting legislation in support of the commercial space launch industry. Commercial expendable launch vehicles (ELVs) have an important contribution to make in maintaining and furthering the nation's leadership in space. They serve as a natural complement to the Shuttle. If future Shuttle availability for commercial and foreign payloads were reduced due to increased needs on the part of the U.S. government or by NASA's inability to achieve projected flight rates, there may be need for additional launch capacity. A viable U.S. commercial ELV industry provides the capability to broaden and deepen our domestic space transportation options, at no direct cost to the taxpayer. Private U.S. launch firms will not only be marketing technology similar to that of their foreign competitors, but can also take advantage of the flexibility uniquely available in the private sector to be fully adaptive to special customer requirements.

The following sections discuss statutes, treaties, regulations, and policies which may have adverse effects on the commercial launch industry. The principal topical areas include treaties, pricing policy, tax and tariff, and the licensing of exports and radio frequencies. For each of these subjects, background, current activity, and recommendations follow. . . .

Document III-14

Document title: President Ronald Reagan, Memorandum for the United States Trade Representative, "Determination Under Section 301 of the Trade Act of 1974," July 17, 1985.

Source: Ronald Reagan Presidential Library, Simi Valley, California.

In July 1984, the company formed to market the Delta launch vehicle in the commercial market, Transpace Carriers, Inc., filed a complaint of unfair trade practices by European governments, through the European Space Agency, with respect to its support of Arianespace, the European marketer of the Ariane launcher. Although not directly addressed in the complaint, the implication was that the U.S. government, by setting a Space Shuttle price low enough to compete with Arianespace, was also unfairly constraining trade. After a year's investigation, this determination by President Reagan dismissed the complaint.

[1]   **THE WHITE HOUSE**
                              WASHINGTON

                              July 17, 1985

MEMORANDUM FOR THE
    UNITED STATES TRADE REPRESENTATIVE

SUBJECT: Determination Under Section 301 of the Trade Act of 1974

Pursuant to Section 301 (a) of the Trade Act of 1974, as amended (19 U.S.C. 2411(a)),
I have determined that the practices of the Member States of the European Space Agency
(ESA) and their instrumentalities with respect to the commercial satellite launching services of Arianespace, S.A. are not unreasonable and a burden or restriction on U.S. commerce. While Arianespace does not operate under purely commercial conditions, this is
in large measure a result of the history of the launch services industry, which is marked by
almost exclusive government involvement. I have determined that these conditions do not
require affirmative U.S. action at this time. But because of my decision to commercialize
expendable launch services in the United States, and our policies with respect to manned
launch services such as the Shuttle (STS), it may become appropriate for the United
States to approach other interested nations to reach an international understanding on
guidelines for commercial satellite launch services at some point in the future.

                          Reasons for Determination

Based on a petition filed by Transpace Carriers, Inc. (TCI), the United States Trade
Representative (USTR) initiated an investigation on July 9, 1984, of the European Space
Agency's policies with respect to Arianespace S.A. Arianespace is a privately owned company, incorporated under the laws of France for the purpose of launching satellites.
Arianespace's shareholders include the French national space agency, and aerospace companies and banks incorporated in the ESA Member States.
[2] The Petitioner alleged that 1) Arianespace uses a two tier pricing policy whereby
Arianespace charges a higher price to ESA Member States than to foreign customers;
2) the French national space agency (CNES) subsidizes launch and range facilities, and
services and personnel provided to Arianespace; 3) the French national space agency subsidizes the administrative and technical personnel it provides to Arianespace; and
4) Arianespace's mission insurance rates are subsidized. In addition to these allegations,
the U.S. also investigated three other areas: government inducements to purchasers of
Arianespace's services; direct and indirect government assistance to Arianespace; and
Arianespace's costs and pricing policies.
Our findings with respect to these allegations are set forth below. Many of the factual
allegations were not supported by evidence on the record. While other allegations were
substantiated, the practices were not sufficiently different from U.S. practice in this field
to be considered unreasonable under Section 301.

Government Inducements:

The investigation uncovered no evidence of offsets or insurance being provided by
ESA or its Member States. Member States of ESA do provide export financing for
Arianespace's customers. However, the terms of the financing are consistent with interna-
tional agreements to which the United States is a party.

Direct Government Assistance:

Administrative Personnel: Arianespace and CNES entered into a Head Office Services Agreement pursuant to which CNES personnel perform certain administrative functions for Arianespace. CNES charges Arianespace a flat percentage of annual turnover for its services. While the fee is arbitrary, we have no reason to question CNES's assertion that the fee, in fact, covers actual wage costs plus fringe benefits. The amounts paid to date seem reasonable.

Range Services: The range facilities at Kourou are operated by CNES. Arianespace pays CNES a fee for the use of the range facilities including personnel services. The fee is arbitrary and it does not cover the full range costs incurred by Arianespace. ESA claims that when the fee is raised Arianespace will pay the full cost of range services attributable to Arianespace's activities. Current U.S. policy offers use of the national ranges and launch support services to commercial ELV's on a direct cost, rather than full cost, reimbursement basis.

[3] Loans and Capital Grants: There is no evidence of direct capital grants or soft loans being given to Arianespace by ESA or the Member States other than CNES, which as a stockholder put up equity capital in Arianespace. Of course, Arianespace stockholders, some of whom, e.g. Aerospatiale, are government-owned, have contributed equity capital to the firm. However, we have no evidence to suggest that such transactions are inconsistent with normal commercial practice.

Hardware: ESA provided a certain amount of hardware to Arianespace at less than its cost of acquisition. ESA claims that the cost was reduced because some of the hardware had been used. ESA estimated the value of this hardware to be $50,000. NASA's agreement with TCI for the transfer of the Delta program also provided for transfer of certain flight hardware at less than the government's cost of acquisition.

Protected Home Market: ESA and its Member States have agreed to give Arianespace a preference over other launch service providers with respect to payloads owned and operated by these government entities. Because of this preference and because almost all European communication satellites are operated by governments, rather than private firms, U.S. ELV's and the Shuttle (STS) have limited opportunities to penetrate the European market. In contrast, much of the U.S. market, which is the major market in the world, is open because communication satellites are owned and operated by private sector firms. However, U.S.G. [U.S. Government] payloads also are carried almost exclusively by U.S. launch service providers. Thus, there is little difference in the respective treatment by ESA and the United States of government payloads. The major difference is in the structure of the market with European communication satellites being operated primarily by government entities.

Indirect Government Assistance: Because Arianespace's major suppliers are also major stockholders and because some of these suppliers are, in turn, owned in whole or part by Member State governments of ESA there is concern that the governments, through their ownership of these supplier companies, can artificially reduce Arianespace's operating costs. However, the investigation uncovered no evidence to suggest that Arianespace is obtaining significant assistance by reason of low-cost inputs from its suppliers.

Costs and Pricing: Under current pricing policies, Arianespace is not recovering its full costs, nor is it likely to do so in the near future. ESA has agreed to long-term, fixed-price contracts for launch services with Arianespace. On the other hand, Arianespace has been quite flexible in its [4] price bids to non-ESA customers, and consistently charges less than the price charged to ESA. But it is not uncommon for firms to discount heavily in order to establish themselves in the market, especially when demand is low. Therefore, it appears that market forces, especially the current excess supply of launch capacity, are primarily responsible for current low launch prices.

Since there are no international standards of reasonableness for launch services, we have compared ESA practices to United States practice, and to reasonable commercial practices. The ESA practices are not sufficiently different from those of the U.S. to be actionable under Section 301. This determination is not an endorsement of ESA practices. Our policies in this area are now undergoing revision, and in the future we may wish to reexamine ESA's practices and their effect on U.S.G. launch services. At that time it may be in our mutual interest to engage in international discussions aimed at establishing appropriate guidelines for the commercial launch industry.

This determination shall be published in the Federal Register.

[signed "Ronald Reagan"]

Document III-15

Document title: Alfred H. Kingon, Assistant to the President, Cabinet Secretary, to James C. Fletcher, Administrator, NASA, Memorandum, "Space Commercialization," September 25, 1986.

Source: Ronald Reagan Presidential Library, Simi Valley, California.

President Ronald Reagan announced on August 15, 1986, that the Space Shuttle would no longer be in the business of launching commercial and foreign satellites except under several restrictive conditions. At the time of this announcement, NASA had contracts to launch forty-four such satellites. NASA argued that, while it would not enter into new launch contracts, it should be allowed to launch as many as possible of those satellites already under contract. The advocates of totally removing the Shuttle from the commercial launch market, to open the way for market entry by U.S. ELVs, argued that even these payloads should not be launched on the Shuttle unless there were compelling reasons to the contrary. Their position prevailed, and the Cabinet-level Economic Policy Council, the framework within which debate had been conducted, recommended to President Reagan that he concur. The President agreed, and this memorandum from White House staff assistant Alfred Kingon communicated his decision to NASA Administrator James Fletcher.

THE WHITE HOUSE
WASHINGTON
September 25, 1986

MEMORANDUM FOR JAMES C. FLETCHER
ADMINISTRATOR
NATIONAL AERONAUTICS,AND SPACE ADMINISTRATION

FROM: ALFRED H. KINGON
ASSISTANT TO THE PRESIDENT
CABINET SECRETARY

SUBJECT: Space Commercialization

Pursuant to the Economic Policy council memorandum of September 11, the President has approved Option 1:

NASA shall no longer provide launch services for commercial and foreign payloads subject to exceptions for payloads that: (1) are Shuttle-unique; or (2) have national security and foreign policy implications.

As was discussed at the meeting with the President, NASA will revise its manifest to include only those payloads that are either Shuttle-unique or have national security and foreign policy implications. The manifest then will be made public, with the expectation that current customers who are not included on the manifest will voluntarily seek launch opportunities elsewhere.

Document III-16

Document title: Commercial Space Launch Act Amendments of 1988, Public Law 100–657, H.R. 4399, November 15, 1988.

Source: NASA Historical Reference Collection, NASA History Office, NASA Headquarters, Washington, D.C.

An ad hoc coalition of companies with an interest in the success of the U.S. commercial space transportation industry was quite effective in convincing Congress and the Department of Transportation that changes in the Commercial Space Launch Act of 1984 were required for such success. Therefore, new legislation was passed in the form of these amendments to the earlier law.

|no page number, but would be "H.R. 4399-1"|

H.R. 4399   PUBLIC LAW 100-657

One Hundredth Congress of the United States of America

AT THE SECOND SESSION

Begun and held at the City of Washington on Monday, the twenty-fifth day of January, one thousand nine hundred and eighty-eight

An Act

To facilitate commercial access to space, and for other purposes.

Be it enacted by the Senate and House of Representatives of the United States of America in Congress assembled,

SECTION 1. SHORT TITLE.
    This Act may be cited as the "Commercial Space Launch Act Amendments of 1988."

SEC. 2. FINDINGS.
    The Congress finds that—
    (1) a United States commercial space launch industry is an essential component of national efforts to assure access to space for Government and commercial users;
    (2) the Federal Government should encourage, facilitate, and promote the use of the United States commercial space launch industry in order to continue United States aerospace preeminence;
    (3) the United States commercial space launch industry must be competitive in the international marketplace;
    (4) Federal Government policies should recognize the responsibility of the United States under international treaty for activities conducted by United States citizens in space; and

(5) the United States must maintain a competitive edge in international commercial space transportation by ensuring continued research in launch vehicle component technology and development.

SEC. 3. DEFINITIONS.

Section 4 of the Commercial Space Launch Act (49 U.S.C. App. 2603) is amended—

(1) in paragraph (10) by striking "and" at the end;

(2) by redesignating paragraph (11) as paragraph (12); and

(3) by inserting immediately after paragraph (10) the following new paragraph:

"(11) 'third party' means any person or entity other than—

"(A) the United States, its agencies, or its contractors or subcontractors involved in launch services;

"(B) the licensee or transferee;

"(C) the licensee's or transferee's contractors, subcontractors, or customers involved in launch services; or

"(D) any such customer's contractors or subcontractors involved in launch services; and."

SEC. 4. PRIVATE ACQUISITION OF GOVERNMENT PROPERTY AND SERVICES.

(a) Section 15(a) of the Commercial Space Launch Act (49 U.S.C. App. 2614(a)) is amended by adding at the end the following: "In taking such actions, the Secretary shall consider the commercial [H.R. 4399-2] availability, on reasonable terms and conditions, of substantially equivalent launch property or launch services from a domestic source."

(b) Section 15(b)(1) of the Commercial Space Launch Act (49 U.S.C. App. 2614(b)(1) is amended by adding at the end the following: "For purposes of this paragraph, the term 'direct costs' means the actual costs that can be unambiguously associated with a commercial launch effort, and would not be borne by the United States Government in the absence of a commercial launch effort."

(c) Section 15 of the Commercial Space Launch Act (49 U.S.C. App. 2614) is amended by adding at the end the following new subsection:

"(d) The head of any Federal agency or department may collect payment for activities involved in the production of a launch vehicle or its payload for launch if such activities were agreed to by the owners or manufacturers of such launch vehicle or payload."

SEC 5. INSURANCE REQUIREMENTS OF LICENSEE.

(a) Section 16 of the Commercial Space Launch Act (49 U.S.C. App. 2615) is amended to read as follows:

"LIABILITY INSURANCE

"SEC. 16. (a)(1)(A) Each license issued or transferred under this Act shall require the licensee or transferee—

"(i) to obtain liability insurance; or

"(ii) to demonstrate financial responsibility,

in an amount sufficient to compensate the maximum probable loss (as determined by the Secretary, after consultation with the Administrator of the National Aeronautics and Space Administration, the Secretary of the Air Force, and the heads of other appropriate agencies) from claims by a third party for death, bodily injury, or loss of or damage to property resulting from activities carried out under the license in connection with any particular launch. In no event shall a licensee or transferee be required to obtain insurance or demonstrate financial responsibility under this subparagraph, with respect to the aggregate of such claims arising out of any particular launch, in an amount which exceeds (I) $500,000,000 or (II) the maximum liability insurance available on the world market at a reasonable cost, if such insurance is less than the amount in subclause (I).

"(B) Each license issued or transferred under this Act shall require the licensee or transferee—

"(i) to obtain liability insurance; or

"(ii) to demonstrate financial responsibility,

in an amount sufficient to compensate the maximum probable loss (as determined by the Secretary, after consultation with the Administrator of the National Aeronautics and Space Administration, the Secretary of the Air Force, and the heads of other appropriate agencies) from claims against any person by the United States for loss of or damage to property of the United States resulting from activities carried out under the license in connection with any particular launch. In no event shall a licensee or transferee be required to obtain insurance or demonstrate financial responsibility under this subparagraph, with respect to the aggregate of such claims arising out of any particular launch, in an amount which exceeds (I) $100,000,000 or (II) the maximum liability [H.R. 4399-3] insurance available on the world market at a reasonable cost, if such insurance is less than the amount in subclause (I).

"(C) Each license issued or transferred under this Act shall require the licensee or transferee to enter into reciprocal waivers of claims with its contractors, subcontractors and customers, and the contractors and subcontractors of such customers, involved in launch services, under which each party to each such waiver agrees to be responsible for any property damage or loss it sustains or for any personal injury to, death of, or property damage or loss sustained by its own employees resulting from activities carried out under the license.

"(D) The Secretary, on behalf of the United States, its agencies involved in launch services, and contractors and subcontractors involved in launch services, shall enter into reciprocal waivers of claims with the licensee or transferee, its contractors, subcontractors, and customers, and the contractors and subcontractors of such customers, involved in launch services, under which each party to each such waiver agrees to be responsible for any property damage or loss it sustains or for an personal injury to, death of, or property damage or loss sustained by its own employees resulting from activities carried out under the license. Any such waiver shall apply only to the extent that claims exceed the amount of insurance or demonstration of financial responsibility required under subparagraph (B). After consultation with the Administrator of the National Aeronautics and Space Administration and the Secretary of the Air Force, the Secretary may also waive, on behalf of the United States and any Federal agency, the right to recover any damages for loss of or damage to property of the United States to the extent insurance is not available by reason of policy exclusions which are determined by the Secretary to be usual for the type of insurance involved.

"(2) Any insurance policy obtained, or demonstration of financial responsibility made, pursuant to a requirement described in paragraph (1) shall protect the United States, its agencies, personnel, contractors, and subcontractors, and all contractors, subcontractors, and customers of the licensee or transferee, and all contractors and subcontractors of such customers, involved in providing the launch services, to the extent of their potential liabilities, at no cost to the United States.

"(3) The Secretary shall determine the maximum probable loss under paragraph (1)(A) and (B) associated with activities under a license, within 90 days after a licensee or transferee has required such a determination and has submitted all information the Secretary requires to make such a determination. The Secretary shall amend such determination as warranted by new information. Within 12 months after the date of enactment of the Commercial Space Launch Act Amendments of 1988, and within each 12-month period thereafter, the Secretary shall submit to the Committee on Commerce, Science, and Transportation of the Senate and the Committee on Science, Space, and Technology of the House of Representatives a report on the current determinations with respect to all issued licenses and the reasons for those determinations.

"(4) Within 6 months after the date of enactment of the Commercial Space Launch Act Amendments of 1988, and within each 12-month period thereafter, the Secretary shall review the amounts specified in paragraph (1) (A) (I) and (B) (I), and shall submit a report to the Congress which, if appropriate, contains a proposed adjustment to such amounts to conform with altered liability expectations [H.R. 4399-4] and availability of insurance on the world market. Such proposed adjustment shall take effect 30 days after the submission of such report.

"(b) (1) To the extent provided in advance in appropriations Acts or to the extent there is enacted additional legislative authority to provide for the payment of claims as submitted in the compensation plan outlined in paragraph (4), the Secretary shall provide for the payment by the United States of successful claims (including reasonable expenses of litigation or settlement) of a third party against the licensee or transferee, or its contractors, subcontractors, or customers, or the contractors or subcontractors of such customers, resulting from activities carried out pursuant to a license issued or transferred under this Act for death, bodily injury, or loss of or damage to property resulting from activities carried out under the license, but only to the extent that the aggregate of such successful claims arising out of any particular launch—

"(A) is in excess of the amount of insurance or demonstration of financial responsibilities required under subsection (a) (1) (A); and

"(B) is not in excess of the level that is $1,500,000,000 (plus any additional sums necessary to reflect inflation occurring after January 1, 1989) above such amount.

The Secretary shall not provide for payment of any part of such claim for which the death, bodily injury, or loss of or damage to property has resulted from willful misconduct by the licensee or transferee. To the extent insurance required pursuant to subsection (a) (1) (A) is not available to cover any such successful third party liability claim by reason of insurance policy exclusions determined by the Secretary to be usual for the type of insurance involved, the Secretary may provide for the payment of such excluded claims without regard to the limitation expressed in subparagraph (A).

"(2) The payment of claims under paragraph (1) shall be subject to—

"(A) notice to the United States of any claim, or suit associated with such claim, against a party described in paragraph (1) for death, bodily injury, or loss of or damage to property;

"(B) participation or assistance in the defense by the United States, at its election, of that claim or suit; and

"(C) approval by the Secretary of that portion of any settlement which is to be paid out of appropriated funds of the United States.

"(3) The Secretary may withhold payment under paragraph (1) if the Secretary certifies that the amount is not just and reasonable, except that the amount of any claim determined by the final judgment of a court of competent jurisdiction shall be deemed by the Secretary to be just and reasonable.

"(4) (A) If as a result of activities carried out under a license issued or transferred under this Act the aggregate of the claims arising out of a particular launch are likely to exceed the amount of insurance or demonstration of financial responsibility required under the license, the Secretary shall (i) make a survey of the causes and extent of damage and (ii) expeditiously submit to the Congress a report setting forth the results of such survey.

"(B) Not later than 90 days after any determination by a court indicating that the liability for the aggregate of claims arising out of a particular launch under such a license may exceed the amount of insurance or demonstration of financial responsibility required [H.R. 4399-5] under the license, the President, on the recommendation of the Secretary, shall submit to the Congress a compensation plan or plans that (i) outlines the aggregate dollar value of such claims; (ii) recommends sources of funding to pay for these claims; and (iii) includes any legislative language required to implement the compensation plan or plans if additional legislative authority is required. No compensation plan for a single event or incident may exceed the aggregate of $1,500,000,000.

"(C) Any compensation plan transmitted to the Congress pursuant to subparagraph (B) shall bear an identification number and shall be transmitted to both Houses of Congress on the same day and to each House while it is in session.

"(D) (i) The provisions of this subparagraph shall apply with respect to consideration in the Senate of any such compensation plan and to Senate action on such compensation plan.

"(ii) Any such compensation plan that requires additional appropriations or additional legislative authority must be considered by the Senate pursuant to this subparagraph within 60 calendar days of continuous session of Congress after the date on which such plan is transmitted to the Congress.

"(iii) For the purposes of this subparagraph, the term 'resolution' means only a joint resolution of Congress the matter after the resolving clause of which is as follows: 'That the [blank space] approves the compensation plan numbered [blank space] submitted to the Congress on [blank space], 19 [blank space].', the first blank space therein being filled with the name of the resolving House and the other blank spaces being appropriately filled; but does not include a resolution which includes more than one compensation plan.

"(iv) A resolution once introduced with respect to a compensation plan shall immediately be referred to a committee (and all resolutions with respect to the same compensation plan shall be referred to the same committee) by the President of the Senate.

"(v) (I) If the committee of the Senate to which a resolution with respect to a compensation plan has been referred has not reported it at the end of 20 calendar days after its referral, it shall be in order to move either to discharge the committee from further consideration of such resolution or to discharge the committee from further consideration with respect to such compensation plan which has been referred to the committee.

"(II) A motion to discharge may be made only by an individual favoring the resolution, shall be highly privileged (except that it may not be made after the committee has reported a resolution with respect to the same compensation plan), and debate thereon shall be limited to not more than one hour, to be divided equally between those favoring and those opposing the resolution. An amendment to the motion shall not be in order, and it shall not be in order to move to reconsider the vote by which the motion was agreed to or disagreed to.

"(III) If the motion to discharge is agreed to or disagreed to, the motion may not be renewed, nor may another motion to discharge the committee be made with respect to any other resolution with respect to the same compensation plan.

"(vi) (I) When the committee has reported, or has been discharged from further consideration of, a resolution, it shall be at any time thereafter in order (even though a previous motion to the same effect has been disagreed to) to move to proceed to the consideration of the resolution. The motion shall be highly privileged and shall not [H.R. 4399-6] be debatable. An amendment to the motion shall not be in order and it shall not be in order to move to reconsider the vote by which the motion was agreed to or disagreed to.

"(II) Debate on the resolution referred to in subclause (I) of this clause shall be limited to not more than 10 hours, which shall be divided equally between those favoring and those opposing such resolution. A motion further to limit debate shall not be debatable. An amendment to, or motion to recommit, the resolution shall not be in order, and it shall not be in order to move to reconsider the vote by which such resolution was agreed to or disagreed to.

"(vii) (I) Motions to postpone, made with respect to the discharge from committee, or the consideration of a resolution or motions to proceed to the consideration of other business, shall be decided without debate.

"(II) Appeals from the decision of the Chair relating to the application of the rules of the Senate to the procedures relating to resolution shall be decided without debate.

"(5) The provisions of paragraphs (1) through (4) shall apply only to each license issued or transferred under this Act for which a complete and valid application has been received by the Secretary prior to the date that is 5 years following the date of enactment of the Commercial Space Launch Act Amendments of 1988.

"(c) The head of any Federal agency or department shall collect insurance proceeds or any other payment owed for the loss of or damage to Government property under its jurisdiction or control resulting from activities carried out under a license issued or transferred under this Act. Such proceeds or other payment shall be credited to the current applicable appropriations, funds, or accounts of that agency or department."

(b) Section 15(c) of the Commercial Space Launch Act (49 U.S.C. App. 2614(c)) is amended to read as follows:

"(c) Consistent with the requirements of this Act, the Secretary shall establish requirements for proof of financial responsibility and such other assurances as may be necessary to protect the United States and its agencies and personnel from liability, death, bodily injury, or loss of or damage to property as a result of a launch or operation of a launch site involving Government facilities or personnel. The Secretary may not under this subsection relieve the United States of liability for death, bodily injury, or loss of or damage to property resulting from the willful misconduct of the United States or its agents."

SEC. 6. UNITED STATES LAUNCH INCENTIVES FOR CERTAIN SATELLITES.

(a) The requirements of subsection (a)(1)(B) of section 16 of the Commercial Space Launch Act (49 U.S.C. App. 2615), as amended by this Act, shall not apply to eligible satellites.

(b) To the extent approved in appropriations Acts, the United States shall not require payment for the provision of launch services in connection with the commercial launch of an eligible satellite.

(c) For purposes of this section, the term "eligible satellite" means a satellite that—

(1) was under construction on August 15, 1986;

(2) was the subject of a launch services agreement or contract with the National Aeronautics and Space Administration, which as of August 15, 1986, was in effect and not yet carried out; and

[H.R. 4399-7] (3) is licensed for launch under the Commercial Space Launch Act.

SEC. 7. PREEMPTION OF SCHEDULED LAUNCHES.

Section 15(b) of the Commercial Space Launch Act (49 U.S.C. App. 2614(b)) is amended by adding at the end the following new paragraph:

"(4)(A) The Secretary, with the cooperation of the Secretary of Defense and the Administrator of the National Aeronautics and Space Administration, shall take steps to ensure that the launches of payloads with respect to which a launch date commitment from the United States has been obtained for a launch licensed under this Act are not preempted from access to United States launch sites or launch property, except in cages of imperative national need. Any determination of imperative national need shall be made by the Secretary of Defense or the Administrator of the National Aeronautics and Space Administration, in consultation with the Secretary, and shall not be delegated. A licensee or transferee preempted from access to a launch site or launch property shall not be required to pay to the United States any amount for launch services solely attributable to the scheduled launch prevented by such preemption.

"(B) The Secretary of Defense or the Administrator of the National Aeronautics and Space Administration, in cooperation with the Secretary, as the case may be, shall report to the Congress within 7 days after any determination of imperative national need under subparagraph (A), including an explanation of the circumstances justifying such determination and a schedule for ensuring the prompt launching of a preempted payload."

SEC. 8. STUDY OF PROCESS FOR SCHEDULING LAUNCHES.

The Secretary of Transportation, in cooperation with the Secretary of Defense and the Administrator of the National Aeronautics and Space Administration, and in consultation with representatives of the space launch and satellite industry, shall study ways and means of scheduling Government and commercial payloads on commercial launch vehicles at Government launch sites in a manner which—

(1) makes the best practicable use of the launch property of the United States; and

(2) assures that the launch property of the United States that is available for commercial use will be available on a commercially reasonable basis,

consistent with the objectives of the Commercial Space Launch Act. The Secretary shall report the results of such study to the Congress within 90 days after the date of enactment of this Act.

SEC. 9. COMMERCIAL SPACE LAUNCH SERVICE COMPETITION.

It is the sense of the Congress that the United States should explore ways and means of developing a dialogue with appropriate foreign government representatives to seek the development of guidelines for access to launch services by satellite builders and users in a manner that assures the conduct of reasonable and fair international competition in commercial space activities.

[H.R. 4399-8] SEC. 10. LAUNCH VEHICLE RESEARCH AND DEVELOPMENT.

The Administrator of the National Aeronautics and Space Administration shall, in consultation with representatives of the space launch and satellite industry, design a program for the support of research into launch systems component technologies, for the purpose of developing higher performance and lower cost United States launch vehicle technologies and systems available for the launch of commercial and Government spacecraft into orbit. The Administrator shall submit a report outlining such program to the Congress within 60 days after the date of enactment of this Act.

SEC. 11. APPLICABILITY TO LICENSES.

This Act, and the amendments made by this Act shall apply to all licenses issued under the Commercial Space Launch Act before, on, or after the date of enactment of this Act.

[signature]
Speaker of the House of Representatives

[signature]
President of the Senate Pro Tempore

APPROVED
NOV 15 1988
[signature of Ronald Reagan]

Document III-17

Document title: Shellyn G. McCaffrey, The White House, Through Eugene G. McAllister, Memorandum for Nancy J. Risque, "Space Launch Insurance," July 1, 1987.

Source: Ronald Reagan Presidential Library, Simi Valley, California.

Even though the Space Shuttle had been barred from launching commercial payloads in August 1986, there remained a number of barriers to the entry of privately owned and operated U.S. ELVs into the commercial launch market. One large obstacle was deciding who had the responsibility for providing insurance against third-party damages resulting from a commercial launch. This memorandum records an initial policy discussion on this issue. Nancy Risque was a senior White House staffer dealing with economic policy issues, and Eugene McAllister and Shellyn McCaffrey were staff support for the White House Economic Policy Council (EPC). The issue was eventually resolved when Congress, in the Commercial Space Launch Amendments of 1988, agreed that the U.S. government would bear the liability for third-party damages above $500 million.

[1]   **THE WHITE HOUSE**
                        WASHINGTON

                                                                July 1, 1987

MEMORANDUM FOR NANCY J. RISQUE
THROUGH: EUGENE J. McALLISTER [initialed "EM"]
FROM: SHELLYN G. McCAFFREY [initialed "SM"]
SUBJECT: Space Launch Insurance [handwritten underlining]

ISSUE: The EPC Working Group on Space Commercialization [handwritten underlining] met today to discuss for the first time potential options for addressing the issue of space launch provider insurance and third-party indemnity. DOT made the presentation. This memo is to brief you on the discussion.

BACKGROUND: Last year, the EPC determined, with the President's concurrence, that the U.S. should not subsidize insurance costs for a private U.S. space launch industry. NSDD 9 appears to reiterate that policy. The issue nonetheless came to the forefront of the development of the private industry when the Air Force, DOT, and launch vehicle manufacturers could not reach agreement on liability/indemnity provisions with a model launch facility agreement. The issue took on commercial significance when Intelsat recently broke off informal negotiations for a satellite launch with Martin-Marietta reportedly because of the USG's failure to indemnify potential third-party liability beyond Martin-Marietta's insurance coverage. The issue for the Administration subsequently has been whether and to what extent USG third-party indemnity represents an actual road-block to a viable U.S. launch industry.

DISCUSSION: The sense of the Working Group today was that there are several questions outstanding as a prelude to any Council consideration of whether to reverse its previous policy against USG indemnity. These include:

1. How much are U.S. launch providers paying for the insurance coverage they now carry? Can they purchase additional increments of coverage in the market? If not, why not?

2. Does the French Government have an explicit, or implicit, agreement indemnifying Ariane Space (the Government owns at least 34 percent of Ariane), the major U.S. competitor?

[2] 3. Is Intelsat bluffing, considering that there is only the one (USSR) third-party liability case on record, and considering Intelsat's very limited options for getting its satellite into space in the near future?

The current consensus of the group seems to be that it is preferable that the U.S. not indemnify, but seek instead to level the international playing field through consultations/negotiations with our competitors. USTR will begin consultations later this month with the French and Europeans on a range of space subsidy issues, including indemnity. The near-term issue remaining, raised by DOT, is whether the EPC should consider "interim" indemnification until bilateral agreements on Government indemnification are achieved. Because several Working Group members expressed concern about such a two-headed strategy, DOT said it would consider further the issue and related questions and return with a report to the group.

Document III-18

Document title: The White House, National Space Policy Directive 2, "Commercial Space Launch Policy," September 5, 1990.

Source: Vice President Dan Quayle, "Final Report to the President on the U.S. Space Program," January 1993, NASA Historical Reference Collection, NASA History Office, NASA Headquarters, Washington, D.C.

The new administration of President George Bush included a re-created National Space Council, chaired by Vice President Dan Quayle and supported by a small staff in the Executive Office of the President. This was the first time that there had been a separate Executive Office body addressing space issues since 1973, when President Richard Nixon had abolished the National Aeronautics and Space Council that had been created by the 1958 Space Act. The National Space Council and its staff provided the mechanism for a series of space policy directives during the Bush administration. Commercial launch issues were an important agenda item for the National Space Council; this directive was the result of its initial deliberations.

[no pagination]

National Space Policy Directive 2
September 5, 1990

Commercial Space Launch Policy

Policy Findings

A commercial space launch industry can provide many benefits to the U.S., including indirect benefits to U.S. national security.

The long-term goal of the United States is a free and fair market in which U.S. industry can compete. To achieve this, a set of coordinated actions is needed for dealing with international competition in launch goods and services in a manner that is consistent with our nonproliferation and technology transfer objectives. These actions must address both the short term (actions which will affect competitiveness over approximately the next ten years) and those which will have their principal effect in the longer term (i.e., after approximately the year 2000).

— In the near term, this includes trade agreements and enforcement of those agreements to limit unfair competition. It also includes the continued use of U.S.-manufactured launch vehicles for launching U.S. Government satellites.

— For the longer term, the United States should take actions to encourage technical improvements to reduce the cost and increase the reliability of U.S. space launch vehicles.

Implementing Actions

U.S. Government satellites will be launched on U.S.-manufactured launch vehicles unless specifically exempted by the President.

Consistent with guidelines to be developed by the National Space Council, U.S. Government agencies will actively consider commercial space launch needs and factor them into their decisions on improvements in launch infrastructure and launch vehicles aimed at reducing cost, and increasing responsiveness and reliability, of space launch vehicles.

The U.S. Government will enter into negotiations to achieve agreement with the European Space Agency (ESA), ESA member states, and others as appropriate, which defines principles of free and fair trade.

Nonmarket launch providers of space launch goods and services create a special case because of the absence of market-oriented pricing and cost structures. To deal with their entry into the market, there needs to be a transition period during which special conditions may be required.

There also must be an effective means of enforcing international agreements related to space launch goods and services.

Document III-19

Document title: The White House, National Space Policy Directive 4, "National Space Launch Strategy," July 10, 1991.

Source: Vice President Dan Quayle, "Final Report to the President on the U.S. Space Program," January 1993, pp. III-25–III-28, NASA Historical Reference Collection, NASA History Office, NASA Headquarters, Washington, D.C.

Continuing policy issues related to commercial space launches were one of the major factors leading to the issuance of this statement of an overall strategy for space launch by the Bush administration.

[no page number, but would be "III-25"]

National Space Policy Directive 4
July 10, 1991

National Space Launch Strategy

I.   Introduction

a.   National space policy provides a framework within which agencies plan and conduct U.S. Government space activities. The National Space Launch Strategy provides guidance for implementation of that policy with respect to access to and from space.
b.   Assured access to space is a key element of U.S. national space policy and a foundation upon which U.S. civil, national security, and commercial space activities depend.
c.   United States space launch infrastructure, including launch vehicles and supporting facilities, should: (1) provide safe and reliable access to, transportation in, and return from space; (2) reduce the costs of space transportation and related services, thus encouraging expanded space activities; (3) exploit the unique attributes of manned and unmanned launch and recovery systems; and (4) encourage, to the maximum extent feasible, the development and growth of U.S. private sector space transportation capabilities which can compete internationally.

II.   Space Launch Strategy

a.   The National Space Launch Strategy is composed of four elements:

   (1) Ensuring that existing space launch capabilities, including support facilities, are sufficient to meet U.S. Government manned and unmanned space launch needs.
   (2) Developing a new unmanned, but man-rateable [sic], space launch system to greatly improve national launch capability with reductions in operating costs and improvements in launch system reliability, responsiveness, and mission performance.
   (3) Sustaining a vigorous space launch technology program to provide cost-effective improvements to current launch systems, and to support development of advanced launch capabilities, complementary to the new launch system.
   (4) Actively considering commercial space launch needs and factoring them into decisions on improvements in launch facilities and launch vehicles.

b.   These strategy elements will be implemented within the overall resource and policy guidance provided by the President.

III. Strategy Guidelines

a.   Existing Space Launch Capability

   (1) A mixed fleet comprised of the Space Shuttle and existing expendable launch vehicles will be the primary U.S. Government means to transport people and cargo to and from space through the current decade and will be important components of the Nation's launch capability well into the first decade of the 21st century.
   (2) To meet U.S. Government needs, agencies will conduct programs to systematically maintain and improve the Space Shuttle, current U.S. expendable launch vehicle fleets, and supporting launch site facilities and range capabilities. Such programs shall be cost-effective relative to current and programmed mission needs and to investments in new launch capabilities.

(3) As the Nation is moving toward development of a new space launch system, the production of additional Space Shuttle orbiters is not planned. The production of spare parts should continue in the near term to support the existing Shuttle fleet, and to preserve an option to acquire a replacement orbiter in the event of an orbiter loss or other demonstrable need. By continuing to operate the Shuttle conservatively, by taking steps to increase the reliability and lifetime of existing orbiters, and by developing a new launch system, the operational life of the existing orbiter fleet will be extended. The Space Shuttle will be used only for those important missions that require manned presence or other unique Shuttle capabilities, or for which use of the Shuttle is determined to be important for national security, foreign policy, or other compelling purposes.

(4) Consistent with U.S. national security and national space policy, the U.S. Government may seek to recover residual value from ballistic missiles which are, or subsequently become, surplus to the needs of the Department of Defense. Prior to any release of such missiles, including components, beyond those already approved for use as space launch vehicles, the Department of Defense will conduct, and the National Space Council and the National Security Council will review, an assessment of alternative disposition options for such missiles. Disposition options will be evaluated in terms of their consistency with U.S. national security and foreign policy interests, available agency resources, defense industrial base considerations, and with due regard to economic impact on the commercial space sector, promoting competition, and the long-term public interest.

[III-26] b. New Space Launch System

(1) The Department of Defense and the National Aeronautics and Space Administration will undertake the joint development of a new space launch system to meet civil and national Security needs. The goal of this launch program is to greatly improve national launch capability with reductions in operating costs and improvements in launch system reliability, responsiveness, and mission performance.

(2) The new launch system, including manufacturing processes and production and launch facilities, will be designed to support a range of medium- to heavy-lift performance requirements and to facilitate evolutionary change as requirements evolve. The design may take advantage of existing components from both the Space Shuttle and existing expendable rockets in order to expedite initial capability and reduce development costs. While initially unmanned, the new launch system will be designed to be man-rateable [sic] in the future.

(3) The new launch system will be managed, funded, and developed jointly by the Department of Defense and the National Aeronautics and Space Administration. The development program will be structured in the near term toward the goal of a first flight in 1999. However, the program should allow for several schedule options for the first flight and should identify key intermediate milestones. Since the new launch system will provide the opportunity for significant long-term benefits to the commercial space launch industry, the agencies should actively explore the potential for U.S. private sector participation. Final decisions on the program schedule, including the date of the first flight, will be made during fiscal year 1993, based on updated requirements and technical and budgetary considerations at that time. A joint program plan will be prepared by the Department of Defense and the National Aeronautics and Space Administration and reviewed by the National Space Council.

(4) The Department of Defense and the National Aeronautics and Space Administration will plan for the transition of selected space programs from current

launch systems to the new launch system at appropriate program milestones to insure [sic] mission continuity and to minimize satellite and other transition costs.

c.   Space Launch Technology

(1) In addition to conducting the focused development program for a new launch system, appropriate U.S. Government agencies will continue to conduct broadly based research and focused technology programs to support long-term improvements in national space launch capabilities. This technology effort shall address launch system components (e.g., engines, materials, structures, avionics); upper stages; improved launch processing concepts; advanced [III-27] launch system concepts (e.g., single-stage-to-orbit concepts, including the National AeroSpace Plane); and experimental flight vehicle programs.

(2) The Department of Defense, the Department of Energy, and the National Aeronautics and Space Administration will coordinate space launch technology efforts and, by December 1, 1991, jointly prepare a 10-year space launch technology plan.

d.   Commercial Space Launch Considerations

(1) In addition to addressing Government needs, improvement of space launch capabilities can facilitate the ability of the U.S. commercial space launch industry to compete. Consistent with U.S. space policy, U.S. Government agencies will actively consider commercial space launch needs and factor them into decisions on existing space launch capabilities, development of a new space launch system, and implementation of space launch technology programs in the following ways:

(a) U.S. Government-funded investments will be consistent with approved budgets and U.S. Government requirements.

(b) U.S. Government agencies, in acquiring space launch-related capabilities, should:

[1] Allow contractors, to the fullest extent feasible, the flexibility to accommodate commercial needs when developing launch vehicles and infrastructure to meet Government needs.

[2] Emphasize procurement strategies which are based on: "best value" rather than lowest cost, performance-based functional requirements, commercial production and quality-assurance standards and techniques, and the use of commercially offered space products and services.

[3] Encourage commercial and State and local government investment and participation in the development and improvement of U.S. launch systems and facilities.

[4] Provide for private sector retention of technical data rights, except those rights necessary to meet Government needs or to comply with statutory responsibilities.

(c) U.S. Government agencies should seek to remove, where appropriate, legal or administrative impediments to private sector arrangements such as industry teams, consortia, cost-sharing, and joint production agreements which may benefit U.S. [III-28] Government needs and economic competitiveness. Agencies should also seek legislative authority for stable long-term commitments to purchase space transportation services.

(d) Within applicable law, U.S. Government agencies are encouraged to use industry advisory groups to facilitate the identification of commercial space launch needs and the elimination of barriers that unnecessarily impede commercial space launch activities. U.S. agencies are also encouraged to consult with State and local governments.

(2) U.S. Government agencies should develop explicit provisions to implement these guidelines for actively considering commercial space launch needs. As appropriate, agencies should solicit public views on these provisions.

IV. Reporting Requirements

U.S. Government agencies affected by these strategy guidelines are directed to report by December 1, 1991, to the National Space Council on their activities related to the implementation of these policies.

Document III-20

Document title: Elizabeth Dole, Secretary of Transportation, Letter to the President, September 30, 1987.

Source: Ronald Reagan Presidential Library, Simi Valley, California.

By November 1983, Secretary of Transportation Elizabeth Dole had secured for her agency jurisdiction over the commercial space launch industry. Over the following four years, she and her staff were strong advocates for that industry in discussions within the executive branch and with Congress; the office also began to develop an effective policy and regulatory framework for the industry's development.

[1]

THE SECRETARY OF TRANSPORTATION
WASHINGTON, D.C. 20590
September 30, 1987

Dear Mr. President:

As I prepare to leave the Cabinet, I want to give you a status report on the success of your commercial space launch policy. This particularly bold initiative will help keep America on the cutting edge in space by directing the energy and creativity of this nation's private sector to reducing the cost of transportation, enabling us to tap the full economic potential of the space environment.

Because the space transportation industry presented an excellent case for privatization and because the consequences for America's space program were so compelling, we argued at every level of government that the federal monopoly in space be ended. I am pleased to inform you that since the U.S. government is no longer competing for commercial satellite launches, America's commercial launch vehicle industry has made significant inroads against Ariane and other foreign competitors. Firm contracts are in place to launch eight payloads in 1989–90. In addition, these companies have reservations for an additional seventeen launches through 1991, a number that grows daily.

The economic benefits are significant. American companies have already invested at least $400 million in private capital to support commercial launch activities. These firms report that their combined efforts will add at least 8,000 new jobs and nearly a billion dollars a year to America's economy. In fact, every time a foreign customer launches on an American rocket, it offsets our balance of payments by $40 million to $100 million each.

As you directed when you signed Executive Order 12465 in February 1984, DOT has taken the lead within the federal government to develop policies and procedures that would ensure safe and responsible conduct of private launch activities, but not impede the growth or vitality of this critical industry. This Department's efforts have had a common goal: to see that America's transportation industries are the world's safest and most efficient. Our commercial space transportation program is fully consistent with that goal.

[2] We've made significant progress, but the continued commitment of other government agencies in this effort is critical to the commercial viability of this industry. A key factor in the emergence of a private sector launch capability was your decision that NASA not maintain its own ELV adjunct to the Shuttle as well as NASA's proposal to purchase commercial launch services to meet, where appropriate, critical government missions that cannot be scheduled in a timely manner on the Shuttle. The Department of Defense's requirement in a recent ELV procurement that manufacturers demonstrate "commercial adaptability" is the type of action that strengthens the production bases of companies offering commercial launch services in the international marketplace.

The Administration must step up its efforts to reform procurement policies and practices to allow federal agencies to minimize the administrative burden currently placed on commercial firms that want to provide launch services to government agencies on a commercial basis. Finally, DOT will continue to work with the Air Force and NASA to improve the terms and conditions that govern commercial launch operations at national ranges.

These efforts, together with an efficient regulatory program, will serve to advance the critical national interests associated with a competitive U.S. commercial launch vehicle industry. With a strong ELV industrial base, our nation should never again be left with a severe shortage of launch capacity. Your vision in privatizing an industry, whose reliability is unmatched by any other launch system in the world, has far-reaching ramifications for America and will result in lasting benefits for all Americans. I am most grateful for the opportunity that you gave me to play a role in the development of this new and dynamic industry.

Respectfully,

Elizabeth Dole

The President
The White House
Washington, D.C. 20500

Document III-21

Document title: Richard E. Brackeen, Chairman, COMSTAC, and President, Martin Marietta Commercial Titan, Inc., to James H. Burnley, Secretary of Transportation, January 29, 1988.

Source: Office of Commercial Space Transportation, Department of Transportation, Washington, D.C.

Richard Brackeen was the individual within the Martin Marietta Company who had pushed the company to attempt to market a variant of its Titan launch vehicle as a commercial launcher. He also served a term as chair of COMSTAC, the Commercial Space Transportation Advisory Committee. In that capacity, he provided an early 1988 status report on the U.S. space launch industry to Secretary of Transportation James Burnley.

[1] MARTIN MARIETTA COMMERCIAL TITAN, INC.
RICHARD E. BRACKEEN
PRESIDENT

P.O. BOX 179
DENVER, COLORADO 80201
TELEPHONE (303) 971-2034

January 29, 1988

The Honorable James H. Burnley
Secretary of Transportation
U.S. Department of Transportation
400 Seventh Street, S.W.
Washington, D.C. 20590

Dear Mr. Secretary:

The Commercial Space Transportation Advisory Committee (COMSTAC) held its sixth meeting on November 12–13, 1987, in Washington, D.C. Members reported considerable progress in achieving the Administration's goal of commercializing the provision of expendable launch services. With the strong support of your Department's Office of Commercial Space Transportation and of the U.S. Air Force, range use agreements have been signed by Martin Marietta Corporation and General Dynamics Corporation. In addition, a NASA range use agreement has been signed by Space Services, Inc. Major investments [totaling] approximately $400 million have been made in this emerging business by the commercial ELV companies, and have resulted in the creation of some 8,000 new jobs. Most importantly, contracts to launch 12 satellites have been awarded to 3 U.S. ELV companies, contributing over $550 million to the U.S. balance of trade.

At our meeting, representatives of key U.S. Government agencies reviewed with COMSTAC members the status of various U.S. Government policy reviews and other governmental activities affecting the U.S. ELV industry. Of particular interest to COMSTAC was the ongoing National Security Council policy review. COMSTAC reiterated strongly its support of Administration policy prohibiting NASA from maintaining an ELV adjunct to the Space Shuttle, and prohibiting the launch by NASA of commercial and foreign payloads on the Shuttle unless these spacecraft must be man-tended or are important for national security or foreign policy purposes. COMSTAC members have asked me to request that, in this review, you strongly support the continuation of these essential policy elements of the Nation's space recovery program.

Despite progress in many areas since its last meeting, COMSTAC found that certain important issues remain to be addressed, and unanimously adopted the following recommendations:
[2]
• The Secretary should urge the Secretary of State to raise the issue of launches of Western satellites on Proton launch vehicles with members of the Coordinating Committee for Multilateral Export Control (COCOM), and to seek independent statements from members of their intention to abide by COCOM principles with respect to the transfer of critical space technologies.

COMSTAC members strongly support the U.S. Department of State's recent reaffirmation of longstanding U.S. policy prohibiting the transfer of sensitive space technologies to the Soviet Union. Both U.S. ELV and satellite manufacturing companies are concerned, however, that foreign satellite manufacturers may purchase Soviet launches at predatory prices, and thus win certain foreign procurements. Statements of support for existing COCOM principles by COCOM members would reaffirm the existing Western consensus in this arena.

- DOT should promptly exercise its full statutory authority to establish allocation of risk principles and insurance requirements covering liability for damage to third parties, as well as to Government property.

DOT has a draft rulemaking pending in this area. The first launch that will be licensed under the Act is currently scheduled in early 1988. Prompt exercise of DOT's authority is required for the smooth implementation of the Nation's commercial ELV policy.

- DOT should push for a national decision on the nature and magnitude of risk that the U.S. Government should bear in order to ensure a competitive U.S. ELV industry, and should support appropriate legislation.

As a signatory to the 1967 Treaty on the Peaceful Uses of Outer Space and the 1971 Liability Convention, the U.S. Government assumed absolute liability for any damage caused to third persons or their property by any object launched from U.S. territory. In turn NASA was authorized to share this risk with industry by paying third-party claims against its commercial customers to the extent that these claims exceeded the liability insurance NASA required them to carry.

DOT, however, has no such authority. In addition, under the USAF range use agreements, U.S. ELV companies are being required to obtain the "maximum available" insurance at a "reasonable price," currently estimated to be $500 million; above this level, the U.S. [3] Government and the ELV companies are responsible under "applicable law." As a result, these companies are facing risks that, while very remote, are potentially larger than the available third-party insurance capacity. In addition, since their major competitor, Arianespace, is indemnified by the French Government against all third-party claims above 400 million French francs (approximately $70 million), they suffer a competitive disadvantage in the worldwide commercial market.

- DOT should expeditiously explore the mechanisms, if any, required to ensure that adequate insurance capacity is available to cover the third-party liability and property damage risks faced by the commercial ELV and satellite manufacturing industries.

With a growing number of commercial ELV launches anticipated over the next few years, the commercial ELV and satellite manufacturers are concerned that adequate insurance capacity be available to cover their third-party liability and property damage risks. DOT's prompt exploration of appropriate insurance capacity mechanisms should increase the likelihood that all required coverages will be available.

- The Secretary should strongly support the efforts of the Office of Federal Procurement Policy (OFPP) and other appropriate U.S. Government agencies to identify and implement the most appropriate approach to be followed by all U.S. Government space programs in complying with Administration policy of procuring commercial launch services rather than hardware.

Under Administration policy, both DOD and NASA are now procuring commercial launch services, as opposed to launch vehicle hardware. Procurement officials are experiencing difficulties in determining which of the Federal Acquisition Regulations (FAR) apply to these procurements. In addition, in the case of U.S. Government turnkey procurements, it is not clear to U.S. satellite manufacturers which of the FAR requirements must be passed on to the ELV subcontractors. For example, ELV companies recently received RFPs [Requests for Proposals] from two satellite manufacturers bidding on the Navy UHF [ultrahigh frequency] procurement that contained a total of 88 FAR requirements, only 12 of which were the same. Responding to such differing RFPs for one launch services contract places an unnecessary financial burden on the ELV companies.

An effort to resolve these problems is underway within the Administration led by OFPP in consultation with the relevant [4] agencies. DOT's strong support of this effort would help to achieve the prompt resolution of this problem.

- The Secretary should urge the U.S. Trade Representative and other appropriate U.S. Government agencies to assess the long-term impact of growing demands for U.S. commercial ELV and satellite companies to provide mandatory trade offsets at a condition of bidding on foreign commercial space launch programs.

A number of foreign commercial space system procurements are underway at this time. In an increasing number of cases, the customer, whether it be a foreign government or a foreign corporation, requires U.S. companies, as a condition of bidding, to provide trade offsets. COMSTAC is aware that trade offsets have long been a practice in military and aviation procurements. However, they have not until now become the practice in commercial launch procurements. COMSTAC believes that one of the advantages inherent in the Administration's commercial launch services policy is the contribution this emerging industry will make to the U.S. balance of trade. Mandatory trade offsets lessen this positive contribution and could, in the long run, even result in a net loss of trade. In addition, their cumulative effect could so lessen the aerospace companies' profit that they decide to withdraw from these lines of business.

- The Secretary, working with NASA and other appropriate U.S. Government agencies, should seek to duplicate, in the ELV arena, NASA's highly successful R&D role in the development of key aeronautical component technologies.

U.S. ELV and satellite companies have expressed considerable concern about the ability of the U.S. to compete over the long run in the international commercial launch services market since it is not large enough to support the R&D budgets required to develop state-of-the-art vehicles optimized for commercial payloads. Advanced vehicles developed for military purposes are not optimized for commercial uses. They note that the European Space Agency recently approved a $3.6 billion budget for the development of the heavy lift Ariane V that is designed significantly to reduce commercial costs to orbit.

COMSTAC has concluded that it is not necessary for the U.S. Government to support the development of new commercial expendable [5] launch vehicles. Rather it proposes that the Administration duplicate the very successful aeronautics R&D model used by NASA and its predecessor NACA. Under this model, NASA, at a fraction of the development cost of a total vehicle, would develop key advanced component technologies. The commercial ELV industry would then incorporate these technologies into advanced, lower cost, commercial vehicles at their expense. DOT's lead in obtaining Administration support for such an approach would provide immeasurable assistance in maintaining U.S. leadership in this critical area of space commercialization.

Mr. Secretary, the record of your Office of Commercial Space Transportation, with the full support of top Departmental officials, in implementing your responsibilities under the Commercial Space Launch Act of 1984 has been outstanding. We welcome your appointment, and look forward to working with you to ensure the continuing success of the new expendable launch vehicle industry.

Sincerely yours,

Richard E. Brackeen
Chairman, COMSTAC

Document III-22

Document title: Office of Commercial Space Transportation, U.S. Department of Transportation, "Office of Commercial Space Transportation; Licensing Regulations," Final Rule (Preamble), *Federal Register* 53 (No. 64 / Friday), April 4, 1988, pp. 11004–11011.

Source: Office of Commercial Space Transportation, U.S. Department of Transportation, Washington, D.C.

After more than three years of discussions, the Office of Commercial Space Transportation made public in April 1988 the details of what it would take for a private company to obtain the government license required to carry out a private-sector space launch. The following is the Preamble to the Final Rule without the appendices.

[11004] **14 CFR Ch. III**

[Docket No. 43810]

Commercial Space Transportation; Licensing Regulations

AGENCY: Office of Commercial Space Transportation, DOT.
ACTION: Final rule.

SUMMARY: The Office of Commercial Space Transportation is publishing final licensing regulations for commercial launch activities. These regulations constitute the procedural framework for reviewing and authorizing all proposals to conduct non-Federal launch activities, including the launching of vehicles, operation of launch sites, and payload activities that are not licensed by other Federal agencies. The Office also is publishing its general administrative procedures and a revised compilation of its information requirements. This final rule replaces all previous guidance, specifically the interim final rule, published February 26, 1986, and the Licensing Policy Statement, published February 25, 1985.

DATE: This rule becomes effective April 4, 1988.

FOR FURTHER INFORMATION CONTACT: Gerald Musarra, Office of the General Counsel, U.S. Department of Transportation, 400 Seventh Street SW., Room 10424. Washington, DC 20590, (202) 366-9305.

SUPPLEMENTARY INFORMATION:

Background

The Commercial Space Launch Act of 1984, Pub. L. 98–575, authorizes the Secretary of Transportation to oversee and coordinate United States commercial launch activities. The Secretary's mandate embraces the authority to license and otherwise regulate such activities, as well as the responsibility to encourage, facilitate and promote establishment of a competitive United States commercial space transportation industry.

The Department of Transportation is currently implementing its authority in this area through interim regulations published by the Office of Commercial Space Transportation on February 26, 1986. The interim regulations built upon the Office's Licensing Policy Statement, published February 25, 1985, which was the Office's initial exposition of the licensing process it had devised as the means for guiding both the planning and conduct of the private launch activities subject to its authority. In particular, the Office's approach to licensing was intended to ensure that certain national interests received appropriate attention when applications are reviewed. These interests are stated explicitly in the Act: Public health and safety, the safety of property, national security interests and foreign policy interests of the United States. The licensing process described in the policy statement involved two reviews designed specifically to address these interests. One focused on the safety operations that would be used to support launch activities, while the other focused on the proposed mission itself. In addition, the policy statement emphasized the need to streamline procedures for consulting with other Federal agencies on specific commercial launch proposals.

The Office received numerous comments on its licensing policy. These comments, as well as its own greater practical experience with the launch industry, were fully considered in the course of drafting proposed licensing regulations. Because the Office concluded that the launch industry required guidance upon which it could immediately rely, these regulations were published on an interim final basis. Although they went into effect immediately upon publication, the Office requested further comment on its licensing regulations in order to identify revisions or clarifications that might be needed to achieve maximum responsiveness to the wide range of launch activities American firms can be expected to propose.

In addition, much progress has been made since the interim regulations were published in developing the contractual arrangements covering access of commercial launch firms to government-developed launch technology and government-provided safety services. The greater definition that now exists in this area has, in turn, made it both necessary and possible to ensure that government range safety functions and launch firm licensing procedures are efficiently integrated.

The regulations published today constitute the administrative framework for according each proposal to conduct a commercial launch activity a prompt, well-defined, and thorough review. They also reflect the Office's on-going efforts to design a licensing program that will provide unqualified assurance to the public that private firms will operate safely and responsibly. This assurance is indispensable to the success of the American commercial launch industry.

The Office will continue to evaluate and, when necessary, re-shape its program in response to growth, innovation and diversity in this critically important industry.

National Space Policy

The interim regulations were published within a month of the Space Shuttle *Challenger* accident, an event which resulted in the temporary grounding of the nation's primary means to space. This situation, combined with the rapidly growing backlog of government and commercial payloads, caused the government to reevaluate its reliance on a single space transportation system as well as its own role as provider of launch services for all the nation's space needs. Instead, the United States private sector would have to assume a new and significant role, alongside the government, in assuring the nation's access to space.

In August 1986, President Reagan announced a new launch policy, set forth in his United States Space Launch Strategy, which limits the Shuttle's role to certain missions and directs the Department of Defense to develop payloads compatible with both expendable vehicles and the Shuttle. Further, the President directed that virtually all routine commercial payloads be launched by commercial launch firms.

On February 11, 1988, the President issued a directive on National Space Policy, which consolidated and updated previous Presidential guidance on space activities. The National Space Policy identifies, for the first time, a separate and distinct commercial space sector. The policy is especially significant because of its emphasis on commercial launch services as an integral element of the robust transportation capability essential for maintaining United States space leadership. Further, the policy reaffirms the role of the Department of Transportation as lead agency for Federal policy and regulatory guidance pertaining to United States commercial launch activities.

National Space Launch Infrastructure

The National Space Policy is the culmination of a series of Presidential Policies aimed at a fundamental redefinition of the traditional role of the Federal Government in space activities. In the past, the nation's space programs were conducted entirely by the Federal Government. Launch firms participated in these programs only as government contractors, operating in complete conformance to government program requirements and launch practices.

Now, however, launch firms will be operating on a commercial basis, in [11005] direct response to the needs of their customers. In doing so, they will rely on the nation's existing launch infrastructure for the support they need to undertake missions vital to the technological and economic well-being of the United States.

The facilities that comprise this infrastructure are resources in which the nation has invested over the course of three decades to ensure United States preeminence in all activities. At present, demand for program support at these facilities is great and the supply, as with all resources, limited. This potential capacity problem highlights the need for management strategies that will maximize access to the national ranges for all sectors of the U.S. space program: Military, civil government, and private commercial. The Department of Defense, the National Aeronautics and Space Administration, and the Department of Transportation are working in concert to develop the means whereby Federal launch property and services can be made available to the commercial launch industry in a manner that enables it to compete effectively in the world market for launch services.

Pursuant to its authority under section 15 of the Commercial Space Launch Act and consistent with the President's directives in the National Space Policy, the Department of Transportation is working to ensure that government launch property and services requested by launch firms are priced in a manner that provides maximum encouragement to the United States commercial launch industry. The Department is also working, in consultation with other Federal agencies, to establish allocation of risk principles and insurance requirements that are appropriate for commercial launch activities conducted at national ranges.

Safety Roles and Responsibilities

The Federal Government plays two distinct roles related to safety in the context of commercial launch activities. The Department of Transportation bears responsibility for ensuring, through its licensing process, that proposed launch activities are not hazardous to public health and safety or the safety of property. The Department's exclusive and continuing Safety authority extends to such activities regardless of whether they are staged at private or government launch facilities.

Before the Department's Office of Commercial Space Transportation can issue a launch license, it must review an applicant's proposed safety operations. In order to secure approval for its safety operations, an applicant must demonstrate that it can marshall [sic] the resources needed to prepare and launch a launch vehicle safely. These resources can be assembled in a number of ways: A company can choose to conduct all safety operations itself; it may rely on government-provided property and services to support its safety operations; or it may choose to perform safety operations through some arrangement whereby private and government resources are combined. In any case, the company must demonstrate that all aspects of its proposed launch activities will be conducted safely.

In addition, the Federal Government also operates, through the Air Force and NASA, a number of launch ranges and related launch facilities. Numerous safety-related operations are conducted at these ranges. Some of these operations, such as those pertaining to flight safety, can be provided under contract as a service to commercial launch firms. Range operators also conduct safety-related operations that derive from their responsibility to protect government property and personnel. These include safety inspections and monitoring, as well as certain other safety functions performed on a mandatory basis for all range users. Most commercial firms have indicated that they plan to contract with national range operators for flight safety support as the means for obtaining safety approval from the Department of Transportation.

Comments on the Interim Regulations

The Office received 13 comments on its interim licensing regulations. Of this total, two were submitted by private individuals, seven from launch firms and other aerospace companies, one from a coalition of media associations, one from a law firm that represents telecommunications clients, and one from a Federal agency. In addition, the Office also received comments from the House Committee on Science and Technology.

Most of the comments received by the Office expressed general support for the licensing policies and procedures articulated in the interim rule. Several commenters, however, raised questions concerning the standard for granting "mission approval," that is, the standard for determining that a proposed launch activity is not objectionable from the standpoint of safety, United States national security or foreign policy interests, or United States international obligations. Specifically, commenters expressed concern that the terms "national security" and "foreign policy" are not defined in the regulations and could be interpreted too broadly.

The Office wishes to emphasize again the guiding principle established by the Commercial Space Launch Act in this area: the "provision of launch services by the private sector is consistent with the national security interests and foreign policy interests of the United States and would be facilitated by stable, minimal and appropriate guidelines that are fairly and expeditiously applied." As the agency charged with implementing the Act, the Department of Transportation views this passage as forming the basis for a presumption that proposed commercial launch activities are consistent with national interests. Thus, the purpose of the licensing process, so far as national security and foreign policy issues are concerned, is to identify and, whenever possible, ameliorate specific problems with a proposal, not to determine that each and every proposal is generally consistent with those interests.

However, the Office also wishes to emphasize again the consideration of national security and foreign policy factors is required in the first instance by the Commercial Space Launch Act, not commercial launch regulations: the Act requires the Office to consult with the Departments of Defense and State on all matters affecting United States national security or foreign policy interests.

The Office also received comments that focused on the treatment accorded payloads in the course of Mission Review. These comments were filed by a coalition of organizations representing entities engaged in news gathering and dissemination ("the Media Parties"), as well as by a law firm specializing in telecommunications matters. Specifically, the commenters expressed some concern that, as drafted, the regulations seemed to suggest the possibility of redundant regulation for payloads that are already subject to payload regulation by other Federal agencies, notably the Federal Communications Commission (FCC) and the National Oceanic and Atmospheric Administration of the Department of Commerce (NOAA). The Office recognizes that some clarification of its policies and procedures concerning approval of proposed missions may be helpful in order to eliminate any confusion concerning the Office's role relative to Federal agencies with exclusive responsibility for regulating satellites or satellite services. This matter is discussed in greater detail in the Section-by-Section Analysis.

The Media Parties also proposed modifications to Mission Review that [11006] are intended to provide procedural safeguards to applicants whose commercial space proposals may involve activities protected by the First Amendment to the Constitution. In the view of the Media Parties, without these modifications, the regulations may impinge on the First Amendment rights of news organizations.

The Office has not adopted these proposed modifications because they would have the effect of distorting the licensing process. To the extent that a proposal to launch a communications or remote sensing satellite raises First Amendment issues, those issues will be addressed by the agencies with exclusive authority for regulating these satellites or the services provided by them: the FCC or NOAA. Such issues do not fall within the scope of the Office's authority for commercial launch activities and, thus, are not addressed in the course of its licensing process. The Office's sole non-safety concern regarding FCC or NOAA regulated payloads is that such satellites not be launched until they are licensed by those agencies.

Another commenter suggested that Mission Review should examine the impact of proposed new payloads on future, as well as current, uses of space. The Office does expect that its review of such a payload would focus on safety, national security or foreign policy implications associated with the payload. In addition, reviews would also focus on those impacts associated with a new payload that may occur in the reasonably foreseeable future. However, the Office does not consider open-ended speculation regarding possible future uses of space by public and private entities, both domestic and foreign, to be consistent with the well-defined and expeditious processing of applications required by the Act.

The Office received comments from the House Committee on Science and Technology that touched on a number of subjects in the regulations. First, the Committee directed the Office's attention to the fact that since "payloads" are defined as "objects," not people, by the Act, there could be a problem with the Office seeking to offer guidance to private entities who may be planning manned launch activities. Indeed, several such entities have consulted with the Office on a number of occasions and a representative of one start-up firm sits on the Department's Commercial Space Transportation Advisory Committee.

With regard to "payloads" as defined by the Act, the Office does not see this term, however defined, as an impediment to exercising its role as the point of contact within the Federal Government for private entities planning manned launch activities. Neither the

Act nor the Report that accompanied the Act at passage indicates that "launch of a launch vehicle" should be read exclusively as launch of an *unmanned* launch vehicle. While it is clear that the Act was drafted primarily for the launch activities most likely to occur in the near term, commercial launches of unmanned rockets, the Report clearly states that "[t]he Act currently provides adequate supervision for all *non-Governmental* (commercial or noncommercial) space launches. . . ." Regardless of the type of launch activity contemplated by a private entity, manned or unmanned, the Federal Government must be prepared to provide effective guidance. Only in this manner can the Government avoid the unsatisfactory administrative response that firms proposing commercial ELV launches experienced prior to issuance of Executive Order 12465 and passage of the Act.

The Committee also asked several questions concerning the Office's research and analysis program, which is intended to enhance the technical resources the Office needs for effective implementation of the Act. This program consists of studies to be conducted over the course of two years. The Committee asked how the Office can handle private launch site proposals on a case-by-case base, as provided in the regulations, within the statutorily prescribed 180 days or how a meaningful rulemaking proceeding on private launch sites can begin if the Office's safety research and analysis will not be completed for two years.

The Office will review proposed private launch site operations on an *ad hoc* basis relying, as an interim measure, on existing government launch expertise, experience, and safety practices as references. In this way reviews will be conducted thoroughly and within the statutory time limits even though there are not now published standards to guide firms planning to conduct private launch site operations. Indeed, such standards cannot be promulgated until adequate data and analysis has been assembled to support a rulemaking.

Any rulemaking initiated in the near-term on private launch site operations will focus on regulatory *policy* issues; that is, the appropriate approach the Office should take in developing policies and procedures for licensing commercial launch site operations. Thus, both review of private launch site operation proposals and pre-rulemaking notice and comment activity focused on licensing issues can be conducted concurrently with ongoing safety research. Further, although the entire safety research effort may take two years to complete, individual studies will be completed throughout that period, some within the next six months to a year. The results of these studies will form the basis for the Office's basic technical capability, including safety evaluation criteria and a data base for future safety standards. It should be noted that safety research is a continuing and critical component of every safety regulatory program, as demonstrated by the extensive on-going research and analysis conducted by other constituent agencies of the Department of Transportation, such as the Federal Aviation Administration or the National Highway Traffic Safety Administration.

In the area of worker safety, the Committee suggested that there is no need to duplicate the requirements of the Department of Labor's Occupational Safety and Health Administration (OSHA) which would apply to worker safety in the context of licensed launch activities. The Office has no intention of doing so. The Act gives the Office comprehensive safety authority for commercial launch operations, thus raising an issue concerning concurrent authority in this area. As in other areas where there is concurrent safety authority, such as aviation, there is a question concerning the more appropriate approach to safety, OSHA's or that of the agency with primary authority for the activity involved. At this time, the Office will not develop safety requirements for the specific purpose of protecting workers involved in commercial launch operations. OSHA requirements will apply to these activities until the Office and OSHA determine that it is appropriate to do otherwise.

The Committee also suggested that the Office prescribe a format for required information and use forms where appropriate. Although the Office has not ruled out adoption of a required format at some future time, it continues to believe that, for the time being, applicants should organize required information in a manner that reflects the organization of their safety operations. In order to encourage innovation, the Office has tried to accord applicants maximum flexibility and to emphasize content, rather than form. The information requested was identified and organized in close cooperation with NASA and the approach was discussed informally with launch companies before rulemaking was initiated. They all supported our approach then and, in [11007] their formal comments on the rule, have continued to do so.

With regard to license fees, the Committee favors incorporating such fees in the regulations to cover the costs associated with processing applications. The Department strongly supports user fees in all transportation modes. The Office intends to consider establishing reasonable fees for licensing processing, balancing the desirability of reasonable fees with its responsibility to encourage and promote a private launch industry.

The committee also alerted the Office to the need for further clarification of some of the definitions contained in the regulations. The Office has made appropriate revisions to its definitions and these, along with other revisions, are discussed in the section-by-section analysis that follows.

Section-by-Section Analysis

Part 400 — Basis and Scope

Section 400.1 indicates that the commercial space transportation regulations derive from both the Federal Government's domestic responsibilities for commercial launch activities as well as the obligations it has assumed under international agreements, particularly the obligation under Article VI of the 1967 Outer Space Treaty to provide authorization and continuing supervision for such activities.

Section 400.2 specifies the launch activities for which the regulations provide guidance: all United States launch activities except amateur rocket activities and the launch activities of the United States Government. As the Office stated in its initial policy statement on licensing, its licensing policies and procedures have been developed primarily for the private commercial launch activities that are currently being proposed: commercial expendable launch vehicle (ELV) launches. However, consistent with the legislative history of the Act, the Office's regulatory guidance also provides adequate supervision for any other non-Federal launch activity. Thus, launch activities failing within the scope of the Office's authority may include activities conducted for experimental, developmental, or research purposes as well as those conducted without any apparent profit motive.

At the same time, neither the Act nor its legislative history evinces an intention to require licenses for small scale rocket launches conducted for recreational or educational purposes at private sites. These launches, which number annually in the millions, are currently subject to state and local regulation, self-regulation by the organizations sponsoring these activities, and Federal airspace requirements. These existing guidelines and requirements have been effective for purposes of protecting public safety and any other national interest that may be associated with these activities.

Part 401 — Organization and Definitions

Section 401.1 identifies the operating unit within the Department of Transportation with primary responsibility for implementing the Department's authority under the Act, the Office of Commercial Space Transportation. Section 401.3 identifies the Director of Commercial Space Transportation as the official within the Department to whom the Secretary's authority for commercial space transportation has been delegated.

Section 401.5 contains definitions of the major terms used in the regulations. The definitions of "launch" and "operation of a launch site" are intended to convey the complementary, but nevertheless distinct, nature of these two activities. A launch centers on the placement, or attempted placement, of a specified launch vehicle and/or its payload in a suborbital trajectory or in space. A launch license authorizes a launch to be conducted in order to achieve certain mission objectives. The license holder is legally responsible for the proper conduct of such a launch. Although a launch license would seem to be oriented toward singular events, one license could cover a specified series of launches where the same safety resources will support several identical or similar missions.

In contrast, the operation of a launch site involves continuing operation at a permanent location. A license covering such operations authorizes a person to operate a launch range facility and to offer approved services to launch companies

The Office has determined that the inclusion of a definition for "commercial launch activities" in the Interim Final Rule was unnecessary and has deleted it.

Part 404 — Regulations and Licensing Requirements

The Commercial Space Launch Act establishes the licensing standards for commercial launch activities. Section 9(b) of the Act directs the Office to issue a license once it has determined that an applicant meets the requirements for a license identified in section 8(a)(1) of the Act. These include current requirements of Federal agencies which apply specifically to the launch of a launch vehicle or operation of a launch site. If, however, the Office determines, in consultation with the appropriate agencies, that any such Federal requirement is not needed to protect public safety, the safety of property or the national security and foreign policy interests of the United States, then section 8(a)(2) permits the Office to eliminate that particular requirement as a requirement for a license. Moreover, section 8(b) authorizes the Office to prescribe new requirements for commercial launch activities. Together these provisions confer broad authority upon the Office to craft efficient regulatory guidance with specific applicability to private launch activities.

If the Office wishes either to eliminate an existing Federal requirement or to prescribe new ones in order to implement the provisions of the Act, a proceeding must be conducted that would involve notice to and comment by the public. Part 404 of the regulations sets out the procedures the Office will follow when conducting rulemaking proceedings and explains how interested parties may participate.

Section 8(c) of the Act gives the Office discretionary authority to waive a licensing requirement for a license applicant if that waiver would be in the public interest and would not jeopardize public health and safety, safety of property, or any national security or foreign policy interest of the United States. Part 404 also establishes procedures for waiver requests by individual applicants.

With regard to existing Federal requirements, the Office has determined that the only provisions with direct applicability to private launches are those of Part 101, Subpart C, of the Federal Aviation Regulations, 14 CFR 101.21-25, regulating all unmanned rocket activities. The Office of Commercial Space Transportation and the Federal Aviation Administration have agreed that, henceforth, requirements pertaining to the use of domestic United States airspace for commercial launch purposes will be handled by the Office as an intradepartmental matter on behalf of licensees.

It should be noted that the Office's safety authority extends to protecting workers at commercial launch sites. For the present, however, the Office will not prescribe any standards or requirements for worker safety in the context of licensed launch activities. Instead, the appropriate requirements of the Department of Labor's Occupational Safety and Health Administration will apply to privately conducted launch activities.

[11008] *Part 405 — Investigations and Enforcement*

The Office will rely on the provisions of Part 405 to ensure compliance with the terms and conditions of licenses. Section 405.1 requires licensees to cooperate with anyone acting on behalf of the Office to monitor licensed activities, including payload-related activities covered by section 6(b)(2) of the Act. Monitoring will be conducted in the least intrusive manner possible and only for the purpose of determining whether such activities conform to applicable requirements.

Section 405.3 deals with modification, suspension or revocation of licenses. The Office may modify a license either on its own initiative or pursuant to a request by the licensee. All modifications must conform to the same standards, identified in the Act, that apply to initial licenses.

Paragraph (b) of § 405.3 indicates that noncompliance with any requirement applicable to a licensed activity is grounds for suspension or revocation of a license. Moreover, § 405.5 provides for emergency orders to halt any launch activity detrimental to national interests, while § 405.7 provides that acts of noncompliance may be punishable by civil penalties.

With regard to the Director's emergency order authority, which is explicitly mandated by section 11 of the Act, the Office is aware of the concern, expressed through the Commercial Space Transportation Advisory Committee, associated with the exercise of this authority. One of the Office's major goals has been to encourage and promote the industry through carefully considered policies and procedures designed to eliminate, wherever possible, regulatory uncertainties. Thus, the Office wishes to emphasize that it views the exercise of this authority as an extraordinary measure to be relied upon in truly emergency circumstances.

Part 406 — Administrative Review

Part 406 describes the Office's procedures for implementing the Act's administrative review provisions. Section 12 of the Act requires that an opportunity for a hearing be accorded persons seeking reconsideration of certain decisions made by the Office. Specifically, persons who have applied for a license may challenge a decision not to issue a license or challenge the conditions attached to a license that has been granted. In addition, a person holding a license may dispute a decision to modify, suspend or revoke that license or to issue an emergency order. Similarly, a payload operator or owner may request a review of the facts or issues pertaining to a payload whose launch the Office has decided to prevent as may a person against whom the Office has assessed a civil penalty. In these circumstances the Office will, if so requested, provide an opportunity for an impartial hearing on the matter at issue. Part 406 sets out the procedures governing initiation and conduct of such proceedings.

Part 411 — Policy

Part 411 establishes the policies of the Office of Commercial Space Transportation for licensing commercial launch activities, including launches, launch site operations, or some combination of the two activities. These policies augment the general application procedures set out in Part 413 of the regulations and the launch license review procedures contained in Part 415 of the regulations.

Section 411.3 identifies the two reviews, Safety Review and Mission Review, through which the Office will evaluate proposed ELV launches. Although the Office will be responsive to proposals involving manned launches, such proposals may involve issues that require reviews different from or in addition to these two reviews.

In order to accord the industry both flexibility and certainty in the course of developing commercial launch proposals, the Office may conduct Safety Review and Mission

Review independent of each other and in the order, sequential or concurrent, appropriate to the applicant's needs. For example, an applicant may secure approval for a proposed mission early in the planning stages of a launch activity and apply later for approval of the safety operations proposed to support an actual launch. The record upon which to base licensing decisions thereby can be developed in a manner that responds to the planning needs of applicants.

Section 411.3 also discusses requests for licenses authorizing the operation of commercial launch sites. Editorial revisions have been made to this section to make it clear that this activity is comparable to the operation of a commercial airport. Although a separate license covering the operation of a launch site is contemplated by the Act, the regulations were not developed specifically for implementing the Office's authority in that area. Devising an appropriate regulatory framework for commercial launch site operations involves careful consideration of a wide range of complex issues, particularly those relating to requirements or standards for implementing the Office's safety authority. The Office has begun investigating these issues as part of its comprehensive research and analysis program.

At the same time, the Office has received a number of inquiries expressing interest in establishing permanent commercial launch sites and wishes to be responsive to any proposal that may be submitted in the near future. In order to do so, the Office will rely on its Safety Review process, discussed below, as an appropriate general framework for initiating an assessment of commercial launch site proposals.

Section 411.5 addresses safety approval, one of two approvals an applicant must secure in order to be granted a license. At present there are no safety standards or requirements that have been developed specifically for commercial launch activities. Therefore, pending completion of efforts to develop these standards and requirements, the Office will make case-by-case determinations regarding safety operations that commercial firms propose to conduct themselves. The Office will supplement the resources available to it, when necessary or appropriate, by relying on the experience and expertise of other Federal agencies. Minor editorial changes have been made to this section in the final rule.

Section 411.7 discusses mission approval. This is the other approval which must be secured in order for an applicant to be granted a launch license. The Office must assess proposed missions from the standpoint of both the national interests and international obligations of the United States. The review will encompass such factors as the nature and purpose of the proposed payload, the impact of the payload on existing uses of space, and the proposed flight plan.

With specific regard to national security and foreign policy interests, the Office is required to consult with the Departments of Defense and State, the Executive Branch agencies with primary responsibility for safeguarding U.S. national security and foreign policy interests, respectively. The Office must ensure that these agencies are apprised of potential commercial launch activities in order for their views to be taken into account. The Office wishes to emphasize again that, as a general matter, Congress has declared privately conducted commercial launches to be consistent with the national security and foreign policy interests of the United States. The Office fully recognizes that the commercial viability of providing such services on a routine basis requires [11009] that review of proposed missions not be encumbered by unnecessary process. Therefore, the Office will seek to identify specific problems associated with a proposed mission, not seek to determine de novo that each launch proposal is consistent with United States interests.

However, the Office has revised § 411.7 of the regulations to correct any impression created in the Interim Rule that the Office was establishing an evidentiary standard for adverse licensing decisions that is higher than or different from that set forth in the Act.

The mission of most proposed orbital launches will be to place a payload in space. Thus, the most significant part of the Office's review of proposed missions will pertain to the payload to be launched. The Office wishes to clarify the nature and scope of its authority with regard to payloads launched by commercial launch firms. A launch license issued

by the Office authorizes the licensee to launch a launch vehicle and any payload to be carried by the launch vehicle. In order to authorize a launch involving a payload, the Office
must first identify the nature of the payload to be launched. This identification is necessary
in order for the Office to determine how to proceed, in practical terms, with a review of a
proposed mission. There are two general options: (1) The payload the applicant proposes
to launch is identified as one which is subject to existing payload regulation. At present, this
category includes only telecommunications satellites licensed by the Federal
Communications Commission (FCC) and remote-sensing satellites licensed by the National
Oceanic and Atmospheric Administration of the Department of Commerce (NOAA). (2)
The payload the applicant proposes to launch is identified as one which is not subject to
existing payload regulation. Only for this latter category will the Office initiate a review, pursuant to its authority under section 6(b)(2) of the Act, in order to determine that the proposed launch of the payload will not jeopardize public health and safety, safety of property
or any national security or foreign policy interest of the United States. The Office does not
conduct such a review for any payload that requries [sic] either an FCC or NOAA license to
launch or operate. Rather, pursuant to section 6(b)(1) of the Act, the Office simply requires
that the appropriate license be secured before the payload can be launched. The Office will
not examine any issues pertaining to payloads licensed by the FCC or NOAA before license
application is made to either of those agencies or during the pendency of any review of a
license application at either agency. Nor will the Office re-examine any matter associated
with a payload that was or could have been subject to FCC or NOAA review during their
respective licensing processes. In order to eliminate any lingering ambiguities in this area,
the policies and procedures in the regulations pertaining to proposed missions have been
revised or clarified, as appropriate. It should be noted, however, that in the course of Safety
Review the Office will seek to ascertain whether all applicants possess the requisite resources
and expertise to conduct safely any planned payload-related operations as part of the
process whereby a launch vehicle is prepared and launched.

Payloads that are subject to review by the Office under section 6(b)(2) of the Act
include all domestic payloads not presently regulated by the FCC or NOAA and all foreign
payloads. The Office is authorized to determine whether the launch of any such payload
would jeopardize public safety, safety of property, or any national security or foreign policy interest of the United States. If necessary, the Office may act to prevent the launch of
the payload in question. As it has done in other areas, the Office has molded its policies
and procedures carefully in this area so that legitimate Federal interests associated with
proposed launches of these payloads are not served at the unnecessary expense of commercial space enterprise. Thus, the Office will exercise its authority under section 6(b)(2)
in a manner that minimizes regulatory uncertainties for those planning or sponsoring
new space applications and missions involving foreign payloads.

Section 411.9 discusses the information the Office will require applicants to submit in
order to initiate review of applications. The Office's approach to this information corresponds to its goal of fostering reliable, low-cost commercial space transportation services.
The Office's information requirements have been organized intentionally into general categories that identify the basic information needed to initiate an appropriate review. However,
although all the requested data must be provided for an application to be considered complete, the Office has not prescribed any particular format for submitting it. Because commercial firms may develop new approaches to the design of launch vehicles, the delivery of
launch services, or the location and organization of launch operations, information submissions may reflect the unique structure or organization of their launch operations.

The Office has made a number of changes to the information requirements identified in the Interim Rule. The Office expects to continue refining these requirements
based on the products of its research program and consultations with other agencies, as
well as formal and informal interaction with the commercial space industry. Therefore,
the Office has concluded that this information should not be included in its published

regulations. So that prospective applicants are assured of having ready access to the most current and accurate version of the Office's information requirements, they will be set out in a separate document that will be available upon request. The first such version of the Office's information requirements is published as an appendix to this preamble.

Part 413 — Applications

Part 413 sets out general license application procedures. These procedures apply to all commercial launch activities, regardless of whether an applicant seeks a license to launch a vehicle, operate a launch site, or for a combination of the two. The application procedures in Part 413 are supplemented by the provisions of Part 415, which contains a detailed description of the review procedures for launch license applications. A separate part has been reserved for future regulations addressing applications for licenses authorizing launch site operations.

Since the nature of a proposed launch activity affects the timing and scope of the Office's review, as well as the degree to which other Federal agencies will be involved, § 413.3 encourages prospective applicants to initiate preapplication consultations with the Office of Commercial Space Transporation [sic].

Section 413.7 contains revised procedures for handling confidential information. These revisions have been made to bring this section into conformity with section 9(c) of the Act, which directs that certain information provided to the office by applicants not be disclosed unless the Secretary determines that withholding such information is contrary to the public or national interest.

Section 413.9 outlines the process for reviewing all applications. Section 413.9(a) has been amended to indicate that information required to initiate a review of an application is available upon request. [11010] Section 413.9(b) states that an application is accepted for review by the Director if it is substantially complete; that is, if it contains sufficient information for a meaningful review. Once an application is accepted for review, § 4119(d) indicates that the Director will initiate an appropriate interagency review. The Office, not the applicant, will assume the burden of shepherding the application through the review process. Additionally, the reference in § 413.9(d) to an "appropriate" review is intended to make clear that the administrative response to an application may not be standard or uniform in all circumstances; the Office has taken great care to insure [sic] that each review is tailored to the application's particular characteristics. In this fashion, the Office intends to avoid any unnecessary regulatory stumbling blocks to proposed launch activities.

Section 413.9(e) indicates that a determination on a license application will be made within 180 days of receipt. As a matter of policy, however, the Office intends to conduct all application reviews on an expedited basis and anticipates that most determinations will be made well before this statutory deadline.

All licenses issued will contain terms defining the activity authorized by the license and the person responsible for conducting that activity. In addition, conditions will be incorporated into all licenses to ensure compliance with statutory and regulatory requirements. Section 413.15 addresses certain standard conditions, including the need for an on-site mechanism to verify that the licensed activity conforms to information that was submitted to and reviewed by the Office during the application review process.

Section 413.17 indicates that a license authorizing a launch activity is separate from the license required for any satellite to be launched. The Act preserves the existing authority of Federal agencies with primary responsibility for payload regulation. At present, this includes only the FCC and NOAA, which are responsible for licensing telecommunications and remote sensing satellites, respectively. Thus, issuance of a launch license has no effect on the exclusive authority of the FCC or NOAA to license such satellites or the services provided by them.

Section 413.19 establishes the applicant's responsibility for the continuing accuracy of information submitted as part of an application review.

Part 415 — Launch Licenses

Part 415 establishes procedures for reviewing launch license applications and the general standards for approving such applications. The provisions of this part apply only to prospective launch license applications and should be read together with the general application procedures in Part 413. A future regulatory proposal addressing commercial launch site operations will establish procedures and standards specifically for license applicants seeking authorization for that activity in a separate part.

Section 415.3 identifies the proposed launch activities that will require a launch license. Any person proposing to launch from U.S. territory must obtain a license authorizing the launch. A U.S. citizen proposing to launch from U.S. territory or from international territory must also obtain such a license, unless (in the case of launches from international territory) another nation has agreed to exercise jurisdiction over the launch. Foreign corporations, partnerships, joint ventures, associations or other entities controlled by U.S. citizens do not need licenses to conduct a launch from foreign territory, unless the foreign nation involved has agreed that the U.S. shall exercise jurisdiction over the launch.

Section 415.5 identifies the two approvals that must be secured in order for a launch license to be issued: safety approval and mission approval. Safety Review and Mission Review are conducted to determine whether these approvals can, in fact, be given. Once secured, no other approval is required from the Office in order for an applicant to be granted a license for an ELV launch.

The Office will accept applications for Safety Review, Mission Review, or for a determination that the launch of a payload covered by section 6(b)(2) of the Act will not be prevented, independent of one another and before submission of an application for a license. Section 415.7 makes clear that any approval or determination made on such applications will be made part of a licensing record. Thus, when an applicant does apply for a launch license, any approval or determination previously made that relates to the activity for which a license is sought remains valid. The Office will not duplicate a relevant review as long as no material changes have been made in matters previously reviewed and approved.

Section 415.9 identifies standard conditions for launch licenses. One of these is securing third-party liability insurance coverage. In exercising its authority in this area, the Office will be looking to set required insurance amounts that accurately reflect the potential losses associated with launch failures. The Office has begun several studies to determine what these amounts should be. For the time being, the Office will prescribe insurance requirements for each licensed activity on a case-by-case basis.

The final regulations include a new provision, § 415.10, which sets out requirements pertaining to the registration of objects launched into space.

Subpart B of Part 415 focuses on Safety Review. Section 415.13 identifies the major elements of Safety Review: the proposed launch site, procedures, personnel and equipment. Section 415.15 notifies applicants that Safety Review can be requested either as part of the license request or before a license request is submitted. This provision responds to the need some prospective licensees may have for explicit approval of their safety operations at an early planning stage.

Section 415.17 of the interim regulations set out the information requirements for Safety Review applicants. This section has been deleted. The information currently required for a Safety Review is contained in the appendix to this preamble. It should be noted that launches from sites with pre-approved safety operations will be treated differently from those occuring [sic] at other sites. At present, the only sites with pre-approved safety operations are Federal launch ranges. In the future, this category would also include commercial launch sites operated under the authority of a license issued by the Office.

Subpart C of Part 415 focuses on Mission Review. Section 415.23 states that for Mission Review, as for Safety Review, applicants may request approval either as part of a license request or before such a request is made. Sections 415.25 and 415.27 of the interim regulations set out the information requirements for applicants seeking mission review or a determinations [sic] on a payload not regulated by FCC or NOAA. These sections have been deleted. Information required for Mission Review, including information pertaining to payloads that are not regulated by the FCC or NOAA, is set forth in the appendix of this preamble. The nature of the proposed mission will affect both the nature and the quantity of information needed by the Office to conduct its review. For |11011| proposals which involve licensed payloads, the payload requirements of Mission Review will be satisfied by the issuance of a license by the responsible Federal agency. Proposals involving other kinds of domestic payloads or foreign payloads must be accompanied by more extensive information, reflecting the more extensive review such proposals must receive from the Office.

Subpart D of Part 415 identifies circumstances wherein applicants may be required to submit information to the Office as part of Safety Review, Mission Review, or both, in order to satisfy the requirements of the National Environmental Policy Act. This information will be needed when some element of a proposal is not covered or addressed by existing environmental documentation on the effects of launch activities.

Executive Order 12291, Regulatory Flexibility Act, and Paperwork Reduction Act

The interim regulations were evaluated under Executive Order [E.O.] 12291, "Federal Regulation," dated February 17, 1981, and the Department of Transportation's Regulatory Policies and Procedures, dated February 26,1979. The regulations were not considered to be "major," as defined by E.O. 12291, because they will not have an annual cost impact exceeding $100 million; they will not cause a major increase in costs or prices for consumers, individual industries, government agencies, or regions; and they will not have a significant adverse impact on competition, employment, investment, productivity, innovation or on the ability of United States-based enterprises to compete with foreign-based enterprises in domestic or export markets. The regulations were considered to be "significant" as defined by the Department's Regulatory Policies and Procedures because of the novelty of space transportation as a private sector activity, the interest of the public and other Federal agencies, and the effect of the regulations on the competitive position of United States launch firms. The Office prepared a Regulatory Evaluation to accompany the interim regulations, which was made available for public review and comment in the rulemaking docket. Since the final regulations are not materially different from the interim ones, the Office considers all regulatory analysis prepared for the interim regulations to be applicable to the final ones. The regulations are largely procedural in nature and are intended to eliminate regulatory obstacles to private launch firms, large or small. Small entities are likely to be involved in launch activities and, as a consequence, affected by the regulations.

The regulations do not impose significant economic costs on them. Therefore, it is certified that the regulations will not have a significant economic impact on a substantial number of small entities.

National Environmental Policy Act

The Office completed an environmental assessment of the commercial space transportation program and made the assessment available for public inspection and comment. The programmatic assessment did not identify any significant impacts that the conduct of commercial launch activities would have on the human environment.

However, certain factors associated with individual launch proposals were not addressed in the assessment and may require further review during the licensing process. These include use of new propellants, new site development, or environmental effects associated with some payloads in the event of a launch accident. Copies of the assessment may be requested from: Office of Commercial Space Transportation, S-50, Washington, DC 20590. Based on the assessment and comments received on it, the Office published a finding of No Significant Impact in the Federal Register on November 19, 1986.

List of Subjects in 14 CFR Parts 400, 401, 404, 405, 406, 411, 413, 415

Administrative practice and procedure, Space transportation and exploration.

(Commercial Space Launch Act of 1984, Pub. L 98–575, October 30, 1984)
Issued in Washington, DC, on March 24, 1988.
Courtney A. Stadd,
Director, Office of Commercial Space Transportation.

Document III-23

Document title: Samuel Skinner, Secretary of Transportation, Letter to Member of Congress transmitting a study by the Office of Commercial Space Transportation on the scheduling of commercial launch operations at Government launch sites, June 1, 1989, with attached: "Executive Summary," pp. iii–vii.

Source: Office of Commercial Space Transportation, Department of Transportation, Washington, D.C.

One issue of concern to both the U.S. commercial space transportation industry and its potential customers was the conditions under which a commercial launch would be scheduled, given that it was using government-owned launch facilities. One concern was that a government launch could preempt a scheduled commercial launch; such a delay could be costly to the commercial launch customer. Reflecting this concern, Congress included in the Commercial Space Launch Act Amendments of 1988 a requirement that the Department of Transportation study the issues associated with launch scheduling to minimize the chances of preemption and other undesirable actions. Below is the standard letter Transportation Secretary Samuel Skinner sent to members of Congress, along with the executive summary of the study report.

THE SECRETARY OF TRANSPORTATION
WASHINGTON, D.C. 20590

June 1, 1989

Dear Member of Congress:

I am pleased to submit to you this study by the Office of Commercial Space Transportation [OCST] on the scheduling of commercial launch operations at Government launch sites. The study, mandated by Congress in Section 8 of the 1988 Amendments to the Commercial Space Launch Act, focuses on the best means of assuring efficient and commercially reasonable access of private sector launch companies to available launch site property and resources.

The scheduling process at Government launch sites is designed to cope with multiple user demands in a dynamic environment. Conflicts between military, civil, and commercial users of launch site resources can and do occur. Significantly, however, the OCST study team found that the principal fear among customers of United States commercial launch providers—that of preemption of a scheduled commercial launch by a military mission—is highly unlikely to occur. In fact, both the Secretaries of Defense and Transportation must concur in such an unlikely event.

The emergence of a new commercial sector has been characterized by rapidly escalating demands for services and by the formation of new working relationships. Inevitably, problems develop that may impede the smooth functioning of commercial operations. Some of these problems are transitional in nature, but others may require corrective actions. The study team was able to identify several approaches that could improve the ability of launch companies to compete more effectively in the international market. These approaches would support the commitment of Government agencies to make significant efforts to meet the requirements of commercial interests in the best manner possible.

The United States private sector has the capability and entrepreneurial spirit needed to expand its role as a major competitor in the world commercial space launch market. The economic, technological, scientific, foreign policy, and national security benefits the Nation would reap from this achievement are great. It is our intent that the information contained in this study provide the basis for continued constructive public policy-making to enhance the commercial environment for the launch industry.

Sincerely,

Samuel K. Skinner

****

[iii]   **EXECUTIVE SUMMARY**

Background

Since 1983 the U.S. government has encouraged the development of a privately-owned, commercial space launch industry. This industry has been especially important since 1986 when, in the wake of the loss of the **Challenger,** many commercial payloads (consisting primarily of communications satellites and occasionally industrial manufacturing experiments) required commercial launch services.

Currently commercial space launch companies provide launch vehicles and the technical support necessary to operate them, and contract from the government the use of national launch ranges, similar to other forms of transportation which rely on government-funded infrastructure. Commercial firms will continue to depend on government-operated ranges regardless of whether private sites are built, and even private launch sites will probably use such government range assets as tracking and telemetry systems. As a result, the ability of U.S. launch companies to compete in the world market will depend heavily on the ability of the national ranges to respond to their needs.

The commercial space launch industry is highly competitive, and the difference between a winning and a losing bid in the competition for a launch contract can be extraordinarily slim. If U.S. launch companies are unable to be fully responsive to the needs of their payload customers, these customers have ample opportunity to take their business elsewhere. The French Ariane program already launches more than half of all orbital commercial payloads, and additional competition is appearing or is expected to appear from China and other foreign countries.

The main factor guiding the decision of a payload owner in the selection of a launch company are price, the ability to launch a payload at a desired time and on schedule, and launch vehicle performance and reliability. While U.S. vehicles are well-poised to compete in terms of their performance and reliability, payload companies have said that U.S. firms are in a less strong position on responsiveness to customer launch date requirements. Customers are also concerned about the effect of using national launch ranges on launch service price.

Delayed launches are very expensive—depending on a company's situation, each month's delay in the use of a communications satellite can cost the owner of the satellite from hundreds of thousands to more than a million dollars in expenses and lost revenue. Therefore, the scheduling of launches and use of facilities at U.S. ranges is critical to the success of the U.S. commercial launch industry. Although launch companies can take steps to improve their competitiveness, the effectiveness of these measures will depend greatly on the support they receive from the national ranges.

Purpose and Scope of the Study

During hearings leading to passage of the Commercial Space Launch Act Amendments of 1988, several Members of Congress and witnesses representing the commercial space industry expressed their concern that private launch firms may encounter difficulties in using government-owned ranges. With this in mind, Section 8 of the Act directed the Secretary of Transportation (in cooperation with the Secretary of Defense and Administrator of NASA) to study ways and means of scheduling government and commercial launches at national ranges "in a manner that would make the best practicable use of U.S. launch property and assures that the U.S. launch property available for commercial use is available on a commercially reasonable basis."

To respond to this directive, the Office of Commercial Space Transportation (OCST) has met with the operators of the national ranges, companies that provide launch services, and the customers that use these services. On the basis of these meetings and documentation provided to the Office, OCST has:

- Identified the policies and procedures range operators follow to schedule launches
- Determined the extent to which these policies and procedures provide a scheduling system that is equitable to commercial users of government ranges
- Documented the ability of government-owned ranges to respond to the requirements of commercial launch firms
- Identified measures that would ensure the best practicable use of government-owned launch property and ensure that this property is made available on a commercially reasonable basis

In carrying out its study, OCST found that the most significant "ways and means" that affect launch [iv] scheduling at national ranges are not the launch scheduling procedures used by the ranges, but rather various factors that affect these procedures. Thus, OCST focused on these factors.

Because of its importance to commercial launch operations, the study focused on operations at the Eastern Space and Missile Center (ESMC) and associated services provided by the nearby Kennedy Space Center (KSC). (Approximately nine-tenths of commercial launches are currently scheduled to originate from ESMC.) OCST did, however, ask other range operators to provide written responses to a series of questions concerning range procedures and operations. The responses did not suggest significant inconsistencies between conditions affecting launch scheduling at ESMC and those at other ranges (although interaction between commercial launch companies and these other ranges has occurred on a much more limited scale to date).

Key Findings

It is important to note that the relationship between operators and commercial users of national ranges is still relatively new, and the details of many procedures, methods, and responsibilities have had to be defined. It is also important to note the environment in which these relationships had to be developed. The rapid growth in demand for commercial launch services was largely unexpected and followed a period when most experts and government officials expected the Space Shuttle would be the primary U.S. space launch system. U.S. capabilities to operate expendable launch vehicles were allowed to decline, so both the launch ranges and the commercial launch firms have had to expend considerable labor and capital to reverse the trend.

OCST believes that government range operators are attempting to provide services to commercial launch companies in the best manner possible. Nevertheless, what is "commercially reasonable" may ultimately depend on what customers—in this case, payload owners—believe best meets their needs, given the alternatives that are available in a highly competitive market. OCST was able to identify several approaches that could improve the ability of U.S. launch companies to compete and demonstrate the commitment of government agencies to the commercial environment required by payload owners.

Moreover, OCST found that, although certain changes in policies and procedures could improve the commercial environment at government ranges, limitations in the physical plant at the ranges would continue to present a problem for the competitiveness of commercial users. There is a class of schedule slippages which remains outside the immediate control of the range operators; historically, most slippages in launch dates have been the result of delays in the preparation of a launch vehicle or a payload, or of system standdowns, which have then resulted in other vehicles being delayed.

The major factors concerning scheduling at government ranges that affect the competitiveness of U.S. launch companies include the following:

1. The availability and capacity of launch pads.

The single most significant factor constraining the capacity of government launch ranges currently is the limited number of launch pads that are available. Other issues, such as the need to share facilities with government missions or the impact of delays on commercial launch schedules, would be much less significant if additional launch pads were available.

Currently ESMC operates six ELV pads at Cape Canaveral Air Force Station (CCAFS) for orbital missions. Most of these pads are fully booked for the next several years. These pads would provide the capacity required for the number of launches currently scheduled for ESMC if all operations took place as planned. However, complications and delays in the preparation of a launch vehicle on the pad are quite common, and can delay later missions scheduled to use the pad. Depending on the length of the delay, it is possible that the effects of the schedule slippage could extend over several subsequent missions.

Some of the factors that cause on-pad delays include:

- Hardware problems, including test anomalies in the launch vehicle or payload while being prepared for launch on the pad, and system standdowns following launch failures
- Payload preparation and on-pad encapsulation
- Launch window constraints and interference from operations at adjacent launch pads

The impact of some of these factors could be reduced by modifying existing hardware and procedures for preparing launches, or by building additional facilities for preparing launch vehicles and payloads at CCAFS; some of these measures, such as those requiring modifications to spacecraft, would require a [v] long-term process. However, even these measures would not in themselves eliminate the basic constraints imposed by the limited number of launch pads.

2. Procedures for scheduling launches.

Several features of the procedures used to assign launch dates to commercial payloads at U.S. ranges appear to reduce the confidence of payload customers that companies operating from U.S. ranges will deliver a payload to orbit on schedule. These include:

- **U.S. launch companies are assigned a 3-month launch slot, rather than a firm launch date before a sale is made.** Unlike foreign companies, U.S. firms, when bidding for a launch contract, cannot contractually commit the ranges they use to launch a mission on a specific date. This is primarily because, whereas foreign launch services are "vertically integrated" and operate from their own ranges, U.S. launch companies depend on launch facilities owned by an independent party—i.e., the U.S. government. Thus, the procedure at U.S. ranges is to provide range users (commercial launch companies and DOD programs alike) a tentative three-month launch slot when an initial request for launch support is made. In the case of commercial launch companies, the assignment of this slot is contingent on a successful sale. A firm launch date is then provided one year prior to launch. Several payload customers indicated that this level of commitment was less satisfactory than that provided by foreign launch services.

- **Allocation of launch opportunities is limited by law and the National Space Policy.** Given current range resources, it is possible that the U.S. launch industry would be unable to obtain launch slots that would be necessary to capture a larger share of the commercial launch market. The National Space Policy gives first priority for the use of national ranges to government payloads. Also, Section 15(A) of the Commercial Space Launch Act authorizes government agencies to provide only launch property that is "excess" or "otherwise not needed for public use," and launch services that are "otherwise not needed for public use." These policies are reflected in the Model Agreement, the basic contractual agreement permitting U.S. companies access to national ranges. Under current procedures, the number of commercial launches that will be supported at a government range is determined in the Model Agreement which, consistent with the policy of first priority[,] allocates launch opportunities to commercial launch companies after government requirements are determined.

- **Procedures for rescheduling delayed launches.** Although range operators assign initial launch dates on a "first come, first served" basis, the procedures for reassigning launch dates if the schedule is disrupted would be handled on a case-by-case basis. As noted above, the National Space Policy and Commercial Space Launch Act give the national government the first priority at national ranges. The Air Force has indicated that, in general, should a commercial launch slip, it would retain its place "in line," unless national security or critical mission requirements required otherwise. However, these conditions are not defined formally at either the national policy level or at the level of the Model Agreement, and the manner in which national needs would be weighed against the need to promote a vigorous commercial launch industry is not documented.

On the other hand, the often-cited possibility of "preemption," in which a scheduled commercial mission being prepared for launch would be removed from line to make way for a government mission, is in reality extremely small, if not negligible. In particular, OCST noted:

- Commercial launch vehicles are owned and titled to private firms and thus cannot be seized by the government. Only the Department of Transportation has the authority to take such action and, even then, only in the event of a national emergency.

- Because launch vehicles are usually designed for specific payloads, it would be extremely difficult from a technical viewpoint to replace a commercial payload on a launch vehicle with a government payload. Moreover, it is technically difficult to prepare a mission quickly enough that it would make sense from the viewpoint of the U.S. government to remove a commercial launch vehicle from a launch pad in order to make room for a government mission.

- Section 7 of the Commercial Space Launch Act Amendments requires the Secretary of Defense or the Administrator of NASA, in consultation with the Secretary of Transportation, to approve any preemption personally, to approve such [vi] actions only when imperative national needs are at stake, and to report to Congress when such action is taken. This elaborate system tends to discourage such actions.

Historically, the main cause for launch date delays has been hardware problems— usually when a component in the launch vehicle or payload fails a prelaunch test on the launch pad, or causes a launch failure, and the anomaly must be identified and corrected. Thus, to a great extent the launch companies themselves have a significant degree of control in maintaining schedules.

3. Potential single point failures at launch ranges.

The loss of any one of several critical facilities for preparing launch vehicles from accident, natural disaster, or attack could bring the system for operating a particular type of launch vehicle to a halt. In the case of ESMC, these include (but are not limited to):

- Launch pads
- Solid rocket storage facilities
- The Titan Solid Motor Assembly Building
- Certain ground equipment used for the movement, inspection, and testing of launch vehicles

Although the probability of a single point failure may be small, and although foreign launch facilities have similar vulnerabilities, it is nevertheless a potential factor in the ability of U.S. launch companies to maintain schedules.

4. Safety review procedures.

Few launch companies expressed dissatisfaction with the range safety requirements themselves or the level of acceptable risk implicit in these requirements. However, several industry representatives indicated that the procedures for implementing these standards are time consuming and sometimes duplicative, with the same data having to be provided to several offices. Further, payload customers remarked that safety regulations were not coordinated with range users, nor were they developed with consideration of their economic impact on commercial operations.

It must be noted that the range safety requirements are intended by the Air Force to protect both public safety and a national resource (i.e., the ESMC launch facilities) and to prevent damage that would affect the ability to conduct both government and commercial launches. Even so, it must also be noted that these procedures were originally designed for government operations. In designing these procedures, the Air Force intended to ensure both the protection of government property and personnel and the success of government missions; however, in a commercial environment, safety is equally important, but the risk of mission success is generally left to the firm undertaking the operation.

The most important reason for delays in safety reviews appears to be a lack of adequate safety staff at ESMC, given the many procedures which must be observed by a range safety representative, and the extensive paperwork which must be reviewed and approved by range safety staff. This is a transitional difficulty, as additional safety personnel are now being put into place. Other factors affecting responsiveness include:

- Most (but not all) ESMC safety personnel are currently located at Patrick Air Force Base, 20 miles from the launch facilities at CCAFS. Onsite safety representatives would improve the accessibility and availability of the safety organization for both commercial and military launches. One launch firm representative suggested using safety representatives trained by and accountable to the agency responsible for safety, but salaried by the commercial firm and located on the firm's premises (implementation of such an approach would, of course, require measures to ensure that conflict of interest situations did not occur).

- Some commercial launch company representatives expressed a concern that ESMC officials had begun to interpret existing safety requirements more narrowly, possibly in response to the loss of the Challenger and ELV mission failures that had occurred at about the same time. For example, in one case the definition of "lifting equipment" that were required to be certified was expanded to include not only cranes and hoists, but also transporters equipped with jacks, trucks equipped with loading platforms, etc. These representatives stated that some mechanism was required to ensure consistency with the intent of the requirements.

[vii] Conclusions

OCST believes that U.S. range operators have made significant progress in supporting commercial launch companies in a manner consistent with the Commercial Space Launch Act and National Space Policy. However, if the U.S. is to be competitive in the years ahead, additional measures are likely to be required to establish a reputation for U.S. ranges being concerned with the demands of the market and the needs of launch customers.

The Department of Defense and the Department of Transportation have forged a close working relationship to address many of the issues raised in this study and to work to resolve them. Together with the cooperation of the private sector, DOD and DOT will continue their efforts to provide an environment conducive to the development of a robust, competitive space launch industry.

In some cases, the options for addressing the issues cited above are straightforward. For example, the constraints created by the limited number of pads could be alleviated by constructing new pads, by refurbishing unused pads, by implementing procedures that minimize on-pad time (such as payload encapsulation at off-pad facilities), or by introducing new technologies (such as off-shore launch systems). The possibilities offered by such potential measures suggests that the national space infrastructure warrants additional study, especially in light of the current interest in commercial and state-operated spaceports.

OCST believes, however, that such specific issues can be addressed only by considering them in the broader context of how the national ranges can meet the commercial requirements of launch companies and their payload customers. Ultimately, whether U.S. ranges provide "commercially reasonable" conditions depends on whether launch companies and payload customers believe conditions at national ranges are commercially reasonable, given the international market for launch services. Thus, OCST notes that any strategy for improving the competitiveness of the U.S. launch industry must be oriented toward providing not just adequate support or support consistent with that provided to U.S. government missions. Rather, such a strategy must be oriented toward providing support that, without direct federal subsidy or adverse affects on national security interests, is fully competitive with the launch support that is available on the world market. . . .

Document III-24

Document title: "Memorandum of Agreement Between the Government of the United States of America and the Government of the People's Republic of China Regarding International Trade in Commercial Launch Services," January 26, 1989.

Source: International Trade Administration, Department of Commerce, Washington, D.C.

The People's Republic of China began to market its Long March space launch vehicle as a commercial space launcher in 1985. It offered potential customers a price considerably lower than that being offered by Arianespace and competitors in the emerging U.S. commercial space launch industry. In both Europe and the United States, space launch providers complained that because China was a non-market economy, this price did not have to reflect actual costs and that China was in essence subsidizing its entry into the commercial launch market. On the other hand, some non-U.S. buyers of communications satellites wanted to take advantage of the Chinese prices, which would lower their costs of doing business. Because U.S. satellite manufacturers were required to get an export license under the International Trade in Arms Regulations to ship a satellite to China for launch, the U.S. government was able to control whether those manufacturers could compete for contracts that specified a Chinese launch.

The U.S. government tried to balance the competing interests of U.S. satellite manufacturers and commercial launch service providers by negotiating this agreement with the Chinese government. It set a quota on the number of Chinese launches carrying an American-built satellite and otherwise specified conditions under which China could enter the global competition for launch contracts. Similar launch trade agreements were signed between the United States and Russia in 1992 and the United States and the Ukraine in 1995.

[1]

Memorandum of Agreement Between the Government of the United States of America and the Government of the People's Republic of China Regarding International Trade in Commercial Launch Services

I.   PURPOSE

The Government of the United States of America (U.S.) and the Government of the People's Republic of China (PRC) have entered into this Memorandum of Agreement

(Agreement), of which the attached Annex is an integral part, to address certain issues regarding international trade in commercial launch services including entry in an appropriate manner of the PRC into the international market for commercial launch services.

II. TRADE ISSUES AND MARKET ENTRY

The Delegation of the People's Republic of China and the Delegation of the United States of America held two rounds of negotiations in Beijing and Washington, D.C. As a result of these discussions, the parties have agreed that certain measures are appropriate to address certain issues regarding international trade in commercial launch services, including entry in an appropriate manner of PRC providers of commercial launch into the international market for commercial launch services. Accordingly, the U.S. and the PRC have agreed as follows:

a.   The U.S. and the PRC support the application of market principles to international competition among providers of commercial launch services, including the avoidance of below-cost pricing, [2] government inducements, and unfair trade practices.

b.   To bring about entry in an appropriate manner, the PRC shall take steps to ensure that providers of commercial launch services controlled by or operating within the territory of the PRC do not materially impair the smooth and effective functioning of the international market for commercial launch services.

(i) Among these steps, the PRC shall ensure that any direct or indirect government support extended to its providers of commercial launch services is in accord with practices prevailing in the international market.

(ii) The PRC shall require that its providers of commercial launch services offer and conclude any contracts to provide commercial launch services to international customers at prices, terms, and conditions which are on a par with those prices, terms, and conditions prevailing in the international market for comparable commercial launch services.

(iii) The PRC agrees that it will prevent its providers of commercial launch services from offering introductory or promotional prices for launch services except for the first or, in extraordinary circumstances, second successful commercial launch of a new launch vehicle. In this regard, promotional prices will not be offered for launches on the Long March IIE or III under any contract other than the contract for the successful launch of the AUSSAT B-1 and B-2 satellites.

(iv) The PRC agrees to require its launch service or [3] insurance providers to offer international customers any insurance or reflight guarantees on a par with prevailing rates and practices in international markets for comparable risk.

c.   In view of the concerns about the launch services market expressed by several countries, the PRC expressed its understanding. The PRC explained that: China has a limited capability of manufacturing launch vehicles. In addition to meeting the needs of domestic Chinese satellite launches, its providers of commercial launch services are only able to offer a limited number of communications satellite launches each year for international customers. Chinese launch services, therefore, are only a supplement to the world market, providing international customers with a new option.

After mutual and friendly consultations, the U.S. and the PRC agreed:

(i) PRC providers of commercial launch services shall not launch more than 9 communications satellites for international customers (including the two AUSSAT and one ASIASAT satellites) during the period of this Agreement, and

(ii) The PRC shall require that any commitments to provide commercial launch services to international customers by PRC launch service providers are proportionately distributed over the period of the Agreement. To this end, the PRC shall prevent a disproportionate [4] concentration of such commitments during any two-year period of the Agreement. The PRC may make commitments in any 3-year period of the Agreement consistent with subparagraph (i) above. The PRC shall also require that PRC launch service providers shall not commit at any time to launch in any calendar year covered by the Agreement more than twice the average annual number of launches permitted under subparagraph (i) above. The PRC shall seek to ensure that PRC launches of communications satellites for international customers are performed as scheduled in the original launch commitment.

    d. The U.S. stated that the U.S. does not provide government inducements of any kind in connection with the provision of commercial launch services to international customers which would create discrimination against launch service providers of other nations and has no intention of providing such inducements in the future. Accordingly, the PRC stated it agreed not to offer inducements of any kind in connection with the provision of commercial launch services to international customers which would create discrimination against launch service providers of other nations.

III. NON-DISCRIMINATION

    1. The U.S. stated that U.S. providers of commercial launch services do not discriminate unfairly against any international customers or suppliers and that it is not U.S. Government policy [5] to encourage any such unfair discrimination by U.S. providers of commercial launch services.

    2. Accordingly, in implementing its commitments under this Agreement, the PRC shall require that its providers of commercial launch services not discriminate unfairly against any international customers or suppliers.

IV. CONSULTATIONS

    1. The PRC and U.S. will consult annually with respect to the obligations in this Agreement and related matters, including the nature and extent of direct and indirect government support provided to commercial launch services providers and developments in the international market for commercial launch services.

    2. In addition, each party undertakes to enter into consultations within thirty (30) days of a request by the other party to discuss matters of particular concern.

    3. During annual consultations, the limitation on the total number of communications satellites that may be launched by PRC providers of commercial launch services may be reconsidered upon request of the PRC in light of unforeseen developments in the commercial launch services market. A U.S. decision on such a request shall be made within thirty (30) days after the completion of the annual consultations.

    4. The U.S. and the PRC agree to work toward a common understanding of the application of market principles to prices, terms, and conditions of commercial launch services for international [6] customers.

    5. To facilitate the annual consultations, the U.S. and the PRC agree to exchange information as follows:

    (a) The U.S. shall each year in advance of such consultations provide to the PRC such publicly releasable information as it possesses with respect to prices, terms and conditions prevailing in the international market for commercial launch services.

    (b) The PRC shall each year in advance of such consultations provide comprehensive information to the U.S. regarding prices, terms, and conditions offered by PRC providers of commercial launch services for the launch of satellites licensed by the U.S. The PRC may also provide other information that it believes may have a material effect on pricing practices of PRC providers of commercial launch services.

(c) The PRC may request that the U.S. provide additional publicly releasable information with respect to international prices, terms and conditions, and may in addition request U.S. views regarding prevailing international market conditions and likely future developments, as well as government supports or inducements. The U.S. shall respond to such requests within thirty (30) days. If such information cannot be provided directly because of business confidentiality, the U.S. shall provide such information in summary form.

(d) The U.S. may request additional information with respect to the prices, terms, and conditions offered by PRC providers [7] of commercial launch services and any PRC government supports or inducements. The PRC shall respond to such requests within thirty (30) days. If such information cannot be provided directly because of business confidentiality, the PRC shall provide such information in summary form.

(e) The U.S. and the PRC shall keep all information received from each other under this paragraph strictly confidential and shall not provide it to any other government or any private person without the written consent of the other.

6.   The U.S. and the PRC shall also provide each year in advance of annual consultations information on a consolidated basis concerning the commitments their launch service providers have undertaken to provide commercial launch services for international customers. This information may be made publicly available.

7.   If a launch of a communications satellite for an international customer will not be performed as scheduled, the PRC shall notify the U.S. regarding the reasons for the delay and the new date for the launch as soon as possible.

8.   It is understood that the U.S. and the PRC will review the information contained in this Article during annual consultations in the context of developments in the international market for commercial launch services.

V.   CLARIFICATION OF RIGHTS AND OBLIGATIONS

1.   If, after friendly consultations with the PRC, the U.S. determines that there is clear evidence that the provisions of [8] this Agreement have been violated, the U.S. reserves its right to take any action permitted under U.S. laws and regulations. The U.S. shall seek to avoid actions inconsistent with this Agreement.

2.   With regard to export licenses, any application for a U.S. export license will be reviewed on a case-by-case basis consistent with U.S. laws and regulations. Nothing in this Agreement shall be construed to mean that the U.S. is constrained from taking any appropriate action with respect to any U.S. export license, consistent with U.S. laws and regulations. Nevertheless, the U.S. will do its utmost to assure, consistent with U.S. laws and regulations, continuity of issued license(s) and the completion of the transactions covered in such license(s).

VI. DISCUSSIONS ON INTERNATIONAL RULES

The U.S. and the PRC are prepared to enter into discussions with other interested parties on comprehensive international rules with respect to government involvement in, and other matters relating to, the international market for commercial launch services. It is understood, however, that nothing in this Agreement shall prejudice any position on any issue that either the U.S. or the PRC may take in those discussions.

VII. COMPREHENSIVE REVIEW

The U.S. and the PRC shall engage in a comprehensive review of the terms and operation of this Agreement beginning in September 1991.

[9] VIII. <u>ENTRY INTO FORCE</u>

This Agreement shall enter into force upon notification by the Government of the United States of America to the Government of the People's Republic of China that a U.S. license for the export of the ASIASAT or AUSSAT satellite(s), or any other satellite, to the People's Republic of China for launch therein, has been approved. Unless extended by agreement of the PRC and the U.S., this Agreement shall terminate on December 31, 1994. It may be terminated at any time by mutual agreement if superseded by an international agreement on government involvement in, and other matters relating to, the international market for commercial launch services or under such other circumstances as may be mutually agreed.

IN WITNESS WHEREOF, the undersigned, being duly authorized by their respective Governments, have signed this Agreement.

DONE at Washington, D.C., in duplicate, in the English and Chinese languages, both texts being equally authentic this twenty-sixth day of January, 1989.

For the Government of the United States
of America:

For the Government of the People's
Republic of China:

[10] <u>ANNEX</u>

The following agreed definitions constitute an integral part of the Memorandum of Agreement Between the Government of the United States of America and the Government of the People's Republic of China Regarding International Trade in Commercial Launch Services of January 26, 1989.

1. The term "commercial launch services" refers to any commercially provided launch of any satellite, including communications satellites, for an international customer.

2. The term "communications satellite" refers to any satellite which is a primary payload of a launch, and which provides telecommunications services. It refers primarily to, but is not limited to, communications satellites in geostationary orbit.

3. The term "international customer" refers to the following:
    (a) any institution or business entity, other than those institutions or entities located within the territory of the PRC and owned or controlled by PRC nationals; or
    (b) any government other than that of the PRC; or
    (c) any international organization or quasi-governmental consortium;
[11] which is the ultimate owner or operator of a satellite or which will deliver the satellite to such ultimate owner or operator.

4. The term "practices prevailing in the international market" in Article II (b)(i) refers to practices by governments of market economies.

5. The term "prices, terms, and conditions prevailing in the international market for comparable launch services" in Article II (b)(ii) includes but is not limited to prices, financing terms and conditions and the schedule for progress payments offered to international customers by commercial launch service providers in market economies.

6. Government "inducements" with respect to particular launch services transactions include, but are not limited to, unreasonable political pressure, the provision of any resources of commercial value unrelated to the launch service competition and offers of favorable treatment under or access to: defense and national security policies and programs, development assistance policies and programs, and general economic policies and programs (e.g., trade, investment, debt, and foreign exchange policies).

7. The term "commitment" means any agreement by an international customer with PRC providers of commercial launch services to launch a communications satellite, which effectively removes the [12] launch from international commercial competition. The term "commitment" does not include reservation agreements.

Chapter Four

Exploring Future Space Transportation Possibilities

by Ivan Bekey

After the 1981 introduction of the Space Shuttle into service, the rest of the 1980s and early 1990s marked one and a half decades for space transportation that might be characterized as mostly running in place. The shortcomings of various U.S. space launch systems became well understood. There were numerous new transportation system concepts generated and clear administration policy statements issued. Also, more than enough major studies of potential new launch systems were carried out. However, except for the introduction of the Titan IV into the U.S. launch fleet, there was little tangible progress until about 1994. The period since then has seen some progress, however, and as the century nears its end, there is hope that the space transportation picture is improving in significant ways. This essay discusses some of the major steps and undercurrents that shaped progress toward developing advanced space transportation systems—or rather the lack of it—in the 1981–1994 time period. It also describes more recent forward movement.

The Background: Air Force-NASA Antagonism and Early Studies of New Space Transportation Systems

During the 1970s, the U.S. Air Force operated and upgraded a fleet of reliable expendable launch vehicles (ELVs), including the Delta, Atlas, and Titan III/34D boosters.[1] NASA successfully developed the partially reusable Space Shuttle, despite its very advanced technologies and inadequate budget. In 1981, the space agency resumed the human spaceflights suspended since the 1975 Apollo-Soyuz mission. As the 1980s began, U.S. space transportation capabilities seemed in good order.

Underneath this apparent progress were turbulent undercurrents that were to shape events in the launch vehicle area for at least fifteen years. These were driven to a considerable degree by Air Force-NASA animosity resulting from a series of decisions during the 1970s, forced on the Air Force by its civilian leaders in the Pentagon and the White House, to end the use of all of its ELVs and commit to flying all its payloads on the Space Shuttle when it entered operational service. Although largely under the surface, the impacts of this antagonism on events in the 1980s and 1990s must not be underestimated. The Air Force had worked very hard and long to develop and refine its expendable vehicle fleets to respectable reliability. Switching all payloads to the expensive-to-operate Space Shuttle, and developing a west coast Shuttle launch facility for launches into polar orbit, imposed unwelcome additional burdens on the Air Force budget. Furthermore, the Air Force, as the service responsible for assuring U.S. access to space, felt very strongly that the classified character and criticality of National Reconnaissance Office (NRO) payloads to national security and the importance of Department of Defense (DOD) payloads to national

---

1. David N. Spires, *Beyond Horizons: A Half Century of Air Force Space Leadership* (Washington, DC: U.S. Government Printing Office, 1997), pp. 113–15.

defense meant that their launches should be under Air Force control; they should not be intermingled with the public limelight associated with the open civil space program.[2]

In addition, senior Air Force officers argued, without success, that the urgent national need for DOD and NRO payloads made it inappropriate to commit in advance to launching them on the untested Space Shuttle, which they felt would have extensive downtimes after inevitable failures, precisely because it was a manned system. NASA countered that the extensive instrumentation on the Space Shuttle would result in quick failure determination and that the use of solid-fueled boosters, which were thought to be more reliable than liquid-fueled alternatives, would mean short downtimes. Furthermore, as long ago as 1971, NASA had made a telling point with the White House: the Space Shuttle needed the DOD and NRO payloads in its manifest if it was to fly frequently enough to become the cost-effective launch vehicle that it was proposed to be.[3]

In hindsight, it can be seen that both sides were right. The Space Shuttle has evolved into the world's most reliable launch vehicle; however, the downtime from its only failure was almost three years—far longer than either the Air Force or NASA had expected. This lengthy absence from service was to a significant degree determined by the media-driven politics of humans in space.[4]

These feelings came to the surface as the Space Shuttle entered operational service in 1982 and culminated with the January 1986 *Challenger* accident. Air Force concerns were intensified by the nearly concurrent but unrelated 1986 launch failures of a Delta and Titan 34D, so that for a period of time the United States was essentially grounded.[5] Even after the Titan returned to flight, it could not launch a number of heavy payloads critical to national security, because they had been redesigned so that only the Space Shuttle could launch them.

Even before the *Challenger* accident, the Air Force leadership had succeeded in convincing the White House that it was unwise to have only one means of getting the most critical national security payloads into orbit. In 1985, the Air Force received approval to develop a new heavy-lift launch vehicle, dubbed the Complementary Expendable Launch Vehicle (CELV), which would have substantially the same capability as the Space Shuttle but be unmanned and expendable. The Air Force preferred to develop a CELV that would be an evolution of the Titan 34D. NASA countered with a proposal for an unmanned cargo derivative of the Space Shuttle, dubbed Shuttle-C, which would have a heavy-lift capability three to four times greater than that of the CELV. The Shuttle-C would thus not only be able to launch critical DOD/NRO payloads, but also the space weapons of the Strategic Defense Initiative (SDI) and NASA's crewed Mars exploration vehicles, both at that time in early planning stages. In addition, an important consideration for NASA was that the increased use of Space Shuttle components would reduce the flight costs of the Space Shuttle for other NASA missions.

Not unexpectedly, the Air Force chose the new expendable vehicle route. This decision was made for a number of reasons, not the least of which was to give back to the Air Force control over its own launches. However, a significant part of the rationale was to enable the nation to have two different heavy launch vehicles, resulting in a more robust

2.   For a discussion of the stresses in the NASA-Air Force relationship and to examine some of the documents that reflect that stress, see the essay by Dwayne A. Day, "Invitation to Struggle: the History of Civilian-Military Relations in Space," and Documents II-29 through II-44 in John M. Logsdon, gen. ed., with Dwayne A. Day and Roger D. Launius, *Exploring the Unknown: Selected Documents in the History of the U.S. Civil Space Program, Volume II: External Relationships* (Washington, DC: NASA Special Publication (SP)-4407, 1996), 2: 233-70, 364-410.

3.   Jerry Grey, *Enterprise* (New York: William Morrow, 1979), pp. 57-88.

4.   For a discussion of the process of bringing the Space Shuttle back into service after the *Challenger* accident, see John M. Logsdon, "Return to Flight: Richard H. Truly and the Recovery from the *Challenger* Accident," in Pamela E. Mack, ed., *From Engineering Science to Big Science* (Washington, DC: NASA SP-4219, 1998), pp. 345-64.

5.   Spires, *Beyond Horizons*, p. 222.

launch capability. The new expendable vehicle was dubbed the Titan IV and is still the largest vehicle in the Air Force fleet. Nevertheless, it has turned out to be as expensive per launch as the Space Shuttle, for reasons having to do as much with manufacturing costs as with the propensity of the Air Force to customize each launcher to its payload and to take up to eighteen months to integrate payloads and vehicles on the pad.

During the 1984–86 period, national space transportation policy deliberations were shaped by a number of studies, conducted by ad hoc groups convened by the White House, the congressional Office of Technology Assessment, and the National Research Council.[6] These studies uniformly restated these conclusions, as follows:

- The U.S. launch capability was broken and needed fixing.
- The needs and characteristics of human spaceflight and national security and other cargo payloads were such that separate launch capabilities should exist for both.
- Costs were excessive for both.
- Technologies could be identified that could make major improvements, but their development required additional funds (which were not forthcoming).
- The needs of commercial programs should be considered to make the U.S. launch industry cost competitive once again.

Despite these perceptions of a problematic reality, little concrete was done to address critical problems for almost a decade.

Lots of Studies, But Little Progress

An initial step in reacting to already perceived problems was President Ronald Reagan's National Space Strategy of 1984; this strategy laid out a set of early remedial steps to be pursued by NASA and the Air Force. [IV-1] These included a call for cooperative study defining desirable options for future space transportation systems. The possibility of launch requirements generated by a space-based ballistic missile defense system was to be a factor in the study, which would be preceded by a cooperatively developed technology plan. This plan, called the Launch Vehicle Technology Study, was issued in December 1984. [IV-2]

That study was followed in 1985–86 by a major joint NASA-Air Force effort to specify the preferred "architecture" (the general characteristics of separate systems and how they would complement one another) of a future U.S. launch capability; this effort was called the National Space Transportation and Support Study. [IV-3] This study developed a number of possible architectures, each of which could, in principle, satisfy the assessed future needs. These included expendable, partially reusable, and fully reusable launch vehicles, as well as a set of upper stages and orbit-to-orbit stages, including both expendable and reusable designs.

The National Space Transportation and Support Study was influential; it solidified White House and congressional support for moving forward with the definition of new space transportation capabilities. Most notable among these were the call for a new unmanned cargo vehicle and the decision to continue to keep manned and unmanned

---

6. Harry S. Dawson, *Review of Space Shuttle Requirements, Operations, and Future Plans* (Washington, DC: U.S. House of Representatives Committee on Science and Technology, 1984); Hearings before the Subcommittee on Space Science and Applications, 98th Cong., 2d sess., *Space Shuttle Requirements, Operations, and Future Plans* (Washington, DC: U.S. House of Representatives Committee on Science and Technology, 1984); Hearings before the Subcommittee on Science, Technology, and Space on 10 September 1984, 98th Cong., 1st sess., *Vandenberg Space Shuttle Launch Complex* (Washington, DC: U.S. Senate Committee on Commerce, Science, and Transportation, 1984); David H. Moore, *Pricing Options for the Space Shuttle* (Washington, DC: U.S. Senate Budget Committee, 1985); *Assured Access to Space During the 1990s* (Washington, DC: U.S. House of Representatives Committee on Science and Technology, 1986); *Assured Access to Space: 1986* (Washington, DC: U.S. House of Representatives Committee on Science and Technology, 1986).

vehicles separate. In addition, the study concluded that there was not an urgent need for an advanced manned vehicle; incremental improvements to the Space Shuttle would suffice.

While the initial study results treated both near-term and longer term options, such as reusable single- and two-stage launch vehicles, the study eventually concluded that a nearer term capability was urgently needed, which could be either expendable or partially reusable. This conclusion, reinforced by the space transportation crisis following the 1986 *Challenger* accident, led in 1987 to a joint Air Force-NASA system definition study of a launcher family called the Advanced Launch System (ALS). [IV-4, IV-5] The ALS was studied extensively, but in 1989, it was abandoned when it became clear that it could not deliver the promised major cost savings that were initially touted. This result was a surprise only to those deeply involved in the ALS study activity, for it was a foregone conclusion to many others from the outset, based on results from a number of other NASA and industry studies.

The state of affairs in space transportation was reviewed again in 1990–91; the result was the Bush administration's National Space Policy Directive 4, "National Space Launch Strategy."[7] This presidential directive was action oriented. It directed NASA and the Air Force to develop a new jointly funded and jointly managed launch vehicle system. It was to be initially unmanned but manned later on. It would have major improvements in reliability, cost, and responsiveness. The directive also tasked the space agency and the Air Force to coordinate their technology programs to enable more advanced and reusable vehicles to eventually complement the jointly developed new system and to actively consider and support commercial industry and its space launch needs in their activities.

The main result from these directives was the start of the National Launch System (NLS) program, a jointly funded and jointly managed NASA-Air Force system with a single program office. Though well intentioned, the NLS program was characterized by two less-than-exemplary major features. As in the previous ALS program, there was consistent over-optimism on the cost savings that could be expected from a new ELV. Cost goals of 100-percent reduction were still being suggested, when many knowledgeable people in industry and government yet again said that 30- to 50-percent savings were probably the most that expendable vehicles could offer. However, these voices were again ignored in the euphoria of possibly initiating the first major new launch vehicle development in the United States since the Space Shuttle in the 1970s.

The second negative feature was that the joint nature of the management did not work well. Although there were some lingering, privately voiced doubts because of the past Air Force-NASA animosities, officially both NASA and Air Force committed to making the joint program work. Most of the management problems could be attributed to the funding uncertainties caused by the different congressional paths for obtaining budget authorizations and appropriations; there were a number of different committees and subcommittees for NASA and DOD, each subject to different priorities and pressures. Furthermore, the management of one program by a joint program office with the different constituencies and orientations of NASA and DOD was proving increasingly difficult.[8] When confidence in the NLS being able to achieve major cost reductions began evaporating, so did its support in Congress, and the program died without having entered the hardware phase.

Additional studies now documented what was clear to even casual observers: despite presidential directives, little progress was being made toward a new launch vehicle capability, not because of a lack of will or ill intent, but because of poor coordination and differing

7. This strategy is also discussed in Chapter Three in this volume of *Exploring the Unknown* and appears as Document III-19.

8. Jerry Grey, "Ups and Downs of the New Space Launcher," *Aerospace America*, June 1992, pp. 26–29.

goals and orientations. [IV-6][9] Furthermore, an important casualty of the lack of progress was the ever more critical capability for the nation to compete in the commercial space launch arena, which was rapidly becoming larger and more visible on the world stage.[10]

While the ALS and NLS activities were in full swing, efforts were made to coordinate the longer term NASA and Air Force technology programs. These activities resulted in many committees and documents, but unfortunately little newly developed technology. This was partly because of the different orientations of the two organizations. NASA was pushing principally those technologies that could improve the Space Shuttle or apply to a new generation of reusable manned vehicles. [IV-7] Meanwhile, DOD was focusing on those technologies that could improve ELVs. One of the few exceptions was in the engine area; new engine technology test beds were pursued on a cooperative basis.

Developing Advanced Technology Systems

Throughout this period, there were a number of programs that attempted to develop breakthrough technologies and embody them in systems that would allow radically lower costs via reusability. These technologies were generally grossly underfunded, if funded at all. There was one exception—the only really major advanced launch technology program prior to 1994—the National Aerospace Plane (NASP). [IV-8, IV-9]

The NASP was touted as a single-stage-to-orbit (SSTO) fully reusable vehicle using air-breathing engines and wings; it was thus often compared to a very high-speed airliner (Figure 4-1). It was first defined at the Defense Advanced Research Projects Agency through a program known as "Copper Canyon." It was sold as a technology demonstration program that would result in a prototype vehicle. It was also promoted as having a capability to fly halfway around the world in a few hours, with the media dubbing it the "Orient Express."[11] Although it advanced many technologies, the NASP program eventually died, after billions of dollars had been spent, because of serious technical problems and overpromises, which derived as much from political desires for rapid progress as from the great demands placed on the technology itself.

Figure 4-1. An artist's conception of the National Aerospace Plane in orbit. (NASA photo HqL-348)

The technologies that would be needed by the NASP were far more demanding than those for pure rocket SSTO fully reusable launch vehicles were. However, because the NASP took off like an airplane, a large number of operational advantages and radically lower operational costs were claimed for it. Unfortunately, these were principally

---

9. Note that the documents following this essay are not necessarily in chronological order.

10. For a review of the launch vehicle situation during this time period, see U.S. Congress, Office of Technology Assessment, *Access to Space: the Future of U.S. Space Transportation Systems*, OTA-ISC-415 (Washington, DC: U.S. Government Printing Office, April 1990), and Vice President's Space Policy Advisory Board, "The Future of the U.S. Space Launch Capability," November 1992 (parts of which appear as Document IV-6 at the end of this chapter).

11. Fred Hiatt, "Space Plane Soars on Reagan's Support," *The Washington Post*, February 6, 1986, p. A4.

undocumented assertions. Nonetheless, being airplane-like, the NASP concept attracted powerful backing because it was intuitively easy to grasp. The nation fooled itself into believing that
because the NASP image was what was desired, the reality itself was therefore attainable.

While the NASP technologies could also be used for single-stage rocket vehicles that
would be far easier to develop and offered the same or even greater cost reduction potential, advocacy politics surrounding the program were fierce—and effectively suppressed all
dissent. For several years, many at NASA and in industry could not publicly voice any
doubts about the NASP or any support for SSTO rockets for fear of losing their jobs.
Dubious security classification existed around some NASP propulsion concepts, which,
while doubtless protecting some U.S. competitive advantages from other nations, also
unquestionably served to diminish public debate over the merits of the concept.

Part of the reason for the acceptance of the NASP was from a misleading "figure of
merit" that was being promulgated by its advocates. This was that because the vehicle
obtained much of its oxygen from the atmosphere, it could be significantly lighter than
an SSTO rocket vehicle, which has to carry all its oxygen in a tank. While this is a true
statement, it is not a meaningful one. The structure and propulsion system of the NASP
had to be considerably heavier than that of the SSTO rocket to survive the much lengthi-
er and higher heat and dynamic loads inherent in a cruise type air-breathing vehicle.
Furthermore, increasing the size of the already large hydrogen tank could only offset the
resulting large drag losses. Thus, the empty weight of the NASP was bound to be considerably greater than that of an SSTO rocket, even though its gross weight was indeed less.

The significance of this is fundamental, because both the development and production costs of any launch vehicle are based mostly on its empty weight, not its gross
weight—the gross weight depending mostly on propellants, which are relatively inexpensive. Therefore, the NASP would inherently be more expensive to develop and build than
an SSTO rocket and would have dubious operations cost advantages. Nonetheless, the
arguments in favor of the NASP were not challenged for years, and this set back progress
on a more achievable SSTO rocket launch vehicle by almost a decade. [IV-10]

During this period, the overriding financial reality was that politically well-supported
space programs such as the NASP, SDI, and Space Station Freedom, along with the need
to maintain a stable of ELVs, were sucking up most of the funds available, leaving little
funding for new activity. The inevitable result was that technology programs suffered the
most. Study after study, such as NASA's internal Space Shuttle-II conceptual definition,
showed that major cost reductions in launch could only be realized by reusing the hardware, not throwing it away after one use.[12] Nonetheless, mainly because of budget pressures, the conclusions of none of these studies were pursued seriously or resulted in
significant technology programs.

There was one exception to this generalization: the Delta Clipper (DC-X) program
undertaken by the Strategic Defense Initiative Organization (SDIO) during the 1990–93
period (Figure 4–2). Visionary advocates of a rocket-based SSTO vehicle succeeded in
1989 in convincing Vice President Dan Quayle, as chair of the newly reconstituted
National Space Council, that such a vehicle was feasible and could be used to deploy key
elements of the then-current SDI system. After an independent review by the Aerospace
Corporation verified the potential feasibility of the concept, the SDIO let contracts, first
for a study and then for a suborbital demonstration of the operational concepts associated with a rocket-powered SSTO vehicle. Plans were to move toward an advanced technology orbital vehicle, but once the Soviet Union collapsed and ballistic missile defense
concepts were revised, there was insufficient funding available to continue the program.

---

12. See, for example, Office of Technology Assessment, *Access to Space*.

Nevertheless, there were a number of sub-orbital demonstration flights that attracted widespread attention. [IV-11, IV-12, IV-13]

With this important exception, there was nothing happening by the early 1990s in the launch vehicle area except a seemingly endless set of starts and stops on ELVs, as well as poorly funded technology programs that would never result in technology mature enough to be taken seriously as the basis for starting a more advanced vehicle. An attempt to increase the payload capability of the Space Shuttle into higher orbits was undertaken, starting with the development of a version of the Centaur upper stage to fit in the Space Shuttle payload bay.[13] This activity was eventually scuttled because of safety concerns stemming from the Centaur's basic common-bulkhead tank design and from the difficulties of dumping propellants in an abort situation.

Figure 4-2. An artist's conception of the McDonnell Douglas DC-XA vehicle. (NASA photo 95-H-672)

A number of upper stage studies followed to define new and better upper stages to be used for orbit insertion and orbit-orbit maneuvering. Chief among them was a proposed joint Air Force-NASA hydrogen/oxygen short upper stage called the High Energy Upper Stage. It was intended to take as little room in the Space Shuttle payload bay as possible, with length equaling flight charges, as well as to be compatible with flying on ELVs. This concept did not progress beyond the early definition stages, partially because of budget woes and partially because of concerns for the ability to manage a joint NASA-Air Force program.

Finally—More Studies, Then Action

By the early 1990s, the realization that something had to change to make major progress in space launch resulted in two seminal studies. The first was the "Access to Space" study conducted by NASA. [IV-14] This was followed shortly by the "Space Launch Modernization" study by the Air Force. [IV-15] Although conducted by their parent organizations, each of these studies had significant participation from the other agency.

The NASA study was the first post-NLS study to seek impartial conclusions on the type of vehicle most appropriate to develop. It set out to develop an "apples-to-apples" comparison of three main space transportation options for at least the next twenty years. These options were as follows:

1. To upgrade the Space Shuttle and to continue to rely on it for the bulk of payloads and missions
2. To develop a new, mostly expendable vehicle with all improvements and techniques to make it as low in cost as possible
3. To develop a new technology, the reusable launch vehicle, which could be either air breathing or rocket powered and with one or two stages

---

13. Spires, *Beyond Horizons*. p. 225.

This study was organized for maximum credibility, with three internal advocacy teams, one for each approach. This resulted in each of the three advocacy teams giving it their "best possible shot" in putting forward arguments in favor of their assigned concept.

The "Access to Space" study was completed in late 1993 and made public in early 1994. Its results were unequivocal. Improving and upgrading the Space Shuttle were costly and unlikely to result in any significant cost savings. Mostly expendable vehicles, no matter how defined, were incapable of reducing the cost of launch by more than about 30 to 50 percent from then-current levels. In contrast, reusable launch vehicles held the promise of being able to reduce the cost of launch by almost an order of magnitude.

Within this reusable launch vehicle category, another apples-to-apples competition was held between NASP-type air breathers and pure rocket SSTOs. The results were also unequivocal. When compared using the same ground rules, pure rockets had about half the cost to develop and produce than air breathers, were considerably less difficult to achieve because they required far less demanding technology, and cost about the same to operate. Given these results and conclusions, the NASA study recommended that a technology maturation program for SSTO rockets be undertaken and that a flight demonstration vehicle be built to validate the technologies acting together in actual flight.

The Air Force "Space Launch Modernization" study, dubbed the "Moorman Study" after its leader, Lt. General Thomas Moorman, was started when NASA's "Access to Space" study was almost complete. It drew on the NASA study and other prior studies and examined a similar spectrum of vehicles. It also included additional DOD-specific requirements. It came to many of the same qualitative conclusions as the NASA study; however, its recommendations were more conservative than those of NASA were. The study recommended that a new generation of ELVs be pursued. These vehicles would evolve from the then-current ELV fleet, with the advanced technology reusable vehicles being relegated to a future technology activity.

The principal reasons given for this conclusion were that the ELV development costs and technical risks were much lower than those of reusable vehicles. While both reasons were true, this result was nonetheless also consistent with still-prevailing Air Force views that expendable vehicles were "the only way to fly." Greater savings in eventual operating costs from reusable vehicles were sacrificed to obtain an earlier and less expensive development program.

The Moorman study recognized that although the cost reductions from upgraded ELVs that could be expected were relatively modest and in line with those forecast by the NASA "Access to Space" study, lower risk and cost and earlier capability were, from a DOD perspective, mandatory and thus deciding factors. Even though the Air Force officially recognized reusable vehicles as eventually being more desirable than expendable vehicles, they were relegated to a future growth capability. The evolved ELV recommended by the study was thus understood to be, in a significant sense, an "interim" capability.

A major outcome of these two studies by DOD and NASA was the issuing of a new National Space Transportation Policy by the White House in August 1994. [IV-16] The most important aspect of this document was to recognize that the different approaches by NASA and DOD came from fundamentally different orientations and constituencies, and thus it was in the national interest to allow each organization to go its own way. The policy assigned to DOD the responsibility of developing evolved ELVs and to NASA the pursuit of reusable launch vehicles. There was to be cooperation in the technology development programs of both agencies.

This resulted in two major and different programs being rapidly started. The Air Force moved to begin the Evolved Expendable Launch Vehicle (EELV) program, with a goal of creating a modular family of new ELVs with the cost goal of 50-percent reduction of launch costs but a minimum requirement of 25-percent reduction.[14] These new EELVs would replace the costlier Delta, Atlas, and Titan vehicles. NASA started a broad ground technology program to develop all the needed technologies for SSTO rockets and a scaled flight demonstration program of SSTO integrated technologies. This latter program centered on using the DC-X (renamed the "Clipper Graham" after retired Air Force General Daniel Graham, who had been a leading advocate of the concept), the vertical takeoff and landing vehicle originally started by the SDIO but taken over by NASA in its new lead role with respect to reusable vehicles. Unfortunately, the modified DC-X crashed during an early flight test, and it could not be returned to service.

In response to the new National Space Transportation Policy, NASA also initiated the X-33 program (Figure 4-3). The X-33 was to be a half-scale demonstrator for an eventual reusable launch vehicle. The X-33 was to be developed as a cooperative venture between NASA and industry; this was a major change from prior launch vehicle developments, which were totally funded by the government. [IV-17] In 1996, NASA chose Lockheed Martin as the industry contractor for the X-33. This selection was in part based on the firm's design and business plan for a full-scale reusable launch vehicle, which is to be a commercially developed and funded vehicle, known as VentureStar, capable of providing launch services to NASA and other customers (Figure 4-4). NASA has also started the X-34 program and awarded a contract to Orbital Sciences Corporation to demonstrate key technologies applicable to future low-cost reusable launch vehicles (Figure 4-5).

A number of privately funded launch vehicle programs have also appeared in recent years, aiming principally at capturing a market share of the smaller satellites characteristic of recent NASA and DOD efforts and of the multisatellite low-orbit communications constellations that have appeared recently on the scene. Two of these private developments use new approaches, if relatively conventional technologies. The first is the Pegasus, an operational air-launched small vehicle; the other is the Kistler K-1, which is a larger two-stage reusable vehicle.[15] While the Pegasus has so far been successful in its intended small payload market niche, it is an extremely expensive vehicle per pound of payload. The Kistler is an ongoing development that aims at halving the cost of ELVs, and it is completely privately funded.

Figure 4-3. An artist's conception of the X-33 Advanced Technology Demonstrator in flight. (NASA photo MSFC 96-1)

Figure 4-4. An artist's conception of the X-33 in preparation for launch. (NASA photo MSFC 96-2).

14. General Accounting Office, *Evolved Expendable Launch Vehicle: DOD Guidance Needed to Protect Government's Interest*, GAO/NSIAD-98-151 (Washington, DC: U.S. Government Printing Office, June 11, 1998), pp. 1-3.

15. For more information on the Kistler project, see "Kistler Aerospace" file, NASA Historical Reference Collection, NASA History Office, NASA Headquarters, Washington, DC.

Figure 4-5. An artist's conception of the X-34 Technology Testbed Demonstrator in flight. (NASA photo MSFC 97-1)

In addition to the above, some other extremely encouraging recent events are under way, with many small entrepreneurs starting to develop small reusable launch vehicles.[16] These were started in the hopes of capturing some of the new low-Earth orbit communications satellite market, but they are also offering many innovative introductory-level potential services, ranging from public space travel (space tourism) to fast package delivery anywhere in the world. A number of different technical approaches are being pursued by these ventures, which include Rotary Rocket, Pioneer, Kelly, and others.[17] This is an extremely encouraging development, as the true era of space entrepreneurship seems to have started.

History teaches that such entrepreneurial involvement is from where the real service improvements and cost reductions come. After a lengthy period of stagnation in new space transportation developments, the outlook for the next decade, with combined government and private-sector involvement, is thus extremely promising.

---

16. An example is Rhoton, Inc., which is developing a revolutionary rotary rocket that will land with helicopter blades rather than a parachute.

17. See, for example, Robert Pearlman, "Space Tourism: A Consumer's Guide," *Ad Astra,* May/June 1998, pp. 22–27; Gregg Maryniak, "X Prize Update," *Ad Astra,* May/June 1998, pp. 30–36; Stewart Taggart, "Rocket Change," *Wired,* October 1998, pp. 139–44, 202.

Document IV-1

Document title: The White House, Fact Sheet, "National Space Strategy," National Security Decision Directive 144, August 15, 1984.

Source: The National Archives, Washington, D.C.

Space policy issues during most of Ronald Reagan's administration were addressed by a Senior Interagency Group (Space) operating within the framework of the National Security Council. Between 1983 and 1998, this group issued a number of policy directives. The major purpose of this directive was to set out the comprehensive principles to govern the Reagan administration's approach to major space issues. The directive also contained the first "post-Shuttle" call for examining the technologies that would be needed for future space launch systems.

|1| FACT SHEET

National Space Strategy

INTRODUCTION

On August 15, 1984, the President approved a National Space Strategy designed to implement the National Space Policy, as supplemented by the President's 1984 State of the Union Address. The strategy identifies selected, high priority efforts and responsibilities, and provides implementation plans for major space policy objectives. This strategy is consistent with other space-related National Security Decision Directives and other Administration policies. A summary of the strategy's contents is provided below.

THE SPACE TRANSPORTATION SYSTEM (STS)

— Insure [sic] routine, cost-effective access to space with the STS. The STS is a critical factor in maintaining U.S. leadership, in accomplishing the basic goals of the National Space Policy, and in achieving a permanent manned presence in space. It is the primary space launch system for both national security and civil government missions. As such, NASA's first priority is to make the STS fully operational and cost-effective in providing routine access to space.

Implementation: The STS program will receive sustained commitments by all affected departments and agencies. Enhancements of STS operational capability, upper stages, and efficient methods of deploying and retrieving payloads will be pursued as national requirements are defined.

NASA and Department of Defense will jointly prepare a report that defines a fully operational and cost-effective STS and specifies the steps leading to that status. This will be prepared and submitted for review by the Senior Interagency Group for Space— SIG(Space)—no later than November 30, 1984.

The STS will be fully operational by 1988. On October 1, 1988, prices for STS services and capabilities provided to commercial and foreign users will reflect the full cost of such services and capabilities. NASA will develop a time-phased plan for implementing full cost recovery for commercial and foreign STS flight operations. At a minimum, this plan will include an option for full cost recovery for commercial and foreign flights which occur after October 1, 1988. OMB [Office of Management and Budget], in consultation with [the Department of Commerce], [the Department of Transportation], DOD, NASA and other agencies[,] will prepare a joint assessment of the ability of the U.S. private sector and the STS to maintain international competitiveness in the provision of launch services. This analysis should include an assessment of all factors relevant to foreign ELVs, U.S. ELVs and the STS. NASA will keep OMB fully apprised of the [2] elements of its time-phased plan as it is being developed. Both the time-phased plan and the OMB analysis will be submitted for review and comment by the SIG (Space) and the Cabinet Council on Commerce and Trade no later than September 15, 1984, and subsequently submitted for the President's approval in order to permit their consideration in the development of the FY 1986 budget.

The Department of Defense and NASA will jointly conduct a study to identify launch vehicle technology that could be made available for use in the post-1995 period. The study should be completed by December 31, 1984.

THE CIVIL SPACE PROGRAM

–   Establish a permanently manned presence in space. NASA will develop a permanent-ly manned Space Station within a decade. The development of a civil Space Station will further the goals of space leadership and the peaceful exploration and use of space for the benefit of all mankind. The Space Station will enhance the development of the com-mercial potential of space. It will facilitate scientific research in space. It will also, in the longer term, serve as a basis for future major civil and commercial activities to explore and exploit space.

Implementation: As a civil program, the Space Station will be funded and executed by NASA beginning in FY 1985 with the goal of the establishment of a permanently manned presence in space within a decade.

–   Foster increased international cooperation in civil space activities. The U.S. will seek mutually beneficial international participation in its civil and commercial space and space-related programs. As a centerpiece of this priority, the U.S. will seek agreements with friends and allies to participate in the development and utilization of the Space Station.

Implementation: NASA and the Department of State will make every effort to obtain maximum mutually beneficial foreign participation in the Space Station program, consis-tent with the Presidential commitment for international participation and other guid-ance. The broad objectives of the United States in international cooperation in space activities are to promote foreign policy considerations; advance national science and tech-nology; maximize national economic benefits, including domestic considerations; and protect national security. The suitability of each cooperative space activity must be judged within the framework of all these objectives. Consistent with these objectives, the SIG (Space) will review all major policy issues raised by proposed agreements for interna-tional participation on the Space Station program prior to commitments by the U.S. Government.

[3] – Identify major long-range national goals for the civil space program. Major long-range goals for the civil space program are essential to meeting the national commitment to maintain United States leadership in space and to exploit space for economic and sci-entific benefit.

Implementation: In accordance with the FY 1985 NASA Authorization Act, the President will appoint a National Commission on Space to formulate an agenda for the United States space program. The commission shall identify goals, opportunities, and policy options for United States civilian space activity for the next twenty years. Upon submission of the Commission report to the President, the Office of Science and Technology Policy, in cooperation with NASA and other appropriate agencies, will review the report and will provide their comments and recommendations to the President through the SIG (Space) within 60 days of the submission of the Commission report.

—   Insure [sic] a vigorous and balanced program of civil scientific research and exploration in space. The U.S. civil space science program is an essential element of U.S. leadership in space, a vehicle for scientific advancement and long-term economic benefits, and a valuable opportunity for international cooperation.

Implementation: NASA and other appropriate agencies will conduct their activities in a manner that will maintain a vigorous and balanced program of civil space research and exploration. NASA will explicitly factor the broad spectrum of capabilities necessary for space science into the planning and development of the manned Space Station and will implement those plans in a manner that will lend stability and continuity to research in the space sciences. Furthermore, the Office of Science and Technology Policy, in conjunction with NASA and other appropriate agencies, will review and define the goals and missions of the various civil agencies in the area of earth sciences research and will provide their recommendations in a report to the SIG (Space) by April 1, 1985.

COMMERCIAL SPACE PROGRAM

—   Encourage commercial Expendable Launch Vehicle activities. The U.S. will encourage and facilitate commercial expendable launch vehicle operations. U.S. Government policies will promote competitive opportunities for commercial expendable launch vehicle operations and minimize government regulation of these activities.

Implementation: The Department of Transportation will carry out the responsibilities assigned by Executive Order 12465 on Commercial Expendable Launch Vehicle Activities. [4] Appropriate agencies will work with [the] Department of Transportation to encourage the U.S. private sector development of commercial launch operations in accordance with existing direction.
The U.S. Government will not subsidize the commercialization of ELVs but will price the use of its facilities, equipment, and services by commercial ELV operators consistent with the goal of encouraging viable commercial ELV launch activities in accordance with existing direction.

—   Stimulate private sector commercial space activities. To stimulate private sector investment, ownership, and operation of civil space assets, the U.S. Government will facilitate private sector access to civil space systems, and encourage the private sector to undertake commercial space ventures without direct Federal subsidies.

Implementation: The U.S. Government will take the following initiatives:

—   Economic Initiatives. Tax laws and regulations which discriminate against commercial space ventures need to be changed or eliminated.
—   Legal and Regulatory Initiatives. Laws and regulations predating space operations need to be updated to accommodate space commercialization.
—   Research and Development Initiatives. In partnership with industry and academia, [the] government should expand basic research and development which may have implications for investors aiming to develop commercial space products and services.

— Initiatives to Establish and Implement a Commercial Space Policy. Since commercial developments in space often require many years to reach the production phase, entrepreneurs need assurances of consistent government actions and policies over long periods.

NASA, [the] Department of Commerce, and [the] Department of Transportation all have roles and will work cooperatively to develop and implement specific measures to foster the growth of private sector commercialization in space. A high level national focus for commercial space issues will be created through establishment of a Cabinet Council on Commerce and Trade (CCCT) Working Group on the Commercial Use of Space. The SIG(Space) will continue its role of coordinating the implementation of policy for the overall U.S. Space Program.

[5] NATIONAL SECURITY SPACE PROGRAMS

— Maintain assured access to space. The national security sector must pursue an improved assured launch capability to satisfy two specific requirements—the need for [a] launch system complementary to the STS to hedge against unforeseen technical and operational problems, and the need for a launch system suited for operations in crisis situations.

Implementation: In order to satisfy the requirement for assured launch, the national security sector will pursue the use of a limited number of ELVs to complement the STS.

— Pursue an [sic] long-term survivability enhancement program. The national security sector must provide for the survivability of selected, critical national security space assets to a degree commensurate with the value and utility of the support they provide. This will contribute to deterrence by helping to ensure that potential adversaries cannot eliminate vital U.S. space capabilities without considerable expenditure of their own resources.

Implementation: The high priority and emphasis on survivability reflected within the Department of Defense space programs will continue.

— Stem the flow of advanced western space technology to the Soviet Union. The U.S. cannot be complacent about the increasing Soviet efforts to erase the U.S. advantage through vigorous Soviet research and development efforts and through technology transfer.

Implementation: All agencies of the Government will cooperate in order to prevent the transfer of space technology to the Soviet Union and to its allies, either directly or through third countries, if such transfer is potentially detrimental to the national security interests of the United States.

— Continue to study space arms control options. The United States will continue to study space arms control options.

Implementation: The Senior Arms Control Policy Group will continue to study a broad range of possible options for space arms control. The studies will be undertaken with a view toward negotiations with the Soviet Union and other nations, compatible with national security interests. All actions will be conducted within the constraints of existing treaty commitments.

— Insure [sic] that DOD space and space-related programs will support the Strategic Defense Initiative. In light of the uncertain long-term stability of offensive deterrence, an effort will be made to identify defensive means of deterring nuclear war. The U.S. has been investigating the feasibility of eventually shifting toward reliance upon a defensive [6] concept. A program has been initiated to demonstrate the technical feasibility of enhancing deterrence through greater reliance on defensive strategic capabilities. The Department of Defense will posture its space activities so as to preserve options to support the demonstration of capabilities as they are defined and become available, and as justified by the state-of-the-art technology.

—   Maintain a vigorous national security space technology program to support the development of necessary improvements and new capabilities. The changing nature of the world environment presents new challenges at the same time as advances in technology present new opportunities.

Implementation: The Department of Defense will provide strong emphasis on advanced technology to respond to changes in the environment, to improve our space-based assets, and to provide new capabilities that capitalize on technological advances.

Document IV-2

Document title: NASA and the Department of Defense, "National Space Strategy—Launch Vehicle Technology Study," December 1984.

Source: Ivan Bekey, Bekey Designs, Bethesda, Maryland.

This study was carried out jointly by NASA's Office of Space Flight and the Office of the Deputy Assistant Secretary of the Air Force for Space Plans and Policy in the fall of 1984. It is in response to a requirement set forth in NSDD-144, "National Space Strategy." It represented the first attempt in many years by the two primary government users of space to define potential future missions, to assess the ability of existing launch systems to meet those requirements, and to identify new technologies needed to development space transportation systems to complement or replace existing launch systems. The appendices to this report are not included here.

National Space Strategy
Launch Vehicle Technology Study

December 1984

[1]   NASA/DoD Space Launch Technology Study
                                      Response to
                            NSDD-144 National Space Strategy

1.   BACKGROUND

The President asked the Department of Defense (DoD) and National Aeronautics and Space Administration (NASA) to conduct a joint study to identify launch vehicle technology that could be made available for use in the post-1995 period. This was one of several actions assigned to both organizations as a part of the National Security Decision Directive-144, National Space Strategy. The following report was jointly worked from the outset, and represents the consensus of both DoD and NASA.

DoD and NASA identified launch needs using representative missions. Neither agency is commited [sic] to carry out these missions; rather, they identified the technology that would give the country the capability to carry out such missions in the future.

The actions taken were started in early October, 1984, when the Office of Space Flight in NASA and the Office of the Deputy Assistant Secretary of the Air Force for Space Plans and Policy were designated as focal points for the study. Shortly thereafter, two joint study groups were formed, with the organizational representation shown in Appendix 1. The first group looked at the missions and their space launch capability requirements for the post-1995 period. The second group examined the availability of new technology during that same period. Both groups consulted heavily with industry and internal planning organizations. The efforts of both groups were melded to form the response contained in the remainder of this report.

The term "technology" as used throughout this report is intended in its broadest sense. That is, "technology" includes existing vehicle systems, classical technology disciplines, and related techniques (e.g. on-orbit assembly). In this context, both NASA and DoD consider that current technology will continue to be available for appropriate use after 1995. For example, the Space Shuttle will remain the primary means of access to space for the Nation until after 2000.

[2] 2. FUTURE LAUNCH NEEDS AND THEIR SYSTEM IMPLICATIONS

In preparing this report, the launch needs of the civil, commercial, and military space programs were examined. The time around 2000 is pivotal, for that is when systems based on technology new in 1985 could become available.

The civilian sector's more demanding missions were defined and are summarized in Appendix 4. These missions rely heavily on the Space Station and include platforms, servicing of satellites and platforms at low and geostationary altitudes, commercial use of microgravity, and several classes of scientific missions. Manned missions to geostationary orbit are projected around 2000. After 2000, NASA foresees manned activity at the Moon and beyond.

DoD's present and projected needs are also outlined in Appendix 4. In the pre-2000 timeframe, they include satellites in geostationary orbit, various polar orbiting satellites, the launch of experiments/prototypes for the technological demonstrations of the Strategic Defense Initiative (SDI), and limited use of the Space Station. Post-2000 needs relate to operational deployment and support of a Strategic Defense system, phase-in of survivable launch capability, and potential manned aerospace planes. The need for assured access to space will remain important to DoD, but the exact means of assuring such access will have to be reassessed as plans for the post-1995 timeframe solidify.

The commercialized launch vehicle programs do not, for the most part, generate nor require new technology. Commercial booster ventures usually minimize cost and schedule risk by using existing technology and previous developments. Therefore, they will tend to take advantage of technology developed by the government or others.

In general, the pre-2000 mission areas identified do not require development of advanced technology. Accordingly, both DoD and NASA consider that the primary means of launch until 2000 will be the Space Shuttle. By the mid- to late-1990s the Shuttle's capabilities will be enhanced by on-orbit assembly techniques, especially in conjunction with the Space Station, Orbital Maneuvering Vehicles (OMVs), and Orbit Transfer Vehicles (OTVs). These missions may require significant growth of the Space Station to support space-based launch vehicle buildup, servicing, [3] and launch and recovery operations.

NASA and DoD have found that, of all the missions examined, the following (not prioritized) appear to drive the need for advanced launch and support systems that would benefit the most from advances beyond current technology:

- Spacecraft Missions Requiring Orbit Transfer Vehicles (Manned or Unmanned)
- Assembly and Construction of Large Spacecraft in Low and Geosynchronous Earth Orbits
- Strategic Defense Space Systems
- Manned Reusable Systems (e.g. Military Aerospace Planes)
- Manned Lunar/Planetary Missions

The long-range and large-scale missions, such as manned exploration of the solar system and the programs generated by the SDI, may require new launch vehicle technology applications for the post-2000 period. In those years, the potential need also exists for a tanker to carry OTV propellants to the Space Station, and some form of a heavy-lift vehicle may be needed to support manned lunar/planetary missions.

NASA believes that since large, unmanned cargo vehicles will probably be available in the post-2000 timeframe, the possible wearout of the Shuttle fleet in this timeframe may make appropriate the development of a second-generation shuttle configured primarily to ferry people to the Space Station and to de-orbit people and cargo.

The launch vehicle needs of the Strategic Defense space systems will probably drive the DoD launch vehicle needs, but these needs are, as yet, largely unknown. The potential needs encompass the spectrum of launch vehicle configurations, such as large vehicles, small vehicles and on-orbit assembly, or a combination of large and small vehicles. The difference between the alternatives will be based on economic tradeoffs and the availability and practicality of on-orbit assembly technology. For the mid- to late-1990s, the plans are for SDI demonstrations to use the Shuttle. These demonstrations would, however, benefit from the enhanced operations and performance resulting from an advanced cryogenic engine. For the deployment and operation of the [4] systems generated as a result of the SDI, the cost of launch will be an important, if not one of the most important, considerations. The solution to the launch cost problem may include using current technologies and techniques in new ways, using new technologies, or using a combination of both. Firm conclusions regarding the Strategic Defense launch requirements await architectural definition and trade studies.

While the pursuit of new technology is important, so too is the exploitation and further refinement of current technology. Using existing technology in new ways and using previous extensive national investments, such as the Space Shuttle with advanced cryogenic engines and the capabilities of the Space Station, may provide cost-effective solutions for meeting future requirements.

3. TECHNOLOGY THRUSTS

Based on the vehicle concepts to support the missions and needs discussed above (Appendix 3), DoD and NASA agree that the new technology areas with the broadest application for accomplishing the missions cited above are the following (not prioritized):

- Advanced Cryogenic Engine (SSME Class)
- Advanced LOX/Hydrocarbon Engine (1-2M Pound Thrust Class)
- Advanced Power Systems
- Advanced Space Engine
- Aerobrakes
- Reentry/Recovery Systems
- Robotics
- Solid Rocket Propulsion

Although the above list is rather short, it should be recognized that while many other technological developments may also apply, their range of application is narrower and more selective. Furthermore, for some specific concepts, there are enabling technologies which, although they do not have broad application, ought to be pursued. Experience in space-based payload buildup, assembly, servicing, and launch and recovery operations in low earth orbit, may identify new areas of development and new technologies to support missions in the post-2000 timeframe. The applicability of all the key technology areas examined is detailed in Appendix 4.

[5] The proper mix of technologies to be pursued depends heavily on which courses the Nation will follow after 1995. Although many other technologies have been identified that might apply to post-1995 launch vehicles, major technological developments cannot be justified without detailed system studies.

An important point needs to be made concerning the Nation's advanced technology base to support future systems. The trend for the past decade and a half has been one of general decline in the national investment in advanced research not related to some specific vehicle development. The current advanced research and technology base is at a minimum subsistence level. The private sector has virtually stopped supporting advanced technology in the areas where there is more than minimal risk. Government investment in basic research not related to ongoing developments is a small fraction of the level it needs to be if breakthroughs in future launch concepts are ever to become a reality.

This study has highlighted several technologies where the payoffs would seem to span broad areas. To preserve the Nation's superiority and world leadership in technology, however, other areas of research, although not required at present for any specific vehicle development, will serve to advantageously broaden the technology base of the Nation. For example, more emphasis on fundamental research and technology is needed to allow the Nation to move away from conventional, chemical-based propulsion.

4. CONCLUSIONS

A. Comprehensive Studies

As a follow-on to this short-term study, a more comprehensive, long-term study should be performed. The study should examine the missions thoroughly to determine specifically the launch vehicle systems needed, including not only those systems centered on new technology, but also those employing existing technology in new ways. This study will provide insight into the payoffs likely to be gained in system operations as a result of investment in new technology. Recognizing that, while mission requirements must be generated independently by DoD and NASA, this study, aimed at technology identification, can and should be conducted jointly.

[6] B. Space Transportation System Enhancements

Since the Shuttle will be the primary means of access to space through the 2000 timeframe, it would be desirable to increase its robustness by taking steps to continue Shuttle availability to the end of the century and by improving the fleet's operations and performance with an advanced cryogenic engine as well as with other possible enhancements.

C. Industrial Base

The currently diminished technology base and the trend that established it cause considerable concern for the future and lead us to conclude that a national commitment to a vigorous research, technology, advanced development and demonstration program, with adequate consideration of provision for major test facilities, is necessary to revitalize the space launch industrial base of the Nation. Such a program must emphasize aspects of technology involving risks that may prevent independant [sic] research in the private sector, but promising high leverage potential for future space launch systems.

Document IV-3

Document title: NASA/DOD Joint Steering Group, "National Space Transportation and Support Study, 1995-2010," May 1986, pp. ii-iii, 1-9, 21-24.

Source: Ivan Bekey, Bekey Designs, Bethesda, Maryland.

National Security Decision Directive 164, "National Security Launch Strategy," signed by President Reagan on February 25, 1985, which appears in Volume II of Exploring the Unknown as Document II-44, directed the Department of Defense and NASA to "jointly study the development of a second-generation space transportation system." This document summarizes the results of the year-long study carried out in response to that directive.

NATIONAL SPACE TRANSPORTATION STRATEGY

National Space Transportation and Support Study 1995–2010

SUMMARY REPORT
PREPARED BY THE
JOINT STEERING GROUP
MAY 1986

|ii| JOINT STEERING GROUP MEMBERSHIP

<u>NASA</u>

- Jesse W. Moore, Associate Administrator of Space Flight (through February 19, 1986)
- R. Adm. Richard H. Truly, Associate Administrator of Space Flight (effective February 20, 1986)
- Dr. William R. Lucas, Director, George C. Marshall Space Flight Center
- Norman E. Terrell, Associate Administrator for Policy
- Dr. Raymond S. Colladay, Associate Administrator for Aeronautics and Space Technology

<u>DoD</u>

- Edward C. Aldridge, Jr., Under Secretary of the Air Force (through April 11, 1986)
- Lt. Gen. Bernard P. Randolph, USAF Deputy Chief of Staff, Research, Development, and Acquisition
- Lt. Gen. James A. Abrahamson, Director, Strategic Defense Initiative Organization
- Dr. Larry L. Woodruff, Office of the Under Secretary for Defense Research and Engineering (Strategic and Theatre Nuclear Forces)

Two Executive Secretaries were appointed to the Joint Steering Group as non-voting members. Ivan Bekey, Director of Advanced Programs, Office of Space Flight, represents NASA, and Dr. Thomas P. Rona, Office of Secretary of Defense, represents DoD.

Codirectors of the Joint Task Team are Paul F. Holloway, Deputy Director, NASA Langley Research Center, and Col. William F. H. Zersen, Assistant for Advanced Launch Systems, Deputy Commander for Launch and Control Systems, USAF's Space Division. Darrell Branscome, NASA Office of Space Flight[,] serves as Mr. Holloway's Deputy.

[iii]   FOREWORD

This report summarizes the findings and recommendations of a year-long cooperative study by the Department of Defense (DoD) and the National Aeronautics and Space Administration (NASA). Detailed data, discussions, and study rationale are presented in the expanded Overview document and its supporting annexes.

[1]   1. BACKGROUND AND PURPOSE

National Security Decision Directive (NSDD) "National Security Launch Strategy," was signed by President Reagan on 25 February 1985. This decision directive presents guidance for near-term implementation of the policies delineated in a prior National Security Decision Directive "National Space Strategy." The latter NSDD stated that the Space Transportation System (STS) will continue as the primary space launch system for both national security and civil government missions and directed that DoD pursue an improved, assured launch capability that will be complementary to the STS to ensure the national security launch requirements are met. The February 1985 NSDD also specified that:

> "DoD and NASA will jointly study the development of a second-generation space transportation system—making use of manned and unmanned systems to meet the requirements of all users. A full range of options will be studied, including Shuttle-derived technologies and others."

To implement the 25 February decision directive, the President signed a National Security Study Directive (NSSD) in May 1985. This document directed that a joint DoD/NASA study be accomplished within one year and delineated four tasks which would provide the basis for a space transportation technology program plan:

Task 1. Compile sets of national security and civil space mission classes for the 1995 period and beyond.

Task 2. Determine space transportation system capabilities which could cost-effectively support the mission needs specified in Task 1.

Task 3. Identify the transportation technologies that are necessary and could be available for the systems to be used in the post-1995 period.

Task 4. Based on the technological needs and opportunities specified in Tasks 2 and 3, identify the technology development programs needed for timely realization.

[2] Four objectives or guiding principles were specified for the joint study:

- Satisfy the future needs of authorized users
- Substantially reduce the cost of space operations to the government
- Develop a flexible and robust space transportation system
- Maintain world leadership in space transportation

In preparing for the study, the existing U.S. space transportation capabilities and related activities were evaluated. The key lessons learned are as follows:

- The current launch systems of the U.S. represent the best technology and the best operations costs available at their individual initial operating dates. Viewed as an architecture, they have kept the U.S. in the dominant leadership role in space transportation. However, in planning for the 1995–2010 time period, achieved plus readily achievable technology advances must be exploited to ensure that the current U.S. leadership posture is maintained.
- The current systems were originally designed to meet space mission planning models which never fully materialized.
- Funding limitations during the development phases of current systems precluded existing national launch systems from realizing full potential for cost effective operations.
- A complementary strategy (i.e., no dependence on a single launch system) must be inherent in the national space policy to increase the probability of continuous access to space.
- Space transportation costs have been substantially driven by both launch system and spacecraft designs which require lengthy manpower-intensive, technically complex, high-cost integration efforts. Too frequently, special spacecraft and launch system modifications and high performance (low margin, experimental-type operations) missions are required to meet spacecraft-unique needs.
- The nation has neither funded nor maintained a vigorous advanced space transportation technology program to improve the operational effectiveness of the existing national launch systems nor provided the appropriate technical foundation for future launch systems.
[3] - The national launch systems industrial base cannot rapidly react to changing space launch requirements and/or adversity.
- Substantial reductions in space transportation costs must be attained if the nation is to meet the demanding needs of the future.
- Foreign space-related developments are beginning to erode the United States' preeminence in space launch activities. The U.S. technology and system development efforts must be made in the context of vigorous and increasing international competition.

2. MISSION NEEDS

Technology initiatives must be defined in terms of potential system concepts within the future robust space transportation architecture for satisfying space mission needs. Therefore, sets of space mission classes have been compiled independently by DoD and NASA which reflect potential national security and civil (government and commercial) space traffic, respectively, for the time period 1995 through 2010. These space mission sets are representative only for the purpose of identifying system capability and related technology needs, and they do not constitute specific plans or requirements for either DoD or NASA. Particular attention was given by DoD to the emerging requirements of the President's Strategic Defense Initiative (SDI), and NASA consulted with the National Commission on Space during the development of these space mission sets.

Because it is not possible to precisely predict the level or nature of future space activities, DoD and NASA each developed four alternative sets of projected mission needs reflecting different space traffic levels. Five combinations of these sets were selected to ensure that mission needs from constrained through aggressive cases were modeled. These five mission model cases are detailed in Table 1.

The Constrained Case, the lowest projected level of national space activity, comprises contemporary national security missions, a low level of SDI experimentation, and a civil core program. The latter includes a permanently manned Space Station, as well as both domestic and foreign components encompassing science and applications, technology developments, and commercial activities. The Normal Growth Case adds some new DoD program starts, has increased levels of SDI experimentation, and includes [4] [original placement of Table 1] [5] manned civil missions in geosynchronous orbit and lunar sorties. The Modest Expansion/Partial-SDI Case is principally characterized by the addition of a representative SDI kinetic energy weapon (KEW) deployment, a second Space Station, and Mars/Asteroid sample returns. The Full SDI Case postulates increased emphasis on national security space activities, including very extensive SDI operational deployments of directed energy weapons (DEWs), and the Aggressive Civil Case postulates substantially expanded civil space activities encompassing a broad spectrum of space exploration and earth-focused programs.

Table 1. National Space Mission Case Construction

Case	Civil Option		DoD Scenario
CONSTRAINED	Core Program – Ongoing Civil Discipline Programs — Science and Applications — Technology Development – LEO Space Station – Polar Platform – GEO Experiments Platform – LEO Space Station Growth	(1992–2010) (1994) (1994) (1998) (1997, 2003)	Constrained Activity – Contemporary Missions/ Spacecraft – Low Level of SDI Experiments
NORMAL GROWTH	Baseline Program Civil Option I Plus: – GEO Sorties and Shack – GEO Manned and Automated Servicing – Lunar Sorties	 (2002, 2004) (2004) (2009)	Normal Growth – Adds Advanced Missions and New Starts (AF, Navy, DNA) – Increased Level of SDI Experiments – Adds Advanced Payload/ Operations/Servicing Capabilities Development
MODEST EXPANSION PARTIAL SDI	Modest Expansion Civil Option II Plus: – Second LEO Space Station – Mars and Asteroid Sample Returns – Commercial Growth – Quarantine Facility	 (2008) (2004, 2008) (1996) (2003)	SDI KEW – Adds Operational KEW, SSTS Deployment – Adds KEW, SSTS Servicing Missions – Reduced Level of SDI Experiments
FULL SDI	Baseline Program Same as Civil Option II above		Full SDI – Adds Operational DEW Deployment, Servicing – Transition to Advanced SSTS

Case	Civil Option		DoD Scenario
	Aggressive Expansion		Normal Growth
	Civil Option III Plus:		Same as DoD Scenario 2 above
	– Earlier Deployment Dates for Most Programs		
AGGRESSIVE CIVIL	– Lunar Surface Camp and Orbit Station	(2006, 2008)	
	– Manned Mars Mission Buildup	(2012)	
	– Nuclear Waste Disposal (Solar Orbits)	(1998)	
	– Third LEO Space Station	(2007)	
	– Space-Based Energy	(2007)	
	– Public Access	(2008)	
	– Extended Communications	(2001)	

Significant transportation requirements derived from the mission models are:

- For any of the five cases, traffic to orbit would be higher than present day levels, both in terms of weight to orbit and frequency of flight. Spacecraft, payloads, flight crews, and servicing materials to be orbited range from approximately 1.25 to 1.75 million pounds annually for the lower activity level cases (Constrained and Normal Growth) to upwards of 5 million pounds annually for the Full-SDI and Aggressive Civil Cases. The orbit transfer system (OTS) weights necessary for transporting selected items beyond low earth orbit (LEO) represent a sizable additional transportation need.
- The national security and civil space missions both involve continued spacecraft and payload placements into the variety of orbits being used by present day U.S. space traffic. However, a new class of orbits (mid inclination, low altitude) not used today would experience the largest traffic levels if operational SDI spacecraft were deployed. Depending on the types of space transportation systems ultimately developed, significant activity increases at the Western Test Range (WTR) or a new launch site could be required to accommodate this SDI traffic.
- Potential SDI architectures involve large satellite constellations to provide continuous coverage of the earth. Establishing such constellations would introduce new requirements for precise timing of launches and orbit transfers to achieve proper orbital plane placements and mission control of multiple spacecraft orbit transfers occurring simultaneously.
[6] - Manned operations in space are a significant element of the civil space program for the first Space Station beginning in the mid-1990s and for geosynchronous earth orbit (GEO) servicing and lunar sorties beyond the turn of the century. In the most aggressive civil option, manned missions to the Moon and to Mars are projected. Potential roles for man in national security space missions are still under study. Assured return from space will be an important transportation requirement.
- Space servicing activities include maintenance, replacement, upgrade, assembly, checkout, retrieval, return, and repair. Approximately half of the civil mass transportation needs are devoted to space servicing, and DoD spacecraft servicing requirements could evolve.
- Some national security space operations must be possible during various conflict levels or natural adversity (e.g., to supplement, redeploy, or replenish space assets), and selected space transportation systems will therefore have to satisfy more stringent functional/operational needs in such areas as availability

(readiness to operate when required, regardless of the circumstances), performance margins, flexible response to changing situations, survivability, and positive control. Specifics of these functional/operational needs are being developed
by the new Unified Space Command in consonance with the evolution of its
strategies and by the SDI Organization.

3. ARCHITECTURES

Space transportation architecture issues for the post-1995 time period derive from the
nature of the existing (pre-1995) space transportation architecture as well as from the
combined DoD/civil mission model sets. Unless there are new initiatives, the U.S. would
enter the post-1995 time period with a relatively high operating cost space transportation
architecture consisting principally of a modest Shuttle fleet with ground processing and
launch facilities at both the Eastern Test Range (ETR) and WTR designed for limited
launch rates. The architecture existing then could also include the Complementary
Expendable Launch Vehicle (CELV) presently planned for launch from ETR to provide
an increased assured access probability for critical DoD spacecraft and possibly other
expendable launch vehicles (e.g., Delta, Atlas, and Titan II). The use of two CELVs per
year is anticipated upon its introduction in 1988, though planned utilization [7] rates and
launch sites are presently undergoing review. The pre-1995 space transportation architecture will have a number of OTSs including Centaur G, Centaur G', the Payload Assist
Module (PAM) Series, the Inertial Upper Stage (IUS), and the Transfer Orbit Stage
(TOS). Orbital Maneuvering Vehicles (OMVs) with a robotic smart front end would be
utilized extensively in any future architecture for payload positioning and for assembly
and servicing operations. The OMV is under development and will be operational prior
to 1995, but the smart front end introduces additional technology requirements.

The existing space transportation architecture would be unable to effectively handle
the increased traffic anticipated for the post-1995 time period. Just to accommodate the
Normal Growth Case flight rates, expenditures for additional Orbiters, expendables, facilities, and operations personnel at both ETR and WTR would be needed. Further substantial expenditures at WTR would be required for an SDI-KEW deployment. Such
expenditures could perpetuate the use of a relatively high operating cost transportation
architecture and preclude the opportunity to exploit technology advances and innovative
operations approaches which can significantly reduce costs.

Space traffic growth beyond the mid-1990s leads to a preferred architecture employing two new launch systems (an unmanned cargo vehicle and a new manned vehicle); a
new, reusable OTS; and new, innovative launch and flight operations approaches (supplementary use of contemporary expendables for selected, specific missions cannot yet be
ruled out). This preferred architectural approach has been shown to be cost effective over
a broad range of mission scenarios. The unmanned cargo vehicle could effectively replace
the CELVs, complement the Shuttle cargo capability, and help to increase probability of
assured access to space.

Even with an unmanned cargo vehicle (UCV) introduced in the mid- to late-1990s, a
new manned vehicle is necessary after the turn of the century for a more cost-effective,
robust space transportation architecture. The then existing Shuttle fleet will be reaching
lifetime limits and would represent 25-year-old technology. Therefore, architectures involving two new vehicles, a UCV followed by a new manned vehicle (and supplemental use of
contemporary ELVs), have become the prime focus of ongoing space transportation activities. For two stage options, common elements, such as the booster first stage for both the
UCV and new manned vehicle, appear effective in terms of life-cycle costs, but [8] this
approach must be assessed for soundness of assured access. (The Full SDI Case would
require substantial development expenditures for an additional, heavy-lift, unmanned
cargo vehicle to launch the large DEWs if their modularization proves impractical.)

There are numerous launch vehicle system options, as well as OTS options, for structuring such architectures (see Figure 1), all of which are currently under study. Additionally, new systems and approaches for launch and flight operations which would significantly lower costs are being identified and assessed. Alternative architectures comprised of various combinations of these system options are being evaluated using cost, performance, operations, operational availability and flexibility, risk, safety, world transportation leadership, and other political/programmatic considerations.

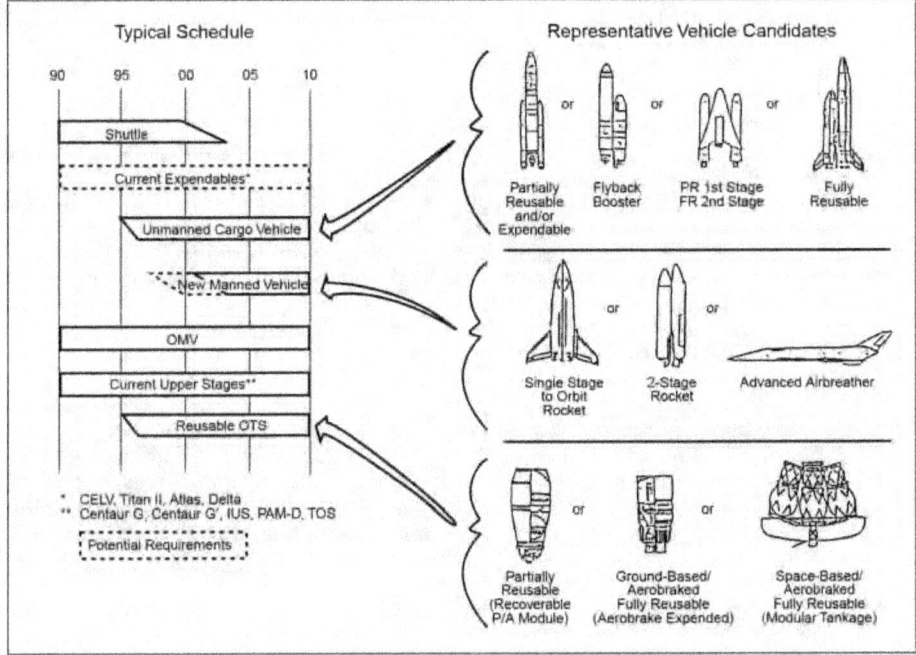

Figure 1. Representative Architectures

[9] The concepts evaluated ranged from single- and two-stage rocket systems to the emerging airbreathing engine technologies. The application of advanced materials, structures, and engines could dramatically increase the performance and decrease the weight and size of rocket vehicles. Simple, highly automated, and airline-like procedures could lead to very low-cost manned or unmanned rocket vehicle operations. A new concept to achieve orbit is the airbreathing National Aero-Space Plane (NASP) Program. The NASP is a focused technology program with an FY 1988–1989 decision date leading toward an FY 1993 technology demonstration research aircraft with a horizontal takeoff and landing single-stage-to-orbit potential. This approach could have a major impact on any future space architecture. Possible operational implications include operation from military airfields, high flight rate, survivability, and flexibility. A NASP vehicle with these characteristics could alter the nature of the entire logistics, operations, and support systems. . . .

[21]   6. FINDINGS

The key findings from this study are:

- The existing Shuttle/CELV/Other ELVs launch vehicle architecture has relatively high operating costs when compared to that achievable in the 1995 to 2010 time period. Although its continued use would require no significant development investments, anticipated traffic growth would necessitate high Shuttle/CELV flight rates and sizable investments in additional orbiters, expendables, ground processing and launch facilities, and operations support.
- Many technologies critical to the future of space transportation are poised for major advances that could greatly benefit both existing and new systems in the post-1995 time period.
[22]- Current funding levels severely inhibit the timely development of a majority of necessary key technologies.
- Facilities in the areas of propulsion, structures, and aerothermodynamics are demonstrably inadequate to cope with development testing requirements inherent in the realization of complex new technologies and systems.
- Future U.S. launch systems design must be driven by operations and support as well as assured access considerations, which may include launch sites within the interior of the United States, in order to achieve operational flexibility and cost effectiveness. A substantial reduction in recurring operations cost is achievable if launch vehicles are designed for operational efficiency rather than maximum performance.
- Mission models should not be evaluated based on total tonnage to orbit alone. Frequency of flight and payload sizes should play a role in the architecture. Modularity of vehicles and payloads may provide increased operational flexibility.
- Preferred architectures employ two new launch systems, an unmanned cargo vehicle and a new manned vehicle (with supplementary use of ELVs for specific missions); a new reusable OTS; and new launch and flight operations approaches. This architectural approach is cost effective across a wide range of mission scenarios and would improve assured access capabilities.
- Integration of payloads with the launch system is a significant operational cost. New processing and integration methods approaching those applied to cargo aircraft and other truly "operational" transportation systems must he developed.
- Numerous system and technology options must be explored in parallel to enable selection of a future U.S. space transportation architecture.
- The generic technology investment plan required to achieve low operations cost, robustness, flexibility, and world leadership in space transportation has been defined. The recommended plan provides a road map with decision dates for final architecture selection.
- Implementation of the recommendations of this report will assure that the U.S. has a solid beginning toward revitalizing its national launch systems technology and industrial base and retaining uncontested leadership in space.

[23]   7. RECOMMENDATIONS

The Joint Steering Group recommends the following:

- If new manned and unmanned launch systems and lower costs for space launch operations are to be attained, the U.S. must commit to implementing the technology plan of this report. This plan, which is complementary to other firmly planned technology activities (e.g., the National Aero-Space Plane Program, ongoing DOD/NASA programs, and industry programs), is focused to provide a base for new systems which can achieve the objective of substantially reduced operations costs. The plan supports the development of both evolutionary and revolutionary technology alternatives necessary to assure continued U.S. world leadership in space transportation.

- Maintain the DoD/NASA Joint Steering Group (JSG) to guide the national efforts toward a second-generation space transportation system. Establish a more permanent organizational structure, which would replace the current Ad Hoc Joint Task Team, to further refine the future space transportation architecture, coordinate technology activities, and coordinate plans for new systems as the need arises and technology becomes available. The JSG must also ensure that close coordination/liaison is maintained with the National Aero-Space Plane Program Office, the Strategic Defense Initiative Organization, and other appropriate DoD and civil offices. Additionally, continued ties must be established with all space transportation users to ensure that transportation-related issues such as spacecraft modularity, standardization, containerization, and servicing are addressed from an overall space program perspective, including identification of needed technologies. Provide for mutual DoD/NASA approval of the structure, staffing, and location for the organization.
- Continue the joint NASA/DoD Space Transportation Architecture Studies to include:
    - Conduct trade studies and sensitivity analyses to refine and confirm the cost beneficial investments which will provide the most efficient operations and vehicle systems for the future.
[24]   - Reassess the transition to the next generation space transportation systems while considering all elements of the architecture and the current Space Shuttle and Titan recovery plans.
    - Develop planning to accommodate the unique military operational and functional needs of Unified Space Command and SDI as these needs evolve.
    - Preserve the option for a near-term space transportation architecture to accommodate potential deployment options for Strategic Defense Initiative systems and meet other increasing civil and DoD launch demands on a cost-effective basis.
- Direct that the study results be reviewed by the space transportation user community for applicability to spacecraft production/operations.

Document IV-4

Document title: Department of Defense, NSDD-261 Report, "Recommendations for Increasing United States Heavy-Lift Space Launch Capability," April 29, 1987, pp. iii–xvi.

Source: NASA Historical Reference Collection, NASA History Office, NASA Headquarters, Washington, D.C.

The Space Transportation Architecture Study identified as the earliest requirement for a new U.S. space transportation capability a vehicle capable of launching heavy national security, missile defense, and civilian payloads. National Security Decision Directive 261 ordered a study to identify the character of such a vehicle, to be called the Advanced Launch System (ALS). This was the Department of Defense response to that order; only the executive summary of the report appears here.

[each page of original marked "SECRET" (crossed out by hand) and "UNCLASSIFIED"]

NSDD-261 REPORT

Recommendations for Increasing United States Heavy-Lift Space Launch Capability

29 APRIL 1987

[iii]   EXECUTIVE SUMMARY

Recent launch failures, a diminishing space transportation technology base, diminishing capacity, lack of flexible launch capability, and growing global space competition have seriously undermined America's traditional leadership in space. To restore and strengthen space leadership, the United States must increase its heavy-lift capability. To achieve this, the DOD, in coordination with NASA, proposes to implement the development of the Advanced Launch System (ALS). The ALS, which includes a new operational launch vehicle, can meet national space launch needs and revitalize the U.S. space transportation capacity and technology base. In addition, the ALS would send the strong message to the Soviets that SDI is viable and could be operational in the mid-1990s. The attached Advanced Launch System Plan is based on two years of substantial, cooperative DOD and NASA studies. The concept is technically sound, affordable, and ready to enter the concept definition phase. This report recommends that the United States immediately commit to the ALS program, and in particular the intensive technology development upon which a successful ALS effort depends.

International Competition and Space Leadership

The United States faces a growing international space challenge. The Soviet Union, European community, Japan, India, and China are all developing space capabilities that challenge U.S. leadership. To compete effectively with emerging foreign space launch capability and regain the [iv] ability to project a U.S. leadership role in space applications during the next decade and into the next century, U.S. space transportation capabilities must move ahead. To do this, the United States will have to modernize its space launch systems to take advantage of advanced technologies and a design philosophy based on operational rather than a research and development orientation.

For almost 30 years the United States has exploited space systems generally to satisfy important scientific and military requirements. We have been the world leader in space exploration due to our technical competence and national commitment. This leadership has been seriously eroded in recent years. If the United States is to regain its space leadership role, it must have the launch capacity, flexibility, and availability to perform ambitious goals. Current launch systems will not be able to satisfy the projected growth of U.S. space launch requirements into the 1990s much less into the next century. Also, these systems do not provide the quick-response and surge capability required of an operational launch system in crisis and conflict situations. Many elements of these systems have not been updated to take advantage of new, more efficient, more capable technology. The languishing U.S. technology base largely depends on 15- to 20-year-old investments. The newest U.S. launch system, the Space Shuttle, is based largely on technologies that were developed in the early 1970s; our expendable launch vehicles are based on even older technologies. While launch vehicles based on those technologies are capable of meeting today's launch requirements and will continue to play an important role in the future, they will not be able to satisfy future operational heavy-lift requirements from the standpoints of launch capacity, assured space access, flexibility, availability, and cost-effectiveness.

To exploit space fully, the cost of space launch must be significantly reduced from today's average of approximately $3,600 per pound. Three decades of U.S. space technology advances and our ability to better understand launch system design, manufacturing, and processing offer the potential of an order of magnitude reduction in operational space launch costs over the next decade. With the exception of communication satellites, which have an economic justification despite their high cost, full commercialization of space has failed to materialize. A primary reason is the prohibitively high cost of space launch. Unless launch costs are [v] reduced, space will be used primarily for government activities and only then for those of the highest national priority. The ALS would provide technology and operational concept advances that will benefit all U.S. space interests.

Incorporation of new technologies and a design philosophy intended to maximize routine operations would provide the first truly operational launch system in the free world. The ALS would significantly enhance the support provided by U.S. military space systems during crisis or conflict situations. Vastly decreased launch vehicle processing times would provide an invaluable quick-reaction capability that currently is available only in Soviet launch systems.

Soviet Threat and Operational Approach to Space Access

The development of this new system becomes particularly important when comparing U.S. and Soviet launch requirements and capabilities. The Soviet Union has multiple launch systems that can perform each of their space launch requirements. They have greatly expanded their launch facilities and manufacturing capability. They have, or will soon have operational, a space station, space shuttle, space plane, and heavy-lift launch vehicle. Why the Soviets are developing such a large, flexible space transportation system capability is not fully understood. However, we are sure that their robust space launch posture provides them with ample opportunity to deploy future Sputnik-like initiatives. One important fact is paramount: whatever the Soviet Union chooses to do in space in the coming decades, it will not be constrained by launch capacity.

[vi]

Figure A
Soviet/U.S. Space Launch Comparison

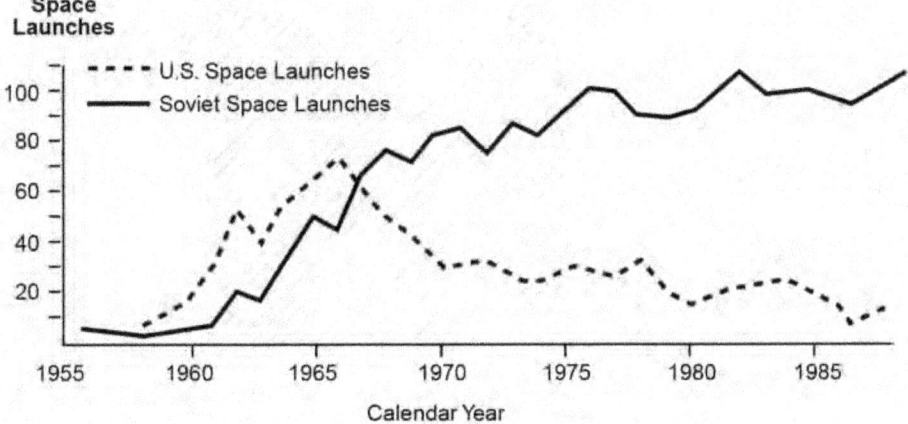

Figure B
Projected Soviet Capability/Requirements Versus U.S. Capability (Figure Secret)

[Figure B omitted in original]

[vii] National Requirements

DOD requires a truly operational launch system (as opposed to current labor-inten-
sive, R&D-oriented systems) to provide additional capacity, availability, flexibility, assured
access, and low cost. Current launch systems do not fully satisfy this need. In addition,
deployment of a strategic defense will require a sixfold increase in annual payload launch
requirements by the late 1990s (see Figure C). This requirement cannot be satisfied with
existing launch systems. In the near term, beginning the development of the ALS would
clearly indicate the seriousness of U.S. resolve to pursue a strategic defense option without violating current U.S. treaty obligations. The beneficial effects of this perception can
be achieved well in advance of a decision to develop a strategic defense.

Figure C
DOD Launch Requirements Versus Capability: SDI + Normal Growth

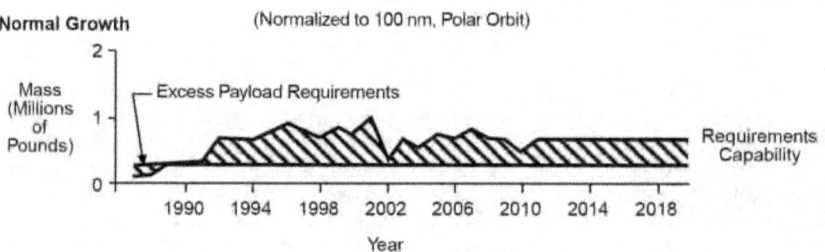

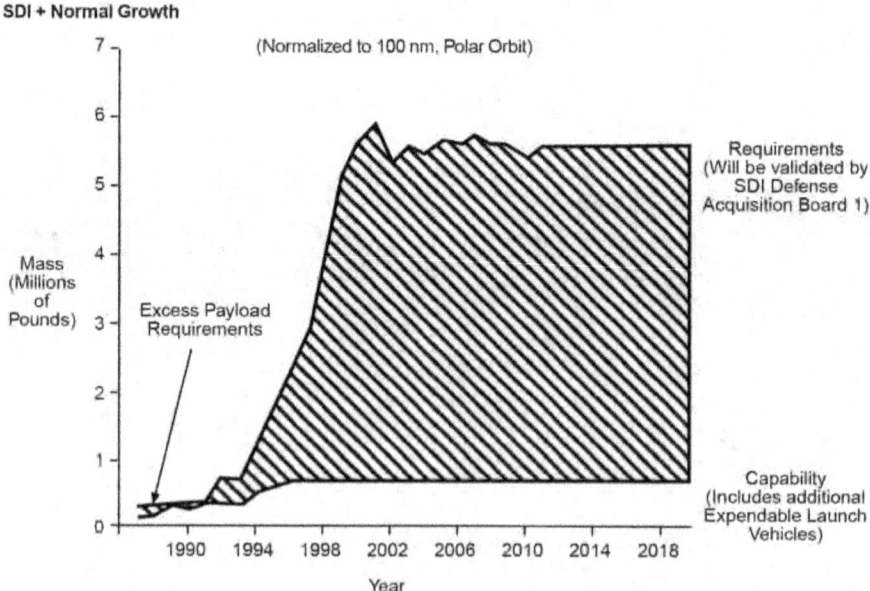

The aggregate civil launch requirement exceeds the current launch system capability. An interim ALS is needed in late 1993 for testing purposes to achieve planned civil objectives such as Space Station deployment in 1994. The system would also enhance the mission capability and scientific returns of many existing civil programs and would make future U.S. civil space leadership initiatives possible (see Figure D). Some specific benefits of the ALS are:

* Assist the Space Station effort by improving assembly and logistics, and crew safety
* Enhance planetary missions through shorter trip times, reduced mission complexity, simplified spacecraft designs, and additional science opportunities
* Support space leadership initiatives which require the use of a heavy-lift capability (Two candidate initiatives are a lunar base manned by 2008 and a manned mission to Mars early in the 21st century.)

[viii] [original placement of Figure C]

[ix]   Figure D
                           Civil Launch Requirements Versus Capability

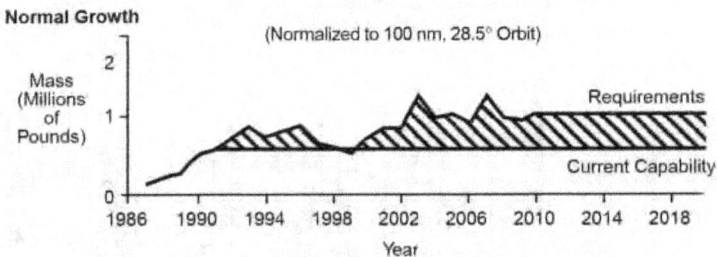

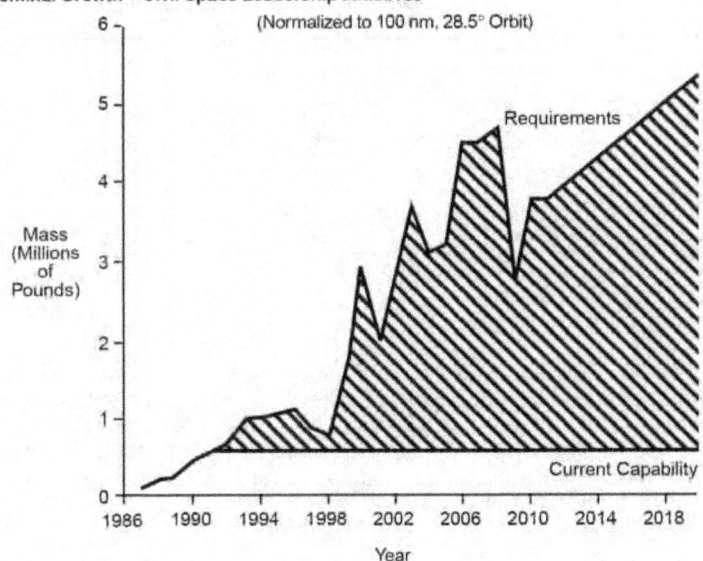

[x] Space Transportation Architecture Study

For nearly two years, the DOD and NASA have been conducting the Space Transportation Architecture Study (STAS) to identify future national space transportation and technology requirements through the year 2010. As a near-term requirement, the STAS identified the need to develop an unmanned cargo vehicle with capability beyond today's systems. It also identified a range of candidate vehicles that could help satisfy that need (see Figure E). It was because of the substantial STAS effort that the requirement for an ALS is well established and the decision can be made to immediately undertake an intensive, focused ALS technology and development program. The STAS also identified the national need for a mixed space transportation fleet (both manned and unmanned systems). The Space Shuttle and Titan accidents that grounded many of our larger and most important spacecraft underline the need for robust national space [original placement of Figure E] [xi] transportation and a mixed fleet of space launch vehicles. Because the final version of the ALS would be largely independent of existing launch systems, it would support the mixed fleet, assured access to space policy. This nation should never be totally committed to reliance on one system for space launch.

Figure E
System Candidates: Unmanned Cargo Vehicles

Type	Large Expendable	Titan Heavy Lift	Two Stage	STS Heavy Lift	Two Stage	Two Stage Down Cargo
Booster	Expendable Liquid	Solids (6)	Partially Reusable Liquid	Solids (2)	Flyback Liquid	Flyback Liquid
Second Stage	Expendable Liquid	Two Liquid Stages	Partially Reusable Liquid	Partially Reusable Liquid	Partially Reusable Liquid	Partially Reusable Glideback
Payload (100 x 100 nm) 28.5° Inclination (lb) Polar (lb)	112,000 90,000	147,000 123,000	190,000 120,000	130,000 112,000	150,000 120,000	140,000 75,000

Advanced Launch System Concept

The ALS concept is dedicated to the development of a new launch system that will reduce substantially the cost of space transportation while enhancing U.S. space launch capacity, availability, flexibility, and reliability (see Figure F). To accomplish this, the ALS program would immediately initiate an extensive space launch technology program to provide the technical advances necessary to support the development of a fully operational ALS. The technology program will address key areas of vehicle performance, ground and launch operations, and manufacturing. Emerging technologies in areas such as state-of-the-art computer systems, manufacturing automation, and data management offer the potential for much higher operational efficiency in use of personnel, facility, and vehicle resources. It is of the utmost importance that these focused technology programs begin immediately in order to realize the full potential of the ALS.

The emphasis of the ALS development program will be on the so-called "objective" ALS vehicle. The vehicle, which would become available in the late 1990s, would incorporate all new technologies to meet national requirements. Recognizing that U.S. launch requirements may dictate the fielding of a heavy-lift capability in the 1993–1994 time frame, the ALS concept also calls for combining new technologies that are central to the objective ALS vehicle with appropriate current technologies. While not as effective as an objective ALS vehicle, such an "interim" vehicle still could reduce current launch costs by as much as a factor of three and substantially increase U.S. launch capability. Reliability features will be incorporated to the maximum extent feasible. The implications of developing a variant of the ALS with requisite high reliability for potential use with manned systems or launching high cost, one of the kind payloads will also be studied. By taking this approach, the United States could provide itself with a near-term, heavy-lift capability without losing its option of continuing the development of the objective vehicle.

[xii]

Figure F
Advanced Launch System (ALS) Concept

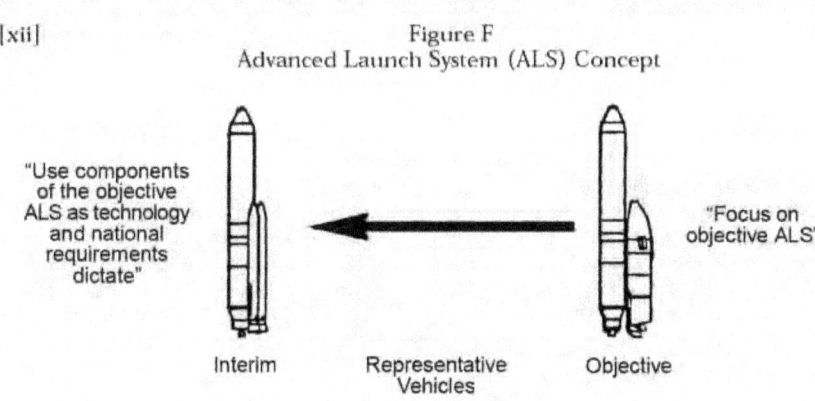

"Use components of the objective ALS as technology and national requirements dictate"

"Focus on objective ALS"

Interim   Representative   Objective
                     Vehicles

Design Philosophy

- Emphasis on operational flexibility
- Reliable with goal of tenfold reduction in operations cost
- High capacity, robust
- Allows for contingencies and assured access to space
- Competitive design and development

- Optimum combination of advanced technology, manufacturing techniques, quality assurance practices, and ground/on-orbit operations concepts
- Revitalize national space transportation technology base
- Thorough demonstration of new technology

[xiii] This approach to ALS development affords great flexibility. If the United States opts for an SDI deployment and/or major civil space initiative (e.g., manned missions to the Moon or Mars) in addition to its other space activities, first the interim, then the objective vehicle would be available. However, if cut-backs in future space activities occur and some heavy-lift capability is still required, but not enough to justify development of an objective vehicle, development of the objective vehicle could be deferred until future requirements dictate. Other alternatives would be assessed such as tailoring the interim system

configuration to the remaining requirements and incorporating emerging technologies into existing systems. In any case, the nation will need to move toward a new launch system to meet the requirements for assured access, availability, flexibility, and low cost to support all space launch users.

ALS Costs

There is no definitive ALS cost estimate since the specific system design has not been selected. Cost estimation is a major task in the concept definition phase of the program planned for the next year. In lieu of an identified systems concept, estimates were made using a representative ALS based on STAS architectures and programs of similar magnitude. Assuming an SDI deployment, and a major civil space initiative (e.g., manned Mars or lunar mission), the total design, development, technology, and evaluation (DDT&E) cost would be approximately $8 billion spread over 11 years. In addition, it would be necessary to substantially augment existing West and East coast facilities at a cost of about $11 billion also spread over 11 years. These costs are comparable to other major DOD and NASA systems acquisitions. The goal for operations cost is a tenfold reduction compared to the cost per pound to orbit of the current systems.

The total cumulative cost of the ALS (including technology development, facilities, production, and operations), satisfying all national requirements, is estimated at about $50 billion through the year 2000. Required funding in FY 87-92 is approximately $8.5 billion (see Figure G). Approved DOD and estimated NASA funding in FY 87-92 is approximately $5.8 billion, leaving a $2.7 billion shortfall. Assuming the estimates for the representative system are correct, this shortfall would have to be funded in order to meet early launch requirements. Failure to fund the program fully would probably delay the ALS program three to five years, which in turn could delay [xiv] [original placement of Figure G] Space Station and SDI deployment. Most of the near-term funding will be invested in critical technology efforts that determine the pace of the program schedule. If the SDIO FY 87 supplemental (which includes a request of $140 million for the ALS) is not approved by Congress, the first operational launch of the objective system could slip by as much as one year. This is due to the fact that these funds are intended to support pacing technology efforts.

[xv] Impact of ALS on Other DOD Programs

The overall impact of the ALS on other DOD programs is positive. In a quantitative economic sense this positive impact is the consequence of the lower space launch operating costs per pound to orbit of an ALS—costs that would significantly drive space transportation costs downward compared to current systems. The cost to DOD of using a current system (e.g., Titan IV) to support the same DOD mission requirements as the ALS through the year 2000 is calculated at over $90 billion, compared with some $40 billion (DOD requirements alone including operations through the year 2000) for the ALS. Just to satisfy estimated SDI launch requirements of nearly five million pounds per year with current launch systems (at approximately $3,600 per pound to orbit) would cost approximately $18 billion annually. Using an objective ALS, these costs could be reduced to $2 billion annually. These savings in operations costs would have a positive impact on future DOD budgets. In none of the FYDP years does the cost of the ALS program exceed 1 percent of the total DOD budget. ALS funding varies from about 0.1 percent at the start of the program to about 1.5 percent in the mid 1990s. In addition, the total development and acquisition cost of the ALS is comparable to other major DOD systems. At a cost of $35 billion (excluding operations cost for DOD), the ALS compares favorably with the Trident II [submarine-launched ballistic missile] at $38 billion and the small ICBM at

$39 billion. While any expenditure of $35 billion would have an impact on other near-term DOD programs over the course of the next several years, the low ALS operations cost would more than pay back the initial investment. In addition to the economic benefits, the ALS will offer the United States a flexibility and operational surge capability in times of crisis with timely delivery of essential on-orbit spacecraft to support military missions.

Figure G
Advanced Launch Systems (ALS) Planned ([Fiscal Year Development Plan] and
[Program Operations Plan]) Versus Required Funding

(Then-Year Dollars in Millions)

	87	88	89	90	91	92	Total
Funding:							
SDIO Budgeted							
Program Specific/Technology	31	289	453	568	546	673	2560
Supplemental	140						140
USAF Budgeted							
Program Specific				100	200	500	800
Relevant Technology	35	39	40	41	42	43	240
NASA							
Program Specific (estimate)	10	(10)	(160)	(365)	(560)	(485)	(1590)
Relevant Technology	63	85	93	94	88	65	488
Total	279	423	746	1168	1436	1766	5818
Required Funding:	279	423	975	1529	2267	3010	8483

NOTE: FY 89–92 funding shortfalls are not firm since there is no definitive ALS cost estimate.

Recommendations

- The United States should immediately embark on a high priority Advanced Launch System program to meet firm DOD and civil requirements.
- A national commitment is needed to initiate and maintain an aggressive technology program to meet the goal of developing U.S. space heavy-lift capability and to enhance existing launch systems.
- The program should be centered on an objective system for the late 1990s.
[xvi]• An interim vehicle should be emphasized to satisfy early Space Station and/or national security needs and to assure compatibility with both near-term requirements and the objective system.

Document IV-5

Document title: James Fletcher, Administrator, NASA, and Frank Carlucci, Secretary of Defense, and approved by Ronald Reagan, President, "Advanced Launch System (ALS) Report to Congress," January 14, 1988.

Source: NASA Historical Reference Collection, NASA History Office, NASA Headquarters, Washington, D.C.

The Space Transportation Architecture Study (which became known as the "STAS study") had suggested the need for a new heavy-lift expendable launch vehicle to carry the heaviest Department of Defense (DOD) and NASA payloads into space. This vehicle became known as the Advanced Launch System (ALS). The ALS program was to be managed by a joint DOD-NASA program office, but all funding was to be provided by DOD except to cover the costs of any NASA-unique requirements. This report to Congress summarizes the program status in its early stages; ultimately, the ALS was not approved for development.

[no page number]

Advanced Launch System (ALS)
Report to Congress

In accordance with the Public Law, PL 100-71, this report is submitted to the Congress. The FY 1987 Urgent Supplemental Report 100-195 was used to guide this report's preparation process. The Advanced Launch System is the next major national initiative in Space Transportation. Both agencies are moving aggressively forward with ALS and are jointly managing the effort. The ALS Program, if approved, will permit this nation to achieve the goal of reduced cost to space.

James C. Fletcher   Frank C. Carlucci
Administrator   Secretary of Defense
National Aeronautics and Space Administration

                                        Approved:

                                        Ronald Reagan
                                        President

[no page number]

I. INTRODUCTION

Public Law 100-71 directed the Secretary of Defense and the Administrator of NASA to submit a plan approved by the President to the Committees on Appropriations which delineates the respective responsibilities of and apportions costs to the Department of Defense and NASA and provides a plan to make maximum use of test facilities relative to the ALS program. This report is in response to the public law.

The report is organized to respond to the requests stated in the 1987 Fiscal Year Supplemental Report 100-195. The major areas presented are program structure, costs, and facilities.

II. PROGRAM STRUCTURE

A. Design Approach

The design approach to be taken for the ALS is based on three principles: develop requirements; use government and industry expertise; and emphasize competition.

1. The basic requirements for the ALS are to provide a launch system that: meets the national launch needs; is flexible, robust, reliable, and responsive; and, significantly lowers the costs of getting payloads to space.

These requirements set the overall guiding philosophy for the ALS. Achieving these goals will require vehicle and operational technology advances, as well as the best abilities of the Department of Defense, NASA, and industry working together. The Defense Acquisition Board will validate the ALS National Security Requirements in a Milestone Zero Review and the NASA Administrator will validate unique civil requirements.

2.   Industry's role is to analyze the government validated requirements and to develop concepts to satisfy those requirements. It is the government's role to manage industry to ensure the concepts fulfill the requirements and to stimulate industry towards the innovations needed to accomplish the necessary technological breakthroughs. By using the expertise of industry, DoD and NASA, this nation can achieve a truly operational system which will lower our costs of getting to space.

3.   The ALS program will emphasize open competition during all phases of the program. The structure to accomplish this is based on three phases:

(a) During the ongoing Phase I Concept Definition, seven contractors are working to define an ALS which achieves the overall requirements. This phase began in July 1987. The end product for each contractor is a System Design Review, that will describe the basic concept and system-level specifications for their concept. DoD and NASA are conducting supporting focused technology efforts, under direction of the program office, in parallel with the contracted studies. Reusable and expendable approaches will be evaluated during the definition and focused technology development phases. During this phase the Defense Acquisition Board (DAB) Milestone Zero Review will be held to approve the mission need and acquisition approach for the ALS and a Milestone One Review will be held to approve entry into Phase II.
[2]   At the end of this phase, the program office will conduct a competitive source selection. This will be an open competition leading to a selection of the contracting teams to conduct Phase II.

(b) The purpose of Phase II is to refine and develop the concepts created in Phase I, to identify the most promising approaches to meeting the ALS requirements, and to conduct preliminary design reviews. During Phase II, focused technology efforts will continue, as required, to provide the necessary technology readiness for [the] ALS. After completion of another open competition, the approaches suitable for full scale development will be selected for Phase III.

(c) Following a successful DAB Milestone Two, Phase III, the Full Scale Development phase, will be initiated Its purpose is to complete development of the ALS leading to an operational capability no later than 1998. The actual Initial Launch Capability (ILC) will depend upon the launch requirements, complexity of the concept, cost, funding availability, and benefits of early availability. These will be assessed at each milestone review.

During all three phases of the program, open competition will be encouraged. A competitor not funded in a phase may compete for the following phase as long as requirements are met for that phase. The acquisition strategy to accomplish these three phases will be determined by the program office and approved through the DAB process.

B. Management Plan

1.   Overall Structure

The ALS is a joint DoD/NASA Program to develop and field this nation's next generation unmanned launch system. It will require the efforts of the finest engineering and management talent available. Therefore, the overall program and sub elements have been structured as joint efforts to allow direct involvement of all appropriate DoD and NASA organizations.

The ALS joint program office will be headed by a Program Manager (PM) appointed by the Air Force. The Deputy PM will be appointed by NASA. The program office will be jointly manned.

The joint program office will have final responsibility for all aspects of the program. Funding, direction, authority to contract; and, cost, schedule, and performance requirements for all program elements will be provided by the program office.

The management elements which make up the joint program office will vary according to the phase of the program. The program manager will determine the structure in order to achieve the overall program mission. Air Force and NASA personnel will head the elements depending on requirements, expertise, and capabilities.

The element managers will be responsible to the program manager for cost, schedule, and technical performance of the element. The managers will still have considerable latitude in managing their respective elements. The location of the management offices will be determined by the PM based on recommendations from the Air Force and NASA.

Work in all areas will be performed by an appropriate mix of DoD and NASA organizations under the direction of the manager, who will have commensurate authority and responsibility to accomplish the program with the available assets. The actual application of resources to tasks within a work element will depend on the particular tasks, available skills and facilities, other requirements for the same assets, and the need to maintain a broad-based national capability for space-related technology.

[3]   2.   Program Guidance

The Defense Acquisition Board will provide major policy and program direction in accordance with [Department of Defense Directive] 5000.1. In addition to the DAB membership specified by [Department of Defense Instruction] 5000.49, NASA will be a participating member for all ALS DAB reviews. Additionally, an Executive Committee, made up of representatives from DoD and NASA, shall provide management oversight.

3.   Management Assignments

Because the ALS program is in its early definition stage and a final concept design has not been selected, definitive management breakout cannot be made at this time. Thus, the management plan which follows is representative and subject to change as the concept development matures. Management responsibilities will be based on experience and expertise, and in all cases will be responsible to the joint program office. Changes to these assignments will be approved by the Executive Committee subject to review by the Secretary of Defense and the Administrator of NASA.

The following is a listing of the ALS program management structure agreed by the DoD and NASA at this time:

ALS Program Management. The overall ALS program management is assigned to a joint DoD/NASA Program Office.

ALS Systems Engineering and Integration (SE&I). The SE&I element will accomplish overall ALS requirements reviews, overall systems coordination and major program oversight. The SE&I element will be managed by DoD with NASA assistance.

ALS Vehicle. The vehicle will be managed by DoD with NASA participation.

ALS Liquid Engine Systems. The liquid engine systems will be managed by NASA with DoD participation.

ALS Solid Motor. Should the concept selected require new solid motor development, the element responsibilities will be determined then.

ALS Flyback Booster. Should the concept selected require a flyback booster development, the element responsibilities will be determined then.

ALS Payload Module. The payload module, including the payload fairing, launch vehicle-to-payload interfaces, and payload handling will be managed by DoD with NASA participation.

ALS Logistics. The logistics package will be managed by DoD with NASA participation.

ALS-Focused Technology. The ALS-Focused Technology Program will be managed by NASA with DoD participation. This effort will be located in the ALS Program Office.

Any other program assignments and structure will be determined by the program office.

[4] III. COSTS

A. Total Costs

DOD and NASA believe an accurate cost estimate is necessary for a successful program and is an essential part of the program approval process. However, the costs for ALS are extremely concept dependent and, since the concepts are still under definition, accurate costs are difficult to estimate at this time. A cost estimate based on preliminary concepts and the overall system requirements will be available in April 1988. DoD and NASA will provide these estimates to the Congress.

B. Cost Sharing

The DoD and NASA recognize the importance of initiating and sustaining the ALS development to meet future national security and other national requirements. DOD will accept full funding responsibilities for developing a national ALS. Those efforts to satisfy unique civil requirements not addressed by the joint ALS baseline design will be funded by NASA.

IV. FACILITIES

The ALS Program will make maximum use of Federal testing facilities, modified as required, to meet ALS requirements. As the concept definitions mature, and test requirements become known, the ALS joint program office will identify respective responsibilities of the Federal entities and facilities to be used.

As an example, the existing rocket propulsion facilities which will be used for ALS testing are located at the National Space Technology Laboratories (NSTL), Bay St. Louis, Mississippi; Air Force Astronautics Laboratories (AFAL), Edwards Air Force Base, California; Marshall Space Flight Center (MSFC), Huntsville, Alabama; Lewis Research Center (LeRC), Cleveland, Ohio; and the Arnold Engineering Development Center (AEDC), Arnold AFB, Tennessee.

As the program matures and concepts are further developed, specific facilities will be selected for testing depending on capability and availability. The program office will produce an interim report at the end of Phase I and a final report on facility use at the end of Phase II. These will be provided to Congress in response to the request for detailed information on facilities and utilization.

V. CONCLUSION

Both DoD and NASA fully support the need for the ALS program and, through the program office, are working to assure a major advance in cost-effective space transportation.

Document IV-6

Document title: Vice President's Space Policy Advisory Board, "The Future of the U.S. Space Launch Capability," November 1992, pp. 3–11, 29–40.

Source: Documentary History Collection. Space Policy Institute. George Washington University, Washington, D.C.

Although it had been in existence since the start of the Bush administration in 1989, the National Space Council did not convene its mandated advisory group, the Vice President's Space Policy Advisory Board, until mid-1992. Then the board spent the remaining months of the Bush administration carrying out several broad policy studies. One of these studies, carried out by a task group headed by former Secretary of the Air Force E.C. "Pete" Aldridge, examined the nation's space transportation situation. The report was issued in November 1992, after George Bush had been defeated for reelection by Bill Clinton. The new Clinton administration dissolved the National Space Council, and it did not implement most of the key recommendations of the Aldridge report. The following are two major excerpts from that report.

VICE PRESIDENT'S
SPACE POLICY ADVISORY BOARD

The Future of the U.S.
Space Launch Capability

A TASK GROUP REPORT

NOVEMBER 1992

[3]   **The New Environment**

The space launch capability of the United States is the most critical aspect of our over-all space program, for without the ability to reliably deliver payloads to orbital velocities, the U.S. space program would not exist. It was only after the United States demonstrated it had the ability to launch payloads, even very modest ones, in the 1958 period that the space program began to emerge and flourish. And only if we have the ability to continue to provide reliable, safe, and relatively inexpensive access to space will technologists, experimenters, and innovators find ways to fully exploit the benefits of space.

We are at a major decision milestone for our future space launch capability. We now have a mixed fleet of space launched vehicles—variants of expendable vehicles that were derived from military ballistic missiles, a manned space transportation system using the technologies of the 1970s, and a new class of small payload launch vehicles using variants and derivatives of existing missiles. These vehicles meet the fundamental lift requirements of the payloads they launch, but the larger vehicles are expensive to operate and do not have the operational flexibility that would otherwise be desirable.

Since approval of the launch strategy in 1991, world events have changed the environment in which the strategy was approved and in which we must implement the strategy. These changes include the intensification of the competitive environment, the realization of the advantages of commercial practices, the availability of excess missile assets for space launch, the reflection of the latest, and reduced, demands on [4] space launch capabilities by a new mission model, and the growing uncertainty of the industrial base that supports production of U.S. space launch vehicles.

Any decisions on the implementation of a space launch capability must be based on the "national" perspective, that is, what is in the best overall interest of the nation rather than the individual interests of the government agencies affected, the programs involved, or the commercial space industry. That was the fundamental objective and focus of this Task Group's review.

Competitive Environment

Changes in the world environment have brought new challenges to the space launch capability of the United States. These challenges exist in the form of a variety of existing and new foreign space launch vehicles, shown in Figure 1, which are priced below comparable U.S. launch vehicles.

While price competition from Ariane has been felt in the United States for years, Ariane could not absorb all commercial payloads being planned around the world. For this reason, and the fact that commercial satellite builders were concerned over a potential monopoly for Ariane, the United States continues to receive launch orders for some of the world's commercial payloads at a rate of three to five per year.

New competition has now emerged which could significantly threaten both the United States and the foreign launch vehicle marketplace. That competition is from the tremendous excess ballistic missile and derived space launch vehicles from the Confederation of Independent States, particularly Russia, and from the very inexpensive launch vehicles in the People's Republic of China. Russia has an impressive space launch infrastructure that could be used to seriously challenge U.S. competitiveness. However, questions exist as to whether we want to take advantage of these new products for U.S. space launches, whether we could rely on these products being in production for long periods, and whether we should place great reliance on the existing but fragile near-term political relationships to commit critical space missions to these components for the long term.

[5]

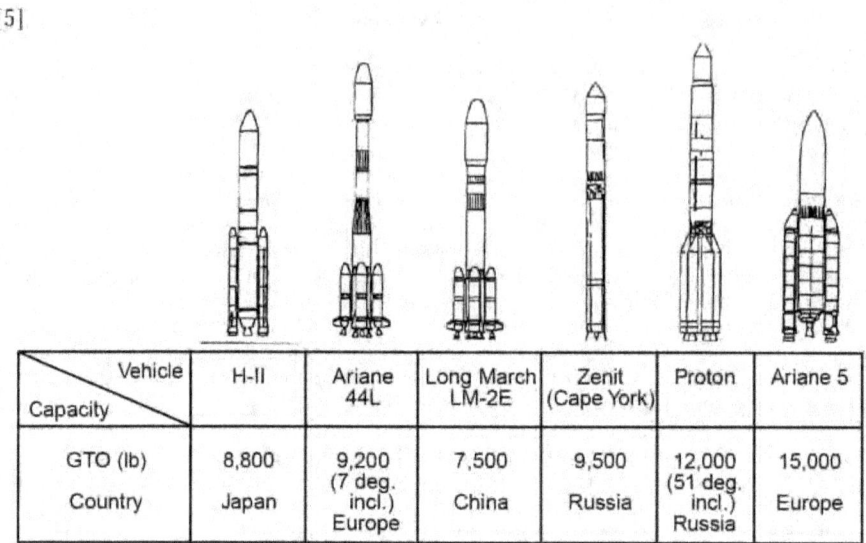

Vehicle Capacity	H-II	Ariane 44L	Long March LM-2E	Zenit (Cape York)	Proton	Ariane 5
GTO (lb)	8,800	9,200	7,500	9,500	12,000	15,000
Country	Japan	(7 deg. incl.) Europe	China	Russia	(51 deg. incl.) Russia	Europe

Figure 1. Foreign Commercial Launch Service Competition

Figure 2 illustrates international launch vehicle competitiveness. If the United States is to remain competitive, it must reduce its cost (and price) to launch payloads by a factor-of-two, as shown by the "Low-cost ELV Goal" line in Figure 2.

Commercial Practices

There have been suggestions by Congress and industry that the government should take advantage of "commercial practices" to reduce the cost of launch vehicles and services. Five distinctions separate commercial from non-commercial practices:

— First, the procurement process, whether the government procures custom-built products priced by negotiation or off-the-shelf products priced by the manufacturers in an open marketplace.

[6] [original placement of Figure 2]

— Second, wide requirement ranges placed on manufacturers by the government with numerous multi-tier design specifications in government procurements versus only end-product or on-orbit performance specifications in commercial procurements.

— Third, the extent of oversight of the manufacturing process, with extensive oversight in government procurements and much less oversight in commercial procurements.

— Fourth, the government limitation on the operating profit of launch vehicle manufacturers under government contracts, which is uncontrolled in commercial contracts.

— Fifth, the financial risks of failure, which are borne by the manufacturer in a commercial contract and are mostly borne by the government in a government contract.

To minimize its risks the government requires more oversight of the launch vehicle manufacturer's processes and specifications. Because mission success is more important in

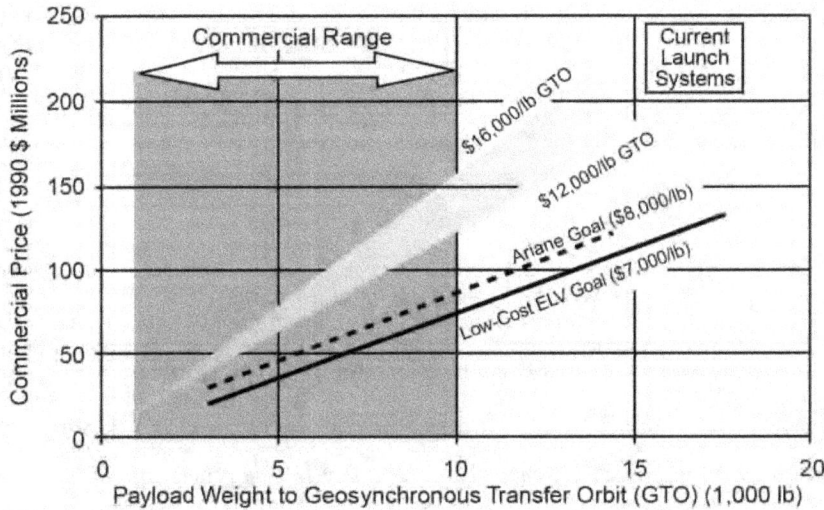

Figure 2. Launch Vehicle Recurring Price versus GTO Payload Weight

government operations than recovery of resources, as is the case in commercial operations, the government is [7] unlikely to accept the full range of commercial practices for space launch operations.

However, U.S. space launch is already "commercial" to some degree. Virtually every U.S. space launch vehicle launching satellites into Earth's orbit is built by a U.S. commercial firm—Martin Marietta, General Dynamics, McDonnell Douglas, Rockwell, LTV, Boeing, or Orbital Sciences—and all of these companies participate extensively in the launch process.

One question that must be addressed is what can the government do, as it works towards its own space launch objectives, to take advantage of the potential cost savings from more application of the commercial practices outlined above and, at the same time, make the U.S. launch vehicle manufacturers more competitive in the commercial world market.

Excess Ballistic Missiles

The phase down of the intercontinental and submarine-launched ballistic missiles (ICBM and SLBM) forces, such as the Titan II, Poseidon, and Minuteman, has provided assets that could and are being used for space launch vehicles. Contracts already exist to convert 15 Titan IIs to space launch vehicles and a contract has been let to begin the conversion of the Minuteman to sub-orbital test vehicles. There is some concern that these "free" vehicles will compete with the production of newer space launch vehicles by reducing the production rate, decreasing the number of production units, and increasing costs. Opponents of using these assets argue that a more efficient, lower cost space launch production program could be built if the government would deny the use of these assets for competition with newly producted [sic] space launch vehicles. In addition, using the excess assets perpetuates a "dead-end" program at the expense of longer range, small payload space launch programs.

Proponents argue that the use of these surplus assets will facilitate lower cost access to space and, in so doing, foster more space-related research and development in both the commercial and university-based sectors than would have been the case without these assets. This additional activity will generate significant and profitable business for the fledg-ling commercial launch industry as it converts surplus assets and provides the [8] associat-ed launch services. Finally, proponents argue that this demonstration of the market for launch services would allow entrepreneurial launch services companies to raise the capital needed for the development of new, more cost competitive launch vehicles and services.

Both positions have merit and a balance between the two points of view must be found.

Future Mission Model Requirements

Projections for the future show a stability in the annual space launch rates for the Department of Defense (DoD), civil, and commercial payloads at about 40 per year (Figure 3). Of these, about seven to eight flights are attributed to the Shuttle and about eight to ten per year are based upon the assumption that commercial satellite manufac-turers, United States and foreign, will continue to rely on U.S. space launch vehicles in the future. The DoD launch rate of 15 to 17 per year is based on a revised estimate of space requirements and funding based on projected future national security needs in a new world environment.

These launch plans are, of course, very dependent on the projected costs of future launch vehicles. U.S. commercial satellite launch rates will either decrease if U.S. launch vehicles can no longer compete financially with foreign launchers or the demand could or might increase if the United States makes a significant reduction in launch costs, thus encouraging the exploitation of space.

Industrial Base

As DoD resources decline, and the industrial organizations that support defense sys-tems shrink and question their future, more and more attention will be placed on options to protect the critical and unique parts of that industry that might be required in the future. The maintenance of a healthy launch industry through the development of new space launch vehicles would appear to be a responsive and efficient way to alleviate the defense conversion problem of our former missile industry. Expansion of [9] [original placement of Figure 3] the space launch vehicle industry through new technology for upgrades to existing vehicles, or the initiation of new vehicle developments to make the U.S. industry more competitive, would be a direct, expeditious, and valuable way to protect this section of the industrial base for future national security requirements. The United States is a world leader in space technology and the conversion of defense resources to pro-tect that leadership would be a valuable way to enhance U.S. competitiveness.

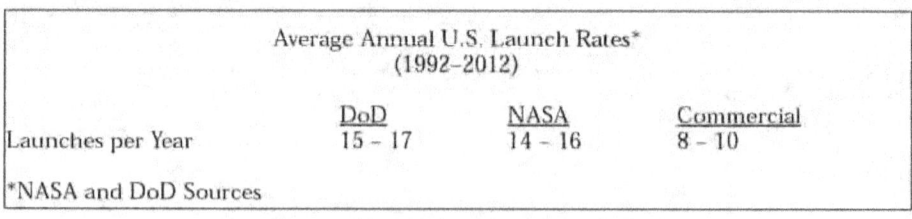

Average Annual U.S. Launch Rates* (1992–2012)		
DoD	NASA	Commercial
Launches per Year 15 – 17	14 – 16	8 – 10
*NASA and DoD Sources		

Figure 3. Average Annual U.S. Launch Rates (1992–2012)

Space launch vehicle contractors have been lacking in incentives to participate actively in, or even argue for, the development of a new launch vehicle. The current contractors
for Titan, Atlas, Delta, and upgrades to these systems are worried about their current business base and are reluctant to abandon near term business for an uncertain future program. Also, they are worried about the potential "winner-take-all" aspects of a future
vehicle competition and the lack of Congressional support for the program. It is understandable that they have a cautious viewpoint and have been somewhat unenthusiastic
about a new system without some changes in the management approach, political support, or investment incentives.

[10]   A recent National Security Industries Association (NSIA) study on the space
transportation system made observations that give a more positive assessment of the industry's perception of the space launch situation. The more pertinent observations from the
NSIA study are as follows:

— A new launch system is required.
— The current fleet does not meet DoD, NASA or commercial cost, responsiveness,
  availability, and operability requirements.
— Some of the present fleet should be retained until a new launch system is proven
  operational and price competitive.
— A new launch vehicle, with performance in the range of 20,000 pounds to low-
  earth orbit is of major interest for DoD, NASA, and commercial users.
— If industry invests in the new program, it will expect an adequate return on
  investment.

Not only did this study indicate a more positive view of a new launch system, it implied
that industry might be willing to share in the development costs.

A New Direction

The 1991 National Space Launch Strategy was based on the conclusion that if the
United States is to compete effectively in the future it must take near-term actions that will
improve the efficiency of its space launch operations, maintain its reputation for reliability, and significantly reduce the cost (and price) to launch. The issue facing the Task
Group was whether the conditions leading to this strategy continue to be relevant in
today's environment.

Developing a "New" or "National" Launch System (NLS) will be relatively expensive
and many related programs are currently underway that will compete for the same scarce
fiscal resources. The Task Group knows that it will be difficult for DoD to step up to a
multi-billion dollar development program when its resources are declining rapidly. DoD
has [11] acceptable alternatives that meet its near-term needs in the Delta, Atlas, and
Titan family of vehicles and its projected launch rates are declining which will extend the
life of this existing fleet. It has been equally difficult for NASA to find the resources to support its share of a new launch vehicle. Congress has been reluctant to give NASA increasing resources and the demands on NASA's budget for Shuttle operations, the Space
Station, Earth observation, and planetary missions will consume the majority of its available resources. So far, there has not been a strong economic imperative or a critical payload requirement to drive the development of a new space launch capability. . . .

[29]   **Recommendations**

Task Group recommendations respond to the findings outlined above and to
Congressional action, which implicitly and explicitly terminates the NLS effort.

1.   **Revalidate the 1991 National Space Launch Strategy and establish a national policy and goal to remain internationally competitive in the space launch marketplace.** The National Space Policy Directive 4, which establishes the National Space Launch Strategy[,] continues to be valid guidance for developing the space launch system for the United States and the implementation of that strategy to remain internationally competitive should continue to receive priority within the affected government agencies. Alternatives to the strategy to either a) forgo new vehicle development and maintain existing launch vehicles, or b) attempt to "leapfrog" existing launch vehicle capability with reusable, and high-risk technology, we reject as inconsistent with maintenance of an effective, competitive, and high confidence space program.

2.   **Create a more formal "national" space launch management arrangement led by an individual with responsibility and authority for the planning and coordination of U.S. space launch capability.** There is a need to provide a more centralized planning, integration, and coordination function for implementing the National Space Launch Strategy and associated programs. Several management models could achieve the desired results. The Task Group recommends the following actions. First, establish an Executive Committee consisting of the heads of major agencies involved in space launch (DoD, NASA, and the Space [30] Council) to provide overall space launch guidance, review and approve plans and program guidance, and adjudicate disputes among agencies involved. Second, designate a single authority (a "space launch authority") responsible to the Executive Committee for planning, coordinating, and integrating U.S. space launch capabilities. This individual should: 1) be an Executive-Level appointee assigned within either NASA or DoD who reports directly to the agency head[;] 2) have the authority to recommend an overall plan and agency funding allocations to the Executive Committee and, within the guidance provided by the Executive Committee, provide program direction to each organization or agency acquiring or operating space launch systems, and oversee program execution[;] 3) be responsible for planning and coordinating space launch technology programs for both existing and new launch vehicles[;] 4) be a focal point for factoring the interests of the U.S. commercial launch industry into government space launch plans[;] and 5) be responsible for government support of a small launch vehicle program.

3.   **The space launch range modernization program being planned in the Air Force, known as the Range Standardization and Automation (RSA) project and related activities, should receive the highest priority in the space launch strategy implementation.** Without the RSA modernization effort and other improvements that will support both the existing and future space launch vehicles, it is doubtful the necessary and desirable safety, reliability, and cost reduction improvements in space launch operations can be achieved. Furthermore, these improvements will enhance the competitiveness of commercial launches that share these facilities.

4.   <u>Terminate the NLS development</u> **within the government agencies and establish a new space launch capability program within the United States, consistent with the revalidated strategy, and under the planning responsibility of the new "space launch authority."** The NLS program was oriented to develop a family of vehicles and design concepts that would lead to an ultimate heavy-lift launch vehicle. The Task Group rejects the near-term requirement for such a vehicle and believes that almost all of the government and commercial space launch requirements for the foreseeable future can be achieved with a vehicle in the lower range of payload performance being considered in the NLS program.

[31] 5. A <u>single</u> "core" space launch vehicle should be pursued that, through modular performance improvements, can meet all the medium and heavier lift requirements (20,000 to 50,000 pounds to low earth orbit [LEO]) of civil, DoD, and commercial users. The new space launch vehicle program, to be known as *"Spacelifter,"* should have the following characteristics:

— employ applicable NLS technology and operational concepts that would reduce its hardware and launch costs and increase its reliability to the maximum extent reasonable and affordable

— compatible with both cargo and manned payloads, and have a performance capability that ranges from 20,000 pounds to 50,000 pounds to LEO with modular concepts (such as strap-on boosters or other innovative modular approaches to achieve the range of performance desired)

— a new high-energy upper stage to satisfy the full range of payload requirements

— a "design-to-launch-cost" goal of a factor-of-two below existing U.S. launch vehicles

— utilize appropriate commercial practices for the acquisition and operation

— extensively instrumented to minimize down-time if failure should occur

— man-rateable [sic]

— a very desirable goal is to be as nearly "environmentally clean" as possible

— Initial Launch Capability planned for the 2000 period to be consistent with depletion of comparable performance launch vehicle inventories and satellite block changes (such as the Follow-on Early Warning System (FEWS), or planned commercial satellites) required at that time

[32] — a transition plan to the new launch vehicle that continues technology applications to improve near-term launch vehicle capabilities, reduces costs, improves reliability, and maintains high confidence in existing launch vehicles and supporting infrastructure until cost and performance of a new space launch vehicle has been demonstrated.

The Spacelifter vehicle will establish U.S. commercial competitiveness, reduce government launch costs, and provide the momentum to move modern technology and operations concepts from the drawing board to real operations. Higher priority should be placed on the design of launch base facilities using improved operational concepts.

If the United States is to depend on the Spacelifter/PLS for all future manned space flight and a majority of the unmanned space missions, the launch vehicle must have attributes that minimize the impact of potential launch failures in the future. The probability of failure must be reduced and the return to operational space flight after the failure must be as quick as possible.

6. **The Air Force should be designated as the manager of the Spacelifter vehicle development and operations.** Since the first payloads to transition to this vehicle will be those produced by DoD, it is more appropriate that the Air Force manage the development of this vehicle. With the termination of NLS, the Air Force should develop a revised acquisition strategy based on performance rather than design specifications. It should encourage the widest application of technology, new contractor arrangements to preserve the space industrial base, and the application of the appropriate commercial practices to the development and operation of the new vehicle.

The acquisition model the Task Group suggests for Spacelifter has three phases. First, competition for Spacelifter would be open to all interested U.S. companies and these companies would be asked to submit conceptual designs, either individually or in teams. Companies would be permitted to incorporate the SSME or any other technologies in their design. Second, the Air Force would select at least two organizations or teams to continue the competition for a short period of time, finalizing their vehicle design and operations concept. Finally, at the competition's conclusion, the Air [33] Force would select the winning concept and industrial organization or team to complete the Spacelifter development and procurement.

    7. **NASA should immediately initiate and manage a two-phased space launch program to deploy and sustain the Space Station.**

        — **The first phase would continue to utilize the Shuttle for the deployment and man-tended phases of the Space Station.** Developing a heavy lift expendable vehicle based on Shuttle components to launch the Space Station would significantly increase the risk to the deployment schedule for the Space Station, divert resources from a more effective long term "national" solution to efficient launch operations, and be "dead-ended" in its application to future manned and unmanned heavy lift requirements. The Task Group questions whether the development of the heavy lift vehicle would be cost effective relative to continuing with the Shuttle to deploy and resupply the Space Station during the early phases of deployment and notes the difficulty and risks of transitioning the Space Station design, optimized for the Shuttle, to a new launch configuration associated with the heavy lift vehicle. Therefore, the Task Group does not recommend the development of a heavy lift launch vehicle based on Shuttle components for deployment of the Space Station. NASA should investigate the feasibility of introducing contingency plans to mitigate the effects of failures during the initial deployment and operation of the Space Station.
        — **The second phase would utilize a man-rated version of the Spacelifter, a Personnel Launch System (PLS), and a Cargo Transfer and Return Vehicle (CTRV) to augment and then replace Shuttle support for the sustained operation of the Space Station.** The Spacelifter/PLS/CTRV would become the primary, long-term support to the Space Station. Funding within NASA for the PLS and CTRV developments needs to be provided immediately if these systems are to be available to support Space Station operations after the year 2000. In order to minimize the negative impact of down-load requirements on CTRV, NASA should undertake a study of options to dispose of non-essential materials from the Space Station.

[34] **8. To offset some of the development costs of the Spacelifter components and vehicles and to demonstrate the commitment to the Spacelifter development, plan for the following changes:**

        — a major near-term reduction in the costs of Shuttle operations by contract incentives, reduction in Shuttle flights at the earliest opportunity, and the reallocation of personnel from Shuttle to the PLS, ACRV [Assured Crew Return Vehicle], and CTRV programs;
        — plan to phase out the Shuttle at the earliest opportunity after the introduction and operational demonstration of the Spacelifter/PLS/CTRV capability;
        — terminate MLV III, avoiding the potential of an additional U.S. launch vehicle, and continuing with the existing medium lift vehicles until Spacelifter becomes available;
        — review the EELV competition and modify it to account for the transition of appropriate NASA payloads to a Spacelifter configuration;
        — slow Titan IV production to about 3 per year and terminating further production upon transition of Titan IV payloads to a Spacelifter configuration;
        — terminate the Advanced Solid Rocket Motor [ASRM] program;

— **terminate the procurement of Shuttle structural spares and mothball the production tooling.**

A substantial part of the near-term investment to develop the Spacelifter vehicle can be offset by these reductions and the redirection of NASA personnel from Shuttle support to planning for the PLS and CTRV. The Task Group recognizes that some of these offsets will be controversial but it believes investments which add only marginally to current capabilities while diverting resources and attention from the required fundamental improvements just cannot be supported. The Task Group also believes MLV III will neither substantially reduce cost nor increase responsiveness and may add to an already overcrowded infrastructure base. With regard [35] to the ASRM program, there is considerable doubt that it will provide significant improvements in safety or reliability. Since Shuttle would be phased out shortly after ASRM became operational, ASRM development costs would not be recovered. Further, ASRM is not environmentally clean. The Task Group also suggests that the existing Shuttle solid rocket motor recovery system and associated refurbishment operations be eliminated at an appropriate point prior to Shuttle system final phase out.

9. **Establish a government-supported, small payload launch program, using low cost launch vehicles, to encourage and promote space research and experimentation that will have a positive long term benefit to the overall national space program.** Military satellite technology, civilian space research, university space research projects, and commercial space applications are focusing more and more on small satellites and associated small launch vehicles. Yet, as in the case of the larger launch vehicles, there is a lack of centralized planning for the use of small launch vehicles resulting in performance gaps and redundancy. The Task Group believes the government should establish a centralized small launch vehicle program that would better plan, integrate, and coordinate government-wide efforts for this class of vehicle. The planning for this program would be the responsibility of the "space launch authority," but the management would remain within the agencies utilizing these capabilities.

10. **To augment the small payload launch program, the Administration should permit the use of excess ballistic missiles for use as space launch vehicles for government sponsored research or commercial applications under specifically controlled conditions.** The Task Group recognizes the controversial nature of this issue but believes that the long-term benefit to the space program and ultimate positive impact on the overall space launch industry in the future justifies use of these assets under certain conditions. Space research and experimentation and new mission concepts will be encouraged and "enabled" by the use of very inexpensive launch vehicles of the class represented by excess ballistic missiles. The use of these assets should be permitted when the following conditions are met: 1) the missions and payloads for such launch vehicles are for government authorized or sponsored research, technology development and test, experimentation and/or education and training, 2) there are no commercially available U.S. space launch vehicles that meet [36] the performance and cost requirements of the mission, 3) the use of more expensive commercially available launch vehicles in lieu of the excess missiles would have precluded the accomplishment of the mission, and 4) the conversion of the excess missiles and all of the launch services are performed by commercial companies selected under competitive processes. The "space launch authority" would determine if these conditions were being met on a case-by-case basis and, if so, recommend that DoD release the assets. The affected government agencies should be encouraged to develop arrangements that would facilitate use of these assets and that would minimize government exposure and liability.

11. **Within the context of the overall approach outlined by these recommendations, the "space launch authority" should continue to plan technology efforts to: 1) improve performance, decrease cost, and improve reliability, safety, responsiveness, and competitiveness of** <u>existing</u> **space launch vehicles ([solid rocket motor upgrade], new low pressure engine concepts, materials, avionics, electronics, testing, etc.), and 2) provide for the next generation of low cost, reliable space launch vehicles that would fully exploit the value of** <u>reusability</u> **(NASP, SSRT, and HSC-F).** Our existing space launch vehicle fleet should continue to receive reliability and cost reduction improvements until the cost and performance goals of Spacelifter are <u>demonstrated</u>. This will provide a hedge against failure to achieve Spacelifter's performance and cost goals and maintain a viable contractor base to support the existing launch vehicle fleet. The Ten Year Space Launch Technology Plan, currently in coordination within the government, would form an acceptable baseline for budget planning and implementing this recommendation. NASA should continue to study heavy lift options for future application to manned and unmanned lunar and planetary missions. The Space Nuclear Thermal Propulsion (SNTP) program is an enabling technology for future manned exploration missions and should be continued to validate the feasibility, cost, and performance consistent with this future requirement.

12. **A vigorous effort must be undertaken to reach a consensus with all government agencies and Congress to pursue and fund the recommended space launch program. If the restructuring efforts, including termination of on-going programs, are accepted without the full commitment to pursue and fund the new Spacelifter efforts, the entire military and civilian space program could be seriously damaged [37] with unacceptable gaps in space system operations.** As stated previously, failure to fund this plan is equivalent to an implicit <u>policy decision</u> to forgo U.S. competitiveness in space launch and increase the long-term cost to the government. Once government funding stability can be achieved, industry will be encouraged to invest its own resources, leveraging government funds and further enhancing launch vehicle capabilities and competitiveness.

13. **While the use of Russian space components might be appropriate on a one-time basis for technology assessment and transfer, or for a very few unique space missions, the Task Group does not recommend the use of Russian manufactured equipment on multiple, routine, or critical space missions.** Russian equipment in the form of engines, space qualified components, and launch vehicles appears to be capable, effective, reliable, and available at competitive prices. This equipment may provide opportunities for positive technology transfer and licensing agreements, and could, in limited situations, advance the U.S. launch industry in technology and capability. However, the uncertainty of a sustained industrial base in Russia and the Ukraine (as well as access to launch facilities in Kazakhstan), the uncertainty of a stable long-term political relationship between the United States and Russia, and the detrimental impact such an arrangement could have on the U.S. industrial base and U.S. competitiveness demand caution and restrictions on cooperative arrangements.

14. **Create a mechanism for downsizing both the space launch industry and supporting government infrastructure while continuing to satisfy future space launch requirements of the United States and taking into account commercial competitiveness of U.S. industry.** Industry has indicated the government has certain impediments to the proper "right-sizing" of U.S. industry (e.g., anti-trust laws) and political pressures will inhibit government from taking necessary steps to reduce or eliminate unnecessary government organizations or facilities that support launch development and operations. Participation of the launch vehicle industry in determining cost-sharing options and unique management arrangements to facilitate a new launch vehicle development should be solicited and

encouraged. Since it is expected that industry would benefit from the introduction of a highly competitive Spacelifter, there should be some incentive for industry to share in the development cost.

[page 38 is blank]

[39]   **Concluding Comments**

The United States is in a very critical period in ensuring continued competitiveness in space launch in both the government and commercial marketplace. The Shuttle program is costing $5 billion per year (absorbing about 35% of the NASA budget) yet is planning to launch only seven to eight flights per year. The government is paying too much to launch government satellites on expendable launch vehicles. U.S. launch vehicles are not competitive with foreign launch vehicles and are receiving market share only because of rate limitations on the current foreign vehicles and fears of a monopoly by commercial satellite customers. New foreign space launch vehicle players have now entered the marketplace with even more competitively priced vehicles. U.S. government launch rates are declining which make U.S. vehicles even less competitive and government cost per launch even higher.

The technology developments in new launch vehicles and revised operational concepts give us confidence that we can produce a space launch vehicle that can save the taxpayer a significant amount in the future and make U.S. space launch vehicles extremely competitive in the world market. The up-front development costs of new launch vehicles and manned spacecraft are high, but we will be able to achieve a very high return on this investment within a reasonable period of time by phasing out obsolete and expensive launch vehicles. Much of the initial cost can be offset with aggressive efforts to reduce current operating costs and termination of those programs that will not be necessary if we initiate the development of a new class of launch vehicles. Other near-term, indirect cost savings, resulting from elimination of launch delays, wasted efforts, [40] and failures resulting from the continued use of older technology vehicles can be achieved.

It is the unanimous view of the Task Group that now is the time to initiate an aggressive effort toward the development of a new generation space launch vehicle that will replace existing manned and unmanned launchers. The cost of this effort will be more than offset with the increased U.S. competitiveness, lower costs to government users, improved reliability, safety, and efficiency, and encouragement of additional research and experimentation to broaden our use of space. It is an essential step to ensure the United States enjoys the benefits of space exploration and exploitation, and it is the manifestation of the U.S. commitment to space leadership.

Document IV-7

Document title: Darrell R. Branscome, Director, Advanced Program Development Division, Office of Space Flight, NASA, "The Next Manned Spacecraft . . . Which Path to Follow?," November 17, 1988.

Source: NASA Historical Reference Collection, NASA History Office, NASA Headquarters, Washington, D.C.

This briefing by NASA engineer Darrell Branscome is representative of the thinking within NASA during the post-Challenger period as the space agency considered what vehicle it might eventually propose to replace the Space Shuttle as the U.S. means for human access to space.

[NASA-001-881028-PA] The Next Manned Spacecraft . . . Which Path to Follow?

November 17, 1988

Darrell R. Branscome
Director, Advanced Program
Development Division
Office of Space Flight

[NASA-002-881028-PA]
* Satisfy People/Payload Requirements
* Improve Cost/Effectiveness
* Increase Reliability
* Increase Margins

Which Path to Follow?

STS Evolution   Simple Rugged People Carrier   Advanced Manned Launch System

[NASA-004-881028-PA] STS Evolution
* Exploit New Technologies
* Build on Existing Engineering Data Base
* Minimize Mold-Line/Configuration Changes
* Counter Obsolescence
* Increased People Carrying Capability

Simple Rugged People Carrier
* Winged or Blunt Body
* Increased Design Margins
* ELV Launched
* Configuration/Size Open
* Limited Return Cargo Capability
* Up Payload on Cargo Vehicle

Advanced Manned Launch System
* Fully Exploit New Technologies
* Improved Design Margins
* Configuration/Size Open
* People Only Option Available

[NASA-066-881028-PA]

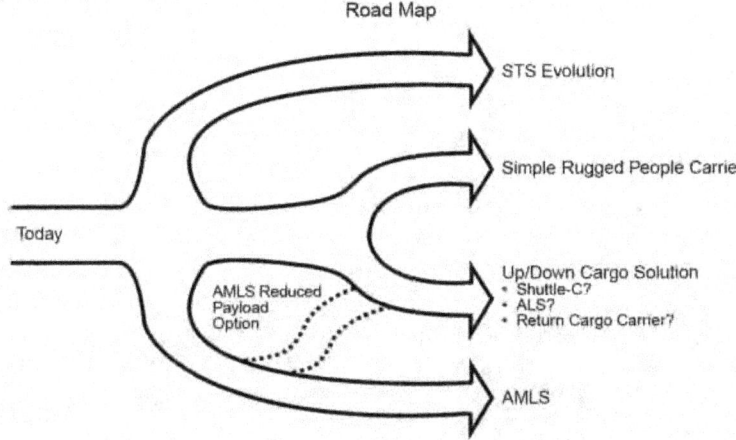

Road Map

STS Evolution

Simple Rugged People Carrier

Today

Up/Down Cargo Solution
• Shuttle-C?
• ALS?
• Return Cargo Carrier?

AMLS Reduced Payload Option

AMLS

[NASA-003-881028-PA]   **Future Requirements**

- Crew Rotation
    - Initial Space Station (8 Persons)
    - Space Station Growth (12–16 Persons)
    - Exploration Missions (? Persons)
- On-Orbit Servicing
    - Scientific Observatories
    - Space Station
    - Polar Operation (?)
- Return Cargo
    - Station Logistics
    - Station Scientific Instruments
    - Manufactured Products

[NASA-032b-881028-PA]

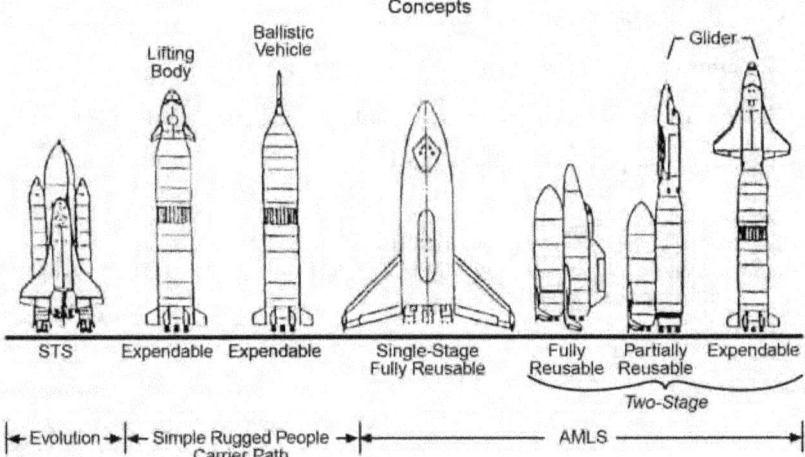

Concepts

Lifting Body

Ballistic Vehicle

Glider

STS Expendable Expendable Single-Stage Fully Reusable Fully Reusable Partially Reusable Expendable

Two-Stage

← Evolution → ← Simple Rugged People Carrier Path → ← AMLS →

[NASA-045-881028-PA] **Low-Cost Ground and Flight Operations**

Vehicle Design Features

- Large Performance Margins/Mature Technology
    - Simplified Flight Planning
    - Minimum Flight to Flight Software Reconfiguration
    - Improved Weather Prediction

- Standard Payload Services and Interfaces
    - Minimum Vehicle/Payload Interface Reconfiguration
    - Payload Containerization

- Autonomy/Automation
    - Paperless Management Information
    - Automated Flight Planning
    - Automated Systems Interface Verification
    - On-Board Hardware Self[-]test/Diagnostics
    - Critical Fault Tolerance/Redundancy

- Maintainability/Operability
    - Non-Hypergolic Reaction Control System
    - Electromechanical (Non-Hydraulic) Actuators
    - Durable Thermal Protection System
    - Total Access to Critical Components

Operations to Drive Vehicle Design

Source: STS Lessons Learned
STAS/KSC Gnd. Ops. Efficiency Study

[NASA-008-881028-PA]   **Higher Reliability**

- Conservative Design Margins
    - Structural
    - Engine Performance
    - Operating Envelope

- Fault Tolerance
    - Engine Out Capability
    - Redundancy (e.g., Electronics, Selected Electromechanical Subsystems)

- Manufacturing/Processing Quality Control

- Engineering Data Base
    - Technology Demonstration/Validation
    - Vehicle System/Subsystem Test and Evaluation
    - Flight Experience/Ground Testing

Document IV-8

Document title: Secretary of Defense and NASA Administrator, "Memorandum of Understanding Between the Department of Defense and the National Aeronautics and Space Administration for the Conduct of the National Aero-Space Plane Program (Revision B)," August 31, 1988.

Source: NASA Historical Reference Collection, NASA History Office, NASA Headquarters, Washington, D.C.

It took more than two years from the time that President Ronald Reagan announced his approval of what became known as the National Aerospace Plane (NASP) program for NASA and the Department of Defense to agree on the program's management structure. Relations between the two agencies with respect to space transportation issues were tense in the years following the Challenger accident, and both had other demands on their advanced technology budgets.

Memorandum of Understanding Between the Department of Defense and the National Aeronautics and Space Administration for the Conduct of the National Aero-Space Plane Program

(Revision B)

[1] PURPOSE

The purpose of this Memorandum of Understanding (MOU) is to establish the mechanisms for the joint conduct of a National Aero-Space Plane (NASP) Program by the Department of Defense (DoD) and the National Aeronautics and Space Administration (NASA).

OBJECTIVE

The National Aero-Space Plane Program is a technology program to provide the basis for hypersonic flight vehicles that will result in space transportation systems, superior U.S. military aircraft, and civil transports that will have technical, cost and operational advantages over existing systems into the next century. The objective of the NASP Program is to develop, and then demonstrate in an experimental flight vehicle, the requisite technologies to permit the Nation to develop both military and civil vehicles capable of operating at sustained hypersonic speeds within the atmosphere and/or as space launch vehicles with the capability of delivering payloads into orbit. The NASP is envisioned to be an airbreathing, hydrogen-fueled, horizontal takeoff and landing vehicle with single-stage-to-orbit capability.

BACKGROUND

During the past decade, substantial progress has been made in hypersonic airbreathing propulsion, advanced materials and structures, and computational technologies contributing to the consensus that operational hypersonic/transatmospheric vehicles may be possible around the turn of the century. However, establishing a valid data base, including the interactive solutions to the propulsion, structures, and aerodynamic problems, is dependent upon accelerating ground-based technology development and verifying the results at sufficient scale in flight over the entire speed range. The DoD and NASA, having engaged in an aerospace plane concept feasibility study in 1984--1985, and having considered the additional technology and operational data needed to support potential future applications, have concluded that the combined objectives and the National interest are best served by a joint program.

PROGRAM DESCRIPTION

The first phase (Phase I) was a Defense Advanced Research Projects Agency (DARPA) concept feasibility study (Copper Canyon) that began in 1982 and concluded in 1985 with NASA, Air Force and Navy participation in the latter part of the study. The study results form a point of departure for the NASP Program.

[2] The NASP Program consists of two phases: Phase II, Technology Development and Application Studies, and Phase III, Experimental Flight Vehicle. The goals of Phase II are to (1) demonstrate the technology maturity, and (2) provide the supporting utility and survivability assessments of potential applications before committing to an experimental flight vehicle. The results of Phase II will be the basis for a decision, prior to the commitment of large resources, on whether to proceed to Phase III. The goal of Phase III is to accomplish sufficient flight demonstration to provide a verified technological basis for future operational vehicles.

PROGRAM DIRECTION

The NASP Program is governed by this MOU between the Secretary of Defense and the NASA Administrator. Under the broad framework of this MOU, the Under Secretary of Defense for Acquisition (USD(A)) and the Associate Administrator, Office of Aeronautics and Space Technology (AA/OAST) are responsible for DoD and NASA participation in the program through a NASP Steering Group. Annex A to this MOU establishes the Terms of Reference and membership for the Steering Group. The Steering Group will provide policy, guidance, and broad programmatic direction and will have issue resolution authority. The Steering Group will make the decision whether to proceed to Phase III, which will be subject to the consent of the Secretary of Defense and the NASA Administrator. The Steering Group will be chaired by USD(A) with AA/OAST as vice-chair.

ORGANIZATION AND RESPONSIBILITIES

The DoD is responsible for overall management of the joint NASP program. Within this program NASA has lead responsibility for civil applications and an integral role in the overall program. Personnel from both agencies will participate in all phases of the technology development, application studies and the design, fabrication, and flight test of experimental flight vehicles. Within the DoD, the Air Force has been assigned the overall responsibility for the NASP program.

The Air Force will maintain the NASP Joint Program Office (JPO) at Wright-Patterson Air Force Base, Ohio with an Air Force Program Manager (PM), a NASA Principal Deputy, and Air Force, Navy and NASA Deputies. The management responsibilities within the JPO will be shared jointly between DoD and NASA. The JPO is responsible for planning and conducting Phase II and Phase III of the NASP program. All JPO Deputies and Directors will be located full time at the JPO.

[3] The Air Force shall establish a NASP Inter-Agency Office (NIO) reporting directly to the Assistant Secretary of the Air Force for Acquisition, with an Air Force Director, a NASA Principal Deputy, and joint Air Force, Navy and NASA staffing. The NIO is responsible for coordination and oversight of policy, budgetary, program progress, congressional and public affairs, and other matters, as required.

SECURITY

The joint DoD/NASA NASP Program Security Guide will be maintained by the JPO; proposed changes will be submitted to the Steering Group for concurrence. Overall security cognizance will be maintained by the Air Force. For NASA, the focal point for security cognizance will be NASA Headquarters. The JPO will have primary responsibility for implementing and managing NASP program security procedures.

PUBLIC AFFAIRS/LEGISLATIVE MATTERS

Guidelines for public affairs activities will be prepared by DoD and NASA, and will be submitted to the Steering Group for approval. Both DoD and NASA will retain the right to release, within these guidelines, information in their respective areas of responsibility. Approval for public affairs activities will be obtained through the normal DoD or NASA channels in accordance with these guidelines.

Although each organization will have specific responsibility for legislative inquiries from their respective oversight committees, prior coordination will be accomplished.

PROGRAM PLANS

The Program Management Plan (PMP) for the overall NASP program will set forth program goals, major tasks and milestones, organization and responsibilities, resources and procurement approach. The PMP will also define a resource allocation and control system. The PMP will be reviewed annually, updated as required, and will be approved by the Steering Group. For substantive changes to the program, approval to proceed must be obtained from the Steering Group prior to change implementation. Updates for the PMP will be the responsibility of the JPO.

[4] RESOURCES

For planning purposes, the funding required for the NASP program is currently estimated to be (then-year dollars in millions):

	Prior	FY88	89	90	91	92	93	94	95	96
DoD	155	183	245	300	390	425	425	413	357	125
NASA	78	70	104	149	119	72	46	46	46	38

DoD and NASA will provide this funding in accordance with the PMP, and their own internal administrative procedures, subject to the availability of funds or other constraints which may be imposed on DoD and NASA. DoD and NASA will be individually responsible for providing accountability for the funds appropriated to their respective agencies by the Congress which are applied to the joint program. Proposed changes in program funding will be reviewed by the Steering Group. Changes endorsed by the Steering Group will be recommended to each agency for approval, allocated on a pro-rata share based on the percentage of each agency's funding for any particular year affected, unless specified otherwise. Each agency will endeavor to provide its share of any recommended increases in program funding. Appropriate institutional support will also be provided by each agency.

DURATION

This MOU will remain in effect when signed by both parties until or unless modified or extended by mutual agreement. Either party to this MOU may terminate its participation upon 120 days written notice to the other party. In the case of such a termination the party terminating will undertake payment of costs incurred up to the point of termination.

MOU REVIEW

If the Steering Group determines the NASP Program is to proceed to Phase III, this MOU will be reviewed for its applicability and revised as necessary.

SIGNATURES

Department of Defense   National Aeronautics and Space Administration

Secretary of Defense   NASA Administrator
Date: 31 AUG 1988   Date: SEP 27 1988

[1 - Annex A]   ANNEX A

TERMS OF REFERENCE

NATIONAL AERO-SPACE PLANE STEERING GROUP

I. PURPOSE

The purpose of this Group is to provide policy, guidance, and broad programmatic direction for the conduct of the National Aero-Space Plane Program.

II. SCOPE

The scope includes current and future phases of the National Aero-Space Plane Program which are concerned with technology development and demonstration, but does not include any subsequent phases which may be devoted to operational systems development.

III. MEMBERSHIP

Chairman: Under Secretary of Defense
                       (Acquisition) (USD(A))

Vice Chairman: National Aeronautics and Space Administration, Associate
                       Administrator for Aeronautics and Space Technology (AA/OAST)

Executive Secretary: Deputy Under Secretary of Defense
                       (Research and Advanced Technology) . . .

Members: Director of Defense Research and Engineering . . .
Assistant Secretary of the Navy (Research, Engineering and Systems) . . .
Assistant Secretary of the Air Force (Acquisition) . . .
Director, Defense Advanced Research Projects Agency (DARPA)
Director, Strategic Defense Initiative Office (SDIO)

Honorary Member: Director, Office of Science and Technology Policy (OSTP)

|2 - Annex A| IV. <u>RESPONSIBILITIES</u>

The Group is responsible for:

1. Ensuring that the Program is conducted in accordance with this MOU.
2. Periodically reviewing progress of the Program.
3. Providing policy, guidance and broad programmatic direction for the conduct of
    the Program, consistent with both the needs of the Program and the other needs
    of the organizations involved.
4. Resolving such policy and guidance issues as may be brought before it.

V. <u>PROCEDURES</u>

1. The Group shall meet, at the call of the Chairman or Vice Chairman, as needed
    to fulfill its responsibilities.
2. Records of such meetings will be maintained by the Executive Secretary, and dis-
    tributed to all Group members.
3. The Executive Secretary is responsible for all administrative and procedural mat-
    ters related to the functioning of the Group.

Document IV-9

**Document title: Department of Defense, "Report of the Defense Science Board Task
Force on the National Aerospace Plane (NASP)," September 1988, pp. 2–25.**

Source: Defense Technical Information Center, Ft. Belvoir, Virginia.

*From almost the beginning of the National Aerospace Plane program, there were doubts about whether
its objectives were technically feasible, particularly in the context of the ambitious schedule set for developing and testing the actual flight hardware. This early independent review of the program by the
Defense Science Board, a top-level external technical advisory group within the Department of Defense,
stressed the demanding technical requirements of the effort and expressed skepticism that the program
could meet its planned schedule. The report's appendices do not appear here.*

Report of the
Defense Science Board Task Force
on the National Aerospace Plane (NASP)

Dr. Joseph F. Shea
Chairman . . .

[stamped "CLEARED FOR OPEN PUBLICATION" and "SEP 29 1988" and "Directorate for Freedom of Information and Security Review (OASD-PA) Department of Defense"]

[2]   SUMMARY

The NASP started in 1984 as a DARPA Program to explore hypersonic air breathing propulsion. It transitioned during 1985 to a program with the dual goals of demonstrating single stage to orbit and hypersonic cruise with the same vehicle. When President Reagan included the NASP in his 1986 State of the Union Message, it became a major national program.

Early estimates of vehicle size, performance, cost and schedule were extremely optimistic. Hypersonic technology had been dormant in the United States for over a decade. It took about a year for both Government and Industry to recognize the technical deficiencies which existed in all the critical technologies, and the lack of ground test facilities to explore the hypersonic environment. Late in 1985, DARPA formed a committee, chaired by Dr. Victor Reis, to review technical and management issues on the program. Among their recommendations was the initiation of a Technology Maturation Program (TMP) to better integrate technology efforts with the design program and to address the most critical technical gaps. Implementation began early in 1987.

This Defense Science Board Task Force was chartered in late 1986 to review the sufficiency of the TMP to support a decision to proceed with detailed design and fabrication of a flight test vehicle by the end of 1989.

When our review began, the program was supported by five airframe and three engine contractors doing Phase 2, Part II configuration studies. The Technology Maturation Program brought in additional contractors [3] and Government Laboratories for specific tasks, and was supplemented by contractor Independent Research and Development efforts.

Late in the Summer of 1987, the planned down select to three airframe and two engine contractors occurred. The program is now in Phase 2, Part II, tentatively scheduled to complete during 1990, at which time one contractor would enter Phase 3 detailed design, fabrication and flight test of the flight test article.

The National Aerospace Plane Program today is significantly different from that envisioned at its outset in 1985. Vehicle weight has grown considerably, as have program cost estimates. Schedules continue to lengthen because of both technical and budgetary issues. We believe more such change can be expected.

The Task Force held four meetings in which the overall program and the Technology Maturation Program were reviewed, four sub-panel meetings on specific technologies and one three day meeting with the contractors. Several of the members have had extensive involvement with the NASP either through membership on the Reis Panel or through consulting assignments directly from NASA or the Air Force.

[4] The recommendations from the Task Force members are unanimous.

Basically, we believe that, as a significant national program, the NASP should be realistically presented to its sponsors within DoD, its supporters in Congress and ultimately, through the White House to the American public. We define "realistic" as a program with a reasonable chance (above 75%, to choose an arbitrary measure) of meeting the performance, schedule and cost goals projected by its proponents. In today's budgetary environment, lack of realism which leads to significant overruns or performance shortfalls can result in loss of program support, and the national embarrassment of a major technical effort poorly executed.

Having looked in some depth into the technologies of importance to the NASP, we are impressed with the progress being made. But we are even more impressed by what has yet to be done to reduce the remaining uncertainties to a reasonably manageable level.

Until these uncertainties are reduced, the NASP should not be a schedule driven program. Rather, it should be paced by events. In particular, we recommend that a set of technical milestones be established which must be demonstrated before a configuration is baselined and Phase 3 detailed design, fabrication, and flight test initiated.

The following sections summarize the Task Force charter and our response to the terms of reference, the major areas of technical concerns, the concerns expressed by the contractors and our conclusions and detailed recommendations. Six appendices [5] discuss the critical technologies in more detail. The seventh appendix summarizes individual contractor comments.

The Task Force strongly supports the overall goals for the National Aerospace Plane Program. We believe our recommendations suggest a realistic path by which those goals can be achieved.

During the period of our review, the program has continued to evolve. This report contains our interpretation of data gathered January–June 1987 and reflects NASP program status and information current as of that time. We believe management has already begun to respond to the recommendations of the Task Force which have been extensively briefed to DARPA, the Air Force, NASA and DoD.

[6]   TERMS OF REFERENCE

The Task Force was chartered to address, but not be limited to, the following issues:

1) The overall sufficiency of the Program's Technology Maturation Plan (TMP).
2) The degree to which the overall program effort adequately supports the achievement of the technical objectives of Phase 2 of the NASP Program.
3) The need for additional technology development efforts which would extend beyond the time frame of the Phase 2 program.
4) The adequacy and viability of criteria to be satisfied in order to justify a decision to proceed to Phase 3 of the NASP Program.
5) The range of missions for the NASP and variants to the degree required to identify technology issues. New capabilities provided by the NASP which offer the potential for new mission possibilities.

[7]   RESPONSE TO ISSUES POSED

Detailed conclusions and recommendations are presented in later sections. This section summarizes the Task Force response to the issues raised in our terms of reference.

1) Although the Technology Maturation Plan is a good start, it is far short of what will be required to enable the NASP Program to enter Phase 3 on the present schedule with any degree of acceptable technical risk.

2) The TMP does not adequately support the objectives of Phase 2. Some tasks provide data too late to help in the configuration decisions which are required to start Phase

3. More importantly, major technology issues in structures and materials, propulsion, aerodynamics, controls, validation of computational aerodynamic codes and ground testing are not being addressed.

3)   To close the risks in the areas indicated above, funding of the Technology Maturation Program should be increased. We estimate that twice as much as presently planned could usefully be invested. Since total program funding is unlikely to increase, this means that the configuration efforts in airframe and propulsion should be scaled back to a level sufficient to provide a focus for the technology effort.

4)   No quantitative criteria have been established to justify a decision to proceed to Phase 3 of the program.

[8] 5) The Task Force did not review the range of missions for NASP. Such studies are still in an embryonic state. However the Task force members believe that the NASP is a vitally important national program because of the missions, both military and commercial, it will enable, and the technology which will be matured.

Hypersonic, air breathing propulsion can attain a Specific Impulse approaching 2000 seconds, compared to about 460 seconds for conventional high energy cryogenic fuel rocket engines. A single stage to orbit, reusable air breathing vehicle is a possibility for low cost to orbit transportation.

Hypersonic cruise vehicles will enable our Military to project American presence anywhere in the world within a few hours, providing timely response for crisis intervention, strategic reconnaissance and terrorist attack. Civilian hypersonic transports will further shrink the world.

The National Aerospace Plane is a necessary precursor to these three classes of vehicles. As an X-airplane it will explore the realm of hypersonic flight, gathering the data necessary to overcome the limitations of analysis and ground test facilities. Of equal importance, the NASP will provide a focus for the development of the six technologies critical to hypersonic vehicle design, aerodynamics, supersonic mixing and fuel-air combustion, high temperature materials, cooled structures, control systems and computational fluid dynamics.

The following sections address the technical concerns encountered in our review.

[9]   DISCUSSION

The technologies critical to the NASP are aerodynamics, propulsion, materials, structures, controls and computational fluid dynamics (which must support several of the disciplines).

The recommendations of the Task Force are based on review of these technologies and the technical and management experience of the Task Force members. This section summarizes the major concerns which shaped our recommendations.

The appendices contain more detailed discussion of each area.

Aerodynamics

The NASP requires an unprecedented degree of integration of the airframe with the propulsion system. Although this is well recognized by program management and the contractors, the problems of integration are formidable. Because of a lack of adequate ground test facilities above about Mach 10, some of the critical design issues may only be resolved by flight test of the vehicle.

The largest uncertainty is the location of the point of transition from laminar to turbulent flow. Estimates range from 20% to 80% along the body span. That degree of uncertainty significantly affects the flow conditions at the engine inlet, aerodynamic heat transfer to the structure and skin friction. These in turn affect estimates of engine performance, structural heating and drag. The assumption made for the point of transition can affect the design vehicle gross take off weight by a factor of two or more.

[10] Computational fluid dynamics cannot predict transition because turbulence must be introduced into the calculations empirically, and no relevant data base exists for the high Mach number flight regime. In addition, while CFD [computational fluid dynamics] is reasonably accurate for two dimensional laminar flows, calculations of three dimensional flow around structural details usually needs to be calibrated by experimental data. Therefore estimates of local heating conditions will be imprecise.

Historically, calculations of aerodynamic performance have been validated in ground test facilities. For Mach numbers between ten and twenty five no ground test facilities exist which can produce true stagnation enthalpy and full scale Reynolds numbers. One or several of the critical parameters can be simulated separately in existing or proposed facilities, and these will provide useful data which may narrow the uncertainties. However there is currently no way to validate methods for combining such partial simulation results to represent the true flight environment.

The uncertainties of aerodynamic performance will affect all aspects of the NASP design.

The NASP program has initiated a major analytic and experimental effort to understand the nature of transition. It would seem prudent to delay initiation of detailed vehicle design until that effort has narrowed the uncertainty in location of the transition point to an acceptable tolerance.

[11] The air breathing propulsion system for the NASP must operate from a standing start to Mach 25. It will consist of three distinct cycles, low speed (up to about Mach 1), ram jet (subsonic combustion), and scram jet (supersonic combustion).

The low speed cycle is a significant design challenge, but can be adequately tested in ground facilities and independent flight, as can the ram jet. Transition from ram jet to scram jet could be the most critical stage of flight, when a normal shock must be forced through the diffuser, combustor and nozzle without flameout or loss of thrust so that the vehicle can continue to accelerate. The system must avoid any strong shock waves that might be caused by fuel injection or details of the variable geometry in the engine flow path required to optimize performance over the wide flight regime. Unwanted shocks could destroy performance or cause unstart which could place heavy demands on the vehicle attitude control system.

Very little is presently known about the mixing and combustion of hydrogen at very high supersonic velocities. It is possible that some of the reactions will not be completed in the combustion chamber, or even in the nozzle, which would result in a loss of performance. Fundamental research in this area has been proceeding slowly because of computational and experimental limitations.

[12] Calculations of flow through the engine will have larger uncertainties than those discussed for aerodynamics because of the uncertainty in inlet conditions, the more complex geometry of the flow path and the introduction of combustion kinetics. Ground test facilities will not provide data much above Mach 8, and full scale testing will probably not exceed Mach 4. Valid testing at higher Mach number will only be done by expanding the flight envelope of the full scale vehicle. It is highly likely that flow anomalies will be encountered in the propulsion system which will require redesign before the flight test program can proceed. Non[-]intrusive instrumentation which can provide the data to

resolve such problems must be developed.

The NASP program should consider conducting the equivalent of a limited pre-flight readiness test . . . for the NASP propulsion system, as is conventional practice for a manned aircraft program. This would require a ground test facility with continuously variable Mach capability to as high a velocity as practical. Ability to demonstrate ram jet to scram jet transition would be very desirable. To this end, modification of the Aeropropulsion Systems Test Facility . . . tunnel at Tullahoma, Tennessee should be studied.

Materials

Based upon the preliminary design and performance estimates presented to the Task Force, surface temperatures of the NASP structure will range from in excess of 3000°F to less than 1200°F. For a typical configuration, some 15% of the wetted [13] area might be exposed to temperatures above 2600°F, 20% to temperatures between 1800°F and 2600°F, about 50% to temperatures between 1200°F and 1800°F, with only 15% below 1200°F where "conventional" materials are available. The higher temperature requirements force the vehicle designer to make a choice between new, promising materials which are in various stages of advanced development (in general, available only in laboratory quantities), or active cooling of a major fraction of the structure.

There appeared to be a discrepancy between consideration of the advanced materials for the high temperature structure and the availability of such material on a schedule compatible with vehicle fabrication. Development of new materials including scaled up production facilities is estimated to take twelve to fifteen years. At the time of our review, the NASP program schedule would have allowed only five to seven years. We also noted that no funds were programmed to facilitate whatever scale up is finally required, although the new materials would not see immediate demand outside the NASP Program and therefore would not be likely to attract private investment.

The lack of scaled up production processes also affects the quality of the material characterization data available to the structural designer. Small quantity lots will not provide the range of material properties required to establish design allowables, damage tolerance and fatigue characteristics for production materials.

[14] The NASP structure will be exposed to high temperature, high enthalpy, disassociated gas. Reusable coatings will be essential to protect the materials.

In areas where the structure is exposed to hydrogen at high temperature and pressure (such as active cooling channels), the hydrogen molecules can penetrate the material and cause embrittlement. The problem is not well understood. The program is raising contractor awareness of the problem, but no funded effort was underway at the time of our review.

It is the opinion of the Task Force that availability of suitable materials in production quantities will be the pacing element in the NASP schedule, and that resources must be identified to fund the necessary scale up and characterization effort.

Structure

The structural designer has the fundamental task of designing an optimum structure to acceptable minimum margins of safety commensurate with man rating the NASP. To do that requires that:

1) The materials to be used must be fully characterized from material reasonably close to or in production, not from small laboratory samples.
2) The complete operating environment must be reasonably known.
[15] 3) The analysis methodology to determine external loads and derive therefrom internal loads must be available, verifiable, accurate and reasonably efficient.

4) The design can be verified through adequate ground and flight test.

Because of the uncertainties noted in earlier sections in aerodynamic loads and heating, materials availability, precision of computation and lack of ground test facilities to replicate thermal and structural flight loads, the current ability to meet the structural designers requirements are marginal to non[-]existent.

To achieve the NASP performance goals, the vehicle structural weight fraction will have to be twenty five to thirty percent less than the Shuttle.

In most conventional aircraft the prime loads are aeroelastic. Environmental loads (thermal, acoustic, dynamic response) may be critical locally, but are not usually coincident with the critical aero loads and are normally analyzed as separate design conditions. But for the NASP the loading is aero thermal elastic acoustic and is coincident at the critical design conditions. Achieving the required structural mass fraction in the face of existing computational capability and uncertainties in the load and material data bases is problematic.

Effort must also be directed at fabrication methods for the new materials. Fastening poses a particular problem because some of the materials demonstrate extreme brittleness in certain temperature ranges, as well as a negative coefficient of thermal expansion.

[16] Because of the lack of structural test facilities, adequate instrumentation with real time data transmission will be a flight safety requirement. Transmission through the plasma sheath which will envelope [sic] the vehicle at the higher Mach numbers presents a severe challenge.

The Task Force believes it would be prudent to establish technical milestones to develop the data bases required for structural design with acceptable tolerances and refine analytic methods. These milestones should be accomplished before proceeding with detailed design.

Controls

The National Aerospace Plane (NASP) Program has some of the most demanding design problems of any flight vehicle development program to date. The extent of coupling between the NASP control system and the vehicle airframe/propulsion system requires that they evolve simultaneously. The degree of uncertainty regarding available component technology and associated performance complicates the task of control system development and mandates early identification of principal design sensitivities and trades. Also, uncertainties regarding environment characteristics demand development of control strategies which maximize available adaptability and authority and minimize the adverse influence of hostile environment effects. All these considerations are as applicable for development and testing of a research vehicle as for an operational system. Less specific knowledge of the environment during early test flights may actually demand more control system adaptability.

[17] To successfully develop the NASP control system, it is necessary to identify the most significant design concerns involving vehicle control and to initiate a technology development plan capable of addressing the issues. The effort should occur early enough to influence overall vehicle design in a manner that will assure successful vehicle and control system integration.

The issues which must be addressed include:

- attitude control (with accuracy to, perhaps, 0.1 degrees while the vehicle undergoes thermoelastic deformation)
- trajectory optimization
- propulsion optimization, including algorithms and sensors to control both throttle and variable geometry

- stability and control with large uncertainties
- sensors and instruments for the high Mach number regime
- handling qualities
- abort scenarios
- integrated guidance and control system

Most of these issues are vehicle design dependent. Therefore, a satisfactory vehicle design cannot be developed independent of the on board control system.
[18] The Task Force found [that] the technology road map developed by the flight systems working group has the elements described to provide an adequate understanding and effective development program for NASP. However, it is not being adequately funded. Controls and flight dynamics optimizations can relieve environments inimical to successful realization of key but very uncertain technologies involved with structures, structural materials, and propulsion systems, for instance. At the levels of program funding currently applied to the flight systems technologies, it is doubtful that these optimizations can be examined adequately and that alternatives will be available on a schedule compatible with the air frame/propulsion developments. As a rough example of the disparity, it now appears that approximately 1 to 2% of the currently identified funding in the program is intended to cover this functional area. It is our experience for aerospace vehicles that avionics represent a much larger percent of the total value of the vehicle. It will not be possible to reach the goals of the program at the level of funding now allocated to the controls and guidance functions.

Computational Fluid Dynamics

The preceding sections highlighted the question of the accuracy of the CFD codes. Much progress has been made in this discipline in recent years, but there is still a long way to go, particularly at the higher Mach numbers. The Task Force found that the CFD team had a realistic view of the limitations of their calculations, and a [19] well thought out plan for improved capability. However the program must guard against exaggerated claims about the efficacy of CFD as a substitute for wind tunnel or flight test data. If expectations are raised too high too soon, CFD could be put in the unfortunate position of losing credibility when, in fact, the community will have been making significant advances that should be recognized as such.

Today, two dimensional calculations are good; three dimensional capability is evolving. But even where the codes are good, they must be calibrated and validated from real world data. This arises from the need to insert certain empirical data such as the onset and length of the transition to turbulence and turbulence characteristic length. In Mach and Reynolds number regimes where no data, or incomplete data exist, the calculations will be precise but not necessarily accurate. The calculations are also strained when all relevant parameters, such as combustion kinetics, must be included.

CFD is essential to the NASP program. But it must be recognized that the accuracy attainable over the next few years will fall short of what is required for vehicle design and performance estimates.

Another potential problem is the computational requirements. Some of the codes take a long time on a powerful computer to converge, on the order of 24 hours. It is likely that several thousand such runs will be required to design the vehicle. Measures should be taken to assure that computer resources will be available, as well as effort directed at reducing execution time.

[20]   CONTRACTOR RESPONSE

The Task Force met with all eight contractors to review their technology efforts and explore their views of the issues critical to the program. Each meeting lasted approximately three hours, thirty minutes of which was a private session with the Task Force.

The following paragraphs summarize the observations and concerns common to most of the eight discussions. (The meetings occurred in late June, 1987, and reflect perceptions of the program at that time.)

- The NASP is truly an experimental vehicle, not a prototype of a space booster or a hypersonic cruise airplane. It will be a success if it achieves high Mach number flight. Design iterations may well be required before orbital insertion is achieved.
- There is little confidence that the aero-breathing propulsion alone will be sufficient to gain orbit in the early phases of the program.
- There are approaches to compensate for the uncertainties in the aero-breathing propulsion, e.g.[,] rockets to help achieve orbital velocity and/or very low drag designs.
- Uncertainties in aerodynamic data, particularly as they affect temperature estimates and propulsion performance, drive the vehicle configuration. Estimates of gross take off weight range [21] from about 300,000 pounds to 500,000 pounds. Confidence in these numbers is not yet high.
- Materials development and manufacturability pace the program. Materials characterization and scale up for production are not adequately funded. The time required for these efforts is too long to support the (then) existing Phase 3 schedule.
- The Technology Maturation Program is a good start, but is not sufficiently focused on the requirements of the most probable configurations. Although information exchange is good, stronger contractor participation in defining the program might help.
- Teaming of airframe and engine contractors would be welcomed. Coordination among several contractors presents a significant burden.
- A variable Mach number wind tunnel is required.
- The (then) scheduled Phase 3 schedule was not realistic.
- The (then) planned Phase 3 funding was not realistic.

The Task Force found these thoughts congruent with our own observations.

[22]   CONCLUSIONS

Based on our review of the NASP program which extended over a six month period, the Task Force reached the following conclusions:

1) The NASP program goals are valid. The technologies which NASP will develop will make significant contributions to our national military and space capabilities and our civilian economy as we enter the twenty first century.
2) The NASP is truly an X-Vehicle. Expectations of short term operational utility should not be raised.
3) Technical uncertainties in all critical disciplines must be narrowed before detailed design is initiated. Uncertainties are too large to estimate with any degree of accuracy the cost, schedule or performance which can be achieved in Phase 3.
4) Readjust the program funding priorities to favor the Technology Maturation effort, while retaining sufficient effort in definition airframe and propulsion configuration to provide focus for the technology work.

5) An experimental program of this type should be event driven, not schedule driven. Demonstration of quantitative technical milestones in all critical disciplines should pace the program.
6) Hypersonic flight will be important to the United States in the decades ahead. Adequate national ground test facilities must ultimately be provided.

[23]   RECOMMENDATIONS

These findings lead the Task Force to make the following recommendations:

1) Maintain the present program objectives. A manned hypersonic vehicle, with the potential of demonstrating a single stage to orbit and extended hypersonic cruise, provides challenging focus for the development of the critical technologies.
2) Complete a rigorous risk identification and closure analysis. Identify the funding, schedule and technical resources required to reduce the risks to a level commensurate with the experimental nature of the vehicle.
3) Establish a quantitative set of technical milestones in all critical disciplines which must be demonstrated before entering Phase 3.
4) In anticipation of the results of the risk closure analysis, begin now to replan the program by making the start of Phase 3 dependent upon demonstration of the technical milestones and by significantly decreasing the portion of program funding devoted to maturing the technology.
5) Emphasize the experimental nature of the program. Once flight test begins, several design iterations may be expected before orbital insertion is achieved. Program planning should anticipate the resources which will be required.
[24] 6) Proceed with the planned down select for both engine and airframe contractors. To reduce the number of design combinations which must be considered, team airframe and engine contractors at an early date.
7) Focus the Technology Maturation Program to support the selected configurations. Strengthen the contractor's input to the definition of Technology Maturation Tasks.
8) Develop a plan to man rate the air breathing engine. Investigate the addition of a variable Mach number nozzle to the Aeropropulsion Systems Test Facility tunnel at Arnold Engineering Development Center to provide a ground test propulsion facility.
9) NASA and DoD should study the possibilities for national hypersonic test facilities for aero-thermal, propulsion and structures.
10) Materials availability will be a pacing item for the program. Develop a plan to scale up to production quantities for the materials selected and to provide characterization data for structural design.
11) Fund the flight control system technology road map tasks to a level commensurate with the importance of integrated flight controls to the program
12) Continue strong support to CFD validation and the narrowing of the uncertainty in location of the point of transition to turbulence.
[25] 13) Identify the computational resources which will be required to support the detailed design phases of the NASP.

We have refrained from making detailed recommendations in each of the technology areas in the belief that the risk closure analysis recommended above will provide the definitive plan required for the program. . . .

Document IV-10

Document title: The White House, Office of the Press Secretary, "Statement by the Press
Secretary," July 25, 1989.

Source: NASA Historical Reference Collection, NASA History Office, NASA
Headquarters, Washington, D.C.

*When the Bush administration began its review of space transportation programs after it took office
in January 1989, a target for cancellation was the NASP program, which was experiencing significant technical problems. Secretary of Defense Richard Cheney moved to cancel the program. He was
overruled by the White House on the advice of the National Space Council. Instead, President George
Bush decided to extend the Phase II technology development phase of the NASP program for an additional three years before a decision on whether to build a flight test vehicle. Ultimately, that decision
was negative, and the program was cancelled.*

THE WHITE HOUSE

Office of the Press Secretary

For Immediate Release   July 25, 1989

Statement by the Press Secretary

The President, acting upon the recommendation of the Vice President, has approved
the continuation of the National Aero-Space Plane (NASP) program as a high priority
national effort to develop and demonstrate hypersonic technologies with the ultimate
goal of single-stage-to-orbit.

The government will complete the Phase II technology development program, and
plans to develop an experimental flight vehicle after completion of Phase II, if technically feasible. The system will be designed to focus on the highest priority research, as
opposed to operational, objectives. Unmanned as well as manned designs will be considered and the program will be conducted in such a way as to minimize technical and cost
uncertainty.

The President also approved an implementation plan to carry out this policy. The
plan extends technology development until early 1993 to reduce technical and cost risks.
It retains an experimental flight vehicle focused on research and technology objectives
and retains a joint program management structure with participation by both the
Department of Defense and NASA.

The Space Council recommendations approved by the President termed the National
Aero-Space Plane a vital national effort which benefits the civil, commercial and national
security interests of the nation. The NASP program promotes industrial competitiveness, fos-
ters U.S. space leadership, and provides the technological basis for greatly expanded access
to space in the 21st century. We call on Congress to join in fully implementing the Space
Council recommendations and in moving forward with the important NASP program.

Document IV-11

Document title: Maxwell W. Hunter, "The Opportunity," April 26, 1987 (revised).

Source: Maxwell W. Hunter (reprinted by permission).

*Max Hunter was a pioneer in innovative space transportation concepts. He was one of the individu-
als whose ideas were incorporated into the Space Shuttle design that NASA decided to develop in 1971.
Hunter worked for Lockheed during the 1980s. In 1987, he began to suggest that there was an
opportunity to develop a new, low-cost launch vehicle, probably based on a single-stage-to-orbit hydrogen-oxygen design. This white paper was a condensation of Hunter's early thinking. When Lockheed
decided that it was not interested in advocating such a vehicle or investing its own funds in its development, Hunter resigned and began individually seeking support for the concept. In December 1988,
he presented his ideas to a group called the "Citizens Advisory Council on National Space Policy."
Attending that meeting was retired U.S. Army General Daniel Graham, who had earlier formed an
organization called High Frontier to support the creation of a defense system against strategic ballistic
missiles, which eventually became the Strategic Defense Initiative (SDI). Both Graham and members of
the council had ties to Vice President-elect Dan Quayle, who would be heading the new National Space
Council once the Bush administration entered the White House. In March 1989, Graham arranged
for Hunter to present a briefing to Vice President Quayle on his SSX concept. Quayle's interest helped
Graham, Hunter, and science fiction author Jerry Pournelle garner support for moving ahead with
what ultimately became known as the Delta Clipper-Experimental (DC-X) program.*

The Opportunity

by

MAXWELL W. HUNTER

Rev 26 April 1987

[1] THE OPPORTUNITY

There exists on this planet today a classical entrepreneurial opportunity. It is in space
commerce, indeed all of space, but as in all great opportunities, it is invisible to most. Else,
it would not be such an opportunity. For such an opportunity to exist, virtually all accepted authorities must be either unaware, or so unperceptive as to be effectively blind.
Historically, the great opportunities flew in the face of accepted authority. The big problem, then, is how to detect the opportunity, especially if you command sufficient resources
for its implementation. They are large, there will be no easy way to feel comfortable, and
many voices will be raised on the side of discomfort. This we discuss herein.

The key to all this is space transportation. Any who believe either that the Space
Shuttle is the final word in space transportation, or that NASA is the ultimate authority,
should read no further. No communication will be possible. Those who remember what
Isabella did to her scientific advisors, and with what results, should read on. The shuttle is
a beautiful flying machine, and still is, even after the accident. It has done much to remove
the mystery from space. With an air transport looking machine going to and from space,
carrying rather normal looking people, the day of the superman in the tin can has been
relegated to history. Due to the accident, it may return, but only briefly. Much technology, e.g.[,] heat shields, has been put into the inventory. So far, good.

The shuttle, however, was supposed to make space transportation very economical. In this, it has failed miserably. The shuttle is a psychological triumph and an operational disaster. It was supposed to have been operated with the culture of an air transport system. If its operational cost were calculated with the methods used by the Air Transport Association, the cost per pound to orbit would be three orders of magnitude lower than it is. No one expected it to be that low, but the idea was to give it a real try and explore the techniques that lead that way. Alas, NASA elected instead to support three massive centers on the program. The shuttle has shown, the hard way, that arbitrarily large bureaucratic expenditures do not create safe flying vehicles. The shuttle does not even attempt to operate in a transportation culture . . . it is still mired in the culture of Apollo.

This situation is what creates the great opportunity. The shuttle will almost certainly always be withheld from being a true space competitor by its primary use as a source of employment for NASA personnel. Now, there may never be enough anyway. A new vehicle could be put on stream which would be a devastatingly effective competitor, both to the shuttle and to the expendable vehicle stable including international entries. With what we now know about space transport design, such a new vehicle should not cost much more per pound to develop than an experimental airplane and possess airplane-class safety. One should not underestimate the problems routinely solved in developing high performance aircraft. They are as high in technology as space vehicles, indeed usually higher. The transports are man, woman and child-rated from the beginning. Millions of people trust their lives to them every day. Their development philosophy, if permitted, could contribute very much to real space transportation.

The cost of propellants is the only fundamentally different price that rockets must pay compared to airplanes. It isn't much. This situation was criminally misunderstood after Sputnik by our mighty scientific community. Rockets, because they must carry all oxidizer on board, were relegated forever to the limbo of massive expense compared to airplanes which get their oxygen "for free" from the atmosphere. It turns out that liquid oxygen is extremely low cost, and carrying it along in a light-weight tank is vastly superior to the frightfully complicated engine cycles, ducting systems and hot, heavy airframes required as the air-breather desperately searches for [2] oxygen instead of proceeding to space on efficient trajectories. The presumed rocket inefficiency has been used in the past to justify all rocket expenses, even those which came from scientific naiveness or bureaucratic technological featherbedding. This is a gross misconception, which has left a lasting terrible weight on our space program.

A relatively small hydrogen-oxygen rocket could be built to place about 20,000 pounds in orbit. It would stand no higher than the tail of a Boeing 747 on the pad. With readily available modern electronics, only a few people would be required for launch and operations (the shuttle electronics are 15 years old—ancient by electronic standards). If the development cost per pound were even five times as high as a modern airplane, it would still cost only several $100 million (not billions) to develop. The cost of the propellants would be less than $5 per pound of payload placed in orbit.

With a few such rockets (followed later, no doubt, by larger sisterships) the current space transportation market could be spirited away. Furthermore future markets, ranging from support for the Strategic Defense Initiative to space tourism, will be vastly larger, and will grow indefinitely. Some, like tourism, will not be dependent on national policy or congressional budgetary outlays. The marketing effort by Society Expeditions, both before and after the Challenger disaster, is most encouraging in this regard. The opportunity may or may not be as great as the New World of 500 years ago, but it has the same flavor, and it could easily be greater.

The possibility even exists for pure glory—the sort of private record setting that so enlivened the airplane scene during the twenties and thirties. Actually its [sic] still going on in airplanes, but spectacular military planes (and expenditures) took the edge off the situation after the real war. The idea of setting space records doesn't even occur to the traditional fly-boys and girls. That's a damn shame. It would be easy for a small manned rocket to generate a higher space speed than any so far. The distance record for a woman away from earth isn't much, and the extra velocity to break that record is trivial. Actually, looping beyond the moon to get further away from earth than anyone has yet gone would be child's play. The possibilities are endless. Fabulous headlines (and history) can be written.

The shuttle has accidentally created this absolutely wonderful opportunity. Its use as a NASA support program has removed it from the competitive market. It cannot easily, if at all, return. It has achieved the untouchable status of "national resource." Destroying its current culture (it is bound to get more expensive for awhile) is unthinkable to its owners, and a massive taxpayer subsidy just to compete with private enterprise is not likely to fly. Fortunately for private enterprise, the most obvious government competitor has declared itself non-combatant.

The opportunity, then, consists of the development, using aircraft-like techniques, of a new launch vehicle. It would be relatively small (so was the Douglas DC-3), use the most modern of operational techniques, and rely heavily on basic technology developed by the shuttle and other programs. It would likely be a single-stage-to-orbit hydrogen-oxygen rocket. It would have a sufficient number of engines that it could stand an engine failure at any time after launch and either complete its mission or successfully abort. This is the key to both operational and test flying with airplane-like techniques. Such rockets can be shown to be extremely competitive, even compared to such exotic devices as scramjets utilizing air-breathing to orbit (and without their vast engine development costs). It thus would be expected to have a very long useful life. It would have both military and civilian application. It would open up space, the New Worlds of today.

[3] The same factors which create the opportunity, also erect the most formidable barriers to it. The shuttle, after all, is most spectacular. How can one expect to do better? The funds required for a new vehicle are of the order of several $100 million. This is not the world of a few $100,000 seed money, or even a few million. This is a classical entrepreneurial opportunity, not an ordinary one. To do it right, greater resources should be available if needed. The program should not be marginally funded. It's hard to overemphasize this point. Multiple flight vehicles should be provided, for flight delays due to equipment unavailability must be avoided. Thus a substantial additional amount should be available, with the objective of using it to start major production if successful, but having it available for contingencies if truly needed.

Private enterprise can supply such funds, but the people who control them already spend them, often on more risky adventures, but adventures with which they are familiar. Conceivably, the operation could be bootstrapped starting with a few $100,000 getting small study contracts (likely military), and by living hand-to-mouth eventually build a vehicle with marginal funds, cutting corners all the way. If necessary, this can be done. It would be far preferable for a real classical entrepreneur to materialize. In fact, its [sic] overdue.

The people who can supply such funds, however, will want independent authoritative opinions as to the risks they will be taking. This is where the situation becomes truly classic, for getting a favorable opinion is likely to be impossible. The greatest experts, at least in the investors['] eyes, will reside at NASA. Any supplementary opinions are likely to be solicited from academia and science. Massive ridicule can be predicted. The only authoritative group claiming low orbital costs today are those government folks promoting air breathers. They automatically assume airplane-like operations, thus agreeing with the basic premise that such things can be applied to space. But they claim that only by breathing air and using horizontal take off can it be achieved. They freely admit to the necessity of massive funding to develop the engines. They do not understand how good rockets can be and actually, rather hate them. They should.

It is this barrier of pseudo-technical opinion which must be surmounted for the opportunity to be exercised. It cannot be surmounted by normal committee action. It can be surmounted because someone has a dream, an innate distrust of "expert" opinion, something to prove, or even an urge to get even. It has to feel right, in spite of expert opinion. Moving out resolutely, with these feelings onboard, is the mark of the classical entrepreneur. Either one arises, or the opportunity goes unfilled. Right now, its [sic] unfulfilled . . . and the ghost of Isabella weeps.

[signature of Maxwell W. Hunter]

Document IV-12

Document title: Gary Hudson, Pacific American, Memo to Thomas L. Kessler, General Dynamics/Space Systems Division, "Comments on SSTO Briefing and a Short History of the Project," December 17, 1990.

Source: Gary Hudson (reprinted by permission)

Gary Hudson was another pioneer in the attempts to develop new launch vehicles, in his case through private funding. This document contains his version of the events leading to the initiation of a single-stage-to-orbit (SSTO) rocket program in 1990 by the Strategic Defense Initiative Organization (SDIO).

[each page marked "Eyes Only" and "Steve Hoeser"]
[1]
Memo

Date: 17 Dec. 90

To: Thomas L. Kessler
General Dynamics/Space Systems Division

From: Gary C. Hudson
Pacific American

Subject: Comments on SSTO Briefing and a Short History of the Project

History of the SDIO SSTO Project

During the 1960s, one man, Phil Bono of Douglas, tirelessly promoted the concept of a fully reusable single stage vehicle which would takeoff and land vertically. Bono's work was essentially ignored by both his management and the aerospace establishment of the day. Frustrated by this reception, and in cooperation with Ken Gatland of the British Interplanetary Society (whose *Journal* had published many of Bono's papers), Bono wrote a book called "Frontiers of Space," which was issued in 1969. Much of the book was an exposition of his VTOL [vertical takeoff and landing] SSTO concepts. This book was my first introduction to the field. In fact, when it came out, I wrote Bono asking if he had ever considered private financing of this idea. He wrote back that he doubted if it was feasible, but urged me to try to secure such interest. Over those early years from 1969 to 1974, I met with him, Ken Gatland, Arthur C. Clarke and others in a futile attempt to promote a private reusable VTOL.

Beginning in 1972, I began to modify Bono's designs. In 1973, through the good offices of a friend, I was invited to meet with John Yardley at his home in St. Louis. Yardley had just been named head of the NASA Office of Manned Spaceflight, and was on his way to DC. He brought several MDAC [McDonnell Douglas Aerospace Corporation] people to the presentation I and a colleague provided. The consensus of the attendees was that it was possible to build a vehicle of the type I described, which was called "Phoenix." The Phoenix was to have a plug cluster powered by RL10 chambers and turbopumps. Most interesting from my perspective was the acceptance of my cost estimates for prototype development, which were in the range of $100 million 1972 dollars. Several people at the meeting said that this was not out of line, if the project was conducted in a "skunkworks" fashion. Among the people agreeing with this were Yardley and the MDAC Chief Engineer. Yardley also said that he would try to have NASA HQ investigate the idea once again (the first study having been the Chrysler SERV Project three years earlier), and he invited me to DC to brief HQ staff and also Langley researchers. [2] Nothing came from these contacts.

For the rest of the 1970s and into the early 1980s, I continued to study the problem of low-cost space transportation. While I worked on designs for several types of expendable rockets, I always kept coming back to the VTOL SSTO. During those years a few others also explored the field, most notably Bob Salkeld, working with Rudi Beichel at Aerojet (mostly on [vertical takeoff and horizontal landing]) and occasionally others such as Boeing in support of Satellite Solar Power studies. But by and large, this was an inactive period.

(During the mid 1970s to mid 1980s, studies were also underway on military manned and unmanned winged vehicles, several of which were single stages or air-launched. These included early an SAMSO [horizontal takeoff and horizontal landing] [1972] which was the forerunner of the 1976 RASV, the TAV/MAV/AMSC studies, and the Have Region activities.)

In 1982, after the failure of investors to finance my company to produce a low-cost modular expendable pressure-fed liquid launch system for Space Services, Inc. of Houston, I founded Pacific American Launch Systems, Inc. to build a small SSTO. My reasoning was that to compete with a government-subsidized space transportation system, it was necessary to operate at a cost at least one to two orders of magnitude lower than [the] Shuttle. This would remove [the] Shuttle from the marketplace and, hopefully due to elasticity of demand, greatly increase the market for space transportation services.

Obviously, it was harder than expected to raise the funds necessary to begin Phoenix development. Even with the assistance of Max Hunter as a Senior Vice President, investors did not believe that the vehicle could be built: everyone [sic] of them seemed to have a brother-in-law in the aerospace industry or NASA who said we were crazy.

When the Shuttle accident occurred, the opportunity was at hand to try once more to find funding. Because the myth of NASA space superiority was shattered by the loss of *Challenger* and subsequent missteps, more people were willing to take the concept of Phoenix seriously. My first attempt at selling the idea was a briefing to the USAF Space Division/XR in early 1986. The idea was ultimately shot down by Aerospace Corporation's negative three-page memo review, but we came close to obtaining a $100K study contract. This was my last attempt to find any government support for the project.

At this point a frustrated Max Hunter got agreement from me to try something new. His plan was to conduct an internal Lockheed study on a vehicle named X-rocket. This "new" concept would not have the rejection issued by Aerospace Corporation, and might, with the Lockheed name, win converts in DC. If that occurred, maybe investors would pay more attention. I readily agreed.

The X-rocket was widely briefed and fairly favorably received over a one year period. [3] Unfortunately, Lockheed support was contingent upon a review by the [Lockheed Missile and Space Company] Missile Systems Division (builders of Trident). Naturally, they didn't know what to make of this wild notion and said it couldn't work. (It was interesting that they concluded the vehicle would have no payload, not negative payload, for a 0.5 million pound takeoff weight. This from a group who was used to building heavy, robust solid launch systems.) At the same time, Aerospace Corporation did another back-of-the-envelope review which also ridiculed the idea. The combination killed the effort and Max Hunter retired at the end of his 20 years in 1987.

During the 1980s, an *ad hoc* group of space professionals and enthusiasts met annually at the home of science fiction writer Larry Niven. This group, chaired by author Jerry Pournelle, was established at the request of the Reagan space transition team following the 1980 election. It was called the "Citizens Advisory Council on National Space Policy." Several times during the decade I presented the Phoenix concept to the assembled group, but never got endorsement of the idea. (Hunter, also a member, was initially negative on the idea, but then became a strong supporter.)

A frustrated council met in December of 1988 to try and forge a consensus regarding what to tell the new Vice President, Dan Quayle, who would be chairing the new National Space Council. The meeting highlight was a presentation on the vehicle known as SSX, or Space Ship Experimental. (Hunter had renamed X-rocket following his departure from Lockheed.) Besides Hunter and myself, Daniel O. Graham was in attendance. He and the rest of the council agreed to endorse SSX, and Graham agreed to take the idea to the Vice President.

Two months later, in February 1989, Graham, Hunter and Pournelle briefed Quayle, who expressed serious interest in the concept. Graham then made the rounds in DC with Hunter and Steve Hoeser (of the SDIO program office) to sell the idea to the community.

In June the Aerospace Corporation was once again tasked with an analysis: Jay Penn performed it on a mixture of Phoenix and SSX vehicle concepts. This time, a one month study was performed with about 10 engineers, and the results were positive. In fact, the July report on Phoenix/SSX was quite favorable; a subsequent official report dated August was still upbeat, but somewhat less so than the July version. The difference was that one of the individuals who had to sign off on the final version was the author of the previous two Aerospace memos which ridiculed the concept, and he had to be placated.

This favorable report allowed SDIO to begin the process of funding study contracts under the management of Col. Gary Payton, later replaced by Lt. Col. Ladner. It is interesting to note that, with the exception of one person in the program office, no one at SDIO seems to be aware of the history of this idea. Ladner once asked me "Just who the hell is this Jerry Pournelle, anyway?"

Document IV-13

Document title: **Department of Defense, Strategic Defense Initiative Organization, "Solicitation for the SSTO Phase II Technology Demonstration," June 5, 1991.**

Source: **NASA Historical Reference Collection, NASA History Office, NASA Headquarters, Washington, D.C.**

After the initial studies of a single-stage-to-orbit (SSTO) concept by the Strategic Defense Organization (SDIO) in 1990, the organization decided to move forward to the next step in developing the concept, a suborbital demonstration of key elements of the concept. Minimal funds were available for this step, and the SDIO emphasized that the winning contractor would operate with minimal government oversight. This procurement was one of the forerunners of the "faster, better, cheaper" approach to developing space systems. McDonnell Douglas won the competition and named its vehicle the Delta Clipper-Experimental (DC-X).

Statement of Work for
Single-Stage-to-Orbit (SSTO)
Phase II
Technology Demonstration

ATTACHMENT 1

[2] 1.1 GOAL

Design, develop, and demonstrate the ability to provide a reusable single-stage-to-orbit and return launch system capable of conducting routine, low cost, and highly reliable space transport.

1.2 BACKGROUND AND PROGRAM PHILOSOPHY

The Single Stage To Orbit (SSTO) rocket concept of launching payloads to orbit dates back to the early 1960s. Further development, in these early efforts, was abandoned primarily due to the lack of supporting technology needed to build this class of launcher.

Within the last 15 years, the Department of Defense (DoD), the National Aeronautics and Space Administration (NASA), the Strategic Defense Initiative Organization (SDIO) and the National Aerospace Plane (NASP) have delivered impressive advances in propulsion, avionics, structures, and materials technologies. Recent Government and private industry assessments indicate that a significant payoff may be available in applying these technologies to a reusable, SSTO rocket-propelled vehicle with significantly reduced servicing, integration requirements and cost.

In general, an SSTO vehicle will operate as a single unit, launched into orbit and returned completely intact. The system will be unique because of its capability to operate over a range of failure modes. The vehicle will be designed to require minimal maintenance between flights, allowing rapid turnaround by today's standards. New user options for deployment and space operations will be provided by the SSTO's extensive schedule and mission flexibility. By balancing design, operational and maintenance factors, the SSTO will drive system costs to their lowest possible level.

The SSTO Phase I concept evaluation contracts provided ample evidence to indicate that the capability now exists to build SSTO vehicles designed to achieve cost-effective operations.

The objective of Phase II of the program is to refine and fully define the concept, and conduct critical hardware and software demonstrations. The program will then proceed into Phase III, the construction and testing of a prototype vehicle. This prototype vehicle will be the equivalent of a Y-class vehicle in aircraft development. It should demonstrate all functional characteristics of the operational vehicle although some performance degradation from the operational vehicle is acceptable [3] if attributable to lower performing components or subassemblies used in the Y-vehicle for schedule or cost reasons only. A clear path to full capability in the operational vehicle is required.

Demonstrations of critical hardware, software, configurations, technology, and capability prior to the Critical Design Review (CDR) forms an integral part of the program concept. The details of these demonstrations are to be defined by the contractor as appropriate to the concept and configuration.

The CDR planned for May 1993 will be the final input in a decision process to proceed with the Phase III construction of the Y-vehicle and ground and flight testing through 1997.

The general schedule for Phases II and III of the SSTO project is:

- Selection of Phase II contractor(s) in AUGUST 1991
- Preliminary Design Review (PDR) 6 months after contract award
- Progress/status reviews, probably bimonthly, between PDR and CDR
- CDR May 1993
- Go/No Go decision in June 1993 for Phase III
- Phase III Y-vehicle first suborbital flight, 1995
- Phase III Y-vehicle first orbital flight, 1997

* Note: Phase III will be a separate procurement subsequent to Phase II.

1.3 SCOPE AND OBJECTIVES

1.3.1 Scope

Phase II of the SSTO program will include all design, analysis, and testing tasks necessary to achieve a successful CDR for a Y-vehicle as defined in section 2.0. A subsequent acquisition for Phase III will follow for the Y-vehicle construction and flight test.

1.3.1.1 Scope of Phase II. Phase II will include completion of all SSTO vehicle design activities, planning for the initiation of Y-vehicle material needs, demonstrations planning for the critical hardware, software, technology, design/operational features, and test activities related to achieving a successful go-ahead for a subsequent Phase III at the 1993 CDR.

[4] 1.3.1.2 Scope of Phase III (For Planning Purposes Only). Phase III will include all activities for procurement, fabrication, testing, construction of a full-scale Y-vehicle, and ground and flight test program from first suborbital test flight (1995) to first orbital test flight (1997).

1.3.2 Objectives

1.3.2.1 Objectives of Phase II. The following objectives are to be accomplished in Phase II:

1.3.2.1.1 Completion of the vehicle system design.

1.3.2.1.2 Completion of design of all support infrastructures.

1.3.2.1.3 Completion of demonstrations of critical hardware, software, technology, design and operational characteristics of the vehicle, and test activities to support the 1993 CDR.

1.3.2.1.4 Initiation of plans for material needs to manufacture the Phase III Y-vehicle.

1.3.2.2 Objectives of Phase III (For Planning Purposes Only). The following objectives will be accomplished in Phase III:

1.3.2.2.1 Complete fabrication of full-scale Y-vehicle

1.3.2.2.2 Complete subsystem and system ground testing and integration in preparation for first flight

1.3.2.2.3 Complete all test site preparations for conduct of flight test program

1.3.2.2.4 Deliver to the test site a flight ready Y-vehicle to begin flight test program in 1995

1.3.2.2.5 Complete all ground and flight testing through orbital test flight in 1997.

2.0 SSTO REQUIREMENTS

2.1 Single Stage to Orbit

In order to meet overall program goals, single stage capability to orbit and return is essential. Some forms of takeoff enhancement may be acceptable as long as inclusion does not seriously compromise turnaround, orbit access, or other major characteristics of the system.

[5] 2.1.1 Manned/Unmanned Operation

The vehicle must be capable of both manned and unmanned operation to and from all normal orbits. Designs must include the ability to conduct normal programmed operations in the absence of direct man-in-the-loop contact. The ability to control the vehicle via a remote pilot (virtual cockpit) is also a design goal. The vehicle will not be designed to be man-rated; instead, after a reasonable number of incremental test flights, combined with the inherent reliability of the system, the vehicle will be manned.

2.1.2   Responsiveness

The vehicle and system must be able to respond to mission requests that fit standard parameters within 30 days from the initial request. Highly nonstandard missions may take longer. Note that this is not turnaround. This reflects the amount of time that a mission must spend in the "queue" between initial request and launch. It is desirable to have a 24-hour launch capability following a 30-day advance notification. (This will accommodate last minute launch delays without causing extensive rescheduling and preparation time once this vehicle is mission scheduled and launch ready.)

2.1.3   All-Weather Capability

The vehicle must be able to take off and land without damage in inclement weather conditions. This capability should be analogous to the weather constraints used by operational airliners. Details as to equivalent capability (e.g., vehicle category 1, 2, or 3) will be resolved based upon improved understanding [of] practical vehicle capability. The vehicle must be capable of take off and landing in a crosswind of 25 Kts [knots] with gusts to 35 Kts.

2.1.4   All-Azimuth Launch

All-azimuth launch capability is highly desirable, although some limitations may be acceptable so long as orbit inclination access is not seriously compromised.

2.1.5   Payload Capacity and Accommodations

The ability of the vehicle to carry Medium Launch Vehicle . . . , payloads is mandatory to capture a reasonable part of the mission model. The goal is payloads as large as 10,000 lb for polar launch azimuth into 100-nmi circular orbit and commensurate lift weights for due easterly launch azimuth into 100-nmi circular orbit. Payload interfaces should be compatible with standard existing reference satellites and should provide both standard and optional services for new payloads. The intent is to maximize use [6] of containerization/encapsulation. A desirable goal is as near as practical to 15 ft diameter by 30 ft length. (Note: This is not a hard requirement—merely an indication of desirable range and potential future missions.)

2.1.6   Payload Access in Space

The current baseline views the payload as a "black box" with minimal interface and no crew interaction other than deployment. It should be recognized, however, that to achieve the full potential of the vehicle or to respond to safety critical events, crew access may be required for some missions either in an operational or contingency mode. Access in this context may mean command, control, and data; actual physical contact; or both. In the case of the latter, EVA (extravehicular activity) or IVA (intravehicular activity) can be options. The vehicle should be designed, as a minimum, to include such access.

2.1.7   Orbital Maneuvering

On-orbit maneuvering velocity change (unrefueled) of 600 ft/sec in addition to reentry and any landing delta-V (if necessary) is required. Designs showing preplanned improvement paths to provide additional cost-effective mission delta-V margin, or the ability to conduct missions beyond the baseline, through refuel, payload offload, or other means are encouraged.

2.1.8   Rendezvous/Docking

Provide standard or optional provisions allowing the operational SSTO to perform Space Transportation System (STS)-like operations for rendezvous, docking, and also propellant refueling to either active (cooperative) or passive (noncooperative) spacecraft. Provisions include low plume impingement separation maneuvers, attitude control with failed on or off thruster, and V-bar (velocity) and R-bar (radius) approaches. Rendezvous radar ([radio frequency] or laser) will be required for non-cooperative targets.

2.1.9   Design for Low Cost

Recent studies indicate that with reusable vehicles, design for operability and reliability is synergistic with cost. The contractor should select appropriate operations, reliability, technologies and design features to ensure minimized life cycle cost.

[7] 2.1.10 Abort/Emergency

Intact abort is desired in all noncatastrophic failure cases. In the event that intact abort is not possible, a high probability of crew escape/survival is desired. Aspects of this area of concern are main engine out capability, Reaction Control System (RCS) failure, Orbital Maneuvering System (OMS) failure (if separate from main propulsion), abort modes, emergency landing capability/sites, and escape mechanisms.

The vehicle must be able to accommodate complete loss of thrust from one or more engines during powered flight such that the probability of catastrophic vehicle loss is minimized. Flight dynamics envelopes will be maintained (demonstrated) to provide sufficiently wide margins of safety to account for the majority of foreseen failure conditions. The propellant feed system shall be designed to detect and isolate a massive leak and provide sufficient propellant flow through redundant feed line(s) to allow safe abort. Means to detect and mitigate failures will be critical to fail-safe operation. Concepts for emergency operations will be provided to handle unforeseen failures. This will include as a minimum concepts to maintain vehicle integrity, contain drastic engine system failures including hydraulic systems (if used), crew safety/escape measures, and crash landing options.

2.1.11   Landing

The operational vehicle must be able to control its landing site to an accuracy such that it can consistently land on a runway capable of handling large commercial jet aircraft, or in the case of a vertical lander in a similar area. Touchdown point within the defined area must be controllable to allow touchdown in the first 1/3 of the runway for horizontal landers or within an area not to exceed 1000 ft in diameter for a vertical lander. Maneuverability during terminal descent should be adequate for obstacle avoidance and touchdown point selection. The vehicle must be able to control touchdown point and drift rate at contact to values consistent with safe landing. Following landing, the vehicle must be transportable to its maintenance/turnaround facility in winds of 25 Kts with gusts to 35 Kts.

2.1.12   Takeoff

The vehicle must be able to maintain attitude and directional control during liftoff and ascent consistent with safe operation. This includes engine out and other emergency situations.

[8] 2.1.13 Turnaround

The vehicle must be capable of being turned around (i.e., serviced as required, reloaded, and prepared for launch) within 7 days of landing with the expenditure of no more than 350 man-days. These values should be viewed as an initial operational capability (IOC) requirement. The Y-vehicle turnaround may require more than 350 man-days. The measure of this objective will also be based on the vehicle's and supporting ground equipment's capability to be maintained and serviced by personnel with a high school education or equivalent plus two years technical training or equivalent. As defined here, turnaround does not include periodic inspection and maintenance, but only those actions required to get ready to fly again, assuming no major problems with the vehicle.

The vehicle will require periodic downtimes for more detailed inspection and maintenance as is done with commercial and military aircraft. This does not form a part of turnaround, but should be analyzed, clearly defined in the maintenance concept, with frequency and duration minimized.

2.1.14   Extra-Vehicular Activity

Provide accommodations to perform a one- and two-person EVA via a separate airlock and/or by depressurizing the cabin. If the depressurization option is pursued, the avionics, life support, and power subsystems must be compatible with a depressurized cabin environment. The vehicle must allow manual payload bay door closure via EVA or IVA and a hatch to egress the vehicle.

2.1.15 On-Orbit Dwell

The vehicle must provide life support consumables, for two-person crew for two days with contingency supplies for an additional two days. Preplanned improvement path providing additional on-orbit dwell of up to 14 days by substituting consumables for payload or by other means is desirable.

2.1.16 Life Support

The designs shall provide for human compatible life support and system interfaces such that a comfortable environment is provided and the highest possible functional efficiency is maintained at all times. Cabin pressure shall be defined by the contractor. Rapid response EVA, docking with NASA space station and Soviet Mir must be considered.

[9] 2.1.17 Surge

Launch rate surge capability shall be provided to double the routine launch rate for a minimum of 30 days. Contractors are encouraged to provide for additional surge duration, capacity or future growth potential.

2.1.18 Environmental Impact

The vehicle must meet all range and overflight environmental requirements including explosion hazard, sonic boom, toxic propellants, and destruct systems.

2.1.19 Advanced Tracking System

The contractor must provide an advanced configuration, maintenance and logistics tracking system for vehicle, subsystem, and component status heritage and tracking.

2.1.20 Reliability

The operational vehicle must demonstrate a (safe recovery) reliability of >.999. Reliabilities approaching those of aircraft (>.999995) are desirable.

2.1.21 Dependability

The operational vehicle should be available to support launch more than 95% of the time on the originally planned day (from the start of processing).

2.1.22 Design Margin

The dependability, reliability, and maintainability of commercial aircraft are due primarily to the margin and robustness incorporated into the design. The following list is provided as a suggested basis (i.e., not a firm requirement) for design margin for SSTO.

–	Structure	margin of safety (goal >= 1.4 with test)
–	Tankage	burst margin of safety (goal >= 1.5 with test)
–	Propulsion	Assume 2% loss of I_{sp} from prediction
		Assume 20% of reduction in engine thrust-to- weight ratio
		Do not normally operate engine above 90% design thrust level
		Do not operate engines above 100% design level in abort contingency

[10] – Control   Assume >6 dB of control margin over worst case conditions
                   Design for [Federal Aviation Administration] wind/gust loads
      – Thermal protection system (TPS) and thermal systems
                   Maintain 100 degrees C of temperature margin on TPS worst-case
                   reentry (function of number of planned thermal cycles)
                   Maintain 100 degrees C thermal limit on all engine components
                   (function of number of thermal cycles)
      – Propellants (cryogens)
                   Allow for RMS propellant trajectory, weather, fuel biases, and loading
                   errors
      – Avionics/subsystems/equipment
                   Design for single fault tolerant (or better) on all mission success
                   systems and two-fault tolerance in avionics strings. Provide maximum
                   protection against anticipated failures.

2.1.23   Example Missions

The need for the SSTO system's capabilities was initially based only on the Strategic
Defense System deployment requirements. However, the SSTO concept definition studies
have indicated greater potential for diverse applications. The example mission areas
described below should be used as a guide for SSTO system design refinement and mission characterization.

2.1.23.1 Payload Placement: This mission area covers normal satellite/payload
deployment missions requiring an orbital range from Low Earth Orbit (LEO) to
Geosynchronous Earth Orbit (GEO).

2.1.23.2 Orbital Platform Support: In general, this mission area is concerned with
on-orbit support of various orbiting platforms. It consists of:

        –   Space Station/Shuttle support (cargo, personnel)
        –   satellite/vehicle on-orbit servicing
        –   Personnel Launch System (PLS) type missions
        –   Assured Crew Recovery Vehicle (ACRV) type missions
        –   satellite recovery/replacement/upgrades

[11]   2.1.23.3 Advanced Mission Support: The Space Exploration Initiative (SEI) provides the opportunities for greatly expanding the SSTO's potential in this mission area.
Contractors are encouraged to determine if, and how, their designs could support SEI
missions to include lunar or Mars landing, and also to determine support to other potential commercial space ventures. . . .

Document IV-14

**Document title: Office of Space Systems Development, NASA Headquarters, "Access to
Space Study—Summary Report," January 1994, pp. i–ii, 1–6, 59–72.**

Source: NASA Historical Reference Collection, NASA History Office, NASA
Headquarters, Washington, D.C.

In 1993, NASA Administrator Daniel Goldin chartered an internal study aimed at identifying and assessing the major alternatives for a long-range direction for space transportation that would satisfy all U.S. needs for several decades into the future. In the same year, Congress also requested NASA to carry out such a study. This report summarizes the results of that study; its recommendation that the United States begin to develop a single-stage-to-orbit reusable space transportation system was incorporated into national policy later in 1994. What follows are major excerpts from the beginning and end of the report.

Access to Space Study
Summary Report

Office of Space Systems Development
NASA Headquarters

January 1994

[i]   **Synopsis**

This study was undertaken in response to a Congressional request in the NASA FY1993 Appropriations Act. The request coincided with an on-going internal NASA broad reassessment of the Agency's programs, goals, and long-range plans. Additional motivations for the study included a recognition that while today's space transportation systems meet current functional needs, they are costly and less reliable than desired, and lack desired operability. This has resulted in increased costs to the government and in severe erosion of the ability of U.S. industry to compete in the international space launch market. A further motivation is the past failure of the Administration and Congress to reach consensus on developing more efficient new launch systems.

This report summarizes the results of a comprehensive NASA in-house study to identify and assess alternate approaches to access to space through the year 2030, and to select and recommend a preferred course of action.

The goals of the study were to identify the best vehicles and transportation architectures to make major reductions in the cost of space transportation (at least 50 percent), while at the same time increasing safety for flight crews by at least an order of magnitude. In addition, vehicle reliability was to exceed 0.98 percent, and, as important, the robustness, pad time, turnaround time, and other aspects of operability were to be vastly improved.

This study examined three major optional architectures: (1) retain and upgrade the Space Shuttle and expendable launch vehicles, (2) develop new expendable vehicles using conventional technologies and transition from current vehicles beginning in 2005, and (3) develop new reusable vehicles using advanced technology, and transition from current vehicles beginning in 2008. The launch needs mission model utilized for the study was based upon today's projection of civil, defense, and commercial mission payload requirements.

Each of the three options resulted in a number of alternative architectures, any of which could satisfy the mission model needs. After comparing designs and capabilities of the alternatives within each of the three options, all defined to an equivalent depth using the same ground rules, a preferred architectural alternative was selected to represent each option. These were then compared and assessed as to cost, safety, reliability, environmental impact, and other factors.

The study concluded that the most beneficial option is to develop and deploy a fully reusable single-stage-to-orbit (SSTO) pure-rocket launch vehicle fleet incorporating advanced technologies, and to phase out current systems beginning in the 2008 time period. While requiring a large up-front investment, this new launch system is forecast to eventually reduce launch costs to the U.S. Government by up to 80 percent while increasing vehicle reliability and safety by about an order of magnitude. In addition, it would place the U.S. in an extremely advantageous position with respect to international competition, and would leapfrog the U.S. into a next-generation launch capability.

[ii] The study determined that while the goal of achieving single-stage-to-orbit fully reusable rocket launch vehicles has existed for a long time, recent advances in technology make such a vehicle feasible and practical in the near term provided that necessary technologies are matured and demonstrated prior to start of vehicle development.

Major changes in acquisition and operations practices, as well as culture, are identified as necessary in order to realize these economies. The study further recognized that the confident development of such a new launch vehicle can only be undertaken after the required technology is in hand. Therefore, the study recommended that a technology maturation and demonstration program be undertaken as a first step. Such a program would require a relatively modest investment for several years.

The study thus recommended that the development of an advanced technology single-stage-to-orbit rocket vehicle become a NASA goal, and that a focused technology maturation and demonstration be undertaken. Adoption of this recommendation could place the U.S. on a path to recapture world leadership in the international satellite launch marketplace, as well as enable much less costly and more reliable future government space activities. . . .

[1]   Introduction

The 1993 NASA Appropriations Act included language that expressed Congress' concern about the rising costs of the Space Station and space transportation, and the likelihood that NASA's program budgets would, at best, be limited in the future. In view of these trends, the Congress' concerns focused on NASA's ability to field a viable space program. Congress requested that a study be performed to recommend improvements in Space Station *Freedom* and space transportation, and to examine and revalidate civilian and defense requirements for space launch. This study was to be done in close cooperation with other agencies.

At about the same time, NASA independently undertook a series of internal studies as part of a reassessment of the Agency's programs, goals, posture, and long-range plans. These studies considered various options for the redesign of Space Station *Freedom*, Space Shuttle safety and reliability improvements, alternative transportation systems, and others. Since the Space Station Redesign Study developed into a full-fledged program reorientation activity during 1993, space transportation emerged as the key remaining area of focus, being at the heart of NASA's ability to support a wide range of national objectives and continue a visionary civil space program.

Another major factor for this study's focus was that NASA, together with the Department of Defense (DOD) and the aerospace industry, had spent nearly a decade defining and advocating a new launch vehicle program (which culminated in the proposed National Launch System), without being able to reach consensus with the Congress that it should be developed.

Yet another factor was the continued erosion of the international market share for U.S. launch vehicles. This market share has dropped from 100 percent to about 30 percent, largely due to the development and fielding of the French-built Ariane system, which targeted and captured at least 50 percent of the world's space launch market. U.S. industry has found itself increasingly unable to effectively compete using the current generation of launch vehicles.

As a result of all these factors and trends, as well as the specific Congressional request, a comprehensive in-house study was undertaken by NASA to identify and assess the major alternatives for a long-range direction for space transportation. The scope of the study was to support all U.S. needs for space transportation—including civilian, commercial, and defense needs—for several decades into the future. This is the Access to Space Study, which was recently completed and is summarized herein.

|2| **Purpose**

The U.S. space transportation architecture meets the current needs for access to space. The Space Shuttle is the world's most reliable launch system, and also functions as a human-tended research laboratory and satellite deployment, retrieval, and repair facility. The expendable launch vehicle fleet and related upper stages can lift all required defense and commercial spacecraft to their required destinations.

While these systems are by no means dysfunctional, they have major shortcomings that will only increase in significance in the future, and thus are principal drivers for seeking major improvements in space transportation. While the launch vehicles differ in their particular characteristics, their aggregate shortcomings are well known. They are too costly, insufficiently reliable and safe, insufficiently operable, and increasingly losing market share to international competition.

This study focused on identifying long-term improvements leading to a space transportation architecture that would reduce the annual cost of space launch to the U.S. Government by at least 50 percent, increase the safety of flight crews by an order of magnitude, and make major improvements in overall system operability (turnaround time, schedule dependability, robustness, pad time, and so forth). The study horizon was set at the year 2030 in order to allow time for new vehicles using advanced technology to fairly demonstrate their potential.

Using these criteria, this study identifies options for a long-term direction for the U.S. to meet government, defense, and commercial needs for space transportation, together with long-range program plans for implementation. While the focus of the study is long term, it recognizes that immediate improvements are needed. Therefore, program recommendations identifying realistic near-term activities for transitioning to the long-term capability are also included.

[3]   **Approach, Ground Rules, and Organization**

Approach

The Access to Space Study team began by recognizing that the Space Shuttle and the expendable launch vehicle fleet represent a very large investment both in vehicles and their supporting infrastructure. It recognized, based on many past studies, that the replacement of the current capability with any new vehicle or vehicles designed to overcome the above named shortcomings is likely to be an expensive and lengthy process.

Thus, the study approach considered, in parallel, a number of alternative approaches that differ in the degree of replacement of current capability, in the pace at which current systems are phased over to the new, and in the degree of utilization of new technologies. Three major alternative options were defined:

1. Provide necessary upgrades to continue primary reliance on the Space Shuttle and the current expendable launch vehicle (ELV) fleet through 2030.

2. Develop a new expendable launch system utilizing today's state-of-the-art technology and transition from the Space Shuttle and today's expendable launch vehicles starting in 2005.

3. Develop a new reusable advanced technology next-generation launch system, and transition from the Space Shuttle and today's expendable launch vehicles starting in 2008.

This strategy and approach is illustrated in figure 1.

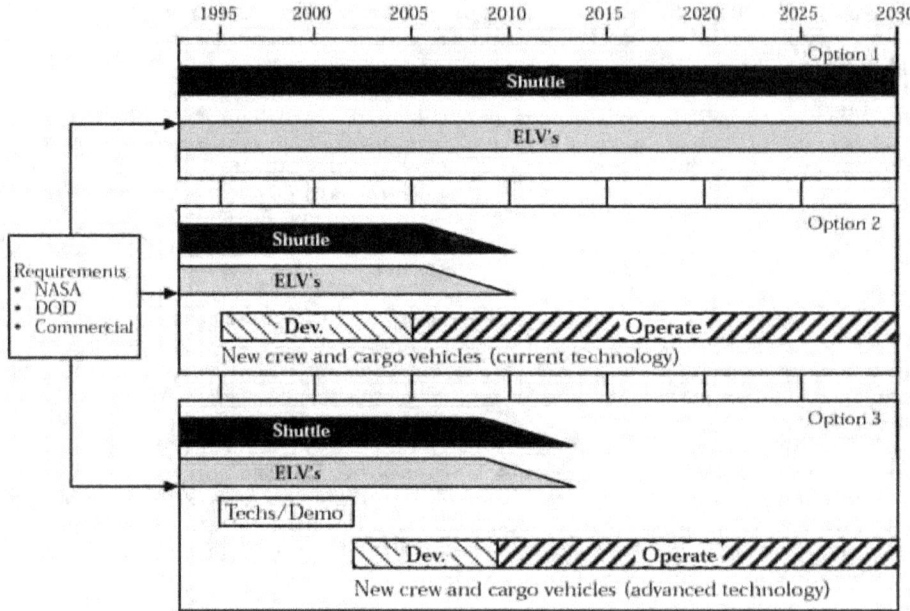

Figure 1.—Study strategy and approach.

[4] Each of the options was to treat the entire architecture of launch vehicles required. Each would be analyzed by a separate study team working independently of the others. The recommendations of these teams would be assessed by a small group reporting to the study director.

Common goals were established, and evaluation criteria were developed based on the goals- against which each of the options could be measured. These included performance and cost goals, operability, growth potential, environmental suitability, and others, as are shown in figure 2. These were organized into three categories in order of priority to facilitate both design selections and eventual comparative evaluation of the recommended architectures.

Fundamental Requirement	Essential Characteristics	Desired Features
1.1 Satisfy the national launch needs • NASA crewed • NASA uncrewed • DOD • Commercial (This includes definition of payloads from small to Shuttle/Titan class, and destinations at all altitudes and inclinations, as well as planetary.)	2.1 Improve crew safety by an order of magnitude (crew survivability >0.999). 2.2 Acceptable life-cycle costs, to include: A. Affordable DDT&E B. Improved operability and annual operating cost reduction over current systems (for STS equivalent <50%). Exclude costs of commercial flights. 2.3 Vehicle reliability of at least 0.98. 2.4 Environmentally acceptable: meet all environmental requirements planned for the year 2002.	3.1 Improve commercial competitiveness of launch vehicles. 2.2 Contribute to industrial economy (dual-use technology and processes). 3.3 Enable incremental development or improvements. 3.4 Improve capability relative to current systems (including STS).

DDT&E—Design, development, test, and evaluation   STS—Space Transportation System

Figure 2.—Access to Space capability goals.

The most beneficial designs that survived elimination within each of the three option teams were to be assessed against these criteria, and a preferred architecture as to be selected from them. An implementation plan and recommended actions were to be the final output of the study. The overall schedule of the study is shown in figure 3.

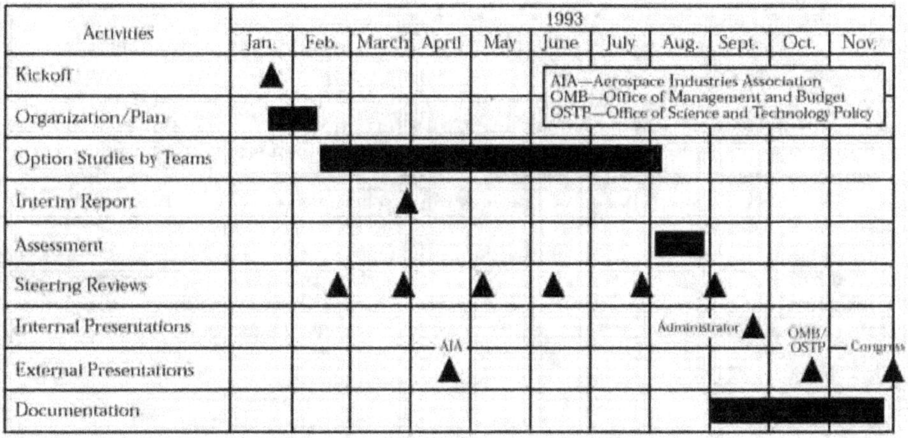

Activities	1993										
	Jan.	Feb.	March	April	May	June	July	Aug.	Sept.	Oct.	Nov.
Kickoff	▲										
Organization/Plan		▬									
Option Studies by Teams			▬▬▬▬▬▬								
Interim Report				▲							
Assessment								▬			
Steering Reviews		▲	▲	▲	▲		▲	▲			
Internal Presentations								▲			
External Presentations				▲						▲	▲
Documentation									▬▬▬		

AIA—Aerospace Industries Association
OMB—Office of Management and Budget
OSTP—Office of Science and Technology Policy

Figure 3.—Access to Space Study schedule.

[5]   **Ground Rules**

A number of ground rules were established for the Access to Space Study. Since a Space Station redesign was in progress, the Space Station *Freedom* design was utilized, but placed into the Mir orbit of 220 nautical miles (nmi) circular altitude at 51.6 degrees inclination. This was done to represent a worst-case scenario for the space transportation systems' requirements.

A common mission model was defined that included all US. defense, civilian, and commercial user elements covering the period from 1995 through 2030. This model was based on conservative extrapolation of current requirements and planned programs, and did not include major future possibilities such as exploration missions to the Moon and Mars. This mission model is shown in figure 4.

Vehicle Class	NASA	Commercial	DOD
Pegasus/Taurus Class	2.0	1 Nominal + 7 Growth	2
Delta Class	3.0	1 Nominal + 2 Growth	6
Atlas Class	2.0	3 Nominal + 0 Growth	3
Titan Class	0.3	—	3
Shuttle Class	8.0	—	—
Total Launches	15.3	5 Nominal + 9 Growth	14

Figure 4.—Annual launch demand mission model from 1995 to 2030.

For lack of solid forecasts of future traffic, the model was assumed to be constant through 2030. It was recognized that such a flat model was unlikely to endure over the long term and that excursions would eventually have to be treated as better models became available, as human exploration or other ambitious missions became better focused, or, hopefully, from additional market demand enabled by future reductions in the costs of access to space.

The annual payload weight to orbit represented by this model and the annual costs for current launch vehicles to launch the model are shown in figures 5 and 6, respectively. The U.S. Government launches 660,000 pounds of payload to space annually at a total cost of $6.7B dollars.

Uniform costing guidelines were developed using conventional weight-based estimating algorithms to allow direct comparison of all alternatives. It was recognized that innovative and potentially lower cost strategies based on major management, contracting, and operating changes might be considered by some, but not all, of the option teams. Therefore, it was decided that these changes were to be treated as excursions to the "business-as-usual" mode.

It was also decided that the commercial traffic estimates of the mission model were to be used for fleet sizing and as a basis for estimating the production base. However, since the principal study aim was to reduce launch costs to the government, the cost projections of the options were to include only government-sponsored missions.

[6]

Vehicle	NASA Plus DOD	Commercial	All
Pegasus/Taurus	4 at 1k = 4k	8 at 1k = 8k	12k
Delta	9 at 10k = 90k	3 at 10 = 30k	120k
Atlas/Centaur	5 at 18k = 90k	3 at 18k = 54k	144k
Titan/SRMU/IUS/Centaur	3.3 at 4k = 156k	0	156k
Shuttle/RSRM			
• S.S. Freedom	5 at 36k = 180k	0	180k
• Low-Earth Orbit	3 at 47k = 141k	0	141k
Totals	661 k	92k	753k

- k = weight in thousands of pounds
- Payload weight expressed in $28°$ low-Earth orbit equivalent, except Space Station (220 nmi at $56°$ inclination)

Figure 5.—Mission model—annual weight to orbit.

Vehicle Class	NASA	DOD	Total
Pegasus/Taurus	2 at 13M = $26M	2 at 13 = $26M	$52M
Delta	3 at 50M = 150M	6 at 50 = 300M	450M
Atlas/Centaur	2 at 115M = 230M	3 at 115 = 345M	575M
Titan/IUS or Centaur	0.3 at 375M = 125M	3 at 375 = 1,125M	1,250M
Shuttle	Annual Program Costs = 3,850M	—	3,850M
Infrastructure	—	526M	526M
Total	$4,381 M	$2,322M	$6,703M

- All costs in FY93 dollars, millions.

Figure 6.—Current fleet launch costs.

Organization

The Access to Space Study was directed by Arnold Aldrich, Associate Administrator for Space Systems Development, NASA Headquarters. The leaders of the three option teams were Bryan O'Connor, NASA Headquarters, and Jay Greene, Johnson Space Center (JSC) for Option 1; Wayne Littles and Len Worlund, Marshall Space Flight Center (MSFC) for Option 2; and Michael Griffin, Headquarters, and Gene Austin, Marshall Space Flight Center, for Option 3.

Mr. Aldrich formed a senior-level steering group to periodically review progress and provide advice. This steering group included members from NASA Headquarters and field installations, as well as representatives from the Department of Defense, the U.S. Air Force, and the Office of Commercial Programs in the Department of Transportation.

A small group of NASA Headquarters staff, reporting to the study director, was to ana-lyze the team reports, make strawman assessments and recommendations, and present them to the steering group and the director. The final study conclusions, presentations, and report were to be prepared by this group. . . .

Option Team Down-Selects

The most beneficial architectures as recommended by the Option teams are shown in the shaded areas in figure 36. These architectures were presented to the study steering group. They were then subjected to comparative analysis from which a preferred architecture was to be selected.

Option 1	Option 2	Option 3
Shuttle-Based	**Conventional Technology**	**New Technology**
• Retrofit: Evolutionary improvements. Keep the current ELV fleet. • New Build: Above changes plus major internal mods; new orbiter. Keep the current ELV fleet. • New Mold Line: Above changes plus major external mods; new orbiters and boosters. Keep the current ELV fleet.	• 84 configurations with differing crew carriers, cargo vehicles, stage configurations, engine types, and number of new vehicles. Reduced to four primary candidate architectures: – (2A): New large vehicle • Keep Atlas, Delta ELVs • HL-42 plus ATV – (2B): New lg. and sm. vehicle • Keep Delta ELV • CLV-P for crew plus cargo – (2C): New lg. and sm. vehicle • Keep Delta ELV • HL-42 plus ATV • Hybrids: STME engines – (2D): New lg. and sm. vehicle • Keep Delta ELV • HL-42 plus ATV • RD180/J2S engines	• 3A: Single-stage-to-orbit all rocket – With Titans • 3B: Single-stage-to-orbit all rocket – No ELVs • Single-stage-to-orbit air-breather/rocket – No ELVs • Two-stage-to-orbit air-breather/rocket – No ELVs

Figure 36.—Architectural alternatives proposed by the teams.

The Option 1 team down-selected to the Retrofit Alternative. This is the alternative that incorporated only internal changes to the Space Shuttle orbiter, retrofitted them into the fleet as the orbiters came in for major maintenance, and replaced orbiters only for attrition. The rationale for the down-select was that this alternative had the lowest design, development, test, and evaluation cost, while enabling about the same level of annual operations cost savings as the other alternatives.

The Option 2 team down-selected to the 2D architecture. This is an architecture that built a new expendable 20k-pound payload launch vehicle to replace the Atlas, a new 85k-pound lift expendable vehicle to replace the Titan and the Shuttle, separate new cargo and crew carriers, and the single-engine Centaur upper stage. It kept the Delta as a cost-effective launcher for smaller payloads. The principal reasons for the down-select were that this alternative did not require new engine development (the RD180 was claimed to be a low-risk modification of the currently operational RD170), had low life-cycle costs, and had the lowest operations costs for the Atlas-class missions, which have a high level of commercial interest. It accepted the limitations inherent in reduced down-mass capability from the Space Station.

[60] The Option 3 team down-selected to an all-rocket, fully reusable single-stage-to-orbit vehicle. The recommended configuration for this vehicle incorporated a tripropellant propulsion system, graphite-composite structure, aluminum-lithium propellant tanks, and an advanced thermal protection system and subsystems. Added margin could be attained by using graphite-composite fuel tanks rather than those made with aluminum-lithium fuel tanks. Rocket vehicles were selected over air-breathing vehicles on the basis that they had lower design, development, test, and evaluation costs; lower technology phase costs; and required less demanding technology that would translate into a more quickly developed and less risky program.

Two versions of the single-stage-to-orbit rocket were recommended. The first (Option 3A) had a transverse payload bay 15 feet in diameter and 30-feet long, which could not accommodate the largest of the Titan-class missions. This architecture thus required continuation of the Titan expendable launch vehicles in parallel with the new vehicle operations. The second version of the single-stage-to-orbit rocket vehicle (Option 3B) had a 45-feet [sic] long longitudinal payload bay that could accommodate all Titan payloads if some were somewhat downsized (a plan which is under serious consideration within the Department of Defense), and thus would not require continuation of expendable launch vehicles as part of the architecture. This version was included because of the high costs of operating the Titan expendable launch vehicle.

[61] **New Operations Concept**

All the option teams recognized that if large savings in annual costs were to be realized, new management, contracting, design, development, and, particularly, operations concepts had to be devised. The fundamental change required was that all phases had to be driven by efficient operations rather than by attainment of maximum performance levels. This, in turn, required maximizing automation and minimizing the number of people in the "standing army" on the ground, as well as requiring redundancy, engine-out capability, and robust margins in all subsystems. In addition, both of the Options 2 and 3 teams recommended avoiding development of new technology in parallel with vehicle development in order to minimize program risks and cost growth.

The Options 2 and 3 teams recommended a streamlined management and contracting approach patterned after the Lockheed "Skunk Works," which features smaller, but dedicated and collocated government oversight, a more efficient contractor internal organization, rapid prototyping, and team continuity from design to flight.

The recommendations also included a number of specific operations-oriented items, some of which are applicable to reusable vehicles and others that apply to both expendable and reusable vehicle operations. They included using well-matured technologies, demonstrated through a number of flights of an experimental vehicle; demonstration and validation of vehicle design via flights of a full-scale prototype, with gradual stretching of the flight envelope; certification of the vehicle design and type-certification of the fleet; avoiding continual engineering changes and long-term development engineering overhead by freezing the design for long periods between block changes; avoiding most detailed inspection and maintenance after each flight unless the need is clearly indicated by an onboard health monitoring and reporting system, or if the immediately previous flight exceeded the flight envelope limits charted in the prototype program; operating the single-stage-to-orbit fleet using a depot maintenance philosophy in which maintenance is only done by exception or every 1 to 2 years; use of small, dedicated ground crews led by a crew chief empowered to make all decisions in operations and maintenance; a reduced ratio of nontouch to touch labor compared to that utilized in today's operations; and much use of automation on the ground, as well as in the vehicle. These amount to a complete change in the way vehicles are developed and operated compared to current practice, and are patterned after several high-performance aircraft programs.

In the aggregate, the above recommendations amount to a "new way of doing business," which was recognized as being essential if low operating costs were to be realized. Its attainment would be a major shift from today's practices in launch vehicle operations.

[62]   **Comparative Analysis**

The down-selected architectures were compared so that a decision could be made on the most attractive option. The major factors considered in the evaluation were design, development, test, and evaluation costs; operations costs; life-cycle costs; and the safety and reliability of the concepts. These and other factors considered followed the major evaluation criteria identified in the Purpose section.

Costs Assessment

The costs presented in this report were developed from a common set of ground rules developed by the Comptroller's Office and are predicated on the technical complexity, operability, and flight-related assumptions of each of the option teams. The costs of the recommended architectures were analyzed, with design, development, test, and evaluation and total program costs treated separately. All cost figures are shown in constant FY94 dollars and in a business-as-usual mode, that is, without incorporation of the operations or management changes discussed in the New Operations Concepts section. This is because the NASA cost models were designed around the historical data base, and NASA does not have a mature basis for estimating costs incurred in a different culture.

The NASA Comptroller assembled a cost team to attempt to estimate the savings that might accrue if new ways of doing business were adopted, and this team concluded that a 30 to 40 percent reduction of the costs shown might be expected operating in such a mode. However, the cost team felt that since each of the options benefited differently from changes in culture, the comparison of the different options would be best served by using the business-as-usual method and then applying estimated reduction factors.

The design, development, test, and evaluation costs of the three options are shown in figure 37. These curves include a technology phase for Option 3. The curves are annotated with a callout indicating the total technology, design, development, test, and evaluation costs, which are $2.4B for Option 1; $11.1B for Option 2; and $17.6 and $18B for Options 3A and 3B, respectively-. These curves do not include facilities, production, or operations. If the new ways of doing business were adopted, these costs could be as much as 30 to 40 percent lower, or $1.5 to $1.7B for Option 1; $6.7 to $7.7B for Option 2; and $10.6 to $12.6B for Option 3.

The profiles of these technology, design, development, test, and evaluation expenditures are very different. Options 1 and 2 require large budgets essentially immediately, while Option 3 has a 4 to 5 year technology phase funded at relatively modest levels before the large budget requirements start. This technology phase requires $900M over 5 years and has an annual peak of about $240M. The profiles of Options 3A and 3B are essentially the same.

The life-cycle cost profiles of the three options are shown through the year 2030 in figure 38. These are total costs for the entire period to deliver the mission model of the Approach, Ground Rules, and Organization section, and include the technology, design, development, test, and evaluation costs of figure 37. A fourth curve is included in figure 38, labeled "current systems," which represents the cost to the U.S. Government if no changes are made and the current systems are operated for the entire period. In 1995, this current systems cost will be comprised of $3.8B for the Space Shuttle, $2.4B for the Department of Defense expendable launch vehicles and infrastructure, and $0.5B for the NASA expendable launch vehicles, totaling $6.7B.

[63]

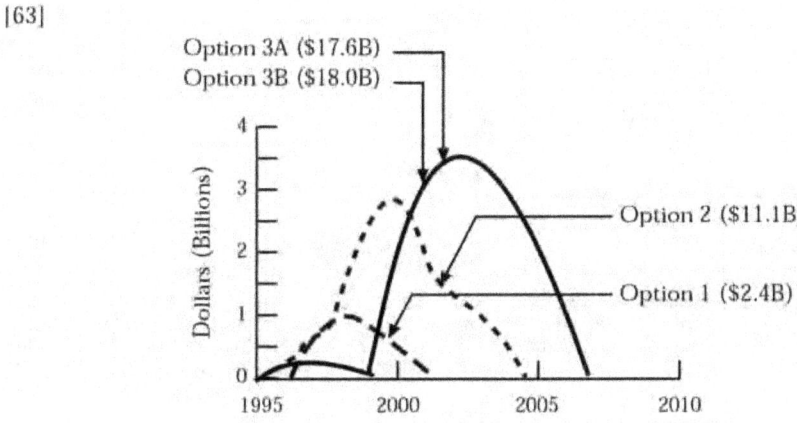

Figure 37.—Design, development, test, and evaluation costs of the options.

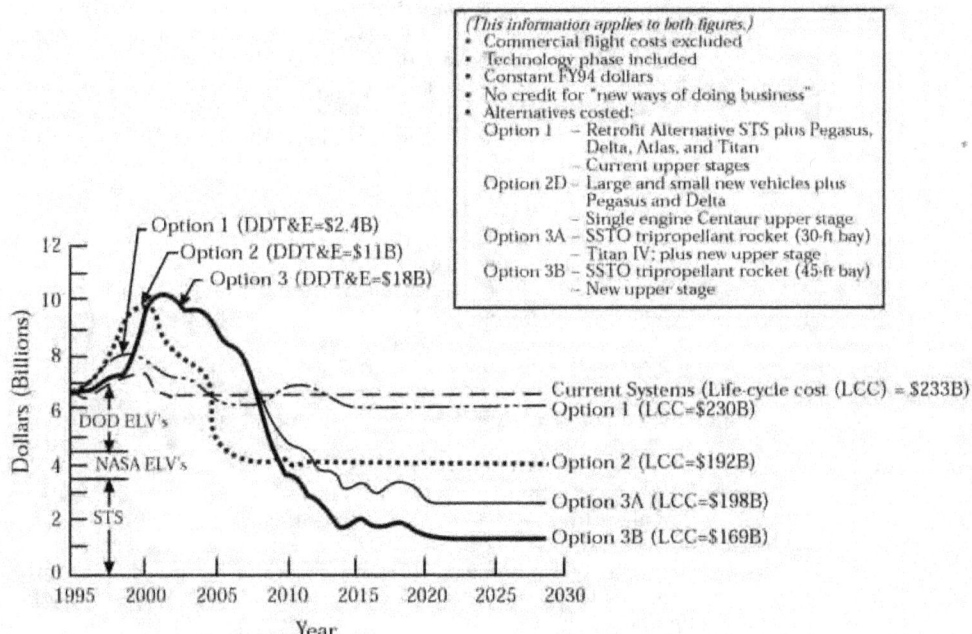

Figure 38.—Total U.S. Government launch costs.

[64] This reference varies somewhat with the expendable launch vehicle annual buys, infrastructure investments, and programmed Shuttle improvements. It was assumed as a point of reference that the expenditures remain essentially fixed after 2000, and that no additional orbiters will be acquired through 2030, even though a replacement orbiter is likely to be needed sometime during that interval. The life-cycle cost of this activity, if nothing is done differently than today, is $233B through 2030.

The cost plot of the architecture of Option 1 shows the increase for the $2.4B investment, followed by the $6.5B to retrofit the fleet, and then by a programmed buy of a replacement orbiter in 2010. The annual realized savings in operations costs is only about $0.25B per year. Its life-cycle costs are $230B. The investment in design, development, test, and evaluation is recovered after 10 years of steady-state operations. The total investment including design, development, test, and evaluation, and the replacement orbiter is recovered in slightly more than 20 years of operation.

The cost plot of the architecture of Option 2 shows the investment of $11.1B in design, development, test, and evaluation costs upon the immediate start of new vehicle development, followed by a rapid reduction in the operations costs starting in 2005 when the new vehicles are introduced and the Shuttle and most expendable launch vehicles are phased out. These vehicles are all phased out over 2 years. The operating costs are reduced to $4B annually beginning in 2006. The life-cycle costs of Option 2 were $192B. The recovery time for the investment in design, development, test, and evaluation is about 4 years of steady-state operation. The recovery of the total design, development, test, and evaluation plus production investment is about 5 years of steady-state operation.

The plot of the architecture of Option 3A shows the investment of $17.6B for technology, design, development, test, and evaluation through 2008, with the start of the development program delayed by about 5 years due to the technology maturation and demonstration phase. This option features the vehicle with the shorter payload bay, which requires continuation of the Titan expendable launch vehicles in parallel.

The Option 3A architecture results in a steady-state operations cost of $2.6B per year. That level is not achieved until after 2020 due to a deliberately slow production phase for the reusable vehicles and upper stages and their spares, which are all purchased continuously and then the production line is shut down. These purchases are stretched over 10 years or more to minimize peak funding needs. The technology, design, development, test, and evaluation investment would be recovered in 4 1/2 years of steady-state operations, while recovery of the total investment, including production of the vehicles, requires 9 years. The life-cycle cost of Option 3A is $198B.

Option 3B has the longer payload bay and could carry all DOD payloads with some downsizing, which the DOD may accomplish at the program's block change time in the first part of the 2000 to 2010 time period. The cost profile for this option follows that of Option 3A during development, but decreases to an annual operations cost of $1.4B since no Titans need to be retained. The life-cycle cost for this option is $169B. The technology, design, development, test, and evaluation investment would be recovered in only 3 1/2 years of steady-state operation, while recovery of the total investment would take only 7 years.

The clear message from figure 38 is that new vehicles are required if substantial savings are desired, and that attaining the greatest savings requires the largest investment.

The most significant aspects of the costs of the three options, and some associated metrics, are shown in figure 39. This figure displays the costs for the technology phase, the design, development, test, and evaluation (including the technology phase), the production of one–time or reusable hardware, the annual operations costs in the out-years, and the life-cycle costs.

[65] In addition to the previous observations, it is important to note that if nothing is done differently, the U.S. Government will spend $233B for space launch through 2030

for the assumed mission model of section 2. Option 1 only reduces that total by $3B over 35 years. Option 2 reduces the life-cycle cost by $41B in non-discounted dollars, or 17.6 percent. Option 3A reduces the life-cycle cost by $35B, or 15 percent. Option 3B reduces the life-cycle cost by $64B, or 27.5 percent.

Thus, the life-cycle cost savings for Option 3B are the greatest of all of the options, averaging a savings of $1.8B per year over the 35 year period through 2030.

| | | Current Program | Option 1 (Retrofit + ELV Fleet) | Option 2 (Lg. + Sm. Veh. + Delta) | Option 3 | |
					(SSTO-R, 30-ft. Bay + Titan)	(SSTO-R, 45-ft. Bay)
Costs	Technology	0	Incl. in DDT&E	$0.4B	$0.9B	$0.9B
	DDT&E (Incl. Technology)	0	$2.4B	$11.1B	$17.6B	$18B
	Production	0	$5.6B	$2.0B	$18.1B	$18.7B
	Operations (Out-Years)	$6.4B/yr	$6.1B/yr	$4.0B/yr	$2.6B/yr	$1.4B/yr
	Life-Cycle Costs	$233B	$230B	$192B	$198B	$169B
Operations Cost Metrics**	Average $/Launch (Shuttle replacement)	$322M (STS)*	$293M (STS)*	$85M (Sm.) $205M (Lg.)	$41M	$38M
	$/lb of Payload (Fleet Average for Mission Model)	$7,488/lb	$6,814/lb	$6,100/lb	$3,900/lb	$2,100/lb
	$/lb of Payload (Full Veh., to LEO, 28°)	$6,850/lb	$6,234/lb	$3,900/lb (Sm.) $1,600/lb (Lg.)	$980/lb	$920/lb
	$/lb of Payload (to the Space Station)	$12,880/lb	$11,720/lb	$3,700/lb (Lg.)	$1,600/lb	$1,500/lb

* Current Space Shuttle capability (no ASRM)
** In the out-years

• Constant FY94 dollars; no "new ways of doing business."

Figure 39.—Summary of option costs.

Referring to the cost metrics portion of figure 39, it is shown that the fleet-average launch costs for the mission model were reduced from the current values of $7,488 per pound to $6,814 per pound for Option 1; $6,100 per pound for Option 2; $3,900 per pound for Option 3A, and to $2,100 per pound for Option 3B. The lowest cost per pound of payload for the new vehicles launching into a 28-degree inclination low orbit were $920 and $980 per pound for the two Option 3 cases. Next higher were the $1,600 per pound to $3,900 per pound for the two different sized vehicles in Option 2, with the commercially significant smaller vehicle having the larger cost per pound. The cost for Option 1 was $6,234 per pound.

The Space Shuttle costs per launch were calculated consistent with the methodology historically presented to [the Office of Management and Budget] and [the General Accounting Office]. While all the costs were lower than the $6,850 to $7,488 per pound for the current Shuttle program when computed the same way, it is clear that the major cost savings targeted as a goal for this study only accrue in architectures employing new vehicles. In addition, it is also clear that Option 3 lowers the launch costs by the largest amount. [66] The cost per launch to a Space Station in a 220 nautical mile circular, 51-degree orbit showed similar trends, the lowest being $38 to $41M for Options 3A and 3B, $85 to $205M for the Option 2 vehicles, and $293M per launch for the Space Shuttle, computed in the same way. The cost per pound of payload to the new Space Station orbit also showed similar trends.

It is possible that the above operations cost metrics might be reduced further by adopting the so-called new ways of doing business, but the savings obtained may be less than the 30 to 40 percent predicted for the design, development, test, and evaluation, and production reduction. This is because the operations costs are already based on stream-lined operations concepts, at least for Options 2 and 3. In addition, further reductions may be possible by buying launch services from the private sector, but the effects have not been well quantified.

It is clear from examination of the cost results that large annual cost savings are pos-sible, but they can only be attained by considerable up-front investment—the larger the investment, the larger the operations cost savings. It is also clear that the attainment of costs substantially below about $900 per pound of payload into a 28 degree low-Earth orbit requires further understanding of the savings obtainable with new ways of doing business, larger mission models requiring more frequent flights, technology beyond that of any alternatives considered in this study, or, most likely, a combination of all these factors.

Other Assessment Factors

Eight major factors were assessed, including a summary of the costs from the previous figure. These assessment factors are displayed in the matrix of figure 40.

	Option 1 Shuttle Retrofit	Option 2 Architecture 2D (Lg. + Sm. + Delta)	Option 3 SSTO Rocket + Titan	Option 3 SSTO Rocket
National Launch Needs	Meets Model	Meets Model Except 125k lb/yr Downmass (Provides 25k lb)	Meets Model	Meets Model (If DOD P/L Shortened)
Vehicle Reliability	Meets 0.98 Goal for Shuttle and Delta	Meets 0.98 Goal for New Vehicles and Delta	Meets 0.98 Goal for New Vehicle	Meets 0.98 Goal
Crew Safety	Does Not Recommend Significant Improvement	Meets 0.999 Goal	Meets 0.999 Goal	Meets 0.999 Goal
Summary Costs	Does Not Approach 50 Percent Reduction Goal	Approaches 50 Percent Reduction Goal	Exceeds 50 Percent Reduction Goal	Far Exceeds 50 Percent Reduction Goal
Operability	Significant Shuttle Improvement; ELV Fleet As Is	New Vehicles: Robust and Highly Operable; Delta, Pegasus As Is	New Vehicles: Robust and Highly Operable; Titan As Is	New Vehicles: Robust and Highly Operable
Technical Risk	Low	New Vehicle–Low; HL-42–Moderate	Moderate-to-High	Moderate-to-High (More Technology Required)
Cost Risk	Low-to-Moderate	Moderate	Moderate-to-High	Moderate-to-High
Other Factors	Additional Orbital Capabilities	Achieves Parity With International Competitors	Major Increase in International Competitiveness	Major Increase in International Competitiveness

Figure 40.—Option comparison.

National Launch Needs

All the options met the requirement to launch the mission model of the Purpose section. The requirement also existed to return all of the mass taken to the Space Station, which was met by Options 1 and 3, but not by Option 2, which returned only approximately 20 percent. This was a feature of the down-selected architecture, and was adopted in order to minimize new vehicle and carrier sizes and costs. The cost of the expended Space Station carriers and racks resulting from this limitation were accounted for in the operations cost analysis.

[67] An additional factor applied to Option 3B, which was able to launch the longest DOD payloads only if the DOD downsized them to 45-feet in length. Preliminary discussions with the DOD indicated that such downsizing was a distinct possibility at the time the payloads were due for a block change, about 2002. Indeed, there has already been some Congressional language urging the DOD in this direction in order to allow retirement of the expensive Titan vehicles. Thus, while the possibility of having shorter payloads might be realistic, nonetheless, the viability of Option 3B rests on this assumption.

Vehicle Reliability

All vehicles except the Atlas and Titan met the goal of having a vehicle reliability greater than 0.98 percent. It was felt that it was unlikely that these two expendable launch vehicles could be upgraded to that reliability in a cost-effective way, while the Delta is almost at this reliability level already. All the new vehicles were designed to exceed this requirement.

Crew Safety

The improvement of crew safety (probability of crew survival) to at least 0.999 from the 0.98 of the Space Shuttle was met or exceeded by the new vehicles of Options 2 and 3. Option 2 had a launch escape propulsion system for the entire crew carrier, while Option 3 adopted escape seats and intact abort of the vehicle into orbit or return to the launch site.

Option 1 did not recommend the addition of escape seats, an escape pod, or liquid boosters to the Shuttle and, thus, did not improve significantly on the current crew safety analysis. The reason for this recommendation was that the analysis showed that the expense for incorporation of additional escape capabilities was high, and that there was a significant impact on current vehicle capabilities due to factors such as a major shift in the orbiter center of gravity.

Summary Costs

The costs discussed with reference to figure 39 indicate that Option 1 did not approach the 50 percent cost savings goal; Option 2 approached it, though it did not meet the goal, reducing operations costs by about 37 percent; and both Option 3 alternatives exceeded that goal—Option 3A reducing costs by 59 percent and Option 3B by 78 percent.

A number of observations were made regarding relative costs. One was the difficulty of reconciling cost estimates for operational systems, which are well understood, with those for new vehicles whose definition is still in the pre-Phase A state.

Compounding that difficulty was an uncertainty in the amount of cost growth margin to include in the estimates, which, in existing systems, was felt to be largely governed by external factors rather than inherent growth due to inadequate definition or design errors. The teams questioned, therefore, whether the historical cost growth allowances using conventional NASA models are too conservative if new management schemes are to be adopted that might better be able to shield the program from external factors.

An additional observation is that the NASA cost models are designed to predict development costs and lack a rigorous process for predicting operations costs. Nevertheless, the estimates developed for the Access to Space Study were made with guidance from experienced costing teams using the best costing tools available.

Operability

Enhancements in the operability of the three options were also assessed. Option 1 improved the Shuttle operability somewhat, but that of the companion expendable launch vehicles was unchanged. Thus, taken as a whole, the operability of Option 1 was not significantly improved over the present situation.

[68] All the new vehicles of Options 2 and 3 had designs, infrastructure, and operations concepts specifically tailored for operability and robustness, and associated significant reductions in operations costs. However, Option 2 retained the Delta and Option 3A retained the Titan, and, thus, their overall operabilities were thus somewhat degraded. Therefore, Option 3B promised the best operability of the three options.

Technical Risk

It is apparent that the technical risk will increase with adoption of new design vehicles, and even more so if new technology is utilized. Thus, the technical risks were assessed as low for Option 1, low for the new vehicles of Option 2 since their designs have been defined in detail under the Advanced Launch System and National Launch Systems programs, moderate for the HL-42 crew carrier vehicle of Option 2, and moderate to high for Option 3 due to the incorporation of new technology. Even though Option 3 incorporates new technology, its risk was felt to be manageable due to the 4 to 5 year technology maturation phase which would develop and demonstrate the needed technologies to at least a level 6 technology readiness level (proven in their operating environment).

Cost Risk

The cost risk was principally due to the schedule impacts of technical uncertainties during development. It was felt to be low to moderate for Option 1, moderate for Option 2, and moderate to high for Options 3A and 3B, the latter driven largely by the presence of new developments and new technology.

There was also a recognition that while the options that had new vehicles incurred greater cost and schedule risk, this risk increased in proportion to the cost savings they would enable.

Other Factors

In addition to the factors assessed above, there are a number of other distinguishing features of the options that should be considered in making an architectural selection.

The first of these is the total capability of the Space Shuttle which, in addition to providing launch and return of payloads, has a capability to capture and repair spacecraft, and is also a crewed orbital research and development facility with an orbital flight duration of at least 2 weeks. These capabilities would not be replicated if Options 2 or 3 were to be selected, as crewed orbital laboratory functions are to be assumed by the Space Station. However, if the Space Station is not available, for whatever reason, this factor could have an overriding importance.

Another such factor is the ability for the U.S. commercial launch industry to compete in the international satellite launch market. Option 1 does nothing to improve the current situation. Option 2 would achieve approximate parity with the projected prices of the Ariane IV and Ariane V, the most efficient of the foreign systems, only after a lengthy development period. Option 3, on the other hand, would lower launch costs so dramatically that U.S. industry could underprice all competitors. The U.S. would likely capture, and once again dominate, the international satellite launch market for a considerable period of time, utilizing these unique advanced technology vehicles.

Lastly, it was recognized that providing two different means for assured access to space for every important payload will be prohibitively expensive, no matter how desirable. One way out of this dilemma is to recognize that the world has changed and that the international space launch community now has the capability and reliability to function as a back-up for launching U.S. payloads in the case of extensive groundings of U.S. launch vehicles. Thus, while some payloads would have to be designed to be compatible with more than one launch vehicle, assured access to space may be attained by any of the options studied, without major additional investment, by proper agreements with other nations.

[69]   **Observations and Conclusions**

Assessment of the characteristics, performance, and costs of the architectures recommended by the option teams led to a number of observations which, in turn, lead to the study conclusions. These are presented below.

Cost Reductions and Safety Increases

The study determined that it is indeed possible to achieve the objectives of large reductions of operations costs and increases in reliability and crew safety at the same time in the same architecture. It did not appear that reasonable modifications to the Space Shuttle could achieve these objectives in a cost-effective manner, though a number of beneficial improvements to the Shuttle system were identified.

New vehicles were required in the architectures to attain these objectives. These vehicles could be constructed using either conventional or advanced technologies, with the conventional technology vehicles approaching the 50 percent desired minimum operations cost reduction (37 percent reduction), and the advanced technology vehicles greatly exceeding it (up to 78 percent operations cost reduction).

Design, Development, Test, and Evaluation Budget

Both current technology and new technology vehicles achieved the targeted operating cost reductions only after sizable design, development, test, and evaluation budget investments. This budget investment was smaller, but immediate, for the Option 2 architecture using current technology new launch vehicles and carriers. Both of the Option 3 architectures required a larger design, development, test, and evaluation budget, but start of their development was delayed 4 to 5 years as a result of the necessity of maturing and demonstrating the required technologies. Thus, Option 3 is more consistent with projected near-term budget availability.

Annual Operations Costs

The annual operations costs of the Option 3B architecture were the lowest of all, since the new vehicle replaced all the current generation launch vehicles which have large operations costs.

The achievement of these low operating costs was completely dependent on making large-scale changes in the way vehicles are designed, developed, managed, contracted for, and operated. It was concluded that associated designs must all be driven by operations, as well as by performance, and that resulting architectures must also entail the major changes in launch infrastructure and operations "culture" referred to as "new ways of doing business."

Most Attractive Option

In view of the above, an architecture featuring a new advanced technology single-stage-to-orbit pure-rocket launch vehicle was recommended as the most attractive option. It has the greatest potential for reducing annual operations costs as well as life-cycle costs, it would develop important new technologies with dual-use in industry (such as composite vehicle structures for cars and airplanes), it would place the U.S. in an extremely advantageous position with respect to international competition, and [it] would leapfrog the U.S. into a next-generation launch capability.

[70] The preferred single-stage-to-orbit rocket alternative is that in which the vehicle is sized so as to accommodate all payloads in the mission model, so as to avoid the need to carry current Titan expendable launch vehicles in parallel. The lowest operations costs resulted from selecting this single-stage-to-orbit pure-rocket vehicle as the focal point of the new launch architecture.

The large development costs associated with this new vehicle would be put off for at least 5 years while the technology was being matured and demonstrated. This would allow at least that time period for measured consideration of the decision to start a new vehicle program.

On the other hand, delaying the decision of which vehicle architecture to select by 4 or 5 years but not funding a focused technology phase will achieve nothing, since the lack of a focused technology program during that period will not reduce the risks of developing an advanced technology vehicle. Therefore, the choices available in 4 to 5 years would be exactly the same as those we face today.

Technology Maturation and Demonstration

The assessment that the best option is to develop a new, fully reusable, advanced technology single-stage-to-orbit rocket launch vehicle is absolutely dependent on maturing and demonstrating the required technologies before initiating development.

Though it is possible to start development right away and perform technology maturation and demonstration concurrently, such an approach carries with it greater technical, schedule, and cost risks. Further, it would immediately require large budgets, precluding the 4 to 5 years of relatively modest budgetary investment. However, once the required technologies are matured and demonstrated at the subsystem/system level in the pertinent environment, the perceived risk is much reduced and should be manageable.

The technologies that require maturation and demonstration include graphite-composite reusable primary structures, aluminum-lithium and graphite-composite reusable cryogenic propellant tanks, tripropellant or [liquid oxygen]-hydrogen engines designed for robustness and operability, low-maintenance intergral [sic] or standoff thermal protection systems, autonomous flight control, vehicle health monitoring, and a number of operations-enhancing technologies.

These technologies must be demonstrated on the ground and through flights of an experimental rocket vehicle. Technologies that interact should be tested together, both on the ground and in the experimental vehicle. A second objective of an experimental vehicle would be to validate the vehicle design models that are used to predict the characteristics and performance of single-stage-to-orbit rocket vehicles.

Technology Applicability

The current expendable launch vehicles and the Space Shuttle will have to be operated for at least another 10 to 15 years before new launch vehicles can be available. Improvements to the fleet vehicles that significantly improve their operability and possibly reduce their operating cost should continue to be considered for implementation.

[71] The technology program for the single-stage-to-orbit rocket would result in the evolution of numerous capabilities and/or components/subsystems that could be directly applied to these current launch vehicle systems. These could improve the operability and, to some degree, the cost performance of the current generation expendable launch vehicle fleet and the Space Shuttle until such time as the new vehicles became available to be phased in. The decision to upgrade the current fleet can be incremental and independent from that to start the technology program.

The new technologies will generally support the development of any type of new generation launch vehicle, even if initiated further in the future. In addition, most of these technologies are highly beneficial in their own right for applications throughout the civilian and defense communities and the commercial marketplace.

Space Shuttle

Even though improvements to the Space Shuttle were identified and new vehicle designs were conceived that potentially could improve its cost and safety, it was clear that the Space Shuttle remains the world's most reliable launcher and is safe to fly utilizing today's rigorous processes until a next generation system becomes available.

The cost savings reported by the Option 1 team did not consider management or contract infrastructure changes. These areas have the potential to offer additional cost reduction benefits; however, considerations such as these were beyond the scope of the Access to Space Study. Such studies may be appropriate and beneficial and, if so, should be undertaken by the Space Shuttle Program. It is recognized that the Space Shuttle Program has already emphasized operational efficiency improvements in its program.

Lastly, the Option 1 team recommended further studies of flyback, fully reusable liquid-fueled boosters for the Space Shuttle in order to increase safety and potentially reduce costs. These studies should be performed to further develop the possible benefits such a configuration might offer.

National Aerospace Plane

The selection of the rocket single-stage-to-orbit over the air-breathing single-stage vehicle by the Option 3 team was done for significant cost, risk, and schedule considerations. The air-breather option was determined to have more difficult technology and, therefore, would be more costly and take longer to develop.

However, air-breathing launchers potentially offer a number of unique mission capabilities in which they may have an advantage. These include launch into orbits with lower inclination than the latitude of the launch site, performing synergetic plane changes in order to over fly a given Earth location on successive orbits, and flexibility to perform single-orbit data collection missions. In addition, their technology is applicable to future hypersonic aircraft, both for civilian and defense applications.

Thus it was concluded that the National Aerospace Plane enabling technology program should continue independently of any decision to proceed with development of a nearer-term low-Earth orbit launch system.

[72]   **Recommendations**

The Access to Space Study makes a number of recommendations. These are summarized below.

1. Adopt the development of an advanced technology, fully reusable single-stage-to-orbit rocket vehicle as an Agency goal.

2. Pursue a technology maturation and demonstration program as a first phase of this activity.

   - The technologies developed should be aimed at a single-stage-to-orbit rocket using tripropellant propulsion and advanced structures and materials. This program would mature and demonstrate the technologies described in the Description of the Option Teams Analysis (Option 3) section and summarized in the Observations and Conclusions section.
   - A complementary experimental rocket vehicle technology demonstration flight program should be pursued in parallel with the technology development activity.
   - These activities should be paced so as to allow the earliest informed decision on development of a full-scale vehicle.

3. The technology, advanced development, and experimental vehicle programs should be coordinated with the Department of Defense.

4. The Space Shuttle and the current expendable launch vehicle programs should be continued. The most beneficial and cost-effective upgrades should be considered for incorporation into these vehicles until the new single-stage-to-orbit vehicle becomes available.

5. Although the focus of these recommendations is a technology maturation and demonstration program, additional studies should be conducted in parallel. They include system trade studies for the single-stage-to-orbit rocket vehicle configuration in order to guide the technology activities, and assessment of a flyback reusable liquid booster concept for the Space Shuttle.

6. The National Aerospace Plane enabling technology program should be continued as a separate and distinct activity, as it contributes to future defense and civilian hypersonic aircraft programs. and it has potentially unique future mission applications.

Document IV-15

Document title: Department of Defense, "Space Launch Modernization Plan—Executive Summary," May 1994, frontmatter and pp. 1–18, 23–30.

Source: Documentary History Collection, Space Policy Institute, George Washington University, Washington, D.C.

The congressional overseers of the Department of Defense (DOD) directed the Secretary of Defense in 1993 to develop a plan for the modernization of DOD (or all U.S.) launch capabilities. The Secretary of Defense assigned this task to the Under Secretary of Defense for Acquisition and Technology, who formed an interagency team led by Lt. General Thomas S. Moorman, Jr., Vice Commander of the U.S. Air Force Space Command, to carry out the study. (The study became widely known as the "Moorman Study.") The study team developed a Space Launch Modernization Plan that, together with the results of NASA's Access to Space Study, formed the basis of many of the Clinton administration's policies set out in the August 1994 statement of National Space Transportation Policy (see Document IV-16). The appendices and annexes that accompanied this executive summary do not appear here.

Space Launch Modernization Plan

Executive Summary

May 1994

[no page number]

DEPARTMENT OF THE AIR FORCE
HEADQUARTERS AIR FORCE SPACE COMMAND

5 May 1994

MEMORANDUM FOR DEPUTY SECRETARY OF DEFENSE

FROM: HQ AFSPC/CV
150 Vandenberg Street, Suite 1105
Peterson AFB CO 80914-4020

SUBJECT: Space Launch Modernization Plan

In December 1993, you directed that a study group be formed to address the FY 94 Defense Authorization Act tasking to develop roadmap options establishing priorities, goals, and milestones for the modernization of US space launch capabilities on behalf of the Secretary of Defense. From January through March 1994, an inter-agency study group with participation from each of the nation's four space sectors—defense, intelligence, civil, and commercial—examined this complex issue.

Primary goals of the study were to investigate all facets of space launch, develop a comprehensive understanding and data base, and foster as much consensus among the government agencies as possible. The attached Executive Summary highlights the findings and recommendations of this group and has been coordinated by your staff through all appropriate executive agencies. In addition, detailed sub-panel annexes are being finalized; they should provide supporting data and rationale for the Executive Summary. Finally, a summary briefing is available for presentation to interested parties.

During the course of this three-month intensive effort, the study team developed a set of roadmap options for modernizing US space launch capabilities. These roadmap options include sustaining current space launch systems, evolving current expendable launch systems, developing a new expendable launch system, and developing a new reusable launch system—all keyed to payload user needs to minimize transition costs. For all roadmap options, we recommend revitalizing the US "core" space launch technology program.

Though this study does not recommend a specific program approach, we believe the roadmap options we have defined will provide the Department of Defense a range of choices to help the United States reduce the cost and improve the operational effectiveness of our space launch capabilities.

THOMAS S. MOORMAN, JR.
Lieutenant General, USAF
Chairman, DoD Space Launch Modernization Study . . .

[no page number]

THE SECRETARY OF DEFENSE
WASHINGTON, D.C. 20301

May 6, 1994

Honorable Al Gore
President of the Senate
Washington, DC 20510

Dear Mr. President:

Section 213 of the National Defense Authorization Act for Fiscal Year 1994, directed the Secretary of Defense to develop, in consultation with the Director, Office of Science and Technology Policy, and submit to Congress, a plan that "establishes and clearly defines priorities, goals, and milestones regarding modernization of space launch capabilities for the Department of Defense or, if appropriate, for the Government as a whole." It also directed the Department to examine requirements for a new launch system, identify the means of reducing production costs for current launch systems, and conduct a comprehensive study of the differences between existing U.S. and foreign expendable space launch vehicles.

This latter study on the differences, which is to be completed by October 1, 1994, will be provided separately and is not addressed by this action.

The Department is not now in a position to submit the plan that establishes priorities, goals, and milestones for modernization, as required by section 213. The Department, however, has developed a plan for modernization of space launch capabilities and is forwarding herewith the Executive Summary of that plan. This summary should be viewed as the first step in complying with section 213. The summary identifies the options for modernizing the current expendable launch vehicle fleet, the milestones for each, and associated development and operations costs. At this time, the Department has not selected a specific option, nor have we chosen to implement any of the recommendations. Those actions will be addressed as we formulate the Department's fiscal year 1996 budget. That budget submission will respond fully to section 213, because we will have chosen a specific plan of action, which, in turn, will establish the goals, priorities, and milestones for implementing that plan.

A similar letter has been sent to the Speaker of the House.

Sincerely,

John M. Deutch
Deputy Secretary of Defense

Enclosure

[no page number]   **Foreword**

Over the past decade, space launch has been a very challenging and unsettled mission for the Department of Defense (DOD). Since the decision in the early 1980s to rely upon the Space Shuttle as the sole access to space for the Nation, there have been costly accidents, significant policy and program changes, and countless studies on future needs and options. In the aftermath of the Challenger accident, the DOD quickly reestablished expendable launch vehicle (ELV) capabilities to regain access to space for critical national security missions. However, these regenerated capabilities were based upon existing launch systems (Titan, Atlas, and Delta) that have significant limitations in terms of cost, operability, and responsiveness. Several efforts have been made in recent years to develop a new ELV system—Advanced Launch System, National Launch System, Spacelifter—but all have been terminated. At the same time, competition is growing for launch systems and services from foreign providers, including Europe, Russia, China, and Japan, which creates further policy and economic issues. Thus, there is a growing sense within the Congress, key agencies and offices within the Executive Branch, and influential industry and public interest circles that while space launch is a critical issue for the America's [sic] future in space, there is no coherent national plan to guide our actions into the next century. . . .

[1]   **A. Tasking**

Section 213 of the National Defense Authorization Act for Fiscal Year 1994 (Appendix 1) directed the Secretary of Defense (SECDEF) to develop, in consultation with the Director, Office of Science and Technology Policy (OSTP), a plan that "establishes and clearly defines priorities, goals, and milestones regarding modernization of space launch capabilities for the Department of Defense or, if appropriate, for the Government as a whole." It also directed that the plan specify whether the SECDEF intends to allocate funds for a new space launch vehicle or other major space launch development initiative in the next Future Years Defense Program (FYDP). For any new non-man-rated expendable or reusable launch vehicle technology development or acquisition identified in the plan, the Act directed exploration of innovative government-industry funding, management, and acquisition strategies to minimize cost and acquisition time. Additionally, the congressional direction specified that the Plan provide a means of reducing the cost of producing existing launch vehicles. Finally, the Act directed a separate report to provide a comparison between U.S. and foreign expendable launch systems. This separate report is to be prepared in consultation with the Administrator of NASA and, as appropriate, the heads of other federal agencies and experts from industry and academia. That report will be provided separately and is not addressed by this action.

Within the Department of Defense (DOD), the task was assigned to the Undersecretary of Defense for Acquisition and Technology, USD(A&T), who in turn approved the Terms of Reference (TOR) for the Space Launch Modernization Plan (SLMP—"the Plan") on 23 December 1993 (Appendix 2). The TOR established an interagency Study Group (Appendix 3) to prepare the plan and a Steering Group (Appendix 4) to oversee and guide the effort. In developing the Plan, the TOR tasked the Study Group to examine space launch systems requirements, past studies, reducing production and operations costs for current systems, space launch technology development efforts being conducted in Government, and innovative funding and management.

In addition, the TOR directed the Study Group to compare U.S. and foreign space launch systems in terms of design, manufacturing, processing, management, and infrastructure to assess their effect on cost, reliability, and operational effectiveness. The TOR directed the Plan be submitted to USD(A&T) within 90 days and the comparison with foreign systems be completed by 1 October 1994.

[2]   **B. Approach**

USD(A&T) appointed Lieutenant General Thomas S. Moorman, Jr., Vice Commander of Air Force Space Command, to lead the study. Both the Study and Steering Groups had broad representation from the National Aeronautics and Space Administration (NASA), the Departments of Commerce and Transportation, the military departments, the Joint Staff, U.S. Space Command, Defense agencies, and the Office of the Secretary of Defense

(OSD), The Study Group worked continuously during the study period, while the Steering Group met periodically to review and guide the effort. The guiding principle throughout the study was to develop consensus among all sectors—defense, intelligence, civil, and commercial—on space launch needs, solutions, and priorities.

The Study Group established a goal to develop a plan to improve the Nation's space mission accomplishment through an integrated, efficient, and balanced space launch capability. The study goal was supported by the following objectives:

- Establish a comprehensive and accessible *database* of program, technology, policy, and budgetary information
- Understand and synthesize *requirements*
- Identify *deficiencies* in current and planned capabilities
- Examine *options* to correct those deficiencies
- Formulate alternative program *roadmaps* and *strategies*
- Develop *findings* and *recommendations*.

The Study Group was organized into five panels: environment, requirements, technical, operations, and business/management (Appendix 5). The Study Group received more than 130 presentations from Government agencies, industry, laboratories, and think tanks. It conducted interviews and roundtable discussions with congressional members and staff, industry executives, and current and past national space leaders. The Study Group developed a detailed understanding of the Nation's launch capabilities and needs and identified "facts of life" that impact future choices. The group then developed four options with associated alternative roadmaps and assessed each one in terms of requirements satisfaction, cost, and risk. Details on the analysis and findings of each panel and the options and roadmaps are contained in Annexes A through E; classified launch requirements for the intelligence sector are documented in a compartmented report (Annex F).

[3]   **C. Background**

The environment within which the national spacelift mission is conducted involves a complex web of actors, objectives, responsibilities, and influences. National security, economic interest, commercial competitiveness, technology excellence, and international relations all drive as well as limit our space launch needs and options. To understand this environment, a broad review of current circumstances and forces is essential.

1. Policy

Past national space policies have emphasized the need for assured access to space. The current national policy context is dominated by the theme of improving the Nation's economy by investing in U.S. industrial competitiveness as well as by encouraging technology transfer from defense to U.S. commercial industry. As this study neared completion, OSTP was in the process of developing the Administration's space launch policy embodying this theme. While past and evolving national policy has included specific direction on modernizing the Nation's space launch capability, little progress has been made due in large part to widely differing views and interests in this area and the inability to maintain consensus within the Executive Branch. To tackle this problem, the Administration's new draft space launch policy addresses DOD and NASA roles and provides guidance for implementation.

2. Prior Studies

The Space Launch Modernization Plan drew extensively from prior launch studies. Highlights and key items from four prior launch studies are included for background.

a. **Report of the Advisory Committee on the Future of the U.S. Space Program (the Augustine Report).** Requested by NASA and completed in December 1990, this study advised the NASA Administrator on the overall approaches NASA management could use to implement a balanced U.S. space program in the future. The committee stated a number of general concerns affecting America's space program, including

- Lack of consensus
- Over commitment of financial and personnel resources
- Program turbulence because of unforeseen technical problems or unrealistic program goals
- Institutional aging and large bureaucracies
- Need to maintain a technically qualified work force
- Declining technology base whose scarce resources are often threatened by mission needs
- Limited resilience of the Space Shuttle.

[4] Not surprisingly, the SLMP identifies some of the same issues today in relation to the U.S. space launch situation. The Augustine Committee recognized that access to space is "the most fundamental building block without which there can be no future space program" and recommended reducing dependence on the Space Shuttle, developing a new, unmanned (but potentially man-rateable [sic]) launch vehicle, and maintaining an advanced launch system technology program to enhance current and evolving capabilities and provide a basis for new and revolutionary launch systems.

b. **The Future of the U.S. Space Launch Capability (the Aldridge Study).** Chartered by the National Space Council and completed in November 1992, this study examined the Nation's spacelift needs and recommended proceeding immediately into the development of a new expendable launch system called Spacelifter—a medium lift vehicle in the 20,000 pounds to low-earth-orbit class with modular growth up to 50,000 pounds to accommodate heavy lift requirements. The report noted that technology efforts such as the National Aero-Space Plane (NASP) and the Single Stage Rocket Technology (SSRT) programs were essential to future generations of fully reusable space launch systems. The report recognized the high costs of the Space Shuttle and suggested that an eventual solution to its high cost must be found. Finally, the report recommended that a new management structure, to include a launch "czar," be created to provide more centralized planning, integration, and coordination for implementing the Nation's launch strategy.

c. **NASA Access to Space Study.** Completed in 1993 in response to tasking in the FY 93 Appropriations Conference Committee language, NASA's *Access to Space Study* examined the Nation's space launch needs. The agency studied three options: Option 1 maintained the Shuttle and current ELV fleet until 2030; Option 2 examined a new expendable launch system using state-of-the-art technology with a transition date of 2005; Option 3 developed a new advanced technology, next-generation reusable launch system with a technology demonstration program and an operational transition date of 2008. NASA recommended adoption of Option 3.

d. **DOD Bottom-Up Review.** The DOD Bottom-Up Review (BUR), completed in 1993, included a review of DOD's space launch program—taking into consideration commercial concerns, the needs of the civil space sector, and impacts on the U.S. industrial base. The BUR examined three alternatives: Alternative 1, a life extension of the current expendable DOD fleet; Alternative 2, the development of a new launch system; and Alternative 3, the development of a "leapfrog" technology launch system. Alternative 3 was eventually eliminated as a viable alternative, but a reusable single-stage-to-orbit (SSTO) rocket was included in Alternative 2. The BUR acknowledged that spacelift modernization was a desirable national goal but concluded that DOD's requirements were being met with the current fleet of expendable boosters. So, Alternative 1 was selected as the most cost-effective option in the near term and as such provided the basis for the DOD space launch program in the FY 95 President's Budget.

[5] **3. Management**

Four major sectors coexist in the national space community: defense, intelligence, civil, and commercial. Each sector has distinct space missions and to a significant degree has developed unique cultures and practices. However, the requirement for space launch is common to all sectors. The first step in developing a modernization plan for space launch is to understand the needs and perspectives of the principal customers and suppliers of spacelift systems and services.

    a.   **Defense Sector.** The defense sector's principal objective is to have efficient and cost-effective space launch capabilities to carry out its warning, surveillance, communication, weather, and navigation missions from space. The evolving National Military Strategy places increased reliance on smaller, more mobile military forces to respond to crises and conflicts around the world. This requires highly capable space force and space launch capabilities with the operability, dependability, and responsiveness to meet operational needs. Because of the increasing costs of launch, the defense sector has generally been pursuing lighter satellites to meet future needs, resulting in a focus on medium lift capabilities.

    b.   **Intelligence Sector.** The intelligence sector provides critical information to national and military decision makers. Their payloads are generally large and expensive, so reliable, heavy lift capability is a top concern. The intelligence sector is also concerned about transition to any new launch vehicle because of experience with transitions from expendable launchers to the Space Shuttle and back to expendables after the Challenger accident. These changes required costly satellite modifications and caused long launch delays.

    c.   **Civil Sector.** Human spaceflight and the need to reduce the costs of Space Shuttle operations dominate NASA's needs. Accordingly, NASA's most important requirement is a more cost-effective reusable space launch system. For the near term, NASA plans to meet its Space Station assembly and resupply requirements with the Space Shuttle and Russian Proton and Soyuz boosters. For its scientific, communications relay, and earth observation missions, NASA will rely on a limited number of medium lift expendable boosters.

    d.   **Commercial Sector.** Today's commercial space launch requirements are dominated by geosynchronous communications satellites. Both commercial satellite builders and launch service providers want low launch service prices and dependable launch schedules, creating a natural synergy between the needs of the defense and the commercial sectors. Although commercial competitiveness characterizes the dialogue in this sector, the Government is the predominant purchaser of launch products and services, and today there are limited opportunities to significantly expand the space launch market.

    e.   **New Management Models.** Many different management schemes have been [6] proposed to deal with the new, more stringent environment. One of particular interest is a proposal to establish a quasi-public launch corporation similar to COMSAT. This corporation would be chartered by Congress to develop, operate, and sell spacelift services to U.S. public and private customers. Such a corporation would provide a national entity that operates on business principles and practices to provide space transportation. As a quasi-public entity, the corporation would deal directly with spacelift users such as NASA, the Air Force, NOAA, and commercial customers. The U.S. Government would invest in the corporation—about $3.5 billion over the first 5 to 7 years of the corporation's existence—and would include a continuing anchor tenancy agreement. While many questions remain concerning implementation, the fundamental concept appears to address many management problems that the Government has found intractable. On the other hand, discussions with a variety of industry leaders as well as those familiar with COMSAT-like activities led the Study Group to conclude that absent a major breakthrough in the commercialization of space, this very innovative approach is not required at this time, but should continue to be examined.

4. Economics

   a.   Space Economics. Roughly 6 percent of the DOD budget is spent on space, of which about 20 percent of this funding is spent on space launch—a figure roughly on the order of $2.5 billion in today's dollars. In contrast, space activities make up about 93 percent of NASA's budget, with aeronautical activities accounting for the remaining 7 percent. Launch costs account for about 31 percent of NASA's budget—about $4.3 billion in today's dollars.

   b.   Hardware Costs. Within defense, hardware costs in the medium (Delta II and Atlas II) and heavy (Titan IV) lift categories are increasing. Atlas costs have risen nearly 50 percent as new capabilities have been added; these are expected to increase again when new contracts are awarded in the late 1990s. Titan costs have been driven up almost 60 percent—approaching $325 million for a Titan IV Centaur. Inefficient production rates primarily account for the increase in Titan IV costs—Titan production was originally sized to support a production rate of 10 per year versus today's rate of 3 per year.

   c.   Failure Costs. Launch accidents are costly. The cost of expendable launch vehicles failures averages roughly $300 million per year and is growing. Failure to achieve predicted reliability and the high costs of boosters and satellites are the principal contributors. Achieving predicted reliability rates could reduce the cost of failure by half, but low launch rates make meeting these performance goals difficult.

   d.   "Niche Markets." While the overall DOD launch demand is decreasing, the division of U.S. launch capability into "niches" with limited ranges of performance—small launchers, Titan II, Delta II, Atlas II, Titan IV and Shuttle—further contributes to the low launch rates. As depicted in Figure 1 below, no single heavy or medium launcher is projected to have a production or launch rate of more than nine per year.

[7] [original placement of Figure 1]

   e.   Competitiveness. The commercial competitiveness of the U.S. fleet has eroded over time. Figure 2 below shows cost per pound of payload to geo-transfer orbit for all launch vehicles. The chart suggests that U.S. systems, in particular Atlas, are generally price competitive with Ariane IV today. However, there is some evidence, anecdotal in nature, which suggests that subsidization may permit competitors to price somewhat lower than the curve shown in Figure 2. Besides pricing, it is clear that other factors are at play

Figure 1: Launch Vehicle "Niche Markets"

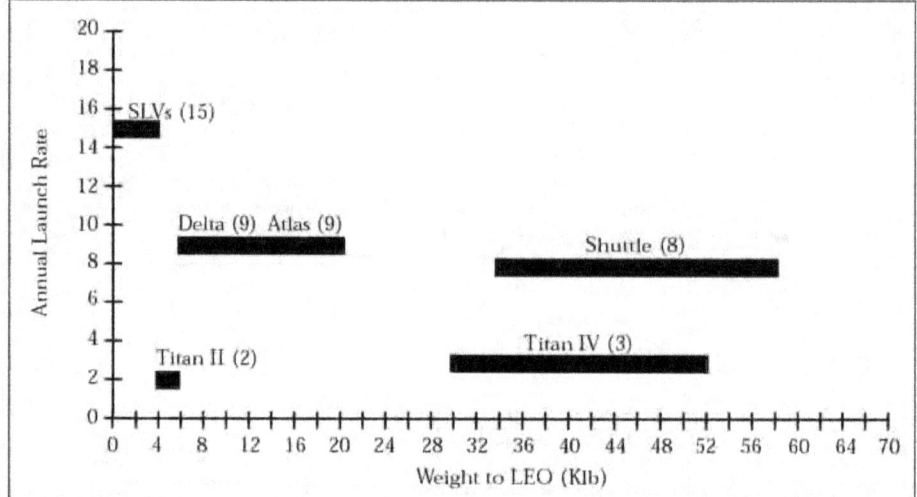

such as international politics, perceptions about U.S. launch systems reliability and sched-ule dependability, and marketing techniques that also contribute to the loss of U.S. mar-ket share. There was general consensus and concern that the U.S. will be even less price competitive with the advent of the new Ariane 5 system and the increasing use of the non-market economy launchers—China's Long March and Russia's Proton and Zenit. A rela-tively new commercial sector—the small communications satellite market—has the potential to drastically change the space launch landscape of all four sectors, but the actu-al size and viability of this new element of the commercial sector are still uncertain. A recent Department of Transportation, Office of Commercial Space Transportation (OCST) study estimated the size of this market for 1994–2005 at between 4 and 10 medi-um launches for constellation deployment and between 8 to 12 small launches for con-stellation sustainment, noting that this estimate is highly uncertain.

[8] [original placement of Figure 2]

    f. Launch Business. The medium/heavy launch market will continue to be domi-nated by Government launches for the foreseeable future. Launch demand has declined as a result of defense reductions, significantly increasing per flight costs. Future Government mission requirements will not likely increase, and the commercial launch market provides little potential for significant growth or economies. From these trends, the Study Group concluded the United States has too many space launch providers with too much production capacity.

    All launch providers are wary of committing any large corporate resources to mod-ernize their product lines and will remain cautious. These companies view the risks as high and the return on investment as low and uncertain. There are indications some pri-vate funding could be made available, given certain guarantees, investment underwriting, and/or anchor tenancy; optimistically, the total would probably be less than $1 billion. This amount would represent a significant downpayment but would not be sufficient to fund a major modernization effort.

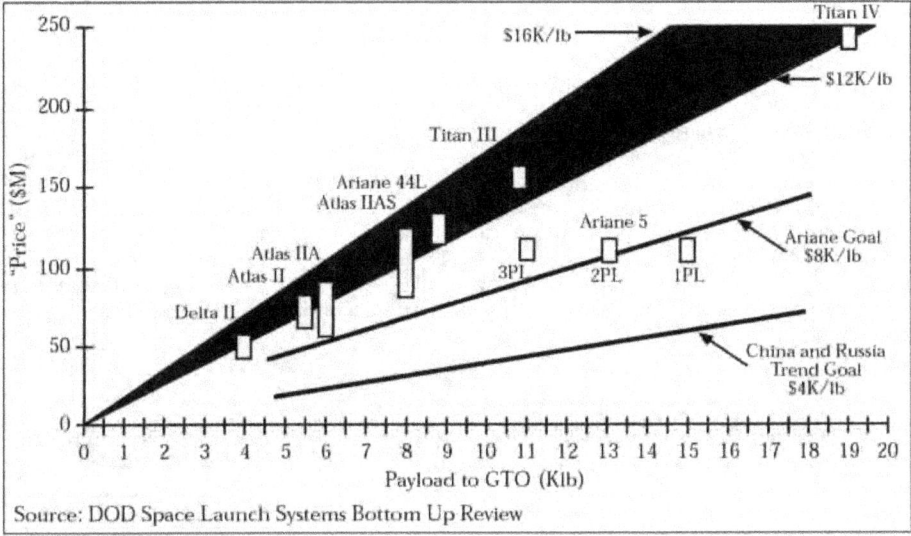

Figure 2. Cost Per Pound

Source: DOD Space Launch Systems Bottom Up Review

5. International Factors

Foreign space launch competition has grown and has become more effective. The European Space Agency (ESA) will remain the principal competitor well into the future. Bilateral agreements limit the purchase of Russian medium/heavy launch services until 2000, while trade with China is limited until 1994 (with a renewal under consideration). Beyond 2000, the Russians and Chinese can be expected to be more competitive. Japan is entering the market with the H-2 booster, but its price and launch [9] base limitations will constrain its market share.

In addition to the competitive landscape described above, the worldwide commercial launch market is influenced by other factors, such as economics and politics. For example, INTELSAT, an international consortium with close to 130 member nations, bases launcher selection primarily on cost but also considers the need to maintain competition among launch providers and the political interests of consortium members. Given the environment, analysis estimates that only 12 to 15 satellites per year are actually open for bid by all launch service providers. Consequently, it is believed that relatively little that can be done in the near term to recapture a significant portion of the market. Hence, the U.S. market share, roughly 30 percent since 1990, will not change significantly absent a modernization initiative.

While the competition for launch services is increasing, there are opportunities for increased cooperation in spacelift. For example, U.S. and Russian cooperation in space endeavors is growing. Changes in foreign policy have encouraged and resulted in significant U.S.-Russian cooperation underscored by the Space Station agreement and trade with Russia in space-related products and technology. Russia possesses highly effective space launch systems and technologies that may provide attractive alternatives to domestic systems or technologies. However, the United States must also be cautious of creating unacceptable dependencies.

6. Technology

The Nation's space launch technology investment—Defense, NASA, and industry—has dropped dramatically in the last 2 years from $570 million in FY 92 to $351 million in FY 94, a decrease of nearly 40 percent. The funding drop is due primarily to major program cancellations including the National Launch System and the Space Nuclear Thermal Propulsion programs, which exposes a weakness in our technology strategy. Dependency on major programs for the technology base provides robust funding while the program is healthy, but the efforts are eliminated as programs are canceled.

Leaving out industry investment, the combined DOD/NASA space launch technology total for FY 94 is $312 million, with much of the funding earmarked for specific developments. Only 14 percent of the total, or $45 million, supports DOD core technology efforts. Without a change in priority, funding will decline in FY 95, leaving a total of about $31 million. These funding levels are insufficient to accomplish a meaningful core space launch technology program.

7. Operations

a. Launch Delays. As a result of system design choices made years ago and the primacy of performance requirements, U.S. launch systems do not have the desired operability characteristics. Delays adversely impact cost, DOD mission performance, and throughput for defense and commercial customers. Delta is the most operable U.S. [10] expendable launch system today with average delays of 22 days. For Atlas, recent statistics show an 88-day average delay. Titan must be considered a system still in development with long on-pad processing times—the average Titan delay is 223 days. Hardware tends to dominate delay statistics, but evidence indicates a significant percentage of the delays are traceable to faulty instrumentation.

b. Manpower. U.S. launch system manufacturing and operations are manpower intensive. Current system designs fundamentally limit processing and operability improvements. U.S. manufacturing processes extend from the plant to the launch pad in increasing degrees from Delta to Titan IV. In contrast, Arianespace, with Ariane 4, has segregated manufacturing from operations. However, when assessed on an equivalent basis by labor category, the launch processing teams for Atlas and Delta are not disproportionately large and compare favorably with Ariane. In the case of Titan IV, the launch team is sized for substantially greater launch activity than is now planned. Misperceptions arise because U.S. *launch bases* are often compared with foreign *launch complexes*. A substantial amount of the activity at the U.S. ranges is not space launch related.

c. Capability. The current U.S. spacelift systems all meet their capability requirements, but often at the price of reduced operating and performance margins. Growth in payload mass typically necessitates expensive increases in space launch vehicle performance. An increase in launch rate would force expensive changes in the ground infrastructure, including launch pads, ranges, and supporting facilities. Without extensive redesign and requalification, virtually no room exists for future payload weight growth in the current fleet.

d. Reliability. Space launch vehicle reliability is inherently dependent on a number of factors including complexity, flight rate, and design stability. The Delta II has quite high reliability rates, while systems that include more stages, hardware, and flight events, such as Atlas and Titan IV, are not as reliable. Likewise, flight rate directly impacts reliability. Systems with high flight rates, such as Delta II, have had more opportunity to identify and correct problems than those with low flight rates, such as Titan IV. Flight rates are tied directly to production rates and the production learning curve and quality. Delta, in contrast to Titan IV, enjoys higher production rates, which help to increase system reliability.

   e. **Responsiveness.** None of the current launch systems were built to be responsive, either in the vehicles or in their associated support launch complex. Small launch vehicles fare best by the very nature of their size. As system size and complexity increase, system responsiveness decreases. One measure is the flight rate for each system. On the Eastern Range, Delta II can launch up to 12 missions per year, if needed. Atlas is limited to eight per year. On the low end, Titan IV can launch four missions per year. Shuttle can launch up to eight missions per year, but at high cost and labor intensive operations. Of the current medium and heavy fleet, the only system with a true launch-on-need (LON) capability is Delta II.

[11]   **D. Requirements**

   There are widely divergent views within the space community on how to define and characterize spacelift requirements. Traditionally, definition has focused on mission models and fundamental performance parameters. Early on, the Study Group concluded that a new method was needed to investigate requirements. Spacelift system requirements were analyzed using a Quality Function Deployment (QFD) process to define, develop, and rank system requirements. This methodology allowed participants of all four space sectors to develop a preliminary set of requirements that represent the "wants" of all the sectors.

   Five top-level requirements were developed—capability, operability, economics, mission success, and responsiveness:

- Capability describes the system's ability to provide accurate, sufficient, predictable, and repeatable performance in operation. It covers access to multiple orbits, crew transport (currently a unique NASA requirement), launch rate, launch system performance, and payload accommodation.
- Operability describes the spacelift system's ability to accomplish the spacelift mission in a timely manner and to support customer needs. It includes supportability, maintainability, operable processes and designs, availability, and schedule dependability.
- Economics describes whether the system is efficient to develop, operate and support. It addresses the entire spectrum of cost-effectiveness and competitiveness.
- Mission success describes the system's ability to satisfy spacelift requirements with a very low incidence of failure. It is characterized by system reliability, crew survival (currently a unique NASA requirement), payload survival, and effectiveness.
- Responsiveness describes the ability of the system to quickly and dependably respond to changing requirements. Responsiveness includes resiliency, ability to launch on need, and flexibility.

[12]   **E. Current System Capabilities**

   Current U.S. spacelift systems share some common characteristics. The expendable systems are all derived, to one degree or another, from ballistic missile systems. All launch systems operate at or very near their maximum performance capability. In many cases, modifications have been made to extend performance capabilities that compromise flight margins, operability, and supportability. Figure 3 summarizes the current spacelift systems in terms of the above requirements.

Figure 3: Characteristics of Current U.S. Space Launch Systems

Payload Class	Spaceflight System	Capability (Performance; Launch Rate)	Operability	Economics	Mission Success for Current Configuration	Responsiveness
Small	Pegasus XL	Less than 1,000 lb to LEO (east or polar): 4 per year	Modern, operable design; maintainable; routine operations; contractor logistics support	$14 million per flight; only flight-proven commercial SLV; very reproducible	1.0 mission success rate	2–4 month call up; standard interface
Medium	Titan II	4,200 lb to LEO polar: 3 per year	Refurbished ICBM—no enhancements; contractor logistics support	$35 million per flight; hand-refurbished from ICBM	0.75 mission success rate; 1.0 launch success rate	90 day call up for [Defense Meteorological Satellite Program]; 66 days on pad
	Delta II	4,010 lb to GTO: 12 per year	Most dependable ELV; some [Air Force] logistics support	$40 million per flight; modern production line	1.0 mission success rate	98 day call up; 28 days on pad
	Atlas I, II, IIA, IIAS	4,970 lb to 8,450 lb to GTO; 8 per year	Contractor logistics support	$90 million per flight; modern production line	0.863 mission success rate for Atlas-Centaur system	No call up [next words illegible] days time on [next words illegible]
Heavy	Titan IV	Up to 10,000 lb to GEO, 49,000 lb to LEO; 4–5 per year (both coasts)	Contractor logistics support; not designed for operability	$250 million to $325 million per flight; very low production rates (3 per year)	0.857 mission success rate; still in development	180+ day call up; 110 days on pad
	Shuttle	Up to 53,500 lb to LEO; crewed; 8 per year	Contractor logistics support; some operability features	$375 million per flight at 8 per year	0.982 mission success rate (ops flights only)	12–33 month call up; 21 days on pad

[13]   **F. Centers of Gravity**

Of the many metrics that could be used to measure improvement in space launch, the Study Group identified five key leverage areas or "centers of gravity." Centers of gravity describe points or elements which when pushed on provide the highest leverage in achieving desired goals. These centers can be mutually independent or highly interdependent and can change in value over time. The centers of gravity for spacelift and the results of improvements in each center are as follows:

- *Production and launch rate and stability*—Reduce the high costs of launch; maintain production, processing, and operations continuity; and improve the ability to meet reliability goals.
- *Reliability*—Control the high costs of failure and thereby improve the availability of resources for investment.
- *Technology availability*—Provide a foundation for force modernization at reasonable cost, schedule, and technical risk.

- *Space launch management*—Achieve and maintain consensus, move from available technologies to fielded capability, and reverse technological and industrial drift and atrophy.
- *Funding commitment*—Move beyond the austere upgrades to current systems that limit the U.S. ability to perform its mission and compete effectively in the international marketplace.

The recommendations of the Study have been assessed using these centers of gravity to ensure that they work these high-leverage areas.

[14]   **G. Options**

The Study Group developed four options for modernizing U.S. spare launch capabilities:

- Option 1: Sustain existing launch systems
- Option 2: Evolve current expendable launch system
- Option 3: Develop a new expendable launch system
- Option 4: Develop a new reusable launch system

Collectively, they represent program "building blocks" from which separate roadmaps were developed. The options generally correlate with those in the *DOD Bottom-Up Review* and NASA's *Access to Space Study.* The individual options describe a range of approaches and costs, not point designs. They were based upon compilations of contractor or system program office estimates plus a management factor applied by the Study Group. Costs are presented to provide relative comparisons between options. In addition to developing the program options, the Study Group defined an enhanced core technology program and examined continued space launch infrastructure sustainment and modernization.

1. Core Technology

A key element of any program for space launch modernization is the "core" space launch technology investment. Currently, DOD core space launch technology is funded at roughly $45 million per year. A time-phased increase from that level to $120 million per year would allow DOD to pursue a coherent strategy for space launch technology development to support a wide range of future launch system and program options. This strategy should begin with an appropriate distribution of the FY 94 Advanced Research Projects Agency (ARPA) funding consistent with congressional direction. Areas for increased technology investment are shown in Figure 4.

2. Sustainment

Spacelift system sustainment covers the launch bases, space launch complexes (SLCs), and the ranges. The majority of sustainment is funded by the Air Force through the Space Launch Infrastructure Investment Plan (SLIIP), an investment strategy that includes both critical upgrades to SLCs and the Range Standardization and Automation (RSA) program. The Air Force's commitment to improving the infrastructure is commendable. SLC sustainment under the SLIIP addresses critical upgrades to launch pads and their associated complexes. When RSA is completed in 2003, it will have brought the ranges' 1950s equipment and methodologies up to the state of the art. Current range equipment and facilities must be sustained until the benefits of RSA are fully realized.

[15]

Figure 4: DOD Core Space Launch Technology

	Propulsion	Vehicle	Operations
Expendable Unique	Low Cost Engine Storable Propellants Clean Solid Propellants Hybrid Propulsion	Low Cost Booster	
Common	Upper Stage Propulsion Russian Engine Test Simple Pumps Chambers/Injectors Test Beds High Energy Fuels	Adaptive [Guidance, Navigation, and Control] [Aluminum-Lithium] Structures Composites Low Cost Mfg Man Tech	Automated Processes Health Management Non Destructive Inspection Leak Free Joints Fault Isolation
Reusable Unique	Linear Aerospike Advanced Propulsion Preburner Turbopumps Tripropellants	Primary Structure Insulation Reliable Sensors Cryo Tanks Aerothermo	Recovery/Refurbishment
Total FYDP Unfunded Core Technology Investment $384M (CY94$)			

94	95	96	97	98	99
$0M	$45M	$89M	$86M	$83M	$81M

3. Transition Windows

Transition costs for new launch systems include those for concurrent operation and maintenance of old and new boosters, infrastructure, and personnel until all payloads are being launched on the new system(s). One way to minimize this cost is to ensure new launch systems are available in time to influence designs for new satellites or planned satellite block changes. Each of the options has been structured to make maximum use of program phasing such that new launch systems are introduced in convenient transition windows.

- Medium lift: 2003–2005
- Heavy lift: 2005–2007, 2009, 2011–2013
- Space Shuttle: 2006–2010.

4. Option Descriptions

a. **Option 1: Sustain Existing Launch Systems.** Option 1 maintains the current fleet of launch systems—Delta, Atlas, Titan, and the Space Shuttle—for the foreseeable future. Funding, based on the FY 95 President's Budget, includes only "austere" upgrades to enable missions, improve reliability and safety, or to address obsolescence.

NASA plans to continue Space Shuttle operations through the early part of the next decade and to continue to use existing ELVs for science missions. The NASA budget funds a focused technology program for reusable launch vehicles accomplished in [16] cooperation with planned DOD technology investments. Tentative plans include conducting flight demonstrations prior to the turn of the century. Such demonstrations could support a Space Shuttle replacement decision in 1999–2000 with credible cost and engineering data. At that point, NASA will either recommend a new start for a Space Shuttle replacement or will program additional safety and reliability upgrades to the existing Shuttle system and procure an additional orbiter.

The FY 95 President's Budget includes money for a competition for a medium class launch vehicle (MLV IV) in FY 96 to support operational Air Force launches. The Request for Proposals (RFP) for MLV IV may contain provisions for support to new DOD on-orbit capabilities: the ALARM early warning satellite and advanced EHF [extremely high-frequency] satellites.

Market-driven industry downsizing may reduce operating costs from current levels. Under Option 1, per flight costs are anticipated to be as follows. The range in costs are due to differences in booster type and configuration (w/ or w/o an upper stage).

- Medium lift: $50-$125 million per flight
- Heavy lift: $250-$320 million per flight
- Space Shuttle: $375 million per flight.

b.   Option 2: Evolve Current Expendable Launch Systems. Key features of Option 2 include flying out current launch vehicles already on contract, evolving a family of launch vehicles from current systems by consolidating medium and heavy lift booster families, and fielding the evolved vehicles to meet payload transition windows. This option would cost between $1.0 billion and $2.5 billion in CY 94 dollars, but would significantly lower operations costs by increasing production rates. Private financing may be available for this option with suitable Government guarantees, such as anchor tenancy or low-interest loans.

As in Option 1, NASA will continue Shuttle operations through the early part of the next decade, continue to use existing ELVs for science missions, and fund a reusable technology program with coordinated DOD investments.

Option 2's acquisition approach includes a competitive procurement with the Request for Proposals (RFP) structured to allow bidders to propose against various sets of payload weight and orbit requirements, launch rates, and operations concepts. Key RFP elements should include firm cost targets, performance-based Government specifications, and strong incentive structures. Recurring costs for this option are estimated at

- Medium lift: $50-$80 million per flight
- Heavy lift: $100-$150 million per flight
[17]• Space Shuttle: $375 million per flight.

c.   Option 3: Develop a New Expendable Launch System. Option 3 would correct deficiencies in current expendable launchers by developing an entirely new launch vehicle family with significantly improved reliability, operability, and cost. This "clean sheet of paper" approach for a new expendable system would use a modular family composed of a common core vehicle and/or common major subsystems—strap-on stages, upper stage(s), payload fairings, and processing and launch facilities. There are two major paths a new expendable system development could follow: (a) replace only the current expendable systems, or (b) replace current ELVs and the Space Shuttle. Replacing the Space Shuttle would require significant additional investment for crew rating enhancements and personnel and cargo transport systems development.

The nonrecurring development cost for the basic new expendable vehicle is estimated to be in the $5 billion to $8 billion range. The crew-rated launcher and associated personnel/cargo vehicles would require an additional $5-$6 billion to develop. The recurring flight costs are estimated to be

- Medium lift: $40–$75 million per flight
- Heavy lift: $80–$140 million per flight
- Personnel launch: $90–$190 million per flight
- Cargo transport: $130–$230 million per flight.

d. Option 4: Develop a New Reusable Launch System. Option 4 would develop a fully reusable space launch system with the objective of substantially reducing flight costs while improving operability and responsiveness. Since a fully reusable system requires significant advances in technology and substantial engineering development, this option is based on a phased development.

The overall approach for Option 4 is to undertake a focused technology development and demonstration effort, followed by a decision as to whether to proceed with development of a prototype system and production of a fleet of operational vehicles. A parallel technology development and flight demonstrator program would be conducted to define technology and engineering feasibility and risks before committing to full-scale system development.

Because of the wide range of technologies, designs, and operating concepts among the various reusable concepts, the cost estimates for a new reusable launch system span a broad range. The technology development and demonstration would require $0.6 billion to $0.9 billion. The cost for engineering development ranges from $6 billion to $20+ billion. This wide range captures the most innovative industry approaches on one end and NASA's estimate from Option 3 of the *Access to Space Study* on the other end. The cost for procuring a four-vehicle fleet ranges from $2.5 billion to $10.5 [18] billion spent beginning in the year 2004 and continuing through 2009. Although the nonrecurring development and procurement investment is relatively high, the annual operational cost of the fleet is estimated to be in the $0.5 billion to $1.5 billion range, compared with today's annual Space Shuttle and expendable launch costs of over $6 billion. . . .

[23] **I. Findings and Recommendations**

The study developed 15 findings and recommendations divided into four groups:

- Fundamental drivers of the space launch industry
- Critical drivers of cost, capability, or operations
- Special focus areas
- Current operations enhancement areas.

1. Fundamental Drivers of the Space Launch Industry

Finding #1: Excess production and processing capacity exist within the space launch industry.

The space launch industry grew up in times of increasing budgets, strong national interest, increasing requirements, and a technology base that produced many satellites with limited lifetimes. The result was a high launch rate and a robust space launch industry. Today, we do more missions with fewer satellites, and the on-orbit lifetimes are very long. The net result is that the launch rate has decreased markedly, yet the industry still has multiple providers with several families of launch vehicles and a capacity to produce more than is needed. Different elements of the industry have developed niches of capability, each of which operates at low, inefficient rates, and none of which remain cost-effective.

Recommendation #1: A major objective of future modernization efforts should be to reduce indus-
trial overhead through downsizing and reduction of niche markets.

Finding #2: Industry is unwilling to fund major space launch modernization alone, but pri-
vate "up front" investment may be available given United States Government guarantees.

Because of high costs and decreasing demand, the space launch industry has little
incentive to make the significant capital investment necessary to modernize its product
lines. Several innovative funding concepts exist, some of which may require special legislation, that could enable the Government to become a partner with industry to encourage modernization, such as off-budget financing (e.g., loan guarantees, tax incentives,
government indemnification), and anchor tenancy (guaranteed minimum launch rates
and prices). Such guarantees would also encourage private investment to levels perhaps
as high as $1 billion.

Recommendation #2: DOD should pursue innovative incentives to encourage private and indus-
trial investment in space launch modernization.

[24] **Finding #3: Driven by user (DOD and National) requirements and current booster**
and spacecraft technology, heavy lift is required for the foreseeable future.

Any restructure of the space launch industry will require a solid understanding of the
range of lift capability required. The number of launches of the Titan IV, today's [heavy-
lift vehicle], has decreased substantially. Therefore, it has been suggested that the Nation
could move all satellites to either medium launch vehicles or to the Space Shuttle, eliminating the need for a heavy lift vehicle. The Study Group examined in detail the user
requirements that drive heavy lift and the technology potential for heavy satellite downsizing to MLV class payloads. These heavy lift requirements are principally intelligence
related, including but not limited to military operational and science and technology
(S&T) intelligence requirements. Intelligence needs and technology limit the potential to
downsize intelligence satellites, and it is unlikely that any known technologies could
enable similar mission success at MLV weights and sizes in the near term.

Recommendation #3A: In the near term, DOD must continue and improve heavy lift capability.

Recommendation #3B: In the longer term, DOD should review and revalidate its intelligence
requirements (both operational and S&T) that drive heavy lift. The NRO should continue to exam-
ine advanced spacecraft technologies that could provide major reductions in payload size and
weight.

Finding #4: Opportunities for payload-booster transition are currently not fully coordi-
nated to maximize the cost-benefit to the Government.

The introduction of new space launch capabilities must be timed properly to realize
cost-effective transitions of spacecraft to the new capabilities. Redesigning satellites to fly
on new boosters is extremely costly, delays the satellite program, and often does not
improve satellite capability. The movement of payloads onto and then off the Space
Shuttle is the case in point, where the payload transition costs were extraordinary. Based
upon current plans for future new starts and/or block changes to satellite systems, windows of opportunity for transition of satellites to new launch vehicles occur for heavy lift
in the years 2005–2007, 2009 and 2011–2012; for medium lift in 2003–2005; and for the
Shuttle in 2006–2010. Any major changes in the industry structure should be timed such

that the initial launch capability (ILC) of new spacelift systems occurs at the satellites' transition points.

Recommendation #4: If a new or evolved space launch system is pursued, the ILC should be planned to coincide with anticipated payload block changes and/or new starts.

[25] 2. **Critical Drivers of Cost, Capability, or Operations**

Finding #5: Increased cost of failure demands [that] greater emphasis be placed [on] improving reliability.

The cost of the vehicles (booster and spacecraft) destroyed in the August 1993 Titan IV failure exceeded $1 billion. Over the past 10 years, the average yearly cost of launch failures has exceeded $300 million and is rising. Such failures directly and substantially impact on-orbit mission capability. Additionally, however, post-accident standdowns for failure resolution create lost opportunity costs that are often hard to quantify. The Nation's fleet of launch vehicles is not as reliable as it should be. As the Nation moved onto the Shuttle, ELV launch rates dropped, production lines slowed, and engineering expertise eroded. Another contributing factor is the lack of sufficient fault tree and failure mode analysis, process control, and instrumentation in the launch system and infrastructure.

Recommendation #5: Support and sustain funding for launch system and infrastructure reliability improvements.

Finding #6: Operations costs per launch for Titan IV are significant and rising.

Although there have been eight Titan IV launches to date, it has not yet reached its full operational capability (FOC) and must be classified as in the development phase. Thus, operation of the system requires more time and people than for a mature system. In 1989, operations cost per launch was $34 million (CY 94 $); by 1994 it increased to $54 million; and by 1999 it is projected to be $72 million. As the launch bases conduct further Titan IV launches and the system approaches FOC, the on-pad time should shrink, and the number of personnel, particularly those involved in Titan [research, development, test, and evaluation], should diminish. If the number of Titan IV launches per year remains very small, it would be appropriate to consider closing or putting into a backup mode one of the East Coast Titan IV launch pads.

Recommendation #6: Aggressively restructure and streamline Titan launch base operations to reduce current and future operations costs.

Finding #7: A cross-sector process to collect, coordinate, and consolidate space launch requirements does not exist.

The most fundamental driver of space launch capability is the set of space launch requirements, yet there are widely differing views and definitions of these throughout the four space sectors. No forum or mechanism has been available to coordinate [26] intersector launch requirements, which has hampered the Executive Branch's ability to articulate needs and sustain support for spacelift modernization. A cross-sector process that balances performance, sustainability, reliability, and cost-effectiveness, such as the Quality Function Deployment used in this study, would greatly facilitate a national consensus on where this country should go in space launch. The results of the QFD process performed during the Study form the basis for follow-on work in this area.

Recommendation #7: Institutionalize a process to gain and sustain community agreement on requirements and associated metrics.

Finding #8: The DOD core space launch technology program is significantly underfunded and externally constrained, which has hindered opportunities for space launch modernization.

Future capability depends on the availability of technology, but space launch technology has suffered in terms of quality and quantity such that current modernization options are limited. Much of the technology work has been accomplished in major programs (ALS, NLS) that no longer exist. Other work is specifically directed such that it cannot be refocused on the most pressing technology issues. Overall space launch technology funding has decreased, and the amount available for core technology, such as engines and structures, is a small fraction of the total. While the emphasis in launch technology has traditionally been on performance, in the future, greater focus on technology to decrease cost is needed. Core technology needs to be increased in the near term; FY 94 ARPA funding should be used to enhance the core DOD launch technology program, consistent with congressional guidance. This includes completion of the Delta Clipper-Experimental (DC-X), investigation of Russian engine technology, and initial work on reusable launch system "long pole" technology and demonstrations, and low cost expendable boosters.

Recommendation #8: Increase funding for a core space launch technology program as an enabler for future investment.

Finding #9: Air Force launch base operations are constrained by antiquated and unsupportable ground systems and facilities.

A critical limit in launch operations is the ground equipment at the launch bases, particularly at the Eastern and Western Ranges, much of which is antiquated and unsupportable. Some range systems average three failures per mission. On 16 Delta missions between February 1992 and September 1993, Eastern Range equipment problems caused 22 delays. In light of those deficiencies, the Air Force has instituted and funded the Range Standardization and Automation (RSA) and launch base infrastructure improvement programs. The RSA program has been a very successful program to date; it requires continued advocacy and support.

[27] **Recommendation #9: Continue funding RSA and launch base infrastructure improvements.**

3.   **Special Focus Areas**

Finding #10: A detailed understanding of Russian engine technology can potentially lead to reduced cost for modernization.

The end of the Cold War and the demise of the Soviet Union create some significant opportunities for cooperation on space launch. Specifically, Russian rocket engines demonstrate high performance, robust margins, and proven ruggedness. Cooperation with Russia has foreign policy benefits; however, at the same time, reliance on Russian engine technology has potential national security implications from a dependency point of view. The prime Russian candidate for cooperation in this area is the RD-170 engine, which the Air Force, in cooperation with NASA and industry, should procure and test. RD-170 testing will give the U.S. Government and rocket engine industry significant insight into alternative design approaches and technical solutions that have apparently enhanced Russian rocket engine performance and durability. Similarly, NASA, with DOD and industry participation, may choose to investigate the use of Russian engine technology applicable to future reusable vehicles.

Recommendation #10: DOD should lead and fund a cooperative effort, with NASA and industry, to investigate the use of Russian engines and engine technology in future ELVs.

Finding #11: There exists general consensus on the potential benefits of a new reusable system; however, there are widely divergent views on timing, approach, cost, and risk.

A fully reusable launch system is an intriguing concept to all the space sectors and industry alike. It offers the potential benefits of responsiveness, reliability, operability, and very low cost per flight, which are universally agreed to be desirable. However, the feasibility of achieving those benefits is uncertain. Based on its needs to continue human spaceflight and provide options to replace the Shuttle, NASA should be assigned the lead for reusables with DOD maintaining a cooperative reusable program. On the other hand, DOD should lead in the ELV arena. Each agency should manage and fund efforts within their respective areas of responsibility. To prove the concept, sustain support, and enable lower risk entry into system development, the reusable technology program should include flight demonstrations.

Recommendation #11: Pursue a cooperative NASA/DOD technology maturation effort that includes experimental flight demonstrations.

[28] **Finding #12: DOD and NASA space launch program coordination needs to be improved.**

While the civil and defense space programs are clearly separate and distinct, space launch is an area of common interest and interdependence that needs interagency coordination. In particular, organizational roles in launch vehicle technology need to be defined and coordinated to avoid confusion and overlap. The Aeronautics and Astronautics Coordination Board (AACB) has been used in the past for high-level DOD/NASA coordination, but in recent years the Board has been used infrequently. In addition to improved DOD/NASA oversight, coordination with other Executive departments is likewise important.

Recommendation #12A: Assign DOD the lead role in expendable launch vehicles and NASA the lead in reusables.

Recommendation #12B: Maintain top-level DOD/NASA oversight and coordination through a mechanism such as the AACB.

Finding #13: The small launch vehicle market is uncertain but could be a major growth area—the key is development of distributed communications and surveillance systems.

An exciting but uncertain trend in the space program is toward small satellites in distributed architectures. Emerging distributed low-earth-orbit constellation concepts for communications and the Brilliant Eyes concept for surveillance in DOD could revolutionize space missions and create a large, new, and different market for small launch vehicles. However, these concepts are not yet proven.

The Government is clearly making progress with its support of the U.S. commercial launch industry and should continue to look for further improvements that would result in enhanced opportunities for commercial launch suppliers, such as improved access to launch facilities and user friendly range services. However, the Study concluded the Government should let commercial market forces function rather than taking a lead role at this time.

Recommendation #13: DOD should continue to monitor development of the small launch vehicle market but not take an active leading role.

4. Current Operations Enhancement Areas

Finding #14: Substantial data on DOD launch operations exist; however, the [29] information is difficult to access and use effectively.

The Air Force routinely collects, maintains, and analyzes operations and maintenance data on its aircraft systems to properly operate and manage its air operations. Similarly, a substantial amount of data is collected and maintained on launch vehicles, equipment, facilities, operations, and processes. This information, however, is scattered, poorly organized, and inconsistently collected and analyzed, which inhibits its use, raises costs, and often results in duplication. Systematic data collection and formatting would allow easier analysis and interpretation of the information to support operations and sustainment decisions.

Recommendation #14: Establish a standardized program for metrics, data collection, and supporting analysis.

Finding #15: There is a lack of standardization within Air Force space launch systems and operations.

Standardization at the launch bases is lacking in areas beyond just data. The launch systems and operations themselves are different at Cape Canaveral Air Station than at Vandenberg Air Force Base. Each launch base developed its own procedures when launch was under R&D management. Notwithstanding the transfer of the launch bases to an operational command, the unique systems and operations remain. Air Force Space Command launch wings, system program offices, and NASA should work together to define and implement a common set of standards.

Recommendation #15: Develop a standard set of procedures, systems, interfaces, processes, and infrastructure across all the launch bases.

[30]   J. Concluding Remarks

While this study makes no recommendation for any one specific program option, or roadmap, 15 recommendations are offered that have common themes—how do we get the maximum payoff for our limited dollars, and how do we create options for the future? These recommendations focus on cheaper approaches, such as using foreign technology; on innovative funding where Government and industry share the risks and rewards; and on preserving future options by investing in enabling technology.

Although the Study Group members received widely differing views and recommendations on launch needs, technologies, programs, and management, one consistent theme pervaded the study. *Space launch is the key enabling capability for the Nation to exploit and explore space.* Serious deficiencies in space launch, if left uncorrected, will have profound impacts on the Nation's future space program. While resources to correct these problems will be limited, a long-term commitment to improve cost and operational effectiveness is essential. Whatever path is chosen must be done as part of a coordinated, time phased, integrated, long term plan. The consensus begun in this study can and should be used to foster Administration and congressional support. The Nation can accept the status quo or choose to establish a future vision and begin to take steps, however bold or measured, towards a more robust and capable space launch future. The choice remains open.

Document IV-16

Document title: The White House, Office of Science and Technology Policy, "Fact Sheet—National Space Transportation Policy," August 5, 1994.

Source: White House Press Office, Washington, D.C.

When it entered office in January 1993, the administration of President Bill Clinton abolished the National Space Council and assigned space policy responsibilities to the Office of Science and Technology Policy in the Executive Office of the President. That office convened an interagency working group to develop a new statement on National Space Transportation Policy. This policy incorporated the recommendations of the NASA Access to Space Study and the DOD Space Launch Modernization Plan, and it provided a comprehensive set of policies to shape future U.S. space transportation activities.

[1]   THE WHITE HOUSE
                        Office of Science and Technology Policy

For Immediate Release   August 5, 1994

Fact Sheet
National Space Transportation Policy

<u>Introduction</u>

The United States space program is critical to achieving U.S. national security, scientific, technical, commercial, and foreign policy goals. Assuring reliable and affordable access to space through U.S. space transportation capabilities is a fundamental goal of the U.S. space program. In support of this goal, the U.S. Government will:

(1) Balance efforts to sustain and modernize existing space transportation capabilities with the need to invest in the development of improved future capabilities;

(2) Maintain a strong space transportation capability and technology base, including launch systems, infrastructure, and support facilities, to meet the national needs for space transport of personnel and payloads;

(3) Promote the reduction in the cost of current space transportation systems while improving their reliability, operability, responsiveness, and safety;

(4) Foster technology development and demonstration to support future decisions on the development of next generation reusable space transportation systems that greatly reduce the cost of access to space;

(5) Encourage the cost-effective use of commercially provided U.S. products and services, to the fullest extent feasible, that meet mission requirements; and

(6) Foster the international competitiveness of the U.S. commercial space transportation industry, actively considering commercial needs and factoring them into decisions on improvements in launch facilities and launch vehicles.

This policy will be implemented within the overall resource and policy guidance provided by the President.

I. Implementation Guidelines

To ensure successful implementation of this policy, U.S. Government agencies will cooperate to take advantage of the unique capabilities and resources of each agency.

This policy shall be implemented as follows:

(1) The Department of Defense (DoD) will be the lead agency for improvement and evolution of the current U.S. expendable launch vehicle (ELV) fleet, including appropriate technology development.

(2) The National Aeronautics and Space Administration (NASA) will provide for the improvement of the Space Shuttle system, focusing on reliability, safety, and cost-effectiveness.

[2] (3) The National Aeronautics and Space Administration will be the lead agency for technology development and demonstration for next generation reusable space transportation systems, such as the single-stage-to-orbit concept.

(4) The Departments of Transportation and Commerce will be responsible for identifying and promoting innovative types of arrangements between the U.S. Government and the private sector, as well as State and local governments, that may be used to implement applicable portions of this policy. U.S. Government agencies will consider, where appropriate, commitments to the private sector, such as anchor tenancy or termination liability, commensurate with the benefits of such arrangements.

(5) The Department of Defense and the National Aeronautics and Space Administration will plan for the transition between space programs and future launch systems in a manner that ensures continuity of mission capability and accommodates transition costs.

(6) The Department of Defense and the National Aeronautics and Space Administration will combine their expendable launch service requirements into single procurements when such procurements would result in cost savings or are otherwise advantageous to the Government. A Memorandum of Agreement will be developed by the Agencies to carry out this policy.

II. National Security Space Transportation Guidelines

(1) The Department of Defense will be the launch agent for the national security sector and will maintain the capability to evolve and operate those space transportation systems, infrastructure, and support activities necessary to meet national security requirements.

(2) The Department of Defense will be the lead agency for improvement and evolution of the current expendable launch vehicle fleet, including appropriate technology development. All significant ELV technology-related development associated with medium and heavy-lift ELVs will be accomplished through the DoD. In coordination with the DoD, NASA will continue to be responsible for implementing changes necessary to meet its mission-unique requirements.

(3) The objective of DoD's effort to improve and evolve current ELVs is to reduce costs while improving reliability, operability, responsiveness, and safety. Consistent with mission requirements, the DoD, in cooperation with the civil and commercial sector, should evolve satellite, payload, and launch vehicle designs to achieve the most cost-effective and affordable integrated satellite, payload, and launch vehicle combination.

    (a) ELV improvements and evolution plans will be implemented in cooperation with the Intelligence Community, the National Aeronautics and Space Administration and the Departments of Transportation and Commerce, taking into account, as appropriate, the needs of the commercial space launch sector.

    (b) The Department of Defense will maintain the Titan IV launch system until a replacement is available.

[3] (4) The Department of Defense, in cooperation with NASA, may use the Space Shuttle to meet national security needs. Launch priority will be provided for national security missions as governed by appropriate NASA/DoD agreements. Launches necessary to preserve and protect human life in space shall have the highest priority except in times of national emergency.

(5) Protection of space transportation capabilities employed for national security purposes will be pursued commensurate with their planned use in crisis and conflict and the threat. Civil and commercial space transportation capabilities identified as critical to national security may be modified at the expense of the requesting agency or department. To the maximum extent possible, these systems, when modified, should retain their normal operational utility.

III. Civil Space Transportation Guidelines

(1) The National Aeronautics and Space Administration will conduct human space flight to exploit the unique capabilities and attributes of human access to space. NASA will continue to maintain the capability to operate the Space Shuttle fleet and associated facilities.

    (a) The Space Shuttle will be used only for missions that requires human presence or other unique Shuttle capabilities, or where use of the Shuttle is determined to be important for national security, foreign policy or other compelling purposes.

    (b) The National Aeronautics and Space Administration will maintain the Space Shuttle system until a replacement is available.

    (c) As future development of a new reusable launch system is anticipated, procurement of additional Space Shuttle orbiters is not planned at this time.

(2) The National Aeronautics and Space Administration will be the lead agency for technology development and demonstration of next generation reusable space transportation systems.

    (a) The objective of NASA's technology development and demonstration effort is to support government and private sector decisions by the end of this decade on development of an operational next generation reusable launch system.

    (b) Research shall be focused on technologies to support a decision no later than December 1996 to proceed with a sub-scale flight demonstration which would prove the concept of single-stage-to-orbit.

    (c) Technology development and demonstration, including operational concepts, will be implemented in cooperation with related activities in the Department of Defense.

(d) It is envisioned that the private sector could have a significant role in managing the development and operation of a new reusable space transportation system. In anticipation of this role, NASA shall actively involve the private sector in planning and evaluating its launch technology activities.

[4] IV. Commercial Space Transportation Guidelines

(1) The United States Government is committed to encouraging a viable commercial U.S. space transportation industry.

   (a) The Departments of Transportation and Commerce will be responsible for identifying and promoting innovative types of arrangements between the U.S. Government and the private sector, as well as State and local governments, that may be used to implement applicable portions of this policy.

   (b) The Department of Transportation will license, facilitate, and promote commercial launch operations as set forth in the Commercial Space Launch Act, as amended, and Executive Order 12465. The Department of Transportation will coordinate with the Department of Commerce where appropriate.

   (c) U.S. Government agencies shall purchase commercially available U.S. space transportation products and services to the fullest extent feasible that meet mission requirements and shall not conduct activities with commercial application that preclude or deter commercial space activities, except for national security or public safety reasons.

   (d) The U.S. Government will provide for the timely transfer to the private sector of unclassified Government-developed space transportation technologies in such a manner as to protect their commercial value.

   (e) The U.S. Government will make all reasonable efforts to provide stable and predictable access to appropriate space transportation-related hardware, facilities, and services; these will be on a reimbursable basis. The U.S. Government reserves the right to use such facilities and services on a priority basis to meet national security and critical civil sector mission requirements.

   (f) U.S. Government agencies shall work with the U.S. commercial space sector to promote the establishment of technical standards for commercial space products and services.

(2) U.S. Government agencies, in acquiring space launch-related capabilities, will, to the extent feasible and consistent with mission requirements:

   (a) Involve the private sector in the design and development of space transportation capabilities and encourage private sector financing, as appropriate.

   (b) Emphasize procurement strategies that are based on the use of commercial U.S. space transportation products and services.

   (c) Provide for private sector retention of technical data rights, limited only to the extent necessary to meet government needs.

   (d) Encourage private sector and State and local government investment and participation in the development and improvement of U.S. launch systems and infrastructure.

[5] V. Trade in Commercial Space Launch Service

(1) A long term goal of the United States is to achieve free and fair trade. In pursuit of this goal, the U.S. Government will seek to negotiate and implement agreements with other nations that define principles of free and fair trade for commercial space launch services, limit certain government supports and unfair practices in the international market, and establish criteria regarding participation by

space launch industries in countries in transition from a non-market to a market economy.

    (a) International space launch trade agreements in which the U.S. is a party must allow for effective means of enforcement. The range of options available to the U.S. must be sufficient to deter and, if necessary, respond to non-compliance and provide effective relief to the U.S. commercial space launch industry. Agreements must not constrain the ability of the United States to take any action consistent with U.S. laws and regulations.

    (b) International space launch trade agreements in which the U.S. is party must be in conformity with U.S. obligations under arms control agreements, U.S. nonproliferation policies, U.S. technology transfer policies, and U.S. policies regarding observance of the Guidelines and Annex of the Missile Technology Control Regime (MTCR).

VI. Use of Foreign Launch Vehicles, Components, and Technologies

(1) For the foreseeable future, the United States Government payloads will be launched on space launch vehicles manufactured in the United States, unless exempted by the President or his designated representative.

    (a) This policy does not apply to use of foreign launch vehicles on a no-exchange-of-funds basis to support the following: flight of scientific instruments on foreign spacecraft, international scientific programs, or other cooperative government-to-government programs. Such use will be subject to interagency coordination procedures.

(2) The U.S. Government will seek to take advantage of foreign components or technologies in upgrading U.S. space transportation systems or developing next generation space transportation systems. Such activities will be consistent with U.S. nonproliferation, national security, and foreign policy goals and commitments as well as the commercial sector guidelines contained in this policy. They will also be conducted in a manner consistent with U.S. obligations under the MTCR and with due consideration given to dependence on foreign sources and national security.

VII. Use of U.S. Excess Ballistic Missile Assets

(1) U.S. excess ballistic missile assets that will be eliminated under the START [Strategic Arms Reduction Treaty] agreements shall either be retained for government use or be destroyed. These assets may be used within the U.S. Government in accordance with established DoD procedures, for any purpose except to launch payloads into orbit. Requests from the Department of Defense or from other U.S. Government agencies to use these assets for launching payloads into orbit will be considered by DoD on a case-by-case basis and require approval by the Secretary of Defense. [6] Mindful of the policy's guidance that U.S. Government agencies shall purchase commercially available U.S. space transportation products and services to the fullest extent feasible, use of excess ballistic missile assets may be permitted for launching payloads into orbit when the following condition are met:

    (a) The payload supports the sponsoring agency's mission.

    (b) The use of excess ballistic missile assets is consistent with international obligations, including the MTCR guidelines and the START agreements.

    (c) The sponsoring agency must certify the use of excess ballistic missile assets results in a cost savings to the U.S. Government relative to the use of available commercial launch services that would also meet mission requirements, including performance, schedule, and risk.

VIII. Implementing Actions

(1) Within 90 days of approval of this directive, United States Government agencies are directed to prepare the following for submission to the Assistant to the President for Science and Technology and the Assistant to the President for National Security Affairs:

(a) The Secretaries of Defense, Commerce, Transportation, and the Administrator of the National Aeronautics and Space Administration, with appropriate input from the Director of Central Intelligence, will provide a report that will include a common set of requirements and a coordinated technology plan that addresses the needs of the national security, civilian, and commercial space launch sectors.

(b) The Secretary of Defense, with the support of other agencies as required, will provide an implementation plan that includes schedule and funding for improvement and evolution of the current U.S. ELV fleet.

(c) The Administrator of the National Aeronautics and Space Administration, with the support of other agencies as required, will provide an implementation plan that includes schedule and funding for improvements of the Space Shuttle system and technology development and demonstration for next generation reusable space transportation systems.

(d) The Secretaries of Transportation and Commerce, with the support of other agencies as required and U.S. industry, will provide an implementation plan that will focus on measures to foster an internationally competitive U.S. launch capability. In addition, the Secretaries will provide recommendations to the Department of Defense and the National Aeronautics and Space Administration that promote the full involvement of the commercial sector in the NASA and DoD plans.

Document IV-17

Document title: NASA, "A Draft Cooperative Agreement Notice—X-33 Phase II: Design and Demonstration," December 14, 1995, pp. A-2–A-4.

Source: NASA Historical Reference Collection, NASA History Office, NASA Headquarters, Washington, D.C.

By the end of 1995, NASA was ready to move forward with the X-33 technology demonstration phase of its Reusable Launch Vehicle (RLV) program. Through this Cooperative Agreement Notice (CAN), the agency invited U.S. industry to participate in the program. The use of a CAN was innovative; prior launch vehicle development programs had always been carried out under contract to the government, rather than as a government-industry cooperative undertaking. This CAN was also one of the first government requests for a proposal for a major undertaking that asked for submissions on CD-ROMs rather than in paper format. Lockheed-Martin, Rockwell International, and McDonnell Douglas responded to the CAN; ultimately, Lockheed-Martin was selected as the industrial partner for the X-33 program. What follows is just the program/technical description that was Appendix A.

December 14, 1995

National Aeronautics and
Space Administration

A Draft
Cooperative Agreement Notice

X-33 Phase II: Design and Demonstration . . .

[A-2]   APPENDIX A

PROGRAM/TECHNICAL DESCRIPTION

Section 1 - Program/Technical Requirements

1.0 Introduction

This program will implement the National Space Transportation Policy, specifically Section III, paragraph 2(a): "The objective of NASA's technology development and demonstration effort is to support government and private sector decisions by the end of this decade on development of an operational next-generation reusable launch system." The objective of this NASA Cooperative Agreement Notice (CAN) is to initiate the final design, construction, flight and ground test of an advanced technology demonstrator vehicle, Experimental-Thirty Three (X-33) as a part of the RLV Technology Program. The X-33 must adequately demonstrate the key design and operational aspects of a Single Stage to Orbit (SSTO) RLV rocket system so as to reduce the risk to the private sector in developing such a commercially viable system.

In order to meet its objectives, the X-33 program must be a very aggressive, focused launch technology development program. It has extremely demanding technical goals and equally demanding business goals. Through ground test and flight test, the program will demonstrate improved operability, safe abort, reusability, and affordability. Technical objectives include improved mass fraction for vehicle structures, and improved thrust to weight for rocket propulsion systems.

The implementation phases of the X-33 are structured as follows:

1.1 X-33 Design/Demonstration Phase (Phase II)

Phase II shall develop necessary data to support an informed program continuation decision at the completion of the Phase. This phase will consist of the final design, fabrication, assembly and test of the X-33 system. The X-33 vehicle will be flight tested using an incremental expansion of the flight envelope to demonstrate "aircraft like" operations. Flight testing will be accomplished at an appropriate test range. Phase II will be completed on or before the end of the decade. This CAN is a solicitation for the X-33 Design/Demonstration Phase (Phase II).

1.2 Commercial RLV Development/Operation Phase (Phase III)

The previous phase is focused towards demonstrating the technology to build reusable launch vehicles with aircraft-like operations. If fully successful, it will enable a low risk, low cost development of a commercially operated RLV system. This final phase will design, manufacture, and operate the RLV system.

[A-3] 1.3 Phase II Scope

The program shall develop necessary data to support a Phase III decision at the end of the decade. The X-33 will demonstrate the critical technologies needed for orbital SSTO rockets in realistic operational environments. To the extent practical, the X-33 will be tested in the ascent and reentry flight environments of a full-scale SSTO rocket. In addition, the X-33 will focus on those operational issues which are critical to the development of reliable low cost reusable space transportation. The X-33 will incorporate more advanced materials with weights and margins traceable and scaleable to those required by an SSTO rocket. The X-33 ground support and flight control systems will be designed to accomplish operations and supportability goals which are key to lower cost system operations. The operability and performance demonstrated by the X-33 will provide the necessary data to establish the detailed requirements for a future operational SSTO. Key technologies required for low cost space access that are not integrated onto the X-33 flight vehicle will also be demonstrated, e.g., ground based technology demonstrations.

1.4 Phase II Technical Objectives

- Technology demonstrations (flight and ground) must be implemented to reduce the business and technical risks which will enable privately financed development and operation of a next generation space transportation system.
- The X-33 flight system, subsystems, and major components shall be designed and tested (in flight and ground) so as to ensure their traceability (technology and general design similarity) and scaleability (directly scaleable weights, margins, loads, design, fabrication methods, and testing approaches) to a full scale SSTO rocket system. At a minimum, key demonstrations should include: structural mass fraction and main engine thrust to weight.
- The X-33 system must demonstrate key "aircraft like" operational attributes required for a cost effective SSTO rocket system. At a minimum, key demonstrations should include: operability (e.g., increased TPS robustness, weather, etc.), reusability, affordability, and safe abort.
- The X-33 system must begin flight testing by March 1, 1999.
- Program must meet X-33-related criteria as specified in Appendix H.

Section 2 - Resources

2.0 Resource Sharing

Significant cost sharing by industry is anticipated during Phase II. For cost sharing purposes, the Government's share is defined as that amount to be funded under the Cooperative Agreement. Industry cost sharing may include cash (profit based or venture capital), Independent Research and Development (IR&D) funds to be expended in performance of the Cooperative Agreement, and non-cash contributions. Industry non-cash contributions are governed by Office of Management and Budget (OMB) Circular A-110, Section 23, entitled "Cost Sharing or [A-4] Matching." Industry's cost share shall not be charged to the Government under the Phase II Cooperative Agreement or any other contract, grant, or Cooperative Agreement, except for allocation as an indirect cost as part of an IR&D program. However, offerors shall not count IR&D funds already allocated as cost sharing to existing or previous Cooperative Agreement efforts. Offerors' proposed cost sharing shall begin upon award.

2.1 Government Budget Information

Any award is subject to the availability of funds. The following funding information is provided as a guide to the potential level of funding available.

Expected real year funding in millions of dollars, by government fiscal year (FY), is as follows:

	1996	1997	1998	1999	Total
Total X-33 Budget	43.00	197.90	340.30	359.90	941.10

The funding included in this table represents NASA's projected funding to complete Phase II. This represents a fixed investment by NASA.

This profile includes funding for all of the following:

a. Funds provided directly to the selected offeror under the resulting Cooperative Agreement, in conjunction with the payment milestones.

b. Funds required to pay for charges relating to the performance of Government responsibilities under the resulting Cooperative Agreement (Government responsibilities may require non-cash resources in the form of personnel, facilities, services, etc., made available through the various installations). These may include charges for program support, materials, facility modifications, etc., but do not include salaries or travel for Government personnel. Offerors are responsible for negotiating and obtaining commitment letters from participating installations and associated task agreements, which will define candidate installation responsibilities/contributions and the charges relating to the performance of these responsibilities (see Appendix B, Section 1.7 and Appendix E). Payment of these charges will be made internal to the Government out of the available program funding.

c. Funds required for term/high payoff technology demonstration and indirect Government program support.

For purposes of planning, each fiscal year budget shall allow the program (Cooperative Agreement and government installation funding) to operate through the month of November of each year. . . .

Biographical Appendix

James A. Abrahamson (1933–   ) was the first director of the Strategic Defense Initiative Organization (SDIO). He received a bachelor's degree from the Massachusetts Institute of Technology (MIT) in 1955 and completed his pilot training in the U.S. Air Force in 1957. After earning his master's degree in aeronautical engineering in 1961, he was assigned as spacecraft project officer on the VELA Nuclear Detection Satellite Program. He later became the program director for the Maverick missile program at Wright-Patterson Air Force Base and, in 1976, the program director of the F-16 Air Combat Fighter program. In 1981, Abrahamson assumed the position of associate administrator for the NASA Office of Space Transportation Systems, responsible for the nation's Space Shuttle program. From 1984 to 1989, he was the chief commander of President Ronald Reagan's Strategic Defense Initiative (SDI) program. Today, he is a director of Orbital Imaging Corporation in Dulles, Virginia, and the chair and chief executive officer of StratCom and Air Safety Consultants. See "Abrahamson, Maj. Gen. James A.," biographical file, NASA Historical Reference Collection, NASA History Office, NASA Headquarters, Washington, DC (hereafter referred to as "NASA Historical Reference Collection").

Sherman Adams (1899–1986) had the title of assistant to the President and served as Dwight D. Eisenhower's chief of staff between 1953 and 1958. Previously, he had been a member of the House of Representatives (R-NH) between 1945 and 1947 and governor of New Hampshire from 1949 to 1953. Adams resigned from the Eisenhower administration in 1958. See Kenneth E. Shewmaker, "The Sherman Adams Papers," *Dartmouth College Library Bulletin* 10 (April 1969): 88–92; John E. Wickman, "Partnership for Research," *Dartmouth College Library Bulletin* 10 (April 1969): 93–97; *Historical Materials in the Dwight D. Eisenhower Library* (Abilene, KS: Dwight D. Eisenhower Library, 1989), pp. 8, 48; *New York Times*, October 28, 1986, p. D28.

Spiro T. Agnew (1918–1998) was elected Vice President of the United States in November 1968, serving under Richard M. Nixon. He served as chair of the 1969 Space Task Group that developed a long-range plan for a post-Apollo space effort. *The Post-Apollo Space Program: Directions for the Future* (Washington, DC: President's Science Advisory Council, September 1969) developed an expansive program that included building a space station, a space shuttle, a lunar base, and a mission to Mars (the last goal had been endorsed by Agnew at the time of the Apollo 11 launch in July 1969). Nixon did not accept this plan, and only the Space Shuttle was approved for development. See Roger D. Launius, "NASA and the Decision to Build the Space Shuttle, 1969–72," *The Historian* 57 (Autumn 1994): 17–34.

Arnold Aldrich (1936–   ) was the associate administrator for the NASA Office of Space Systems Development from 1991 to 1994. In 1959, soon after graduating from Northeastern University, he joined NASA. Working at the Johnson Space Center, he held a number of key positions in the Mercury, Gemini, Apollo, Skylab, Apollo-Soyuz, and Space Shuttle programs. In 1986, he was named the director of the Space Shuttle program and led the recovery activities after the *Challenger* accident. After 35 years of service to the agency, he retired from NASA in 1994 and joined Lockheed Missiles and Space Company as vice president for commercial space programs. See "Aldrich, Arnold," biographical file, NASA Historical Reference Collection.

David E. Aldrich headed the F-1 program since its inception in the late 1950s. As program manager of the Engine Program Office at Marshall Space Flight Center, he directed F-1 engine's development from its early design through development, qualification, and successful flight record. Before this program, he was a program engineer with Rocketdyne. See "Aldrich, David E.," biographical file, NASA Historical Reference Collection, Marshall Space Flight Center, Huntsville, AL (hereafter referred to as "Marshall's NASA Historical Reference Collection").

Edward C. ("Pete") Aldridge, Jr. (1938–   ), spent his entire career in the aerospace community as a corporate and government official. He served as under secretary and then secretary of the Air Force during the Reagan administration. Before then, he was educated at Texas A&M University and the Georgia Institute of Technology (Georgia Tech). He entered the Department of Defense as assistant secretary for systems analysis from 1967 through 1972. He then went to LTV Aerospace Corporation for a year and in 1973 was named senior management associate in the Office of Management and Budget. Returning to the Department of Defense in 1974, Aldridge served as assistant secretary for strategic programs until 1976. He then moved back to private industry until reentering government service with the Air Force in 1981. See "Aldridge, Edward C.," biographical file, NASA Historical Reference Collection.

H. Julian Allen (1910–1977) was an eminent space pioneer. Upon graduation from Stanford University in 1936, he joined the Langley Memorial Laboratory of the NACA in Hampton, Virginia. In 1941, when the Ames Research Laboratory was established, he moved to California and spent the rest of his career at Ames. His most outstanding engineering achievement was as the originator of the concept of bluntness as an aerodynamic technique for reducing the heating of spacecraft reentry into Earth's atmosphere. This design was successfully used in the Mercury, Gemini, and Apollo programs. See "Allen, H. Julian," biographical file, NASA Historical Reference Collection.

Milton B. Ames, Jr. (1919–1978), graduated from Georgia Tech with a degree in aeronautical engineering in 1936. He then joined the NACA as a research engineer before becoming chief of its aerodynamics division. When NASA was born in 1958, he became chief of its Aerodynamics and Flight Mechanics Division. Before retiring in 1974, he also served in a variety of advanced space research positions. He received numerous awards from NASA. See "Ames, Milton B.," biographical file, NASA Historical Reference Collection.

William A. Anders (1933–   ) was a career U.S. Air Force officer, although a graduate of the U.S. Naval Academy. Chosen with the third group of astronauts in 1963, he was the backup pilot for Gemini XI and the lunar module pilot for Apollo 8. He resigned from NASA and the Air Force (active duty) in September 1969, when he became executive secretary of the National Aeronautics and Space Council. He joined the Atomic Energy Commission in 1973 and became chair of the Nuclear Regulatory Commission in 1974. Anders was named U.S. ambassador to Norway in 1976. Later, he worked as a vice president of General Electric and then as senior executive vice president of operations for Textron, Inc. He retired as chief executive officer of General Dynamics in 1993, but remained chair of the board. See "Anders, W.A.," biographical file, NASA Historical Reference Collection.

Norman C. Appold (1917–   ) was born in Detroit, Michigan, on April 3, 1917. He attended Valparaiso University and graduated with a master's degree in chemical engineering from the University of Michigan. From mid-1942 to September 1944, he flew B-24s in combat. In 1947, after graduating with a master's degree in aeronautical engineering from the California Institute of Technology (Caltech), he worked for the U.S. Air Force specializing in aeronautical and aircraft propulsion. Appold served as chief of special projects for the deputy commander, weapons systems, at the Air Research and Development Command's headquarters from 1956 to 1957. See "Appold, Col. N.C.," biographical file, NASA Historical Reference Collection.

Neil A. Armstrong (1930–   ) was the first American to set foot on the Moon on July 20, 1969. He had become an astronaut in 1962, after having served as a test pilot with both the NACA (1955–58) and NASA (1958–62). He flew as command pilot on Gemini VIII in March 1966 and commander of Apollo 11 in July 1969. During 1970–71, he was deputy associate administrator for the Office of Advanced Research and Technology at NASA Headquarters. In 1971, he left NASA to become a professor of aerospace engineering at the University of Cincinnati and to undertake private consulting. He also served as vice chair of the Presidential Commission on the Space Shuttle *Challenger* Accident in 1986. See Neil A. Armstrong, *et al.*, *First on the Moon: A Voyage with Neil Armstrong, Michael Collins and Edwin E. Aldrin, Jr.* (Boston: Little, Brown, 1970); Neil A. Armstrong, *et al.*, *The First Lunar Landing: 20th Anniversary as Told by the Astronauts, Neil Armstrong, Edwin Aldrin, Michael Collins* (Washington, DC: NASA EP-73, 1989).

J. Leland Atwood (1904–1999) was the president and chief executive officer of Rockwell International Corporation. He began work as an aeronautical design engineer for the Douglas Aircraft Corporation in 1930 and moved to North American Aviation in 1934. He became assistant general manager in 1938 and was named North American's first vice president in 1941. He became president in 1948, gaining the title of chief executive officer in 1960, then chairman of the company two years later; he served until 1970, when he retired. In 1969, he was awarded NASA's Public Service Award, the highest honor for a nongovernmental employee. See "J.L. Atwood," biographical file, NASA historical Reference Collection; Associated Press article, "Ex-Rockwell CEO Atwood Dies at 94," Sunday, March 7, 1999.

Robert E. ("Gene") Austin (1939–   ) was the X-33 program manager assigned to NASA's Marshall Space Flight Center. Before being selected to this position in 1996, he headed up the Advanced Technology Team for three years. In that role, he was part of NASA's "Access to Space" study. Prior to that, he served as the deputy director of the Advanced Transportation Technologies Office at Marshall. He worked with senior NASA management to change existing policies to permit the innovative new "partnering" approach used with the X-33 program. See "Gene Austin," biographical file, Marshall's NASA Historical Reference Collection.

B

James E. Beggs (1926–) served as NASA administrator between July 10, 1981, and December 4, 1985, when he took an indefinite leave of absence pending disposition of an indictment from the Justice Department for activities taking place prior to his tenure at NASA. This indictment was later dismissed, and the U.S. Attorney General apologized to Beggs for any embarrassment. His resignation from NASA was effective on February 25, 1986. Prior to serving at NASA, he had been executive vice president and a director of General Dynamics Corporation in St. Louis, Missouri. Previously, he had served with NASA in 1968–69 as associate administrator for the Office of Advanced Research and Technology. From 1969 to 1973, he was under secretary at the Department of Transportation. He went to Summa Corporation in Los Angeles, California, as managing director of operations and joined General Dynamics in January 1974. Before joining NASA, Beggs had been with Westinghouse Electric Corporation in Sharon, Pennsylvania, and Baltimore, Maryland, for thirteen years. A 1947 graduate of the U.S. Naval Academy, he served with the Navy until 1954. In 1955, he received a master's degree from the Harvard Graduate School of Business Administration. See "Beggs, James E.," Administrator files, NASA Historical Reference Collection.

Rudi Beichel (1913–) was assigned to test the A-4/V-2 turbopump and its hydrogen-peroxide power supply system during World War II. Immediately after the war, he came with Wernher von Braun and 117 other specialists to the United States. Before being transferred to the Redstone Arsenal, he assisted the U.S. Army at the White Sands Proving Ground. After leaving the von Braun team, he joined the Aerojet Corporation in Sacramento, California. While there, he became a leading specialist for liquid propellant systems. In 1997, he founded Beichel Technologies, International, which applies rocket technologies to other world problems in the areas of combustion and power generation. Though retired now, he is still a consultant for Aerojet. See "Rudi Beichel," biographical summary, Marshall's NASA Historical Reference Collection.

Roger M. Boisjoly (1939–) worked as an engineer for Morton Thiokol, Inc., the maker of the solid rocket boosters for the Space Shuttle. He warned of a possible problem with the O-rings before the *Challenger* accident. He became involved with several lawsuits related to this incident. See "Boisjoly, Roger," biographical file, NASA Historical Reference Collection.

Philip Bono (1921–1993) was a distinguished aerospace design engineer. He was the investigator of novel reusable launch systems that embodied the principle of vertical takeoff and landing examined at the McDonnell Douglas Corporation in the 1960s. A fuller explanation of his ideas appears in his and K.W. Gatland's book *Frontiers of Space* (London: Blanford Press; New York: Macmillan Publishing Co., 1969). See "Bono, Philip," biographical file, NASA Historical Reference Collection.

Frank Borman (1928–) was the commander of the December 1968 Apollo 8 circumlunar flight. He had been chosen as a NASA astronaut in the early 1960s and had been on the Gemini VII mission in 1965. After leaving the astronaut corps, he became president of Eastern Airlines. See Andrew Chaikin, *A Man on the Moon: The Voyages of the Apollo Astronauts* (New York: Viking, 1994); Frank Borman, with Robert J. Serling, *Countdown: An Autobiography* (New York: William Morrow, 1988).

Karel J. Bossart (1904–1975) was a pre–World War II immigrant from Belgium, who was involved early on in the development of rocket technology with Convair Corporation. In the 1950s, he was largely responsible for the design of the Atlas ICBM booster with a very thin, internally pressurized fuselage instead of massive struts and a thick metal skin. See Richard E. Martin, *The Atlas and Centaur "Steel Balloon" Tanks: A Legacy of Karel Bossart* (San Diego: General Dynamics Corp., 1989); Robert L. Perry, "The Atlas, Thor, Titan, and Minuteman," in Eugene M. Emme, ed., *A History of Rocket Technology* (Detroit: Wayne State University Press, 1964), pp. 143–55; John L. Sloop, *Liquid Hydrogen as a Propulsion Fuel, 1945–1959* (Washington, DC: NASA SP-4404, 1978), pp. 173–77.

Darrell R. Branscome (1944–   ) began his career in 1966 at NASA's Langley Research Center. He started as a space technologist before serving as manager of advanced engineering systems analysis and economic studies. In 1973, he moved to the Management Development Program at Headquarters, also serving as a technical assistant in the Manned Space Technology Program Office and as acting program manager of the Space Technology Shuttle Payloads Program. He returned to Langley in 1974 as technical project manager responsible for systems engineering support for spacecraft systems and space technology experiments. In 1975, he went back to NASA Headquarters to serve as staff director for the Subcommittee on Space Science and Applications, House of Representatives Committee on Science and Technology. In 1985 Branscome became special assistant to the associate administrator for the Office of Space Flight. He also served as NASA deputy co-director of the joint NASA-DOD National Space Transportation and Support Study. During 1986–87, he was director of special programs, Office of Space Flight, as well as NASA co-director of the National Space Transportation and Support Study and the NASA Mixed Fleet Study. He then became director of the Advanced Program Development Division in the Office of Space Flight. During 1991–92, he was technical assistant to the associate administrator for the Office of Space Flight and then deputy associate administrator for that office (management). He finally returned to Langley after being named chief engineer. He has received the NASA Exceptional Service Medal and several group and special achievement awards. See "Branscome, Darrell R.," biographical file, NASA Historical Reference Collection.

Wernher von Braun (1912–1977) was the leader of what has been called the "rocket team," which had developed the German V-2 ballistic missile during World War II. At the conclusion of the war, von Braun and some of his chief assistants—as part of a military operation called Project Paperclip—came to America and were installed at Fort Bliss in El Paso, Texas, to work on rocket development and use the V-2 for high-altitude research. They used launch facilities at the nearby White Sands Proving Ground in New Mexico. In 1950, von Braun's team moved to the Redstone Arsenal near Huntsville, Alabama, to concentrate on the development of a new missile for the Army. They built the Army's Jupiter ballistic missile and before that the Redstone, used by NASA to launch the first Mercury capsules. Eventually, the team became part of NASA and developed the powerful Saturn rockets, and von Braun became director of NASA's Marshall Space Flight Center. The story of von Braun and the "rocket team" has been told many times. See David H. DeVorkin, *Science With a Vengeance: How the Military Created the US Space Sciences After World War II* (New York: Springer-Verlag, 1992); Frederick I. Ordway III and Mitchell R. Sharpe, *The Rocket Team* (New York: Thomas Y. Crowell, 1979); Erik Bergaust, *Wernher von Braun* (Washington, DC: National Space Institute, 1976).

Harold Brown (1927–   ) was director of defense research and engineering at the Pentagon from 1961 to 1965 before becoming secretary of the Air Force from 1965 to 1969. After spending eight years as the president of Caltech, he returned to Washington to serve as the secretary of defense under President Jimmy Carter from 1977 to 1981. He currently works at the Center for Strategic and International Studies in Washington, D.C. See "Brown, Harold," biographical file, NASA Historical Reference Collection.

Wilber M. Brucker (1894–1968) was secretary of the Army between 1955 and 1961. An attorney, he had also held a number of important government positions, including governor of Michigan (1930–32), prior to becoming secretary. Brucker had also served with the Army in World War I. After leaving federal service, Brucker returned to his law practice in Detroit. See William Gardner Bell, *Secretaries of War and Secretaries of the Army: Portraits & Biographical Sketches* (Washington, DC: Center of Military History, 1982), p. 140; *New York Times*, October 29, 1968, p. 41.

Zbigniew Brzezinski (1928–   ) served as President Carter's national security advisor from 1977 to 1981. See "Brzezinski, Zbigniew," biographical file, NASA Historical Reference Collection.

James H. Burnley IV (1948–   ), with degrees from Yale and Harvard, became an associate in the firm Brooks, Pierce, McLendon, Humphrey & Leonard in 1973. From 1975 through 1981, he was a partner with Turner, Enochs, Foster, Sparrow & Burnley. He went on to be director of VISTA for a year, then deputy attorney general for the Department of Justice for another year. In 1983, he became general counsel for the Department of Transportation, a position he held until 1987, when he became secretary of transportation until 1989. See *Who's Who in America 1988–1989*, 45th edition, vol. 1 (Wilmette, IL.: Marquis Who's Who, 1989).

George H.W. Bush (1924–   ) served as the forty-first U.S. President between 1989 and 1993 and as Vice President under Ronald Reagan (1981–89). His career began after high school when he enlisted in the Navy and was a pilot during World War II. During this time, he was awarded the Distinguished Flying Cross and three Air Medals. Following an unsuccessful bid for a Senate seat, he was elected to the U.S. House of Representatives in 1966 from Texas. In 1971, he was named U.S. ambassador to the United Nations. He held that position until 1973, when he became chair of the Republican National Committee. In 1976, he was appointed director of the Central Intelligence Agency. See "Biography of George Bush," George Bush Presidential Library, Texas A&M University, College Station, TX.

C

Howard W. Cannon (1912–   ) was a first lieutenant in a Combat Engineers unit at the start of World War II, but shortly switched over to the Army Air Corps. During the war, he was shot down and spent forty-four days behind enemy lines. He retired from the Air Force as a major general, having flown almost every fighter in the Air Force inventory. Cannon was first elected to the Senate as a Democrat from Nevada in 1958, shortly after the creation of NASA, and immediately became involved with the space program. He pursued all aspects of the relationship between aviation and national policy, ultimately stepping in for the ailing Senator Clinton Anderson, chair of the Senate Aeronautical and Space Sciences Committee, to prevent the Space Shuttle from being cancelled. He also served as the ranking Democrat on the Senate Commerce Committee, which oversees NASA. In 1971, he received the Wright Brothers Memorial Trophy for outstanding service to U.S. aviation. He remained in the Senate until 1983, after being defeated for reelection. See "Cannon, Howard W.," biographical file, NASA Historical Reference Collection.

Frank Carlucci (1930–   ) was the national security advisor for President Reagan. He was appointed secretary of defense in November 1987, a position he held for fourteen months. After leaving the Pentagon, he joined the Carlyle Group, a Washington investment partnership, as vice president and managing director; he later became chairman. See "Carlucci, Frank C.," biographical file, NASA Historical Reference Collection.

Jimmy Carter (1924–   ) was the thirty-ninth U.S. President between 1977 and 1981. He graduated from the Naval Academy in 1946. After seven years as a naval officer, he returned to Georgia. In 1962, he entered state politics in the Georgia State Legislature, and eight years later, he was elected governor of Georgia, a position he held until 1975 when he began campaigning for the presidency. After leaving the White House in 1981, Carter returned to Georgia, where in 1982 he founded the nonprofit Carter Center in Atlanta to promote peace and human rights worldwide. See "Biography of Jimmy Carter," The White House, Washington, D.C.

James Chamberlin (1915–1981) was born in British Columbia and held aerodynamics engineering jobs throughout Canada. He was chief of design for the C102 jetliner and for the CF100 and CF105 all-weather fighters at AVRO Canada. He played a leading role in the design of the Mercury capsule and was project manager for the Gemini spacecraft, under the direction of the Space Task Group at NASA's Langley Research Center. He also held the position of director of their Space Shuttle study. See "Chamberlin, J.A.," biographical file, NASA Historical Reference Collection.

Richard Cheney (1941–   ) was born in Lincoln, Nebraska, on January 30, 1941. He attended Yale University, Casper College, and the University of Wyoming, earning B.A. (1965) and M.A. (1966) degrees. In March 1989, he was appointed secretary of defense under President Bush, a position he held until January 1993. In October 1995, he became the president and chief executive officer of the Halliburton Company in Dallas, Texas. See "Cheney, Dick," biographical file, NASA Historical Reference Collection.

Arthur C. Clarke (1917–   ) is one of the most well-known science fiction authors. He has also been an eloquent writer on behalf of the exploration of space. In 1945, before the invention of the transistor, Clarke wrote an article in *Extraterrestrial Relays* describing the possibility of geosynchronous orbit and the development of communications relays by satellite. He also wrote several novels, the best known being *2001: A Space Odyssey*, based on a screenplay of the same name he prepared for film director Stanley Kubrick. The movie is still one of the most realistic depictions of the rigors of spaceflight ever to be filmed. See "Clarke, Arthur C.," biographical file, NASA Historical Reference Collection.

William J. ("Bill") Clinton (1946–   ) became the forty-second U.S. President in 1993, a position he held for two terms. While attending Georgetown University, he interned for Senator J. William Fulbright of Arkansas. After graduating in 1968, he won a Rhodes Scholarship, studying at Oxford University for two years. He went on to receive a law degree from Yale University in 1973 and then returned to his home state of Arkansas to teach law at the University of Arkansas and to prepare to enter politics. Clinton was elected Arkansas attorney general in 1976 and then went on to win the governorship in 1978, becoming the youngest U.S. governor at age thirty-two. He lost in his try for a second term, but he regained the office later and served as governor until 1992, becoming the second person in Arkansas history to be reelected to a fifth gubernatorial term. See "Biography of William J. Clinton," The White House.

Aaron Cohen came to NASA in 1962 and played a key role in the Apollo program, where his efforts were critical to the success of all six lunar landings. He later became manager for the Space Shuttle orbiter, directing the orbiter's design, development, production, and initial flight-testing. During this period, he worked at the Johnson Space Center. In 1986, he became Johnson's director, from which he was called to Headquarters as acting NASA deputy administrator in 1992. He then retired from NASA in 1993. Cohen was educated at Texas A&M University, and upon his retirement on August 20, 1993, he became the Zachry professor of mechanical engineering at his alma mater. See "Cohen, Aaron," biographical file, NASA Historical Reference Collection.

Raymond S. Colladay (1943–   ) was named associate administrator for the NASA Office of Aeronautics and Space Technology on June 14, 1985, after having served as the deputy associate administrator since April 1982. He is a graduate of Michigan State University and began his career at NASA's Lewis Research Center in 1969 as a research engineer for space propulsion. He has written more than twenty NASA technical reports and articles relating to aeronautical research. In 1988, he left NASA to direct the Defense Advanced Research Projects Agency. Today, he is the vice president of advanced technology for Martin Marietta, an aerospace and defense contracting firm. See "Colladay, Raymond S.," biographical file, NASA Historical Reference Collection.

Eugene E. Covert was an aerospace engineer and professor of aeronautics and astronautics at MIT beginning in 1952. He also served as the Air Force's chief scientist, on a number of prestigious national science advisory boards, and on the Rogers Commission investigating the Space Shuttle *Challenger* accident. See "Covert, Eugene E.," biographical file, NASA Historical Reference Collection.

Laurence C. Craigie (1902–1994), was a career Air Force officer and the first U.S. military jet pilot in 1942 when he flew the Bell XP-59. A graduate of the U.S. Military Academy at West Point, in 1923, he joined the Army Air Corps and became a pilot. During World War II, he served in a variety of weapons development programs, as well as in a combat role in North Africa and Corsica. After the war, he directed the Air Force's research and development programs, serving as deputy chief of staff for development (1951–54) and commander of the Allied Air Force in Southern Europe before his retirement following a heart attack in 1955. See "Lieut. Gen. Laurence Craigie, 92; First Military Jet Pilot for the U.S.," *New York Times*, March 1, 1994.

Robert L. Crippen (1937–   ) was selected as an astronaut for the Manned Orbiting Laboratory program in 1966 and transferred to the NASA astronaut program in 1969. He commanded the Skylab Medical Experiments Altitude Test and was part of the support crew for the Skylab 2, 3, and 4 missions and the Apollo-Soyuz Test Project. He piloted STS-1, the first orbital flight test of the shuttle *Columbia*, in April 1981. Following retirement from the astronaut corps, he served as the deputy director of the Flight Crew Operations Directorate at Johnson Space Center and then deputy director of NSTS Operations at NASA Headquarters, stationed at Kennedy Space Center. He served as director of Kennedy from 1992 to 1995. See "Crippen, Robert L.," biographical file, NASA Historical Reference Collection.

John W. ("Gus") Crowley, Jr. (1899–1974), joined the Langley Aeronautical Laboratory in 1921 after earning his mechanical engineering degree from MIT the year before. He became head of the research department at Langley in 1943 and then transferred to the NACA's Washington headquarters in 1945 to become acting director of research there. He assumed the post of associate director for research in 1945, and when NASA replaced the NACA, he became director of aeronautical and space research. He retired in 1959. See "John W. Crowley, Jr.," biographical file, NASA Historical Reference Collection.

Philip E. Culbertson (1925–   ) served the American space exploration program for twenty-two years in high-level technical and policy positions. He was appointed to the position of associate administrator for NASA's Office of Policy and Planning in January 1987, where he remained until his retirement in January 1988. After his retirement, he became president of the Lew Evans Foundation. See "Culbertson, Philip E.," biographical file, NASA Historical Reference Collection.

Robert Cutler (1895–1974) was admitted to the Massachusetts bar in 1922 and became an associate with the Boston firm of Herrick, Smith, Donald & Farley, becoming partner in 1928. In 1946, he became president and director of the Old Colony Trust Company, eventually becoming its chairman for the next several years. He served as special assistant for National Security Affairs for President Eisenhower and later chair of the National Security Council Planning Board (1953–55 and 1957–58). From 1960 to 1962, he served as special assistant to the secretary of the U.S. Treasury. See *Who Was Who in America 1974–1976*, vol. 6 (Chicago: Marquis Who's Who, 1976).

D

Richard Darman (1943–   ) served as a deputy assistant to President Reagan and deputy to the chief of staff. He also served as the director of the Office of Management and Budget for the Bush administration. He received a master's in business administration from Harvard University in 1967. See "Darman, Richard," biographical file, NASA Historical Reference Collection.

Edward E. David, Jr. (1925–   ), served as science advisor to President Richard M. Nixon in 1970 and then as director of the Office of Science and Technology. Previously, he had served between 1950 and 1970 as executive director of research at Bell Telephone Laboratories. For a discussion of the President's Science Advisory Committee, see Gregg Herken, *Cardinal Choices: Science Advice to the President from Hiroshima to SDI* (New York: Oxford University Press, 1992).

LeRoy Day received a bachelor's degree from Georgia Tech in aeronautical engineering and then a master's degree in industrial management from MIT. He worked briefly for the Navy on missile development in the early 1960s and then joined NASA in the Gemini Program Office in 1962. After being acting deputy director of the Gemini program, he became director of the Apollo test program. See "Day, LeRoy," biographical file, NASA Historical Reference Collection.

Kurt H. Debus (1908–1983) earned a B.S. in mechanical engineering (1933) and then an M.S. (1935) and Ph.D. (1939) in electrical engineering, all from the Technical University of Darmstadt in Germany. He became an assistant professor at the university after receiving his degree. During the course of World War II, he became an experimental engineer at the A-4 (V-2) test stand at Peenemünde (see Wernher von Braun above), rising to become superintendent of the test stand and test firing stand for the rocket. In 1945, he came to the United States with a group of engineers and scientists headed by von Braun. From 1945 to 1950, the group worked at Fort Bliss, Texas, and then moved to the Redstone Arsenal in Huntsville, Alabama. From 1952 to 1960, Debus was chief of the missile-firing laboratory of the Army Ballistic Missile Agency (ABMA). In this position, he was located at Cape Canaveral, Florida, where he supervised the launching of the first ballistic missile fired from there, an Army Redstone. When the ABMA became part of NASA, Debus continued to supervise missile and space vehicle launchings, first as director of the Launch Operations Center and then of Kennedy Space Center as it was renamed in December 1963. He retired from that position in 1974. See "Kurt H. Debus," biographical file, NASA Historical Reference Collection.

John Deutch (1938–   ) served as deputy secretary of defense (1994–95) and then as director of the Central Intelligence Agency (1995–97). He received a Ph.D. from MIT and also served as that school's dean of science and provost. See "Deutch, John M.," biographical file, NASA Historical Reference Collection.

Thomas F. Dixon (1916–1998) was appointed deputy associate administrator of NASA, in charge of the planning and construction of space vehicles and research programs. He held this position from 1961until 1963. He was schooled at Vanderbilt University and received his master's degree in chemical engineering from the University of Michigan in 1940, as well as a master's in aeronautical engineering from Caltech in 1945. Later positions included president and chairman of the board of Airtronics, Inc., and president of Teledyne McCormick/Selph. He was affiliated with the College of William and Mary at the time of his death. See "Thomas F. Dixon," biographical file, NASA Historical Reference Collection.

Elizabeth H. Dole (1936–   ) was admitted to the D.C. bar in 1966 and practiced law in Washington from 1967 to 1968 before becoming associate director of legislative affairs and then executive director of the President's Committee for Consumer Interests until 1971. She was deputy director of the Office of Consumer Affairs at the White House for the next two years before becoming commissioner of the Federal Trade Commission from 1973 to 1979. She held various other government and public service positions before becoming the secretary of transportation, the first woman to hold that position, for the Reagan administration (1983–87). Two years later, she became secretary of labor for the Bush administration, leaving that position in 1990. The following year, she became president of the American Red Cross until she resigned in 1999. See *Who's Who in America 1992–1993*, 47th edition, vol. 1 (New Providence, NJ: Marquis Who's Who, 1993).

Charles J. Donlan served as NASA's deputy associate administrator (technical) for the Office of Manned Space Flight in the late 1960s and early 1970s and participated in Space Shuttle planning. See "Space Shuttle (1969–72) Charles Donlan," file, NASA Historical Reference Collection.

Walter R. Dornberger (1895–1980) was Wernher von Braun's military superior during the German rocket development program of World War II. He oversaw the effort at Peenemünde to build the V-2, fostering internal communication and successfully advocating the program to officials in the German army. He also assembled the team of highly talented engineers under von Braun's direction and provided the funding and staff organization necessary to complete the technology project. After World War II, Dornberger came to the United States and assisted the Department of Defense with the development of ballistic missiles. He also worked for the Bell Aircraft Company for several years, helping develop hardware for Project BOMI, a rocket-powered spaceplane. See Walter R. Dornberger, *V-2*, trans. by James Cleugh and Geoffrey Halliday (New York: Viking, 1958); Gerald L. Borrowman, "Walter R. Dornberger," *Spaceflight* 23 (April 1981): 118–19.

Hugh L. Dryden (1898–1965) was a career civil servant and an aerodynamicist by discipline who had also begun life as something of a child prodigy. He graduated at age fourteen from high school and went on to earn an A.B. in three years from Johns Hopkins University (1916). Three years later, he earned his Ph.D. in physics and mathematics from the same institution, even though he had full-time employment at the National Bureau of Standards since June 1918. His career there, which lasted until 1947, was devoted to studying airflow, turbulence, and particularly the problems of the boundary layer—the thin layer of air next to an airfoil that causes drag. In 1920, he became chief of the bureau's aerodynamics section. His work in the 1920s on measuring turbulence in wind tunnels facilitated research within the NACA that produced the laminar flow wings used in the P-51 Mustang and other World War II aircraft. From the mid-1920s to 1947, his publications became essential reading for aerodynamicists around the world. During World War II, his work on a glide bomb named the Bat won him a Presidential Certificate of Merit. He capped his career at the Bureau of Standards by becoming its assistant director and then associate director during his final two years there. He then served as director of the NACA from 1947 to 1958, after which he became deputy administrator of NASA under T. Keith Glennan and James E. Webb. See Richard K. Smith, *The Hugh L. Dryden Papers, 1898–1965* (Baltimore, MD: The Johns Hopkins University Library, 1974).

E

A.J. Eggers, Jr. (1922–   ), worked for NASA as the deputy administrator for advanced research and technology from 1964 to 1968. Prior to that, he served as assistant director of the Ames Research Center. In 1968, he became the assistant administrator for policy of NASA, and in 1971, he left NASA for a job with the National Science Foundation. Eggers was later named director of the Palo Alto Research Laboratory of the Lockheed Missiles and Space Company in 1977. See "Eggers, Dr. Alfred," biographical file, NASA Historical Reference Collection.

Dwight D. Eisenhower (1890–1969) was the thirty-fourth U.S. president between 1953 and 1961. Previously, he had been a career U.S. Army officer and, during World War II, was supreme allied commander in Europe. As president, he was deeply interested in the use of space technology for national security purposes and directed that ballistic missiles and reconnaissance satellites be developed on a crash basis. See Rip Bulkeley, *The Sputniks Crisis and Early United States Space Policy* (Bloomington, IN: Indiana University Press, 1991); R. Cargill Hall, "The Eisenhower Administration and the Cold War: Framing American Astronautics to Serve National Security," *Prologue: Quarterly of the National Archives* 27 (Spring 1995): 59–72; Robert A. Divine, *The Sputnik Challenge: Eisenhower's Response to the Soviet Satellite* (New York: Oxford University Press, 1993).

John C. Evvard (1915–) was the assistant director of NASA's Lewis Research Center (renamed Glenn Research Center in 1999) in the early 1960s. He received his M.S. from Caltech in 1940. He originated simple methods to evaluate the load distributions on wings and to extend the range of supersonic jets. See "Evvard, John C. (Dr.)," biographical file, NASA Historical Reference Collection.

F

Maxime A. Faget (1921–), an aeronautical engineer with a B.S. from Louisiana State University (1943), joined the staff at Langley Aeronautical Laboratory in 1946 and soon became head of the Performance Aerodynamics Branch of the Pilotless Aircraft Research Division. There, he conducted research on the heat shield of the Mercury spacecraft. In 1958, he joined NASA's Space Task Group, forerunner of NASA's Manned Spacecraft Center that became the Johnson Space Center, and he became its assistant director for engineering and development in 1962 and later its director. He contributed many of the original design concepts for Project Mercury's manned spacecraft and played a major role in designing virtually every U.S. crewed spacecraft since that time, including the Space Shuttle. He retired from NASA in 1981 and became an executive for Eagle Engineering, Inc. In 1982, he was one of the founders of Space Industries, Inc., and became its president and chief executive officer. See "Maxime A. Faget," biographical file, NASA Historical Reference Collection.

Richard P. Feynman (1918–1988) was a brilliant, iconoclastic physicist who shared a Nobel Prize in 1965 for work in quantum electrodynamics. He taught theoretical physics at Caltech from 1950 until his death. Feynman participated in the Manhattan Project and was a member of the Rogers Commission that investigated the Space Shuttle *Challenger* accident. He wrote several popular books, including the best-seller *Surely You're Joking Mr. Feynman*. See "Feynman, Richard," biographical file, NASA Historical Reference Collection.

Peter M. Flanigan (1923–) was an assistant to President Nixon from 1969 to 1974. Previously, he had been involved in investment banking with Dillon, Read, and Co. He returned to business when he left government service. His position in the White House involved him in efforts to gain approval to build the Space Shuttle during the 1969–72 period. See "Miscellaneous Other Agencies," biographical file, NASA Historical Reference Collection.

James C. Fletcher (1919–1991) received an undergraduate degree in physics from Columbia University and a doctorate in physics from Caltech. After holding research and teaching positions at Harvard and Princeton Universities, he joined Hughes Aircraft in 1948 and later worked at the Guided Missile Division of Ramo-Wooldridge Corporation. In 1958, Fletcher co-founded Space Electronics Corporation in Glendale, California, which after a merger became Space General Corporation. He was later named systems vice president of the Aerojet General Corporation in Sacramento. In 1964, he became president of the University of Utah, a position he held until he was named NASA's administrator in 1971. He served until 1977. He also served as NASA administrator a second time, for nearly three years following the loss of the Space Shuttle *Challenger*, from 1986 until 1989. During his first administration at NASA, Fletcher was responsible for beginning the Shuttle effort. During his second tenure, he presided over the effort to recover from the *Challenger* accident. See Roger D. Launius, "A Western Mormon in Washington, D.C.: James C. Fletcher, NASA, and the Final Frontier," *Pacific Historical Review* 64 (May 1995): 217–41.

Robert A. Frosch (1928–) was NASA administrator throughout the Carter administration from 1977 to 1981. He earned undergraduate and graduate degrees in theoretical physics at Columbia University, and from September 1951 to August 1963, he worked as a research scientist and director of research programs for Hudson Laboratories of Columbia University. Until 1953, he worked on problems in underwater sound, sonar, oceanography, marine geology, and marine geophysics. Thereafter, he was first associate and then director of the laboratories. In September 1963, Frosch came to Washington to work with the Advanced Research Projects Agency of the Department of Defense, serving as director for nuclear test detection (Project VELA), and then as deputy director of that agency. In July 1966, he became assistant secretary of the Navy for research and development, responsible for all Navy programs of research, development, engineering, test, and evaluation. From January 1973 to July 1975, he served as assistant executive director of the United Nations Environmental Programme. While at NASA, Frosch was responsible for overseeing the continuation of the development effort on the Space Shuttle. During his tenure, the project underwent testing of the first orbiter, *Enterprise*, at NASA's Dryden Flight Research Facility in southern California. The orbiter made its first free flight in the atmosphere on August 12, 1977. He left NASA with the change of administrations in January 1981 to become vice president for research at General Motors Research Laboratories. See "Frosch, Robert A.," Administrator files, NASA Historical Reference Collection.

Craig L. Fuller (1951–   ) worked for a corporate public affairs firm in California prior to joining the new administration in 1981 as President Reagan's assistant for cabinet affairs, arranging for NASA's space station proposal to be discussed at a meeting of the Cabinet Council for Commerce and Trade. In 1985, he became chief of staff to Vice President Bush. After Bush became President, Fuller served as co-chair of his transition team, going on to be a member of the advisory committee for the Economic Summit held in Houston, Texas, and chair of the 1992 Republican National Convention. After leaving the White House in 1989, he was president of the Washington firm of Wexler, Reynolds, Fuller, Harrison and Schule and then president of Hill and Knowlton USA. He later served as senior vice president for corporate affairs at Philip Morris Companies, Inc. See *Who's Who in America 1992-1993*, 47th edition, vol. 1 (New Providence, NJ: Marquis Who's Who, 1993).

Clifford C. Furnas (1900–1969) earned his Ph.D. from the University of Michigan in 1926 and served as a chemist with the U.S. Bureau of Mines from 1926 to 1931. He then taught chemical engineering at Yale from 1931 to 1942. He became director of research at Curtiss-Wright Airplane Division (1943–46) and served as vice president for Cornell Aeronautical Laboratory (1946–54). After serving as chancellor at the University of Buffalo from 1954 to 1962, he became president of the State University of New York at Buffalo. See "Furnas, Dr. Clifford C.," biographical file, NASA Historical Reference Collection.

G

Yuri Gagarin (1934–1968) was the Soviet cosmonaut who became the first human in space with a one-orbit mission aboard the spacecraft Vostok 1 on April 12, 1961. The great success of that feat made Gagarin a global hero, and he was an effective spokesman for the Soviet Union until his death in an aircraft accident. See "Gagarin, Yuri," biographical file, NASA Historical Reference Collection.

Ken Gatland (1924–1997) was educated at Hawker Aircraft Technical School in 1941. In the astronautics field, he became an internationally recognized author of many books and articles that enjoyed a world wide distribution. He was a distinguished member of the British Interplanetary Society and a major contributor to its development. See "Gatland, Ken," biographical file, NASA Historical Reference Collection.

Hugo Gernsback (1884–1967) was a U.S. publisher known as one of the fathers of science fiction because he founded the magazine *Amazing Stories* in 1926. He designed the world's first home radio set, the Telimco Wireless. His *Telimco* catalog evolved into science magazines such as *Modern Electronics and Electrical Experimenter.* Gernsback was an instrumental force in the formation of the Science Fiction League in 1936. He patented more than eighty inventions and published hundreds of works. See "Gernsback, Hugo," biographical file, NASA Historical Reference Collection.

Robert R. Gilruth (1913–   ) was a longtime NACA engineer working at the Langley Aeronautical Laboratory from 1937 to 1946. He then was chief of the Pilotless Aircraft Research Division at Wallops Island from 1946 to 1952. He had been exploring the possibility of human spaceflight before the creation of NASA. He served as assistant director at Langley (1952–59) and as assistant director (manned satellites) and head of Project Mercury (1959–61), technically assigned to the Goddard Space Flight Center but physically located at Langley. In early 1961, T. Keith Glennan, NASA's first administrator, established an independent Space Task Group (already the group's name as an independent subdivision of Goddard) under Gilruth at Langley to supervise the Mercury program. This group moved to the Manned Spacecraft Center in Houston, Texas, in 1962. Gilruth was then director of the Houston operation from 1962 to 1972. See Henry C. Dethloff, *"Suddenly Tomorrow Came . . .": A History of the Johnson Space Center* (Washington, DC: NASA SP-4307, 1993); James R. Hansen, *Engineer in Charge: A History of the Langley Aeronautical Laboratory, 1917–1958* (Washington, DC: NASA SP-4305, 1987), pp. 386–88.

John H. Glenn, Jr. (1921–   ), was selected with the first group of astronauts in 1959. He was the pilot for the February 20, 1962, Mercury-Atlas 6 (*Friendship 7*) mission, the first American orbital flight. He made three orbits on this mission. He left the NASA astronaut corps in 1964 and later entered politics as a senator from Ohio. He returned to space in 1998 as a Space Shuttle payload specialist aboard STS-95. He was a test subject for specific investigations that mimic the effects of aging, including loss of muscle mass and bone density, disrupted sleep patterns, a depressed immune system, and loss of balance. See Loyd S. Swenson, Jr., James M. Grimwood, and Charles C. Alexander, *This New Ocean: A History of Project Mercury* (Washington, DC: NASA SP-4201, 1966); Shuttle Press Kit—STS-95.

T. Keith Glennan (1905–1995) served NASA's first administrator, while on leave from Case Institute of Technology, from August 7, 1958, to January 20, 1961. He was educated at Yale University and worked in the sound motion picture industry with the Electrical Research Products Company. He was also studio manager of Paramount Pictures and Samuel Goldwyn Studios in the 1930s. Glennan joined Columbia University's Division of War Research in 1942, serving through World War II, first as administrator and then as director of the U.S. Navy's Underwater Sound Laboratories at New London, Connecticut. In 1947, he became president of the Case Institute of Technology in Cleveland. During his administration, Case rose from a primarily local institution to rank with the top engineering schools in the nation. From October 1950 to November 1952, Glennan served as a member of the Atomic Energy Commission. After leaving NASA, he returned to Case, where he continued to serve as president until 1966. See J.D. Hunley, ed., *The Birth of NASA: The Diary of T. Keith Glennan* (Washington, DC: NASA SP-4105, 1993).

Robert H. Goddard (1882–1945) was one of the three most prominent pioneers of rocketry and spaceflight theory. He earned his Ph.D. in physics at Clark University in 1911 and went on to become head of the Clark physics department and director of its physical laboratories. He began to work seriously on rocket development in 1909 and is credited with launching the world's first liquid-propellant rocket in 1926. He continued his rocket development work with the assistance of a few technical assistants throughout the remainder of his life. Although he developed and patented many of the technologies later used on large rockets and missiles—including film cooling, gyroscopically controlled vanes, and a variable-thrust rocket motor—only the last of these contributed directly to the furtherance of rocketry in the United States. Goddard kept most of the technical details of his inventions a secret and thus missed the chance to have the kind of influence his real abilities promised. At the same time, he was not good at integrating his inventions into a workable system, so his own rockets failed to reach the high altitudes he sought. See Milton Lehman, *Robert H. Goddard: A Pioneer of Space Research* (New York: Da Capo, 1988).

Daniel S. Goldin (1940–) became the ninth NASA administrator in April 1992 and immediately began to earn a reputation as an agent of change in bringing reform to America's space agency. In addition to implementing many management changes, Goldin negotiated with his Russian counterpart, Yuri Koptev, the head of the Russian Space Agency, to construct the International Space Station with a partnership of European countries, Canada, and Japan. Before coming to NASA, Goldin was vice president and general manager of the TRW Space & Technology Group in Redondo Beach, California. During a twenty-five-year career at TRW, he managed the development and production of advanced spacecraft, technologies, and space science instruments. Goldin began his career as a research scientist at NASA's Lewis Research Center in Cleveland in 1962 and worked on electric propulsion systems for human interplanetary travel. See "Daniel S. Goldin," Administrator files, NASA Historical Reference Collection.

Nicholas E. Golovin (1912–1969), born in Odessa, Russia, but educated in the United States (with a Ph.D. in physics from George Washington University in 1955), worked in various capacities for the government during and after World War II, including the Naval Research Laboratory (1946–48). He held several administrative positions with the National Bureau of Standards from 1949 to 1958. In 1958, he was chief scientist for the White Sands Missile Range and then worked for the Advanced Research Projects Agency in 1959 as director of technical operations. He became deputy associate administrator for NASA in 1960. He joined private industry before becoming, in 1961, the director of the NASA-DOD Large Launch Vehicle Planning Group. He joined the Office of Science and Technology at the White House in 1962 as a technical advisor for aviation and space and remained there until 1968. He was then a research associate at Harvard and a fellow at the Brookings Institution. See obituaries in *Washington Star,* April 30, 1969, p. B-6; *Washington Post,* April 30, 1969, p. B14.

Andrew Jackson Goodpaster, Jr. (1915–), was a career Army officer who served as defense liaison officer and secretary of the White House staff from 1954 to 1961, being promoted to brigadier general during that period. He later was deputy commander of the U.S. forces in Vietnam (1968–69), and commander-in-chief of the U.S. Forces in Europe (1969–74). He retired in 1974 as a four-star general but returned to active duty in 1977 and served as superintendent of the U.S. Military Academy, a post he held until his second retirement in 1981. See "Goodpaster (Andrew J., Jr.)," biographical file, NASA Historical Reference Collection.

Mikhail S. Gorbachev (1931–) became leader of the Soviet Union in 1985 and restructured the nation, presiding over the demise of the communist state and the end of the Cold War in 1989. In the process, he opened negotiations with the United States for significant international cooperation in space exploration. See Thomas G. Butson, *Gorbachev: A Biography* (New York: Stein and Day, 1985); "Gorbachev, Mikhail Sergeyevich," biographical file, NASA Historical Reference Collection.

Albert A. Gore (1907–1998), was admitted to the bar in 1936, after being in the field of education, and began practicing law. As a Democrat from Tennessee, he was first elected to the U.S. House of Representatives in 1938 and reelected for two more terms before resigning in 1944 to join the U.S. Army. He returned to Congress for four more terms from 1945 to 1953. He then won election to the Senate and served from 1953 to 1971, after an unsuccessful reelection in 1970. He then returned to Tennessee and resumed the practice of law, becoming vice president and member of the Board of Directors of the Occidental Petroleum Company. Gore also taught law at Vanderbilt University until 1972 and was a member of the Board of Petroleum and Coal Companies. See "Gore, Albert Arnold," biographical summary, U.S. Senate, Washington, DC.

Daniel O. Graham (1926–1996) was a space expert who pioneered the concept of the Strategic Defense Initiative (SDI), also known as "Star Wars." A staunch conservative, he was appointed director of the Defense Intelligence Agency in 1974. In 1981, he became the director of the High Frontier organization, becoming a major voice in support of SDI. In 1990, Graham founded the Space Transportation Agency to support the development of launch vehicles that embodied the single-stage-to-orbit concept. He died on December 31, 1996, after a long battle with cancer. See "Graham, Lt. Gen. Daniel," biographical file, NASA Historical Reference Collection.

William R. Graham (1937–) served three years of active duty as a project officer with the Air Force Weapons Laboratory at Kirtland Air Force Base in New Mexico. He then went on to spend six years with the Rand Corporation in Santa Monica, California, before becoming a founder and executive of R&D Associates, based Marina Del Rey, California. In 1980, he served as an advisor to presidential candidate Reagan and was a member of the President-elect's transition team. He then served as chair of the General Advisory Committee on Arms Control and Disarmament for three years. Nominated as NASA's deputy administrator by President Reagan on September 12, he was confirmed by the Senate on November 18, 1985. Graham left NASA in October 1986 to become director of the White House Office of Science and Technology Policy, eventually becoming science advisor to Reagan, a position he held until June 1989 when he left government service to join Jaycor, a high-technology company. See "Graham, Wm.," Deputy Administrator files, NASA Historical Reference Collection.

Jay Greene (1942–) joined the Johnson Space Center in 1965. From 1974 to 1980, he headed the Flight Dynamics Section and served as a flight dynamics technician for the Apollo program. From 1980 to 1982, he was chief of Mission Operations Branch. In 1987, he became chief of the Safety Division at Johnson and served in that capacity until he became deputy manager of the Space Shuttle program in 1989. In 1991, he was appointed deputy associate administrator for exploration. See "Greene, Jay H.," biographical file, NASA Historical Reference Collection.

Michael Griffin is the senior vice president and chief of technology at Orbital Sciences Corporation. Previously, he was NASA's chief engineer from 1993 to 1994. Griffin received his master's degrees in aerospace science from Catholic University and in electrical engineering from the University of Southern California. He later received his Ph.D. in aerospace engineering from the University of Maryland. See "Griffin, Michael," biographical file, NASA Historical Reference Collection.

H

John P. Hagen (1908–1990) was director of the Vanguard program during the 1950s. He had been an astronomer at Wesleyan University (1931–35) before working for the Naval Research Laboratory (1935–58). With the creation of NASA, he became the assistant director of spaceflight development (1958–60), and in 1962, he returned to higher education, becoming a professor of astronomy at Pennsylvania State University. See obituary in *New York Times*, September 1, 1990, p. 25.

James C. Hagerty (1909–1981) was on the staff of *The New York Times* from 1934 to 1942, the last four years as legislative correspondent at the paper's Albany bureau. He served as executive assistant to New York Governor Thomas Dewey from 1943 to 1950 and then as Dewey's press secretary for the next two years before becoming press secretary for President Eisenhower from 1953 to 1961. See "Miscellaneous Other Agencies," biographical file, NASA Historical Reference Collection.

Eldon W. Hall (1919–) joined NASA as chief of analysis and requirements in October 1958. He joined the NACA in 1943 at the Lewis Research Center in Ohio when it was known as the Aircraft Research Laboratory and served as head of the Propulsion Systems Analysis Section there from 1946 to 1951. With NASA, he served as assistant director of systems from July 1962 to December 1963, director of Gemini systems engineering from December 1963 to November 1966, and director of advanced manned mission systems engineering from November 1966 to April 1968. Hall later worked for General Electric as the manager of advanced technology projects. In 1996, he wrote the book *Journey to the Moon: The History of the Apollo Guidance Computer.* See "Hall, Eldon W.," biographical file, NASA Historical Reference Collection.

Bryce N. Harlow (1916–1987) was a part of the congressional staff from 1938 to 1951, rising to be chief clerk in 1950. From 1951 to 1952, he was vice president of Harlow Publishing Corporation in Oklahoma City. He returned to Washington and held positions on the White House staff, beginning in 1953, becoming deputy assistant to the President for congressional affairs in 1959. In 1961, he became director of governmental relations for Proctor and Gamble Manufacturing Company until 1969, when rejoined the White House as assistant to the President for legislative and congressional affairs. He became counselor to the President from 1969 to 1970 and then served as vice president of Proctor and Gamble (1970–73) before returning again to the White House as counselor to President Nixon at the height of the Watergate scandal, remaining until April 1974, when he resigned and returned to private life. See *San Francisco Chronicle,* obituaries, February 18, 1987, p. 32; Allan Cromley, "Sooner Presidential Advisor Bryce Harlow Dies," *Daily Oklahoman,* February 18, 1987.

Klaus P. Heiss (1942–) is an Austrian-born economist who prepared a major economic feasibility study for the Space Shuttle program in 1971. He later worked with Econ, Inc., and founded and headed Space Transportation Corporation in Princeton, New Jersey. See "Heiss, Klaus P.," biographical file, NASA Historical Reference Collection.

Lawrence F. Herbolsheimer joined NASA in June 1986, having previously served in various senior management positions in the public and private sectors, including being a corporate planning/business development manager in the International Division of Container Corporation of America in Chicago. He was a co-founder of Vertechs Corporation, Montgomery Foods, Inc., Apex Corporation, and Middle West Consultants, Ltd. He also served as associate director in the White House Office of Cabinet Affairs, was a presidential appointee to the Commercial Space Working Group under the Cabinet Council on Commerce and Trade, and was a representative to executive branch committees. Herbolsheimer was appointed deputy assistant administrator for the Office of Commercial Programs at NASA Headquarters. In this position, he was responsible for advancing the interests and participation of the private sector in the U.S. space program. He was named by NASA Administrator Dr. James C. Fletcher to serve as the acting assistant administrator until a permanent replacement could be selected. See "Herbolsheimer, Lawrence," biographical file, NASA Historical Reference Collection.

William M. Holaday (1901–) was special assistant to the secretary of defense for guided missiles between 1957 and 1958. He then was Department of Defense director of guided missiles in 1958 and chairman of the Civilian-Military Liaison Committee from 1958 to 1960. Previously, he had been associated with a variety of research and development activities, notably as director of research for the Socony-Mobil Oil Company (1937–44). See "William M. Holaday," biographical file, NASA Historical Reference Collection.

John K. Holcomb (1920–1989) was a NASA official and Navy captain. In 1962, he was assigned to NASA as the assistant director of launch operations in the Office of Manned Space Flight. He worked at NASA Headquarters throughout the Apollo program before retiring in 1976. After retirement, he served as docent at the National Air and Space Museum in Washington until 1986. See "Holcomb, John K. Capt.," biographical file, NASA Historical Reference Collection.

D. Brainerd Holmes (1921–) was involved in the management of high-technology efforts in private industry and the federal government. He was on the staff of Bell Telephone Laboratories (1945–53) and at RCA (1953–61). He then became deputy associate administrator for the NASA Office of Manned Space Flight from 1961 to 1963. Thereafter, he assumed a series of increasingly senior positions with Raytheon Corporation and, since 1982, has been chairman of Beech Aircraft. See "D. Brainerd Holmes," biographical file, NASA Historical Reference Collection; "Holmes, D(yer) Brainerd," *Current Biography Yearbook 1963*, pp. 191–92.

Paul F. Holloway (1938–) was the director of NASA's Langley Research Center from 1991 to 1996. He began his NASA career at Langley in 1960 as an aerospace research engineer. In 1969, he was appointed head of the Systems Analysis Section and became the head of that branch in 1971. He continued his service to NASA and was named deputy director of Langley in 1985, until he became director in 1991. Holloway retired in 1996. See "Holloway, Paul," biographical file, NASA Historical Reference Collection.

George W. Hoover was an early space enthusiast who entered the Navy in 1944 and became a pilot. He moved to the Office of Naval Research to conduct a program in all-weather flight instrumentation. Later, he helped originate the idea of high-altitude balloons, used in a variety of projects, such as Skyhook, which supported cosmic-ray research and served as a research vehicle for obtaining environmental data relevant to supersonic flight, among other uses. In 1954, he was project officer in the field of high-speed, high-altitude flight, with involvement in the Douglas D-558 project leading to the X-15. Hoover was also instrumental in establishing Project Orbiter with von Braun and others, resulting in the launch of Explorer I, the first American satellite. See "George W. Hoover," biographical file, NASA Historical Reference Collection.

S. Neil Hosenball joined NASA in 1961 as an attorney. He served as the agency's general counsel from 1975 to 1985. Upon his retirement from NASA, he became director of the University of Colorado's Center for Space Law and Policy. See "S. Neil Hosenball," biographical file, NASA Historical Reference Collection.

Robert B. Hotz served as a member of the Rogers Commission investigating the Space Shuttle *Challenger* accident. He was the editor of *Aviation Week & Space Technology* from 1953 to 1980. See "Hotz, Robert," biographical file, NASA Historical Reference Collection.

Gary Hudson (1950–) attended the University of Minnesota and was an *Aviation Week* "Laurel" award recipient in 1994 for outstanding achievement in the field of space, having been working in the field of commercial space for more than twenty-five years. From 1980 to 1981, he was president of GCH, Inc., and chief systems designer of the Percheron 055 experimental launch vehicle, the first private launcher built in the United States. He then co-founded Pacific American Launch Systems, Inc., becoming president and chief executive officer. Also during this time, he was a consultant to the U.S. Air Force's Project Forecast II. In 1994, Hudson co-founded HMX, Inc., which developed a monopropellant rocket engine propulsion system for Kistler Aerospace Corporation. He is currently a founder and member of the Board of Directors of the Rotary Rocket Company, also currently serving as president and chief executive officer, as well as a board member of the Space Transportation Association. See "Gary C. Hudson," biographical file, Rotary Rocket Company, Redwood City, CA.

Maxwell W. Hunter II (1922–) was the principal designer of the Civilian Delta Rocket. He joined the National Aeronautics Space Council in the early 1960s, helping decision-makers during the Apollo years. In the late 1960s, he returned to designing and advocating advanced space systems, including the Hubble Space Telescope. In recent years, he has originated the SSX, a single-stage-to-orbit launch system. See "Hunter, Maxwell W., II," biographical file, NASA Historical Reference Collection.

Abraham Hyatt (1910–) earned a B.S. in aeronautical engineering from Georgia Tech in 1933. After working for the U.S. Geodetic Survey and private industry, he became head of the Design Research Branch for the Navy's Bureau of Aeronautics in 1948 and advanced to chief scientist and research analysis officer (1956–58). In 1959, he became assistant director for propulsion in NASA. The following year, he became director of NASA's Office of Program Planning and Evaluation. He remained in that position until 1964, when he became a professor at MIT and then, in 1965, executive director for corporate planning at North American Aviation, Inc. See "Abraham Hyatt," biographical file, NASA Historical Reference Collection.

J

George Jeffs was president of Rockwell International's North American Aerospace Operations, as well as president of the Space Systems Group, in the early 1980s. As such, he was heavily involved with the Space Shuttle's development. See "Jeffs, George," biographical file, NASA Historical Reference Collection.

Roy W. Johnson (1906–1965) was named the first director of the Department of Defense's Advanced Research Projects Agency and served from 1958 to 1959. As such, he was head of Defense Department's initial space efforts. Prior to joining the government, he worked for General Electric and retired as an executive vice president. See "Johnson, Roy W.," biographical file, NASA Historical Reference Collection.

J. Wallace Joyce (1907–1970) was an engineer and the head of the International Geophysical Year Office of the National Science Foundation. A few years after graduating school, he worked for the Navy as an electric engineer (1937–42). In 1949, he was employed by the Department of Defense in the mutual defense assistance program. That same year, he also worked for the Department of State (1949–50) as the deputy science advisor before becoming the head of the International Geophysical Year Office in 1955. See *Who's Who in America 1958-1959*, 30th edition, vol. 1 (Chicago: Marquis Who's Who, 1959); *Who Was Who in America 1969-1973*, vol. 5 (Chicago: Marquis Who's Who, 1973).

K

Joseph Kaplan (1902–1991) was born in Tapolcza, Hungary, and came to the United States in 1910. He trained as a physicist at Johns Hopkins University and worked on the faculty of the University of California at Berkeley from 1928 until his retirement in 1970. He directed the university's Institute of Geophysics (later the Institute of Geophysics and Planetary Physics) from the time of its creation in 1944. Kaplan was heavily involved in efforts in the 1950s to launch the first artificial Earth satellite, serving as the chair of the U.S. National Committee for the International Geophysical Year (1953–63). See "Kaplan, Joseph," biographical file, NASA Historical Reference Collection; Joseph Kaplan, "The Aeronomy Story: A Memoir," in R. Cargill Hall, ed., *Essays on the History of Rocketry and Astronautics: Proceedings of the Third Through the Sixth History Symposia of the International Academy of Astronautics* (Washington, DC: NASA Conference Publication 2014, 1977), 2: 423–27; Joseph Kaplan, "The IGY Program," *Proceedings of the IRE* (June 1956): 741–43.

Theodore von Kármán (1881–1963) was a Hungarian aerodynamicist who founded the Aeronautical Institute at Aachen before World War I and achieved a world-class reputation in aeronautics through the 1920s. In 1930, Robert A. Millikan and his associates at Caltech lured von Kármán from Aachen to become the director of the Guggenheim Aeronautical Laboratory at Caltech (GALCIT). There, he trained a generation of engineers in theoretical aerodynamics and fluid dynamics. With its eminence in physics, physical chemistry, and astrophysics as well as aeronautics, it proved to be an almost ideal site for the early development of U.S. ballistic rocketry. See Judith R. Goodstein, *Millikan's School: A History of California Institute of Technology* (New York: W.W. Norton, 1991); Clayton R. Koppes, *JPL and the American Space Program: A History of the Jet Propulsion Laboratory* (New Haven, CT: Yale University Press, 1982); Michael H. Gorn, *The Universal Man: Theodore von Kármán's Life in Aeronautics* (Washington, DC: Smithsonian Institution Press, 1992).

John F. Kennedy (1916–1963) was the thirty-fifth U.S. president from 1961 to 1963. A senator from Massachusetts between 1953 and 1960, he successfully ran for president as the Democratic candidate, with party wheelhorse Lyndon B. Johnson as his running mate. Using the slogan "Let's get this country moving again," Kennedy charged the Republican Eisenhower administration with doing nothing about the myriad social, economic, and international problems that festered in the 1950s. He was especially hard on Eisenhower's record in international relations, taking a "cold warrior" position on a supposed "missile gap" (which turned out not to be the case), wherein the United States lagged far behind the Soviet Union in intercontinental ballistic missile technology. On May 25, 1961, President Kennedy announced to the nation the goal of sending an American to the Moon before the end of the decade. The human spaceflight imperative was a direct outgrowth of it; Projects Mercury (at least in its latter stages), Gemini, and Apollo were each designed to execute it. On this subject, see Walter A. McDougall, . . . *The Heavens and the Earth: A Political History of the Space Age* (New York: Basic Books, 1985); John M. Logsdon, *The Decision to Go to the Moon: Project Apollo and the National Interest* (Cambridge, MA: MIT Press, 1970).

George A. Keyworth II (1939–   ) was director of the Office of Science and Technology Policy and science advisor to President Ronald Reagan between 1981 and 1986. Formerly the head of the Los Alamos Scientific Laboratory, Keyworth received a Ph.D. in nuclear physics from Duke University in 1968. He began work at Los Alamos after graduation and remained there until 1981. See "Keyworth, George A(lbert), 2d," *Current Biography Yearbook 1986*, pp. 265–68.

James R. Killian, Jr. (1904–1988), was president of MIT between 1949 and 1959, on leave between November 1957 and July 1959 when he served as the first presidential science advisor. President Dwight D. Eisenhower established the President's Science Advisory Committee, which Killian chaired, following the Sputnik crisis. After leaving the White House staff in 1959, Killian continued his work at MIT but in 1965 began working with the Corporation for Public Broadcasting to develop public television. Killian described his experiences as a presidential advisor in *Sputnik, Scientists, and Eisenhower: A Memoir of the First Special Assistant to the President for Science and Technology* (Cambridge, MA: MIT Press, 1977). For a discussion of the President's Science Advisory Committee, see Gregg Herken, *Cardinal Choices: Science Advice to the President from Hiroshima to SDI* (New York: Oxford University Press, 1992).

George B. Kistiakowsky (1900–1982) was a pioneering chemist at Harvard University, associated with the development of the atomic bomb, and later an advocate of banning nuclear weapons. He served as science advisor to President Eisenhower from July 1959 to the end of the administration. He later served on the advisory board to the U.S. Arms Control and Disarmament Agency from 1962 to 1969. See *New York Times*, December 9, 1982, p. B21; "George B. Kistiakowsky," biographical file, NASA Historical Reference Collection.

S. J. Kline (1922–   ) is a mechanical engineer specializing in fluid mechanics. Educated at Stanford University and then MIT, he has been recognized through a number of awards for his work in his field. He has spent most of his career as a professor at Stanford is noted for his work on internal flow and zonal modeling of turbulent flows. See *American Men and Women of Science: 1998–1999*, 20th edition, vol. 4 (New Providence, NJ: RR Bowker, 1999), p. 465.

Hugh J. Knerr (1887–1971) is recalled as "a man of courage, vision, and organizational genius whose contributions to the establishment and shaping of the United States Air Force are a legacy to be treasured by those who have followed after him." Knerr graduated from the Naval Academy in 1908 and went through the programs at the U.S. Army Staff School, the U.S. Army War College, and various special schools in the Air Corps. His military career was spent fighting to get close to airplanes and, once there, fighting to prove their worth. Leaving the Army Air Corps in 1942, he returned to service and rose to brigadier general in the U.S. Strategic Air Forces in Europe. See John L. Frisbee, *Makers of the United States Air Force* (Washington, DC: U.S. Government Printing Office, 1996); *Who Was Who in America 1982–1985*, vol. 8 (Chicago: Marquis Who's Who, 1985); John. S. Bowman, ed., *The Cambridge Dictionary of American Biography* (Cambridge, Eng.: Cambridge University Press, 1995).

Sergei P. Korolev (1907–1966) was the chief designer of spacecraft and rockets for the Soviet space program. Imprisoned by Stalin for many years, he was finally "rehabilitated" after Stalin's death in 1953. By 1945, he had developed the long-range missile. In the 1950s he headed a design team that made an intercontinental ballistic missile. He later directed the launches of the first satellite, the first single and multi-crew missions, and the Soviet human lunar program. He died of complications during surgery in 1966. See "Sergei Korolev," biographical file, NASA Historical Reference Collection.

Christopher C. Kraft, Jr. (1924–   ), was a long-standing official with NASA throughout the Apollo program. He received a bachelor of science degree in aeronautical engineering from Virginia Polytechnic and State University in 1944 and joined the Langley Aeronautical Laboratory of the NACA the next year. In 1958, still at Langley, he became a member of the Space Task Group developing Project Mercury and moved with the group to Houston in 1962. He was flight director for all of the Mercury and many of the Gemini missions and directed the design of Mission Control at the Manned Spacecraft Center, which was renamed the Johnson Space Center in 1973. He was named the Manned Spacecraft Center's deputy director in 1970 and its director two years later, a position he held until his retirement in 1982. Since then, he has remained active as an aerospace consultant. See "Kraft, Christopher C., Jr.," biographical file, NASA Historical Reference Collection.

Donald J. Kutyna (1933–   ) was the commander of the Air Force Space Command from November 1987 through March 1990. A West Point graduate, in 1982, Kutyna became the deputy commander for space launch and control systems at the Space Division of the Air Force Systems Command. In that position, he managed the military's participation in the Space Shuttle program and assumed responsibility for the Air Force's expendable launch vehicles. He has achieved the rank of lieutenant general. See "Kutyna, Donald," biographical file, NASA Historical Reference Collection.

L

Chester M. Lee (1919–   ) retired from the U.S. Navy in 1965. His Navy career included a variety of guided missile programs such as the Polaris submarine-launched ballistic missile. He joined NASA as the chief of plans at the Mission Operations Directorate and later served as assistant mission director for Apollo 11 and mission director for Apollo 12 and 13. He was then named program director at NASA Headquarters for the Apollo-Soyuz Test Project. These assignments were followed by director of Shuttle operations and then director of Shuttle customer utilization and services. He retired from NASA in 1987 as assistant associate administrator for policy, planning, and DOD affairs in the Office of Space Flight. He joined SPACEHAB, Inc., as executive vice president and later became president of the company. See "Lee, Chester M.," biographical file, NASA Historical Reference Collection.

Lyman L. Lemnitzer (1899–1988) was a career Army officer who served as Army vice chief of staff (1957–59), Army chief of staff (1959–60), chairman of the Joint Chiefs of Staff (1960–62), commanding general of U.S. Forces in Europe (1962–69), and supreme allied commander in Europe (1963–69). See William Gardner Bell, *Commanding Generals and Chiefs of Staff: Portraits & Biographical Sketches* (Washington, DC: Center of Military History, 1982), p. 132; *New York Times*, November 13, 1988, p. 44.

Charles A. Lindbergh (1902–1974) was an early aviator who gained fame as the first pilot to fly solo across the Atlantic Ocean in 1927. His public stature following this flight was such that he became an important voice on behalf of aerospace activities until his death. He served on a variety of national and international boards and committees, including the central committee of the NACA. He became an expatriate living in Europe, following the kidnapping and murder of his two-year-old son in 1932. In Europe, during the rise of fascism, Lindbergh assisted American aviation authorities by providing them with information about European technological developments. After 1936, he was especially important in warning the United States of the rise of Nazi air power. He assisted with the war effort in the 1940s by serving as a consultant to aviation companies and the government, and after the war, he lived quietly in Connecticut and then Hawaii. See Walter S. Ross, *The Last Hero: Charles A. Lindbergh* (New York: Harper & Row, 1967).

Robert Lindstrom joined NASA's Marshall Space Flight Center in 1960 at its establishment and became the program manager for the Saturn I/IB launch vehicle. He was named deputy manager of the Space Shuttle Project Office at Marshall in 1972 and was named manager in 1974. See "Lindstrom, Robert," biographical file, NASA Historical Reference Collection.

J. Wayne Littles (1939–   ) received his bachelor's degree in mechanical engineering from Georgia Tech in 1962. That same year, he became an aerospace engineer with Rocketdyne in California. While there, he worked on the Saturn launch vehicle engines. After receiving his master's, in the same field, from the University of Southern California, he worked for Teledyne Brown Engineering in Huntsville, Alabama, as a research engineer. He joined NASA in 1967, moving rapidly through the ranks. Beginning as an engineer in the former Propulsion and Vehicle Engineering Directorate, he was appointed deputy director of the Marshall Space Flight Center in 1989. In 1994, he was assigned to NASA Headquarters as chief engineer. Later that year, he was named associate administrator for the Office of Space Flight. In early 1996, Littles returned to Marshall as the center's eighth director. He retired from NASA in 1997. See "Littles, J. Wayne," biographical file, NASA Historical Reference Collection; "Dr. J. Wayne Littles," biography, Marshall's NASA Historical Reference Collection.

Alan M. Lovelace (1929–) was born in St. Petersburg, Florida, and was educated at the University of Florida in Gainesville, receiving a bachelor of science in chemistry in 1951, a master of science in organic chemistry in 1952, and a Ph.D. in organic chemistry in 1954. Shortly after the end of the Korean Conflict, he served in the U.S. Air Force from 1954 to 1956. Thereafter, he began work as a government scientist at the Air Force Materials Laboratory at Wright-Patterson Air Force Base in Dayton, Ohio. In January 1964, he was named chief scientist of the laboratory and then its director in 1967. In October 1972, Lovelace was named director of science and technology for the Air Force Systems Command at Headquarters, Andrews Air Force Base, Maryland. In September 1973, he became the principal deputy to the assistant secretary of the Air Force for research and development. In September 1974, he left the Department of Defense to become the associate administrator of the NASA Office of Aeronautics and Space Technology. With the departure of George Low as NASA deputy administrator in June 1976, Lovelace became deputy administrator, serving until July 1981. He retired from NASA to accept a position as corporate vice president—science and engineering with the General Dynamics Corporation in St. Louis, Missouri. See "Lovelace, Alan M.," Deputy Administrator files, NASA Historical Reference Collection.

George M. Low (1926–1984), a native of Vienna, Austria, came to the United States in 1940 and received an aeronautical engineering degree from Rensselaer Polytechnic Institute in 1948 and a master of science in the same field from that school in 1950. He joined the NACA in 1949, and at Lewis Flight Propulsion Laboratory, he specialized in experimental and theoretical research in several fields. He became chief of manned spaceflight at NASA Headquarters in 1958. In 1960, Low chaired a special committee that formulated the original plans for the Apollo lunar landings. In 1964, he became deputy director of the Manned Spacecraft Center in Houston, the forerunner of the Johnson Space Center. He became deputy administrator of NASA in 1969 and served as acting administrator from 1970 to 1971. He retired from NASA in 1976 to become president of Rensselaer, a position he held until his death. In 1990, NASA renamed its quality and excellence award after him. See "Low, G.M.," Deputy Administrator files, NASA Historical Reference Collection.

W. R. Lucas was the director of NASA's Marshall Space Flight Center from 1974 to 1986. He played a key role in the development of the Saturn V rocket and an even greater role in the development of the Space Shuttle. Lucas was with NASA for more than thirty years before his retirement in 1986. See "Lucas, W.R.," biographical file, NASA Historical Reference Collection.

M

Richard C. McCurdy (1909–), an engineer specializing in petroleum, was associate administrator for organization and management at NASA Headquarters in Washington, D.C., from 1970 to 1973 and a consultant to the agency from 1973 to 1982. See "McCurdy, Richard," biographical file, NASA Historical Reference Collection.

Robert McFarlane (1937–) was President Reagan's national security advisor from October 1983 to December 1985. Prior to that position, McFarlane had been deputy to his predecessor William Clark, counselor to Alexander M. Haig, Jr., when he was secretary of state, a member of the staff of the Senate Armed Services Committee, and military aide to Henry Kissinger when he was national security advisor to President Nixon. An Annapolis graduate, he commanded the first U.S. Marine battery to land in the Republic of South Vietnam, and he completed two tours there. Along with Oliver North, McFarlane is considered a key figure in the Iran-Contra scandal of the early 1980s, in which arms were secretly traded to Iran in exchange for hostages, and resulting monies were funneled into the illegal support of Nicaraguan Contras. After a lengthy investigation, McFarlane was convicted of unlawfully withholding information from Congress about his and North's Contra-support activities and about the solicitation of foreign funding for the Contras, although he was later pardoned by President Bush. See *Who's Who in America 1984–1985*, 43rd edition, vol. 2 (Chicago: Marquis Who's Who, 1985).

Robert S. McNamara (1916–) was secretary of defense during the Kennedy and Johnson administrations from 1961 to 1968. Thereafter, he served as president of the World Bank, where he remained until retirement in 1981. As secretary of defense in 1961, McNamara was intimately involved in the process of approving Project Apollo by the Kennedy administration. See "McNamara, Robert S(trange)," *Current Biography Yearbook 1987*, pp. 408–13; John M. Logsdon, *The Decision to Go to the Moon: Project Apollo and the National Interest* (Cambridge, MA: MIT Press, 1970).

Frank J. Malina (1912–1981) was a young Caltech Ph.D. student in the mid-1930s when he began an aggressive rocket research program to design a high-altitude sounding rocket. Beginning in late 1936, Malina and his colleagues started the static testing of rocket engines in the canyons above the Rose Bowl, with mixed results, but a series of tests eventually led to the development of the WAC-Corporal rocket during World War II. After the war, Malina worked with the United Nations and eventually retired to Paris to pursue a career as an artist. See "Malina, Frank J.," biographical file, NASA Historical Reference Collection.

Myron Malkin (1924–1994) was director of the Space Shuttle development program from 1973 to 1980. After serving in the Marine Corps during World War II, he earned M.S. and Ph.D. degrees in nuclear physics from Yale University. During the 1960s, he worked for General Electric on missile and launch vehicle programs, as well as on the Manned Orbiting Laboratory, which was never built. In 1972, he was named deputy secretary of defense for intelligence. After he retired from NASA in 1980, he worked for Fairchild Industries and established his own aerospace consulting firm. See "Malkin, Myron," biographical file, NASA Historical Reference Collection.

Hans Mark (1929–   ) became NASA deputy administrator in July 1981. He had previously served as secretary of the Air Force from July 1979 until February 1981 and as under secretary of the Air Force since 1977. In February 1969, Mark became director of NASA's Ames Research Center in Mountain View, California, where he managed the center's research and applications efforts in aeronautics, space science, life science, and space technology. Born in Mannheim, Germany, he came to the United States in 1940 and became a citizen in 1945. He received a Ph.D. in physics from MIT in 1954. Upon leaving NASA, he became chancellor of the University of Texas at Austin. See "Mark, Hans," Deputy Administrator files, NASA Historical Reference Collection.

Robert P. Mayo (1916–   ) was an economist and President Nixon's first director of the Bureau of the Budget. On July 1, 1970, when the Bureau of the Budget was replaced with the Office of Management and Budget, Mayo was shifted to the White House as a presidential assistant. Shortly thereafter, he left Washington to assume the presidency of the Federal Reserve Bank of Chicago. See "Mayo, Robert P(orter)," *Current Biography Yearbook 1970*, pp. 282–84.

Elliott Mitchell earned a B.S. in chemistry from William and Mary in 1941 and served from 1942 to 1950 as a physical chemist and chemical engineer in the Department of the Navy. From then until 1958, he was physical sciences administrator and then chief of propulsion research and development in the Navy's Bureau of Ordnance. In 1958, he joined NASA as chief of the solid rocket development program. When he left NASA in 1961, he was assistant director of manned spaceflight programs for propulsion. Thereafter, he became a consultant. See "Elliott Mitchell," biographical file, NASA Historical Reference Collection.

Walter F. Mondale (1928–   ) was the U.S. Vice President under President Jimmy Carter (1977–81). He ran for President himself in 1984 but lost to incumbent Ronald Reagan. Mondale served in the Senate as a Democrat from Minnesota from 1964 to 1977 and was considered a harsh critic of large technology programs such as the Space Shuttle. He also served as the Clinton administration's ambassador to Japan. See "Mondale, Walter," biographical file, NASA Historical Reference Collection.

Jesse W. Moore joined NASA's Jet Propulsion Laboratory in 1966 and worked in a variety of areas. He went to NASA Headquarters in 1978 as deputy director of the Solar Terrestrial Division in the Office of Space Science. In June 1979, he was appointed director of the Space Flight Division, and in December 1981, he became director of the Earth and Planetary Exploration Division. In April 1984, he assumed the position of the associate administrator for the Office of Space Flight. He was named director of the Johnson Space Center on January 23, 1986, where he remained until reassignment to the general manager in October 1986. After the *Challenger* accident, he resigned from NASA to become the director of program development with Ball Aerospace Systems. In August 1993, he became the vice president of Washington operations at Ball. See "Moore, Jesse W.," biographical file, NASA Historical Reference Collection.

Thomas S. Moorman, Jr., commanded the Air Force Space Command from 1990 to 1992 and then assumed the position of vice commander until 1994. From then until his retirement in 1997, Moorman was vice chief of staff at the U.S. Air Force Headquarters. A driven individual, Moorman has exhibited great leadership, vision, and commitment to advancing all levels of U.S. presence in space. He attended Dartmouth College, received his MBA from Western New England College, and his M.A. from Auburn University. Moorman's vision for the future of the Air Force is one totally integrated with spaced-based systems, from communications to mapping to eventual space weaponry. Before serving at the Space Command, he worked as an intelligence officer and an operations officer. See "Moorman, Thomas S. Jr.," biographical file, NASA Historical Reference Collection.

Oskar Morgenstern (1902–1977) was a German-born and -trained economist. He came to the United States in 1925 and worked at Princeton University after 1938, becoming the first director of the Econometric Research Program there in 1948. He founded and headed Mathematica, Inc., which provided economic analyses to government and industry.

Gerald J. Mossinghoff (1935–   ) joined NASA's Office of General Counsel in 1963. He transferred to the U.S. Patent Office for a brief time and then returned to NASA in 1967 as the director of the Congressional Liaison Division. In 1971, he became deputy assistant administrator (policy) in the Office of Legislative Affairs, and in 1974, he was appointed assistant general counsel of NASA. From 1976 to 1981 he was NASA's deputy general counsel. Mossinghoff left NASA in 1981 to join the U.S. Patent Office as patent commissioner. He resigned in 1984 to take over the presidency of the Pharmaceutical Manufacturers Association. See "Mossinghoff, Gerald J.," biographical file, NASA Historical Reference Collection.

George E. Mueller (1918–   ) was the associate administrator for NASA's Office of Manned Space Flight from 1963 to 1969. As such, he was responsible for overseeing the completion of Project Apollo and beginning the development of the Space Shuttle. He moved to General Dynamics as senior vice president in 1969, where he remained until 1971. He then became president of Systems Development Corporation (1971–80) and then its chair and corporate executive officer (1981–83). See "Mueller, George E.," biographical file, NASA Historical Reference Collection.

Dale D. Myers (1922–   ) served as NASA deputy administrator from October 1986 until 1989. He was also the associate administrator for NASA's Office of Manned Space Flight from 1970 to 1974. From 1974 to 1977, he was a vice president at Rockwell International and president of North American Aircraft Group in El Segundo, California. Then he was under secretary of the U.S. Department of Energy from 1977 to 1979. Earlier, Myers was vice president and program manager, Apollo Command/Service Module Program, North American-Rockwell, from 1964 to 1969. From 1969 to 1970, he served as vice president and program manager, Space Shuttle Program, Rockwell International. After leaving NASA in 1989, he returned to private industry. See "Myers, Dale D.," Deputy Administrator files, NASA Historical Reference Collection.

N

J.V. Naish (1929–   ) was the president of Convair and the senior vice president of General Dynamics Corporation. He was an airplane executive who had also been the director of McDonnell Aircraft. See *Who's Who in America 1964–1965*, 33rd edition, vol. 2 (Chicago: Marquis Who's Who, 1965).

John von Neumann (1903–1959) was a famous, brilliant mathematician and head of Princeton University's Institute for Advanced Study. He received a diploma in chemical engineering from Zurich's Federal Institute of Technology and a Ph.D. in mathematics from the University of Budapest, both in the same year. During World Ware II, he helped build the atomic bomb. He later envisioned the computer as a highly flexible logic machine, helping pave the way for present-day computers. Von Neumann invented at least three new fields: cellular automata theory (showing how inanimate cells can be made to behave as if they were alive), game theory (math in the art of decision-making), and the study of the similarity between minds and computers. See "Von Neumann, John," biographical file, NASA Historical Reference Collection.

John S. Newton (1908–   ) is an electrical and mechanical engineer and president of Newton Engineering. He previously worked for Westinghouse Electric as well as Baldwin Locomotive Works, for which he became vice president of the Locomotive Division in 1951. His work centers on future technologies and applications, such as alternate sources of energy or the manufacturing of hybrid diesel-electric cars. See *American Men and Women of Science: 1998-1999*, 20th edition, vol. 5 (New Providence, NJ: RR Bowker, 1999), p. 823.

Richard M. Nixon (1913–1994) was the thirty-seventh U.S. President between January 1969 and August 1974. Early in his presidency, he appointed a Space Task Group under the direction of Vice President Spiro T. Agnew to assess the future of spaceflight in the nation. Its report recommended a vigorous post-Apollo exploration program culminating in a human expedition to Mars. Nixon did not approve this plan, but he did decide in favor of building one element of it, the Space Shuttle, which was approved on January 5, 1972. See Roger D. Launius, "NASA and the Decision to Build the Space Shuttle, 1969-72," *The Historian* 57 (Autumn 1994): 17–34.

Robert G. Nunn, Jr. (1917–1975), earned a law degree from the University of Chicago in 1942. After four years in the Army during World War II, then private practice of law for eight years in Washington, D.C., and in his hometown of Terre Haute, Indiana, he joined the Air Force Office of General Counsel in 1954. He became NASA's assistant general counsel in November 1958 and then special assistant to T. Keith Glennan in September 1960. He helped draft many legal and administrative regulations for NASA, and then he went to work for the Washington law firm of Sharp and Bogan. Later, he formed the firm of Batzell and Nunn, specializing in energy legislation and administrative law. See "Nunn, R.G., Jr.," biographical file, NASA Historical Reference Collection.

O

Hermann J. Oberth (1894–1989) is one of the three recognized fathers of spaceflight. A Transylvanian by birth but a German in his family heritage, he was educated at the Universities of Klausenburg, Munich, Göttingen, and Heidelberg. His doctoral dissertation being rejected because it did not fit into any established scientific discipline, he published it privately as *Die Rakete zu den Planetenräumen* (*The Rocket into Interplanetary Space*) in 1923. It and its expanded version, titled *Ways to Spaceflight* (1929), set forth the basic principles of spaceflight and directly inspired many subsequent spaceflight pioneers, including Wernher von Braun. See his "Hermann Oberth: From My Life," *Astronautics*, June 1959, pp. 38–39, 100–06; Frank Winter, *Rockets into Space* (Cambridge, MA: Harvard University Press, 1990), pp. 17–25; Helen B. Walters, *Hermann Oberth: Father of Space Travel* (New York: Macmillan, 1962).

Bryan O'Connor (1947–   ) headed the Space Shuttle program at NASA Headquarters from 1994 to 1996. He was an astronaut from 1980 to 1991, commanding a Space Shuttle mission in 1991 and piloting another in 1985. Prior to that, he was a Marine Corps pilot. In 1996, NASA awarded him the Exceptional Service Medal. He resigned from NASA that same year. See "O'Connor, Major Brian D.," biographical file, NASA Historical Reference Collection.

Verne Orr (1916–   ) served as secretary of the Air Force from 1981 to 1985. He received a master's degree in business administration from Stanford University in 1939. After serving in World War II and working in the family car dealership, he served in California state government from the mid-1960s to the mid-1970s. From 1975 to 1980, he taught government finance courses at the University of Southern California. See George M. Watson, Jr., *Secretaries and Chiefs of Staff of the United States Air Force: Biographical Sketches and Portraits* (Washington, DC: Air Force Historical Support Office, 1999).

P

Thomas O. Paine (1921–1992) was appointed deputy administrator of NASA on January 31, 1968. Upon the retirement of James E. Webb on October 8, 1968, he was named acting administrator of NASA. He was nominated as NASA's third administrator on March 5, 1969, and confirmed by the Senate on March 20. During his leadership, the first seven Apollo manned missions were flown, in which twenty astronauts orbited Earth, fourteen traveled to the Moon, and four walked upon its surface. He resigned from NASA on September 15, 1970, to return to the General Electric Company in New York City as vice president and executive of the Power Generation Group, where he remained until 1976. In 1985, the White House chose Paine as chair of a National Commission on Space to prepare a report on the future of space exploration. Since leaving NASA fifteen years earlier, Paine had been a tireless spokesperson for an expansive view of what should be done in space. The Paine Commission took most of a year to prepare its report, largely because it solicited public input in hearings throughout the United States. The report, *Pioneering the Space Frontier*, was published in a lavishly illustrated, glossy format in May 1986. It espoused a "pioneering mission for 21st-century America . . . to lead the exploration and development of the space frontier, advancing science, technology, and enterprise, and building institutions and systems that make accessible vast new resources and support human settlements beyond Earth orbit, from the highlands of the Moon to the plains of Mars." The report also contained a "Declaration for Space" that included a rationale for exploring and settling the solar system and outlined a long-range space program for the United States. See Roger D. Launius, "NASA and the Decision to Build the Space Shuttle, 1969–72," *The Historian* 57 (Autumn 1994): 17–34.

Gary Payton has been deputy associate administrator for what is now known as NASA's Office of Aero-Space Technology since 1996. In this position, he oversees a variety of technology development activities for future space transportation systems. He joined NASA in the spring of 1995. Prior to that, He served for more than 23 years in the U.S. Air Force. In 1985, he flew aboard the Space Shuttle *Discovery* as a payload specialist and later became the deputy for technology in the Ballistic Missile Defense Organization. See "Payton, Gary," biographical file, NASA Historical Reference Collection.

G. Edward Pendray (1901–1987) had been a proponent of peaceful uses of rocket power since the 1930s. He joined the staff of the *New York Herald Tribune* in 1925 as a reporter, later becoming science editor. In 1936, he joined the Westinghouse Electric and Manufacturing Company as assistant to the president, where he remained until 1945. He then opened his own industrial public relations firm, Pendray & Company. He was senior partner there until 1971. Pendray was the founder of the American Rocket Society and wrote several books, including his 1947 work, *The Coming Age of Rocket Power.* See "Pendray, G. Edward," biographical file, NASA Historical Reference Collection.

Wilton B. Persons (1896–1977) was a career Army officer who had entered the U.S. Army Coast Artillery in 1917 and advanced through the ranks to major general in 1944. He had served in the Allied Expeditionary Force in World War I and in Europe in World War II. He headed the Office of Legislative Liaison for the Department of Defense between 1948 and his retirement in 1949. He was called back to active duty as a special assistant to General Dwight D. Eisenhower at Supreme Headquarters of Allied Powers in Europe from 1951 to 1952 and was active on behalf of Eisenhower's presidential campaign in 1952. He became a deputy assistant to Eisenhower in 1953 and then was made an assistant to the president in 1958. He served throughout the Eisenhower presidency, handling congressional liaison before he replaced Sherman Adams in 1958 as Eisenhower's chief of staff.

Rocco Petrone (1926–   ) was heavily involved at NASA with the development of the Saturn V booster used to launch Apollo spacecraft to the Moon in the 1960s and early 1970s. He worked at the Marshall Space Flight Center and became its director in 1973. He left Marshall in 1974 for a position at NASA Headquarters in Washington, D.C., in 1974 and retired from the agency in 1975. He then became president and chief executive officer of the National Center for Resource Recovery. See "Petrone, Lt. Col. Rocco A.," biographical file, NASA Historical Reference Collection.

Samuel C. Phillips (1921–1990) was trained as an electrical engineer at the University of Wyoming, but he also participated in the Civilian Pilot Training Program during World War II. Upon his graduation in 1942, he entered the Army infantry but soon transferred to the air component. As a young pilot, he served with distinction in the Eighth Air Force in England—earning two distinguished flying crosses, eight air medals, and the French croix de guerre—but he quickly became interested in aeronautical research and development. He became involved in the development of the incredibly successful B-52 bomber in the early 1950s and headed the Minuteman intercontinental ballistic missile program in the latter part of the decade. In 1964, Phillips, by this time an Air Force general, was lent to NASA to head the Apollo lunar landing program, which, of course, was unique in its technological accomplishment. He went back to the Air Force in the 1970s and commanded the Air Force Systems Command prior to this retirement in 1975. See "Gen. Samuel C. Phillips of Wyoming," *Congressional Record,* August 3, 1973, S-15689; Rep. John Wold and Gen. Sarah H. Turner, "Sam Phillips: One Who Led Us to the Moon," *NASA Activities,* May/June 1990, pp. 18–19; obituary in *New York Times,* February 1, 1990, p. D1.

William H. Pickering (1910–   ) obtained his bachelor's and master's degrees in electrical engineering and then a Ph.D. in physics from Caltech before becoming a professor of electrical engineering there in 1946. In 1944, he organized the electronics efforts at the Jet Propulsion Laboratory (JPL) to support guided missile research and development, becoming project manager for Corporal, the first operational missile that JPL developed. From 1954 to 1976, he was director of JPL, which developed the first U.S. satellite (Explorer I), the first successful U.S. circumlunar space probe (Pioneer IV), the Mariner flights to Venus and Mars in the early to mid-1960s, the Ranger photographic missions to the Moon in 1964–65, and the Surveyor lunar landings of 1966–67. See "Pickering, William H.," biographical file, NASA Historical Reference Collection.

Richard W. Porter was an electrical engineer who worked on missile programs with the General Electric Company before working on Earth sciences programs at the National Academy of Sciences. In 1964, he was the academy's delegate to the Committee on Space Research (COSPAR). See "Assorted Government Officials," biographical file, NASA Historical Reference Collection.

Frank Press (1924-   ) served as President Carter's science advisor. From 1981 to 1993, he served as president of the National Academy of Sciences. He received a Ph.D. in geophysics from Columbia University in 1949. See "Press, Frank," biographical file, NASA Historical Reference Collection.

Howard Pyle (1906-   ) was the Republican governor of Arizona from 1951 to 1954. He then became an assistant to President Eisenhower in 1955. During World War II, he was a Pacific war correspondent for the American Broadcasting Station. See *Who's Who in America 1958–1959*, 30th edition, vol. 2 (Chicago: Marquis Who's Who, 1959).

Q

Donald A. Quarles (1894-1959) was deputy secretary of defense between 1957 and 1959. Just after World War II, he had been a vice president first at the Western Electric Company and later at Sandia National Laboratories, but in 1953, he accepted the position of assistant secretary of defense (research and development). He was also secretary of the Air Force between 1955 and 1957. See "Quarles, Donald," biographical file, NASA Historical Reference Collection.

J. Danforth ("Dan") Quayle (1947-   ) served in the Indiana National Guard from 1969 to 1975. In 1974, he was admitted to Indiana's Bar and began practicing in Huntington. In 1976, he was elected as a Republican to the ninety-fifth Congress and was reelected to the ninety-sixth Congress. In 1980, he was elected to the U.S. Senate and then reelected in 1986, serving until 1989. He resigned to become U.S. Vice President under George H.W. Bush from 1989 to 1993. As Vice President, he chaired the National Space Council and had significant involvement with the development of the International Space Station, Space Shuttle replacement options, the Space Exploration Initiative, and NASA management. See *Who's Who in America 1988–1989*, 45th edition, vol. 2 (Wilmette, IL: Marquis Who's Who, 1989); Dan Quayle, *Standing Firm: A Vice-Presidential Memoir* (New York: Harper Collins Publishers, 1994).

R

M.L. Raines was director of the Safety, Reliability, and Quality Assurance Division at Johnson Space Center in the 1970s.

Bernard P. Randolph (1933-   ) has degrees in chemistry, electrical engineering, and business administration and has completed the programs at the Squadron Officer School, Air Command and Staff College, and Air War College. He held various assignments at Lincoln Air Force Base, Nebraska, and Los Angeles Air Force Station, California. In 1969, he served in Vietnam as an airlift operations officer at Chu Lai and airlift coordinator at Tan Son Nhut Air Base. Returning to the United States in 1970, he was assigned to Air Force Systems Command headquarters. In 1974, he returned to Los Angeles Air Force Station. In 1978, he assumed responsibility for space defense systems at Space Division headquarters. From 1980 to 1981, he served as vice commander of the Warner Robins Air Logistics Center, Robins Air Force Base, Georgia. He then moved to the Office of the Deputy Chief of Staff, Research, Development, and Acquisition, at the U.S. Air Force Headquarters in Washington, D.C. He returned to Los Angeles Air Force Station in 1983; the following year, he became vice commander of Air Force Systems Command. In 1985, Randolph returned to Washington and served as deputy chief of staff for research, development, and acquisition. In 1987, he became commander, Air Force Systems Command, Andrews Air Force Base, Maryland, at the same time attaining the rank of general. See "General Bernard P. Randolph," U.S. Air Force biography, Secretary of the Air Force, Office of Public Affairs, Washington, DC.

Ronald Reagan (1911–) was the fortieth U.S. President from January 1981 until 1989. During his presidency, the maiden flight of the Space Shuttle took place. In 1984, he mandated the construction of an orbital space station. Reagan declared that "America has always been greatest when we dared to be great. We can reach for greatness again. We can follow our dreams to distant stars, living and working in space for peaceful, economic, and scientific gain. Tonight I am directing NASA to develop a permanently manned space station and to do it within a decade." See Sylvia D. Fries, "2001 to 1994: Political Environment and the Design of NASA's Space Station System," *Technology and Culture* 29 (July 1988): 568–93.

Eberhard F.M. Rees (1908–) was deputy director for technical and scientific matters at the Marshall Space Flight Center. A graduate of the Dresden Institute of Technology, he began his career in rocketry in 1940 when he became technical plant manager of the German rocket center at Peenemünde. He came to the United States in 1945 with von Braun's rocket team and worked with von Braun at Fort Bliss, Texas, moving to Huntsville in 1950 when the Army transferred its rocket activities to the Redstone Arsenal. He served as deputy director of development operations at the Army Ballistic Missile Agency from 1956 to 1960. In 1970, he succeeded von Braun as director of the Marshall Space Flight Center. He retired in 1973. See "Eberhard Rees," biographical file, NASA Historical Reference Collection.

Donald B. Rice (1939–) received a Ph.D. in economics from Purdue University in 1965. In the mid- to late 1960s, Rice served in various staff positions at the Department of Defense. From 1970 until 1972, Rice served as assistant director of the Office of Management and Budget, with responsibility for science, technology, and space programs, among others. He then became president and chief executive officer of the Rand Corporation, a national security think tank. From 1989 to 1993, Rice served as secretary of the Air Force. See George M. Watson, Jr., *Secretaries and Chiefs of Staff of the United States Air Force: Biographical Sketches and Portraits* (Washington, DC: Air Force Historical Support Office, 1999).

R. H. Rice (1904–) was an aerospace engineer. He was president of the Los Angeles division of North American Aviation from 1935 until 1961, later becoming the chief engineer. See *Who's Who in America 1964–1965*, 33rd edition, vol. 2 (Chicago: Marquis Who's Who, 1965).

Sally K. Ride (1951–), the first American woman to fly in space, was chosen as an astronaut in 1978 and served as a mission specialist on STS-7 in 1983 and on STS 41-G in 1984. She also served as a member of the Presidential Commission on the Space Shuttle *Challenger* Accident in 1986, and from 1986 to 1987, she chaired a NASA task force that prepared a report on the future of the civilian space program, titled *Leadership and America's Future in Space* (Washington, DC: U.S. Government Printing Office, 1987). She resigned from NASA in 1987 to join the Center for International Security and Arms Control at Stanford University. She left Stanford in 1989 to assume the directorship of the California Space Institute, part of the University of California at San Diego. See "Ride, Sally K.," biographical file, NASA Historical Reference Collection.

William P. Rogers (1913–) was chair of the presidentially mandated blue ribbon commission investigating the *Challenger* accident in January 1986. It found that the failure had resulted from a poor engineering decision—the use of an O-ring to seal joints in the solid rocket booster that was susceptible to failure at low temperatures, introduced innocently enough years earlier. Rogers kept the commission's analysis on a technical level and documented the problems in exceptional detail. The commission, after some prodding by Nobel Prize–winning scientist Richard P. Feynman, did a credible job of grappling with the technologically difficult issues associated with the accident. See *Report of the Presidential Commission on the Space Shuttle Challenger Accident, Vol. I* (Washington, DC: U.S. Government Printing Office, June 6, 1986).

Milton W. Rosen (1915–), an electrical engineer by training, joined the staff of the Naval Research Laboratory in 1940, where he worked on guidance systems for missiles during World War II. From 1947 to 1955, he was in charge of Viking rocket development. He was technical director of Project Vanguard, the scientific Earth satellite program, until he joined NASA in October 1958 as director of launch vehicles and propulsion in the Office of Manned Space Flight. In 1963, he became senior scientist for NASA's deputy associate administrator for defense affairs. He later became deputy associate administrator for space science (engineering). In 1974, he retired from NASA to become executive secretary of the Space Science Board at the National Academy of Sciences. See "Milton W. Rosen," biographical file, NASA Historical Reference Collection; Milton W. Rosen, *The Viking Rocket Story* (New York: Harper, 1955).

S

Robert M. Salter, Jr. (1920-   ), was a physicist who worked with North American Aviation (1946–48), the Rand Corporation (1948–54), Lockheed Aircraft Company (1954–59), Quantatron, Inc. (1960–62), and Xerad, Inc., since 1962. He was responsible for much of the early thinking at Rand on the possibility of an artificial Earth-orbiting satellite. See "Rand Corp. No - 0262," file, NASA Historical Reference Collection.

Eugen Sänger (1905–1961) was an Austrian scientist whose ideas about reusable spacecraft were commemorated in a 1980s German design for a two-stage launch system that carries his name. See E. Sänger, *Raketenflugtechnik* (1933), whose English version is *Rocket Flight Engineering* (Washington, DC: NASA TT F-223,1965); "Eugen Sanger," biographical file, NASA Historical Reference Collection.

Bernard A. Schriever (1910-   ) earned a B.S. in architectural engineering from Texas A&M University in 1931 and was commissioned in the Army Air Corps Reserve in 1933 after completing pilot training. Following broken service, he received a regular commission in 1938. He earned an M.A. in aeronautical engineering from Stanford in 1942 and then flew sixty-three combat missions in B-17s with the 19th Bombardment Group in the Pacific Theater during World War II. In 1954, he became commander of the Western Development Division (soon renamed the Air Force Ballistic Missile Division), and from 1959 to 1966, he was commander of its parent organization, the Air Research and Development Command (renamed the Air Force Systems Command in 1961). As such, he presided over the development of the Atlas, Thor, and Titan missiles, which served not only as military weapon systems but also as boosters for NASA's space missions. In developing these missiles, Schriever instituted a systems approach, whereby the various components of the Atlas and succeeding missiles underwent simultaneous design and testing as part of an overall "weapons system." Schriever also introduced the notion of concurrency, which has been given various interpretations but essentially allowed the components of the missiles to enter production while still in the test phase, thereby speeding up development. He retired as a general in 1966. See Jacob Neufeld, "Bernard A. Schriever: Challenging the Unknown," *Makers of the United States Air Force* (Washington, DC: Office of Air Force History, 1986), pp. 281–306; Robert L. Perry, "Atlas, Thor . . .," in Eugene M. Emme, ed., *A History of Rocket Technology* (Detroit: Wayne State University Press, 1964), pp. 144–60; Robert A. Divine, *The Sputnik Challenge: Eisenhower's Response to the Soviet Satellite* (New York: Oxford University Press, 1993), p. 25.

George P. Shultz (1920-   ) served as director of the Office of Management and Budget after 1970, during the Nixon administration. Before that time, he had been Nixon's secretary of labor. During the Reagan administration (1981–89), he served as secretary of state. See "Shultz, George P.," *Current Biography Yearbook 1988,* pp. 525–30.

Glenn T. Seaborg (1912-   ) earned a Ph.D. in physics from the University of California at Berkeley in 1937 and worked on the Manhattan Project in Chicago during World War II. Afterward, he became associate director of Berkeley's Lawrence Radiation Laboratory, where he and associates isolated several transuranic elements. For this work, Seaborg received the Nobel Prize in 1951. He also served as chair of the Atomic Energy Commission between 1961 and 1971, and then he returned to the faculty of the University of California at Berkeley. See David Petechuk, "Glenn T. Seaborg," in Emily J. McMurray, ed., *Notable Twentieth-Century Scientists* (New York: Gale Research Inc., 1995), pp. 1803–06.

Robert C. Seamans, Jr. (1918-   ), had been involved in aerospace issues since he completed his Sc.D. degree at MIT in 1951. He was on the faculty at MIT's department of aeronautical engineering from 1949 to 1955, when he joined the Radio Corporation of America (RCA) as manager of the Airborne Systems Laboratory. In 1958, he became the chief engineer of the Missile Electronics and Control Division and joined NASA in 1960 as associate administrator. In December 1965, he became NASA's deputy administrator. He left NASA in 1968, and in 1969, he became secretary of the Air Force, serving until 1973. Seamans was president of the National Academy of Engineering from May 1973 to December 1974, when he became the first administrator of the new Energy Research and Development Administration. He returned to MIT in 1977, becoming dean of its School of Engineering in 1978. In 1981, he was elected chair of the board of trustees of Aerospace Corporation. See "Seamans, Robert C., Jr.," Deputy Administrator files, NASA Historical Reference Collection; Robert C. Seamans, Jr., *Aiming at Targets: The Autobiography of Robert C. Seamans, Jr.* (Washington, DC: NASA SP-4106, 1996).

Willis H. Shapley (1917–   ), the son of famous Harvard astronomer Harlow Shapley, earned a bachelor of arts degree from the University of Chicago in 1938. From that point until 1942, he did graduate work and performed research in political science and related fields at the University of Chicago. He joined the Bureau of the Budget in 1942 and became a principal examiner in 1948. From 1956 to 1961, he was assistant chief (Air Force) in the bureau's military division, becoming progressively deputy chief for programming (1961–65) and deputy chief (1965) in that division. He also served as special assistant to the director for space program coordination. In 1965, he moved to NASA as associate deputy administrator, with his duties including supervision of the public affairs, congressional affairs, Department of Defense and interagency affairs, and international affairs offices. He retired in 1975 but rejoined NASA in 1987 to help it recover from the *Challenger* disaster. He served as associate deputy administrator (policy) until 1988, when he again retired but continued to serve as a consultant to the administrator. See "Shapley, W.H.," biographical file, NASA Historical Reference Collection.

Joseph F. Shea (1926–1999) joined the Office of Manned Space Flight at NASA Headquarters in 1962. The next year, he was named the Apollo program manager at NASA's Manned Spacecraft Center in Houston. In 1967, he moved to NASA Headquarters as deputy associate administrator for the Office of Manned Space Flight. He joined the Raytheon Company in 1968 and served on the NASA Advisory Council for several years. Shea returned to NASA as head of space station redesign efforts in the early 1990s and also served as chair of a task force that reviewed plans for the first servicing mission of the Hubble Space Telescope. He was an adjunct professor of aeronautics and astronautics at MIT. After returning to Raytheon, he worked his way up to senior vice president. See "Shea, J.F.," biographical file, NASA Historical Reference Collection; "Obituary for Joseph F. Shea," *Boston Globe*, February 16, 1999.

Alan B. Shepard, Jr. (1923–1998), was a member of the first group of seven astronauts chosen in 1959 to participate in Project Mercury. He was the first American in space, piloting Mercury-Redstone 3 (*Freedom 7*), and he was backup pilot for Mercury-Atlas 9. He was subsequently grounded because of an inner ear ailment until May 7, 1969 (during which time he served as chief of the Astronaut Office). Upon returning to flight status, Shepard commanded Apollo 14, and in June 1971, he resumed duties as chief of the Astronaut Office. He retired from NASA and the U.S. Navy on August 1, 1974, to join the Marathon Construction Company of Houston, Texas, as partner and chairman. See Alan Shepard and Deke Slayton, *Moonshot: The Inside Story of America's Race to the Moon* (New York: Turner Publishing, 1994); The Astronauts Themselves, *We Seven* (New York: Simon and Schuster, 1962).

Milton A. Silveira (1929–   ) was a longtime NASA employee, who worked at the agency's Lewis Research Center (1955–63) and Manned Spacecraft Center in Houston (1963–67). He also served as deputy manager of the orbiter project at Johnson Space Center (1967–81), assistant to the deputy administrator at NASA (1981–83), and NASA chief engineer (1983–86).

Abe Silverstein (1908–   ), who earned a B.S. in mechanical engineering (1929) and an M.E. (1934) from Rose Polytechnic Institute, was a longtime NACA manager. He had worked as an engineer at the Langley Aeronautical Laboratory between 1929 and 1943 and had moved to the Lewis Laboratory (later Research Center) in a succession of management positions, the last (1961–70) as director of the center. Interestingly, in 1958, the Case Institute of Technology had awarded him an honorary doctorate. When T. Keith Glennan arrived at NASA from Case, Silverstein was on a rotational assignment to the Washington headquarters as director of the Office of Space Flight Development (later the Office of Space Flight Programs) from the position of associate director at Lewis, which he had held since 1952. During his first tour at Lewis, he had directed investigations leading to significant improvements in reciprocating and early turbojet engines. At NASA Headquarters, he helped create and direct the efforts leading to the spaceflights of Project Mercury that established the technical basis for the Apollo program. As Lewis's director, he oversaw a major expansion of the center and the development of the Centaur launch vehicle. He retired from NASA in 1970 to take a position with Republic Steel Corporation. See Virginia P. Dawson, *Engines and Innovation: Lewis Laboratory and American Propulsion Technology* (Washington, DC: NASA SP-4306, 1991); "Abe Silverstein," biographical file, NASA Historical Reference Collection.

Samuel Knox Skinner (1938–   ) was admitted to the Illinois Bar in 1966 and was an assistant U.S. attorney in Illinois from 1968 to 1974. He became a U.S. attorney in 1975, going on to be a partner of Sidley & Austin in Chicago from 1977 to 1989. In 1984, he became chairman of Regional Transportation Authority in Chicago. He then served as President George Bush's secretary of transportation from 1989 to 1991, when he became chief of staff for the same administration. See "Samuel Skinner," biographical file, NASA Historical Reference Collection; *Who's Who in America 1992–1993*, 47th edition, vol. 2 (New Providence, NJ: Marquis Who's Who, 1993).

Murray Snyder (1911–1969) began his career as a reporter for the *San Antonio Light* and then moved to New York to be a political writer. He wrote for the *New York Post* and the *New York Herald Tribune* before becoming assistant press secretary for the White House between 1953 and 1957. Snyder then became assistant secretary of defense for public affairs for the next four years, followed by president of Murray Snyder Associates, a public relations firm. See *Who Was Who in America 1969–1973*, vol. V (Chicago: Marquis Who's Who, 1973).

Athelstan F. Spilhaus (1911–   ) is a prominent meteorologist at New York University in the meteorology department that he founded. From 1949 to 1966, he was the dean of the Institute of Technology at the University of Minnesota. In September 1969, he was elected president of the American Association for the Advancement of Science. He later returned to New York University. See "Athelstan Spilhaus," biographical file, NASA Historical Reference Collection.

C. Starr (1912–   ) is a physicist and engineer and is noted for his work on the Manhattan Project. Educated at Rensselaer Polytechnic Institute, he has worked in both industrial and scholarly research. After his work at Oak Ridge, he went on to become an expert in atomic energy and nuclear reactors, as well as other fields, and he currently directs the Atomic Industry Forum. See *American Men and Women of Science: 1998–1999*, 20th edition, vol. 6 (New Providence, NJ: RR Bowker, 1999), p. 1201.

H. Guyford Stever (1916–   ) was a major contributor not only to U.S. Air Force scientific and technical progress, but to the administration of American science as a whole. A distinguished professor of aeronautical engineering at MIT from the 1940s through the 1960s, he made vital discoveries relating to aerospace science. While undertaking these responsibilities, he also established himself as a prime technical advisor to the Army Air Forces and the Air Force on questions relating to radar, guided missiles, and space. He helped found the Air Force science organization, and he served with Dr. Theodore von Kármán in compiling the seminal report, *Toward New Horizon*. Noted for the report that bears his name and that served to transform the Air Research and Development Command to the Air Force Systems Command, Stever spent twenty-one years on the Air Force Scientific Advisory Board. He later worked as the president of Carnegie Mellon University and, finally, as the White House science advisor under the Nixon and Ford Administrations. He also served as the head of the National Science Foundation. See "Stever, H. Guyford," biographical file, NASA Historical Reference Collection.

Homer J. Stewart (1915–   ) earned his doctorate in aeronautics from Caltech in 1940, joining the faculty there two years before that. In 1939, he participated in pioneering rocket research with other Caltech engineers and scientists, including Frank Malina, in the foothills of Pasadena. Out of their efforts, the Jet Propulsion Laboratory (JPL) arose, and Stewart maintained his interest in rocketry at that institution. He was involved in developing the first American satellite, Explorer I, in 1958. In that year, on leave from Caltech, he became director of NASA's Office of Program Planning and Evaluation, returning to JPL in 1960 in a variety of positions, including chief of the Office of Advanced Studies from 1963 to 1967 and professor of aeronautics at Caltech. See "Stewart, Homer," biographical file, NASA Historical Reference Collection; Clayton R. Koppes, *JPL and the American Space Program: A History of the Jet Propulsion Laboratory* (New Haven, CT: Yale University Press, 1982), pp. 23, 32, 44, 47, 79–80, 82.

Ernst Stuhlinger (1913–   ) was a member of von Braun's Peenemünde rocket team who came to the United States following World War II under Project Paperclip. During his tenure at Marshall Space Flight Center, he directed the early planning for lunar exploration and the Apollo Telescope mount that was flown on Skylab. He was also responsible for the early planning on the High Energy Astronomy Observatory and the contributed to the initial phases of the Space Telescope Project. After retiring from NASA in 1975, he continued an Earthbound extension of his work on interplanetary electric propulsion by researching and advocating electric cars. Stuhlinger has received numerous prestigious awards, authored numerous book and technical articles related to manned and unmanned space activities, and belonged to a number of noted scientific societies. See "Stuhlinger, Ernst," biographical file, NASA Historical Reference Collection.

George P. Sutton (1920–   ) was a rocket engineer. Born in Austria, he came to the United States in 1938 and was naturalized in 1944. He was a research engineer at Aerojet Engineering Corporation from 1943 to 1946. Sutton then became the manager of advanced design for the Rocketdyne division of North American Aviation from 1946 to 1958. For a two-year interim, he worked at the Department of Defense as the chief scientist of the Advanced Research Projects Agency before returning to the North American Aviation in 1960. He won many awards, including the Pendray award, and has written numerous works. See *Who's Who in America 1964–1965*, 33rd edition, vol. 2 (Chicago: Marquis Who's Who, 1965).

Norman E. Terrell (1933–   ) was appointed the associate administrator for policy at NASA in 1984. He has held foreign service and domestic career positions since 1963. He was assistant director of the Arms Control and Disarmament Agency, deputy assistant secretary of state for science and technology, director of international affairs at NASA, and a member of the staff at the Nuclear Regulatory Commission. See "Terrell, Norman E.," biographical file, NASA Historical Reference Collection.

James R. ("J.R.") Thompson, Jr. (1937–   ), became deputy administrator of NASA in 1989. Beforehand, he had served as director of NASA's Marshall Space Flight Center in Huntsville, Alabama. He had assumed his position at Marshall on September 29, 1986, after having served three years as deputy director for technical operations at Princeton University's Plasma Physics Laboratory. From March to June 1986, he was vice-chair of the NASA task force inquiring into the cause of the Space Shuttle *Challenger* accident. He began his career in 1960 as a development engineer with Pratt & Whitney Aircraft in West Palm Beach, Florida. He joined the research and development team at Marshall in 1963 as a liquid propulsion system engineer responsible for component design and performance analysis associated with the J-2 engine system on the Saturn launch vehicle. In 1966, he joined the Space Engine Section in the former Propulsion and Vehicle Engineering Laboratory at Marshall and became chief of the section in 1968. In that capacity, he was responsible for the design and test evaluation of auxiliary space engine propulsion systems for the Saturn and experimental small interplanetary propulsion systems. In 1969, Thompson transferred to Marshall's Astronautics Laboratory, where he served as chief of the Man/Systems Integration Branch from 1969 to 1974. In September 1974, he was named manager of the Main Engine Projects Office at Marshall, where he was responsible for the development and operation of the most advanced liquid propulsion rocket engine ever developed. He served in that position almost from the beginning of early development testing on the Space Shuttle main engine through the initial Shuttle flights. In February 1982, he was named associate director for engineering in Marshall's Science and Engineering Directorate. Born in Greenville, South Carolina, he graduated from Druid Hills High School in Atlanta in 1954. He was awarded a bachelor of science degree in aeronautical engineering from Georgia Tech in 1958 and a master of science degree in mechanical engineering from the University of Florida in 1963. He has completed all course work at the University of Alabama toward a Ph.D. in fluid mechanics. He served as a lieutenant in the U.S. Navy from 1958 to 1960 and was stationed at Green Cove Springs, Florida, as an administrative officer in the Atlantic Fleet. Upon leaving NASA in 1991, he entered private business. See "Thompson, J.R.," Deputy Administrator files, NASA Historical Reference Collection.

Robert F. Thompson (1925–   ) began his career as an aeronautical engineer with the NACA in 1947. He held a series of increasingly responsible jobs at NASA, including manager of the Apollo Applications program and of the Space Shuttle program. He retired from NASA in 1981 and then accepted a job with McDonnell Douglas Technical Services Company in Houston. See "Thompson, Robert," biographical file, NASA Historical Reference Collection.

A.O. Tischler joined the NACA as a chemical engineer on the staff of the Lewis Laboratory in 1942. He remained there until his transfer to NASA Headquarters in Washington, D.C., in 1958. From November 1961 to January 193, he served as the assistant director for propulsion in NASA's Office of Manned Space Flight. In January 1964, he was appointed director of the Chemical Propulsion Division in the Office of Advanced Research and Technology. See "Tischler, A.O.," biographical file, NASA Historical Reference Collection.

Richard H. Truly (1937–   ) became NASA's associate administrator for the Office of Space Flight on February 20, 1986. In this position, he led the painstaking rebuilding of the Space Shuttle program after the *Challenger* accident. This was highlighted by NASA's celebrated "return to flight" on September 29, 1988, when the Space Shuttle *Discovery* lifted off from Kennedy Space Center in Florida on the first Shuttle mission in almost three years. He later served as NASA's eighth administrator from May 1989 to March 1992. Before returning to NASA, the former Shuttle astronaut served as the first commander of the Naval Space Command in Dahlgren, Virginia, established on October 1, 1983. His career in the U.S. Navy began in 1959, when he was commissioned an ensign. This coincided with his graduation from Georgia Tech, which he attended as a Naval ROTC midshipman and earned a bachelor's degree in aeronautical engineering. Following flight school, he was designated a naval aviator in 1960. His initial tour of duty, Fighter Squadron 33, was aboard the *USS Intrepid* and the *USS Enterprise*, and he made more than 300 carrier landings. From 1963 to 1965, he was a student and then instructor at the U.S. Air Force Aerospace Research Pilot School at Edwards Air Force Base in California. In 1965, Truly became one of the first military astronauts selected to the Air Force's Manned Orbiting Laboratory program in Los Angeles, California, and he transferred to NASA as an astronaut in August 1969. He served as capsule communicator for all three Skylab missions in 1973 and the Apollo-Soyuz Test Project in 1975. As a naval aviator, test pilot, and astronaut, Truly logged more than 7,500 hours in numerous military and civilian jet aircraft. He was pilot for one of the two-astronaut crews that flew the 747/Space Shuttle *Enterprise* approach and landing test flights during 1977. He then was backup pilot for STS-1, the first orbital test of the Shuttle. His first spaceflight was November 12–14, 1981, as pilot of *Columbia* (STS-2), significant as the first piloted spacecraft to be reflown in space. His second flight (STS-8) on August 30–September 5, 1983, was as commander of *Challenger*, the first night launch and landing by the Shuttle program. After leaving NASA, he became vice president and director of the Georgia Tech Research Institute in Atlanta. See "Truly, Lt. Cdr. Richard H.," Administrator files, NASA Historical Reference Collection.

Konstantin E. Tsiolkovskiy (1857–1935) became enthralled with the possibilities of interplanetary travel as a boy and, at age fourteen, started independent study using books from his father's library on natural science and mathematics. He also developed a passion for invention, and he constructed balloons, propelled carriages, and other instruments. To further his education, his parents sent him to Moscow to pursue technical studies. In 1878, he became a teacher of mathematics at a school north of Moscow. Tsiolkovskiy first started writing on space in 1898, when he submitted for publication to the Russian journal, *Nauchnoye Obozreniye* (*Science Review*), a work based on years of calculations that laid out many of the principles of modern spaceflight. The article, "Investigating Space with Rocket Devices," presented years of calculations that laid out many of the principles of modern spaceflight and opened the door to future writings on the subject. In it, Tsiolkovskiy described in depth the use of rockets for launching orbital spaceships. There followed a series of increasingly sophisticated studies on the technical aspects of spaceflight. In the 1920s and 1930s, he proved especially productive, publishing ten major works, elucidating the nature of bodies in orbit, developing scientific principles behind reaction vehicles, designing orbital space stations, and promoting interplanetary travel. He also furthered studies on many principles commonly used in rockets today: specific impulse to gauge engine performance, multistage boosters, fuel mixtures such as liquid hydrogen and liquid oxygen, the problems and possibilities inherent in microgravity, the promise of solar power, and spacesuits for extravehicular activity. Significantly, he never had the resources—nor perhaps the inclination—to experiment with rockets himself. After the Bolshevik revolution of 1917 and the creation of the Soviet Union, Tsiolkovskiy was formally recognized for his accomplishments in the theory of spaceflight. Among other honors, in 1921, he received a lifetime pension from the state that allowed him to retire from teaching at the age of sixty-four. Thereafter, he devoted full time to developing his spaceflight theories studies. His theoretical work greatly influenced later rocketeers both in his native land and throughout Europe. While less well known in the United States during his lifetime, Tsiolkovskiy's work enjoyed broad study in the 1950s and 1960s, when Americans sought to understand how the Soviet Union had accomplished such unexpected success in its early spaceflight efforts. See "Tsiolkovskiy, K.E.," biographical file, NASA Historical Reference Collection.

Stansfield Turner (1923–   ) was a four-star admiral who was director of the Central Intelligence Agency from 1977 to 1981. Educated at Amherst College starting in 1941, he transferred to the Naval Academy after two years and took a year at Oxford on a Rhodes Scholarship. He has become an advocate for the reduction of nuclear arms stockpiles. See "Stansfield Turner," *Who's Who in America 1994* (New Providence, NJ: Marquis Who's Who, 1994).

V

James A. Van Allen (1914–   ) was a pathbreaking astrophysicist best known for his work in magnetospheric physics. Van Allen's January 1958 Explorer 1 experiment established the existence of radiation belts—later named for the scientist—that encircled Earth, representing the opening of a broad research field. Extending outward in the direction of the Sun approximately 40,000 miles, as well as stretching out with a trail away from the Sun to approximately 370,000 miles, the magnetosphere is the area dominated by Earth's strong magnetic field. See James A. Van Allen, *Origins of Magnetospheric Physics* (Washington, DC: Smithsonian Institution Press, 1983); David E. Newton, "James A. Van Allen," in Emily J. McMurray, ed., *Notable Twentieth-Century Scientists* (New York: Gale Research Inc., 1995), pp. 2070–72.

W

L.L. Waite (1907–   ) was the chief of aerodynamics at Berliner-Joyce Aircraft (then North American Aviation). He held many positions there, including assistant chief of aerodynamics, thermodynamics and flight test, assistant to the president to organize guided missiles operation, vice president in charge of guided missiles, control equipment and atomic energy research, and senior vice president of the organization. See *Who's Who in America 1964–1965*, 33rd edition, vol. 2 (Chicago: Marquis Who's Who, 1965).

Alan T. Waterman (1892–1967) was the first director of the National Science Foundation from its founding in 1951 until 1963. Waterman received his Ph.D. in physics from Princeton University in 1916 and then served with the Army's Science and Research Division during World War I. He was on the faculty of Yale University between the two world wars. He was with the War Department's Office of Scientific Research and Development during World War II and then with the Office of Naval Research between 1946 and 1951. He and NASA leaders contended over control of the scientific projects to be undertaken by the space agency, with Waterman's National Science Foundation being used as an advisory body in the selection of space experiments. See "Waterman, First NSF Head, Dies at 75," *Science* 158 (December 8, 1967): 1293; Norriss S. Hetherington, "Winning the Initiative: NASA and the U.S. Space Science Program," *Prologue: The Journal of the National Archives* 7 (Summer 1975): 99–108; John E. Naugle, *First Among Equals: The Selection of NASA Space Science Experiments* (Washington, DC: NASA SP-4215, 1991).

James E. Webb (1906–1992) was NASA's administrator between 1961 and 1968. Previously, he had been an aide to a congressman in New Deal Washington, an aide to Washington lawyer Max O. Gardner, and a business executive with Sperry Corporation and the Kerr-McGee Oil Company. He had also been director of the Bureau of the Budget between 1946 and 1950 and under secretary of state from 1950 to 1952. See W. Henry Lambright, *Powering Apollo: James E. Webb of NASA* (Baltimore, MD: Johns Hopkins University Press, 1995).

Caspar W. Weinberger (1917–   ), a longtime Republican government official, was a senior member of the Nixon, Ford, and Reagan administrations. For Nixon, he was deputy director (1970–72) and then director (1972–76) of the Office of Management and Budget. In this capacity, had a leading role in shaping the direction of NASA's major effort of the 1970s, the development of a reusable Space Shuttle. For Reagan, he served as secretary of defense, in which he also oversaw the use of the Shuttle in the early 1980s for the launching of classified Department of Defense payloads into orbit. See "Weinberger, Caspar W(illard)," *Current Biography Yearbook 1973*, pp. 428–30.

Albert D. (Bud) Wheelon has spent his life in the world of science and advanced technology. His first work was focused on guidance systems for long-range ballistic missiles and early space projects at TRW, Inc. He joined the Central Intelligence Agency in 1962 and served as the deputy director for science and technology until 1966. During that time, he received the Distinguished Intelligence Medal for his work in the collection of technical intelligence. He began working at the Hughes Aircraft Company in 1966 and four years later was given responsibility for building Hughes Space and Communications Group. In 1987, he was named chief executive officer and chairman of the board at Hughes. He retired in May 1988. His public service work includes time on the Defense Science Board, the President's Foreign Intelligence Advisory Board, and the Presidential Commission on the Space Shuttle *Challenger* Accident. He received a B.S. in engineering from Stanford University in 1949 and a Ph.D. in physics from MIT in 1952. See "Wheelon, Albert," biographical file, NASA Historical Reference Collection.

Walter C. Williams (1919–1995) earned a B.S. in aerospace engineering from Louisiana State University in 1939 and went to work for the NACA in 1940, serving as a project engineer to improve the handling, maneuverability, and flight characteristics of World War II fighters. Following the war, he went to what became Edwards Air Force Base to set up flight tests for the X-1, including the first human supersonic flight by Captain Chuck Yeager in October 1947. He became the founding director of the organization that became Dryden Flight Research Facility. In September 1959, he assumed associate directorship of the new NASA Space Task Group at Langley, created to carry out Project Mercury. He later became director of operations for the project and then associate director of NASA's Manned Spacecraft Center in Houston (subsequently renamed Johnson Space Center). In 1963, Williams moved to NASA Headquarters as deputy associate administrator for the Office of Manned Space Flight. From 1964 to 1975, he was a vice president for Aerospace Corporation. Then from 1975 to until his retirement in 1982, he served as NASA's chief engineer. See "Walter C. Williams," biographical file, NASA Historical Reference Collection.

Charles E. Wilson (1890–1961) was an electrical engineer with the Westinghouse Electric and Manufacturing Company from 1909 to 1919, leaving that position to become chief engineer and factory manager of the Delco Remy Company until 1926, when he became president. In 1929, he became vice president of General Motors Corporation for ten years, moving up to executive vice president and then president from 1941 to 1953, becoming chief executive officer in 1946. In 1953, Wilson became the fifth secretary of defense, serving during the Eisenhower administration until 1957. He retired from the Pentagon in 1957. See *Who Was Who in America 1961–1968*, vol. IV (Chicago: Marquis Who's Who, 1968).

Robert G. Wilson (1934–) is a nuclear physicist who has worked at North American Aviation and Rocketdyne. Also specializing in electronics, Wilson attended Ohio State University, where he received his Ph.D. in physics. His work has focused on semiconductors, ion implantation, electron and ion emission, and experimental low-energy nuclear physics. See *American Men and Women of Science: 1998–1999*, 20th edition, vol. 7 (New Providence, NJ: RR Bowker, 1999), p. 812.

Y

John F. Yardley (1925–) was an aerospace engineer who worked with McDonnell Aircraft Corporation on several NASA human spaceflight projects between the 1950s and the 1970s. He also served as NASA's associate administrator for the Office of Space Flight between 1974 and 1981. Thereafter, he returned to McDonnell Douglas as president (1981–88). See "Yardley, John F.," biographical file, NASA Historical Reference Collection.

Charles E. ("Chuck") Yeager (1923–) was the U.S. Air Force test pilot who piloted the X-1 research aircraft on the first supersonic powered flight in 1947. Thereafter, he served in several Air Force positions, retiring as a brigadier general. He also served as a member of the Presidential Commission on the Space Shuttle *Challenger* Accident in 1986. See Chuck Yeager, *Yeager* (New York: Bantam Books, 1982).

Clayton Yeutter (1930–) started out as a farmer and rancher in Nebraska in 1957. While maintaining this job, he became a member of the faculty of the University of Nebraska in the Department of Agricultural Economics. He was admitted to the Nebraska Bar in 1963. Three years later, he received his Ph.D. in agricultural economics from the University of Nebraska. He held several government positions in the following years, becoming senior partner of Nelson, Harding, Yeutter & Leonard from 1977 to 1978. He then went on to become president and chief executive officer of the Chicago Mercantile Exchange, a position he held until 1985 when he became the U.S. trade representative. In that capacity, he led the American team in negotiating the U.S.-Canada Free Trade Agreement and helped launch the 100-nation Uruguay Round of GATT negotiations. In 1989, Yeutter was appointed secretary of agriculture under President George Bush. He held that title until 1991. The following year, he became advisor to the president for domestic policy. See *Who's Who in America 1990–1991*, 46th edition, vol. 2 (Wilmette, IL: Marquis Who's Who, 1991).

Herbert F. York (1923–   ) had been associated with scientific research in support of national defense since World War II. He was director of the Livermore Radiation Laboratory for the University of California before moving to the Department of Defense in March 1958 as chief scientist of the Advanced Research Projects Agency. He became the Department of Defense's director of research and engineering in December 1958 during a departmental reorganization; this was the third-ranking civilian office after the secretary and deputy secretary of defense. He served as director of defense research and engineering until 1961. He then moved to the University of California at San Diego as chancellor and professor of physics. He also served as a member of the President's Science Advisory Committee under both Eisenhower and Johnson and was later chief negotiator for the comprehensive test ban during the Carter administration. See "Dr. Herbert F. York," biographical file, NASA Historical Reference Collection; Herbert F. York, *Making Weapons, Talking Peace: A Physicist's Odyssey from Hiroshima to Geneva* (New York: Basic Books, 1987).

John W. Young (1930–   ) served as a fighter pilot and test pilot before being chosen with the second group of astronauts in 1962. He was pilot of Gemini 3, backup pilot of Gemini VI, command pilot for Gemini X, backup command module pilot for Apollo 7, command module pilot of Apollo 10, backup commander for Apollo 13, commander for Apollo 16 (the ninth to walk on the Moon), and backup commander of Apollo 17. He retired from the Navy on September 30, 1976 and served as the chief of the Astronaut Office. He then commanded the first Space Shuttle orbital flight test (STS-1) and then STS-9 (Spacelab 1), becoming the first person to fly in space six times. Currently, he serves as special assistant to the director of Johnson Space Center for engineering, operations, and safety, and he remains an active member of the astronaut corps. See "Young, John W.," biographical file, NASA Historical Reference Collection.

Index

A

Abrahamson, James A., 521

"Access to Space" study, 509–10, 587, 589, 600, 603–04, 617, 620, 626

Adams, Sherman, 52–53

Advanced Launch System (ALS), 189, 506, 529–30, 533–42, 555, 600, 607, 623

Advanced Research Projects Agency (ARPA), 17, 20, 94–97, 109–11, 617, 623; and one million pound thrust engine, 105–09; also see Defense Advanced Research Projects Agency

Advanced Solid Rocket Motor (ASRM), 550

Aerobee rocket, 8, 40–43

Aerojet General Corporation, 10, 13, 18, 26, 576; and one million pound thrust engine, 105–09

Aeronautics and Astronautics Coordination Board (AACB), 624

Aerospace Corporation, 10, 576–77

Agena rocket, 11

Agnew, Spiro T., 165

Air Force, U.S., 408, 418–21, 459–60, 465, 472–73, 479, 494, 496, 503–12, 517–18, 540, 542, 548–50, 558–70, 576, 591, 611, 623, 625; and Air Force Space Command, 604–05, 607, 625

Air Force Astronautics Laboratories (AFAL), 542

Air Force Ballistic Missile Division, 10

Air Force Strategic Missiles Evaluation Committee, 9–10

Air Transport Association, 573

Aldrich, Arnold, 591

Aldrich, David E., 25, 134–35

Aldridge, Edward ("Pete"), 420, 521, 542

Aldridge Study, 609

Allen, H. Julian, 6, 34–38

American Rocket Society, 1, 2

American Telephone and Telegraph (AT&T), 405

Anders, William A., 24, 212, 283–86

Anderson, Martin, 329–33

Apollo, Project, 18, 23–29, 113–14, 573

Apollo 4, 28

Apollo 6, 29

Apollo 7, 24

Apollo 8, 24, 30

Apollo 9, 30

Apollo 11, 30

Apollo 17, 30

Apollo-Soyuz Test Project, 177, 503

Appold, Norman C., and 1959 progress report on Saturn development, 116–19

Ariane European launch vehicle, 186, 406–09, 412–13, 420–22, 445, 447–48, 450, 466, 491, 543–45, 600, 611–14

Arianespace organization, 407–08, 412–13, 419, 421, 454–56, 466, 474

Arms Control and Disarmament Agency, 428

Arms Control Policy Group, Senior, see Senior Arms Control Policy Group

Army Ballistic Missile Agency (ABMA), 8, 20, 49–51, 64, 97–104; and early development of Saturn, 111–15; and 1959 progress report on Saturn development, 116–19; and transfer to NASA, 115–16

Arnold Air Force Base, 542

Arnold Engineering Development Center (AEDC), 542

ASIASAT satellites, 501

Assured Crew Return Vehicle (ACRV), 550

Atlantis orbiter, 179, 187, 191, 192, 425

Atlas launch vehicle (Project MX-774), 7, 10, 11, 12, 13, 14–15, 16, 64–68, 69–70, 75–81, 75–81, 95, 96, 181, 405, 408, 410, 420, 503, 526–27, 547, 592, 599, 607, 611, 613–18

Atomic Energy Commission (AEC), 18, 88–91, 91–95
Atwood, J. Leland ("Lee"), 27
Auburn, MA, 2
"Augustine Report," 609
Austin, Gene, 591
AUSSAT B-1 and B-2 satellites, 498, 501

B

Ballistic Missile Command, 12
Bay St. Louis, Mississippi, 542
Beggs, James M., 407, 412, 414, 446
Beichel, Rudi, 576
Beijing, 498
Bekey, Ivan, 503–12, 517, 521–22
Bell Aircraft Corporation, 1; and one million pound thrust engine, 105–09
Bell Telephone Laboratories, 405
Bilstein, Roger E., 22
Boeing Company, 422, 545
Boeing 747 airliner, 573
Bono, Phil, 575–76
Borman, Frank, 24
Borrasca, Barton, 428
Bossart, Karel J., 10
Brackeen, Richard E., 472–76
Branscome, Darrell, 522, 553–54
Braun, Wernher von, 3, 6, 8, 14, 19, 20, 25, 32–33, 56–63; and all-up testing, 142–43; and early development of Saturn, 111–15; and F-1 engine instability, 135–42; and "National Integrated Missile and Space Vehicle Development Program, A," 97–104; and 1959 progress report on Saturn development, 116–19; and Nova rocket, 20, 25, 96, 129–43; and "pogo effect," 144–47
Brilliant Eyes concept, 624
British Interplanetary Society, 202–05, 575
Brown, Harold, 294–305
Brucker, Wilber, 62
Buran Soviet orbiter, 188
Burnley, James, 472
Bush, George, 19, 186, 416–17, 466–67, 542, 571
Bussard, Robert W., 17

C

Cabinet Council on Commerce and Trade (later renamed the Economic Policy Council), 411, 414, 441, 448, 514, 516
Cannon, Howard, 91–94
Cape Canaveral Air Force Station (CCAFS), 493–96, 625
Cargo Transfer and Return Vehicle (CTRV), 550–51
Carlucci, Frank, 537–38
Carter, Jimmy, 176, 290–91
Centaur Upper Stage, 17, 22, 96, 420–21, 509, 611
Central Intelligence Agency (CIA), 412, 428, 631
Centre Nationale d'Études Spatiales (CNES), 413, 455–56
Challenger orbiter, 175, 176, 179, 183–88, 425; and accident, 413–15, 419–22, 431, 478, 491, 496, 504, 506, 553, 573, 576, 606
Chamberlin, James, 428
Cheney, Richard, 571
China, 405, 417, 422, 491, 497–502, 530, 543, 607, 612–13

Citizens Advisory Council on National Space Policy, 572, 577
Clarke, Arthur C., 575
Cleveland, Ohio, 542
Clinton, William J. ("Bill"), 418, 542, 626
Cohen, Aaron, 282-83
Colladay, Raymond S., 521
Columbia orbiter, 175-79, 323-29, 334-37, 425
Commerce, Department of, 411, 428, 441-42, 486, 497, 513, 516, 607, 628-29, 631
Commercial Space Initiative, 415
Commercial Space Launch Act (of 1984), 410, 413, 415-16, 431-40, 453, 458-65 *passim,* 476-97 *passim,* 629
Commercial Space Launch Amendments of 1988, 458-65
Commercial Space Launch Policy, 416
Commercial Space Transportation Advisory Committee (COMSTAC), 442-43, 472-76, 480; also see Transportation, Department of
Communications Act (of 1934), 434
Complementary Expendable Launch Vehicle (CELV), 181-83, 504, 526-28
COMSAT, 611
Conatec, Inc., 418
Conestoga, 409
Confederation of Independent States (CIS), 543
Congress, 407, 410, 413, 415-16, 420, 431-40 *passim,* 452, 479-80, 490-97 *passim,* 536, 538, 542, 544, 547, 585-86, 607, 611
Convair Corporation, 10, 17, 22, 26, 64-68, 75-81
Coordinating Committee for Multilateral Export Control (COCOM), 473-76
"Copper Canyon" program, 507, 558
Covert, Eugene F., 174-75, 275-82
Crippen, Robert, 177
Cutler, Robert, 53

D

Debus, Kurt H., 147-53
Defense Acquisition Board (DAB), 539-40
Defense, Department of (DOD), 417, 422, 425, 428, 430, 439, 445, 451, 453, 463-64, 469-72, 475, 478, 480, 485, 494-96, 503-634 *passim;* and DOD Bottom-Up Review (BUR), 609-10, 613, 617
Defense Advanced Research Projects Agency (DARPA), 507, 558, 561-63; also see Advanced Research Projects Agency
Defense Science Board, 561-62
Delta Clipper (DC-X) program (renamed "Clipper Graham"), 508, 511, 572-77, 623
Delta launch vehicle, 8, 11, 19, 95, 180, 405-06, 408, 410, 418-22, 424, 454, 503-04, 526-27, 547, 592, 598, 600, 607, 611, 613-18
Deutch, John M., 606
Discoverer I, 11
Discovery orbiter, 179, 186, 425
Dole, Elizabeth, 411-12, 441-42, 444-45, 471-72
Donlan, Charles J., 211-14, 234-38
Dornberger, Walter R., 3
Douglas Aircraft Company, 22, 26
Dryden, Hugh L., 70-72
Dyna-Soar program, 20, 113, 162

E

Eastern Space and Missile Center (ESMC), 492-93, 495-96
Eastern Test Range (ETR), 526, 615, 623
Economic Policy Council (EPC), 414, 457, 465-66; and Working Group on Space Commercialization, 465

Edwards Air Force Base, 542
Eggers, A.J., Jr., 34–38
Eisenhower, Dwight D., 8, 18, 19, 38, 49–50, 51–53, 56–63
Endeavour orbiter, 186, 191
Energiya Soviet launch vehicle, 188, 422
Energy, Department of, 470
Enterprise orbiter, 173–74
Europe, 405–06, 530, 607
European Space Agency (ESA), 407, 413, 416, 455–57, 467, 613
Evolved Expendable launch Vehicle (EELV), 510–11
Explorer I, 8, 54–55, 56–63

F

Faget, Maxime A., 166, 215–21
Federal Acquisition Regulations (FAR), 475
Federal Aviation Administration (FAA), 430
Federal Communications Commission (FCC), 430, 434, 480, 486–89
Flanigan, Peter, 249–52
Fletcher, James C., 91–94, 170–72, 183–85, 212, 414, 457, 537–38; and Senator Walter F. Mondale, 256–61; and Space Shuttle contractor selection, 262–68; and Space Shuttle cost overruns, 286–88; and Space Shuttle decision process, 223–31, 245–49, 252–55; and Space Shuttle economic analysis, 239–44; and Space Shuttle safety, 352–54; and Stever Panel on Redesign of Space Shuttle Solid Rocket Booster, 385–93; and United States Space Launch Strategy, 382–85
Follow-on Early Warning System (FEWS), 549
France, 406, 466, 474, 491
Frosch, Robert A., 283–86, 288–89, 290–91, 294–305, 305–06
Fuller, Craig L., 441, 448–49

G

Gatland, Ken, 575
Gemini, Project, 15–16, 81–85
General Accounting Office (GAO), 22, 597
General Dynamics Corporation, 75–81, 410, 419–20, 422, 450, 473, 545, 575
General Electric Company, 5; and one million pound thrust engine, 105–09; and Space Shuttle, 164–73
Gernsback, Hugo, 1
Gilruth, Robert R., 119–22
Glenn, John, 15
Glennan, T. Keith, 22, 26, 75–81
Goddard, Robert H., 2, 3, 17, 21
Goldin, Daniel S., 585
Golovin, Nicholas E., 119–22
Goodpaster, Andrew J., 49–50, 51–53, 56–63, 69–70
Gorbachev, Mikhail, 188
Gore, Albert, 606
Graham, Daniel, 511, 577
Greene, Jay, 591
Griffin, Michael, 591
Grissom, Virgil I. ("Gus"), 15
Grumman Aerospace Corporation, 166; and Space Shuttle contractor selection, 262–68
Guggenheim Aeronautical Laboratory of the California Institute of Technology (GALCIT), 4
Gunn, Charles, 428
GTE Spacenet, 407
Guggenheim Fund for the Promotion of Aeronautics, 2

H

H-2 booster, 613
Hagen, John P., 7
Hagerty, James C., 56–63
Hall, Eldon, and 1959 progress report on Saturn development, 116–19; and "Recommendations for NASA Manned Space Flight Vehicle Program," 122–29
Harshbarger, James, 428
Hartsfield, Hank, 334–37
Heiss, Klaus, and Space Shuttle decision process, 223–31; and Space Shuttle economic analysis, 239–44
Herbolsheimer, Lawrence F., 448–49
Hermes project, 6
High Energy Upper Stage, 509
Hoeser, Steve, 577
Holloway, Paul F., 522
Holmes, D. Brainerd, 25; and F-1 engine instability, 135–42; and "Recommendations for NASA Manned Space Flight Vehicle Program," 122–29
Hudson, Gary, 575
Huntsville, Alabama, 542
Hunter, Maxwell W., 572, 575–77
Hyatt, Abraham, and 1959 progress report on Saturn development, 116–19

I

India, 530
Indian National Satellite (INSAT), 424
Inertial Upper Stage (IUS), 526
Intelsat VI, 191, 426
International Geophysical Year (IGY), 7, 38–43, 44–45, 52–53, 54–56
International Scientific Radio Union, 7
International Telecommunications Satellite (INTELSAT) consortium, 405, 466, 613
International Trade Administration (ITA), 497; see also Commerce, Department of
International Trade in Arms Regulations, 497
International Union of Geodesy and Geophysics, 7
Isabella, queen, 572, 575

J

Jackass Flats, New Mexico, 18
Japan, 405, 422, 530, 607, 613
Jet Propulsion Laboratory (JPL), 4
Johnson & Johnson, 448
Johnson, Lyndon B., 28
Johnson, Roy W., 109–11
Johnson Space Center (JSC), 591
Joint Chiefs of Staff, 428
Juno I, 8
Juno II, 95
Juno V, 20, 96
Jupiter ballistic missile, 6, 7, 9, 11, 49–51, 54–56, 70, 95; and "National Integrated Missile and Space Vehicle Development Program, A," 97–104
Jupiter Composite Re-entry Test Vehicle (Jupiter C), 6, 7, 8, 12, 49–51, 54–56, 56–63, 95

K

Kazakhstan, 552
Kelly Corporation, 512
Kennedy, John F., 14, 18, 19, 23, 70-72, 163
Kennedy Space Center (KSC), 492
Kessler, Thomas L., 575
Killian, James R., Jr., 69-70
Kingon, Alfred H., 457
Kistiakowsky, George B., 69-70
Kistler Corporation, 511; and Kistler K-1, 511
Kiwi experimental nuclear reactor, 18, 89-91
KIWI reactor series, 18, 89-91
Knerr, Hugh J., 32-33
Korolev, Sergei P., 2
Kraft, Christopher C., Jr., 305-06, 399-404
Krunichev, 422

L

LTV Corporation, 545
Labor, Department of, 481, 483
Land Remote-Sensing Commercialization Act (of 1984), 434
Langley Research Center, 522, 576
Launch Vehicle Technology Study, 505
Lawrence Livermore National Laboratory, 17, 18
Lee, Chester, 344-46
Lemnitzer, Lyman, 60
Lenz, Allen J., 329-33
Lewis Research Center (LeRC), 542
Lindbergh, Charles A., 2
Littles, Wayne, 591
Lockheed Aircraft Corporation, 13, 64-68, 75-81, 422, 572, 576-77, 593, 631; and "Skunk Works," 593; and Space Shuttle, 164-73; and Space Shuttle contract protest, 268-72; and Space Shuttle contractor selection, 262-68
Lockheed Missiles and Space Company, 18; and Space Shuttle contractor selection, 262-68
Long March Chinese launch vehicle, 422, 497-98, 544, 612
Los Alamos Scientific Laboratory, 17, 18, 91, 94
Lovelace, Alan M., 183, 305-06, 352-54
Lovell, James A., Jr., 24
Low, George M., 171-72; and all-up testing, 142-43; and early development of Saturn, 111-15; and Space Shuttle contractor selection, 262-68; and Space Shuttle decision process, 223-34
Lucas, William R., 305-06, 521
Lynn, James T., 286-88

M

McAllister, Eugene G., 465
McCaffrey, Shellyn G., 465
McCarthy, John, 428
McCartney, Forrest, 187
McCurdy, Richard C., 172; and Space Shuttle contractor selection, 262-68
McDonnell Aircraft/McDonnell Douglas Aircraft Company, 27, 410, 418, 448-50, 545, 576, 631; and Space Shuttle, 164-73; and Space Shuttle contractor selection, 262-68
McElroy, Neil H., 13
McFarlane, Robert ("Bud"), 412, 444-46, 449-51

McIntyre, James T., Jr., 288–89

McNamara, Robert S., and NASA-DOD launch agreement, 147–53

Mardell, A.D., 75–81

Mars, 525, 535–36

Marshall Space Flight Center (MSFC), 542, 591

Martin, Glenn L., Company, 7

Martin Marietta Corporation, 172, 183, 410, 419–22, 449–51, 472–76, 545; and Martin Marietta Commercial Titan, Inc., 472–76

Matagorda Island, Texas, 409

Mathematica, Inc., and Space Shuttle economic analysis, 239–44

Mattingly, T.K., 334–37

Maultsby, Thomas, 428

Mayo, Robert P., 222

Mercury, Project, 14–15, 73–75, 75–81, 81–85

Miller, Donald, 428

MIDAS Project, 75–81

Missile Technology Control Regime (MTCR), 630

Minuteman ballistic missile, 13, 69–70, 70–72, 142, 545

Mir Soviet space station, 191, 192, 590

Mondale, Walter F., 256–61

Moon, 525, 535–36

Moore, Jesse W., 521

"Moorman Study," 509–10, 604–26

Moorman, Thomas, 510, 604–05, 607

Morgenstern, Oskar, and Space Shuttle economic analysis, 239–44

Morrell, Jimmey, 428

Morton Thiokol Corporation, 172, 184

Moscow Group for the Study of Reactive Motion (MosGIRD), 1–2

Mossinghoff, Gerald G., 272–74, 442–43

Mueller, George E., 28, 81–84, 164; and all-up testing, 142–43; and presentation of Space Shuttle, 202–05

Murphree, E.V., 49–51

Muse, T.C., and 1959 progress report on Saturn development, 116–19

Musarra, Gerald, 477

Myers, Dale D., and cancellation and disposal of Saturn items, 155–60; and Shuttle-C, 394–99

Myers, Joseph C., 45–49

N

Naish, J.V., 75–81

National Advisory Committee for Aeronautics (NACA), 6, 34

National Aeronautics and Space Administration (NASA), and ABMA transfer to NASA, 115–16; and Ad Hoc Committee for Review of Space Shuttle Main Engine Development, 275–82; and Ad Hoc Committee on Hypersonic Lifting Vehicle with Propulsion, 163; and Ad Hoc Subpanel on Reusable Launch Vehicle Technology, 194–201; and Advanced Launch System (ALS), 189; and Advanced Solid Rocket Motor (ASRM), 190; and all-up testing, 142–43; and Anders study, 283–86; and *Atlantis* orbiter, 179, 187, 191, 192, 425; and cancellation and disposal of Saturn items, 155–60; and *Challenger* accident, 354–75, 378–82; and *Challenger* orbiter, 175, 176, 179, 183–88, 354–75; and characteristics of Space Shuttle, 234–38; and *Columbia* orbiter, 175–79; and DOD-NASA Astronautics and Aeronautics Coordinating Board, 164; and *Discovery* orbiter, 179, 186; and early development of Saturn, 111–15; and early Space Shuttle concepts, 161–63; and *Endeavour* orbiter, 186, 191; and *Enterprise* orbiter, 173–74; and F-1 engine instability, 135–42; and fifth orbiter, 347–52; and first Space Shuttle flight, 177–79, 323–29; and ICBM program, 9–14; and launch agreement with DOD, 147–53; and Launch Complex 39, 154–55; and Lockheed contract protest, 268–72; and Mueller Space Shuttle presentation, 202–05; and National Security Decision Directive 8, 333–34; and National Space Transportation System analysis, 337–44; and 1959 progress report on Saturn development, 116–19; and one million pound thrust engine, 105–09; and orbiter names, 274–75; and O-rings, 183–85, 354–75, 378–82; and "pogo effect,"

144-47; and Reagan Space Shuttle policy, 329-37; and reasons for Space Shuttle, 202-03; and "Recommendations for NASA Manned Space Flight Vehicle Program," 122-29; and Saturn V, 1-160; and Senator Walter F. Mondale, 256-61; and Shuttle-C, 189, 394-99; and space access, 1-160; and Space Launch Complex-6, 181; and Space Shuttle, 161-404; and Space Shuttle classification as space vehicle, 272-74; and Space Shuttle contractor selection, 262-68; and Space Shuttle cost overruns, 286-88, 288-89, 290-91, 294-305, 305-06; and Space Shuttle decision process, 222-34, 245-49, 252-53; and Space Shuttle definition, 211-14; and Space Shuttle economic analysis, 239-44; and Space Shuttle fleet attrition, 393-94; and Space Shuttle fundamental design considerations, 215-21, and Space Shuttle guidelines for development of the flight assignment baseline, 344-46; and Space Shuttle Main Engine (SSME), 174-76, 190-91; and Space Shuttle Management Independent Review, 399-404; and Space Shuttle OMB review, 294-305; and Space Shuttle operational uses, 179-81; and Space Shuttle organization, 249-52; and Space Shuttle return to flight, 185-88, 375-78; and Space Shuttle safety concerns, 306-23, 352-54, 375-78, 378-82, 393-94; and Space Shuttle study contract, 164-73; and Space Shuttle Task Group, 164-66; and Space Task Group, 165-66, 206-10; and Stever Panel on Redesign of Space Shuttle Solid Rocket Booster, 385-93; and TPS inspection and repair on-orbit, 282-83; and United Space Alliance, 191; and United States Space Launch Strategy, 382-85

National Aeronautics and Space Council, 466
National Aerospace Plane (NASP), 470, 507-10, 527-29, 552, 557-71, 578, 603-04, 609
National Commission on Space, 514
National Environmental Policy Act, 430, 489
"National Integrated Missile and Space Vehicle Development Program, A," 97-104
National Launch System (NLS), 506, 509, 547-49, 600, 607, 614, 623
National Oceanic and Atmospheric Administration (NOAA), 480, 486-89, 611
National Reconnaissance Office (NRO), 420, 503-04, 621
National Research Council, 505
National Science Foundation, 52-53
National Security Council, 414, 444, 449-50, 469, 473, 631
National Security Decision Directive, 428, 446, 452, 522, 529
National Security Industries Association (NSIA), 547
National Security Launch Strategy, 521
National Space Council, 416, 466-71, 508, 542, 548, 571, 609, 626
National Space Launch Vehicle Program, 91-97
National Space Launch Vehicle Strategy, 468, 506, 547-48
National Space Policy, 407-10, 415, 426-27, 445, 452, 478, 494, 496, 548
National Space Strategy, 446, 450, 505, 513-21
National Space Technology Laboratories (NSTL), 542
National Space Transportation and Support Study, 505
National Space Transportation Policy, 510, 604, 626
Navaho cruise missile, 5-6, 67-68
Naval Research Laboratory (NRL), 7-8, 41-43, 45-49
Navy, U.S., 475, 558, 561
Neumann, John von, 9
Niven, Larry, 577
Nixon, Richard M., 466
North American Aviation/North American Rockwell, Inc., 5, 17, 23, 26-29, 64-68, 74-75, 85-88, 134-35; and F-1 engine instability, 135-42; and Space Shuttle, 164-73; and Space Shuttle contractor selection, 262-68; also see Rockwell International, Inc.
Nova rocket, 20, 25, 96, 129-43

O

Oak Ridge National Laboratory, 17, 91
Oberth, Hermann, 21
O'Connor, Brian, 591
Occupational Safety and Health (OSHA), 481, 483
Office of Commercial Space Transportation (OCST), see Transportation, Department of
Office of Federal Procurement Policy (OFPP), 474-75

674

Office of Management and Budget (OMB), 411, 428, 446, 448, 453, 513–14, 597
Office of Science and Technology Policy (OSTP), 428, 515, 561, 606–08, 626, 631
Office of Technology Assessment, 505
Ojalehto, George, 428
Orbital Maneuvering Vehicle (OMV), 518, 526
Orbit Transfer System, 525
Orbit Transfer Vehicle (OTV), 518–19
Orbital Sciences Corporation, 511, 545
Ordnance Guided Missile Center, Alabama, 6

P

Pacific American Launch Systems, Inc., 575
Paine, Thomas O., 222
Paperclip, Project, 33–34
Patent and Trademark Office, 442; also see Commerce, Department of
Patrick Air Force Base, 496
Payload Assist Module (PAM), 526
Payton, Gary, 577
Pegasus launch vehicle, 511, 598, 616
Pendray, G. Edward, 1
Penn, Jay, 577
People's Republic of China (PRC), see China
Percheron launch vehicle, 409
Pershing missile, 70
Personnel Launch System (PLS), 550–51
Peterson Air Force Base, 605
Phillips, Samuel C., 27, 28
Phoebus, program, 18, 89–91
Phoenix vehicle, 576
Pickering, William, 61
Pioneer Corporation, 512
Pioneer, program, 9
Pluto, Project, 18
Polaris ICBM, 13, 69–70
Poindexter, John, 449
Porter, Richard H., 61
Poseidon ballistic missile, 545
Pournelle, Jerry, 572, 577
Pratt & Whitney Company, 17; and *Enterprise* orbiter, 173–74; and one million pound thrust engine, 105–09;
Proton launch vehicle, 422, 544, 610, 612

Q

Quarles, Donald A., 38–43, 51–53, 54–56
Quayle, J. Danforth ("Dan"), 466–67, 508, 572, 577

R

Ramo, Simon, 64–68
Ramo-Wooldridge Corporation, 10
Randolph, Bernard P., 521
Range Standardization and Automation (RSA) project, 548, 617, 623
Reaction Motors, Inc., 1, and one million pound thrust engine, 105–09

Reagan, Ronald, 179, 183, 407, 409–14, 428, 441, 446, 453–54, 457, 478, 505, 513, 537–38, 557, 562, 577; and *Challenger* accident, 354–75; and fifth orbiter, 347–52; and National Security Decision Directive 8, 333–34; and National Space Transportation System analysis, 337–44; remarks on completion of fourth Space Shuttle flight, 334–37; and Space Shuttle policy, 329–33; and Space Shuttle return to flight, 185–88, 375–78; and Stever Panel on Redesign of Space Shuttle Solid Rocket Booster, 385–93; and United States Space Launch Strategy, 382–85, 478

Redstone Arsenal, Alabama, 6, 49–51

Redstone rocket, 6, 11, 14–15, 19, 40–43, 49–51, 51–53, 54–55, 56–63, 73–75; and "National Integrated Missile and Space Vehicle Development Program, A," 97–104

Regulus II rocket, 13

Reis, Victor, 562

Report of the Advisory Committee on the Future of the U.S. Space Program, see "Augustine Report"

Rice, Donald B., and Space Shuttle decision process, 223–34

RIFT, project, 18

Risque, Nancy J., 465

Robertson, Reuben B., Jr., 43–45

Rocketdyne Division, North American, 5, 23, 24, 25, 29, 74–75; and F-1 engine instability, 135–42; and one million pound thrust engine, 105–09; and Space Shuttle Main Engine (SSME), 174–76

Rockwell International, Inc., 179, 545, 631

Rogers, William P., and *Challenger* accident, 183–85, 354–75, 414

Rona, Thomas P., 522

Rosen, Milton, and F-1 engine instability, 135–42; and "Recommendations for NASA Manned Space Flight Vehicle Program," 122–29

Rosenberg, Robert, 202–03

Roswell, New Mexico, 2

Rotary Rocket Corporation, 512

Rover/NERVA nuclear rocket, 17–19, 85–91, 91–95, 97

Russia, 191, 422, 497, 543, 552, 607, 612–13, 623–24

Rye, Gilbert D., 449–50

S

Salkeld, Bob, 576

SAMOS, program, 75–81

Sänger, Eugen, 1, 161

Saturn V, 1–160, 18, 19–29, 96; and all-up testing, 142–43; and cancellation and disposal of Saturn items, 155–60; and early history, 111–15; and F-1 engine instability, 135–42; and Launch Complex 39, 154–55; and launch operations, 29–31; and NASA-DOD launch agreement, 147–53; and 1959 progress report on Saturn development, 116–19; and "pogo effect," 144–47; and "Recommendations for NASA Manned Space Flight Vehicle Program," 122–29

Schriever, Bernard A., 64–68, 75–81

Schwenk, F.C., 111–15

Science Wonder Stories, 1

Scout rocket, 8, 13

Seaborg, Glenn T., 19

Sea Launch Company, 422

Seamans, Robert C., Jr., 25; and F-1 engine instability, 135–42

Senior Arms Control Policy Group, 516

Senior Interagency Group (SIG) (Space), 428, 430, 444, 450, 453, 513–15

Sharrard, John, 428

Shea, Joseph F., 562

Shepard, Alan B., Jr., 15

Shultz, George, and Space Shuttle decision process, 223–31

Silveira, Milton A., 215–21

Silverstein, Abe, 20, 23, 25–26; and 1959 progress report on Saturn development, 116–19; and one million pound thrust engine, 105–09

Single Stage Rocket Technology (SSRT) Program, 609
Skinner, Samuel, 490–91
Skylab, 23, 30
Smith, Richard, 305–06
Snyder, Murray, 61
Soviet Union, 188, 405, 417, 466, 473, 508, 516, 530–32, 623
Space Launch Infrastructure Investment Plan (SLIIP), 617
Soyuz launch vehicle, 610
"Space Act" (of 1958), 466
Space Services, Inc., 409–10, 473, 576
Space Launch Complex (SLC)-4, 421
Space Launch Complex (SLC)-6, see National Aeronautics and Space Administration (NASA) and Space Shuttle
"Space Launch Modernization" study, see "Moorman Study"
Space Launch Policy Working Group, 426, 428
Space Launch Strategy, U.S., 414–15, 478
Spacelifter, 549–50, 552, 607, 609
Space Nuclear Propulsion Office, AEC, 18
Space Nuclear Thermal Propulsion, program, 614
Space Policy Advisory Board, Vice President's, see Vice President's Space Policy Advisory Board
Space Ship Experimental (SSX), 577
Space Shuttle, 161–634 passim; and Ad Hoc Committee for Review of Space Shuttle Main Engine Development, 275–82; and Ad Hoc Committee on Hypersonic Lifting Vehicle with Propulsion, 163; and Ad Hoc Subpanel on Reusable Launch Vehicle Technology, 194–201; and Advanced Launch System (ALS), 189; and Advanced Solid Rocket Motor (ASRM), 190; and Anders study, 283–86; and Atlantis orbiter, 179, 187, 191, 192, 425; and Challenger accident, 354–75, 378–82; and Challenger orbiter, 175, 176, 179, 183–88, 354–75; and characteristics, 234–38; classification as space vehicle, 272–74; and Columbia orbiter, 175–79; and Complementary Expendable Launch Vehicle (CELV), 181–83, 504; and contractor selection, 262–68; and cost overruns, 286–88, 288–89, 290–91, 294–305, 305–06; and decision process, 222–34, 245–49, 252–53; and definition, 211–14; and Discovery orbiter, 179, 186; and DOD-NASA Astronautics and Aeronautics Coordinating Board, 164; and early concepts, 161–63; and economic analysis, 239–44; and Endeavour orbiter, 186, 191; and Enterprise orbiter, 173–74; examination, 305–06; and fifth orbiter, 347–52; and first flight, 177–79, 323–29; and fleet attrition, 393–94; and fundamental design considerations, 215–21; and guidelines for development of the flight assignment baseline, 344–46; and Lockheed contract protest, 268–72; and Mueller presentation, 202–05; and National Security Decision Directive 8, 333–34; and National Space Transportation System analysis, 337–44; and OMB review, 294–305; and operational uses, 179–81; and organization, 249–52; and orbiter names, 274–75; and O-rings, 183–85, 354–75, 378–82; and Reagan policy, 329–37; and reasons for, 202–03; and return to flight, 185–88, 375–78; and safety concerns for, 306–23, 352–54, 375–78, 378–82, 393–94; and Senator Walter F. Mondale, 256–61; and Shuttle-C, 189, 394–99; and Soviet Union, 188; and Space Launch Complex-6, 181; and Space Shuttle Main Engine (SSME), 174–76, 190–91; and Space Shuttle Management Independent Review, 399–404; and Space Shuttle Task Group, 164–66; and Space Task Group, 165–66, 206–10; and Stever Panel on Redesign of Space Shuttle Solid Rocket Booster, 385–93; and study contract, 164–73; and TPS inspection and repair on-orbit, 282–83; and United Space Alliance, 191; and United States Space Launch Strategy, 382–85
Space Station, 191, 451, 508, 514–15, 518–19, 524, 533, 536–37, 550, 586, 590, 593, 597, 599–600, 610
Space Task Group, 165–66
Space Technology Laboratories, 10
Space Transportation Architecture Study (STAS), 529, 534, 538, 556
Sputnik 1, 6, 8, 51–53, 97, 98, 531
State, Department of, 428, 430, 439, 473, 480, 485, 514
Stever, H. Guyford, 385–93
Stewart, Homer J., 38–43, 45–49
Storms, Harrison, 28
Strategic Air Command, 10–11
Strategic Arms Reduction Treaty (START), 630
Strategic Defense Initiative (SDI), 451, 504, 516, 518, 523–25, 532, 535, 573; and Strategic Defense Initiative Organization (SDIO), 508, 511, 526, 529, 536, 561, 575–78
Sutton, George P., and 1959 progress report on Saturn development, 116–19
Symphonie communications satellite, 406

T

Terrell, Norman E., 521

Thiokol Corporation, and one million pound thrust engine, 105–09

Thor ballistic missile (Thor-Able), 10, 12, 13, 19, 94

Thor Delta launch vehicle, 405

Titan launch vehicle, 10, 11, 12, 13, 14, 15–16, 19, 69–70, 81–85, 164, 186, 405, 409, 419–22, 472, 495, 503–05, 526–27, 534, 545, 547, 592–93, 596, 598–600, 607, 611, 613–18, 621–22, 628

Tischler, A.O., 134–35; and F-1 instability, 135–42; and "Recommendations for NASA Manned Space Flight Vehicle Program," 122–29

Tracking and Data Relay Satellite (TDRS), 186–87

Trade Representative, Office of the U.S. (USTR), 413, 454–55, 466, 475

Transpace Carriers, Inc. (TCI), 410, 413, 419, 454–55

Transportation, Department of, 411–13, 418, 431, 440, 444, 450, 452, 466, 471–97 *passim*, 513–16, 591, 607, 612, 628–29, 631; and Commercial Space Transportation Advisory Committee (COMSTAC), see Commercial Space Transportation Advisory Committee (COMSTAC); and Office of Commercial Space Transportation (OCST), 411–13, 418, 476–97 *passim*, 612

Treaty on the Peaceful Uses of Outer Space (1967), 474

Trident ballistic missile, 536, 576

Truly, Richard H., 185–88, 375–78, 521

Tsiolkovskiy, Konstantin E., 17, 21

U

Ukraine, 422, 497, 552

Union of Soviet Socialist Republics (U.S.S.R.), see Soviet Union

United Nations, 444; and Committee on the Peaceful Uses of Outer Space, 444

United Space Alliance, 191

Upper Atmosphere Rocket Research Panel, 4–5

U.S. Air Force (USAF), see Air Force, U.S.

U.S. Navy (U.S.N.), see Navy, U.S.

V

V-2 rocket, 3, 4, 5, 11

Van Allen, James A., 4–5, 56–63

Vandenberg Air Force Base, 11, 421, 625

Vanguard, Project, 7, 8, 40–49, 50–51, 54–56, 95

Vega rocket, 96

Verein für Raumschiffahrt (VfR), 3

Vertical Assembly Building (VAB), 30–31

Vice President's Space Policy Advisory Board, 542

Viking, project, 16–17

Viking rocket, 7, 40–49, 54–56

W

WAC Corporal rocket, 4, 5

Waterman, Alan T., 52–53, 62, 63

Webb, James E., 26; and NASA-DOD launch agreement, 147–53; and Launch Complex 39, 154–55

Weber, Arnold R., 249–52

Weinberger, Caspar, 182, 211; and Space Shuttle decision process, 223–31, 245–49, 252–55

Western Development Division, 10–11, 64–68

Western Test Range (WTR), 525–26, 623

Westinghouse Electric Corporation, 18

White House, 407, 410, 412, 414, 420, 426, 441, 503, 505

White House Cabinet Council on Commerce and Trade, see Cabinet Council on Commerce and Trade (later renamed the Economic Policy Council)
White Sands Proving Grounds, New Mexico, 4, 5
Williams, Walter A., 352–54
Wilson, Charles, 10
Woodruff, Larry L., 521
Wooldridge, Dean, 64–68
Worlund, Len, 591
Wright Aeronautical Division, Curtiss-Wright Corporation, and one million pound thrust engine, 105–09
Wright-Patterson Air Force Base, 558

X

X-1 aircraft, 1
X-15 aircraft, 26
X-33 Advanced Technology Demonstrator, 511, 631–34
X-34 Technology Testbed Demonstrator, 511–12

Y

Yanagida, Joy, 428
Yardley, John F., 274–75, 282–83, 305–06, 576
Yates, D.N., 55
Yeutter, Clayton, 413
York, Herbert, 60
Young, John W., 177, 185; and *Challenger* accident, 378–82; and Space Shuttle decision process, 223–31; and STS-1, 323–29

Z

Zenit launch vehicle, 544, 612
Zersen, William F. H., 522

The NASA History Series

Reference Works, NASA SP-4000

Grimwood, James M. *Project Mercury: A Chronology* (NASA SP-4001, 1963).

Grimwood, James M., and Hacker, Barton C., with Vorzimmer, Peter J. *Project Gemini Technology and Operations: A Chronology* (NASA SP-4002, 1969).

Link, Mae Mills. *Space Medicine in Project Mercury* (NASA SP-4003, 1965).

Astronautics and Aeronautics, 1963: Chronology of Science, Technology, and Policy (NASA SP-4004, 1964).

Astronautics and Aeronautics, 1964: Chronology of Science, Technology, and Policy (NASA SP-4005, 1965).

Astronautics and Aeronautics, 1965: Chronology of Science, Technology, and Policy (NASA SP-4006, 1966).

Astronautics and Aeronautics, 1966: Chronology of Science, Technology, and Policy (NASA SP-4007, 1967).

Astronautics and Aeronautics, 1967: Chronology of Science, Technology, and Policy (NASA SP-4008, 1968).

Ertel, Ivan D., and Morse, Mary Louise. *The Apollo Spacecraft: A Chronology, Volume I, Through November 7, 1962* (NASA SP-4009, 1969).

Morse, Mary Louise, and Bays, Jean Kernahan. *The Apollo Spacecraft: A Chronology, Volume II, November 8, 1962-September 30, 1964* (NASA SP-4009, 1973).

Brooks, Courtney G., and Ertel, Ivan D. *The Apollo Spacecraft: A Chronology, Volume III, October 1, 1964-January 20, 1966* (NASA SP-4009, 1973).

Ertel, Ivan D., and Newkirk, Roland W., with Brooks, Courtney G. *The Apollo Spacecraft: A Chronology, Volume IV, January 21, 1966-July 13, 1974* (NASA SP-4009, 1978).

Astronautics and Aeronautics, 1968: Chronology of Science, Technology, and Policy (NASA SP-4010, 1969).

Newkirk, Roland W., and Ertel, Ivan D., with Brooks, Courtney G. *Skylab: A Chronology* (NASA SP-4011, 1977).

Van Nimmen, Jane, and Bruno, Leonard C., with Rosholt, Robert L. *NASA Historical Data Book, Volume I: NASA Resources, 1958-1968* (NASA SP-4012, 1976, rep. ed. 1988).

Ezell, Linda Neuman. *NASA Historical Data Book, Volume II: Programs and Projects, 1958-1968* (NASA SP-4012, 1988).

Ezell, Linda Neuman. *NASA Historical Data Book, Volume III: Programs and Projects, 1969-1978* (NASA SP-4012, 1988).

Gawdiak, Ihor Y., with Fedor, Helen. Compilers. *NASA Historical Data Book, Volume IV: NASA Resources, 1969-1978* (NASA SP-4012, 1994).

Rumerman, Judy A. Compiler. *NASA Historical Data Book, Volume V: NASA Launch Systems, Space Transportation, Human Spaceflight, and Space Science, 1979-1988* (NASA SP-4012, 1999).

Astronautics and Aeronautics, 1969: Chronology of Science, Technology, and Policy (NASA SP-4014, 1970).

Astronautics and Aeronautics, 1970: Chronology of Science, Technology, and Policy (NASA SP-4015, 1972).

Astronautics and Aeronautics, 1971: Chronology of Science, Technology, and Policy (NASA SP-4016, 1972).

Astronautics and Aeronautics, 1972: Chronology of Science, Technology, and Policy (NASA SP-4017, 1974).

Astronautics and Aeronautics, 1973: Chronology of Science, Technology, and Policy (NASA SP-4018, 1975).

Astronautics and Aeronautics, 1974: Chronology of Science, Technology, and Policy (NASA SP-4019, 1977).

Astronautics and Aeronautics, 1975: Chronology of Science, Technology, and Policy (NASA SP-4020, 1979).

Astronautics and Aeronautics, 1976: Chronology of Science, Technology, and Policy (NASA SP-4021, 1984).

Astronautics and Aeronautics, 1977: Chronology of Science, Technology, and Policy (NASA SP-4022, 1986).

Astronautics and Aeronautics, 1978: Chronology of Science, Technology, and Policy (NASA SP-4023, 1986).

Astronautics and Aeronautics, 1979–1984: Chronology of Science, Technology, and Policy (NASA SP-4024, 1988).

Astronautics and Aeronautics, 1985: Chronology of Science, Technology, and Policy (NASA SP-4025, 1990).

Noordung, Hermann. *The Problem of Space Travel: The Rocket Motor.* Stuhlinger, Ernst, and Hunley, J.D., with Garland, Jennifer. Editors (NASA SP-4026, 1995).

Astronautics and Aeronautics, 1986–1990: A Chronology (NASA SP-4027, 1997).

Management Histories, NASA SP-4100

Rosholt, Robert L. *An Administrative History of NASA, 1958–1963* (NASA SP-4101, 1966).

Levine, Arnold S. *Managing NASA in the Apollo Era* (NASA SP-4102, 1982).

Roland, Alex. *Model Research: The National Advisory Committee for Aeronautics, 1915–1958* (NASA SP-4103, 1985).

Fries, Sylvia D. *NASA Engineers and the Age of Apollo* (NASA SP-4104, 1992).

Glennan, T. Keith. *The Birth of NASA: The Diary of T. Keith Glennan.* Hunley, J.D. Editor (NASA SP-4105, 1993).

Seamans, Robert C., Jr. *Aiming at Targets: The Autobiography of Robert C. Seamans, Jr.* (NASA SP-4106, 1996).

Project Histories, NASA SP-4200

Swenson, Loyd S., Jr., Grimwood, James M., and Alexander, Charles C. *This New Ocean: A History of Project Mercury* (NASA SP-4201, 1966; rep. ed. 1998).

Green, Constance McL., and Lomask, Milton. *Vanguard: A History* (NASA SP-4202, 1970; rep. ed. Smithsonian Institution Press, 1971).

Hacker, Barton C., and Grimwood, James M. *On Shoulders of Titans: A History of Project Gemini* (NASA SP-4203, 1977).

Benson, Charles D. and Faherty, William Barnaby. *Moonport: A History of Apollo Launch Facilities and Operations* (NASA SP-4204, 1978).

Brooks, Courtney G., Grimwood, James M., and Swenson, Loyd S., Jr. *Chariots for Apollo: A History of Manned Lunar Spacecraft* (NASA SP-4205, 1979).

Bilstein, Roger E. *Stages to Saturn: A Technological History of the Apollo/Saturn Launch Vehicles* (NASA SP-4206, 1980, rep. ed. 1997).

SP-4207 not published.

Compton, W. David, and Benson, Charles D. *Living and Working in Space: A History of Skylab* (NASA SP-4208, 1983).

Ezell, Edward Clinton, and Ezell, Linda Neuman. *The Partnership: A History of the Apollo-Soyuz Test Project* (NASA SP-4209, 1978).

Hall, R. Cargill. *Lunar Impact: A History of Project Ranger* (NASA SP-4210, 1977).

Newell, Homer E. *Beyond the Atmosphere: Early Years of Space Science* (NASA SP-4211, 1980).

Ezell, Edward Clinton, and Ezell, Linda Neuman. *On Mars: Exploration of the Red Planet, 1958–1978* (NASA SP-4212, 1984).

Pitts, John A. *The Human Factor: Biomedicine in the Manned Space Program to 1980* (NASA SP-4213, 1985).

Compton, W. David. *Where No Man Has Gone Before: A History of Apollo Lunar Exploration Missions* (NASA SP-4214, 1989).

Naugle, John E. *First Among Equals: The Selection of NASA Space Science Experiments* (NASA SP-4215, 1991).

Wallace, Lane E. *Airborne Trailblazer: Two Decades with NASA Langley's Boeing 737 Flying Laboratory* (NASA SP-4216, 1994).

Butrica, Andrew J. Editor. *Beyond the Ionosphere: Fifty Years of Satellite Communication* (NASA SP-4217, 1997).

Butrica, Andrew J. *To See the Unseen: A History of Planetary Radar Astronomy* (NASA SP-4218, 1996).

Mack, Pamela E. Editor. *From Engineering Science to Big Science: The NACA and NASA Collier Trophy Research Project Winners* (NASA SP-4219, 1998).

Reed, R. Dale. With Lister, Darlene. *Wingless Flight: The Lifting Body Story* (NASA SP-4220, 1997).

Heppenheimer, T.A. *The Space Shuttle Decision: NASA's Search for a Reusable Space Vehicle* (NASA SP-4221, 1999).

Hunley, J.D. Editor. *Toward Mach 2: The Douglas D-558 Program* (NASA SP-4222, 1999).

Center Histories, NASA SP-4300

Rosenthal, Alfred. *Venture into Space: Early Years of Goddard Space Flight Center* (NASA SP-4301, 1985).

Hartman, Edwin, P. *Adventures in Research: A History of Ames Research Center, 1940–1965* (NASA SP-4302, 1970).

Hallion, Richard P. *On the Frontier: Flight Research at Dryden, 1946–1981* (NASA SP- 4303, 1984).

Muenger, Elizabeth A. *Searching the Horizon: A History of Ames Research Center, 1940–1976* (NASA SP-4304, 1985).

Hansen, James R. *Engineer in Charge: A History of the Langley Aeronautical Laboratory, 1917–1958* (NASA SP-4305, 1987).

Dawson, Virginia P. *Engines and Innovation: Lewis Laboratory and American Propulsion Technology* (NASA SP-4306, 1991).

Dethloff, Henry C. *"Suddenly Tomorrow Came . . .": A History of the Johnson Space Center* (NASA SP-4307, 1993).

Hansen, James R. *Spaceflight Revolution: NASA Langley Research Center from Sputnik to Apollo* (NASA SP-4308, 1995).

Wallace, Lane E. *Flights of Discovery: 50 Years at the NASA Dryden Flight Research Center* (NASA SP-4309, 1996).

Herring, Mack R. *Way Station to Space: A History of the John C. Stennis Space Center* (NASA SP-4310, 1997).

Wallace, Harold D., Jr. *Wallops Station and the Creation of the American Space Program* (NASA SP-4311, 1997).

Wallace, Lane E. *Dreams, Hopes, Realities: NASA's Goddard Space Flight Center, The First Forty Years* (NASA SP-4312, 1999).

General Histories, NASA SP-4400

Corliss, William R. *NASA Sounding Rockets, 1958–1968: A Historical Summary* (NASA SP-4401, 1971).

Wells, Helen T., Whiteley, Susan H., and Karegeannes, Carrie. *Origins of NASA Names* (NASA SP-4402, 1976).

Anderson, Frank W., Jr. *Orders of Magnitude: A History of NACA and NASA, 1915–1980* (NASA SP-4403, 1981).

Sloop, John L. *Liquid Hydrogen as a Propulsion Fuel, 1945–1959* (NASA SP-4404, 1978).

Roland, Alex. *A Spacefaring People: Perspectives on Early Spaceflight* (NASA SP-4405, 1985).

Bilstein, Roger E. *Orders of Magnitude: A History of the NACA and NASA, 1915–1990* (NASA SP-4406, 1989).

Logsdon, John M. Editor. With Lear, Linda J., Warren-Findley, Jannelle, Williamson, Ray A., and Day, Dwayne A. *Exploring the Unknown: Selected Documents in the History of the U.S. Civil Space Program, Volume I: Organizing for Exploration* (NASA SP-4407, 1995).

Logsdon, John M. Editor. With Day, Dwayne A., and Launius, Roger D. *Exploring the Unknown: Selected Documents in the History of the U.S. Civil Space Program, Volume II: External Relationships* (NASA SP-4407, 1996).

Logsdon, John M. Editor. With Launius, Roger D., Onkst, David H., and Garber, Stephen E. *Exploring the Unknown: Selected Documents in the History of the U.S. Civil Space Program, Volume III: Using Space* (NASA SP-4407, 1998).